Advanced Antenna Systems for 5G Network Deployments

Bridging the Gap Between Theory and Practice

Advanced Antenna Systems for 5G Network Deployments

Bridging the Gap Between Theory and Practice

Henrik Asplund

David Astely

Peter von Butovitsch*

Thomas Chapman

Mattias Frenne

Farshid Ghasemzadeh

Måns Hagström

Billy Hogan

George Jöngren

Jonas Karlsson

Fredric Kronestedt

Erik Larsson

Peter von Butovitsch has served as the driver and main editor throughout the development of this book.

Academic Press is an imprint of Elsevier
125 London Wall, London EC2Y 5AS, United Kingdom
525 B Street, Suite 1650, San Diego, CA 92101, United States
50 Hampshire Street, 5th Floor, Cambridge, MA 02139, United States
The Boulevard, Langford Lane, Kidlington, Oxford OX5 1GB, United Kingdom

British Library Cataloguing-in-Publication Data
A catalogue record for this book is available from the British Library

Library of Congress Cataloging-in-Publication Data
A catalog record for this book is available from the Library of Congress

ISBN: 978-0-12-820046-9

For Information on all Academic Press publications
visit our website at https://www.elsevier.com/books-and-journals

Publisher: Mara Conner
Acquisitions Editor: Tim Pitts
Editorial Project Manager: John Leonard
Production Project Manager: Kamesh Ramajogi
Cover Designer: Greg Harris

Typeset by MPS Limited, Chennai, India

Contents

Authors

Henrik Asplund

Henrik Asplund received his MSc degree from Uppsala University, Sweden, in 1996 and joined Ericsson Research, Stockholm, Sweden, in the same year. Since then he has been working in the field of antennas and propagation supporting predevelopment and standardization of all major wireless technologies from 2G to 5G. His current research interests include antenna techniques, radio channel measurements and modeling, and deployment options for 5G including higher frequencies.

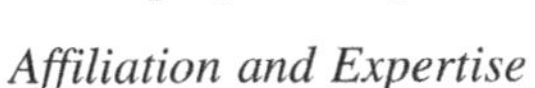

Affiliation and Expertise

Master Researcher in Antennas and Propagation, Ericsson AB, Stockholm, Sweden

David Astely

David Astely is currently a Principal Researcher with Ericsson Research in the radio area. He received his PhD in signal processing from KTH Royal Institute of Technology in 1999 and has been with Ericsson since 2001, where he has held various positions in both research and product development.

Affiliation and Expertise

Principal Researcher, Ericsson AB, Stockholm, Sweden

Peter von Butovitsch

Peter von Butovitsch joined Ericsson in 1994 and currently serves as Technology Manager at Systems & Technology. He has held various positions at Ericsson Research and in RAN system design over the years, and from 1999 to 2014 he worked for Ericsson in Japan and China. He holds both an MSc in engineering physics and a PhD in signal processing from KTH Royal Institute of Technology in Stockholm, Sweden. In 2016 he earned an MBA from Leicester University in the United Kingdom.

Affiliation and Expertise

Technology Manager, Ericsson AB, Stockholm, Sweden

Thomas Chapman

Thomas Chapman is currently working within the radio access and standardization team within the Standards and Technology group at Ericsson. He has been contributing into 3GPP standardization since 2000 to the whole portfolio of 3GPP technologies including UTRA TDD, WCDMA, HSPA, LTE, and NR, and has been deeply involved in concept evaluation and standardization of AAS in RAN4. He holds an MSc (1996) and PhD (2000) in electronic engineering and signal processing from the University of Manchester, UK.

Affiliation and Expertise
3GPP Standardisation Delegate, Ericsson AB, Stockholm, Sweden

Mattias Frenne

Mattias Frenne is currently a Principal Researcher in multi-antenna standardization in Ericsson. He holds an MSc (1996) and a PhD (2002) in engineering physics and signal processing respectively, both from Uppsala University, Sweden. Mattias has contributed to the physical layer concept development for both LTE and NR and is acting as a 3GPP Standardization Delegate in 3GPP RAN WG1 since 2005, mainly covering topics in the multi-antenna area. He was named Ericsson Inventor of the Year in 2016.

Affiliation and Expertise
Principal Researcher, Ericsson AB, Stockholm, Sweden

Farshid Ghasemzadeh

Farshid Ghasemzadeh received an MSc degree in electrical engineering from the Chalmers University of Technology, Gothenburg, Sweden in 1994. He joined Ericsson in 1999 and currently has a position as "Expert in Radio Performance" within the Department of "Standards & Technology." He has held various positions in Ericsson and worked with RAN system design and standardization. Prior to joining Ericsson in 1999, he worked for LGP telecom as specialist in RF and microwave design developing radio products for various standards and technologies.

Affiliation and Expertise
Expert in Radio Performance, Ericsson AB, Stockholm, Sweden

Måns Hagström

Måns Hagström has worked with Radars and Radios for the last 20 years, doing both hardware and software development for real-time applications. He joined Ericsson in 2011. He is currently a systems architect and involved in the evolution of AAS radios. In various roles he has been involved in both the high-band mmW and mid-band TDD AAS development at Ericsson. He holds an MSc in computer science from the University of Gothenburg, Sweden.

Affiliation and Expertise
Senior Radio Systems Architect, Ericsson AB, Stockholm, Sweden

Billy Hogan

Billy Hogan joined Ericsson in 1995. Currently he is the Principal Engineer for AAS technology and strategies within Development Unit Networks where he drives solutions and strategy for AAS in 4G and 5G. Previously he has held various technical and leader positions in Core and Radio Access Network systemization and design. He holds a BE in electronic engineering from the National University of Ireland, Galway, and an ME in electronic engineering from Dublin City University, Ireland.

Affiliation and Expertise
Principal Researcher in AAS Technology Strategies, Ericsson AB, Stockholm, Sweden

George Jöngren

George Jöngren is currently an expert in adaptive multi-antenna technologies at Ericsson. Starting with his PhD studies in 1999, he has two decades of experience working with state-of-the-art techniques in the multi-antenna field. He joined Ericsson in 2005 and has over the years had various roles, including working for 8 years as a 3GPP delegate driving Ericsson's efforts on physical layer multi-antenna standardization. He holds a PhD in signal processing and an MSc in electrical engineering from the KTH Royal Institute of Technology, Stockholm, Sweden.

Affiliation and Expertise
Expert in Adaptive Multi-Antenna Technologies, Ericsson AB, Stockholm, Sweden

Jonas Karlsson

Jonas Karlsson joined Ericsson in 1993. Since then he has held various technical and leader positions in Ericsson covering both radio access research and system management in product development. He is currently an Expert in Multi-Antenna Systems at Development Unit Networks. He holds an MSc in electrical engineering and engineering physics from Linköping University, Sweden, and a PhD in electrical engineering from the University of Tokyo, Japan.

Affiliation and Expertise
Expert in Multi-Antenna Systems, Ericsson AB, Stockholm, Sweden

Fredric Kronestedt

Fredric Kronestedt joined Ericsson in 1993 to work on RAN research. Since then he has taken on many different roles, including system design and system management. He currently serves as Expert, Radio Network Deployment Strategies, at Development Unit Networks, where he focuses on radio network deployment and evolution aspects for 4G and 5G. He holds an MSc in electrical engineering from KTH Royal Institute of Technology, Stockholm, Sweden.

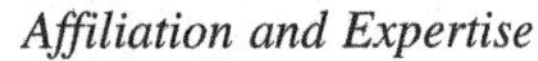

Affiliation and Expertise

Expert in Radio Network Deployment Strategies, Ericsson AB, Stockholm, Sweden

Erik Larsson

Erik Larsson joined Ericsson in 2005. He is currently a researcher working with concept development and network performance for NR with a focus on advanced antenna systems. He holds both an MSc in engineering physics and a PhD in electrical engineering, specializing in signal processing, from Uppsala University, Sweden.

Affiliation and Expertise

Researcher in Multi-Antenna Systems, Ericsson AB, Stockholm, Sweden

Contributors

Bo Göransson

Senior Expert, Multi Antenna Systems, Ericsson AB, Stockholm, Sweden

Jacob Österling

Senior Expert, Radio Base Station Architecture, Ericsson AB, Stockholm, Sweden

Preface

INTRODUCTION

Multi-antenna technologies that exploit the spatial domain of the wireless channel have been available for many decades and have over time been developed to the point that they are now quite sophisticated. Together with rapid advancement of hardware and software technology, this development has recently been embraced by the mobile network industry and an explosion is now seen in the number of products exploiting multi-antenna technologies to improve coverage, capacity, and end-user throughput. In earlier generations of mobile communication standards, coverage and capacity have been enhanced primarily by other means than utilizing the spatial domain. Although 4G contains a rich multi-antenna toolbox of features, it is not until the introduction of 5G that a broader adoption of more advanced solutions is expected.

With the introduction of 5G, the interest in multi-antenna technologies has increased rapidly. The industry, which previously has been cautious in the approach to multi-antenna solutions, has changed attitude and it has become generally accepted to deploy advanced antenna systems (AAS), where advanced is referring to both the multi-antenna feature domain and the corresponding hardware solution, in the mobile networks. It is therefore expected that all 5G network deployments will include AAS to some degree and deploying these in well-planned manner will significantly enhance network performance. It is also believed that both the total volume of multi-antenna solutions in the market and the ratio of sites using AAS in the networks will increase over time as multi-antenna technologies are further enhanced.

Substantial performance gains can indeed be achieved when using AAS compared to conventional antenna systems, if deployed and dimensioned correctly. To get attention towards AAS, the communications community has been keen to show the great technology potential of AAS. This is backed up by much of the theoretical research on so-called massive multiple input multiple output (MIMO) that may be interpreted as showing a huge potential for performance improvements. This book gives insights into the factors that impact performance and what levels of performance can be achieved in different real network deployments.

Multi-antenna technology is a multidisciplinary field to which there are inputs and contributions from different scientific and industry communities, for example, wave propagation, antenna theory, massive MIMO, traffic patterns, etc. To understand the real achievable AAS performance in mobile networks, knowledge from all these different areas needs to be combined. A deeper understanding of AAS, particularly from an interdisciplinary perspective, has not yet been successfully conveyed to the broader communication community. It is therefore believed that there is a need to spread deeper knowledge of AAS technologies concerning what performance they offer in different scenarios and how they can be successfully used in 5G networks in order to facilitate a healthy network evolution and adapt to future needs.

PURPOSE

The purpose of this book is to

1. provide a holistic view of multi-antenna technologies;
2. describe the key concepts in the field, how they are used, and how they relate to each other;
3. synthesize the knowledge of the related disciplines;
4. provide a realistic view of the performance achievable in mobile network deployments and discuss scenarios where AAS add most value;
5. describe the standardized features in the 3GPP specifications of long-term evolution (LTE) and new radio (NR) that most closely relate to multi-antenna operation;
6. clarify some common misconceptions.

To summarize, the essential contribution of this book is to combine the knowledge from the related disciplines, do the analysis from an interdisciplinary perspective, and to put AAS into a mobile network context.

OUTLINE

The outline of the book is as follows. Chapter 1, Introduction, and Chapter 2, Network Deployment and Evolution, provide a general introduction to the field of multi-antenna technologies, including AAS and mobile network deployments. Chapters 3–5 describe different basic technology building blocks underlying the understanding of AAS. Chapter 6, Multi-antenna Technologies, and Chapter 7, Concepts and Solutions for High-Band Millimeter Wave, describe the key concepts of multi-antenna technologies for mid-band and high-band. Chapters 8–10 describe the standard support in 3GPP and some examples of how end-to-end features can be designed based on the 3GPP standard. Chapter 11, Radio Performance Requirements and Regulation, describes the AAS impacts on radio product performance requirements, as specified in standards and regulations, and Chapter 12, Architecture and Implementation Aspects, describes impacts on radio products and sites with respect to architecture and implementation choices that follow from introducing AAS. Chapter 13, Performance of Multi-antenna Features and Configurations, and Chapter 14, Advanced Antenna System in Network Deployments, discuss the radio network performance impact of different AAS features and configurations in different scenario deployments, as well as how AAS can be used in mobile networks. Finally, Chapter 15, Summary and Outlook, summarizes the book. Some specifics follow below (Fig. 1).

- *Chapter 1,* Introduction, provides a background to AAS, how AAS has been used in the past, the introduction of AAS in the mobile industry, and some of the developments in academia that have been important for the adoption of AAS in mobile networks.
- *Chapter 2,* Network Deployment and Evolution, outlines the status of the operator network deployments, the requirements on network evolution, and what role AAS can have in that process. The purpose is to show the typical status of commercial mobile network deployments and the need for evolution of these to meet future requirements with respect to increasing capacity, coverage, and end-user throughput. Specifically, the possibilities to evolve the networks are described and the possibility of using AAS to meet future requirements is discussed.

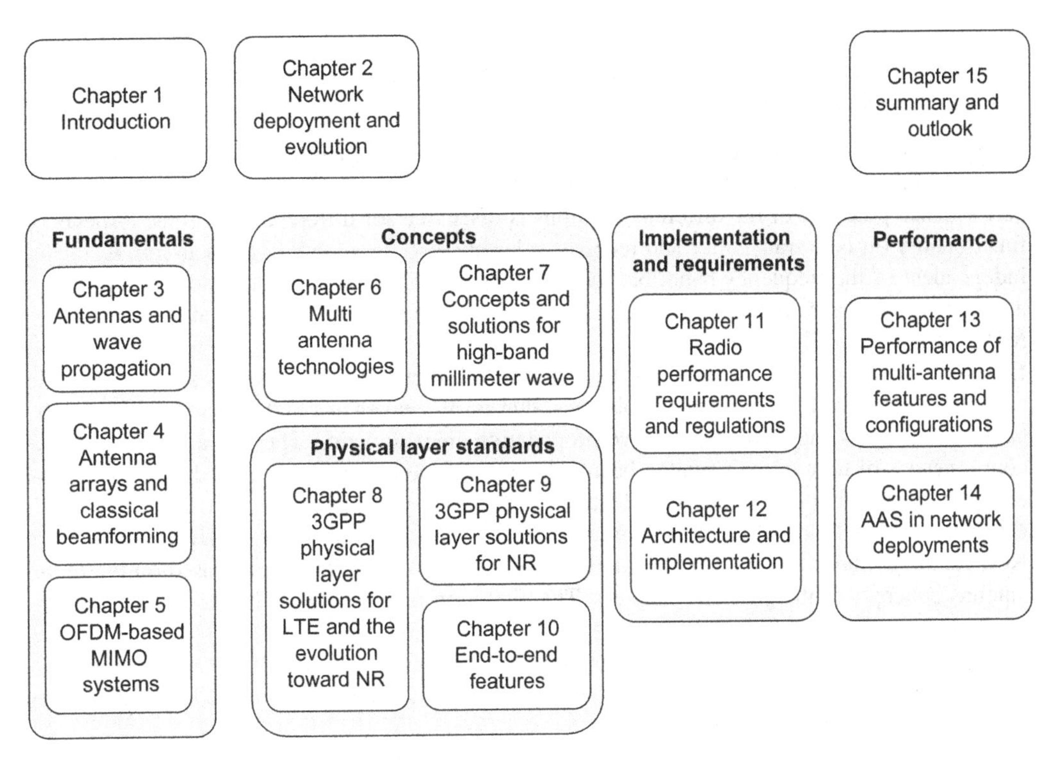

FIGURE 1

Outline of the book where chapters are grouped into five categories.

The purpose of Chapters 3–5 is to give the reader an understanding of the technology elements that influence the performance of AAS or the way AAS are built.

- In *Chapter 3,* Antennas and Wave Propagation, some of the basic antenna and wave propagation properties relevant for mobile communications are explained. The focus is on topics that are needed to understand later chapters in the book.
- *Chapter 4,* Antenna Arrays and Classical Beamforming, describes classical beamforming with antenna arrays, starting from a single antenna element, and then introducing uniform linear and planar antenna arrays.
- *Chapter 5,* Orthogonal Frequency Division Multiplexing–Based Multiple-Input Multiple-Output Systems, introduces some of the fundamentals of *orthogonal frequency division multiplexing*–based MIMO systems, that is, the access technology used for both 4G and 5G, that will be used in later chapters. More specifically, based on results from Chapter 3 & 4, Chapter 5 outlines the equivalent channel model to aid the discussion of multiple antenna techniques in Chapter 6, Multi-antenna technologies.

Chapters 6–10 explain multi-antenna concepts, how these are supported in the standard, and how the standardized hooks can be used to make end-to-end features. Some of the key results in the book are developed in this part.

- *Chapter 6,* Multi-antenna Technologies, describes multi-antenna technologies, for example, beamforming, single-user MIMO, multiple-user MIMO (MU-MIMO), and cell shaping. The performance potential of the different solutions is discussed for different conditions. Aspects of functionality on both transmitter and receiver sides are also discussed. The discussion is independent of the frequency band, but the solutions described are currently most widely adopted on mid-band frequencies, that is, 1–7 GHz. Some misconceptions related to MU-MIMO are specifically addressed.
- *Chapter 7,* Concepts and Solutions for High-Band Millimeter Wave, discusses solution specifics relating to high-band (or mm-wave) solutions, that is, at frequencies higher than 24 GHz, with short wavelengths and challenging wave propagation characteristics. The usage and consequences of time-domain analog beamforming at these bands are discussed, and some examples illustrating impacts on both features and implementation are outlined.
- *Chapter 8,* 3GPP Physical Layer Solutions for Long-Term Evolution and the Evolution Toward New Radio, contains the 3GPP LTE standard support for AAS technologies based on the multi-antenna concepts of the previous chapters. The evolution of LTE toward NR is outlined, where shortcomings of LTE are discussed.
- *Chapter 9,* 3GPP Physical Layer Solutions for New Radio, contains the 3GPP NR standard support for AAS technologies and describes how bottlenecks identified from LTE standard were utilized in designing NR in a better way.
- *Chapter 10,* End-to-End Features, explains how the 3GPP toolbox of Chapter 9 can be used to design system solutions for NR AAS features, and examples of a selection of features from Chapter 9 are presented.
- In *Chapter 11,* Radio Performance Requirements and Regulation, the requirements on radio equipment performance are outlined. This work is based on the work in 3GPP RAN4 where requirements on performance of the radio equipment and the corresponding test procedures are specified. The purpose of this chapter is to explain the rather extensive set of requirements that must be fulfilled and what that means in terms of product design.
- *Chapter 12,* Architecture and Implementation Aspects, outlines some of the base station architecture and implementation considerations that follow from the introduction of AAS features. The relation to feature distribution, location of processing, the use of interfaces, etc. are discussed. Additional aspects specific to high-band solutions are also considered.
- *Chapter 13,* Performance of Multi-antenna Features and Configurations, outlines the radio network performance of different features under given conditions. From this discussion, it follows why some features and AAS configuration choices are favorable. This chapter refers to the results developed in Chapters 3–7 and some 3GPP-related aspects.
- *Chapter 14,* Advanced Antenna System in Network Deployments, outlines how AAS is used to support the network evolution in different typical network deployment scenarios. It will be illustrated that there is a strong dependency between deployment scenario and required AAS characteristics. Hence different scenarios call for different AAS solutions. This chapter refers to Chapter 2, Network Deployment and Evolution, and addresses some of the network

evolution–related issues raised there. A few case studies for how different operators can evolve their networks in terms of spectrum usage, site aspects, choice of antenna solution, etc. are also provided.
- *Chapter 15*, Summary and Outlook, summarizes the key takeaways of the book and provides an outlook of how AAS may develop in the future.

READING GUIDELINES

The main target audience of this book are key stakeholders of AAS within the field of mobile telecommunications, for example, stakeholders in technology, strategy, network planning, network design, and engineering divisions within mobile network operators, and other stakeholders in the industry who want to understand how AAS works and how it can be used.

Another target audience is academia, where the research scope can be expanded or enriched by providing insights from industry and wireless channel modeling community that can be used to create more realistic modeling of network and channel aspects.

No particular prior knowledge is assumed to read the book. Visual and textual examples and explanations are provided, where possible, to give a better intuitive feeling for the mechanisms of antenna arrays and AAS features. A background in science, technology, engineering, or mathematics is however useful. Some basic physics, specifically wave theory, will be useful. For some parts, mathematical descriptions will offer a deeper understanding and are therefore used. For example, elementary matrix algebra, theory of complex numbers, and basic communication theory will help the understanding of the theoretical parts of multi-antenna functionality. The reader who is unfamiliar with the mathematics can however get the essence by only reading the text.

The text aims to be self-contained in the sense that all info relevant for the readers that relate to AAS is included in the book, except for basic mathematics. The reader shall thus not be dependent on other literature. For an in-depth treatment of a specific technology area, the established literature is referred to. Forward and backward references between different chapters are provided to show the connections between the different technology areas, respectively. Each chapter is concluded with a summary covering the key points and the aspects that are used in later chapters.

The book covers a broad range of topics and it may be advisable to digest the material in smaller chunks. Some guidance is provided below.

- The fundamentals in Chapters 3–5 provide a background for the reader who is interested in getting an in-depth understanding, but may be omitted by the reader who wants to get quickly to the essence of multi-antenna functionality.
- Chapter 6, Multi-antenna Technologies, and Chapter 7, Concepts and Solutions for High-Band Millimeter Wave, are essential for most of the remaining chapters.
- Chapter 13, Performance of Multi-antenna Features and Configurations, and Chapter 14, Advanced Antenna System in Network Deployments, can be read directly after Chapter 6, Multi-antenna Technologies, and Chapter 7, Concepts and Solutions for High-Band Millimeter Wave, for those who are mainly interested of AAS performance in mobile networks. Most of the findings in Chapter 13, Performance of Multi-antenna Features and Configurations, and

Chapter 14, Advanced Antenna System in Network Deployments, can be well understood even without following the mathematics in the earlier chapters.

- Chapters 8–12 are very useful for readers who want to understand the multi-antenna aspects of the 3GPP standard and the impacts of AAS on the system architecture and implementation. These chapters can however be read selectively and separately from the other chapters with limited loss of flow. Chapter 8, 3GPP Physical Layer Solutions for Long-Term Evolution and the Evolution Toward New Radio, and Chapter 9, 3GPP Physical Layer Solutions for New Radio, should however be read in succession.

Some chapters are structured to help the reader to get a shorter overview or to obtain a deeper understanding. Chapter 8, 3GPP Physical Layer Solutions for Long-Term Evolution and the Evolution Toward New Radio, and Chapter 9, 3GPP Physical Layer Solutions for New Radio, contain both a shorter basic description and a more extensive part where all the details are explained. Chapter 13, Performance of Multi-antenna Features and Configurations, also contains a shorter overview that outlines the main feature performance results. The details can then be read selectively.

As AAS is a multidisciplinary field, there is a contribution of terminology from different fields, which is partially overlapping. In this book, the aim is to use an aligned terminology. In certain areas, the terminology is adapted to that which is commonly used in that specific area. The specific terminology used is then defined in the context it appears.

Acknowledgments

A project of this magnitude relies on support and understanding from the people around us. The knowledge reflected in this book has been accumulated over many years and a large number of people have made both direct and indirect contributions. Some groups have made contributions in specific areas and have thereby had a greater influence on this book.

We would firstly like to express our most sincere gratitude to all colleagues at Ericsson who have supported us by contributing with text proposals, material and reviews of parts of the manuscript.

The information on performance of advanced antenna systems is largely based on results, learnings, and discussions developed within several Ericsson performance evaluation teams. The work represents a collective effort of significant magnitude and the contributors are too numerous to mention explicitly. Nevertheless, we are deeply grateful for the knowledge and insights we have gained from many studies and hard work conducted in the area, knowledge that made it possible to write this book.

The standardization of long-term evolution and new radio relies on contributions and good ideas from a very large group of people in the academia and in the industry, some of them within our company, whose collective work is fundamental to the technology on which the book is based.

Finally, we would like to express our deepest gratitude to our families for their patience and support.

Abbreviations

1D	one dimension
2D	two dimensions
3D	three dimensions
2G	second-generation mobile system
3G	third-generation
3GPP	3rd Generation Partnership Project
4G	fourth-generation mobile system
5G	fifth-generation mobile system
A/D	analog-to-digital converter
AAS	advanced antenna system
ACK	acknowledgment (Positive)
ACLR	adjacent channel leakage ratio
ACS	adjacent channel selectivity
ADSL	asymmetric digital subscriber line
AFE	analog front-end
AI	artificial intelligence
AIR	antenna integrated radio
AMPS	advanced mobile phone service
AOSA	array of subarrays
ARP	antenna reference point
ARPU	average revenue per user
AWS	advanced wireless services
BCH	broadcast channels
BM	beam management
BPSK	binary phase shift keying
BS	base station
BSC	base station controller
BTS	base station transceiver
BW	bandwidth
BWP	bandwidth part
C-MTC	critical machine-type communication
CA	carrier aggregation
CAPEX	capital expenditure
CAT	category (in LTE)
CBRS	citizens broadband radio system
CCH	control channel
CCE	control channel element
CDF	cumulative density function
CDMA	code division multiple access
CE	control element
CFR	crest factor reduction
CoMP	coordinated multipoint
CORESET	control resource set
COTS	commercial off-the-shelf

CP	cyclic prefix
CPE	customer premises equipment
CPRI	common public radio interface
CPU	CSI processing unit
CQI	channel quality indicator
CRAN	centralized RAN
CRI	CSI-RS Resource Indicator
CRS	cell-specific reference signal
CS	cell shaping
CS	cyclic Shift
CSI	channel-state information
CSI-IM	CSI—interference measurement
CSI-RS	CSI—reference symbol
CSR	codebook subset restriction
CSS	common search space
CW	code word
CWDM	coarse wavelength division multiplexing
D/A	digital-to-analog converter
DAS	distributed antenna system
DC	dual carrier
DC	dual connectivity
DCI	downlink control information
DFE	digital front-end
DFT	discrete Fourier transform
DL	downlink
DM-RS	demodulation reference symbol
DPD	digital predistortion
DPS	dynamic point selection
DRAN	distributed RAN
DRX	discontinuous reception
DSL	digital subscriber line
DSP	digital signal processor
DWDM	dense wavelength division multiplexing
DwPTS	downlink pilot time slot
eCPRI	Evolved CPRI
EF	element factor
EIRP	equivalent isotropic radiated power
EMC	electromagnetic compatibility
EMF	electromagnetic field
EN	E-UTRAN new radio
EN-DC	EN dual connectivity
eNB	evolved node B
EPDCCH	enhanced PDCCH
E-UTRA	evolved UTRA
EVM	error vector magnitude
FB	feedback
FCC	Federal Communications Commission
FDD	frequency division duplex

FDM	frequency division multiplexing
FDMA	frequency division multiple access
FFT	fast Fourier transform
FPGA	field-programmable gate array
FR1	frequency range 1 as Defined in 3GPP TS 38.104
FR2	frequency range 2 as Defined in 3GPP TS 38.104
FS	frequency selective
FS 1	frame structure 1
FS 2	frame structure 2
FTP	file transfer protocol
FWA	fixed wireless access
GAA	general authorized access
GoB	grid of beams
gNB	generalized node B
GP	guard period
GSM	global system for mobile communications
HARQ	hybrid automatic repeat request
HD	high definition
HPBW	half-power beamwidth
HSDPA	high speed downlink packet access
HSPA	high speed packet access
HSUPA	high speed uplink packet access
HW	hardware
IAB	integrated access backhaul
ICS	in-channel sensitivity
IFDMA	interleaved FDMA
IFFT	inverse fast Fourier transform
IODT	interoperability and device testing
IoT	Internet of things
IQ	in-phase and quadrature components
IRC	interference rejection combining
IRR	infrared reflective
ISD	intersite distance
KPI	key performance indicator
L1-RSRP	Layer 1 RSRP
LAA	license assisted access
LI	layer indicator
LNA	low noise amplifier
LO	local oscillator
LTE	long-term evolution
LTE-M	LTE-machine-type communication
MAC	medium access control
MBB	mobile broadband
MBSFN	multimedia broadcast multicast service single-frequency network
MCS	modulation and coding scheme
MIMO	multiple-input multiple-output
MISO	multiple-input single-output
MMSE	minimum mean square error

mm-Wave millimeter wave
MR measurement restriction
MRC maximum ratio combining
MRT maximum ratio transmission
MS mobile station
MU-MIMO multiple-user MIMO
NACK negative ACK
NB-IoT narrowband Internet of things
NC-JT noncoherent joint transmission
NMT Nordic mobile telephony
NR new radio
NZP Nonzero power
OBUE operating band unwanted emission
OCC orthogonal cover code
OFDM orthogonal frequency-division multiplexing
OPEX operational expenditure
OSS operation and support subsystem
OTA over the air
PA power amplifier
PAL priority access license
PAPR peak-to-average power ratio
PBCH physical broadcast channel
PCB printed circuit board
PCFICH physical control format indicator channel
PCS personal communications service
PDC personal digital cellular
PDCCH physical downlink control channel
PDCP packet data convergence protocol
PDSCH physical downlink shared channel
PHICH physical hybrid ARQ indicator channel
PHY physical layer
PMCH physical multicast channel
PMI precoding matrix indicator
PN pseudo noise
PRACH physical random access channel
PRB physical resource block
PRG precoding resource block group
PSS primary synchronization signal
PT-RS phase tracking reference signal
PUCCH physical uplink control channel
PUSCH physical uplink shared channel
QAM quadrature amplitude modulation
QCL quasi co-location
QoS quality of service
QPSK quadrature phase-shift keying
RAN radio access network
RAT radio access technology
RB resource block

RBS radio base station
RE resource element
RF radio frequency
RET remote electrical tilt
RI rank indicator
RLC radio link control
RMa 3GPP rural macro channel model
Rmin minimum data rate
RNC radio network controller
RLM radio link monitoring
RRC radio resource control
RRM radio resource management
RRU remote RU
RS reference symbol
RSRP reference signal received power
RU radio unit
Rx radio receiver
RX receive
SA subarray
SD standard definition
SDL supplementary downlink
SFP small form-Factor pluggable Transceivers
SI study item
SINR signal-to-interference-and-noise ratio
SMa 3GPP suburban macro channel model
SNR signal-to-noise ratio
SOC system-on-chip
SR scheduling request
SRI SRS resource indicator
SRS sounding reference signal
SS synchronization signal
SSB SS/PBCH block
SSS secondary synchronization signal
SUL supplementary uplink
SU-MIMO single-user MIMO
TACS total access communication system
TB transport block
TCI transmission configuration indicator
TCO total cost of ownership
TCP transmission control protocol
TDD time division duplex
TDM time division multiplexing
TDMA Time division multiple access
TM transmission mode
TMA tower mounted low-noise amplifier
TPMI transmit PMI
TRP total radiated power; transmission point
TRS tracking RS

TRX transceiver
TTI transmission time interval
Tx radio transmitter
TX transmit
Tx/Rx radio transmitter/radio receiver
UCI uplink control information
UE user equipment
UESS UE-Specific Search Space
UHD ultra-high definition
UMi 3GPP urban micro channel model
UMa 3GPP urban macro channel model
UMTS universal mobile telecommunications service
UL uplink
ULA uniform linear array
UPA uniform planar array
URLLC ultra-reliable low-latency communication
UTRA UMTS terrestrial radio access
VoLTE voice over LTE
VRB virtual resource block
W_1 wideband, long-term precoding matrix
W_2 per subband or wideband, short-term precoding matrix
WCDMA wideband code division multiple access
WCS wireless communications service
WDM wavelength division multiplexing
WI work item
xDSL DSL family (e.g., ADSL)
ZF zero-forcing
ZP zero power

CHAPTER

INTRODUCTION

1

1.1 MULTI-ANTENNA TECHNOLOGIES AND ADVANCED ANTENNA SYSTEMS

Multi-antenna technologies can be applied at the transmitter, the receiver, or on both sides of the wireless communication link and explore temporal and spatial properties of the radio channel to enhance performance. It allows sharing of communication resources not only in time and frequency as in conventional wireless communication, but also in the spatial domain. The objective when multi-antennas are applied to mobile communication systems is to improve the network performance in terms of coverage, capacity, and end-user throughput.

An *advanced antenna system* (AAS) is one solution to implement multi-antenna technologies. In this book, AAS is referred to as an antenna system comprising an AAS radio and associated AAS features, where the latter comprises various multi-antenna techniques and algorithms. An AAS radio is a hardware unit consisting of an antenna array with a large number of radio chains and possibly parts of the baseband functionality. Furthermore, a distinguishing aspect of an AAS is that the radio and the antenna are tightly integrated.

The AAS radio facilitates AAS features such as beamforming and spatial multiplexing. The AAS features can be executed by algorithms in the AAS radio, in the base station baseband unit or both. These concepts will be defined and discussed in detail later in the book, for example, in Chapters 6 and 12.

To distinguish an AAS from a *conventional system*, the conventional (non-AAS) system consists typically of a passive antenna and remote radio unit comprising a low number of radio chains. Hence the antenna and radio are typically not integrated. There is however no single common and industry-wide definition of AAS, as different industry players have used this term and/or similar terms in different and often overlapping ways. The reason for this lies partially in the fact that the concept of a base station and related terms are intrinsically difficult to define and partially because of differing ideas of what should be encompassed within the AAS definition.

As there are many conventional systems with 2, 4, and 8 radio chains already deployed, the number 8 has commonly been used to define the boundary between AAS and a conventional system, that is, above 8 is typically an AAS. The reason to distinguish AAS from conventional systems is that AAS is associated with a new integrated building practice that has an impact on the whole antenna/radio/baseband architecture and hence also the deployment in mobile networks. However, in this context it should be noted that the building practices of AAS could also be used for 8 or fewer radio chains.

Advanced Antenna Systems for 5G Network Deployments. DOI: https://doi.org/10.1016/B978-0-12-820046-9.00001-0

1.2 BRIEF HISTORY OF MULTI-ANTENNA TECHNOLOGIES AND ADVANCED ANTENNA SYSTEM

1.2.1 BEFORE MOBILE COMMUNICATION SYSTEMS

The use of antenna arrays to direct radio signals is not new and is not restricted to the field of telecommunications. The technique was used by Guglielmo Marconi in 1901 to increase the gain of the Atlantic transmissions of Morse codes [1]. Marconi used four 61 m high tower antennas arranged in a circular array in Poldhu, England, to transmit the Morse signal for the letter "S," a distance of 3425 km to Signal Hill, St. John, Newfoundland, Canada. Another early attempt to use multi-antenna techniques was made by Karl Ferdinand Braun who demonstrated the gains achievable by phased array antennas in 1905. Marconi and Braun received the Nobel Prize in physics 1909 for "recognition of their contributions to the development of wireless telegraphy" [2].

Antenna diversity techniques to overcome fading were developed in the 1940s [3]. The use of antenna array-based beamforming was also developed to steer the power in a certain direction as to improve coverage of the transmitted or received signals. Radar systems were developed that make intrinsic use of phased arrays for direction finding. Radio astronomy also makes use of antenna arrays. For this purpose, the antenna arrays are very large scale; in some cases, the elements of the array are tens of thousands of kilometers apart in order to be able to directionally detect very long-wavelength signals from outer space.

The concept of steering signals based on arrays of transmitters or sensors is not restricted to the electromagnetic domain; arrays are also deployed in sonar systems for directional processing. In fact, the two ears on a human or an animal, spaced apart, utilize the time difference of the reception of an audio signal to determine the direction of the sound source. Such *binaural* information can also be used to separate sound from background noise.

1.2.2 INTRODUCTION OF MULTI-ANTENNA TECHNOLOGIES TO TELECOM

In mobile communication, fixed, directional sector antennas were used already in the first analog mobile communication networks, AMPS, TACS, and NMT, in the early 1980s. These antennas were implemented as columns of antenna elements and were designed to maximize the area coverage. Such antennas with fixed coverage area have been, and are still being, used in all cellular mobile communication systems. They are, by far, the most common antenna type in use.

The telecom industry has acknowledged the potential of multi-antenna systems for a long time and signs of multi-antenna interest for mobile communication can be traced at least back to the early 1990s. Antenna systems allowing for dynamic, steerable beamforming were conceptualized at the same time as the advent of digital cellular systems with GSM and D-AMPS. At that time, requirements on capacity and coverage were still modest and network equipment was relatively expensive and thus a prohibitive factor for large-scale adoption. It is, however, in more recent years with the introduction of 4G (LTE) a decade ago that the use of multi-antenna techniques exploiting antennas arrays became ubiquitous both for transmission and receive purposes. The number of phase adjustable antennas in the array on the base station side was, however, for long kept at a modest level of 2, 4, or 8 in commercial networks. With the advent of AAS and spurred by coming introduction of 5G, antenna arrays with substantially more elements and radio chains have received significant industry interest and are now seen as a powerful and commercially viable tool for evolving the telecommunications environment. Such AAS are thus already playing a key role in both 4G and 5G.

1.2.2.1 2G—Early attempts

In GSM, there was no standard support for multi-antenna technologies. Trials were made by some network vendors, for example, Ericsson [4–6] and Nortel, and mobile network operators (MNOs) and academia to evaluate the technology potential [7]. The Buzzword at that time was *adaptive antennas*, to emphasize that the antenna gain pattern could be modified based on traffic conditions. The installations were however physically large and expensive. The initial focus of GSM multi-antennas was on improving capacity at 900 MHz but as the 1800 MHz band became available, the need for multi-antenna solutions as the capacity booster was reduced. Deploying additional 1800 MHz carriers was a much more cost-efficient and practical solution compared to increasing the number of antennas. The performance potential versus the size and cost for multi-antenna solutions at that time did not provide enough incentive to drive the industry toward large-scale multi-antenna deployments.

1.2.2.2 3G—Introduced but not widely used

In 3G, support for multi-antenna features in the standard was initially very limited. The focus for the first release of 3G was mainly on voice and packet data at modest rates (384 kbps). In order to increase throughput, 2×2 downlink multiple-input multiple-output (MIMO) was introduced in a later release. The observed gains in field were however limited, as the vast majority of mobile terminals already present in the 3G networks, did not have MIMO capability. The new multi-antenna features even had an initial negative impact on those legacy terminals, and hence there was great reluctance among the network operators to enable MIMO functionality.

Despite several efforts from terminal and network vendors, downlink MIMO functionality in 3G did not take-off in practice. Another basic multi-antenna feature, four-way receive (RX) diversity was shown to have excellent gains in uplink for HSUPA operation, allowing doubling of uplink capacity and enhanced uplink coverage. But similar to the fate of downlink MIMO in 3G, that feature also had limited uptake mainly due to the need for costly site visits to upgrade from older 2 RX antennas to new antennas that could support four-way RX.

In contrast, the multicarrier feature that was first introduced in the 3G standard just after MIMO was introduced became successful as it supported increase of peak rates, improved spectral efficiency, and gave capacity gains. Multicarrier had the advantage that it gave gains with new terminals but also worked seamlessly in networks with large populations of legacy terminals as there were no backward compatibility issues and it was relatively easy to deploy.

A lesson learned from 3G was thus that MIMO functionality needs to be supported from the first release in the next generation, to avoid the issues with legacy terminals in the network.

In China, a TDD-based 3G system, time division synchronous code division multiple access (TD-SCDMA), was introduced that included beamforming functionality from start. TD-SCDMA was included in the 3GPP standard as one of the 3G solutions. It was commercially used in China, but, largely because of the late introduction, the spread outside China was limited.

1.2.2.3 4G—Intrinsic, initially limited but gradually evolving

Already the first release of LTE supported basic MIMO techniques; for example, downlink spatial multiplexing with up to four layers to the mobile terminal as well as support for multi-user MIMO (MU-MIMO), see Section 8.2 for an in-depth survey of LTE history and evolution. The spatial domain was further explored in the following LTE releases with more advanced features being added to the standard. For TDD, reciprocity-based AAS solutions were possible already from start since reference signals in the uplink were defined, however, the main purpose of those were not reciprocity based operation.

The first step toward AAS support in standardization came in LTE release 10, as spatial multiplexing of up to eight layers was introduced. The feature was never completed, since the associated radio requirements were not introduced. However, this was the beginning of an expansion of support for MIMO-related functionality over time by the introduction of new enhanced MIMO functionality in every coming 3GPP release.

During Release 11, the industry realized that advanced, integrated base stations with large numbers of phase and amplitude adjustable antennas were on the horizon and that the existing framework for radio performance requirements and evaluations, which was based on the classic single antenna base station architecture, was insufficient. Therefore, 3GPP began studying solutions for AAS radio requirement specification. This led to a process over several years during which the over-the-air (OTA) AAS radio specification was developed. Simultaneously, a new channel model suitable for AAS was developed and features were specified for AAS to enhance MU-MIMO and terminal measurements for base stations with up 32 antennas. Another addition was the introduction of feedback-based two-dimensional beam steering (horizontal and vertical), also a feature enabled by AAS.

1.2.2.4 5G—Intrinsic and advanced

Just as LTE supported MIMO from the first release to alleviate legacy terminal issues, advanced beamforming functionality, and support for AAS base stations have been included as an integral part of the first 5G release, see Chapter 9, on 5G NR specifications. Support for reciprocity-based operation for TDD and UE measurements of up to 32 base station antennas was introduced. The need for solutions for bands above 3 GHz calls for massive MIMO AAS implementations, and especially at millimeter-wave (mm-wave) frequency bands there is a need for massive MIMO AAS implementations and advanced management of beamforming to provide sufficient link budget for operation.

1.3 WHY ADVANCED ANTENNA SYSTEMS NOW?

AAS is a solution that requires a large amount of integrated electronics to achieve the best performance. The initial multi-antenna solutions were physically large and costly compared to conventional solutions. Also, the scenarios for maximizing the benefits of multi-antennas were not fully understood. The early attempts, during the mid-1990s, of introducing multi-antenna solutions in GSM networks were therefore discontinued due to the relatively high cost versus performance achieved.[1]

[1] In Japan, however, some success was reached. For the personal handy system (PHS), circular arrays were used to provide service on a larger scale. This was however a relatively isolated success that did not spread widely in the mobile community.

Recently, however, multi-antenna solutions have become an increasingly attractive solution. There are several reasons for that.

First, most of the newly available spectrum, especially spectrum with high bandwidths, tends to be at higher frequencies where propagation conditions are more challenging. Increased antenna area is needed to compensate the propagation losses, but increased antenna area leads to narrower beam widths that cannot cover the full area of cells. Therefore, AAS solutions supporting dynamically steerable beams are needed to provide coverage in the whole cell. AAS thus unlocks the potential of new spectrum allocations.

Second, the traffic in the networks has increased rapidly ever since the mobile networks were launched and is expected to continue to grow for many years to come, see further discussion in Section 2.2.1. This is mainly due to increasing requirements on higher performance from end-users who gradually adopt more advanced applications, but also due to an increasing number of subscribers. Therefore, the requirements on the networks to support higher capacity have increased continuously.

There are strong incentives for the mobile network operators to reuse existing sites when upgrading the technology. Radio base station sites are often costly and difficult to acquire, particularly in urban areas where the capacity needs are the greatest. In addition to this, many operational costs are also associated with each site. Therefore, many mobile network operators try to exploit the existing sites as much as possible before trying to add new sites. AAS enables increasing coverage, capacity, and end-user performance, and is therefore an attractive solution for network improvements using existing sites.

Third, the cost of hardware is gradually going down. The accumulated effects of Moore's law [8], that is, that the number of transistors per area unit doubles roughly every 18 months, have over time reduced hardware costs to a level that the price/performance ratio has now made AAS deployments commercially attractive.

Fourth, as hardware components have become smaller, it is now also possible to integrate the antenna array, the radio equipment, and parts of the baseband tightly, hence increasing performance and reducing form factors.

At the present time, commercial AAS solutions are reaching the marketplace. AAS base stations with typically 16–64 transceivers have emerged for the 3–6 GHz range, while for mm-wave, 128 transceivers or more is the norm. As with any new technology, an evolution toward reduced cost and weight, and improved performance is expected as AAS base stations become mature and mainstream.

The operation and deployment of AAS are different to that of previous generations of base stations. An understanding of the most optimal deployment scenarios is still in its early stages. Furthermore, the industry has not yet developed a complete means to evaluate and compare the performance of different types of AAS solutions due to the complexity and interaction of factors such as the spatial distribution of signal power and interference, traffic patterns, user behavior, inter-cell interactions, etc. Also, some types of AAS base stations benefit from a different approach to site planning and installation, challenging the established principles for network rollout. It is to be expected that the coming years will witness a learning curve as the industry increases its understanding of the potential and efficient usage of the technology.

1.4 ACADEMIC WORK

Academic research has contributed significantly to the development of multi-antenna technologies. Some milestones in the academic work related to array antennas for wireless communications are briefly described here.

There is a long and gradually evolving research on antenna array processing for wireless communication dating back to at least the 1970s. For example, a paper [9] with the title "Adaptive Arrays" describes a system with multiple antennas and antenna weights updated in real-time as early as 1976. Concepts exploiting adaptive arrays were then further developed during the 1980s [10,11]. The research interest increased substantially in the decade to follow when focus was both on parametric methods for estimation of physical parameters (such as direction of arrival) [12] and space-time processing for more abstract channel properties [13], combined with a popularization of the use of convenient and powerful linear algebra to analyze the systems.

From the middle of the 1990s, theories for spatial multiplexing of several data streams to the same user using so-called MIMO systems started to receive much interest [14–16]. Spatial multiplexing had been a well-known concept for a long time even before that, but then the focus was on multiplexing multiple substantially differently located users, and it could easily be understood how it worked based on pure geometrical considerations using classical narrow beam shapes. Single-user MIMO seemed in contrast much more mysterious, almost defying intuition as all the multiplexed data streams where directed to the same geographical location. In hindsight, however, the concept of multiplexing multiple independent data streams on a channel with cross-talk and letting a receiver separate these signals via filtering was a technique described as early as 1970 [17].

In the early years of the new century, the main parts of the most popular techniques still used today had all been described and thoroughly analyzed in research. The focus was however on antenna arrays with a modest number of elements, not much beyond ten. It was considered unrealistic from a practical point of view to go further even though it was well known that the algorithms as such were completely general and could handle any number of elements and that gains would generally increase with more elements. The fear of assuming unrealistic parameter settings by going for a massive number of elements in the analysis was finally swept away once the massive MIMO research era started around 2010 [18], where the most basic assumption was a very large number of dynamically adaptable antenna elements [19]. This in turn, combined with a general trend of integrated antenna solutions, eventually also pushed the industry to look at techniques to make it more practically feasible to substantially increase the number of elements in real deployments and thus we stand today with multi-antenna systems with a large number of adaptable antennas as a fundamental component of 5G.

Massive MIMO relates to features employing a large number (i.e. massive) of phase and/or amplitude controllable antennas and massive MIMO is often associated with AAS. In academic work, the focus of massive MIMO has been slightly different from what is common in the mobile communication industry. When discussing massive MIMO, the academic literature commonly assumes an *extremely large* number of transmit and/or receive antennas, often together with the concept of "dirty RF," see [20], which assumes that the radio requirements can be relaxed. Within the context of this book, systems with more than eight dynamically adaptable antennas (or transceivers) are considered massive MIMO, including possibility to use frequency domain MIMO algorithms applied per OFDM subcarrier (without the "dirty RF" association), since this number of adaptable antennas is a considerably larger number than has been common in deployed base stations until recent times. Another difference is that the academic literature primarily has considered reciprocity-based beamforming and TDD in the context of massive MIMO, whereas the industry also considers feedback-based beamforming and FDD-based systems since this use of massive MIMO has great commercial interest.

In academic studies, some underlying assumptions are often too simplified to be applicable for commercial mobile networks. Hence, this simplification may lead to discrepancies between results in those studies and the results or performance from real mobile communication networks. Examples on assumptions that may be different relate to radio channel models and traffic models. Effects of such discrepancies will be occasionally discussed in this book, for example, in Section 6.7. It should thereby be emphasized that it is important to combine knowledge from all relevant fields in industry and academia to get a deeper understanding of AAS performance in real networks. Hopefully, this book can provide such background knowledge and insights to benefit both our industry and the academia.

REFERENCES

[1] P.K. Bondyopadhyay The first application of array antenna, in: Proceedings 2000 IEEE International Conference on Phased Array Systems and Technology, 2000.

[2] Nobel Media AB. The Nobel Prize in Physics 1909. NobelPrize.org. <https://www.nobelprize.org/prizes/physics/1909/summary/>, 2019 (accessed 08.10.19).

[3] F.A. Bartlett, A dual diversity preselector, QST XXV (1941) 37–39.

[4] S. Andersson, B. Hagerman, H. Dam, U. Forssen, J. Karlsson, F. Kronestedt, et al., Adaptive antennas for GSM and TDMA systems, IEEE Personal. Commun. (1999) 74–86.

[5] S. Andersson, U. Forssen, J. Karlsson, T. Witzschel, P. Fischer, A. Krug, Ericsson/Mannesmann GSM field-trials with adaptive antennas, 1997 IEEE 47th Vehicular Technol Conf Technology in Motion, 3, 1997, pp. 1587–1591.

[6] H. Dam, M. Berg, S. Andersson, R. Bormann, M. Frerich, F. Ahrens, et al., Performance evaluation of adaptive antenna base stations in a commercial GSM network, in: Gateway to 21st Century Communications Village. VTC 1999-Fall. IEEE VTS 50th Vehicular Technology Conference (Cat. No.99CH36324), 1999, vol. 1, pp. 47–51.

[7] J. Strandell, M. Wennström, A. Rydberg, T. Öberg, O. Gladh, L. Rexberg, et al., Experimental evaluation of an adaptive antenna for a TDMA mobile telephony system, in: Proceedings of 8th International Symposium on Personal, Indoor and Mobile Radio Communications – PIMRC '97, 1997.

[8] G.E. Moore, Cramming more components onto integrated circuits, Proc. IEEE 86 (1) (1998) 82–85. Reprinted from G.E. Moore, Cramming more components onto integrated circuits, Electronics (1965) 114–117.

[9] S. Applebaum, Adaptive arrays, IEEE Trans. Antennas Propag. 24 (5) (1976) 585–598.

[10] J.H. Winters, Optimum combining in digital mobile radio with cochannel interference, IEEE Trans. Vehicular Technol. 33 (3) (1984) 144–155.

[11] J. Salz, Digital transmission over cross-coupled linear channels, AT&T Technical J. 64 (6) (1985) 1147–1159.

[12] M. Viberg, H. Krim, Two decades of array signal processing research: the parametric approach, IEEE Signal. Process. Mag. 13 (4) (1996) 67–94.

[13] A.J. Paulraj, C.B. Papadias, Space-time processing for wireless communications, IEEE Signal. Process. Mag. 14 (6) (1997).
[14] E. Telatar, Capacity of Multi-Antenna Gaussian Channels, Bell Laboratories Technical Memorandum, October 1995.
[15] G.J. Foschini, M.J. Gans, On limits of wireless communications in fading environments when using multiple antennas, Wirel. Personal. Commun. 6 (3) (1998) 311–335.
[16] G.G. Raleigh, J.M. Cioffi, Spatio-temporal coding for wireless communications, IEEE Trans. Commun. 46 (3) (1998) 357–366.
[17] A. Kaye, D. George, Transmission of multiplexed PAM signals over multiple channel and diversity systems, IEEE Trans. Commun. Technol. 18 (5) (1970) 520–526.
[18] T.L. Marzetta, Noncooperative cellular wireless with unlimited numbers of base station antennas, IEEE Trans. Wirel. Commun. 9 (11) (2010) 3590–3600.
[19] T.L. Marzetta, Massive MIMO: an introduction, Bell Labs Technical J. 20 (2015) 11–12.
[20] G. Fettweis, M. Lohning, D. Petrovic, M. Windisch, P. Zillmann, W. Rave, et al., A new paradigm, IEEE 16th International Symposium on Personal, Indoor and Mobile Radio Communications, 2005, pp. 2347–2355.

CHAPTER

NETWORK DEPLOYMENT AND EVOLUTION

2

The goal of this chapter is to outline basic network design principles and key radio network evolution steps. The intention is to provide a background on how the networks are built, the principles for how they are evolved, and the methods used to achieve that. Traditional methods to increase network performance are discussed to provide some context for how advanced antenna systems (AAS) can be used and where AAS will be an effective solution.

2.1 CELLULAR NETWORKS

2.1.1 CELLULAR NETWORK BASICS

A cellular network is designed around the need to serve multiple users over a large geographical area. A cellular network consists of multiple fixed network sites, where users connect and communicate with the sites that provide the best radio signal. There is a limited amount of frequencies in the network for conveying information and these frequencies need to be reused between sites to get enough capacity. The reuse of frequencies means that neighboring sites often interfere with each other, but if the interference is managed properly then users can connect. The users are then divided over a number of sites and can be efficiently served. In Ref. [1], it was shown that interference is minimized by deploying sites on a hexagonal grid using omnidirectional antennas or three sectors with 120-degree antennas (see Fig. 2.1).

The area covered by an antenna is traditionally referred to as a cell (the concept of a cell has then evolved, see, for example, Sections 6.2.1 and 6.6.1). Further, a radio base station is associated to a site. The base station may serve multiple cells or sectors in case the site has multiple sectors,

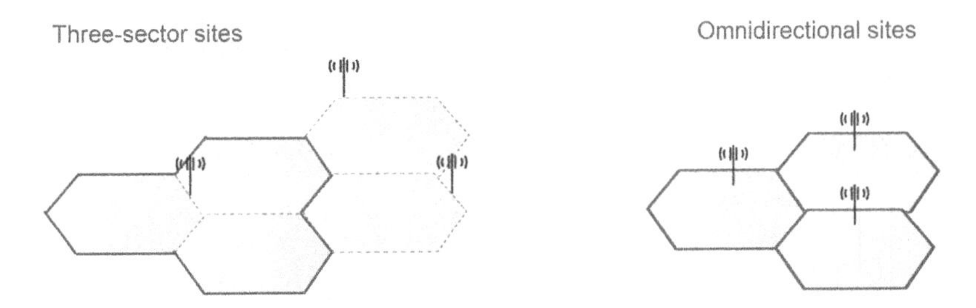

FIGURE 2.1

Base station radio sites ("Ⲯ") on a hexagonal grid. Three-sector site configurations are shown to the left, while omni configurations are to the right.

Advanced Antenna Systems for 5G Network Deployments. DOI: https://doi.org/10.1016/B978-0-12-820046-9.00002-2

while a site with omnidirectional antenna includes only one cell. The terms node B, evolved node B (eNB) (for 4G), and generalized node B (gNB) (for 5G) are also commonly used to refer to a base station, although in principle these terms refer to logical entities and not physical ones. User phones, tablets, etc., are generically referred to as user equipment (UE).

2.1.1.1 Radio communication procedures

In order to communicate, UEs establish a radio connection with a base station. The base station sends and receives data from the UE and forwards the data to a core network. The core network routes data to/from the nearest base station to a recipient, which in turn maintains a radio link in order to create a complete end-to-end connection. The link between the base station and the core network is termed backhaul (see Fig. 2.2). A description of how the underlying network connecting individual base stations is built up is not within the scope of this book, which is focused on AAS base stations. Further details on AAS architecture can be found in Chapter 12.

It is impractical to continuously maintain a radio link between each of the UEs and the base station since this would both use up radio capacity and drain the UE battery. Instead, the base station only sets up a connection to the UE when a call or data transmission is ongoing. When the base station has an ongoing connection to the UE, then the UE is said to be in *connected mode.*

When the UE is not in connected mode, it does not continuously transmit or receive information from the base station. A means is needed for the network to reach the UE and vice versa. The network reaches the UE via a procedure known as paging. A UE that is not connected periodically activates its receiver. During these occasions, UE is able to receive signaling from the network, known as paging signaling that indicates to the UE that there is an incoming session and that it should establish a radio connection with the base station. Connections to the network can also be initiated from the UE side, for example, when a user makes a call. In either case, the UE is able to initiate a radio connection to the base station by using a signaling procedure known as random access.

When first switched on, UEs need to be able to detect the presence of a network and establish contact with it. To facilitate detection of a network, each base station transmits a set of signals across the whole area that the base station covers. The first of these signals is a synchronization signal that uses one out of a set of known sequences (e.g., in Long-term evolution (LTE) there are 504 unique sequences in the set). When a UE is switched on, it can use a correlation process to search for synchronization signals across potential frequencies where it expects to detect a network. When the UE detects a positive correlation with a known correlation sequence, it has detected the presence and frequency of a network and can take further steps to connect to the network. The

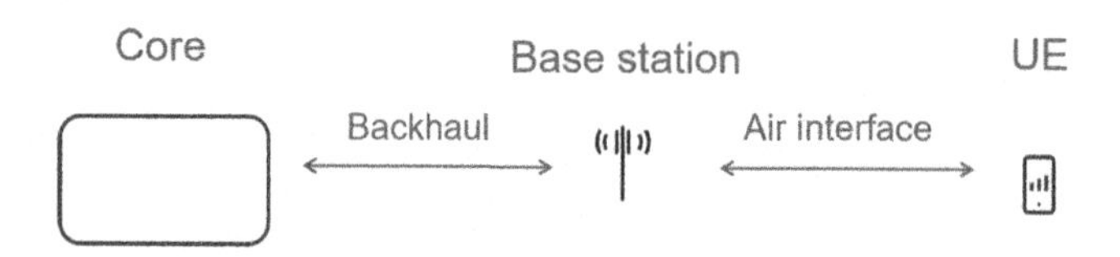

FIGURE 2.2

The UE communicates to another recipient through the core network via the base station over the air interface and backhaul links.

synchronization sequence is generally base station specific. In addition to the synchronization sequence, each base station transmits encoded information containing additional parameters needed to synchronize to the network in a so-called broadcast channel. Once a UE has detected the synchronization sequence, the UE reads the contents of the broadcast channel. This provides sufficient information for the UE to complete synchronization to the base station and transmit to the base station using the random access channel to inform the base station of its presence. An overview of the process is found in Fig. 2.3.

A key aspect of the operation of a mobile network is that users can seamlessly continue calls whilst on the move. This means that the UE maintains a connection when it moves out of the coverage area of one base station and into the coverage area of another. The process of transferring the connection from one base station to another is known as handover. To facilitate handover, UEs that have connections make measurements of the relative strength of the reference signals from different base stations and report them to the network. Based on the reports, the network may decide to transfer the connection from one base station to another.

Also, when UEs do not have active connections, it is important for the network to keep track of the cells within which UEs are present in order that calls can be transferred to the right place. To facilitate updating of the UE position, UEs also make measurements of base stations synchronization signal strength and update the network as they move between base stations. The measurements are highly optimized in order to minimize the need for activation of the UE receiver and help to prolong battery life.

2.1.1.2 Scheduling

Scheduling is another key functionality of a base station to enable user multiplexing. All connected UE served by a cell share a common pool of resources. In an orthogonal frequency division multiplexing (OFDM)–based system, this resource can be represented by a time–frequency grid where time is divided into OFDM symbols and frequency is divided into subcarriers. See also Chapter 5. Scheduling is the operation performed by the network to decide which resources are allocated to a UE for either downlink reception of data or uplink transmission of data. A control channel is transmitted to the UE to indicate the allocated resources for a data transmission/reception. See an example in Fig. 2.4 where two users are frequency multiplexed in an OFDM-based system as LTE,

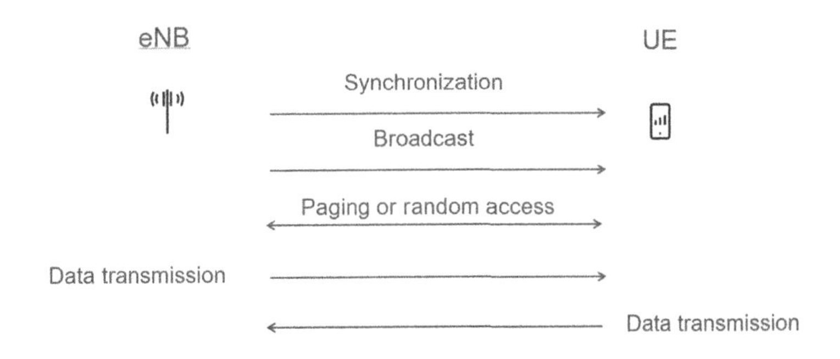

FIGURE 2.3

The process of how a UE synchronizes to eNB after it is, for example, turned on and a communication session is initiated from the eNB or UE.

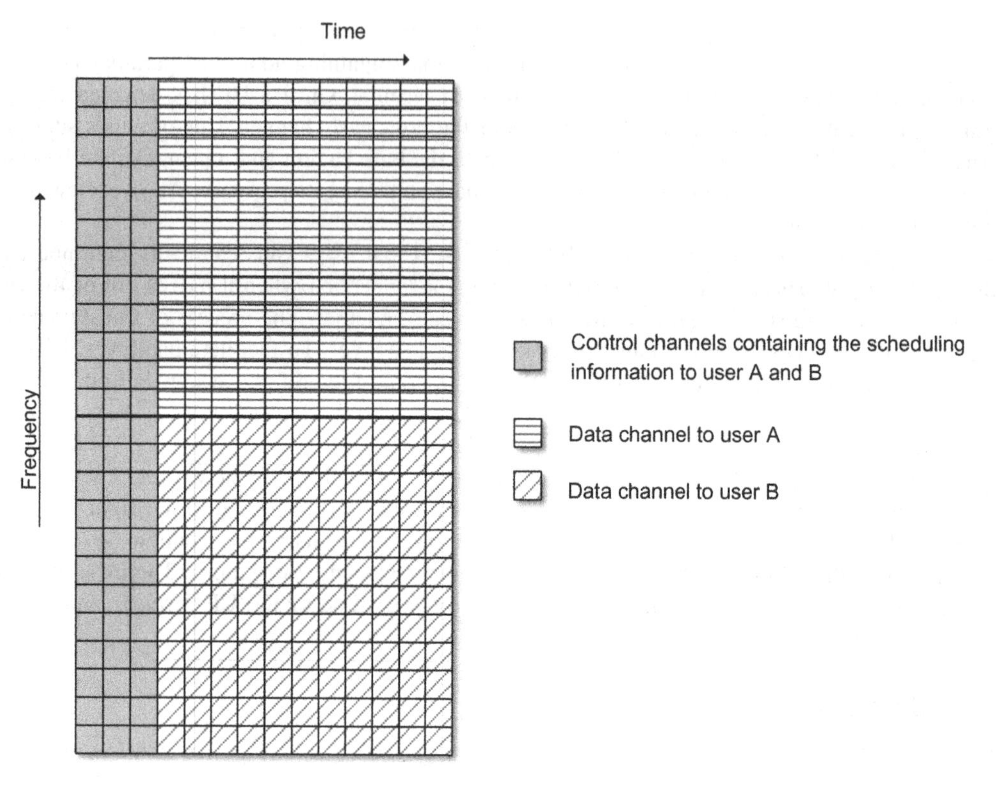

FIGURE 2.4

Scheduling of resources in a time–frequency grid of an OFDM system, to two different users.

where they first receive their control channels in a common region and then receive the data channel in their respective data regions.

The scheduler complexity has increased over the years of cellular communication. In global system for mobile communications (GSM), users are time-multiplexed and have periodically occurring resources, which implies a very simple scheduling task. In code division multiple access (CDMA)–based networks of 3G, both a time allocation and a spreading code are assigned to the UE by a scheduler. In OFDM-based 4G and 5G systems, a resource is instead a time and frequency resource as mentioned above. In addition, the spatial domain, facilitating multiple MIMO (multiple-input multiple-output) layers, constitutes an important third dimension of the shared pool of resources. The fine granularity of the addressable resources in OFDM allows for the introduction of *opportunistic scheduling* in frequency, where the scheduler also considers knowledge of the channel. For example, the scheduler can choose to transmit data to a certain UE only in those selected frequency subbands where the channel provides a high signal-to-noise ratio at the UE. This is known as frequency selective scheduling and is supported in LTE and new radio (NR). By the same principle, the scheduler can decide to transmit data selectively in the spatial domain, using the beam provided by a multi-antenna array.

Moreover, as channel quality information becomes available (how this is achieved in practice, please refer to Section 6.4), scheduling also performs a dynamic adjustment of the targeted spectral efficiency of the transmission over the wireless link, known as *link adaptation*. Hence, the scheduler decides on a code rate for the channel encoder, the modulation constellation to use [e.g., quadrature phase shift keying (QPSK), 16-quadrature amplitude modulation (QAM), etc.], and in the MIMO case, the MIMO precoders and the number of MIMO layers.

Hence, the trend over the years has moved toward a scheduler that considers very recent and accurate time–frequency–spatial channel information for each UE, plus takes into account the amount of data in buffers (and its latency tolerance), waiting to be transmitted to/from each UE. As the channel, interference, and traffic are constantly varying, the scheduler must make scheduling decisions on a short time scale on which UE to prioritize based on many different factors. Hence, the scheduler is the "brain" of the base station, which solves a multidimensional resource assignment problem, see Fig. 2.5.

There is a need to transmit signals that can be reached by terminals in the entire cell as well as an opportunity to optimize communications to individual links. This will motivate both the need for UE–specific beamforming as well as cell-specific beamforming, see Sections 6.1 and 6.2.

2.1.2 NETWORK HISTORY

Analog mobile systems were introduced in the early 1980s using the 450–900 MHz bands. Examples include *Advanced Mobile Phone Service (AMPS), Total Access Communication System (TACS), and Nordic Mobile Telephone (NMT).* These systems are often referred to as the first-generation mobile systems, that is, 1G. The systems were built with the purpose to provide voice

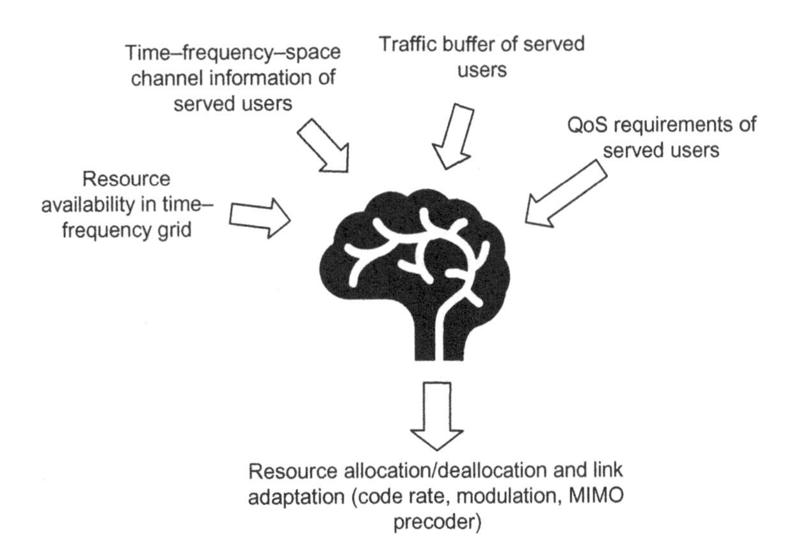

FIGURE 2.5

The scheduler is the "brain" of the base station, taking inputs from various sources and decides on a millisecond basis which users get data and what rates and resources they are allocated.

services. The main design criterion was to provide ubiquitous coverage over wide areas. Radio equipment and site construction were expensive, so the radio networks were designed to maximize coverage with a minimum number of radio sites using the hexagonal pattern (see Section 2.1.1.1).

The baseline network deployment choice was typically to use hexagonal three-sector sites. In practice, it is difficult to maintain hexagonal grids. The desired and optimal site locations cannot always be secured due to commercial reasons or deployment restrictions. For example, varying building heights make sectors in certain directions less useful. As traffic demand increased, the mobile network operators added new sites, that is, made denser deployments where the subscriber density was higher.

2.1.2.1 Introduction of 2G

In the beginning of the 1990s, 2G digital mobile systems such as GSM, time division multiple access (TDMA), and personal digital cellular (PDC) were introduced in bands around 900 MHz. They were all designed for circuit-switched voice, short messages, and low-rate data services (initially circuit-switched, later packet-switched). These networks were built on the same site grid as the analog systems where such existed. In many countries, GSM was the first system deployed. Either way, the networks were mainly built to provide ubiquitous coverage for voice services in the same way as the analog networks. GSM was quickly adopted globally, which led to an increase in volumes of network equipment and UEs. In developing countries, there was also a rapid subscriber uptake, although somewhat later in time.

To meet the increasing demand, new frequency bands were also made available. For example, for GSM, 1.8 GHz was added outside the Americas. Other 2G technologies followed a similar path of adding new frequency bands as a step in the capacity evolution. Coverage was still the most important network design criteria, but capacity became increasingly important.

Multi-antenna for uplink receiver diversity was commonly applied in 2G. Multi-antenna for downlink was used in some rare cases (e.g., antenna hopping and transmitter diversity).

2.1.2.2 Introduction of 3G and high-speed packet access

When developing 3G, the use of higher-rate mobile data was envisioned. 3G did not trigger any great change in mobile phone usage initially. A few years after the initial 3G deployments, HSPA was introduced as a method for packet data access as well as providing an order of magnitude higher data rates, which was a system enabler for technology change. HSPA together with the introduction of smartphones in 2007 had a strong catalytic effect on the consumer market and a whole new ecosystem started to grow. As a result, the mobile Internet era with ever-increasing data usage took off.

The traffic characteristics changed rapidly from being voice centric to being packet data centric. Apart from Internet access and web browsing, many new applications for social networking, gaming, video, and music were introduced with the new smartphones. The new applications generated much more traffic than the early voice and low data rate services. Smartphones were also used as the main Internet access tool in many countries with underdeveloped fixed broadband infrastructure. The result was a rapid shift to smartphones and increased traffic demand relative to GSM and the initial 3G introduction.

3G was generally introduced in frequency bands around 2 GHz. As a result, a denser site grid was required to provide coverage relative to the low-band 2G technologies (<1 GHz). Due to the many sites deployed, a good network capacity baseline was achieved from start and it was easily improved by adding more 3G carriers when the traffic demand from smartphones increased. Low-band (e.g., 850–900 MHz) 3G variants were later introduced as a means to primarily enhance coverage (especially deep indoor).

Multi-antenna for receiver diversity at base station was commonly applied in 3G.

2.1.2.3 Introduction of 4G

The work on a new wireless access technology, 4G, started after HSPA with targets such as simplicity, higher speeds, and enhanced capacity. As opposed to 2G and 3G, 4G was designed to use larger bandwidths and operate in many frequency bands and included multi-antenna technologies already from the start. This facilitated large bandwidths that lead to increased capacity and higher speed compared to 3G. 4G was initially designed to be used both in existing 2G and 3G bands and in new frequency bands, both low bands below 1 GHz (e.g., 700–800 MHz) and frequency bands around 2 GHz (e.g., 2.6 GHz). In America, LTE was successfully introduced at 700 MHz, which made a perfect coverage fit to the GSM/CDMA in 850 MHz on existing sites. In many countries outside the United States, LTE was deployed in 1800 MHz to a large extent, with better coverage than 3G at 2.1 GHz.

The introduction of 4G accelerated the use of smartphones, which leads to further need for capacity. In the early era of LTE, 2G and 3G were used as a voice complement as the voice capabilities in LTE (voice over LTE) were deployed later compared to its data operation.

LTE is often being migrated into the low-frequency bands intended for 2G and 3G such as 850 and 900 MHz in order to improve LTE coverage, especially in markets lacking a new coverage band intended for LTE. The need for increased LTE capacity as well as the need for increasing bandwidth to provide higher user throughput by means of carrier aggregation is another driver for migrating LTE into existing 2G and 3G bands around 2 GHz, in particular when the population of UEs with only 2G and 3G support was becoming small enough.

In most cases, LTE inherited its site deployment baseline from 2G and 3G operations. So-called green field operators with only 4G are very few. Lately 4G has experienced a shift toward higher frequency bands and larger bandwidths, but apart from that, deployments remain basically the same.

Multi-antenna for uplink receiver diversity and downlink MIMO were commonly applied in 4G

2.1.2.4 Introduction of 5G

In 2019, 5G was launched in several markets around the world. In Europe and Asia, 5G deployments based on 3.5 GHz are available, for example, in United Kingdom, Germany, Switzerland, and South Korea. In the United States, there are 5G currently (2019) networks based on both mm-wave and 2.5 GHz bands.

2.1.3 SPECTRUM

Spectrum is the key asset used in wireless communication for electromagnetic transmission over the air. Each transmission occupies a channel bandwidth that depends on the data rate and the carrier modulation method used. The channel bandwidth has a center frequency referred to as carrier frequency.

The spectrum in a country is shared among all users of the spectrum, for example, land mobile radio, satellite communication, radio, TV, mobile communication, Wi-Fi, military use, etc. Spectrum is in most countries treated as a national asset that is managed by authorities on a national level. Over time, the need for spectrum has increased and, in most countries, spectrum is a scarce resource and the national authorities often charge substantial amounts of money to issue licenses for commercial services. Therefore, spectrum is one key asset for network operators.

Mobile communication makes use of spectrum of different bandwidths in various frequency ranges for different mobile technologies. In the beginning, the bandwidths were limited as the usage at the time was limited. As the load and the user experience expectations in the networks have increased, the need for more spectrum has dramatically increased and there is a continuous search for new available spectrum.

2.1.3.1 Global alignment

Since the spectrum allocation and management are handled on a national level, there are differences in how different countries have allocated spectrum. It was however observed early that there are great benefits for the mobile communication industry from the harmonizing of spectrum between different countries to enable international roaming and better economy of scale. Therefore, processes for aligning spectrum for mobile use globally have been established. The *World Radiocommunication Conference* (WRC), which is a forum arranged by the *International Telecommunications Union* (ITU), is held every 4 years and has, as one of its main purposes, to align spectrum to be allocated for new generations of mobile systems. There are also other initiatives to support spectrum alignment, whereof some are on regional level.

2.1.3.2 Characteristics

In lower spectrum, coverage will be easier to achieve. Therefore, lower spectrum is considered valuable to get good coverage. For higher bands, the coverage is more challenging to provide. The characteristics of wave propagation will be discussed in detail in Chapter 3.

The reason to move toward higher frequency bands is that there is a limited availability of spectrum in the lower band range, especially as each band is typically shared by several users or operators. Hence it is important for the operators to acquire large amounts of spectrum to be able to provide good services to many subscribers at the same time. In the higher spectrum range, the spectrum in many countries is not fully utilized and there are larger chunks of spectrum available.

For convenience the range of spectrum has been categorized as low, mid, and high band. The definitions of these ranges are not precise and have changed over time as the target spectrum for mobile communication has changed. In this book, the following definitions are used (see also Fig. 2.6):

- Low band: <1 GHz
- Mid-band: 1–7.125 GHz
- High band: >24 GHz

2.1.3.3 Spectrum evolution from 1G to 5G

For the early mobile system deployments in 1G, the most important aspect was to provide coverage. Therefore some narrow bands at 450 MHz [e.g., 3 MHz frequency division duplex (FDD) per

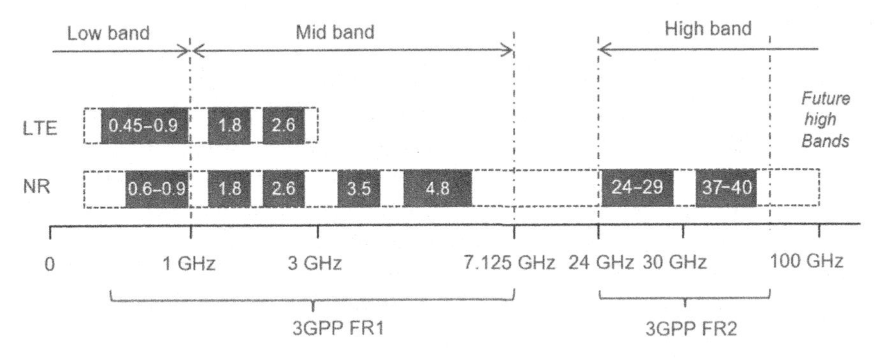

FIGURE 2.6

Illustration of frequency band definitions. FR1 and FR2 are used in 3GPP. Some example bands such as 1.8 and 2.6 are also included.

operator in Sweden] and later 900 MHz were allocated. That was sufficient at the time, since the number of subscriptions was limited.

For GSM, the 900 MHz band was initially used in most countries. Later, when more capacity was needed, the 1.8 GHz band was also allocated.

In order to get a global standard for 3G, the organization *3rd Generation Partnership Project* (3GPP) was formed, which among other things specified the spectrum to be used. The main 3G spectrum was to become Band 1, that is, 2.1 GHz (Downlink) and 1.9 GHz (uplink). 3GPP has since then specified all frequency bands, also for other technologies.

For LTE (4G), the main new bands were allocated around 2.6 GHz both for time division duplex (TDD) and FDD, but also bands around 700–800 MHz were used. With the introduction of LTE, a more rapid re-farming process of GSM spectrum also occurred. For example, Band 3 (around 1.8 GHz) was re-farmed for LTE in many countries.

When starting the spectrum planning for 5G, the goal was to find new large chunks of spectrum to support the evolution of mobile broadband (MBB) and introduction of other new services, for example, see Section 2.1.6.2. The mid-band spectrum range offers bandwidths in the order of 100 MHz, while the mm-wave range provides bandwidths of hundreds of MHz.

In 3GPP rel. 16, spectrum is divided into two frequency ranges (see Fig. 2.6):

- Frequency range 1 (FR1): Existing and new bands in 0.41–7.125 GHz
- Frequency range 2 (FR2): New bands from 24.25 to 52.6 GHz

FR2 is usually referred to as "high band" or "mm-wave" as the wavelength is in a range from about 5–12 mm. In this range, more directional antennas are required to maintain coverage, see Chapters 3, and 7. This requires substantially different solutions, both with respect to functionality and implementation. The suitable use cases are also very different.

As older network generations, 1G, 2G, and 3G, become less popular, the corresponding spectrum is gradually re-farmed for use by later generations, 4G and 5G, which are soon to become the main technologies in most bands. Some spectrum is however usually kept for the remaining users of the older technologies.

2.1.4 CURRENT NETWORK DESIGN

The current radio network deployments are largely built on the earlier radio network designs, that is, 2G and 3G. In some countries, 3G frequency band licenses came with regulatory requirements on coverage, which resulted in many new sites compared to 2G and thereby a high site density. In other markets (such as United States), the site design originates back to a voice-only deployment at 850–900 MHz in many cases. Both 3G and 4G were designed to operate under these circumstances but the result is a lower site density and a lower area capacity baseline.

The basis of most networks today is three-sector macro site deployments on rooftops in urban areas. Mast and tower deployments are typical in residential suburban and rural areas. The site density follows the traffic demands and hence in very crude terms the subscriber density during the busy hour. The site density is the highest in urban areas and typically decreases in suburban and rural areas. Urban and suburban/rural scenarios typically have different radio propagation challenges. The urban environment is characterized by relatively low antennas, many large propagation obstacles, and large, highly attenuating buildings, while the suburban and rural environments have taller antenna installations, fewer propagation obstacles, and buildings with small footprint and wall types that are easier to penetrate, see Sections 3.5 and 14.2.

Due to several factors, the networks have evolved in a less than systematic way, and today there is often only an approximate resemblance to hexagonal site grids. There are many practical limitations, for example, availability of locations for new sites, possibility to build three-sector sites, topography, and uneven area capacity needs that make the site plan to deviate from the ideal hexagonal appearance. Requirements on limited visual impact of large antenna installation also affect the site deployments especially in urban areas.

As a complement to the basic macro site grid, small cells and dedicated indoor deployments provide additional coverage and capacity where needed, see the following sections.

2.1.4.1 Small cells

The capacity demand in most dense urban areas is generally very high due to high concentration of users in geographically limited areas such as central business districts with high-rise buildings or busy areas that attract a lot of people, such as areas with shopping, restaurants, and nightlife. In such areas the site-to-site distance often goes below 200 m. Adding more macro sites can be problematic in these areas from the perspective of finding feasible sites. Instead, a proposed solution is to deploy small cells on street level or near street level to add coverage and capacity. The small cell is typically a small form factor base station that often uses lower power intended to cover a limited area compared to a traditional macro base station. A small cell can be very effective in offloading a macro site if it is placed in the correct location. However, this can be difficult to achieve in practice since the location of subscribers is generally unknown on a detailed level.

2.1.4.2 Indoor systems

In-building deployments play an important role in providing good indoor performance in many parts of the world today. Large footprint buildings with high building entry losses that are difficult to cover from outdoor macro sites is an example of a coverage-driven deployment suitable for in-building solution. Crowded public venues such as a train station or a stadium are capacity-driven examples of where indoor deployments fit well. *Distributed Antenna Systems* (DASs) are currently the most common solution used for indoor deployments. DASs also facilitate multioperator use. In case of passive DAS, the radio equipment is located far away from the antennas, so the cable losses can be significant. In recent years, moving the radio equipment to the antenna location has become more popular, since this almost eliminates cable losses.

Other examples of areas targeted by indoor solutions are underground/subways and road tunnels. Leaky cables and repeaters are solutions that are commonly used in these scenarios. These solutions are similar to passive DAS.

Today, very high capacity demand in limited areas is often present in, for example, stadiums, at music festivals and train stations with high subscriber density. 5G based on mm-wave bands may be a good fit for such venues as the expected operator allocations in this band (up to 1000 MHz) can provide a future proof capacity solution (see Section 7.2).

2.1.5 CURRENT USE OF MOBILE BROADBAND

Since the early 2000s, the cellular traffic has shifted from predominantly voice to being predominantly packet data and smartphone based, often referred to as mobile broadband (MBB). It consists of many different application types such as email, web browsing, messaging, social networking, gaming, music streaming, and video. The MBB traffic is people centric and therefore the spatial traffic distribution typically follows the population density in, for example, suburban and urban areas or daytime population in dense urban areas where few people actually live. Due to costs, mobile networks are often tailored to provide population coverage as well as covering areas that are frequently visited by people (roads) rather than ubiquitous area coverage. Moreover, people tend to spend most of the time indoors (e.g., at work or in school and at home) [2]. This means that the majority of traffic is generated indoors, hence the ability to provide indoor coverage is particularly important for MBB deployments.

2.1.6 USE CASES OTHER THAN MOBILE BROADBAND

LTE (later releases) and 5G are designed for many other types of traffic than MBB, for example, fixed wireless access (FWA), *critical machinetype communication* (C-MTC), and *massive machine-type communications* (M-MTC). The uptake of such use cases is expected to grow over the years to come.

Solutions for FWA will be discussed in Section 14.3.4 and it will be shown that AAS can be very useful in FWA. C-MTC and M-MTC are currently under study and AAS can especially be considered for C-MTC due to high performance of AAS.

2.1.6.1 Fixed wireless access

FWA is a way to provide broadband access to households through use of wireless systems (see Fig. 2.7). FWA is a cost-efficient way of providing Internet access in areas with underdeveloped or no fixed broadband infrastructure, especially if existing sites, radio equipment, and radio spectrum deployed for MBB are reused also for FWA [3]. FWA can be based on 3G, 4G, and 5G depending on the FWA needs. An FWA use case example is that FWA based on LTE can provide much higher speeds than DSL with poor connection speed at long distances from the exchange. The need for high speed is important as web content is getting more data heavy. Another FWA use case is to

FIGURE 2.7

Fixed wireless access is when a cellular system is used for providing broadband access to households in an area. Outdoor deployed customer premises equipment (CPE) is shown in the figure.

connect households in an area totally lacking fixed infrastructure and there is no other option for Internet connectivity. LTE-based FWA is a then a good option due to low costs relative to fixed broadband technologies.

FWA plays an important role also in developed broadband markets where it can be a substitute, for example, for fiber [4]. Fiber is costly to deploy in residential areas with single-family buildings.

Often fixed broadband subscriptions are offered with unlimited data, while MBB subscriptions often use data buckets. FWA must be able to handle much higher capacity and data consumption than regular MBB services. FWA should be dimensioned in the same way as fixed broadband, which is different from MBB generally. An FWA connection to a household is typically shared between several users and can be used for delivering TV/video over the top. As a result, the requirement of demand per subscription and data rates for FWA can be much larger compared to MBB. 5G and AAS are hence means that are particularly important for FWA.

2.1.6.2 New use cases

Mobile communications have up to, and including 4G, predominantly been serving the general public with MBB services. The mobile network operators are in almost all countries the only holders of licensed spectrum for mobile communication and hence the only provider of such services.

When preparing for the standardization of 5G, it was early recognized that 5G not only should be delivering MBB services better but also should address new areas of use. Whereas MBB services cover human needs very well, the new, identified services of potential interest were largely in the nonhuman domain, that is, MTC. Two main categories of MTC were identified. M-MTC services, which are supposed to serve very large numbers of nonhuman devices, for example, sensors and meters. The characteristics of such devices are that the number of devices is potentially very large (many devices per human), do often not require high volumes of data and are often located in places with poor coverage. The other category, C-MTC services need to provide critical communication services with very high quality, for example, very high availability ($>99.999\%$), very high reliability, very low (and stable) latency, and possibly very high bit rates. The intention is to provide services of business-critical use, for example, process control and mission-critical use, for example, *automatic guided vehicles*.

The intention is that mobile systems should be designed with the performance and stability to provide various kinds of industries with high-quality communication systems that can serve as generic communication systems for industries [5]. Through the design to cater for a very wide variety of requirements, 5G can be widely applied as the communication technology of preference in society at large, covering essentially all fields that can make use of wireless communication.[1]

Later releases of LTE, that is, advanced LTE++ (3GPP Rel. 14, 15, and 16), have also addressed M-MTC and C-MTC in various ways and do support these services to the greatest extent possible. In 5G, however, the standard has accommodated these services from start and can hence support them to a much higher degree.

2.2 NETWORK PERFORMANCE

The reason for introducing AAS is to improve network performance, which will be further explained in Section 2.3.4. Therefore relevant aspects of network traffic characteristics and performance will be introduced.

2.2.1 NETWORK TRAFFIC

The MBB network traffic growth depends on two main factors, the number of smartphone subscriptions and the average data consumption per smartphone. The latter is often expressed in monthly data per subscriber, for example, GB per month, and is often used to classify smartphone subscriptions by mobile operators. Today, the smartphone penetration is close to 100% in many mature markets and hence the traffic growth is driven mainly from an increased monthly usage in these markets.

The increasing data demand per smartphone is due to several reasons. The growth of the data per smartphones is stimulated by increased use of video and that the video resolution offered by applications is improved, which increases the data volume for a given video clip length. There is a trend toward better and better video resolution, for example, moving from standard definition (SD) to high definition (HD), and 4K ultrahigh definition (UHD), that also increases the data volumes. For example, Netflix states that SD video consumes 0.7 GB/hour, while HD and UHD video is in the range of 3–7 GB/hour for their movie offerings [6]. Another reason for increased data demand is that the data size of web pages is getting larger due to more page details with more pictures as well as embedded with videos and ads [7]. An implication from these effects is that also the speed or data rate to deliver the data must be increased in order to have the same waiting time. There are generally expectations by the consumers that the networks should provide higher speed and shorter delay in addition to higher capacity.

The global yearly traffic growth has been in the range of 50%–80% in the last 5 years according to Ericsson Mobility Report [4] (see Fig. 2.8). This means that network deployments should

[1]To be noted, some fixed communication technologies, for example, optical fiber communication, still have higher capabilities w.r.t. bit rates, latency, and quality, and therefore there will always be a room use of these where the requirements are truly extreme. The main application for these technologies will however be in the backhaul network rather than communication to the end user devices.

FIGURE 2.8

Measured global data and voice traffic per quarter and year-on-year growth from 2013 to 2019. The voice traffic is insignificant relative to data.

Data from Ericsson Mobility Report.

typically be prepared to meet this level of traffic growth. Note also from the figure that the circuit-switched voice traffic is very small relative to the smartphone data traffic (including VoIP).

The data traffic per smartphone differs also depending on the region in the world. Ericsson Mobility Report [4] makes future projections on this (see Fig. 2.9). It can be seen that the traffic per device is highest in mature markets such as North America (e.g., United States) followed by Western Europe (e.g., United Kingdom).

2.2.2 TRAFFIC PATTERNS

MBB is the traffic that dominates in mobile networks today. It typically originates from many different application types such as email, web browsing, messaging, social networking, gaming, music streaming, and video.

In detail, most MBB traffic types are often very bursty and consist of many small data packets. For a user there are typically many short packages in a burst (e.g., protocol messages and user

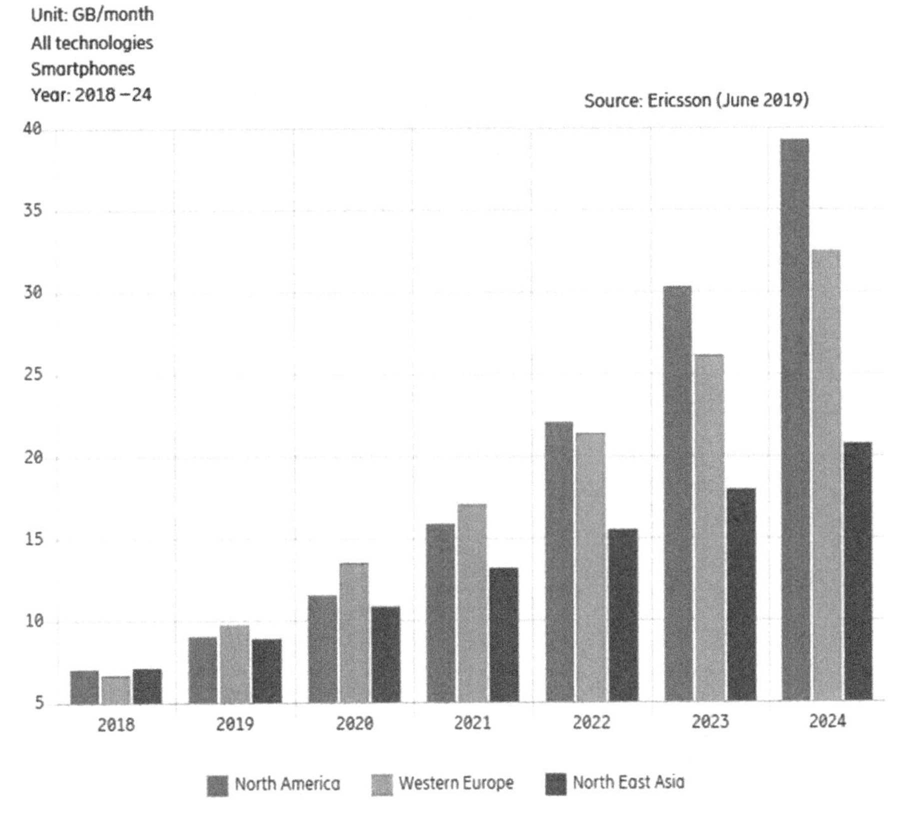

FIGURE 2.9

Monthly data traffic volume (GB per month) per smartphone and region, 2018–24 prediction.

Data from Ericsson Mobility Report.

data) followed by a longer period of silence (see Fig. 2.10). The result is typically a large number of users connected to a cell but the number of actively transmitting users is small in most cases.

In Ref. [8] it was reported that most data come in burst sizes of 250 bytes or smaller when measured on a network level. This is also supported from, for example, Ref. [9], showing packet sizes of 100–400 bytes mostly in the downlink with longer packet interarrival times for background, messaging, and gaming applications. Interactive services such as web browsing and video include more data [8,9], for example, downlink packet sizes of 1500 bytes are not uncommon, which is the maximum Ethernet transmission unit size with shorter packet interarrival times. As a result, video can be expected to have a less bursty behavior than other typical MBB applications. Traffic sessions can take several forms and it is difficult to predict the required data volume (capacity) from the number of connected users and vice versa. There are many dependencies affecting the relation, for example, object sizes, session timers, and traffic profiles. Nevertheless, by classifying sessions

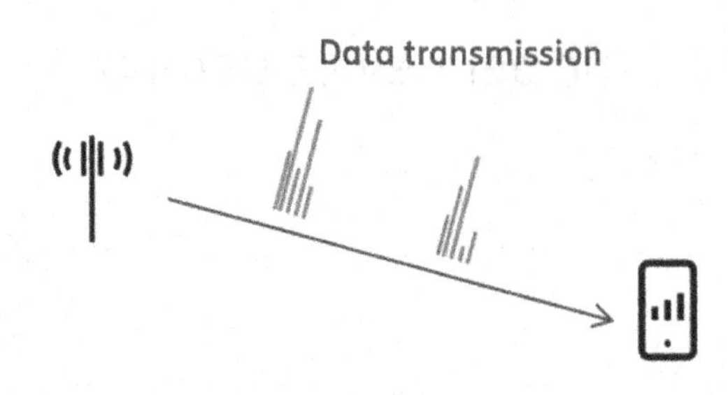

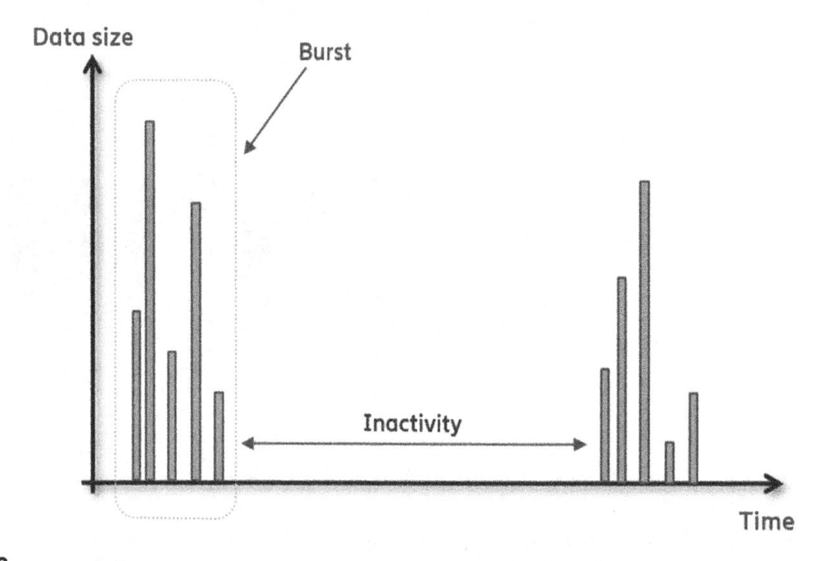

FIGURE 2.10

Download data transmission from a base station to a UE where the typical MBB user traffic behavior is characterized by that some data packets are generated followed by a period of silence.

into three groups referred to as *system information* (<100 kB), *web browsing* (100 kB to 10 MB), and *video* (>10 MB), MBB network measurements typically depict a pattern according to Fig. 2.11. Most of the sessions consist of small packets but their contribution to the total data volume is small. Most of the data volume is consumed by very few but large sessions.

Measurements have shown that typically 80%–90% of MBB traffic goes in the downlink in many networks today. This can be explained by that video and web browsing are typically data heavy and downlink centric. Due to this, it is common that TDD deployments employ downlink heavy slot formats. However, good uplink performance is still important in order to support the downlink data delivery. For example, downlink packets must be acknowledged via the uplink. This is especially important for the transmission control protocol.

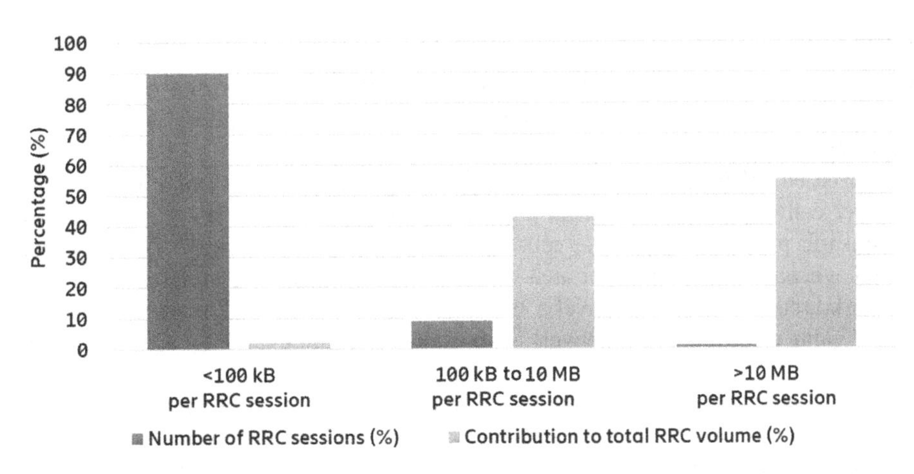

FIGURE 2.11

Example of relations between number of sessions and traffic volume.

2.2.3 PERFORMANCE METRICS

Mobile network performance is often expressed in terms of *coverage*, *capacity*, and *UE (or user) throughput*. These concepts are defined and discussed in Sections 2.2.3.1–2.2.3.3, followed by an illustration in Section 2.2.3.4.

2.2.3.1 Coverage

A UE is said to have coverage if it can maintain a connection with the network. This means that both control information and UE data can be reliably exchanged between the network and the UE.

Mobile networks are often designed so that the control channel coverage is not limiting the coverage and therefore network studies focus mostly on data channel coverage. Coverage is a key performance metric when dimensioning a network as it relates to the number of required sites and therefore also directly affects the total network roll-out cost. Coverage matters for all UEs from day one when deploying a new network, while network capacity can grow gradually as the network load increases.

Different metrics reflecting coverage can be envisioned, but they all try to quantify how reliably UE data and control information can be transmitted.

A common measure of coverage is to look at the achievable UE throughput for the worst UEs in the network, the so-called cell-edge UEs. The definition of cell-edge depends on how good coverage the operator wants to dimension the network for, but a typical definition is the 5:th percentile worst UEs in the network. Combining multiple bands may be used for providing coverage in a network.

Another related coverage indicator is to assess how many of the UEs in the network that satisfy a minimum UE throughput requirement at low network load. The minimum UE throughput requirement depends on end-user expectations and the supported services. For example, a video service would typically require a speed of at least 10 Mbps. This is often referred to as App coverage.

2.2.3.2 Capacity

Capacity is a key metric to decide how many UEs or how much traffic the network can sustain with maintained coverage. The capacity is often of high interest for operators as it reflects the production cost (cost per bit).

For packet-based MBB services, capacity is often expressed as the average served traffic (bits per second) per cell or per area unit given a certain resource utilization or UE throughput requirement at a specific percentile, often the cell-edge user percentile. Average served traffic per cell (bps/cell) and average served traffic per area unit (e.g., bps/km^2) are often referred to as cell capacity and network area capacity, respectively. It is also common to normalize these metrics with the network bandwidth to estimate the network's spectral efficiency expressed as, for example, bps/cell/Hz.

Like for the coverage discussion, the choice of UE throughput requirement where capacity is measured depends on the desired end-user experience. The resource utilization, typically defined as the percentage of used resource blocks averaged over time and frequency relative to the total number of resource blocks, is preferably chosen to reflect the conditions of normal network operation, but a typical choice in performance evaluations is $\sim 50\%$.

2.2.3.3 UE throughput

The UE throughput reflects the user experience and is closely related to coverage and capacity. The throughput will differ between UEs and depends on aspects such as their respective link quality, total cell or network traffic load, scheduling, system assumptions, etc.

It is common to assess UE throughput gains for different user percentiles, for example, the 5:th (cell-edge), 50:th, and 95:th percentiles, at specific network loads. An alternative is to analyze the UE throughput distribution for different levels of network load or resource utilization. Data coverage is then often assessed at low or normal load and data capacity at high load.

2.2.3.4 Illustration of performance metrics and their relations

The absolute performance measured by different key performance indicators is important when evaluating different network evolution and migration paths to understand whether the network can satisfy future needs, while relative performance gains are more useful when comparing different AAS solutions. Relative performance evaluations tend to be more accurate and robust against various system and modeling imperfections.

The performance metrics discussed in Sections 2.2.3.1 to 2.2.3.3 are illustrated in Fig. 2.12, showing UE throughput as a function of served traffic for a relevant UE percentile and for two different network configurations. The aim is to assess the performance gains of a more capable network relative to a reference network in terms of capacity and UE throughput. Capacity gains are read from the x-axis and UE throughput gains are measured from the y-axis.

2.2.4 PERFORMANCE EVALUATIONS

Link budget analysis is a commonly used tool for evaluating network coverage or the maximum cell range. It basically involves calculating all gains and losses from the transmitter to the receiver, for example, from eNB to UE, for determining the maximum signal loss that can be translated into

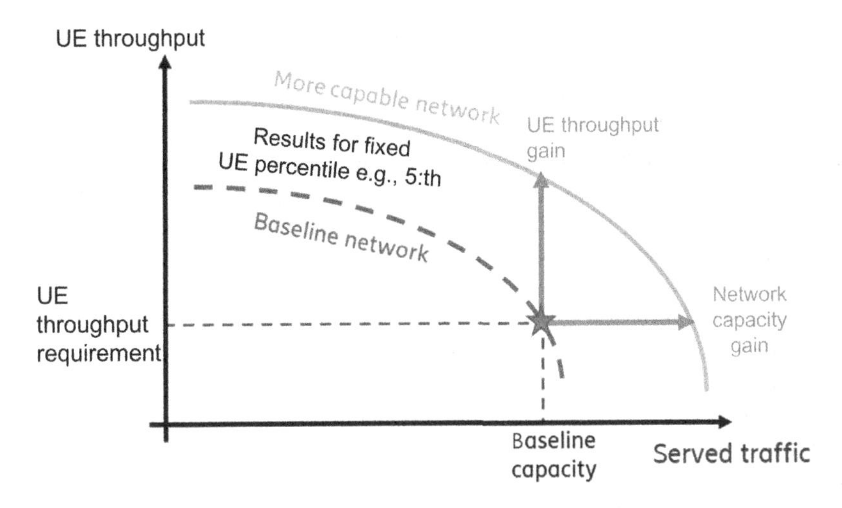

FIGURE 2.12

Illustration of capacity and UE throughput gains for a more capable network relative to a baseline network. The capacity gain is obtained by considering the cell-edge performance (5:th user percentile) and is measured from the x-axis relative to a specific load or throughput requirement for the baseline network. UE throughput gains are measured from the y-axis relative to the same baseline point. UE throughput gains are often given for 5:th (cell-edge) and 50:th user percentiles.

range using a propagation model. The maximum signal loss is usually a function of the desired data rate at cell-edge for, for example, MBB services. The link budget analysis only tells about coverage and not about capacity or distribution of user data rates.

Simulation is another tool for evaluating performance with different radio network technologies or features. Throughout the development of technologies for mobile networks, simulations have been very valuable.

Simulations can be essentially made on two levels. There are link level simulations, considering the radio link between one base station and one UE. These are used to evaluate the link performance using very detailed models of the transmitter, receiver, and radio channel. Link performance results are used as input to link budget analysis used for understanding coverage. In addition, there are radio network simulations, which evaluate the performance of networks with many base stations and many UEs taking the impact of interference and traffic load into consideration. AAS performance evaluations in this book focus on network simulations (see Chapters 13 and 14).

2.2.4.1 Network simulations

Results from network simulations are used for performance evaluations and will be frequently referred to throughout this book.

Network simulations can be time dynamic or time static depending on what is targeted to be studied or important to consider for the analysis. Time dynamic simulations typically model time explicitly and show performance over time. This is required when time aspects are an important part of the analysis, for example, when studying the impact from different traffic models where traffic intensity varies over time, or when comparing scheduling approaches.

Time static simulations analyze the network under a specific load or interference situation, that is, performance is assessed by taking a snapshot of the network under a specific interference situation. It can be used for estimating coverage or user performance in the network as a function of traffic load or utilization of the radio resources. For this, simplifications of the traffic and scheduling models are made but the advantage is often much shorter computational time. The static simulations are often used to determine network capacity or spectral efficiency.

2.2.4.2 Modeling assumptions

Simulations are to a large extent based on modeling in order to reduce complexity and increase computational speed.

Detailed modeling of the radio propagation is needed. The radio propagation is very complicated but relevant radio channel properties need to be modeled. Short-term characteristics such as fading, multipath, and other propagation-related properties need to be modeled according to the adequate environmental conditions. Network simulations usually use a simple model of the layer 1; often it is based on link level simulations that include multipath. For AAS, spatially related propagation phenomena are very important. This will be described in Section 3.6.

Performance results will also depend on the geographical location of the users and these are generally not known in detail. For example, MBB users are located indoors more often than outdoors and there is a larger density of users in urban office areas during day times. A typical choice is to assume 70%–80% indoor users in urban evaluations. This assumption can impact the AAS performance significantly. For example, a macro scenario with many users indoor in high-rise buildings will favor an AAS capable of vertical beamforming. Also, coverage is generally substantially reduced for indoor users compared to outdoor users in macro deployments.

Characteristics of the traffic generated by a specific user needs also to be modeled. A simple traffic model is the "full buffer" model, where the user is constantly transmitting data over the air interface. The traffic pattern of real users is very different. For most MBB applications, the amount of data to be transmitted is small, hence access to the data channel is only required for a fraction of the time. Therefore the performance of more realistic models is quite different from the full buffer model. For example, the cell capacity of full buffer traffic is dominated by the users with the best radio links. For a case where users want to transmit the same amount of data (i.e., equal buffer), the cell capacity is determined by the users with poor radio links since they will spend a longer time in the scheduler consuming more radio resources. See also Sections 6.7.2, 6.7.4.1, and 13.2.5.

Moreover, the interference characteristics are also strongly dependent on the traffic model, see Fig. 2.13, which will affect performance.

Another important aspect of the results is the network deployment model. In many network simulations, hexagon network models are applied. The hexagon models usually use cells of equal size and antenna deployed at the same height as well as uniform traffic demand over the area. This is generally optimistic as in reality there are often limitations on where the operators can put the sites and the user distribution is uneven. However, in analyses where different technical solutions to a problem are to be compared, hexagon models are still useful. 3GPP has decided on some deployment scenarios for use in standardization work [10]. For advanced deployment analysis, real map data are often utilized including details on buildings, roads, and clutter types to improve

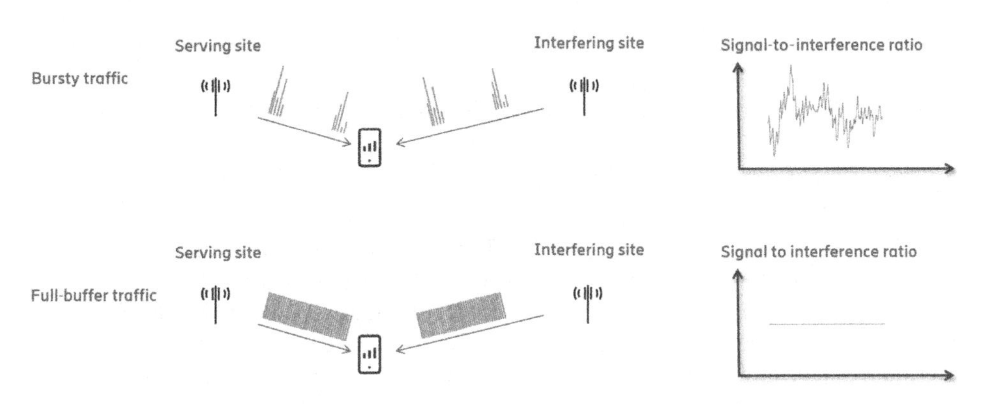

FIGURE 2.13

Illustration of signal-to-interference characteristics for bursty MBB-like and full buffer traffic cases. The bursty traffic signal quality is varying a lot, while full buffer signal quality is more constant.

accuracy using more advanced propagation models. For example, width of streets, building footprints, and heights, as well as building characteristics such as wall material, glass types, and indoor wall structures. For best accuracy, these aspects should be modeled.

In summary, there are many model parameters that can affect the simulation results. Simulations are useful but not precise in an absolute sense. They are good for building up an understanding of trends, that is, how different phenomena influence performance. They can also help to evaluate relative differences between different technical solutions. The performance of networks that use AAS includes many additional parameters and models compared to networks using classic radios. Therefore the need for accurate models and assumptions becomes even more important when analyzing AAS. This will be further elaborated in later chapters, particularly in Section 13.2.

2.3 NETWORK EVOLUTION

2.3.1 NEED FOR CONTINUOUS EVOLUTION

As mentioned already, mobile operators evolve their networks to be able to offer increased capacity, better coverage, and higher end-user throughput.

The network area capacity is on a high-level dependent on the site density (number of sites and hence cells per km^2) and the capacity per cell (or site). The capacity per cell does in turn depend on the amount of spectrum and the radio link performance.

Area coverage depends on the site density but also what frequency band is deployed. The cell coverage is frequency dependent. Low-frequency bands have generally better coverage properties than higher frequency bands as will be further discussed in Chapter 3.

The user throughput is generally determined by the transmission bandwidth (Hz) and radio link performance (bps/Hz). The link performance depends on the transmission technology including coding, modulation, and overhead, as well as the signal quality level, that is, signal-to-interference-and-noise ratio (SINR).

All strategies for network evolution will rely on these fundamental techniques.

2.3.2 STRATEGY

There are several strategies for network evolution. The strongest rationale for the network evolution strategy is usually trying to find the most cost-efficient way of expanding the network.

A network evolution strategy can be summarized as: improve—densify—add. By this, it is meant that first the existing sites are *improved* to maximize cell capacity, for example, by adding more bandwidth and/or improving cell spectral efficiency. *Densify* is the step to introduce new sites to improve area capacity and coverage. *Add* is the last step where site density is increased by adding a new category of cells, small cells below rooftop, or indoor to improve coverage or capacity in local areas with high needs and where macro sites are difficult to deploy.

The reason for this strategy is that improving performance of existing sites provides the lowest cost. Existing site infrastructure investments such as mast, shelter, and backhaul are then reused. Finding new sites may take time and will incur costs in terms of acquisition, permits, build and further operational costs.

There can be deviations from this strategy. One example is in dense urban areas with high-rise buildings that can be difficult to cover from rooftop macro sites. Moreover, these buildings often have a high subscriber density at daytime (office buildings) or night time (residential buildings), which generate very high traffic. An alternative to macro densification is then to deploy dedicated indoor systems to improve performance.

2.3.3 OPTIONS FOR CAPACITY EXPANSION

More details on the network evolution strategy steps are described below.

2.3.3.1 Improving existing sites

Increasing bandwidth is a simple and straightforward way to increase link performance and cell capacity. The radio link performance is proportional to the bandwidth assuming that the transmit power increases proportionally. In 2G and 3G, the bandwidth was initially small but it has been gradually increased. In LTE, the carrier bandwidth is 20 MHz and for 5G it is even larger.

Mobile operators have typically got access to new frequency bands over time. New spectrum is usually sold or allocated when a new mobile system generation is released. Additional bands are sometimes made available between the new generations as well. The network coverage is frequency dependent as will be further explained in Section 3.5.1. Low-bands below 1 GHz have typically been used as coverage bands in rural areas but also in urban areas to provide deep indoor coverage. The higher frequency bands have often been used as supplementary capacity bands, that is, added where higher capacity is needed. The new higher frequency bands, above 2.6 GHz intended for 5G come with larger propagation challenges, as will be further explained in Chapter 3. When deploying new 5G bands at higher frequencies, it is desirable to adopt coverage-enhancing techniques to compensate for the lower coverage, see, for example, Sections 7.2.2 and 7.2.3.

The radio link performance is a measure of the data rate that can be achieved for a user and it can be improved in multiple ways. It has been gradually improved in 2G and 3G. At this point, the radio link performance of both 4G and 5G is close to what is achievable, for single or a few antennas on transmit and receive sides. Improved modulation schemes have been applied as a tool to increase link performance in both 3G and 4G together with link adaptation. Link adaptation involves selecting modulation and coding schemes to maximize link performance given the signal-to-interference ratio. For example, 64 QAM and 256 QAM have been gradually implemented and deployed to improve peak speeds as well as cell capacity in uplink and downlink. These tools are dependent on the capabilities of the UE fleet, that is, sufficiently many subscribers have a device that supports the desired modulation scheme. The remaining tool for improving link performance is to add antennas. For LTE, MIMO was introduced (see Section 8.2 for how LTE evolved and MIMO support was enhanced), which improves the link performance. Using advanced receivers is also a method to improve link performance using multiple antennas.

There is a strong connection between the radio link performance and cell spectral efficiency. Improving the radio link performance will result in that users will finish their data transmissions faster (if users are sending the same amount of data) and there will be time available to serve new users. Hence improved link performance will improve cell spectral efficiency.

Network optimization and tuning are tools for making the cells operate at a higher spectral efficiency by reducing interference. Antenna tilting is one example of how to reduce interference (see Fig. 2.14 and Section 13.3.2).

Increasing output power is another option that could be used to increase performance up to a certain point when the network becomes interference limited. The definition of an interference limited scenario is a scenario where more power does not lead to higher capacity, which is the opposite case to the coverage limited scenario where more power is beneficial. It is important to understand that interference limitation is frequency band dependent, a site grid designed for a low-frequency band, for example, 900 MHz, might become noise-limited for a higher frequency band, for example, 3.5 GHz. There are also practical limits to the levels of output power. Output power has to be limited on both base station and UE side due to requirements on *electromagnetic field* radiation (see Section 11.8). The base station power is needed for serving users all over the

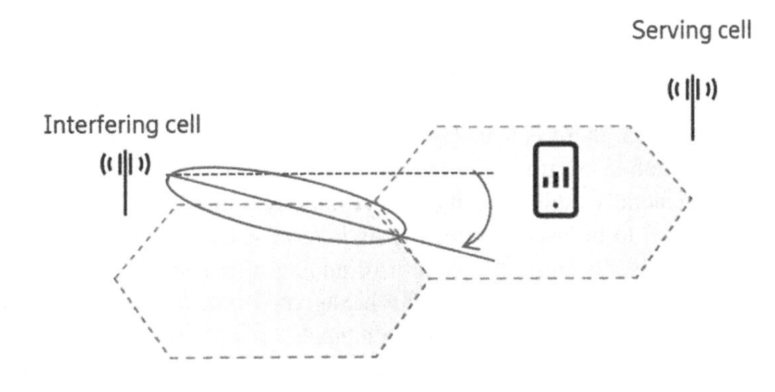

FIGURE 2.14

Illustration of antenna tilting as a means to reduce interference.

intended coverage area and there is a benefit of high output power for the downlink. Due to regulations on maximum output power, the base station antennas are usually at least several meters away from the public reducing the radiation strength. The base station output power can be up to 6–8 W/MHz (see also analysis in Section 13.7.5). On the UE side, where the antenna is located closer to humans and often placed directly against the head, the output power must be lower than on the base station side. There are limited possibilities to increase output power in UE due to battery life, standard health limitations, and regulatory considerations. For 3G, LTE, and NR, the UE output power level is limited to 23 dBm (0.20 W).

Increasing number of sectors per site, that is, high-order sectorization, is a special case of improving existing sites by increasing the number of cells per site. The use of six-sector sites has been commonly discussed, but the actual deployments in the field are rather few so far (see also Section 13.3.4).

The main techniques for improving existing sites and cells are hence increasing bandwidth and cell spectral efficiency. These techniques are often cost-efficient since building new sites, macro, small, or indoor comes with higher costs. Therefore these techniques are often exploited as the first step in the network evolution toward higher capacity and user throughput.

The introduction of 5G at new frequency bands is a natural part of the increasing bandwidth tool. Another way is to re-farm spectrum from an old wireless technology to a newer one. The mobile network operators typically maintain multiple wireless technologies, for example, 2G, 3G, and 4G today and their network and spectrum can be re-farmed from the old (2G and 3G) to a more modern technology (4G and 5G). Typically, when there are enough subscribers having UEs supporting the new technology and the number of subscribers with handsets supporting only the old technology is small. If this is not the case, the operator can stimulate the adoption of the new technology by providing handset subsidies to move subscribers from the old to the new technology. In this way, it is possible to minimize the bandwidth for the old technology and introduce the new technology in the same frequency band. Often a thin layer must be maintained for the old technology since all subscribers will not upgrade in a short time and there might also be a need for roaming reasons, which is a large revenue source for the operators. In 5G, spectrum sharing is introduced to overcome the spectrum migration from 4G (see also Section 13.7.6).

2.3.3.2 Densification of existing macro grid

Generally, when an operator has evolved the network up to the situation that all spectrum assets have been deployed in an area, there is usually no other option than to add new sites, that is, apply site densification. This situation occurs often locally in dense urban areas where capacity demand is highest. Adding sites generally adds capacity almost linearly with the number of sites, and therefore this method is expected to be used continually as long as it is meaningful to further add sites. When the network becomes very dense, the benefit of adding a new site is smaller since the intercell interference is harder to manage. Maintaining a hexagonal network layout is often not possible in practice. Densification can be prolonged using techniques for improving or maintaining cell isolation such as antenna tilting and antenna azimuth or antenna beamwidth changes. The limit on site densification is dependent on the environment and deployment but as a rule of thumb inter-site distances below 100 m are typically difficult to maintain.

2.3.3.3 Adding small cells and indoor systems

When adding more rooftop or tower-based macro sites does not provide cost-efficient capacity expansions, small (street level) cells and indoor systems can be used. For best efficiency in terms of off-loading macro sites, the new small and indoor system should be deployed where macro provides poor performance.

2.3.4 WHY MULTI-ANTENNA AND ADVANCED ANTENNA SYSTEMS

Deploying multi-antenna and AAS techniques provide ways to increase the spectral efficiency and are therefore instrumental tools for improving existing sites in the network evolution toolbox. They will improve both coverage and capacity. This is done by making use of the spatial domain, exploiting that power is transferred from the transmitter to the receiver through many physical paths. This will be illustrated for the case with free-space propagation in Chapter 5, and explained in detail in Chapter 6 for the case with multipath propagation. The benefit in terms of improved SINR stems from the fact that power is directed along the paths that most efficiently transfer power from the transmitter to the receiver rather than transmitting power widely over the sector. It is also possible to avoid transmitting powers along certain paths to avoid generating interference. These benefits will also be present at the base station reception side as well as in the UE reception/transmission for high bands especially.

When exploiting new high-frequency bands, multi-antenna and AAS offer the ability to get higher antenna gain by means of beamforming. In many cases this makes it possible to reuse existing sites and maintain similar coverage of both control signals and data rates at the new higher frequency band as for existing lower frequency bands. This is particularly important for 5G where most new frequency bands are in the 3–6 GHz region or in the mm-wave range (FR2) where the coverage is considerably worse than for the existing mid-band range (below 3 GHz typically). An illustration of this is shown in Fig. 2.15. In particular, AAS is essentially a prerequisite for exploiting FR2 at existing sites. AAS technologies are typically not practical for frequency bands below 1 GHz as the antenna size becomes very large, at least when considering beamforming in the horizontal domain. Multi-antenna deployment at base station side seems to be limited to four radio branches at these bands in practice today.

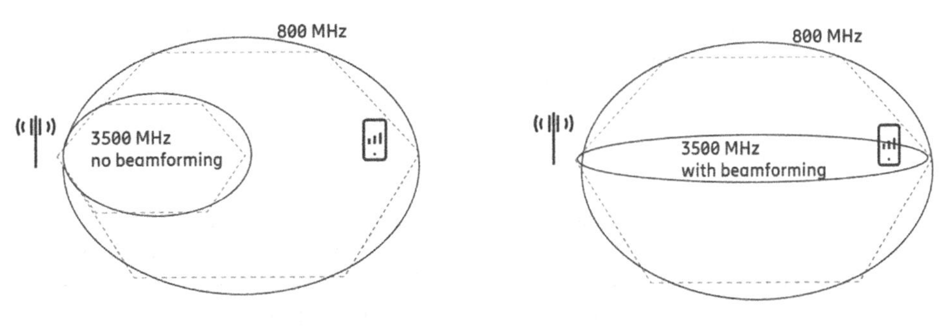

FIGURE 2.15

Illustration of coverage benefits from beamforming at 3.5 GHz relative to 800 MHz coverage.

In summary, with multiple antennas and AAS it is possible to exploit new and most often higher frequency bands with sufficient coverage. Another key advantage is that AAS can offer higher spectral efficiency and thereby provide higher capacity at existing sites. The latter is especially important in markets with smaller spectrum allocations. In other deployments such as FWA, AAS may be required and especially useful due to the need for very high capacity.

The technologies applied by AAS are explained in detail in Chapters 6, and 7. AAS performance results are found in Chapter 13, and Section 14.3.

2.3.4.1 Why now

AAS technologies, for example, beamforming, have been known and used in other fields, for example, radar and satellite communication, with proven performance. This has been recognized by the mobile industry for more than 25 years.

The main reasons why AAS has not been widely deployed earlier in mobile networks are that the implementations have been costly, large, and heavy, and that there have been other solutions that have met the requirements in a simpler and more cost-efficient way, for instance, adding spectrum and additional carriers. Moreover, each mobile system generation has provided improvements in spectrum efficiency and opened up for gradually larger bandwidths in new spectrum bands. The governments, standardization bodies, and spectrum authorities have realized the need for spectrum by the mobile industry and have therefore gradually released new frequency bands. The capacity needs up to now have essentially been met through use of new bands, and therefore there has not been an imminent need to exploit AAS to great extent earlier.

Another key reason for considering AAS is to increase coverage. New frequency bands are usually available on higher frequencies, where the coverage is worse. AAS can be used in this case to improve coverage and thereby increase the usability of the higher frequency band in a way that non-AAS solutions cannot do.

For 5G, the frequency bandwidths have evolved further and allocations of 50–100 MHz or more per operator are desired in FR1. For FR2, the bandwidth allocation can be up to 1 GHz. Larger bandwidths are typically found only in the higher frequency regimes for which AAS is required to provide coverage in a cost-efficient way with small or moderate densification needs. Moreover, there is also an increased pressure of maximizing the use of available spectrum, that is, pushing the industry to further optimize the spectrum efficiency. AAS is the next tool for increasing spectral efficiency since the radio link performance has more or less been fully exploited up to 4G.

Lower hardware cost as well as tighter integration due to smaller components is another reason for the increased focus on AAS now. It is becoming possible to build AAS solutions, which are rather hardware intense, in ways that are more cost-efficient than before. Given the performance benefits, AAS is an option competing with the existing network evolution alternatives. Traditional network evolution methods, see Section 2.3.2, are still expected to be used where they are more cost efficient. The choice for capacity expansion depends on both practical conditions, for example, availability of sites and spectrum, and economical factors, for example, *total cost of ownership* (TCO). For AAS, there are also new factors that are important to consider, for example, multi-antenna/multiband support, visual impact, and site solution impact to mention a few. As the cost of AAS will continue to go down and the adopted AAS technologies will be more efficient, it is expected that the use of AAS will increase gradually.

2.4 SUMMARY/KEY TAKEAWAYS

In this chapter, network deployment and network evolution aspects have been described. The purpose has been to provide rationale for why networks have evolved the way they have up to now and to show how AAS can contribute in the network.

The traffic in all mobile networks is growing at significant pace and the requirements from the end-user to use the mobile network for more advanced applications require continuous network evolution. The networks are evolved to meet the objectives specifically in terms of coverage, capacity, and end-user throughput in the most cost-efficient way. In reality, the most cost-efficient alternative for network evolution is to improve existing sites, for example, by adopting new frequency bands when possible.

AAS is a means of providing further performance improvements by using the spatial domain and thus more efficient use of the characteristics of the radio channel between the transmitter and receiver. AAS comes with an additional cost because it requires more hardware than classical radios. However, it is still a viable solution and more cost-efficient than other means available for the new high-frequency band applications such as 5G, for example, providing coverage at high bands through AAS to avoid building many new sites.

Some additional topics are also introduced to provide the context of AAS. Since the main reason for introducing AAS is to improve network performance, the basic definitions for the network performance measurements were introduced (see Section 2.2.3).

Key takeaways that will be referred to in later chapters are the following.

- Rapid growth of traffic demand and expectations on end-user throughput and coverage will require continuous network evolution (Chapter 14);
- Mobile network operator cost structures usually benefit maximum use of existing sites and hence motivate the introduction of AAS as a means of providing capacity, coverage, and end-user performance improvements (Chapters 13, and 14);
- MBB traffic is built up of many subscribers, each having a bursty traffic profile. Hence the number of connected users is often much larger than the number of active users. This has an impact on the efficiency of AAS features, particularly multiple-user MIMO (Sections 6.7.2, 6.7.4.1, and 13.6);
- AAS is necessary to improve coverage of higher frequency bands (Section 7.2.2);
- AAS requires more hardware, for example, more antenna elements and radio chains, than for conventional systems. This will impact which AAS solutions will be cost-efficient in different deployment scenarios and also the size and weight and hence the deployability (Chapter 14).

REFERENCES

[1] V.H. MacDonald, The cellular concept, Bell Sys. Tech. J. 58 (1) (1979).

[2] Ericsson ConsumerLab, 2014. Liberation from location. <https://www.ericsson.com/assets/local/news/2014/10/liberation-from-location-ericsson-consumerlab.pdf> (accessed 17.06.19).

[3] H. Olofsson et al., 2018. Leveraging LTE and 5G for fixed wireless access, Ericsson Technology Review. <https://www.ericsson.com/en/ericsson-technology-review/archive/2018/leveraging-lte-and-5g-nr-networks-for-fixed-wireless-access> (accessed 17.06.19).
[4] Ericsson, Ericsson mobility report – November 2018. <https://www.ericsson.com/en/mobility-report/reports/november-2018/mobile-traffic-q3-2018> (accessed 17.06.19).
[5] J. Sachs et al., February 2019. Boosting smart manufacturing with 5G wireless connectivity, Ericsson Technology Review. <https://www.ericsson.com/en/ericsson-technology-review/archive/2019/boosting-smart-manufacturing-with-5g-wireless-connectivity> (accessed 17.06.19).
[6] C. Summerson, How much data does Netflix use? <https://www.howtogeek.com/338983/how-much-data-does-netflix-use/> (accessed 17.06.19).
[7] T. Everts, The average web page is 3MB. How much should we care? <https://speedcurve.com/blog/web-performance-page-bloat/> (accessed 17.06.19).
[8] W. Carney et al., Simplified traffic model based on aggregated network statistics, IEEE 802.11ax WG contribution, 11-13-1144.
[9] 3rd Generation Partnership Project; Technical Specification Group Radio Access Network; LTE RAN enhancements for diverse data (Rel. 11), 3GPP, 3GPP TS 36.822.
[10] 3rd Generation Partnership Project; Technical Specification Group Radio Access Network; Study on channel model for frequencies from 0.5 to 100 GHz (Rel. 15), 3GPP, 3GPP TS 38.901.

CHAPTER

ANTENNAS AND WAVE PROPAGATION

3

This chapter treats the fundamentals of radio waves and antennas, their properties and characteristics, and provides some insights into how these affect wireless communication. Challenges and opportunities associated with multi-path propagation and higher frequencies are treated. The key learnings of this chapter will be reused throughout the book when discussing how more capable antennas are utilized to mitigate these challenges and make the most of the opportunities.

The present chapter is structured as follows: an introduction to electromagnetic fields and waves is given in Section 3.1, while general properties of electromagnetic waves are discussed in Section 3.2. Section 3.3 introduces antennas as the interface between conducted and radiated electromagnetic energy. Fundamental properties of transfer of energy from one antenna to another in free space are given in Section 3.4, and in real-world conditions in Section 3.5. Section 3.5.3 introduces the reader to multi-path propagation and all its associated challenges for wireless communication. Section 3.6 introduces a mathematical framework for multi-path propagation, which will be used in the treatment of array antennas (Chapter 4), orthogonal frequency-division multiplexing (OFDM)-based transmission (Chapter 5) and multi-antenna technologies (Chapter 6). Furthermore, Section 3.6 describes stochastic and site-specific models for radio wave propagation. Such models will be used later in the book for evaluations of multi-antenna and advanced antenna systems (AAS) through simulations (see, e.g., Chapters 13 and 14). Finally, a summary of some key points is provided in Section 3.7.

The reader who wishes to follow the mathematical treatment in this chapter is expected to have some familiarity with complex numbers, vector fields and vector analysis, and Cartesian and spherical coordinate systems (a brief overview of these topics is provided in Appendix 1).

3.1 INTRODUCTION

More than 150 years ago, the mysteries of electromagnetic forces and fields were revealed through scientific theory building and experiments. The progress led James Clerk Maxwell to formulate his famous equations that describe how electric and magnetic fields interact with charges and currents (see Fig. 3.1). This set of coupled differential equations can be rearranged in the form of the wave equation, which has solutions for the electric field $\boldsymbol{E}$ (unit: V/m) and the magnetic field $\boldsymbol{B}$ (unit: T) on the form

$$\boldsymbol{E} = f(\omega t - \boldsymbol{k} \cdot \boldsymbol{r}) \quad (3.1)$$

$$\boldsymbol{B} = g(\omega t - \boldsymbol{k} \cdot \boldsymbol{r}) \quad (3.2)$$

Advanced Antenna Systems for 5G Network Deployments. DOI: https://doi.org/10.1016/B978-0-12-820046-9.00003-4

$$\nabla \cdot \boldsymbol{D} = \rho$$
$$\nabla \cdot \boldsymbol{B} = 0$$
$$\nabla \times \boldsymbol{E} = -\frac{\partial \boldsymbol{B}}{\partial t}$$
$$\nabla \times \boldsymbol{H} = \boldsymbol{J} + \frac{\partial \boldsymbol{D}}{\partial t}$$

FIGURE 3.1

The famous equations that take their name from James Clerk Maxwell, here presented in their differential form. The reader is reassured that familiarity with these equations is not a prerequisite for understanding the present chapter. However, for the interested reader, Appendix 1 gives a short introduction to vector fields and the divergence and curl of these.

where the functions f and g can be arbitrary[1] functions, $\boldsymbol{r}$ is the spatial (three-dimensional, 3D) position, t is the time, and ω is in general a proportionality constant (for time-harmonic waves ω is the angular frequency, that is, radians per second). These solutions represent a wave traveling in the direction of the *wave vector* $\boldsymbol{k} = k\widehat{\boldsymbol{k}}$ with speed $v = \omega/k$ (see Fig. 3.2).

The general solution to the wave equation is a linear superposition of such waves; however, for the initial sections of this chapter a single wave is considered. The electric field $\boldsymbol{E}$ and the magnetic field $\boldsymbol{B}$ are coupled and interact with each other as specified by Maxwell's third and fourth equations; hence they are commonly referred to as *electromagnetic* fields and waves.

In other words, Maxwell provided the mathematical basis for the description of electromagnetic waves, which are time-varying fluctuations in the electric and magnetic fields. These waves can be generated by accelerating electric charges such as a time-varying current in a transmitting antenna. Similarly, the time-varying fields exert forces on electric charges, such as electrons within a receiving antenna, thereby generating a current in the antenna. Hence, electromagnetic waves can be used to transport energy and information from one antenna to another.

One important class of waves is those that are sinusoidal or time-harmonic, that is, where the electric field vector $\boldsymbol{E}$ along its direction $\widehat{\boldsymbol{r}}$ is proportional to

$$\mathrm{Re}\{\exp(j\omega t)\} = \cos(\omega t). \tag{3.3}$$

Note that the field is represented by the real part of a complex harmonic rather than as a sine wave. While these two representations are equivalent, it is usually more convenient to work with the complex representation, not the least since the time derivatives in Maxwell's equations can then

[1]Boundary conditions and initial value conditions will also influence the solutions f and g.

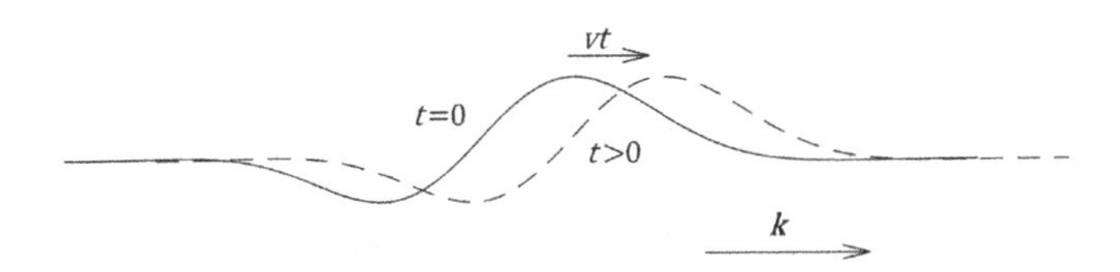

FIGURE 3.2

An arbitrary wave traveling in the direction $\boldsymbol{k}$ with speed v.

be replaced with multiplication by a constant. Often the Re{} is omitted and it is implicitly understood that only the real part of the solution should be used.

Most waves generated by natural or artificial processes are time-harmonic. Additionally, from Fourier analysis it is known that any wave form can be represented by a linear superposition of time-harmonic waves. Therefore in the following we will exclusively consider time-harmonic fields and waves.

3.2 PROPERTIES OF ELECTROMAGNETIC WAVES

The electric and magnetic fields are vector-valued, that is, in each location in space the fields have a strength and a direction in 3D space. Such a vector can be represented by three coordinates in a Cartesian coordinate system collected into a column vector, for example, $\boldsymbol{v} = \begin{bmatrix} v_x & v_y & v_z \end{bmatrix}^{\mathrm{T}}$. This is however not the only way to represent a vector (see Appendix 1.5 for further details about coordinate systems).

The equation for the electric field of a sinusoidal wave propagating in the direction $\widehat{\boldsymbol{k}}$ can be written as

$$\boldsymbol{E}(\boldsymbol{r},t) = \mathrm{Re}\left\{\boldsymbol{E}_0\exp\left(j(\omega t - \boldsymbol{k}\cdot\boldsymbol{r})\right)\right\} \tag{3.4}$$

where $\boldsymbol{k}$ is the wave vector with $\boldsymbol{k} = k\widehat{\boldsymbol{k}} = 2\pi/\lambda\widehat{\boldsymbol{k}}$ and ω now represents the angular frequency in radians/second with $\omega = 2\pi f$. $\boldsymbol{E}_0$ is a vector with complex-valued elements, *phasors*, that represents the amplitude and phase of the electric field in different directions in space (e.g., $\widehat{\boldsymbol{x}},\widehat{\boldsymbol{y}},\widehat{\boldsymbol{z}}$ or $\widehat{\boldsymbol{r}},\widehat{\boldsymbol{\theta}},\widehat{\boldsymbol{\varphi}}$) at a reference time and position.

A time-harmonic wave has a frequency $f = \omega/2\pi$ and a wavelength λ and propagates at the speed $v = \omega/k = \lambda f$ in the direction given by the vector $\boldsymbol{k}$. In vacuum, $v = c = 299{,}792{,}458$ m/s, which is the speed of light. Indeed, light is just electromagnetic waves within a certain range of wavelengths (compare Fig. 3.7).

For the sinusoidal wave in Fig. 3.3, the wave is propagating in the direction $\widehat{\boldsymbol{k}} = \widehat{\boldsymbol{x}}$ and the electric field vector is aligned in the z-direction, that is,

$$\boldsymbol{E}_0 = E_z\exp(j\phi)\widehat{\boldsymbol{z}} \tag{3.5}$$

where E_z is the amplitude and ϕ is the phase of the wave at $t = 0$ and $\boldsymbol{r} = \boldsymbol{0}$.

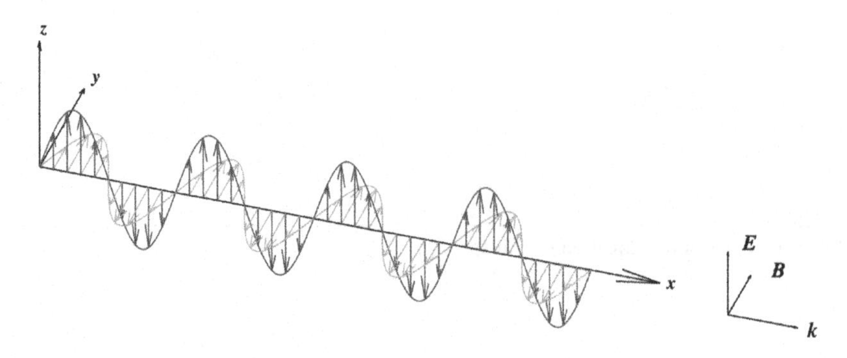

FIGURE 3.3

A sinusoidal transverse electromagnetic wave propagating in the x-direction.

A particular characteristic of electromagnetic waves is that they are *transverse*, meaning that the electric field vector $\boldsymbol{E}$, the magnetic field vector $\boldsymbol{B}$, and the direction of propagation given by the wave vector $\boldsymbol{k}$ are all mutually orthogonal (see Fig. 3.3). The magnetic field is given by

$$\boldsymbol{B} = \frac{1}{\omega}\boldsymbol{k} \times \boldsymbol{E}, \tag{3.6}$$

hence it oscillates in phase with the electric field but in a direction orthogonal to the electric field and the direction of propagation. For the wave in Fig. 3.3 the magnetic field is

$$\begin{aligned} \boldsymbol{B} &= \frac{1}{\omega}\boldsymbol{k} \times \mathrm{Re}\{\boldsymbol{E}_0 \exp(j(\omega t - \boldsymbol{k}\cdot\boldsymbol{r}))\} \\ &= -\frac{1}{c_0}\mathrm{Re}\{E_z \exp(j\phi)\exp(j(\omega t - \boldsymbol{k}\cdot\boldsymbol{r}))\}\hat{y} \end{aligned} \tag{3.7}$$

It is common to only describe and discuss the electric field, with the implicit understanding that the magnetic field can be determined using the above vector cross product. In the remainder of this book we will therefore focus on the electric field.

3.2.1 INSTANTANEOUS AND AVERAGE POWER FLOW

An electromagnetic wave represents a power flow in space, characterized in direction and magnitude by the Poynting vector (W/m^2):

$$\boldsymbol{S} = \frac{1}{\mu_0}\boldsymbol{E} \times \boldsymbol{B} \tag{3.8}$$

where μ_0 is the vacuum permeability ($\mu_0 = 1.26 \times 10^{-6}$ H/m).

The time-average power flow or *power density* $\boldsymbol{S}_{\text{av}}$ for a sinusoidal wave propagating in the direction $\hat{\boldsymbol{k}}$, determined by integrating the power flow $\boldsymbol{S}$ over one period $T = 1/f$, is then

$$\boldsymbol{S}_{\text{av}} = \frac{1}{2}\mathrm{Re}\left\{\frac{1}{\mu_0}\boldsymbol{E} \times \boldsymbol{B}\right\} = \frac{1}{2}\mathrm{Re}\left\{\frac{1}{\mu_0}\boldsymbol{E}_0 \times \boldsymbol{B}_0^*\right\} = \frac{1}{2\eta}|E_0|^2\hat{\boldsymbol{k}} = W_{\text{av}}\hat{\boldsymbol{k}} \tag{3.9}$$

where $W_{av}=\frac{1}{2\eta}|E_0|^2$ is the average power density (W/m^2), and $\boldsymbol{B}_0^*$ denotes the complex conjugate of $\boldsymbol{B}_0$. The constant $\eta = \mu_0 c \approx 377\ \Omega$ is the *intrinsic impedance of free space*.

3.2.2 WAVE FRONTS

The electric field of the transverse wave of Fig. 3.3 that was introduced in the previous section can be rewritten by combining (3.4) and (3.5) as

$$\boldsymbol{E}(\boldsymbol{r},t) = E_z \mathrm{Re}\left\{\exp\left(j(\omega t - \boldsymbol{k}\cdot\boldsymbol{r} + \phi)\right)\right\}\widehat{z} \tag{3.10}$$

As can be seen from this equation, the phase $\omega t - \boldsymbol{k}\cdot\boldsymbol{r} + \phi$ varies both in time and in space. However, around a specific location in space the spatial component $\boldsymbol{k}\cdot\boldsymbol{r}$ of the phase is constant in directions perpendicular to the propagation direction $\boldsymbol{k}$. In general, the *wave front* is a surface in three dimensions along which the phase is constant at a particular time instant (see Fig. 3.4). In our example in Fig. 3.3 with $\widehat{\boldsymbol{k}} = \widehat{x}$ the wave front is planar; hence this is referred to as a *plane wave*. A wave originating from a point-like source, such as the fields generated by a transmitting antenna when observed at a sufficient distance (see Section 3.3) will expand spherically and hence create a

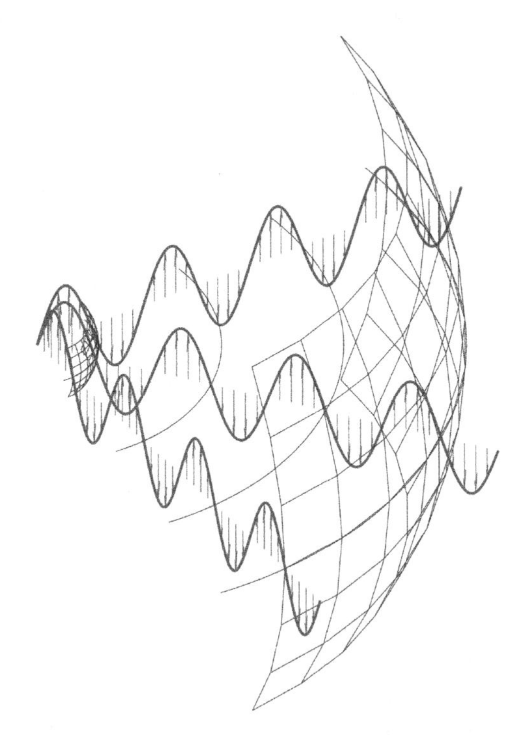

FIGURE 3.4

Illustration of the wave front of spherical waves originating from a point source. The phase is constant along the lines and surfaces indicated in the figure. Locally, the wave front becomes approximately plane at larger distances from the source.

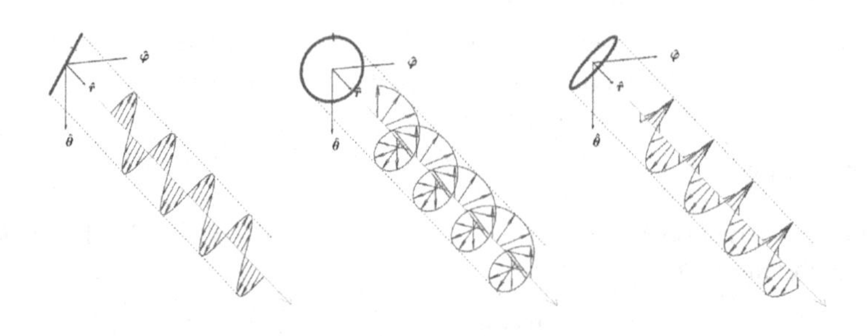

FIGURE 3.5

Different types of polarizations as illustrated by the trace of the electric field vector when looking along the direction of propagation of the wave: linear polarization (left), circular polarization (middle), and elliptical polarization (right).

spherical wave front. However, the wave front is often conveniently treated as being locally plane, which is a reasonable approximation at larger distances where the radius of curvature of the wave front is very large.

3.2.3 POLARIZATION

As described above, radiated electromagnetic waves are transverse, meaning that the electric and magnetic field vectors are orthogonal to the direction of propagation. Hence, in 3D space there are locally two dimensions available for the fields to oscillate in, that is, it is said that the wave has two degrees of freedom. We define the *polarization* of a wave as the orientation of the oscillations of the electric field in these two dimensions as observed when looking along the direction of propagation. When discussing polarization in this book we will utilize spherical coordinates rather than Cartesian coordinates since the former are more suitable for characterizing the fields and polarizations of waves transmitted by antennas.

Mathematically, the electric field of a spherical wave propagating in the $\widehat{\boldsymbol{r}}$ direction can be expressed as

$$\boldsymbol{E}_0 = E_\theta \exp(j\phi_\theta)\widehat{\boldsymbol{\theta}} + E_\varphi \exp(j\phi_\varphi)\widehat{\boldsymbol{\varphi}} \tag{3.11}$$

where E_θ and E_φ are the amplitudes and ϕ_θ and ϕ_φ are the phases of the phasors representing the transverse electric field components along the basis unit vectors $\widehat{\boldsymbol{\theta}}$ and $\widehat{\boldsymbol{\varphi}}$, respectively. Note that these vectors can be represented as 3×1 vectors of coordinates in 3D space (see Appendix 1.5 for how to transform between different coordinate systems).

The relation between the phasors determines the *polarization* of the wave, that is, the time-varying direction and relative magnitude of the electric field vector. In this book, the wave polarization is defined as observed along the direction of propagation. Some particular types of polarizations are the following (see also Fig. 3.5):

Linear polarization occurs when the electric field vector oscillates in a constant direction, that is, when $\phi_\theta = \phi_\varphi$ or when either of E_θ or E_φ is zero. The amplitude relations between E_θ and E_φ

then determine the slant angle of the electric field vector. In this book, *vertical polarization (VP)*[2] is defined as the polarization when the electric field is oscillating along the $\widehat{\theta}$ axis, that is, when $E_\varphi = 0$. Similarly, *horizontal polarization (HP)* is defined as the polarization when the electric field is oscillating along the $\widehat{\varphi}$ axis ($E_\theta = 0$). Other linear polarizations include *+45° polarization* ($E_\theta = E_\varphi$) and *−45° polarization* ($E_\theta = -E_\varphi$).

Circular polarization occurs when the electric field vector rotates in the transverse plane with constant amplitude, that is, when $E_\theta = E_\varphi$ and $\phi_\theta = \phi_\varphi \pm 90°$. The direction of rotation can be either clockwise or counterclockwise, resulting in *right-hand circular polarization (RHCP)* or *left-hand circular polarization (LHCP)*.

Other relations between the phasors result in *elliptical polarization*, that is, where the electric field vector both oscillates in amplitude and rotates.

The polarization of a wave can be generally represented by a two-element complex-valued 2×1 unit vector $\widehat{\underline{\psi}}$, used to rewrite the 3×1 electric field vector as

$$\begin{aligned} \boldsymbol{E}_0 &= E_\theta \exp(j\phi_\theta)\widehat{\theta} + E_\varphi \exp\left(j\phi_\varphi\right)\widehat{\varphi} \\ &= \sqrt{E_\theta^2 + E_\varphi^2}\,[\widehat{\theta} \quad \widehat{\varphi}]\,\widehat{\underline{\psi}} \triangleq \sqrt{E_\theta^2 + E_\varphi^2}\,\widehat{\psi} \end{aligned} \tag{3.12}$$

where

$$\widehat{\underline{\psi}} \triangleq \frac{1}{\sqrt{E_\theta^2 + E_\varphi^2}} \begin{bmatrix} E_\theta \exp(j\phi_\theta) \\ E_\varphi \exp\left(j\phi_\varphi\right) \end{bmatrix} \tag{3.13}$$

contains the coordinates of the polarization vector

$$\widehat{\psi} = [\widehat{\theta} \quad \widehat{\varphi}]\,\widehat{\underline{\psi}} = \left[\widehat{\underline{\psi}}\right]_1 \widehat{\theta} + \left[\widehat{\underline{\psi}}\right]_2 \widehat{\varphi} \tag{3.14}$$

in the basis $\widehat{\theta}$ and $\widehat{\varphi}$. Here, $[\widehat{\theta} \quad \widehat{\varphi}]$ is a 3×2 real-valued matrix containing the basis vectors $\widehat{\theta}$ and $\widehat{\varphi}$ as columns.

In general, this chapter uses the notation $\underline{\boldsymbol{x}}$ to represent the vector $\boldsymbol{x}$ with coordinates in a different basis. The basis is given by the context, usually in terms of $\widehat{\theta}$ and $\widehat{\varphi}$ in either a transmitter or receiver coordinate system. The two different ways to represent the same vector will be used interchangeably in the following. The vector $\underline{\boldsymbol{x}}$ may be referred to as the *Jones vector*[3] corresponding to $\boldsymbol{x}$.

With this notation, the following polarization vectors can be defined:

$$\widehat{\underline{\psi}}_{\mathrm{VP}} = \begin{bmatrix} 1 \\ 0 \end{bmatrix}, \widehat{\underline{\psi}}_{\mathrm{HP}} = \begin{bmatrix} 0 \\ 1 \end{bmatrix} \tag{3.15}$$

$$\widehat{\underline{\psi}}_{\mathrm{LHCP}} = \frac{1}{\sqrt{2}} \begin{bmatrix} 1 \\ j \end{bmatrix}, \widehat{\underline{\psi}}_{\mathrm{RHCP}} = \frac{1}{\sqrt{2}} \begin{bmatrix} 1 \\ -j \end{bmatrix} \tag{3.16}$$

---

[2]In the literature, horizontal and vertical polarizations are sometimes defined with respect to the horizon in a Cartesian coordinate system. However, that definition would later lead to issues when describing antenna properties since the antenna directivity and polarization would become nonseparable, compare Eq. (3.30).

[3]This representation of the polarization state and of the polarization scattering matrix $\boldsymbol{\Psi}$ introduced later in this chapter were introduced by R.C. Jones and the associated mathematical operations to determine the polarization are therefore called Jones calculus.

$$\widehat{\psi}_{+45^\circ} = \frac{1}{\sqrt{2}}\begin{bmatrix}1\\1\end{bmatrix}, \widehat{\psi}_{-45^\circ} = \frac{1}{\sqrt{2}}\begin{bmatrix}1\\-1\end{bmatrix} \tag{3.17}$$

For any polarization $\widehat{\psi}_1$ there is always an orthogonal polarization $\widehat{\psi}_2$ for which $\widehat{\psi}_1 \cdot \widehat{\psi}_2 = \widehat{\psi}_1^* \widehat{\psi}_2 = \underline{\widehat{\psi}}_1^* \underline{\widehat{\psi}}_2 = 0$, for example, $\widehat{\psi}_{\mathrm{VP}}$ is orthogonal to $\widehat{\psi}_{\mathrm{HP}}$, $\widehat{\psi}_{+45^\circ}$ is orthogonal to $\widehat{\psi}_{-45^\circ}$, and $\widehat{\psi}_{\mathrm{LHCP}}$ is orthogonal to $\widehat{\psi}_{\mathrm{RHCP}}$. Any elliptical polarization will have an orthogonal polarization that is also elliptical, but with opposite sense of rotation and orthogonal slant angle. An arbitrary polarization can be described as the linear combination of two orthogonal polarizations, for example, of vertical and horizontal or of $+45^\circ$ and -45° polarizations.

$$\underline{\widehat{\psi}} = a\underline{\widehat{\psi}}_{\mathrm{VP}} + b\underline{\widehat{\psi}}_{\mathrm{HP}} = c\underline{\widehat{\psi}}_{+45^\circ} + d\underline{\widehat{\psi}}_{-45^\circ} \tag{3.18}$$

As will be described in Section 6.3 (cf. Fig. 6.21), the two degrees of freedom of the wave polarization can be exploited for communication purposes to achieve more robust signal transmission or even by transferring different information on different polarizations.

3.2.4 SUPERPOSITION

In linear media,[4] the solutions to Maxwell's equations obey the important principle of superposition, which specifies that any linear combination of solutions (waves) is also a solution to Maxwell's equations. In other words, any number of electromagnetic waves of the same or different frequencies, polarizations, or propagation directions can be superimposed. The local electromagnetic field is then the linear vector sum of the fields contributed by each wave, for example,

$$\boldsymbol{E}(\boldsymbol{r},t) = \sum_n \boldsymbol{E}_n(\boldsymbol{r},t) \tag{3.19}$$

$$\boldsymbol{B}(\boldsymbol{r},t) = \sum_n \boldsymbol{B}_n(\boldsymbol{r},t) \tag{3.20}$$

An illustration of the superposition of two waves with opposite propagation directions along the x-axis and electric field vector along the z-axis is shown in Fig. 3.6. Mathematically, we can represent the electric field of the two waves in the figure as

$$\boldsymbol{E}_1(\boldsymbol{r},t) = \mathrm{Re}\{\exp(j(\omega t - \boldsymbol{k}_1 \cdot \boldsymbol{r}))\}\widehat{z} = \cos(\omega t - kx)\widehat{z} \tag{3.21}$$

$$\boldsymbol{E}_2(\boldsymbol{r},t) = \mathrm{Re}\{\exp(j(\omega t - \boldsymbol{k}_2 \cdot \boldsymbol{r}))\}\widehat{z} = \cos(\omega t + kx)\widehat{z} \tag{3.22}$$

The superposition of these two waves is then

$$\begin{aligned} \boldsymbol{E}(\boldsymbol{r},t) &= \boldsymbol{E}_1(\boldsymbol{r},t) + \boldsymbol{E}_2(\boldsymbol{r},t) \\ &= (\cos(\omega t - kx) + \cos(\omega t + kx))\widehat{z} \\ &= 2\cos(\omega t)\cos(kx)\widehat{z} \end{aligned} \tag{3.23}$$

This represents a *standing wave* since the position of the peaks ($kx = n\pi, n = 0, \pm 1, \pm 2, \ldots$) and nulls ($kx = \pi/2 + n\pi$) do not change over time.

The superposition of multiple waves with possibly different directions of propagation, amplitude, and phase relations and different polarizations will create a much more complex standing

[4] Air behaves as a linear medium except for extremely high field strengths that cause electrical discharge.

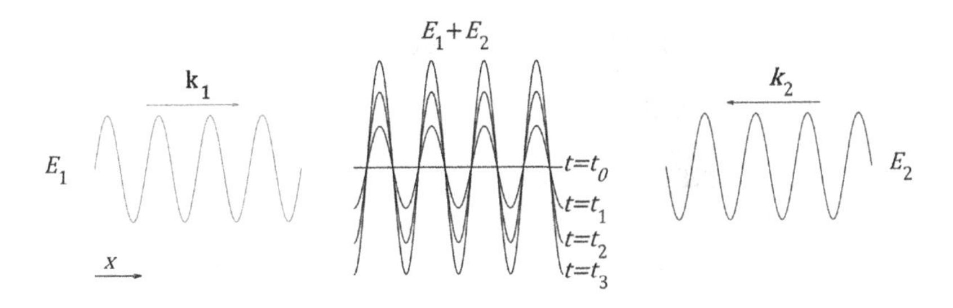

FIGURE 3.6

Illustration of the standing wave pattern (center) of the superposition of two waves (left and right) with opposite propagation directions.

wave pattern (see, e.g., Fig. 3.19). A sensing object such as an antenna traveling through this wave pattern will experience field strength (current amplitude) variations that are commonly referred to as signal *fading*, which can be a challenge to reliable communication. This is discussed in more detail in Section 3.5.3. Superposition is also beneficial as it allows antenna arrays to use constructive superposition of waves to focus radiation in certain directions, or a pair of antenna elements with different polarizations to generate an arbitrarily polarized wave (more on this in Chapter 4).

3.2.5 THE ELECTROMAGNETIC SPECTRUM

Time-harmonic electromagnetic waves come in many forms that have been given various descriptive names such as radio waves, infrared light, visible light, and gamma rays. All of these are simply sinusoidal electromagnetic waves with different frequencies or wavelengths. The range of frequencies over which electromagnetic waves appear is referred to as the *electromagnetic spectrum* (see Fig. 3.7 for an overview). The frequency bands used for mobile wireless communication that were introduced in Section 2.1.3 are highlighted. For systems up to and including 4G, these are primarily in the range of a few hundred MHz to several GHz or equivalently with wavelengths ranging from several decimeters to several centimeters.

Why are these bands used for mobile wireless communication rather than, say, infrared or VLF (very low frequency, kHz range)? The answer to this question is that the range between a few hundred MHz and several GHz strikes a good balance between antenna design (see Section 3.3), signal coverage (radio wave propagation, see Section 3.5), challenges of multi-path propagation (Section 3.5.3), and available communication bandwidth (as was discussed in Chapter 2).

AAS, the topic of this book, is the technology for creating more efficient and capable antennas. One of the main benefits of AAS is that they enable the use of higher frequencies,[5] thereby giving access to much higher communication bandwidths. Opportunities and challenges associated with the use of higher frequency bands are treated in Chapter 7.

[5]Including bands above 30 GHz with wavelengths less than a centimeter, commonly referred to as millimeter-wave bands.

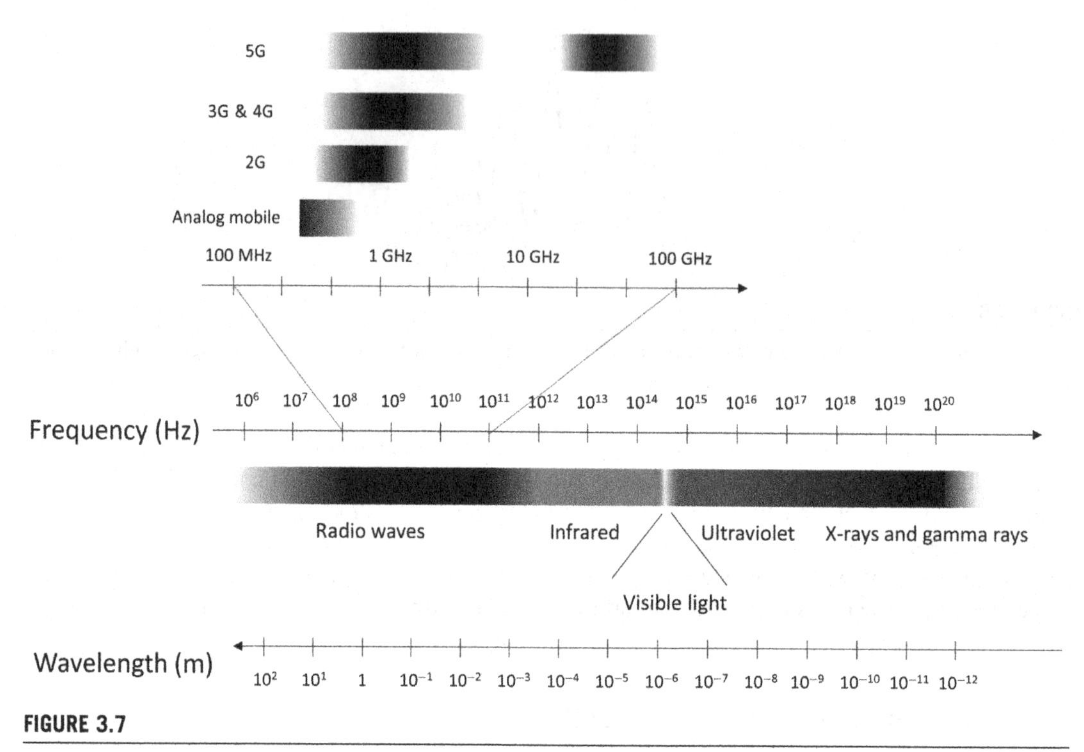

FIGURE 3.7

The electromagnetic spectrum.

3.3 TRANSMISSION AND RECEPTION OF ELECTROMAGNETIC WAVES: BASIC ANTENNA CONCEPTS

The time-varying electric and magnetic fields in an electromagnetic wave will exert forces on any electric charges it encounters, for example, those bound within a conductor, thereby inducing currents in these conductors. Equivalently, time-varying currents will induce time-varying electric and magnetic fields. This forms the basis of radiative transmission of energy and information, where currents may be induced in an antenna at one location that generates electromagnetic waves that propagate in space. These waves may be received — converted into currents — in another antenna at another location. An antenna thus converts conducted power into radiated power or vice versa.

In other words, the antenna is the interface between the electric currents and the electromagnetic waves. An antenna is designed[6] so that the electric currents are guided along suitable paths, where the strength of the current and its geometry decide the properties of the radiated waves (compare the current density $\boldsymbol{J}$ in Maxwell's equations that represents the strength and direction in space of the electric current).

[6]Antenna design is considered by some to be a form of art where the designer uses materials and feeding points to shape current distributions that create the desired radiation characteristics.

3.3.1 INFINITESIMAL DIPOLE

To illustrate some principles, let us first consider a very locally confined sinusoidally time-varying current with amplitude I_o, that is, the current

$$\boldsymbol{J} = I_o \exp(j\omega t)\hat{\boldsymbol{z}} \tag{3.24}$$

only exists along the z-axis over the length l ($l \ll \lambda$) close to the origin. An antenna producing such a current is sometimes referred to as an infinitesimal or Hertzian[7] dipole. If the current, that is, the direction in which the electric charges are moving, is aligned in the $\hat{z}$ direction, the electric field strength of the radiated field will at a sufficient distance r from the antenna be approximately

$$\boldsymbol{E}(\boldsymbol{r},t) \approx j\eta \frac{kI_0 l\sin(\theta)}{4\pi r} \exp(j(\omega t - \boldsymbol{k}\cdot\boldsymbol{r}))\hat{\boldsymbol{\theta}}. \tag{3.25}$$

This represents a transverse spherical wave with vertical polarization propagating radially outward from the antenna (see Fig. 3.9 for a representation of the spatial variation of the field strength for the infinitesimal dipole). In other words, the electric field strength is proportional to the amplitude I_0 of the feeding current, inversely proportional to the distance r, and varies in strength with the sine of the angle θ. The region at which the approximation in (3.25) holds is referred to as the *far-field* of the antenna. In the far-field, the antenna can be considered to behave as a point source, that is, the propagation direction is purely radial.

3.3.2 NEAR-FIELDS AND FAR-FIELDS

For the infinitesimal dipole antenna, the far-field boundary occurs at $r \gg \lambda$, while for larger antennas an additional condition is that $r > 2D^2/\lambda$, where D is the largest dimension (length, diameter, etc.) of the antenna. At shorter distances from an antenna, that is, in the *near field*, the distance and angular dependence of the electric and magnetic fields are much more complex. An intuitive explanation for this is that in the near field different parts of the antenna are in different directions with respect to the observation point and hence wave components with different (non-radial) propagation directions are superimposed. Very close to the antenna in the *reactive near field*, the energy transfer is not strictly outwards, instead some energy is oscillating back and forth between the antenna and the nearby space.

3.3.3 ANTENNA POLARIZATION

The *polarization of an antenna*, $\hat{\psi}_\mathrm{a}$, is defined as the polarization of a wave transmitted by the antenna. As an example, the infinitesimal dipole in Section 3.3.1 is referred to as being vertically polarized (since $E_\varphi = 0$). In general, the antenna polarization may be a function of the observation angle as $\hat{\psi}_\mathrm{a}(\theta, \varphi)$.

From reciprocity considerations (see also Section 3.3.5), the polarization of an antenna also applies at reception and is then understood as the polarization of an incident plane wave for which the received power is maximized. It is important to note that the receive polarization of an antenna is a mirror image of its transmit polarization. To understand this, assume that an antenna radiates

[7]Using dipole antennas, Heinrich Hertz was the first to demonstrate the existence of radio waves in 1887.

outgoing waves with the polarization $\widehat{\psi}_{\mathrm{a}} = a_\theta \widehat{\boldsymbol{\theta}} + a_\varphi \widehat{\boldsymbol{\varphi}}$. However, the *wave polarization* is defined when looking along the direction of propagation; hence an incident wave with the same wave polarization will be oscillating in the direction $a_\theta \widehat{\boldsymbol{\theta}} - a_\varphi \widehat{\boldsymbol{\varphi}}$ in the coordinate system of the receive antenna. Hence, to maximize the received power, the antenna receive polarization $\widehat{\psi}_{\mathrm{a,r}}$ should be equal to $a_\theta \widehat{\boldsymbol{\theta}} - a_\varphi \widehat{\boldsymbol{\varphi}}$. Using Jones vectors for the transmit and receive polarizations, this can be expressed as

$$\underline{\widehat{\psi}}_{\mathrm{a,r}} = \begin{bmatrix} 1 & 0 \\ 0 & -1 \end{bmatrix} \underline{\widehat{\psi}}_{\mathrm{a}} \tag{3.26}$$

As the mirror image relation between transmit and receive polarization is far from intuitive, an illustrative example is shown in Fig. 3.8.

In the case that an incident wave has a different polarization $\underline{\widehat{\psi}}_{\mathrm{w}} = \left[a_{\theta,\mathrm{w}} \quad a_{\varphi,\mathrm{w}}\right]^{\mathrm{T}}$ from that of the antenna, $\underline{\widehat{\psi}}_{\mathrm{a,r}} = \left[a_{\theta,\mathrm{r}} \quad a_{\varphi,\mathrm{r}}\right]^{\mathrm{T}}$, not all of the energy in the wave will be captured by the antenna. In fact, only the component of the wave that has the same polarization $\widehat{\psi}_{\mathrm{a,r}}$ as the receive antenna will induce any current in the antenna. The component with orthogonal polarization $\widehat{\psi}_{\mathrm{a,r}}^{\perp}$ will not induce any current in the antenna, since from reciprocity a current in the antenna will only produce a radiated wave of the first polarization (recalling the mirror image between transmit and receive polarization discussed above).

Hence, an incident wave with electric field $\boldsymbol{E}_0 = E_0 \widehat{\psi}_{\mathrm{w}}$ will produce the same current in the antenna as an incident wave with amplitude χE_0 that is perfectly aligned with the antenna polarization, $\boldsymbol{E}_0' = \chi E_0 \widehat{\psi}_{\mathrm{a,r}}$. Here χ is a complex constant that relates the amplitude of the arbitrarily polarized incident wave with the amplitude of the polarization-matched incident wave giving the same receive antenna response. Using the Jones vector representations, χ can be determined as

$$\chi(\theta,\varphi) = a_{\theta,\mathrm{r}}(\theta,\varphi) a_{\theta,\mathrm{w}} + a_{\varphi,\mathrm{r}}(\theta,\varphi) a_{\varphi,\mathrm{w}} = \underline{\widehat{\psi}}_{\mathrm{a,r}}^{\mathrm{T}}(\theta,\varphi)\underline{\widehat{\psi}}_{\mathrm{w}} \tag{3.27}$$

In the above expression, the angular dependence of the antenna polarization and consequently of χ is explicitly written out.

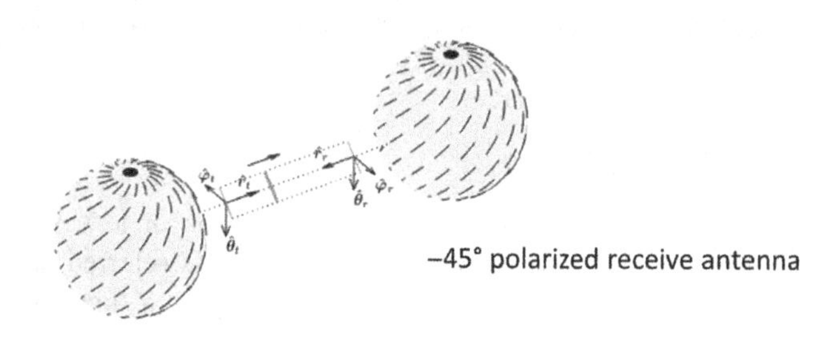

FIGURE 3.8

Illustration of the mirror image relation between the transmit and receive polarization of an antenna. Both antennas in the figure have a −45° (transmit) polarization, yet a −45° polarized wave impinging on the receive antenna is orthogonal to the −45° polarized receive antenna due to the mirror image relation.

3.3.4 RADIATION PATTERN

The absolute value of the complex amplitude of a spherical wave in free space is inversely proportional to the distance (r) from the source as $E_0 \sim 1/r$ (compare Eq. (3.25)). Therefore the time-averaged power density $\boldsymbol{S}_{\text{av}} = W_{\text{av}}\widehat{\boldsymbol{r}}$, which is proportional to the square of the electric field [see Eq. (3.9) in Section 3.2.1], will be inversely proportional to the square of the distance. This relation follows intuitively from energy conservation principles, for example, the power emanating from a source is distributed over a sphere enclosing the source where the area of the sphere is proportional to r^2 (see also Fig. 3.14). A related measure is the *radiation intensity* $U = r^2 W_{\text{av}}$ given as the power radiated per solid angle:

$$U(\theta, \varphi) = \frac{r^2}{2\eta}\left(|E_\theta|^2 + |E_\varphi|^2\right) \tag{3.28}$$

where E_θ and E_φ are the transverse electric field components in the $\widehat{\boldsymbol{\theta}}$ and $\widehat{\boldsymbol{\varphi}}$ directions as in (3.11) and η is the intrinsic impedance of free space (see Section 3.2.1).

The total radiated power P_{rad} is the integral of the radiation intensity over all solid angles, that is, the full sphere:

$$P_{\text{rad}} = \oiint_\Omega U d\Omega = \int_0^{2\pi}\int_0^{\pi} U(\theta, \varphi)\sin\theta d\theta d\varphi \tag{3.29}$$

The radiation properties as a function of angle at a fixed (large) distance (e.g., in the far-field) is commonly referred to as the *radiation pattern*[8] of the antenna. In the literature, a radiation pattern can refer to radiation intensity, field strength, polarization, phase, etc. In this book, the radiation pattern is denoted as $\boldsymbol{g}(\theta, \varphi)$, and is composed of the complex amplitude pattern $g(\theta, \varphi)$ and the antenna polarization $\widehat{\boldsymbol{\psi}}_{\text{a}}(\theta, \varphi)$ introduced in Section 3.3.3:

$$\boldsymbol{g}(\theta, \varphi) \triangleq g(\theta, \varphi)\widehat{\boldsymbol{\psi}}_{\text{a}}(\theta, \varphi). \tag{3.30}$$

The radiation pattern is related to the transmitted spherical electric wave $\boldsymbol{E}(\boldsymbol{r},t)$ via

$$\boldsymbol{E}(\boldsymbol{r},t) \sim \frac{1}{r}\exp(j(\omega t - \boldsymbol{k}\cdot\boldsymbol{r}))\boldsymbol{g}(\theta, \varphi). \tag{3.31}$$

Here the normalization of $\boldsymbol{g}(\theta, \varphi)$ is done such that $U(\theta, \varphi) = P_{\text{t}}/4\pi|g(\theta, \varphi)|^2$, where P_{t} represents the transmit power fed into the antenna. The total radiated power is in turn given by $P_{\text{rad}} = \varepsilon P_{\text{t}}$, where ε represents losses in the antenna (see also Sections 3.3.7.3 and 3.3.7.4) and the reason for the chosen normalization is that $g(\theta, \varphi) = 1$ then corresponds to a lossless isotropic antenna (see also Section 3.3.6).

Hence, while the radiation intensity $U(\theta, \varphi)$ is proportional to the transmitted power and characterizes the absolute radiated power per solid angle, the radiation pattern $\boldsymbol{g}(\theta, \varphi)$ is independent of the transmit power and characterizes the spatial and polarization distribution of the radiated field in a relative sense.

Returning to the example of an infinitesimal dipole in Section 3.3.1, it follows from comparison of (3.31) with (3.25) that the following components of the radiation pattern are identified for the infinitesimal dipole:

---

[8] *Antenna pattern* is often used synonymously with radiation pattern.

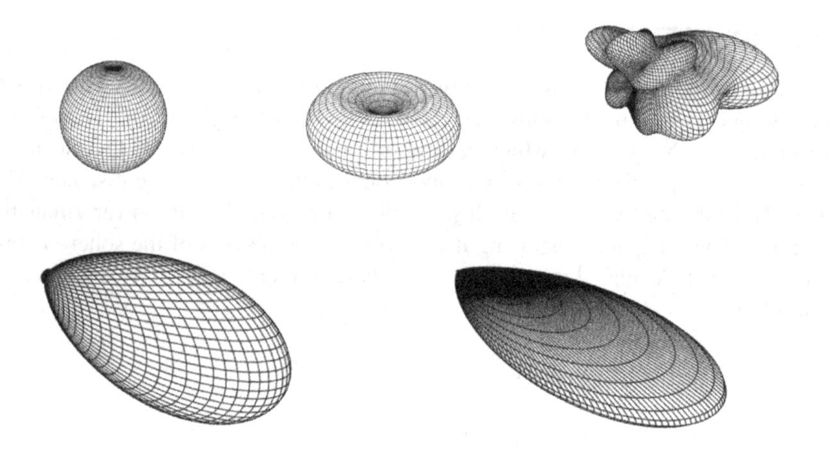

FIGURE 3.9

Some examples of radiation patterns (radiation intensity): isotropic antenna (upper left), infinitesimal dipole (upper middle), a GSM phone (upper right), circular aperture antenna (lower left), and a macro base station antenna for sector coverage (lower right).

$$g(\theta, \varphi) \sim \sin(\theta) \tag{3.32}$$

$$\hat{\psi}_{\mathrm{a}} = \hat{\boldsymbol{\theta}} \tag{3.33}$$

Similarly, the radiation intensity of the infinitesimal dipole is

$$U(\theta, \varphi) \sim \sin^2(\theta) \tag{3.34}$$

Most importantly, it can be seen that the complex amplitude pattern $g(\theta, \varphi)$ and radiation intensity $U(\theta, \varphi)$ are different in different directions. For the infinitesimal dipole antenna, the radiation intensity is maximized along the horizon where $\theta = 90°$ and is zero for $\theta = 0°$ or $\theta = 180°$, that is, directly above and below the antenna (see Fig. 3.9).

More generally, antennas may be designed to create more complicated current distributions, which in turn results in more complicated radiation patterns where both $g(\theta, \varphi)$ and $\hat{\psi}_{\mathrm{a}}(\theta, \varphi)$ can have large and rapid variations with respect to the observation angle. In addition, interaction between the radiating antenna and objects nearby can further distort the radiation pattern, as in the case of mobile phone radiation properties being affected both by the metallic chassis of the phone and by the user holding the phone. Some examples of radiation intensity patterns for different antennas are shown in Fig. 3.9 and for the antenna polarization in Fig. 3.10.

3.3.5 RECIPROCITY OF ANTENNA CHARACTERISTICS

The radiation pattern $\boldsymbol{g}(\theta, \varphi)$ provides the relation between the current amplitude I_o at the antenna terminals and the field strength $\boldsymbol{E}(\boldsymbol{r},t)$ of the radiated electromagnetic waves in the direction (θ, φ). With a few exceptions involving the use of nonlinear materials, most antennas are *reciprocal*, meaning that a plane wave with field strength $\boldsymbol{E}$ impinging from direction (θ, φ) will induce a current I_0 in the antenna. Thus the radiation pattern $\boldsymbol{g}(\theta, \varphi)$ applies both for transmission and for

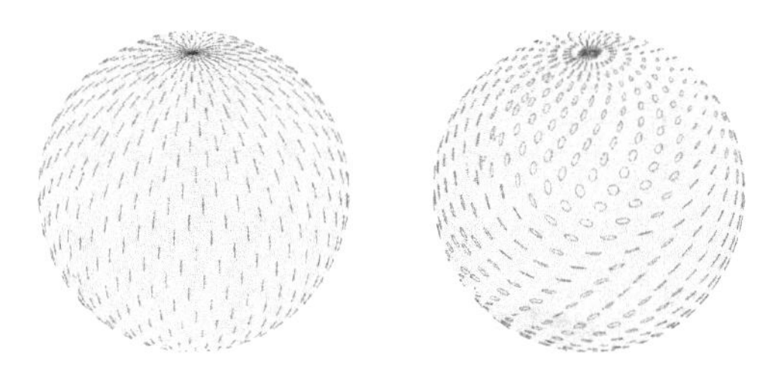

FIGURE 3.10

Some examples of radiation patterns (antenna polarization): infinitesimal dipole (left) having a constant polarization and GSM phone (right) where the polarization varies as a function of direction.

reception. However, recall that the receive antenna polarization is a mirror image of its transmit polarization as was discussed in Section 3.3.3.

In the following sections some further characteristics of the radiation pattern will be discussed and metrics such as directive gain, half-power beamwidth, efficiency, antenna gain, and effective antenna area will be explained. All of these characteristics are equally valid for describing the antenna in transmission and in reception.

3.3.6 ISOTROPIC ANTENNA

When describing the radiation characteristics of different antennas, it is convenient to relate the performance to that of an *isotropic* antenna (see Fig. 3.9), which radiates with the same intensity in all directions: which means that $g(\theta, \varphi)$ is constant independent of the direction. Unless otherwise mentioned, we will in the book assume that an isotropic antenna is also lossless, which corresponds to $g(\theta, \varphi) = 1$. An isotropic antenna is a theoretical concept—all practical antennas will radiate unequally with larger intensity in some directions and smaller (even zero) in other directions. The existence of at least one *null*, a direction in which zero energy is radiated, is a result of the fact that the electromagnetic field is transverse to the direction of propagation; therefore the fields will be tangential along the surface of a sphere enclosing the antenna. Such tangential fields are impossible to arrange without the field becoming zero in at least one point on the sphere, which is proven by the "hairy ball theorem," which in layman terms states that it is impossible to comb a hairy ball without creating a cowlick.

3.3.7 ANTENNA PROPERTIES

3.3.7.1 Directivity

The *directivity*[9] $D(\theta, \varphi)$ of an antenna is defined as the ratio of the radiation intensity $U(\theta, \varphi)$ (power per solid angle, see Eq. (3.28)) in a given direction to the radiation intensity averaged over all directions, that is, the total radiated power P_{rad} divided by the solid angle of the full sphere, 4π.

[9]The term *directive gain* is also used.

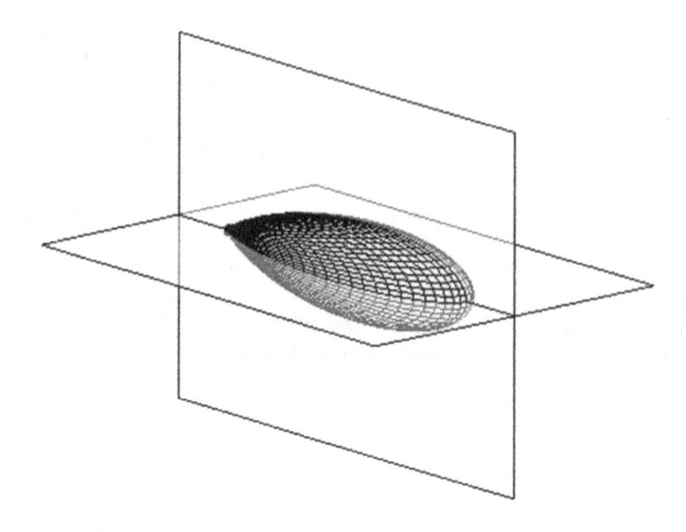

FIGURE 3.11

Illustration of the two main cuts or planes for specifying the antenna directivity. See examples of plots in these planes in Fig. 3.12.

$$D(\theta, \varphi) = \frac{U(\theta, \varphi)}{P_{\text{rad}}/4\pi} \tag{3.35}$$

This power ratio is usually specified in decibels with the unit dBi ("decibels over isotropic," that is, compared to an isotropic antenna that radiates the same total power). The isotropic antenna has a directivity of 0 dBi in all directions, while the directivity of the vertical infinitesimal dipole varies from $-\infty$ dBi when looking straight up or down to 1.76 dBi along the horizontal plane (compare Fig. 3.9).

For antennas with static radiation patterns the directivity is often specified only in the horizontal and vertical planes (see Figs. 3.11 and 3.12).

3.3.7.2 Lobe width and sidelobe levels

In general, the radiation pattern of an antenna may have large variations with respect to the observation angle. A *lobe*[10] can be loosely defined as an angular interval where the directive gain is higher relative to the gain in surrounding directions. By this definition, the infinitesimal dipole has a single lobe, but it is not uncommon for antennas to have many different lobes. Some antennas are designed to produce one *main lobe* with high directive gain (see an example in Fig. 3.12). It is almost unavoidable that such a design has some weaker *side lobes* in other directions and even a *back lobe* in the opposite direction of the main lobe. Other antennas may have multiple lobes of similar strength. A lobe can be characterized by its *half-power beamwidth* (HPBW)[11] that specifies

[10]Traditionally, these have also been referred to as "beams." However, with the use of AAS the term beam has in this book been taken to include the whole radiation pattern (associated with a single stream of information-carrying symbols) and not only a particular lobe of it.

[11]More correctly "half-power *lobe* width."

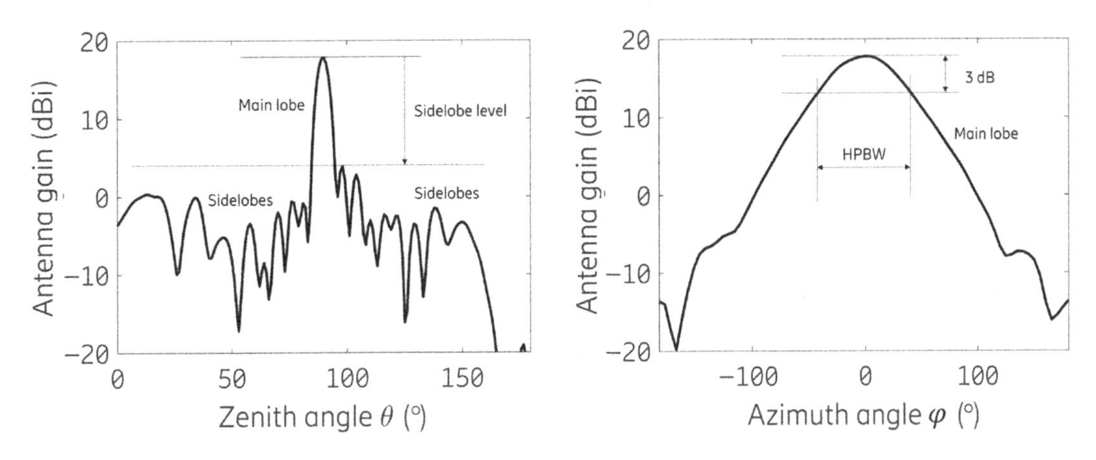

FIGURE 3.12

Plots of antenna gain (directivity times efficiency) in the vertical (left) and horizontal (right) planes for a macro base station antenna.

the angular interval over which the lobe has a directive gain within 3 dB of the peak directive gain of the lobe. The relative levels of the side lobes and back lobes can usually also be found in the specifications of an antenna.

3.3.7.3 Efficiency

Practical antennas always incur some losses as some energy may be converted to heat instead of electromagnetic waves due to finite electric conductivity. The *antenna efficiency* ε characterizes the fraction of the supplied energy that is converted into radiation. It is always lower than 1 and can be as low as 0.1–0.2 for antennas that are small in relation to the wavelength or are exposed to lossy materials in their near field. A user holding a mobile phone is an example of the latter.

3.3.7.4 Antenna gain

The antenna gain $G(\theta, \varphi)$ is the product of the directive gain and the efficiency, $G(\theta, \varphi) = \varepsilon D(\theta, \varphi) = |g(\theta, \varphi)|^2$; hence losses in the antenna are also taken into account. If a single value for the antenna gain is specified, this is understood to be the maximum of the gain over all directions. The directive gain and the antenna gain both describe how power is radiated into space, the difference is the reference which is the total power fed into the antenna for G and the total power radiated from the antenna for D. This means that the antenna gain in a certain direction is the ratio of the radiation intensity to the radiation intensity that would be obtained using an isotropic lossless antenna with the same total power fed into it.

A typical antenna used at a macro base station can have an antenna gain of 15 dBi or more within its main sector of coverage (compare Fig. 3.12), while a mobile phone antenna may have an antenna gain closer to 0 dBi or even negative in dBi. One extreme example at radio frequencies is the radio telescope at the Arecibo Observatory in Chile, which is equipped with a 305 m diameter reflector antenna. At 3 GHz the peak antenna gain of this antenna is roughly 80 dBi and the beamwidth is slightly less than 0.03°.

3.3.7.5 Effective antenna area

One convenient measure is the *effective antenna area* A_{eff}, which relates the average power density S (in power per unit area) of an electromagnetic wave incident from a certain direction to the power P_{r} that is captured by the antenna as

$$P_{\mathrm{r}} = A_{\mathrm{eff}} S \tag{3.36}$$

In other words, the effective antenna area is the size of an area through which the same power flows as is collected by the antenna. As such, it is convenient metric for determining the received power using the very simple Eq. (3.36). From reciprocity considerations it can be shown that the effective antenna area can be expressed using the antenna gain G as

$$A_{\mathrm{eff}}(\theta, \varphi) = \frac{\lambda^2}{4\pi} G(\theta, \varphi) |\chi(\theta, \varphi)|^2 \tag{3.37}$$

The factor $|\chi(\theta, \varphi)|^2$ which is often referred to as the *polarization loss factor* represents the fraction of the incident power of the wave that has a polarization that is aligned with the antenna receive polarization, as discussed in Section 3.3.3. For a lossless isotropic antenna with $G(\theta, \varphi) = 1$ and a receive polarization $\widehat{\psi}_{\mathrm{a,r}}$ that is perfectly matched to the polarization $\widehat{\psi}$ of the impinging wave, that is, where $|\chi(\theta, \varphi)|^2 = 1$, the effective antenna area is $\lambda^2/4\pi$. Note that the effective antenna area is in general a function of both direction and wave polarization. As with the antenna gain, if a single value for the effective area is given, this is understood to refer to the value in the maximum direction.

The effective area is different than the physical area, though there exist relations between the two depending on the antenna type. Some examples from Ref. [1] are summarized in Table 3.1. For small antennas, the effective antenna area is of the order $(\lambda/2)^2$, while for larger antennas it approaches the physical area.

3.3.8 ANTENNA SIZE, GAIN, LOBE WIDTH, AND FREQUENCY DEPENDENCE

The infinitesimal dipole introduced in Section 3.3.1 creates a radiation pattern that is omnidirectional, that is, it has a constant antenna gain as a function of the azimuth angle φ but varying antenna gain in the elevation domain. More directive radiation patterns with higher antenna gain in the maximum direction can be created with electrically[12] larger antennas (see some examples in Fig. 3.9 and Table 3.1). As can be seen from Table 3.1, two antennas of the same type that have equal gains but operate at different frequencies f_1 and f_2 will tend to have physical and effective areas that are related as

$$\frac{A_2}{A_1} = \frac{A_{\mathrm{eff},2}}{A_{\mathrm{eff},1}} = \left(\frac{f_1}{f_2}\right)^2 \tag{3.38}$$

Hence, if the antenna gain is independent of frequency, then the antenna will become physically smaller with frequency and also capture less power in receiving mode due to the reduced effective antenna area. An illustration of the size versus frequency for a set of equal gain horn antennas is

[12]Electrically large means large in relation to the wavelength. A 1 m tall antenna is electrically small for a 1 MHz frequency (wavelength is 300 m) but electrically large for a 10 GHz frequency (wavelength is 3 cm).

Table 3.1 Relations Between Physical Area, Effective Antenna Area, and Antenna Gain.

Antenna Type	Physical Area	Effective Antenna Area (A_{eff})	Antenna Gain (G)
Isotropic antenna	N/A	$\frac{\lambda^2}{4\pi}$	1
Small dipole or half-wavelength dipole	$\ll \lambda^2$	$\approx \frac{\lambda}{2} \cdot \frac{\lambda}{4}$	$\approx \frac{\pi}{2}$
Parabolic reflector antenna	A_{parab}	$\approx \frac{2}{3} A_{\text{parab}}$	$\approx \frac{8\pi}{3\lambda^2} A_{\text{parab}}$
Various horn antennas	A_{horn}	$\approx 0.5 A_{\text{horn}}$ to $0.81 A_{\text{horn}}$	$\approx \frac{2\pi}{\lambda^2} A_{\text{horn}}$ to $\frac{3.24\pi}{\lambda^2} A_{\text{horn}}$
Rectangular array of n half-wave dipoles with half-wavelength spacing in front of a ground plane	$\approx n(0.5\lambda)^2$	$\approx n(0.5\lambda)^2$	$\approx n\pi$

From H.T. Friis, A note on a simple transmission formula, Proc. IRE 34 (5) (May 1946) 254–256, with antenna gain estimated using (3.37).

FIGURE 3.13

10 dBi horn antennas at different frequencies from 700 MHz at the extreme right to 37 GHz at the extreme left.

given in Fig. 3.13. All antennas have the same gain and a similar electrical size of $\sim \lambda \times \lambda$ but evidently very different physical sizes.

Conversely, if the physical antenna area is kept constant while the frequency is increased, the antenna gain will tend to increase at the rate

$$\frac{G_2}{G_1} = \left(\frac{f_2}{f_1}\right)^2 \tag{3.39}$$

At the same time, the lobe width becomes narrower since it is inversely proportional to the directive gain. For a planar array, an approximate relationship between the horizontal HPBW φ_{HPBW} and the vertical HPBW θ_{HPBW} (both in degrees) and the directive gain D is

$$D \approx \frac{32{,}400}{\varphi_{\text{HPBW}} \theta_{\text{HPBW}}} \tag{3.40}$$

Returning to the horn antennas in Fig. 3.13, if the higher frequency horns would be as big as the 700 MHz horn (or rather use the same effective antenna area), the antenna gain would be 19 dBi at 2 GHz, 33 dBi at 10 GHz and a staggering 44 dBi at 37 GHz. Simultaneously, the lobe widths, assuming for simplicity that (3.40) is applicable also for a horn antenna, would go from 57° at 700 MHz to 19° at 2 GHz, 4° at 10 GHz and 1° at 37 GHz.

Increasing the physical antenna size or increasing the frequency[13] (thereby increasing the electrical size) is hence effective means for making an antenna focus its radiation into a narrower angular interval and hence increase the antenna gain in the corresponding directions. This can be very beneficial for communication purposes as more energy can reach a receiver in a certain direction and less energy is transmitted in unwanted directions where it would interfere with other communication equipment. However, high directivity is of little or no use if it cannot be controlled so that the energy is steered in the desired direction to reach the intended receiver. Physical rotation of the antenna is sometimes used, for example, for some radar transmissions. However, electrical steering is much more suited to the high demands of mobile communications where the same antenna is to be used to communicate with one or more mobile users, and transmission directions can change drastically on a millisecond time scale. Such steering is achieved using *antenna arrays*, where an effectively large antenna is synthesized using several individually controlled smaller antennas. Antenna arrays will be discussed in detail in Chapter 4, while methods for determining how to spatially distribute the energy will be treated in Chapter 6.

3.4 TRANSMISSION AND RECEPTION OF ELECTROMAGNETIC WAVES: FREE-SPACE PROPAGATION

3.4.1 THE RECIPROCAL RADIO CHANNEL

The relation between the signal[14] transmitted from a first antenna and the signal received at a second antenna is characterized by the *radio channel*. If the medium in which the radio waves propagate between the antennas is linear and isotropic, one can show using Maxwell's equations that the radio channel is unchanged even if the second antenna is transmitting and the first antenna is receiving. We say that the radio channel is reciprocal. The condition of reciprocity has been found to apply to almost[15] all radio channels experienced by mobile communication systems.

Reciprocity allows information about the radio channel in one direction to be inferred from measurements of the radio channel in the opposite direction, which is of great importance for some advanced antenna transmission schemes as discussed further in Section 6.4.2.

[13]Since the radio frequency spectrum is regulated this is seldom an option.

[14]Signals can be represented by currents or voltages at the antenna terminals that drive the current distribution on the antenna and hence the radiated electromagnetic waves.

[15]This is mostly true for the desired signal; however, nonlinearity in the transmitter or in the antenna can create interfering signals due to *passive intermodulation*. This is a well-known problem in mobile communication systems.

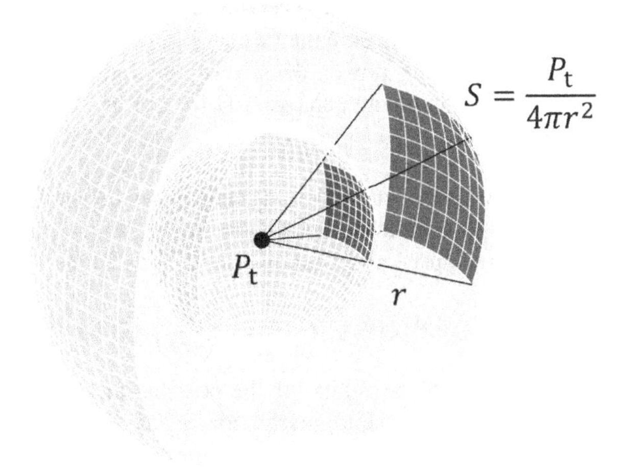

FIGURE 3.14

An isotropic antenna radiates the power P_t. This power is distributed uniformly over the area of a sphere. At a radius r from the antenna, the power density S of the electromagnetic waves is $P_t/4\pi r^2$.

3.4.2 FRIIS' EQUATION

Assume that a lossless isotropic antenna radiates electromagnetic waves with power P_t equally in all directions (see Fig. 3.14). In free space, the power density (power per unit area) of these waves at a distance of r equals

$$S = \frac{P_t}{4\pi r^2} \tag{3.41}$$

since the area of a sphere enclosing the antenna is $4\pi r^2$. Note that the power density is the same for any transmitted frequency. Now insert a lossless isotropic receiving antenna at distance r with a polarization that is matched to the field. Utilizing the effective antenna area of the isotropic antenna that was introduced in Section 3.3.7.5, the received power by this antenna can be determined to be

$$P_r = SA_{\text{eff,isotropic}} = \frac{P_t}{4\pi r^2}\frac{\lambda^2}{4\pi} = P_t\left(\frac{\lambda}{4\pi r}\right)^2. \tag{3.42}$$

In logarithmic units,

$$P_{r,\text{dBm}} = P_{t,\text{dBm}} - 20\log_{10}\left(\frac{4\pi r}{\lambda}\right), \tag{3.43}$$

which can be reformulated as

$$\text{FSPL}_{\text{dB}} = P_{t,\text{dBm}} - P_{r,\text{dBm}} = 20\log_{10}\left(\frac{4\pi r}{\lambda}\right), \tag{3.44}$$

where the right-hand side is often referred to as the *free-space path loss between isotropic antennas* (FSPL). It should be emphasized that the frequency dependence of the FSPL is due to the frequency-dependent effective antenna area of the receiving antenna and *not* due to any losses that occur between the transmitting and receiving antenna (which is a common misconception).

For an arbitrary transmit antenna, the power density of the radiated waves in the direction (θ_t, φ_t) from the antenna will depend on the antenna gain $G_t(\theta_t, \varphi_t)$ as

$$S = S_{\text{isotropic}} G_t(\theta_t, \varphi_t). \tag{3.45}$$

Similarly, the effective antenna area of the receiving antenna is dependent on the antenna gain $G_r(\theta_r, \varphi_r)$ and the receive antenna polarization according to Eq. (3.37).

Hence, the received power in free-space propagation conditions will be

$$P_r = SA_{\text{eff,isotropic}} = P_t G_t(\theta_t, \varphi_t) G_r(\theta_r, \varphi_r) \left(\frac{\lambda}{4\pi r}\right)^2 |\chi(\theta, \varphi)|^2. \tag{3.46}$$

The polarization loss factor $|\chi(\theta, \varphi)|^2$ accounts for the possible mismatch between the transmit antenna polarization $\widehat{\psi}_t$ and the receive antenna polarization $\widehat{\psi}_r$ as discussed in Section 3.3.7.5. In the worst case, when the transmit and receive polarizations are orthogonal, no power is received at all. Antenna systems employing dual polarizations in transmission and/or reception are an efficient way to mitigate this loss, since the power that is lost by using one polarization can be recovered using the orthogonal polarization.

Considering only co-polarized antennas and antennas that are aligned such that the maximum antenna gain is in the direction of the other antenna, Eq. (3.46) can be simplified to

$$P_r = P_t G_t G_r \left(\frac{\lambda}{4\pi r}\right)^2. \tag{3.47}$$

This equation is attributed to Harald T. Friis [1] and tells us that the received power in free space drops with the inverse square of the distance, but that it can be increased using directive antennas at the transmitter or receiver. We can also see that if the directive gains of the antennas are constant, the received power reduces with increasing frequency (the wavelength λ becomes shorter) at a rate of $20\log_{10}(f)$ in dB. Unless we can increase the directive gains of the antennas, the coverage will become worse the higher the frequency becomes. Fortunately, as was described in the previous section, higher frequencies allow the antennas to become electrically larger for a given physical size, and hence the antenna gains will grow with frequency as well. For a fixed physical area, the antenna gain of an antenna grows as $20\log_{10}(f)$ according to Eq. (3.39). Using a fixed size transmitting antenna (such as a parabolic reflector antenna) and an isotropic receiving antenna (or vice versa) results in a frequency-independent path loss, since the reduced power due to the effective antenna area is exactly canceled by the increased directivity. Utilizing fixed size antennas for both transmission and reception results in a received power that *increases* with increasing frequency. The caveat is of course that the directivity also increases with frequency and hence two highly directive antennas need to be pointed toward each other, with a direction error that is less than the HPBW. This can be achieved with little effort for a fixed transmission link but can be very challenging for mobile communication systems, particularly in the presence of multipath propagation, which will be discussed in the next section. Here AAS are required to steer the high directive gains in the desired directions in space. Methods for adapting the radiation pattern in order to utilize the antenna area and directivity are the basis for AAS and will be discussed throughout this book, in particular in Chapter 6.

3.5 WAVE PROPAGATION IN REAL-WORLD ENVIRONMENTS

Real-world wireless communication scenarios involve propagation in complex environments where the radio waves interact with various objects through processes such as reflection, transmission, diffraction, and scattering. These interactions are well understood for simpler canonical scenarios such as those shown in Fig. 3.15 where Maxwell's equations in combination with appropriate boundary conditions can be used to analytically or numerically compute the resulting electromagnetic fields. However, for radio wave propagation in, for example, a building or a city, the scale of the problem makes it challenging or even impossible to accurately calculate the electromagnetic waves and fields using such methods or even to accurately describe the environment with subwavelength accuracy. Instead empirical models and engineering rules of thumb have been developed through measurements of wireless propagation.

In this section, we will discuss how the radio waves interact with the environment creating variations on different scales. We speak about *large-scale channel variations* when discussing characteristics such as path loss, shadow fading, and time and angular dispersion, which vary on the meter to kilometer scale. The large-scale variations can be understood as determining the average power and distribution of multi-path. The constructive and destructive superposition of multi-path creates *small-scale channel variations*, or fast fading, which occur on the scale of the wavelength.

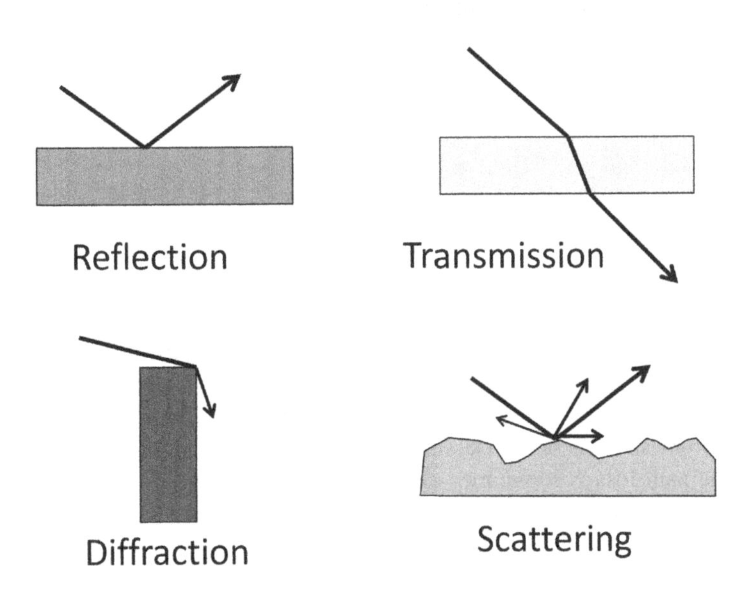

FIGURE 3.15

Schematic illustrations of reflection, transmission, diffraction, and scattering that represent basic types of interactions of electromagnetic waves with the environment.

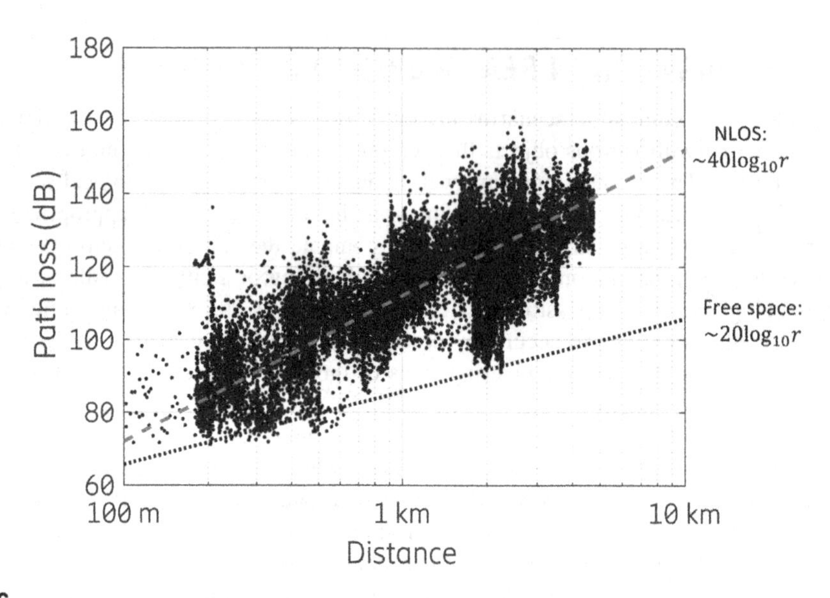

FIGURE 3.16

Measured path loss versus distance (*black dots*) between a macrocell and a user equipment (UE) on ground. The dotted line represents free space path loss and the dashed line represents a slope proportional to r^4.

3.5.1 PATH LOSS AND SHADOW FADING

We define the *path loss between isotropic antennas* ("path loss" in short) as the ratio between the transmitted power and the locally averaged[16] received power where both transmitter and receiver are using lossless isotropic antennas:

$$\mathrm{PL} = \frac{P_\mathrm{t}}{P_\mathrm{r}} \tag{3.48}$$

or, in logarithmic units[17]:

$$\mathrm{PL(dB)} = P_{\mathrm{t(dB)}} - P_{\mathrm{r(dB)}}. \tag{3.49}$$

When there is *line of sight (LOS)* between the transmitter and receiver, the path loss is usually similar to that in free space, that is, proportional to r^2 (or $20\log_{10}r$ in dB units). In *non-line of sight (NLOS)* conditions, the attenuation caused by obstructions such as terrain, buildings, trees, and walls will cause greater path loss variations, including an increase with distance at a more rapid rate. One example from an outdoor measurement in a macrocell scenario is shown in Fig. 3.16. It can be seen that the path loss is sometimes close to the reference FSPL curve. This is typical for locations in LOS. In NLOS the path loss tends to increase faster with increasing distance due to the

[16]The local average is done to remove small-scale variations due to constructive or destructive superposition of multipath. A suitable averaging length is $>10\lambda$ (compare Fig. 3.19).

[17]Radio engineers prefer to use dB values since the path loss conditions under which a communication system typically operate span many orders of magnitude.

presence of obstructions such as terrain, buildings, and trees, and also experience more variability around the mean value for a specific distance. Empirical fitting of a curve with a constant slope against these results produces a path loss proportional to approximately r^4 (or $40\log_{10}r$ in dB units). The variability of observations around this linear model is commonly normal distributed in logarithmic units with typical standard deviations in the range of 6–10 dB. These variations are referred to as *log-normal fading* or *shadow fading* and are caused by local variations in the amount of obstructions that cause variations in how much the radio waves are "shadowed" by the obstructions, hence the name shadow fading. Shadow fading tends to be spatially correlated, that is, in locations near each other a similar offset from the mean path loss is experienced.

A simple empirical one-slope path loss model is the ABG model, defined by:

$$\mathrm{PL(dB)} = A + B\log_{10}r + G\log_{10}f + \sigma, \tag{3.50}$$

where A is an offset, B is the slope with respect to distance, G represents the frequency dependence of the path loss, and σ represents the log-normal fading. Different values for these parameters have been observed for different scenarios (urban, suburban, rural, indoor) and for different antenna heights (see, e.g., the Hata model [2] or Table 3.2). For outdoor deployments, increasing the antenna height tends to increase the coverage but also the interference to neighboring cells.

As seen in the previous section, the frequency-dependent effective antenna area of the isotropic receive antenna causes a $20\log_{10}(f)$ dependence for the path loss in free space. In outdoor NLOS propagation conditions, the frequency dependence is similar but in certain conditions somewhat stronger due to losses incurred by diffraction, attenuation by trees and foliage, etc. In addition, atmospheric losses due to oxygen absorption (around 15 dB/km in a band around 60 GHz) or rain may cause additional losses at higher frequencies; however, these are seldom significant at the frequencies and cell sizes used in cellular networks.

Table 3.2 Typical Distance and Frequency Dependence of the Path Loss in Different Common Scenarios.

Scenario		Typical Distance Dependence (B)	Typical Frequency Dependence (G)
Free space and LOS		20	20
NLOS	Urban macrocell	20 at short range, increasing to ~40 at longer range or for lower frequencies	20–23
	Suburban macrocell		23
	Rural macrocell		20–23
	Urban microcell	35	21–23
	Indoor cell	38	23–25

Based on 3GPP TR 38.901, Study on channel model for frequencies from 0.5 to 100 GHz, 2019 [3], ITU-R Report M.2412, Guidelines for evaluation of radio interface technologies for IMT-2020, 2017 [4]. M. Riback, J. Medbo, J.-E. Berg, F. Harrysson, H. Asplund, Carrier frequency effects on path loss, in: 2006 IEEE 63rd Vehicular Technology Conference, Melbourne, Victoria, 2006, pp. 2717–2721 [5]. Millimetre-wave based mobile radio access network for fifth generation integrated communications (mmMAGIC), Measurement Results and Final mmMAGIC Channel Models, Deliverable 2.2, H2020-ICT-671650-mmMAGIC/D2.2, May 2017 [6].

3.5.2 BUILDING PENETRATION LOSS

Path loss models such as the ABG model have been instrumental in dimensioning and deploying earlier generation cellular communication systems, where the design target was outdoor (vehicular) coverage. However, due to the successful introduction of handheld mobile phones there has been an increased demand for indoor coverage. As was discussed in Section 2.1.5, the majority of network traffic originates indoors. Hence, the additional loss that the radio waves experience when propagating into a building needs to be characterized and taken into account in the radio network design and deployment. This loss obviously depends on the materials used to construct the building. Brick or concrete usually provides larger attenuation that increases rapidly with frequency, while regular glass is quite transparent also to radio waves. On the other hand, metal-coated glass that is being increasingly used for energy-saving purposes incurs a high loss to the radio waves. Since the radio waves may take many different paths into the building, the effective loss L_{eff} through the external building wall can be approximated by a weighted average of the losses $L_n, n = 1 \ldots N$ of the N different materials of the building façade taking into account the relative proportions p_n of the surface area of the external façade composed of the different materials

$$L_{\text{eff}} = \frac{1}{\sum_n p_n \frac{1}{L_n}} \tag{3.51}$$

The intuition behind this expression is illustrated in Fig. 3.17 (left), where the loss is often determined by the most transparent material even if this material only makes up a smaller fraction of the total façade area.

Empirical models for the building penetration loss have been developed and calibrated against measurements, for example, by 3GPP [3] as shown in Fig. 3.17 (right) and by the ITU-R [7]. These

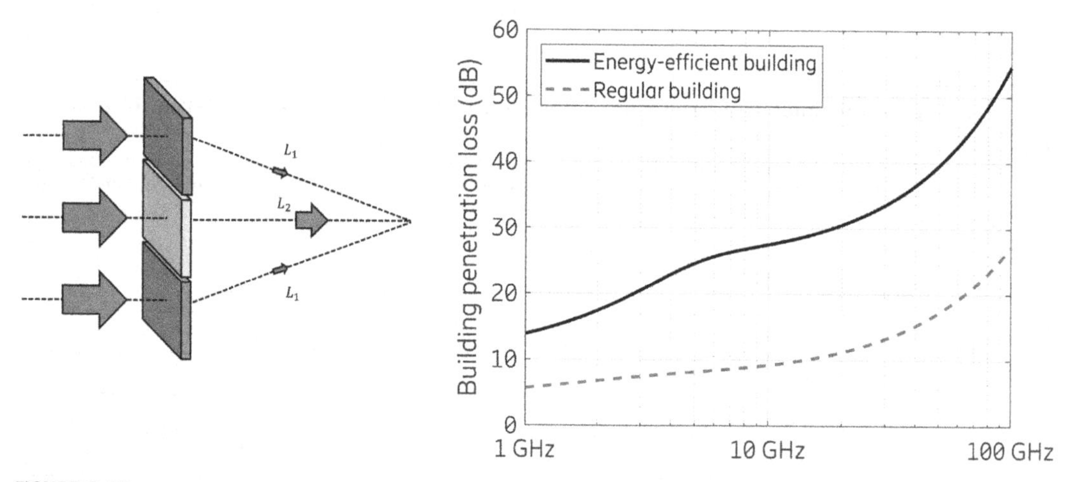

FIGURE 3.17

The effective building penetration loss is often determined by the losses of the different façade materials weighted by their relative area (left). Building penetration loss at perpendicular incidence to the external wall according to a model developed by 3GPP (right).

models distinguish between two classes of buildings, "traditional" buildings having regular glass where the penetration loss is limited even for mm-wave frequencies, and "energy-efficient" buildings equipped with metal-coated glass where the penetration loss can easily reach 20 dB even at low frequencies.

Further losses are experienced due to internal walls and furniture when propagating deeper into the building. As a rule of thumb, the additional loss in a building with an open floor plan can be modeled by 0.2–0.5 dB/m extra loss, while the presence of thicker internal walls may result in 0.5–2 dB/m extra loss.

A practical consequence of the frequency dependence of the NLOS path loss and building penetration loss is the challenge in providing similar coverage at higher frequencies to that of current cellular networks in the 1–2 GHz range. The use of antenna directivity becomes even more important for ensuring high-frequency coverage in real-world propagation environments including outdoor to indoor, where penetration and diffraction losses are higher than in free space. However, the principles outlined for antenna gain as a function of frequency in the previous section are still valid; hence increased directivity and beamforming is the method of choice to ensure high-frequency coverage.

3.5.3 MULTI-PATH PROPAGATION

An electromagnetic wave that is reflected, scattered, etc., in a complex environment will give rise to many secondary electromagnetic waves of varying strength and propagation directions. A number of these waves will reach the receiving antenna which will experience a multi-path propagation channel (see Fig. 3.18). If a certain radio wave is traced back from the receiver to the transmitter, this wave can be seen to have propagated along a certain *propagation path*, as illustrated in the figure. Multi-path propagation has been one of the main challenges to radio wave communications but has in recent years also been understood to provide additional opportunities for efficient communication. This will be further elaborated in the coming chapters, especially in Chapter 6, and in particular when it comes to exploiting multi-path for spatially multiplexing several data streams to the same user as described in Section 6.3.

3.5.3.1 Fast fading

Consider one of the propagation paths in Fig. 3.18. If the length of this propagation path is d_n, the wave front will reach the receiver after a delay $\tau_n = d_n/c$. For a time-harmonic wave with wavelength λ this time delay is equal to a phase shift $e^{-jd_n 2\pi/\lambda}$. If the transmitter or receiver position is (locally) shifted by $\boldsymbol{r}$, each individual propagation path will become shorter or longer depending on the propagation direction $\widehat{\boldsymbol{k}}_n$ in relation to the position shift, that is, as $\Delta d_n = \widehat{\boldsymbol{k}}_n \cdot \boldsymbol{r}$. The superposition of many waves with different directions of propagation $\boldsymbol{k}_n$ and hence different phase shifts Δd_n will result in constructive or destructive superposition depending on the phase relations between the waves. As the spatial position $\boldsymbol{r}$ is varied, the relative phases also change. This will cause a standing wave pattern in space with dips and peaks (see Fig. 3.19).

An extreme example is the superposition of two equal strength waves with opposite propagation directions, that is, $\boldsymbol{k}_2 = -\boldsymbol{k}_1$ (see Fig. 3.6). If these two waves are in phase at a certain point, then they will be 180° out of phase a quarter of a wavelength away along the propagation direction and hence completely cancel each other. Such nulls will occur every half-wavelength. However, if the

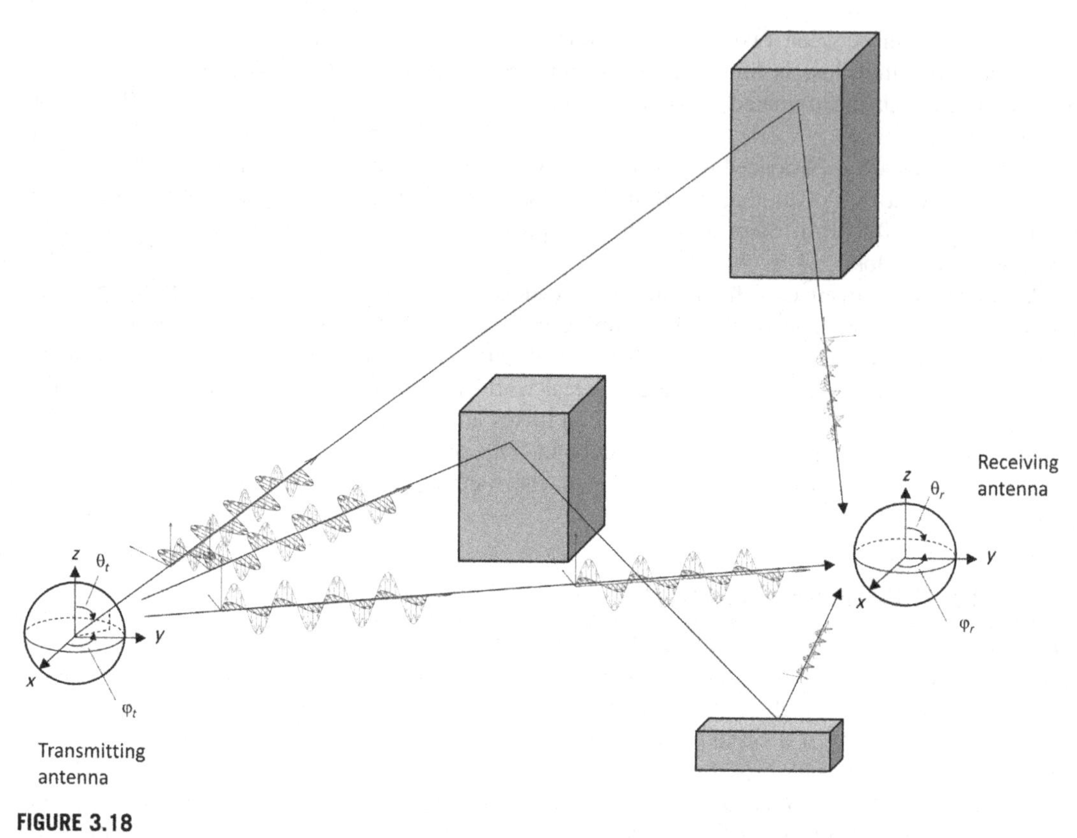

FIGURE 3.18

A multi-path propagation channel, containing a number of propagation paths characterized by their angles of departure (θ_t, φ_t) and arrival (θ_r, φ_r), path lengths, and relations between transmitted and received amplitudes and polarizations.

two waves have almost the same propagation direction, the distance between peaks and nulls will be much longer, since the two waves need to propagate much longer distance before the difference in phase shifts becomes noticeable. This exemplifies the relation between the *angular dispersion* of the waves and the rate of fading and will be discussed more in the next subsection.

An antenna moving through the pattern will experience signal fading due to the varying field strength. The faster the antenna (or objects in the environment) moves, the faster the signal variations. Since these variations occur on the scale of a fraction of the wavelength (decimeter to millimeter range for mobile communication systems) they are typically referred to as *fast fading*, as opposed to the slower variations of the path loss and shadow fading.

Two nearby but not colocated antennas will experience similar but not identical fading variations. As the separation of the antennas increases, the similarities decrease. This is exploited in antenna diversity schemes, where the use of multiple antennas can mitigate the fading variations since the probability that two or more antennas simultaneously experience a fading dip is low.

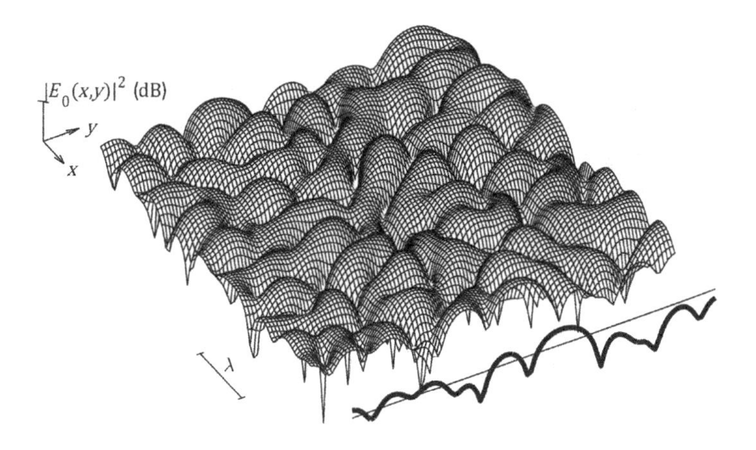

FIGURE 3.19

The constructive and destructive superposition of multiple waves impinging from all directions in the x-y plane creates a standing wave pattern with peaks and dips, here shown as the logarithmic magnitude of the field strength in two dimensions. The black thick line represents the field strength that is experienced along a specific trajectory through this standing wave pattern.

In a rich scattering environment where many waves of similar strength but different propagation directions are superimposed, the real and imaginary parts of the elements of $\boldsymbol{E}_0$ tend to become complex Gaussian as a result of the law of large numbers. Hence, the amplitude of the phasors becomes Rayleigh[18]-distributed. If one wave is much stronger than the others, this results in a non-zero mean of the phasor and hence the variation of the amplitude becomes more shallow. The resulting amplitude distribution is Ricean,[19] characterized by the power ratio K between the strong wave and the sum power of all other waves. *Rayleigh fading* is commonly observed in NLOS conditions, while *Rice fading* is more likely in LOS conditions. However, exceptions to this may occur as sometimes there is a dominant propagation path also in NLOS conditions, or there could be multiple similar strength paths in LOS. Examples of Rayleigh and Ricean fading and their distributions are given in Fig. 3.20.

3.5.3.2 Angular dispersion

As mentioned above, the spatial separation of the fading dips is a function of the angular dispersion of the waves. When waves are incident from all possible directions in two or even three dimensions the fading dips are spaced by approximately $\lambda/2$, as in Fig. 3.19, while a distribution in a narrower angular interval results in much larger separation between the fading dips as along the y-direction in Fig. 3.21.

In a cellular communication scenario, more reflections and scattering occur around the user equipment (UE) than around the base station, since the latter is usually installed with a less

[18]Lord Rayleigh who has given name to the fading and random distribution was a Nobel Prize winner who made many crucial contributions to science including providing the theoretical explanation for why the sky is blue.

[19]Stephen Rice was one of the pioneers of communication theory.

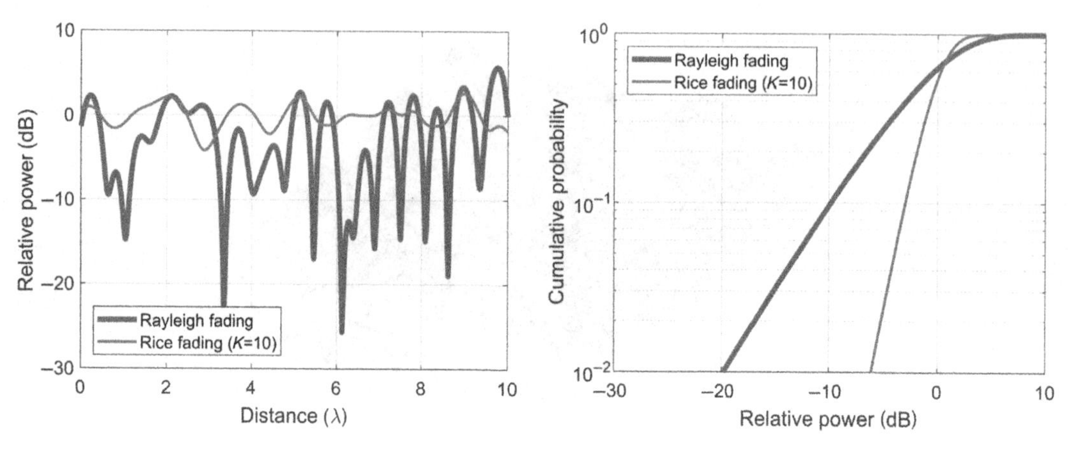

FIGURE 3.20

Examples of Rayleigh and Ricean fading in space (left) and the cumulative distribution functions of these two distributions (right).

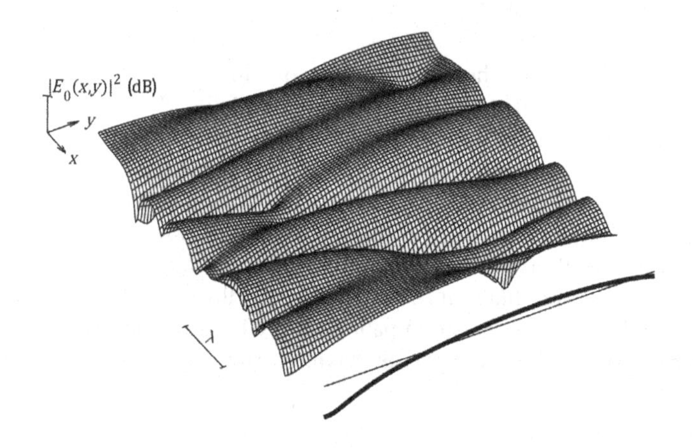

FIGURE 3.21

Example of a standing wave pattern due to waves impinging from a limited angular interval, here within 45° of the positive x-axis. The black thick line represents the field strength that is experienced along a specific trajectory.

obstructed view to provide good signal coverage in an extended area, as in the case with an above-rooftop macrocell base station (compare Fig. 3.22).

An example of the measured angular distribution of multi-path at a macro site is shown in Fig. 3.23. The circles illustrate the directions and relative powers of the multi-path components arriving at the base station from transmission by a single UE. By comparison of the directions with the underlying panoramic photo, the origin of the different paths can be identified as reflections

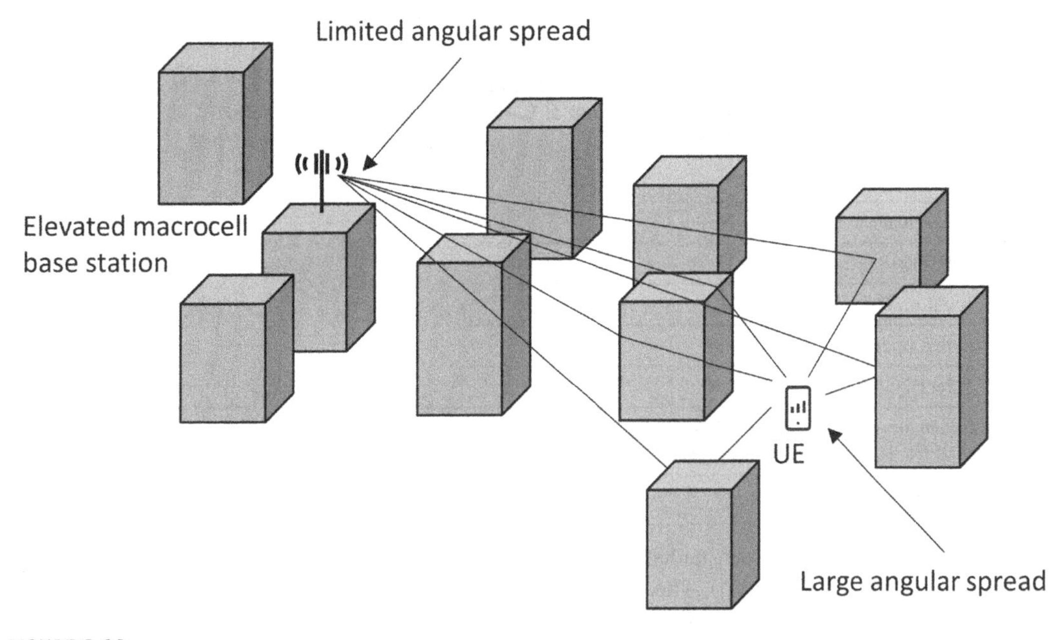

FIGURE 3.22

Sketch of difference in angular spread between an elevated macrocell base station and a user equipment (UE) that is surrounded by scattering objects.

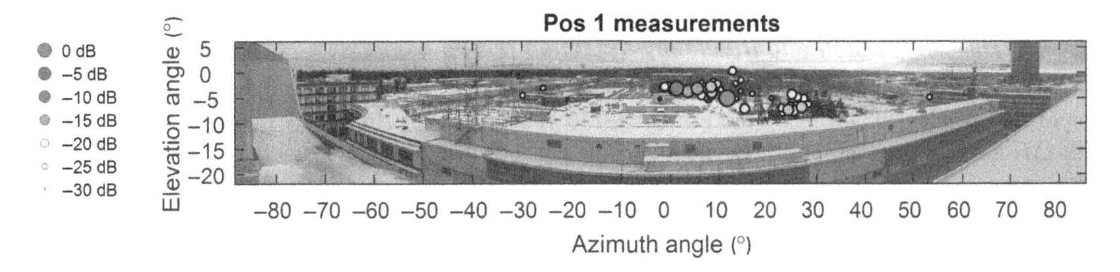

FIGURE 3.23

A panoramic view from a macro base station, with circles representing the measured directions of arrival and relative power of different multi-path components reaching this base station from a single mobile terminal located in the approximate direction (12°, − 5°). These measurements have been reported in Ref. [8].

from buildings, diffraction over roof edges, scattering from trees, etc. The paths are distributed over a wider angular range in the horizontal direction than in the vertical direction. This is a result of the distribution of scattering objects around the UE that is confined in height by the difference between the ground level and the highest objects (buildings, tree tops) but can span a large horizontal area.

Table 3.3 Typical Angular and Time Dispersion in Different Common Scenarios.

Scenario		Typical Horizontal (Azimuth) rms Angular Spread at Base Station (degrees)	Typical Horizontal (Azimuth) rms Angular Spread at the UE (degrees)	Typical rms Delay Spread (ns)
LOS		0–5	0–50	0–100
NLOS	Urban macrocell	5–25	20–90	50–500
	Suburban macrocell	3–15		
	Rural macrocell	3–10		
	Urban microcell	10–30		50–300
	Indoor cell	10–40		20–100

Note: *The angular spread in the elevation dimension is typically an order of magnitude lower than in the azimuth dimension. Based on 3GPP TR 38.901, Study on channel model for frequencies from 0.5 to 100 GHz, 2019 [3], ITU-R Report M.2412, Guidelines for evaluation of radio interface technologies for IMT-2020, 2017 [4].*

Since the UE is typically surrounded by scatterers, the angular spread as seen from it tends to be much larger (see, e.g., Ref. [9]). The resulting fading pattern at the UE side therefore resembles the example in Fig. 3.19, while at the base station side the fading pattern tends to vary at a slower rate as in the example in Fig. 3.21.

The angular dispersion can be quantified by the rms angular spread (definition in Section 3.6.1.2) and typical values are provided in Table 3.3. Angular dispersion is of crucial importance for AAS, as will be further elaborated in Section 6.3.5.

3.5.3.3 Time dispersion and frequency-selective fading

The phase $e^{-jd_n 2\pi/\lambda}$ of a time-harmonic wave propagating along a path with length d_n is also a function of the wavelength. Changing the wavelength or equivalently the frequency of the wave, say by a shift Δf, also changes the phase. In multi-path propagation where the propagation paths of different waves have different lengths, the superposition of the waves also becomes frequency-dependent, which is referred to as *frequency-selective fading*. An example of frequency-selective fading is shown in Fig. 3.24 (right). In analogy to the spatial fading, the rate of fading in frequency is proportional to the spread of propagation path lengths of the radio waves. Since the path length d can be converted into a time delay τ using the speed of light as $\tau = d/c$, the spread is referred to as the *time dispersion*.

Some further intuition can be achieved with the help of a simple example with two propagation paths having equal amplitude and phase at a certain carrier frequency f_c but different propagation lengths according to $d_2 = d_1 + \Delta d$. In this case, the phase difference between the two paths as a function of the frequency $f = f_c + \Delta f$ is $2\pi\Delta\tau\Delta f$. The two paths are in phase when $2\pi\Delta\tau\Delta f = n\pi, n = 0, \pm 1, \pm 2, \ldots$ and 180° out of phase when $2\pi\Delta\tau\Delta f = n\pi + \pi/2$. Thus frequency-selective fading will be observed, where the maxima (or minima) will occur regularly with $0.5/\Delta\tau$ spacing in the frequency domain. As a numerical example, a 250 ns delay difference will result in nulls every 2 MHz, while a 25 ns delay difference creates nulls every 20 MHz.

The time dispersion can be directly observed in the channel impulse response (formally defined in Section 3.6.2.6), which characterizes the relative path powers and times of arrival (see Fig. 3.24,

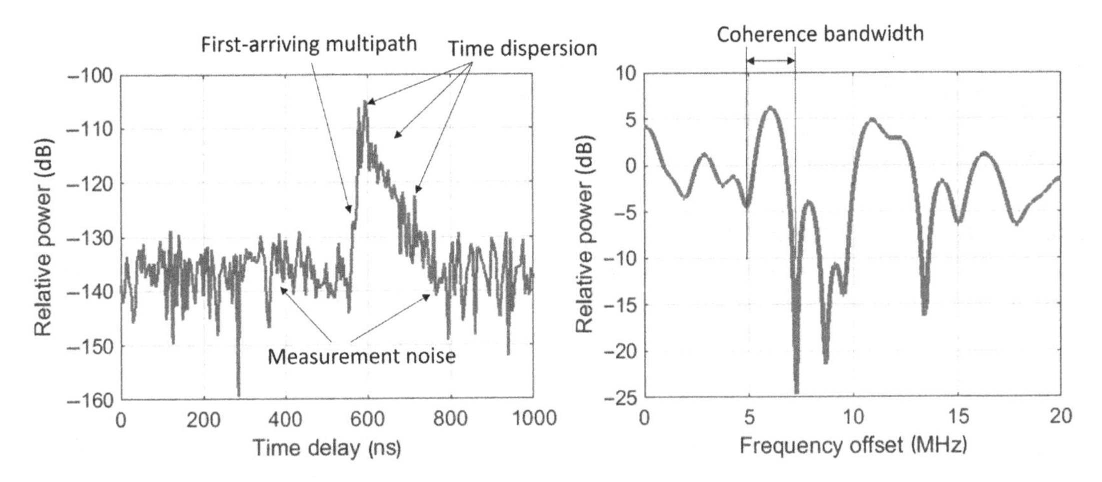

FIGURE 3.24

A measured channel impulse response that shows the power reaching the receiver as a function of the time delay that is proportional to the path length (left), and the same channel as a function of frequency relative to the carrier frequency (right).

left). The rate of variability in frequency, for example, the distance between the fading dips in frequency, is commonly characterized by the *coherence bandwidth*,[20] which is inversely related to the time dispersion of the propagation paths. Similarly, the time dispersion is usually characterized using the *rms delay spread*.[21] Typical values for the rms delay spread in different scenarios are provided in Table 3.3.

Time dispersion has been a constant nemesis in the design of wireless communication system, since it makes it challenging to transmit information at a high rate without encountering intersymbol interference, where previously transmitted symbols interfere with the present one after having propagated through the channel. Chapter 5, will discuss how OFDM-based transmission as used in 4G and 5G systems can mitigate the impact of the time dispersion.

3.5.3.4 Doppler spread

For a moving receiver antenna, the incoming waves will appear to be shifted in frequency due to the Doppler effect. The size of this Doppler frequency shift is

$$f_\mathrm{D} = \frac{v}{\lambda}\cos(\vartheta), \tag{3.52}$$

where v is the speed and ϑ is the angle between the direction of movement and the incident wave. Hence, a moving antenna in a multi-path channel will experience a Doppler spread of the channel due to different Doppler shifts of different incident waves. Note that by reciprocity the same effect occurs for a moving transmitting antenna in relation to the outgoing waves. The Doppler spread is

[20]Different definitions of coherence bandwidth are in use, but it can generally be understood as the frequency range over which the channel variations are smaller than some threshold.

[21]A mathematical definition of the rms delay spread is provided in Section 3.6.1.1, equation (3.56).

related to the rate of change of the channel in time, which is typically quantified by the coherence time. A higher Doppler spread results in more rapid time variations. Increasing the speed of the moving antenna increases the Doppler, as does increasing the frequency. These rapid variations can be challenging for providing a robust communication link, particularly for high-speed mobility and/or higher carrier frequencies.

3.5.3.5 Polarization scattering

Interactions with the environment may also affect the polarization of the wave. Reflection or transmission coefficients can be different for different polarizations resulting in a polarization change. An everyday example of this is when randomly polarized sunlight reflects off wet asphalt or water and becomes more predominantly horizontally polarized such that the reflections can be suppressed using polarized sunglasses. Reflections also involve a direction change that causes the sense of rotation of the electric field vector to change, thereby translating a RHCP to a left-hand circular, or a $+45°$ to a $-45°$, or vice versa. Diffraction and rough surface scattering coefficients may also be polarization-dependent.

When a single wave undergoes multiple polarization-affecting interactions or when multiple waves with dissimilar polarization scattering are superimposed, the resulting polarization will deviate from that of the transmitted wave. It has been found [10] that this deviation is usually the smallest if the wave had been transmitted with vertical or horizontal polarization, most likely due to the fact that the ground and most man-made surfaces are horizontal or vertical. The power in the orthogonal polarization component, for example, horizontal for a vertically polarized transmitted wave, is typically 6–10 dB weaker than that received in the original polarization. The power ratio between the original and orthogonal polarization is commonly denoted as the *cross-polarization ratio* (XPR).

Furthermore, the polarization-dependent scattering coefficients typically cause the fast fading variations to be more or less independent between different combinations of transmit and receive polarizations, for example, the H-to-V channel experiences fast fading variations that are uncorrelated to those on the V-to-H, V-to-V, or H-to-H channel. Here it is important to note that the XPR values and lack of correlation are specific to the VP and HP. Other (orthogonal) polarization pairs may result in other XPRs and correlations with results that can sometimes be a bit surprising. As an example, a $+45°$ transmitted wave will typically be received with similar power on a $+45°$ and $-45°$ polarized antenna in NLOS conditions; hence the XPR for a $+45°$-polarized wave is often close to 0 dB.

3.5.3.6 Summary of multi-path propagation

To summarize, signal fading in space and in frequency occurs due to multi-path dispersion in angle and delay. Multi-path has traditionally been one of the main challenges of wireless communication due to the associated time dispersion and signal fading; however, in more recent years advances in receiver design, link adaptation, frequency-selective scheduling, channel coding, and antenna diversity schemes have mitigated the impact and made communication systems such as 3G and 4G more resilient to multi-path. OFDM-based transmission schemes as used in 4G and 5G are particularly robust to time dispersion.

Moreover, by making observations of the local field strength for multiple spatial positions and/or polarizations through the use of multiple transmit and/or receive antennas, it is possible to distinguish different waves from each other and thereby achieve multiple parallel communication channels. This forms the foundation of multi-antenna and MIMO (multiple-input, multiple-output) communications in which different waves are modulated to carry different information (more on this in Chapter 6).

It should be noted that time dispersion, angular dispersion, and Doppler spread are not always independent of each other. Enhancing some propagation paths while suppressing others using directional antennas or beamforming can have an impact also on the time dispersion and the Doppler spread. For instance, the propagation paths shown in Fig. 3.23 also have different propagation delays. An antenna that focuses the radiation along some of these paths may then also change the experienced time dispersion.

3.6 MODELING OF WAVE PROPAGATION AND THE TRANSMISSION OF COMMUNICATION SIGNALS

Models for the wireless wave propagation are instrumental when developing and evaluating communication technologies including advanced antenna schemes. A model is expected to capture key characteristics such as the path loss and the multi-path and associated channel dispersion. Such models may be used in research, standardization, and predevelopment of wireless communications, sometimes years or even decades before actual products are available. Typical and representative, sometimes even challenging propagation conditions are expected to be reproduced by the model so that the communication can be ensured to be robust against challenging conditions and optimized for typical propagation conditions.

Different ways of representing the wireless propagation of signals between the transmitting and receiving antenna can be used. Two main classes of representations are the *radio channel* and the *directional propagation channel* (compare Fig. 3.25), where the radio channel relates the transmitted and received signals (e.g., voltages or currents at the antenna feeds or antenna ports[22]) while the directional propagation channel relates the transmitted and received radio waves [11]. The main difference between these two representations is that the radio channel includes the impact of the transmitting and receiving antenna through their radiation patterns $\mathbf{g}_\mathrm{t}(\theta, \varphi)$ and $\mathbf{g}_\mathrm{r}(\theta, \varphi)$, while the directional propagation channel is independent of the antennas. Models for the radio channel are also needed for the design and evaluation of the communication system, for example, how information should be converted into signals and how these signals should be decoded into information at the receiver. Earlier generations of communication systems were developed mainly using channel models of the wireless radio channel, typically consisting of a path loss model together with a stochastic model for the channel coefficients $H(f, t)$ over a limited frequency range.

However, the advent of multi-antenna technologies together with advances in propagation measurement capabilities has led to a better understanding of the directional properties of the radio channel and how to utilize these for more efficient communication. The use of more directive and steerable antennas creates a need for more accurate modeling of the directional properties that has led to the development of more advanced directional propagation models.

To summarize, the directional propagation channel describes the propagation paths while the radio channel describes the relation between the transmitted and received signals. In the following we will describe the directional propagation channel and the radio channel in more detail, followed by an overview of different models for these two.

[22]Formal definitions of the antenna port will be introduced in Section 8.3.1.

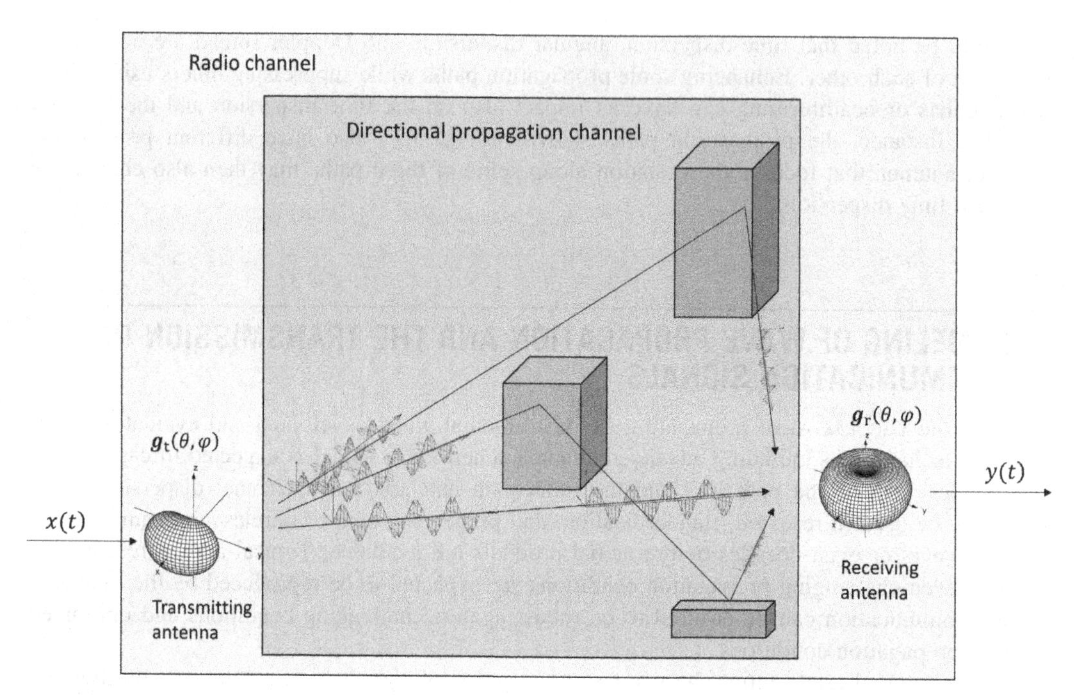

FIGURE 3.25

Illustration of the radio channel and the double-directional propagation channel.

3.6.1 DIRECTIONAL PROPAGATION CHANNEL

The directional propagation channel is most readily described by the propagation paths. With reference to Fig. 3.26, the n:th propagation path can be characterized by a set of properties $\{\tau_n, \theta_{t,n}, \varphi_{t,n}, \theta_{r,n}, \varphi_{r,n}, a_n, \underline{\boldsymbol{\Psi}}_n\}$ defined as follows:

- A time delay τ_n that is proportional to the propagation path length d_n as $\tau_n = d_n/c$;
- A direction of departure from the transmitter, typically specified with respect to a local horizon-aligned coordinate system centered on the transmit antenna as an azimuth angle $\varphi_{t,n}$ and zenith elevation angle $\theta_{t,n}$. This direction of departure can also be expressed using the unit vector $\widehat{\boldsymbol{r}}_{t,n} = \widehat{\boldsymbol{k}}_{t,n}$ pointing in the direction $(\theta_{t,n}, \varphi_{t,n})$, that is, in the propagation direction of the departing wave;
- A direction of arrival at the receiver, typically specified with respect to a local horizon-aligned coordinate system centered on the receive antenna as an azimuth angle $\varphi_{r,n}$ and zenith elevation angle $\theta_{r,n}$. This direction of arrival can also be expressed using the unit vector $\widehat{\boldsymbol{r}}_{r,n} = -\widehat{\boldsymbol{k}}_{r,n}$ pointing in the direction $(\theta_{r,n}, \varphi_{r,n})$, that is, *opposite* to the propagation direction of the arriving wave;
- A relation between the electric field strength $\boldsymbol{E}_{t,n}$ of the transmitted wave along the n:th propagation path and the electric field strength $\boldsymbol{E}_{r,n}$ of the wave as it reaches the receiver. This relation can be expressed with Jones vectors as follows:

$$\underline{\boldsymbol{E}}_{r,n} = a_n e^{-j2\pi f_c \tau_n} \underline{\boldsymbol{\Psi}}_n \underline{\boldsymbol{E}}_{t,n} \quad (3.53)$$

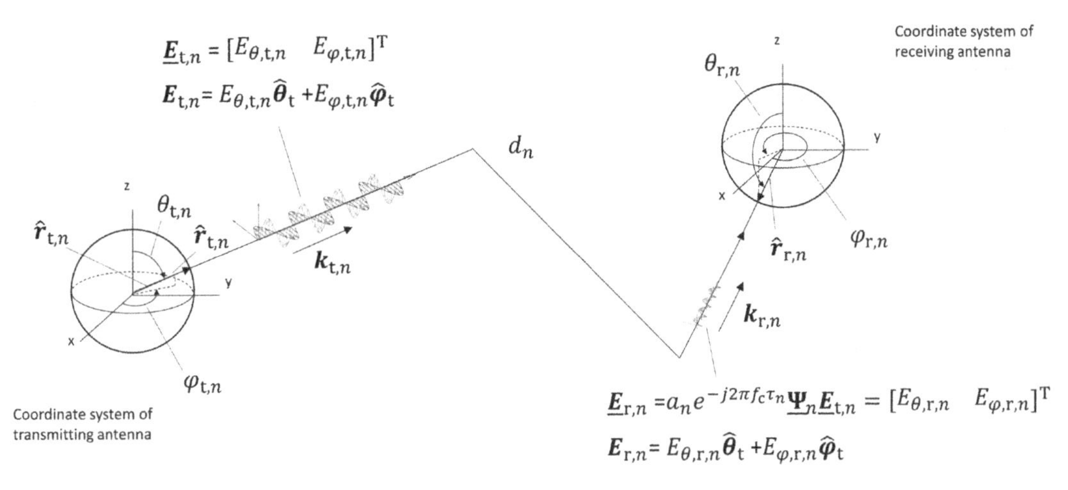

FIGURE 3.26

Illustration of parameters describing the n:th propagation path in a directional propagation channel.

where $e^{-j2\pi f_c \tau_n}$ represents the phase shift of the wave due to the propagation path length, a_n represents additional attenuation and phase shift due to interactions between the wave and the environment (compare Fig. 3.15), and $\underline{\Psi}_n$ accounts for the relative attenuations and phase shifts between the different transverse field components of $E_{t,n}$ and $E_{r,n}$ and is given by the 2×2 *polarization scattering matrix* in a Jones representation form

$$\underline{\Psi}_n \triangleq \begin{bmatrix} \psi_{\theta_t\theta_r,n} & \psi_{\varphi_t\theta_r,n} \\ \psi_{\theta_t\varphi_r,n} & \psi_{\varphi_t\varphi_r,n} \end{bmatrix}. \tag{3.54}$$

The coefficient $\psi_{\theta_t\varphi_r,n}$ contains the relative amplitude and phase shift of the $\hat{\varphi}_r$ component of the wave at the receiver if the wave at the transmitter was polarized in the $\hat{\theta}_t$ direction and similarly for the other elements in the polarization scattering matrix $\underline{\Psi}_n$. Note that the polarization scattering matrix can also be written on a non-Jones form.[23]

Since $\underline{\Psi}_n$ contains the relative amplitudes and phases of the different field components it is customary to normalize it such that one of the co-polarized coefficients equals unity, for example, $\psi_{\theta_t\theta_r,n} = 1$. The relative power of the corresponding co-polarized component of the wave is

[23]The polarization matrix can also be represented in a real-word 3D coordinate system as a 3×3 *matrix*

$$\Psi_n = \left[\hat{\theta}_r \quad \hat{\varphi}_r\right]\underline{\Psi}_n\left[\hat{\theta}_t \quad \hat{\varphi}_t\right]^T$$

and the electric field corresponding to (3.52) but using a 3D coordinate system can then be written as $E_{r,n} = a_n e^{-j2\pi f_c \tau_n}\Psi_n E_{t,n}$. When here evaluating $\Psi_n E_{t,n}$ it is seen how $\left[\hat{\theta}_t \quad \hat{\varphi}_t\right]^T E_{t,n} = \underline{E}_{t,n}$ expresses the electric field on the transmit side in coordinates with basis vectors $\hat{\theta}_t$ and $\hat{\varphi}_t$ while left multiplication with $\left[\hat{\theta}_r \quad \hat{\varphi}_r\right]$ transforms from the Jones vector $\underline{E}_{r,n}$ in basis $\hat{\theta}_r$, $\hat{\varphi}_r$ to 3D coordinates producing electric field vector $E_{r,n}$. This shows the equivalence between $\underline{\Psi}_n$ and Ψ_n, which can be used interchangeably, just like Jones and non-Jones representations for vectors representing transverse electric fields can be interchanged.

$p_n = |a_n|^2$. This power can be specified in relation to a unit power transmitted wave and include the path loss along the propagation path, or it may be normalized so that $\sum p_n = 1$.

Note that for a free-space channel where there is a single propagation path, the polarization scattering matrix for that single path is given by

$$\underline{\Psi}_1 = \begin{bmatrix} 1 & 0 \\ 0 & 1 \end{bmatrix}$$

Based on these characteristics, different metrics for the dispersion of the channel are defined in the following subsections.

3.6.1.1 Delay dispersion

The *mean delay* of the channel is

$$\tau_{\text{mean}} = \frac{\sum p_n \tau_n}{\sum p_n}. \tag{3.55}$$

The *rms delay spread* is

$$\tau_{\text{rms}} = \sqrt{\frac{\sum p_n (\tau_n - \tau_{\text{mean}})^2}{\sum p_n}}. \tag{3.56}$$

As can be seen, the mean delay and rms delay spread take into account both the relative delays τ_n and the relative strengths p_n of the propagation paths. Hence, a later arriving path has a much stronger impact on the rms delay spread if it is strong than if it is weak. For early digital receivers, the rms delay spread was found to be quite well correlated with the bit error rate of the communication link. For OFDM-based transmission, the time dispersion influences the choice of cyclic prefix length. The rms delay spread will vary depending on the scenario, but some typical ranges are provided in Table 3.3.

3.6.1.2 Angular dispersion

The *mean azimuth* and *mean elevation* angles of departure and arrival and the corresponding *rms angular spreads* can be determined in a similar way as the rms delay spread, though one must be a bit careful when determining these measures due to the cyclic nature of angles. Hence, the mean azimuth angle of departure can be determined as

$$\varphi_{\text{t,mean}} = \arg \sum p_n e^{j\varphi_{\text{t},n}} \tag{3.57}$$

and the rms azimuth angular spread of departure as

$$\varphi_{\text{t,rms}} = \sqrt{\frac{\sum p_n (\varphi_{t,n} - \varphi_{t,mean})^2}{\sum p_n}} \tag{3.58}$$

where the difference $\varphi_{\text{t},n} - \varphi_{\text{t,mean}}$ should be determined as the smallest difference in the unit circle. The same definitions can be used for all the angles of arrival and departure by just replacing $\varphi_{\text{t},n}$ with another angle.

Typical ranges of the rms angular spreads are provided in Table 3.3. The rms delay spread and rms angular spread represent convenient metrics for characterizing the time and angular dispersion.

However, the reader should be cautioned that these metrics do not capture all the characteristics of the dispersion. Very different distributions in delay or angle can still result in the same rms spread, for example, two equal amplitude paths with little separation can result in the same rms spread as two widely spaced paths with unequal amplitude.

3.6.1.3 Polarization cross-scattering

The cross-polarization ratio (XPR) κ characterizes the relative power that is scattered from one polarization to the orthogonal polarization and can be determined as

$$\kappa_{\mathrm{VH}} = \frac{\sum p_n \left|\psi_{\theta_t\varphi_r,n}\right|^2}{\sum p_n \left|\psi_{\theta_t\theta_r,n}\right|^2} \tag{3.59}$$

$$\kappa_{\mathrm{HV}} = \frac{\sum p_n \left|\psi_{\varphi_t\theta_r,n}\right|^2}{\sum p_n \left|\psi_{\varphi_t\varphi_r,n}\right|^2} \tag{3.60}$$

Typically, $\kappa_{\mathrm{VH}} \approx \kappa_{\mathrm{HV}}$ and assumes values in the range -6 to -10 dB [10].

3.6.2 RADIO CHANNEL

In practice, it is impossible to directly observe all the propagation paths in the directional channel. Instead these can be estimated from finite resolution measurements of delay-sensitive (i.e., wideband) signals transmitted or received using directionally sensitive antenna arrays. In the following, a model for the radio channel, that is, the relation between transmitted and received signals, will be derived using the propagation paths characterizing the propagation channel. This model forms the basis for developing and evaluating multi-antenna transmission schemes and will be utilized for the mathematical treatment of array antennas, OFDM transmission, and multi-antenna technologies in the coming chapters of this book.

3.6.2.1 Baseband signal representation

With reference to Fig. 3.27, the radio frequency signal $x_{\mathrm{RF}}(t)$ that is the input to the transmit antenna is generated by letting a complex-valued low-pass baseband signal, $x(t)$, modulate a carrier with frequency f_c,

$$x_{\mathrm{RF}}(t) = \mathrm{Re}\left\{x(t)e^{j2\pi f_c t}\right\} \tag{3.61}$$

The signal transmitted at radio frequency is a real-valued passband signal and the baseband signal is often referred to as the *complex baseband equivalent* or *equivalent low-pass signal* since it contains the same information as the radio frequency signal. In particular it can be noted that a single-frequency sinusoid, $x_{\mathrm{RF}}(t) = A\cos(2\pi f_c t + \phi)$, can be represented with the equivalent complex baseband signal $x(t) = Ae^{j\phi}$ which is a complex constant whose magnitude and argument represents the amplitude and phase of the sinusoid. From Fourier analysis it is known that an arbitrary periodic complex signal $x(t)$ can be expressed as a sum of complex sinusoids

$$x(t) = \sum_{n=-\infty}^{\infty} b_n e^{j2\pi nft} \tag{3.62}$$

where b_n are the Fourier series coefficients and nf are the baseband frequencies that can be both positive and negative.

In the same way, the received signal can also be represented by a complex equivalent baseband signal

$$y_{\mathrm{RF}}(t) = \mathrm{Re}\{y(t)e^{j2\pi f_c t}\} \tag{3.63}$$

Analysis and modeling can thus be done using the baseband equivalents and the channel model presented in this section provides a relation between the complex-valued signals $x(t)$ and $y(t)$ rather than between the real-valued signals $x_{\mathrm{RF}}(t)$ and $y_{\mathrm{RF}}(t)$.

Complex-valued baseband signals are also used in practice, and the process of generating a radio frequency signal from a baseband signal is referred to as *upconversion*, whereas the process of generating baseband signals from a received radio frequency signal is referred to as *downconversion*. The principle is illustrated in Fig. 3.27 and it may be noted that in the model presented, the impact of the low-pass filter is not included, mainly for simplicity.

3.6.2.2 Channel coefficient for single-frequency sinusoid

For the sake of simplicity, assume that a constant baseband signal $x(t) = Ae^{j\phi}$ is transmitted from a stationary antenna at the origin of the local coordinate system with the radiation pattern $\mathbf{g}_t(\theta, \varphi)$. This signal could represent the Fourier component b_0 with baseband frequency $f = 0$ of a more general signal. More general signals can be considered through the superposition of such single-frequency sinusoids.

The signal $x = Ae^{j\phi}$ is carried by the electromagnetic waves radiated from the transmitting antenna and will reach a receiver antenna at the origin of the local coordinate system with radiation pattern $\mathbf{g}_r(\theta, \varphi)$ along N different propagation paths. The received baseband signal y will be the transmitted signal x multiplied by the *channel coefficient* H (where the baseband equivalents x, y, and H are all constants),

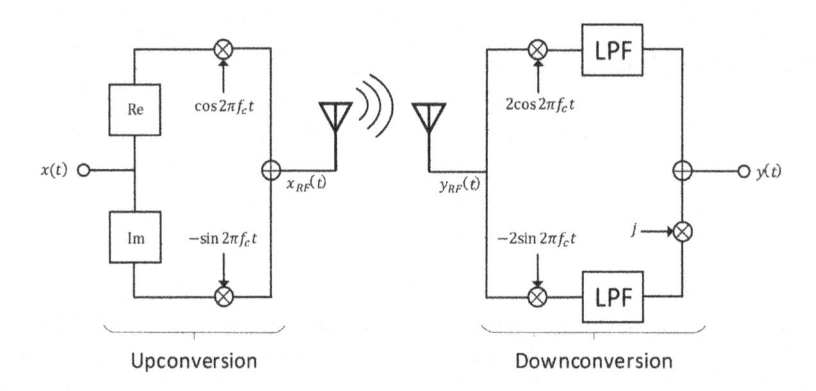

FIGURE 3.27

Illustration of upconversion in a transmitter and downconversion in a receiver using a low-pass filter (LPF).

$$y = Hx = \sum_{n=1}^{N} H_n x, \tag{3.64}$$

where

$$H_n = a_n e^{-j2\pi f_c \tau_n} a_{tr}\left(\theta_{r,n}, \varphi_{r,n}, \theta_{t,n}, \varphi_{t,n}\right) \tag{3.65}$$

and

$$a_{tr}\left(\theta_{r,n}, \varphi_{r,n}, \theta_{t,n}, \varphi_{t,n}\right) \triangleq g_r\left(\theta_{r,n}, \varphi_{r,n}\right)\widehat{\underline{\psi}}_r^T\left(\theta_r,n, \varphi_{r,n}\right)\underline{\Psi}_n\widehat{\underline{\psi}}_t\left(\theta_{t,n}, \varphi_{t,n}\right)g_t\left(\theta_{t,n}, \varphi_{t,n}\right) \tag{3.66}$$

Here H_n can be seen as a variant of Friis' equation (3.46) specifying a complex amplitude gain rather than real-valued power gain. The path loss, for example, $\left(\lambda/4\pi r\right)^2$ in free space, is implicitly included in a_n, while $a_{tr}\left(\theta_{r,n}, \varphi_{r,n}, \theta_{t,n}, \varphi_{t,n}\right)$ captures the impact on the received field strength due to the transmit and receive radiation patterns $g_t\left(\theta_{t,n}, \varphi_{t,n}\right)$ and $g_r\left(\theta_{r,n}, \varphi_{r,n}\right)$ in combination with impact due to the possible misalignment between the incident wave polarization $\underline{\Psi}_n\widehat{\underline{\psi}}_t$ and the receive antenna polarization $\widehat{\underline{\psi}}_r$ as was discussed in Section 3.3.3. The polarization scattering matrix $\underline{\Psi}_n$ was introduced in Section 3.6.1. The phase rotation, $\exp(-j2\pi f_c \tau_n)$, comes from the fact that $e^{-j2\pi f_c \tau_n} x$ is the equivalent complex baseband signal of $x_{RF}(t - \tau_n)$ as can be seen by exchanging t by $t - \tau_n$ in (3.61).

3.6.2.3 Time and spatial variations of the channel coefficient

Next, the variation of the propagation delays τ_n with transmitter and receiver positions and with time will be considered. The motivation for this is to be able to characterize the channel variations over time and with respect to antenna position, for example, within an antenna array. In the general case, the position of a moving receive antenna relative to the local origin can be described in terms of the vector $\boldsymbol{r}_r + \boldsymbol{v}_r t$. By assuming that the wave fronts at the receiver are (locally) planar, it follows from Section 3.5.3.1 that the path length becomes shorter or longer by $\Delta d_{r,n} = -\widehat{\boldsymbol{r}}_{r,n}\cdot(\boldsymbol{r}_r + \boldsymbol{v}_r t)$, where $\widehat{\boldsymbol{r}}_{r,n}$ is the unit vector pointing in the direction $(\theta_{t,n}, \varphi_{t,n})$ as was defined in Section 3.6.1. Similarly, if $\boldsymbol{r}_t$ describes the position of the transmitting antenna with respect to the local origin, then the path length difference is $\Delta d_{t,n} = -\widehat{\boldsymbol{r}}_{t,n}\cdot\boldsymbol{r}_t$, where a stationary transmit antenna has been assumed for simplicity. The total path length difference is thus

$$\Delta d_n = \Delta d_{r,n} + \Delta d_{t,n} = -\widehat{\boldsymbol{r}}_{r,n}\cdot(\boldsymbol{r}_r + \boldsymbol{v}_r t) - \widehat{\boldsymbol{r}}_{t,n}\cdot\boldsymbol{r}_t, \tag{3.67}$$

which corresponds to a delay difference

$$\Delta\tau_n = -\frac{\widehat{\boldsymbol{r}}_{r,n}\cdot\boldsymbol{r}_r}{c} - \frac{\widehat{\boldsymbol{r}}_{r,n}\cdot\boldsymbol{v}_r t}{c} - \frac{\widehat{\boldsymbol{r}}_{t,n}\cdot\boldsymbol{r}_t}{c}. \tag{3.68}$$

By replacing τ_n with $\tau_n + \Delta\tau_n$ in (3.65) and using $2\pi f_c/c\widehat{\boldsymbol{r}} = \boldsymbol{k}$, H_n can be expressed[24] as

$$H_n(\boldsymbol{r}_t, \boldsymbol{r}_r, t) = a_n a_{tr} e^{-j2\pi f_c \tau_n} e^{j\boldsymbol{k}_{t,n}\cdot\boldsymbol{r}_t} e^{j\boldsymbol{k}_{r,n}\cdot\boldsymbol{r}_r} e^{j\boldsymbol{k}_{t,n}\cdot\boldsymbol{v}_r t} \tag{3.69}$$

Variations of the spatial positions $\boldsymbol{r}_t, \boldsymbol{r}_r$ or of time t create phase rotations that can be different for the different propagation paths since the propagation directions, given by $\boldsymbol{k}_{t,n}$ and $\boldsymbol{k}_{r,n}$, are

[24]By convention, $\boldsymbol{k}_{r,n}$ is defined here as pointing outwards from the receiver, that is, opposite of the direction of propagation.

different due to the angular dispersion in the channel. The sum of these per-path channel coefficients according to (3.64) is what creates the fast fading variations seen in Figs. 3.19–3.21.

3.6.2.4 Frequency variation of the channel coefficient

Now, the variation of the channel coefficient with baseband frequency will be considered, assuming fixed antenna positions and a fixed time. Consider a complex baseband signal with a non-zero but single baseband frequency f, that is, $x(t) = b_f e^{j2\pi ft}$. The real-valued passband signal is $x_{\mathrm{RF}}(t) = \mathrm{Re}\{x(t)e^{j2\pi f_c t}\} = \mathrm{Re}\{b_f e^{j2\pi(f_c+f)t}\}$; hence by replacing f_c with $f_c + f$ in (3.65) it can be written as

$$H_n(f) = a_n a_{\mathrm{tr}} e^{-j2\pi f_c \tau_n} e^{-j2\pi f \tau_n} \tag{3.70}$$

In analogy with the time and spatial dimensions, variations of the baseband frequency f creates phase rotations $e^{-j2\pi f \tau_n}$ that are different due to the time dispersion in the channel; hence the sum of these per-path channel coefficients according to (3.71) creates fast fading variations in frequency (compare Fig. 3.24, right).

$$H(f) = \sum_{n=1}^{N} H_n(f) \tag{3.71}$$

We refer to the function $H(f)$ as the *channel transfer function*. If the input signal $x(t)$ is written as a sum of sinusoids, for example, through a Fourier transform with respect to t according to (3.62) as $X(f) = \mathrm{F}_t x(t))$, then the received signal $Y(f)$ is

$$Y(f) = H(f)X(f) \tag{3.72}$$

3.6.2.5 The time- and space-dependent transfer function

In general, the channel coefficient is a function of both the frequency (Section 3.6.2.4) and time and space (Section 3.6.2.3), resulting in a phase rotation of the per-path channel coefficients (3.65), according to

$$e^{-j2\pi(f_c+f)(\tau_n+\Delta\tau_n)} \tag{3.73}$$

where $\Delta\tau_n$ is determined according to (3.68).

Communication signals $x(t)$ are band-limited, that is, they can be expressed as Fourier series with frequencies $|f| < f_{\max}$. Typical communication signals have bandwidths up to a few hundred MHz, that is, $f_{\max} < 10^8$. Therefore the radio channel needs to be described only for baseband frequencies f in the same range. In a local spatial region with radius r_{local} around the origin of the coordinate system for the transmitting (or receiving) antenna, the delay differences $\Delta\tau_n$ will be limited by $\Delta\tau_n < 2r_{\mathrm{local}}/c$, which in turn means that the product $f\Delta\tau_n$ will be bounded by $f\Delta\tau_n < 2f_{\max} r_{\mathrm{local}}/c$. In many cases, such as when observing the fast fading in a local area (compare Fig. 3.19 or 3.20) or over an antenna array, and further assuming a limited bandwidth, the phase rotation due to the term $f\Delta\tau_n$ will be very small and can be neglected. The approximation $f\Delta\tau_n \approx 0$ is often referred to as the *narrowband approximation* in array processing.

Using the narrowband approximation, (3.73) can be written as

$$e^{-j2\pi(f_c+f)(\tau_n+\Delta\tau_n)} \approx e^{-j2\pi f_c\tau_n}e^{-j2\pi f\tau_n}e^{-j2\pi f_c\Delta\tau_n} \tag{3.74}$$

The benefit of this approximation is that the baseband frequency f and the time- and space-dependent delay difference $\Delta\tau_n$ become decoupled and can be treated as independent variables. Hence, applying the same steps as when deriving (3.69), the *time- and space-dependent transfer function* can be written as

$$H(f,t,\boldsymbol{r}_t,\boldsymbol{r}_r) = \sum_n a_n a_{tr} e^{-j2\pi f_c\tau_n}e^{-j2\pi f\Delta\tau_n}e^{j\boldsymbol{k}_{t,n}\cdot\boldsymbol{r}_t}e^{j\boldsymbol{k}_{r,n}\cdot\boldsymbol{r}_r}e^{j\boldsymbol{k}_{t,n}\cdot\boldsymbol{v}_r t} \tag{3.75}$$

This function characterizes the radio channel as a function of baseband frequency, time, spatial position in the transmitter coordinate system, and spatial position in the receiver coordinate system.

3.6.2.6 Channel impulse response

An equivalent way of representing the radio channel is with the *channel impulse response* $h(\tau,\ldots)$, which is the inverse Fourier transform of the transfer function $H(f,\ldots)$ with respect to f, $h(\tau,\ldots) = \mathrm{F}_f^{-1}H(\mathrm{f},\ldots)$. By applying the Fourier transform identity $\mathrm{F}\{\delta(t-\tau_n)\} = e^{-j2\pi f\tau_n}$ the time- and space-dependent impulse response can be written as

$$h(\tau,t,\boldsymbol{r}_t,\boldsymbol{r}_r) = \sum_n a_n a_{tr} e^{-j2\pi f_c\tau_n}e^{j\boldsymbol{k}_{t,n}\cdot\boldsymbol{r}_t}e^{j\boldsymbol{k}_{r,n}\cdot\boldsymbol{r}_r}e^{j\boldsymbol{k}_{r,n}\cdot\boldsymbol{v}_r t}\delta(t-\tau_n) \tag{3.76}$$

A special case of the impulse response is the free-space channel consisting of a single propagation path and assuming stationary transmit and receive antennas at the respective origins,

$$h_{\text{free space}}(\tau) = ae^{-j2\pi f_c\tau}\underline{\mathbf{g}}_r^{\mathrm{T}}(\theta_r,\varphi_r)\begin{bmatrix}1 & 0\\ 0 & 1\end{bmatrix}\underline{\mathbf{g}}\mathrm{t}(\theta_t,\varphi_t)\delta(t-\tau) \tag{3.77}$$

The free-space channel impulse response will be utilized when discussing antenna arrays in Chapter 4. The received signal can then be expressed as a convolution between the channel impulse response and the transmitted signal:

$$y(t) = h(\tau,\ldots) * x(t) \tag{3.78}$$

The channel impulse response is one of the most common ways of representing the radio channel and is used extensively for simulations of wireless communication.

3.6.2.7 Other representations of the channel

The channel transfer function $H(f,t,\boldsymbol{r}_t,\boldsymbol{r}_r)$ is dependent on the time, frequency, and the (local) relative positions and radiation patterns of the transmitter and receiver antennas. Therefore it describes the spatial (and time-selective) and frequency-selective fading. This particular representation of the radio channel can be transformed through Fourier analysis into other representations. The relation between the frequency-dependent channel $H(f,\ldots)$ and the delay-dependent channel impulse response $h(\tau,\ldots)$ was derived in the previous section, and both representations are shown in Fig. 3.24. Similar Fourier transform relations exist between the time t and Doppler shift f_D, and between the spatial locations $\boldsymbol{r}_t,\boldsymbol{r}_r$ and spatial frequencies ω_t,ω_r (which relate to the "beam space" as sometimes used in the literature). These relations are summarized in Table 3.4.

Table 3.4 Multi-path Channel Relations and Representations.

	Dual Domain
Time delay (τ)	Baseband frequency (f)
Time dispersion	Frequency-selective fading
Channel impulse response	Channel transfer function
Delay spread	Coherence bandwidth
Doppler frequency (f_D)	Time (t)
Doppler dispersion	Time-variant fading
Doppler spectrum	Coherence time
Doppler spread	
Spatial frequencies, "beam space" (ω_t, ω_r)	Spatial positions ($\boldsymbol{r}_t, \boldsymbol{r}_r$)
Angular dispersion	Spatial fading
Angular spectrum	Coherence length
Angular spread	Antenna correlation

The radio channel can be described equally well using any of these mathematical representations, so the choice of which to use depends on other aspects such as implementation efficiency or characteristics of the problem to be studied.

3.6.2.8 Correlation

Channel properties are commonly expressed using different correlations, which are metrics that capture how similar the fast fading variations are for different times, frequencies, and positions. In particular the spatial correlations, that is, the correlation as a function of different spatial (antenna) separations, are key metrics for array processing and multi-antenna technologies and will be extensively utilized in Chapter 6 (see, e.g., Section 6.3.5). In the abstract vector space used for multi-antenna signal processing, the spatial correlation captures the directional information in the channel and reflects the impact of angular spread on the fading signals.

We define the correlation between the complex random variables X and Y as

$$\rho_{XY} \triangleq \frac{\mathrm{E}[XY^*]}{\sqrt{\mathrm{E}[XX^*]\mathrm{E}[YY^*]}} \tag{3.79}$$

The correlation between two channel transfer functions $H_1(f_1, t_1, \boldsymbol{r}_{t,1}, \boldsymbol{r}_{r,1})$ and $H_2(f_2, t_2, \boldsymbol{r}_{t,2}, \boldsymbol{r}_{r,2})$ can for a wide range of transfer functions be shown to be dependent only on the difference in frequency, delay and position as $\rho_{H_1H_2}(\Delta f, \Delta t, \Delta \boldsymbol{r}_t, \Delta \boldsymbol{r}_r)$. With the help of (3.75) and by varying only one of the degrees of freedom, different correlation functions can be determined using the properties p_n, τ_n, $\boldsymbol{k}_{t,n}$, $\boldsymbol{k}_{r,n}$, and $\boldsymbol{\Psi}_n$ of the propagation paths in the directional channel model from Section 3.6.1. Under the assumption that all antennas are isotropic and polarization-matched, that is, if $a_{tr} = \mathbf{g}_r^T(\theta_r, \varphi_r)\boldsymbol{\Psi}\mathbf{g}_t(\theta_t, \varphi_t) = 1$, the following correlation functions can be derived:

The *frequency correlation function*

$$\rho(\Delta f) = \frac{\sum_n p_n e^{-j2\pi\Delta f \tau_n}}{\sum_n p_n} \tag{3.80}$$

The *time correlation function*

$$\rho(\Delta t) = \frac{\sum_n p_n e^{j\boldsymbol{k}_{r,n} \cdot \boldsymbol{v}_r \Delta t}}{\sum_n p_n} \tag{3.81}$$

The *spatial correlation function* at the transmitter (receiver) side

$$\rho(\Delta \boldsymbol{r}_t) = \frac{\sum_n p_n e^{j\boldsymbol{k}_{t,n} \cdot \Delta \boldsymbol{r}_t}}{\sum_n p_n} \tag{3.82}$$

$$\rho(\Delta \boldsymbol{r}_r) = \frac{\sum_n p_n e^{j\boldsymbol{k}_{r,n} \cdot \Delta \boldsymbol{r}_r}}{\sum_n p_n} \tag{3.83}$$

Examples of spatial correlation functions for different angular distributions of the incoming waves, that is, different sets of $p_n, \boldsymbol{k}_{r,n}$, are shown in Fig. 3.28. As can be seen in these figures, the correlation tends to reduce with increasing position shift $\Delta \boldsymbol{r}$ although not necessarily in a monotonic fashion. Also note that a higher angular spread leads to a lower correlation.

3.6.3 STOCHASTIC MODELS

Stochastic models represent models where the statistical properties of the propagation paths (Section 3.6.3.2) or of the channel coefficients (Section 3.6.3.1) are determined and fitted to results from propagation measurements. These models can be used in link simulations or radio network simulations of the wireless communication system including the advanced antennas. One of the main benefits of stochastic modeling is that the model descriptions are quite generic and straightforward to implement and therefore suitable for use when comparability is paramount such as in

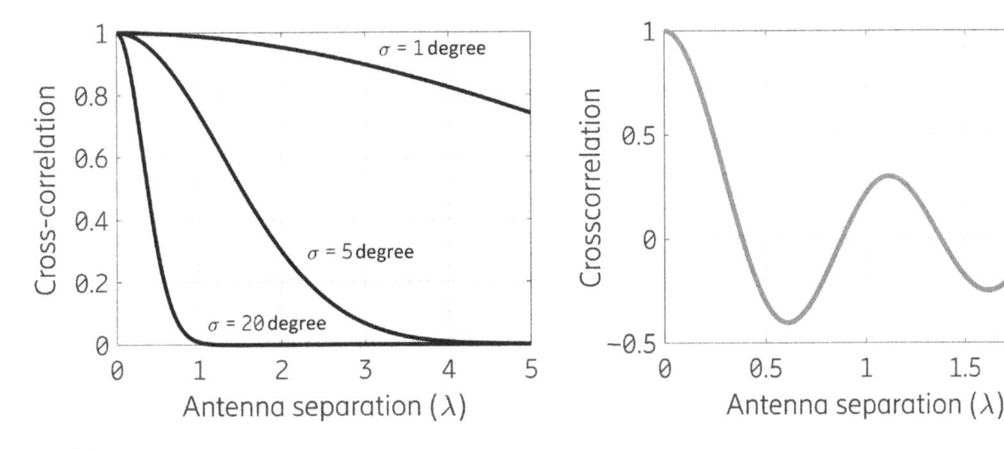

FIGURE 3.28

Correlation as a function of antenna separation in wavelengths assuming that the propagation paths are Gaussian distributed in angle with a mean that is perpendicular with respect to the antenna separation and different standard deviations (angular spreads) (left) and for propagation paths that are uniformly distributed in the horizontal plane according to the Jakes model (right).

standard-setting organizations where many different participants contribute results from proprietary simulators.

3.6.3.1 Stochastic channel models

A *stochastic channel model* of the radio channel can be formed using separate, sometimes even independent, models of the following components:

- Average power, for example, the expectation of the norm of H. This depends on the path loss and shadow fading, but also on the transmit and receive antenna gains. The ABG model or the Hata model [2] are examples of simple stochastic path loss models that can be used to capture the dynamics of path loss and interference levels in stochastic radio network simulations. For simulations of a single communication link it is common to normalize $E|\boldsymbol{H}|^2$ to unity and instead adjust the signal-to-noise ratio by adding white noise of varying power;
- The variability of the signal strength, for example, the fading statistics, where Rayleigh or Rice fading are commonly used models;
- The rate of change of the signal strength versus time, specified using the coherence time or Doppler spread. This is a function of the mobility of the transmitting and/or receiving antenna through space, that is, its speed v. In some cases, variations also occur due to mobility in the environment, for example, reflections off moving cars. Also, the angular spread of the multi-path around a moving antenna affects the rate of change. A common model is the one by Jakes [12], which is derived under the assumption that the multi-path is incident on the moving receiver antenna from directions that are uniformly distributed in the horizontal plane. This model leads to a "bath tub"-shaped Doppler spectrum and an autocorrelation function of the fading with respect to time, $\rho(\Delta t)$, that has the shape of a zero-order Bessel function of the first kind: $\rho(\Delta t) = J_0\left(\frac{2\pi}{\lambda} vt\right)$ (see Fig. 3.28 (right));
- The rate of change of the signal strength versus frequency, specified using the coherence bandwidth or the rms delay spread. This is a function of the multi-path dispersion in delay (propagation distance). Hence, the most common way of modeling this is by representing the channel impulse response or power delay profile using a *Tapped Delay Line* (TDL) model that represents the channel impulse response in (3.76) using a finite set of "taps." An example of a TDL model is given in Fig. 3.29. For TDL models representing NLOS conditions, each tap is usually assumed to experience Rayleigh fading in time with a Jakes-type Doppler spectrum, while a first tap with Ricean fading is used to represent LOS conditions;
- The degree of similarity, that is, correlation between the channels representing different combinations of transmit and receive antennas. This is often specified using models for the spatial cross-correlation between the fading variations experienced between different pairs of transmit or receive antennas, which can be determined using (3.82) and (3.83). As we saw in Section 3.6.2.8 the cross-correlations depend on the angular spread, the spatial antenna separations, and the antenna polarizations. As can be seen in Fig. 3.28, increasing the antenna separation reduces the correlation. One popular assumption is to use completely uncorrelated (Rayleigh) fading on all elements of $\boldsymbol{H}$, which is referred to as the *IID Rayleigh* (independently identically Rayleigh distributed) model. This assumption can be motivated in the case of spatially well-separated antennas in a rich scattering environment or for some specific combinations of angular distribution and spatial separation where the correlation is zero, but as

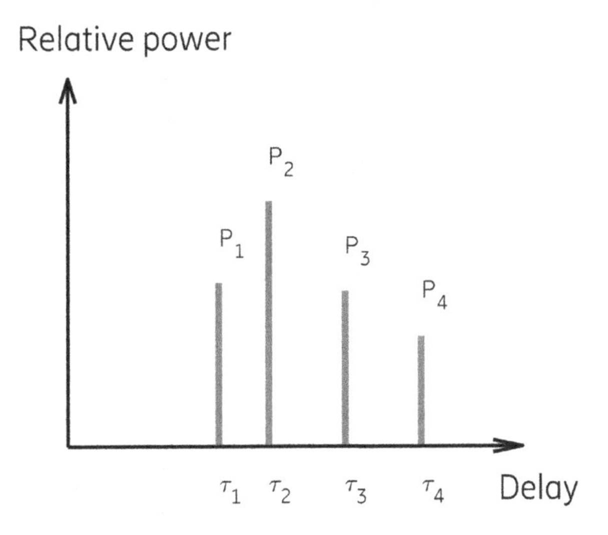

FIGURE 3.29

Example of a tapped delay line (TDL) model representation of a channel impulse response.

can be seen in Fig. 3.28 it is often unrealistic for the case of more closely spaced antennas, particularly when the angular spread is lower such as for an elevated base station (compare Table 3.3). Hence, for the typical antenna arrays used in 4G and 5G systems the correlation properties need to be taken into account.

3.6.3.2 Stochastic geometric models

Stochastic models of the directional propagation channel, or as they are often called *stochastic geometric channel models* or *spatial channel models (SCM)*, have also been developed to support more comprehensive radio network simulations involving multiple communication links spanning both serving and interfering links in a multicell scenario within a geographical area. In such models, the number of paths and their distributions in amplitude, delay, angle, and polarization are described by stochastic distributions that have been characterized using high-resolution spatial channel measurements such as those shown in Fig. 3.23. A common observation from measurements is that the observed propagation paths tend to be grouped into several *clusters* in the angular and delay domains. The preferred modeling methodology is hierarchical, where the mean angles and initial propagation delay of each cluster are first specified or generated, and in a secondary step individual propagation paths within each cluster are generated. Clustered stochastic geometric models such as [13] have become the main tool for simulations involving AAS, and such models have been developed for 5G new radio (NR) standardization by 3GPP [3] and for IMT-2020 evaluation by ITU-R [4]. In these models, different scenarios such as Urban Macro (UMa), Urban Micro (UMi), Rural Macro (RMa), Indoor Hotspot (InH), and Indoor Factory (InF) are captured by different parameterizations of the propagation paths according to corresponding measurements, for example, as outlined in Table 3.3. In radio network simulations, the stochastically generated propagation paths are used to calculate the channel impulse response as in Eq. (3.76).

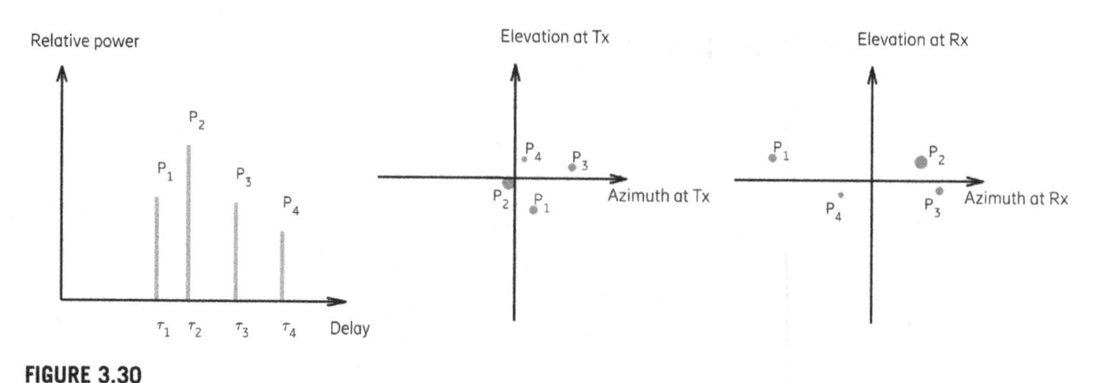

FIGURE 3.30

Schematic example of a CDL model representation of a directional propagation channel.

Through the use of stochastic distributions for the large-scale parameters such as mean angles, angular spread, and delay spread, an SCM model can generate channel realizations that span the range of variability encountered in a cell or a whole wireless communication network. Hence, such a model is well suited for exercising different multi-antenna algorithms and establishing their robustness with respect to both typical and outlier channel conditions.

For simulations of a single communication link, a *clustered delay line model* (CDL) is sometimes used [13], which represents a geometrical extension of the TDL described in the previous section. For the CDL, the mean directions of departure and arrival of each tap are specified. Furthermore, each tap is represented by a cluster of multi-path components (plane waves) that are distributed in angle around the mean direction. A schematic example of a CDL is shown in Fig. 3.30. Note that while very detailed, such a CDL represents one single possible realization of, for example, the time delays and directions of departure and arrival in the radio channel. Hence it is not recommended to optimize, for example, an adaptive beamforming scheme toward a single CDL since a trivial solution with a fixed, preselected, beam shape can perform equally well. In contrast, in an SCM model an unlimited set of different CDLs can be generated.

3.6.4 SITE-SPECIFIC PROPAGATION MODELS

Since the early days of cellular communication, operators have been using *site-specific propagation models* when planning their networks by evaluating how certain key performance indicators such as coverage or capacity are affected by deploying base stations at different potential sites. Such propagation models need to take into account details of the propagation environment around the site, including terrain height variations, building positions and shapes, and indoor floor plans. Radio network planning requires efficient propagation models that can provide predictions of the area coverage from a specific site in seconds or minutes. Also, the models need to be sufficiently accurate so that unpleasant surprises at the end of a costly and lengthy site acquisition process are minimized.

For early voice-based communication systems it was typically sufficient to predict the path loss, as the coverage and interference working points could then be assessed. Link budgets determined using single link calculations or simulations (including margins for shadow fading and inter-cell interference) could then be used to assess the expected radio link quality at a certain location.

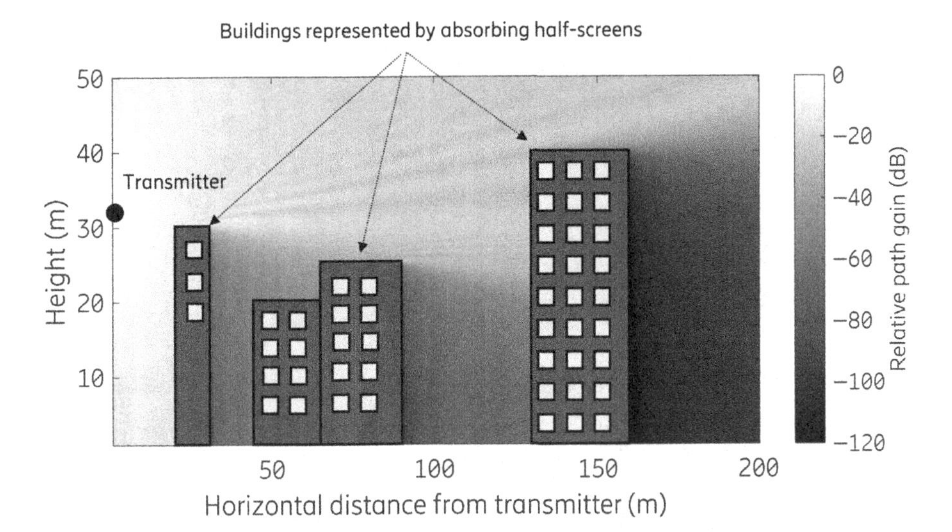

FIGURE 3.31

Prediction of the path loss using a height profile represented by absorbing or reflecting half-screens.

Correspondingly, the most successful site-specific propagation models utilized efficient and often semi-empirical models, for example, by determining the terrain and building profile along a line connecting the transmitter and receiver and applying some semi-analytic or heuristic method for calculating the loss over such a profile [14] (see Fig. 3.31 for one example).

However, as will be seen in Chapters 5, and 6, communication systems are today much more capable and able to adapt to and indeed utilize multi-path propagation through beamforming and multi-stream (MIMO) transmission schemes. Hence, there is a greater need to also predict the main characteristics of the multi-path in the channel so that the operator may assess where more advanced base stations are required and where simpler and less capable base stations provide sufficient performance. This can require a full 3D site-specific propagation model where both the propagation modeling and the environment description requires much higher detail. A schematic illustration of a 3D channel prediction using a site-specific propagation model can be seen in Fig. 3.32. Commercial radio network planning tools are increasingly capable of providing such 3D channel predictions via, for example, ray-tracing techniques, though the computational time remains a challenge.

Another challenge is the need for predictions at the new and higher frequencies being allocated to 5G deployments. Due to the novelty of mobile communication at these frequencies there are fewer propagation measurements available against which the models can be validated.

3.7 SUMMARY AND DISCUSSION

In this chapter, some basic properties of antennas and wave propagation have been described, with the following key takeaways:

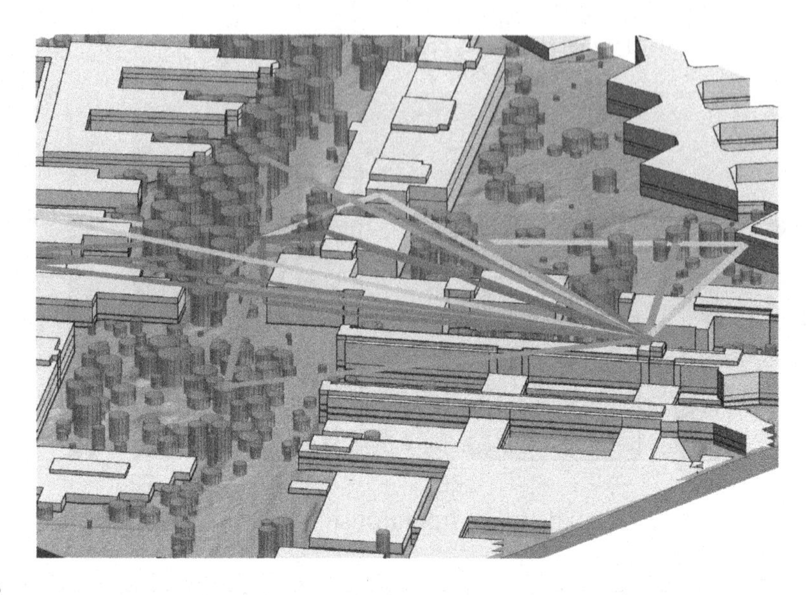

FIGURE 3.32

Illustration of site-specific propagation modeling in an urban environment using the model in Ref. [15]. The different lines represent examples of propagation paths experiencing different types of interactions, for example, reflections, diffuse scattering from buildings and from trees, diffraction, and street canyon wave-guiding.

- Electromagnetic waves can be transmitted by and received by antennas, that is, the antenna is a converter between conducted and radiated energy;
- The antennas can shape how the power is distributed in different directions;
- Larger antennas in relation to the wavelength give more degrees of freedom for spatially shaping the energy. As will be seen in Chapter 4, antenna arrays are one method for constructing large antennas with the added benefit that the spatial shaping can be adaptive.
- Increasing distance and addition of obstructions between transmitter and receiver reduces the energy of the received signals/waves;
- Multi-path propagation causes rapid and seemingly random fluctuations in the received signal as a function of time, frequency, and spatial position.

These properties need to be taken into account or should even be actively exploited when designing AAS solutions and features for mobile communication. Some of these properties are summarized below.

First and foremost, the path loss as discussed in Section 3.5.1 represents a fundamental limit to the quality of the communication link. Too large path loss and any communication signals modulated on the radio waves will be too weak to detect compared to the noise. A low path loss can mean a strong desired signal, but that in itself is not a guarantee of good communication quality as interference from other uses of the same frequencies, for example, for communication links in other cells or even to other users in the same cell, may restrict the information that can be reliably transferred. Clever shaping of transmitting and receiving antenna patterns (including polarization) using antenna arrays (Chapter 4) through AAS solutions aim to mitigate the path loss and hence maximize the desired signal, while also minimizing the interference by suppressing transmission or reception in directions where interfered or interfering devices are located (see Chapter 6).

Of fundamental importance are the relations between the path loss and achievable antenna gains and the frequency. As discussed in Section 3.4 the path loss (between isotropic antennas) increases with frequency, while the antenna gain can also increase with frequency as outlined in Section 3.3.8. These are factors of major importance whenever new and usually higher frequency bands are taken into use (see Chapter 14), but become critically important for high-band where the frequency is one order of magnitude higher than for mid-band (see Chapter 7). These properties also affect the architecture and design decisions (see Chapter 12).

For AAS solutions to be effective the transmitter and/or receiver needs to know how to shape the radiation pattern for good effect, that is, what directions and polarizations to enhance and what directions and polarizations to suppress. Such knowledge can be gained by simple means in a static free-space propagation scenario. However, due to multi-path propagation as discussed in Section 3.5.3 the radio channel will be affected by angular and delay dispersion that causes the channel to vary in time, frequency, and with spatial position. The rapid variation of the channel with respect to time, frequency, and different antenna positions may require additional resources to be devoted to transmitting known signals, pilots, for estimating the channel, rather than for the transmission of payload data. Section 9.3.1.2.1 discusses how different densities of reference signals in time and frequency, used for channel estimation, can be configured to a UE in NR. The selected density depends on how fast the channel is expected to vary for the UE in the given deployment. The angular spread has a crucial impact on AAS solutions, limiting the benefit transmission or reception using a single, well-defined beam. Instead, the use of generalized beamforming may be required if the angular spread is large (see Section 6.1.4). AAS can also exploit multipath propagation for diversity or even to send several data streams simultaneously (cf. Section 6.3), often using antenna arrays and signal processing algorithms that result in radiation patterns with much more complex shapes (see Section 6.3.5.6 and in particular Fig. 6.31).

The fact that electromagnetic waves and antennas are polarized (Sections 3.2.3 and 3.3.3) means that a propagation path may carry two independent and orthogonally polarized waves that can be separately sensed by a pair of orthogonally polarized antennas. Such dual-polarized antenna elements are commonly used in classical antennas and AAS to provide diversity by matching the polarization to that used by the antenna at the other end of the link, or to transmit and receive separate data streams simultaneously. This is also covered in Sections 4.5.2.1 and 6.3.

Other properties are also exploited by AAS and will be referred in later chapters.

REFERENCES

[1] H.T. Friis, A note on a simple transmission formula, Proc. IRE 34 (5) (1946) 254–256.

[2] M. Hata, Empirical formula for propagation loss in land mobile radio services, IEEE Trans. Veh. Technol. 29 (3) (1980) 317–325.

[3] 3GPP TR 38.901, Study on Channel Model for Frequencies from 0.5 to 100 GHz, 2019.

[4] ITU-R Report M.2412, Guidelines for Evaluation of Radio Interface Technologies for IMT-2020, 2017.

[5] M. Riback, J. Medbo, J.-E. Berg, F. Harrysson, H. Asplund, Carrier frequency effects on path loss, in: 2006 IEEE 63rd Vehicular Technology Conference, Melbourne, Vic., 2006, pp. 2717–2721.
[6] Millimetre-wave based mobile radio access network for fifth generation integrated communications (mmMAGIC), Measurement Results and Final mmMAGIC Channel Models, Deliverable 2.2, H2020-ICT-671650-mmMAGIC/D2.2, May 2017.
[7] ITU.R Recommendation P.2109-0, Prediction of Building Entry Loss, June 2017.
[8] J. Medbo, H. Asplund, J. Berg, N. Jalden, Directional channel characteristics in elevation and azimuth at an urban macrocell base station, in: 2012 6th European Conference on Antennas and Propagation (EUCAP), Prague, 2012, pp. 428–432.
[9] C. Larsson, B. Olsson, J. Medbo, Angular resolved pathloss measurements in urban macrocell scenarios at 28 GHz, in: 2016 IEEE 84th Vehicular Technology Conference (VTC-Fall), Montreal, QC, 2016, pp. 1–5.
[10] H. Asplund, J. Berg, F. Harrysson, J. Medbo, M. Riback, Propagation characteristics of polarized radio waves in cellular communications, in: 2007 IEEE 66th Vehicular Technology Conference, Baltimore, MD, 2007, pp. 839–843.
[11] M. Steinbauer, A.F. Molisch, E. Bonek, The double-directional radio channel, IEEE Antennas Propag. Mag. 43 (4) (2001) 51–63.
[12] W.C. Jakes (Ed.), Microwave Mobile Communications, John Wiley & Sons Inc, 1974.
[13] P. Kyösti, et al., WINNER II Channel Models, IST-4-027756 WINNER II Deliverable D1.1.2 V1.2, February 2008.
[14] C. Phillips, D. Sicker, D. Grunwald, A survey of wireless path loss prediction and coverage mapping methods, IEEE Commun. Surv. Tutor. 15 (1) (2013) 255–270.
[15] H. Asplund, M. Johansson, M. Lundevall, N. Jaldén, A set of propagation models for site-specific predictions, in: 12th European Conference on Antennas and Propagation (EuCAP 2018), London, 2018, pp. 1–5.

CHAPTER

ANTENNA ARRAYS AND CLASSICAL BEAMFORMING 4

The purpose of this chapter is to explain the basic concept of classical beamforming using an antenna array, where, like a flashlight beam, signal power can be directed in a desired direction. This is accomplished by a beamformer that adjusts the signals transmitted from the elements of an array so that they combine constructively at an intended receiver. The corresponding gain pattern, sometimes referred to as a beam or beam gain pattern, is derived and illustrated and some properties like maximum gain and width of the main lobe are described for commonly used configurations such as uniform linear and planar arrays.

The focus is on the free-space single-path propagation scenario introduced in Chapter 3. Results in this chapter will then serve as inputs to Chapter 5, where the setup will be extended by considering multipath propagation, multiple orthogonal frequency division multiplexing (OFDM) subcarriers, and multiple antennas at the receiver side as well. More specifically, Chapter 5 will formulate a multiple-input, multiple-output (MIMO) OFDM channel model based on the so-called array response vector introduced in the present chapter. This in turn will allow exploration of various multi-antenna techniques in Chapter 6.

Finally, a uniform planar array (UPA) of dual-polarized element pairs and the partitioning of such an array into an array of subarrays are introduced. The latter partitioning facilitates consideration of how to find a suitable number of radio chains as a function of the deployment, as described in Chapters 13 and 14.

4.1 INTRODUCTION

An antenna array is set of antennas commonly organized in a structure such as an array of rows and columns. An antenna array can be used for both transmission and reception. By transmitting different versions of the same signal from all the antennas, the signal's amplification, referred to as gain, can be controlled so that the gain is different in different directions. Rather than spreading the signal power over an entire cell, the signal power can be directed to where the intended user is located and even be reduced in other directions to avoid interfering other users. Similarly, when the antenna array is used for reception, the gain in the direction of where the user is located can be increased so that the received signal power is increased.

The signals transmitted from different antennas are converted into electromagnetic waves that combine in the air. Considering a receiver located at some distance right in front of the transmitting antenna array, the propagation path lengths from the different antennas to the receiver are essentially the same and if the same signal is transmitted from all the antennas, the waves will add constructively at the receiver. With N antennas, the amplitude of the resulting wave is N times higher

Advanced Antenna Systems for 5G Network Deployments. DOI: https://doi.org/10.1016/B978-0-12-820046-9.00004-6

than the amplitude for a single antenna. For the same total transmission power, the amplitude per antenna is however downscaled with a factor $1/\sqrt{N}$ and the gain in terms of power is then N as compared to a single antenna. In other directions, the contributions from the different antennas can add destructively and may even completely cancel each other so that the amplitude is zero. A receiving antenna converts the incident wave into a signal, whose amplitude therefore depends on its position as illustrated in Fig. 4.1.

The maximum gain can be increased by having more antennas in the array. However, for cellular network, it is typically of interest to have a high antenna gain in all the directions where there are users, and this is enabled by a beamformer that adjusts the amplitude and delay of the signal transmitted from each antenna. Compared to the case when exactly the same signal is transmitted from all antennas, the direction of the maximum gain can thereby be changed so that it is obtained in a desired direction rather than right in front of the antenna array (see the example in Fig. 4.2).

In the present chapter, the opportunities offered by an array of antennas will be demonstrated, primarily by deriving the gain as a function of direction. The case with an antenna array used for transmission of a narrowband bandpass signal to a single receiving antenna is considered and rather than using formulations based on the electric fields and deriving radiation intensities, the radio channel model in Section 3.6.2 is used for the free-space single-path case in Section 3.6.2.6. The reason for taking the approach based on the channel model is to prepare for the MIMO OFDM channel model in Section 5.2 which is an extension of the model presented in the present chapter. The MIMO channel model will include multiple antennas at both the transmitting and receiving

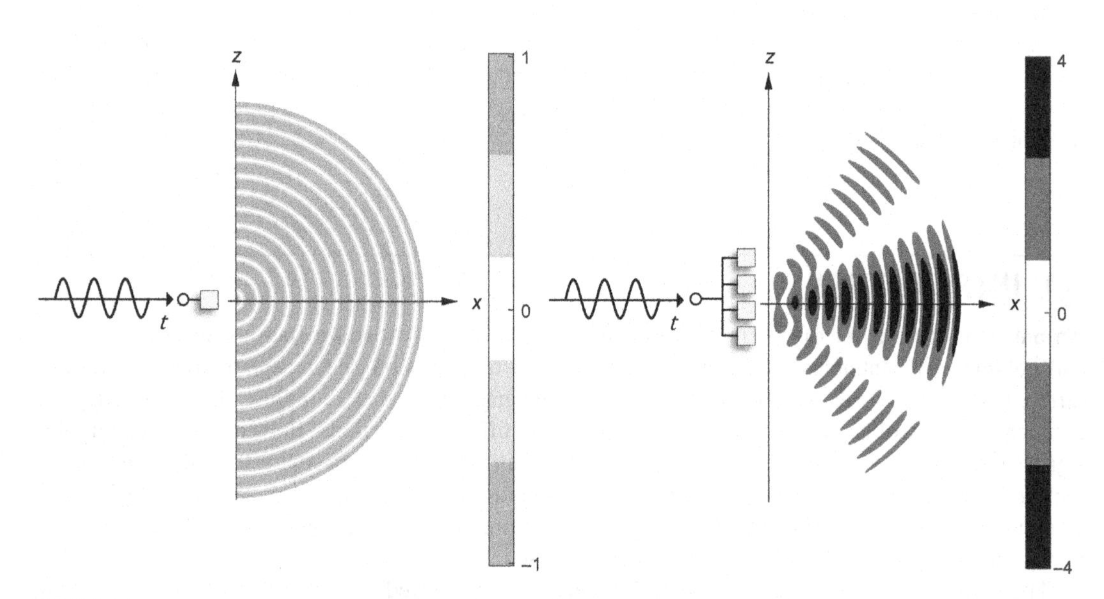

FIGURE 4.1

Illustration of the signal amplitude for different receiver positions in the xz-plane for a fixed time when a sinusoid is transmitted using one antenna (left) or four antennas (right). The amplitude of the signal is given by the bar legends.

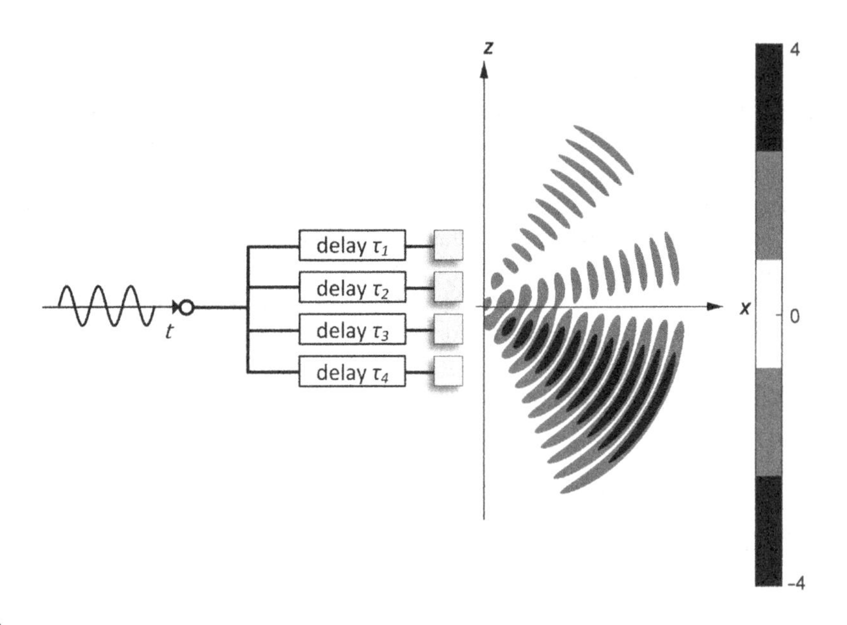

FIGURE 4.2

By transmitting the same signal from all antennas with appropriate delays, $\{\tau_k\}$, the directions where the signals add constructively can be controlled.

sides for the case with wideband OFDM-based transmission and serve as a basis for Chapter 6 on multiple antenna techniques.

Similar to Refs. [1,2], the starting point taken in Section 4.2 is the case with two identical transmitting antennas in free space. The signal received by an isotropic antenna in a certain direction is derived for the case that the same signal is transmitted from both antennas. As will be seen, the signal's amplitude is proportional to a product of the transmitting antennas' amplitude pattern and an array factor. Furthermore, some basic properties of the gain will be described and illustrated. Then, in Section 4.3, a more general antenna array with an arbitrary number of elements is considered, where *element* refers to an antenna in the array. The array response vector is introduced and related to the free-space single-path channel. Thereafter, it is shown that by also introducing a beamforming weight vector that represents the signals transmitted, the received signal's complex amplitude and consequently also the gain, is given by a product between the two vectors. The uniform linear array (ULA) is defined and it is shown that the maximum gain increases with the number of elements while the width of the main lobe decreases.

Classical beamforming, introduced in Section 4.4, allows changing the direction in which the maximum gain occurs by transmitting the same signal from all the elements, but with delays, or equivalently phase shifts chosen so that the signals add constructively in the desired direction. ULAs are then considered, and some remarks are made on element spacing as well as on so-called grating lobes. Some other forms of beamforming techniques are also illustrated to make the point that the properties of the gain pattern depend not only on the array but also on the beamformer used.

The UPA and the dual-polarized UPA are defined in Section 4.5. These configurations represent commonly used special cases of the general array introduced in Section 4.3 and some basic properties of the gain patterns are examined. With a UPA consisting of dual-polarized element pairs, not only the direction of the main lobe but also the polarization can be controlled.

Section 4.6 introduces so-called arrays of subarrays. A dual-polarized UPA can be partitioned into subarrays and a benefit of this that fewer radio chains are needed. At the same time, this constrains the set of beamforming weight vectors that can be applied to the antennas of the underlying array. For classical beamforming, the dependency between the subarray size and the range of angles with high gain is illustrated; the larger the subarray, the smaller the range of angles. This builds intuition for why different deployments call for different subarray sizes as will be illustrated in Sections 13.4 and 14.3.

Finally, in Section 4.7, the key findings of the chapter are summarized.

4.2 ARRAYS WITH TWO ELEMENTS

Before addressing arrays with arbitrary many antennas, the case with two identical antennas transmitting the same signal is considered. If the signal transmitted is a sinusoid of a certain frequency, the antennas convert the signal to sinusoidal electromagnetic waves. A receiving antenna at a certain location will convert the sum of the two incident waves into a received signal. This is illustrated in Fig. 4.3.

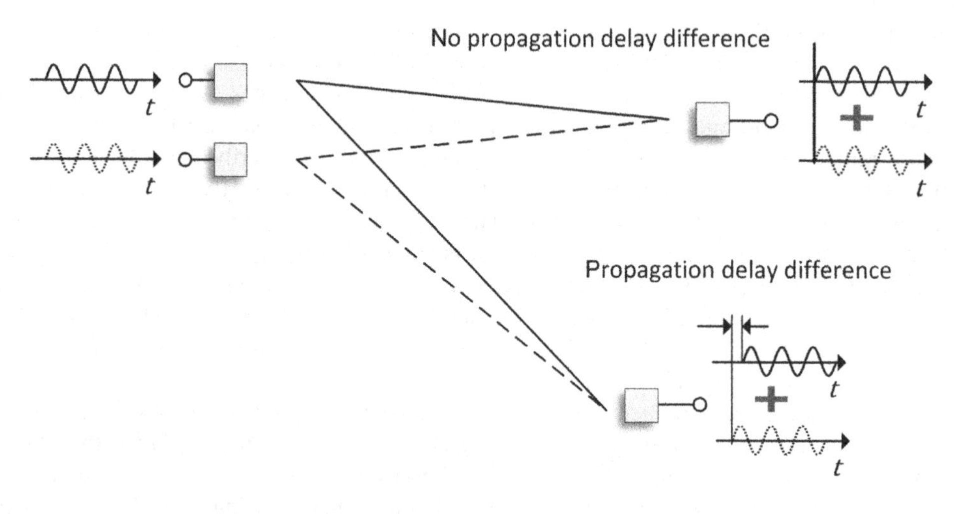

FIGURE 4.3

The received signal is a sum of the signals transmitted from the two antennas. The two signals may experience different propagation delays depending on the receiver's position.

The received signal's amplitude will depend on the difference in propagation delays for the signals transmitted from the two transmitting antennas which in turn depends on the direction from the transmitting antennas to the receiver. If the receiver is right in front of the two transmitting antennas so that the propagation delays are the same, then two signals will add constructively so that the amplitude is twice that of the signal received from a single antenna. If, however, the two signals have experienced different propagation delays, then the amplitude will depend on the time delay difference. In fact, the signals may completely cancel each other. For the case of a sinusoid signal, complete cancelation occurs if the delay difference equals half a period that corresponds to a propagation path length difference of half a wavelength.

From a received signal perspective, the array with two identical transmitting antennas appears as a single antenna with a (complex) amplitude pattern (see Section 3.3.4) that thus depends on the direction to the receiver. The amplitude of the received signal will also depend on the amplitude patterns of the individual transmitting antennas, which as mentioned above are referred to as *elements* in the context of the array. More specifically, the amplitude pattern for the array antenna can for the case with identical elements be expressed as a product of the element amplitude pattern and an array factor, which captures the summation of the signals from the individual elements. Note that the amplitude pattern also determines the gain pattern of the antenna array, that is, the power received as a function of direction relative to a single isotropic transmitting antenna.

Next, the amplitude pattern of the two-element array is derived and used to illustrate some properties of the gain pattern, such as a 3 dB increase in maximum gain as well as a narrower main lobe width relative to a single element.

4.2.1 ASSUMPTIONS

The transmitting and receiving antennas are in free space as depicted in Fig. 4.4. On the transmitting side, a vertical array with two identical elements placed along the z-axis with separation d is considered. The receiver antenna is placed at a distance R from the origin at a position determined by zenith and azimuth angles θ and φ in a right-handed spherical coordinate system (see also Appendix 1). The position of the receiver relative to the origin is given by the vector $\boldsymbol{r}$,

$$\boldsymbol{r} = R\hat{\boldsymbol{r}}, \tag{4.1}$$

where the spherical unit vector $\hat{\boldsymbol{r}}$ in Cartesian coordinates is described by

$$\hat{\boldsymbol{r}} = \begin{bmatrix} \sin\theta\cos\varphi & \sin\theta\sin\varphi & \cos\theta \end{bmatrix}^{\mathrm{T}}. \tag{4.2}$$

The first transmitting element is located at the origin and therefore $\boldsymbol{r}$ also represents the propagation path from the first element to the receiving antenna. The propagation path from the second element to the receiving antenna is represented with the vector $\boldsymbol{r}_2$. To isolate the impact of the transmitting elements, the receiver antenna is finally assumed to be isotropic, lossless and aligned in terms of polarization to the incident waves.

A key assumption that will be used in the derivation of the gain below is that the receiver is located at a large enough distance so that the propagation paths from the two transmitting elements, $\boldsymbol{r}$ and $\boldsymbol{r}_2$, appear (approximately) parallel (see also [1–3]).

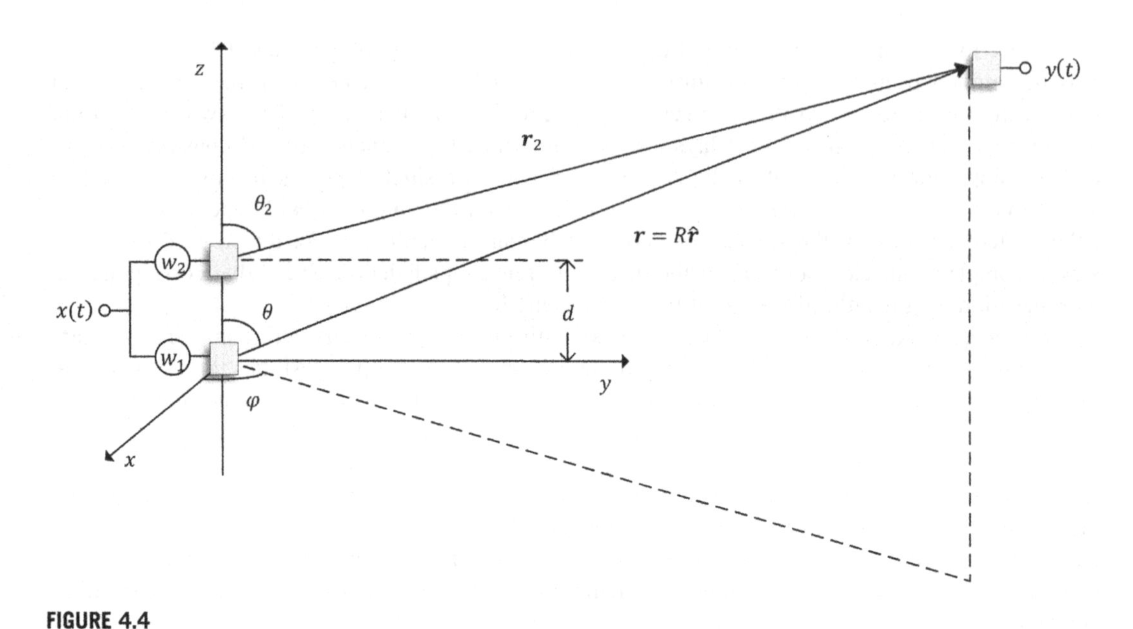

FIGURE 4.4

A transmitting two-element vertical array with element separation d

The assumption is illustrated in Fig. 4.5, together with the vector $\boldsymbol{d}_2$ that represents the second element's position,

$$\boldsymbol{d}_2 = \begin{bmatrix} 0 & 0 & d \end{bmatrix}^{\mathrm{T}}, \tag{4.3}$$

and the propagation path from the second element

$$\boldsymbol{r}_2 = R_2 \hat{\boldsymbol{r}},$$

where R_2 denotes the corresponding path length. The propagation path length of the second element, R_2, can be expressed as the projection of $\boldsymbol{r}_2$ onto $\hat{\boldsymbol{r}}$

$$R_2 = \hat{\boldsymbol{r}}^{\mathrm{T}} \boldsymbol{r}_2 = \hat{\boldsymbol{r}}^{\mathrm{T}} (\boldsymbol{r} - \boldsymbol{d}_2) = R - \hat{\boldsymbol{r}}^{\mathrm{T}} \boldsymbol{d}_2 = R - d\cos\theta, \tag{4.4}$$

where the second equality follows from that the propagation path for the second element can be expressed as $\boldsymbol{r}_2 = \boldsymbol{r} - \boldsymbol{d}_2$. Since the paths $\boldsymbol{r}$ and $\boldsymbol{r}_2$ are assumed parallel, the angular direction to the receiver is furthermore equal to θ for both elements. An interpretation of the approximation of parallel paths is that the waves incident on the receiver from the two transmitting elements can be seen (locally) as plane waves propagating in the same direction $\hat{\boldsymbol{r}}$ (see also the discussion of wave fronts in Section 3.2.2).

Some assumptions are also needed on the signals transmitted and received. More specifically, as shown in Fig. 4.4, the same signal is transmitted from both elements, and for a fair comparison to the case with a single element, the total transmit power is kept the same. This means in this case with two elements that each element uses half the power, and the signals transmitted are taken as

$$x_1(t) = w_1 x(t), \tag{4.5}$$

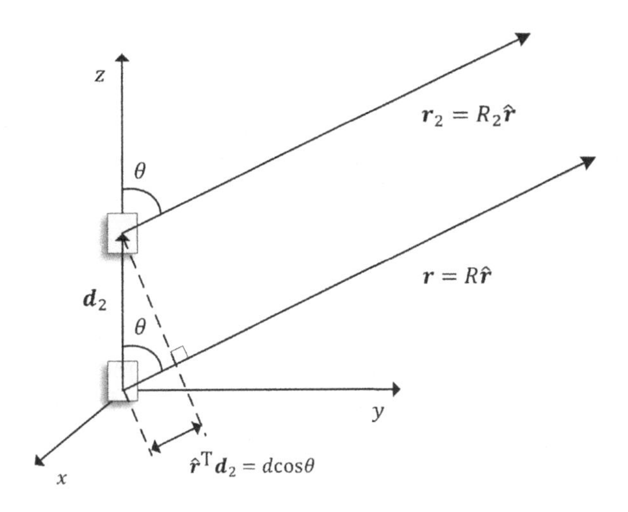

FIGURE 4.5

The distance to the receiver is large so that the two paths to the receiver (not shown in the figure), $\boldsymbol{r}$ and $\boldsymbol{r}_2$, appear parallel.

$$x_2(t) = w_2 x(t), \tag{4.6}$$

with amplitude weights chosen as

$$w_1 = w_2 = 1/\sqrt{2}. \tag{4.7}$$

The transmitted signal $x(t)$ as well as the received signal, $y(t)$, are complex-valued low-pass basebands equivalents [4,5], and the actual signals at radio frequency are narrowband signals given by

$$x_{\mathrm{RF}}(t) = \mathrm{Re}\{x(t)e^{j2\pi f_c t}\}, \tag{4.8}$$

$$y_{\mathrm{RF}}(t) = \mathrm{Re}\{y(t)e^{j2\pi f_c t}\}, \tag{4.9}$$

where f_c is the carrier frequency (see also Section 3.6.2.1 and Fig. 3.27). The special case that the transmitted radio frequency signal is a single-frequency sinusoid means that

$$x(t) = X \rightarrow x_{\mathrm{RF}}(t) = |X|\cos(2\pi f_c t + \arg\{X\}), \tag{4.10}$$

and the magnitude and argument of X thus represent the amplitude and phase of the sinusoid.

4.2.2 RECEIVED SIGNAL IN A FREE-SPACE CHANNEL

Due to the superposition principle (see Section 3.2.4), the received signal $y(t)$ will be a sum of the contributions from the two transmit elements. To determine the contribution from each transmitting element, the multipath propagation model in Section 3.6.2 is used for the special case with a single path to represent free-space propagation. Omitting for simplicity any thermal noise, the received signal can then be written as

$$y(t) = h_1 x_1(t - \tau_1) + h_2 x_2(t - \tau_2) \tag{4.11}$$

where h_n and τ_n represent the complex amplitude and propagation delay for the propagation path associated with transmit element n. More specifically, it can be shown using the definition of the antenna radiation pattern in (3.30) and the free-space radio channel model defined in Section 3.6.2.6 that the complex amplitudes h_n can be expressed as

$$h_n = \alpha e^{-j2\pi f_c \tau_n} g_r(\theta_r, \varphi_r) g_t(\theta, \varphi) \chi. \tag{4.12}$$

Here, α represents the path loss between isotropic lossless antennas, θ_r and φ_r represent the direction at the receiver, $g_r(\theta, \varphi)$ and $g_t(\theta, \varphi)$ are the (complex) amplitude patterns for the receiver antenna and the transmitting elements, respectively. The term χ captures the alignment between the polarization of the receiver antenna, $\hat{\psi}_r(\theta_r, \varphi_r)$, and the polarization of the transmitting elements, $\hat{\psi}_t(\theta, \varphi)$ (see also Section 3.4.2).

As a side note, for the case that the propagation paths $\boldsymbol{r}$ and $\boldsymbol{r}_2$ are not parallel, (4.11) still holds but the complex amplitude gains in (4.12) would need to be modified since the angles $\theta_r, \varphi_r, \theta$, and φ are different for the two elements. Another implicit and reasonable assumption is that the distance between the two elements, d, is much smaller than the propagation path lengths so that the path loss in terms of α is the same for both elements.

To demonstrate the gain of using two transmit elements, the expression in (4.12) is next further simplified using the assumption that the receiver antenna is isotropic, lossless and aligned in terms of polarization so that

$$g_r(\theta_r, \varphi_r) = 1, \tag{4.13}$$

$$\chi = 1. \tag{4.14}$$

By combing the four last equations, together with the definitions of the transmitted signals in (4.5) and (4.6) the received signal becomes

$$y(t) = \alpha e^{-j2\pi f_c \tau_1} g(\theta, \varphi) w_1 x(t - \tau_1) + \alpha e^{-j2\pi f_c \tau_2} g(\theta, \varphi) w_2 x(t - \tau_2), \tag{4.15}$$

where $g(\theta, \varphi) = g_t(\theta, \varphi)$ is the complex amplitude pattern for a transmitting element. Thus, the received signal is as expected a sum of two versions of the transmitted signal with different propagation delays, and the factor $e^{-j2\pi f_c \tau_n}$ represents the phase shift caused by the propagation delay of element n. The propagation delays can in turn using (4.4) be expressed as

$$\tau_1 = \frac{R}{c}, \tag{4.16}$$

$$\tau_2 = \frac{R_2}{c} = \frac{R}{c} - \frac{d\cos\theta}{c} = \tau_1 - \frac{d\cos\theta}{c}, \tag{4.17}$$

where c is the wave propagation speed.

In the next section, the last three expressions will be combined to establish a gain relation between the transmitted and received signals as a function of direction. This is possible since the propagation delay difference does depend on θ and φ in the general case, and on θ in the particular case.

4.2.3 GAIN AND ARRAY FACTOR

The expression for the received signal is further refined using properties of the transmitted signal. For the case that the signal transmitted is a sinusoid as defined in (4.10), also the received signal, as obtained by substituting $x(t) = X$ in (4.15), is a sinusoid at radio frequency,

$$x(t) = X \rightarrow y(t) = Y \leftrightarrow y_{\mathrm{RF}}(t) = |Y|\cos(2\pi f_c t + \arg\{\mathrm{Y}\}). \tag{4.18}$$

Next, the expressions for propagation delays in (4.16), and (4.17), and the fact that $x(t)$ corresponds to a sinusoid at radio frequency is again used in (4.15). Then, by defining the array factor, $\mathrm{AF}(\theta, \varphi)$, as

$$\mathrm{AF}(\theta, \varphi) = w_1 + w_2 e^{jkd\cos\theta}, \tag{4.19}$$

using the relation between carrier frequency f_c, wave propagation speed c, and the magnitude of the wave vector defined in Section 3.2,

$$k = \|\boldsymbol{k}\| = \frac{2\pi}{\lambda} = \frac{2\pi f_c}{c}, \tag{4.20}$$

the amplitude and phase of the received sinusoid can be expressed as

$$Y = \alpha e^{-j2\pi f_c \tau_1} \underbrace{\mathrm{AF}(\theta, \varphi) g(\theta, \varphi)}_{= g_{\mathrm{AA}}(\theta,\varphi)} X \tag{4.21}$$

As can be seen from (4.21), the received signal amplitude, $|Y|$ depends on the product of the elements' amplitude pattern as well as the array factor which thus describes how the contributions from the two elements sum up. In fact, by comparing (4.21) with using only the first element ($w_1 = 1, w_2 = 0 \leftrightarrow \mathrm{AF}(\theta, \varphi) = 1$), the received signal appears to be transmitted from a single antenna with (complex) amplitude pattern

$$g_{\mathrm{AA}}(\theta, \varphi) = \mathrm{AF}(\theta, \varphi) g(\theta, \varphi), \tag{4.22}$$

rather than from an antenna with amplitude pattern $g(\theta, \varphi)$. This explains why an antenna array may also be referred to as an antenna. The amplitude pattern for the array transmission, $g_{\mathrm{AA}}(\theta, \varphi)$, will just as any other amplitude pattern characterize radiation properties such as its gain pattern which relates to the transmitted and received power rather than amplitude. By squaring the magnitude of both sides of (4.21), the relation between transmitted and received power is

$$|Y|^2 = \alpha^2 G_{\mathrm{AA}}(\theta, \varphi)|X|^2,$$

where the gain as a function of direction has been defined as

$$G_{\mathrm{AA}}(\theta, \varphi) = |g_{\mathrm{AA}}(\theta, \varphi)|^2 = |\mathrm{AF}(\theta, \varphi)|^2 G(\theta, \varphi). \tag{4.23}$$

Here, $G(\theta, \varphi) = |g(\theta, \varphi)|^2$ is the (power) gain pattern for one element in the array. Thus, it is assumed that the complex amplitude pattern $g(\theta, \varphi)$ is scaled so that $G(\theta, \varphi)$ is the (antenna) gain as a function of direction relative to an isotropic lossless antenna for one element (see also Section 3.3.7.4). Before examining and illustrating the gain pattern defined by (4.23), the following is noted:

- The amplitude pattern in (4.22) can be expressed as a product of the array factor and the element pattern. This will hold also for the larger arrays considered in later parts of the chapter, including both uniform linear and UPAs as well as cases with beamforming;
- For the gain pattern in (4.23), the factor $|\mathrm{AF}(\theta,\varphi)|^2$ can be recognized as the gain pattern for an array of isotropic elements in terms of received power relative to a single isotropic antenna under free-space propagation conditions. This gain will be sometimes be referred to as the *(free-space) array gain*. The gain pattern for the antenna array in (4.23) is thus the product of this array gain and the element gain patterns.

Having made these observations, some properties of the gain pattern are explored and illustrated in Section 4.2.4 before showing in Section 4.2.5 that the findings apply not only to single-frequency signals but also to narrowband signals.

4.2.4 PROPERTIES OF THE GAIN

To further understand the properties of the (power) gain defined in (4.23), Euler's formula is used together with the assumption on transmitting the same signal from both elements. Combining (4.7) and (4.19) gives

$$|\mathrm{AF}(\theta,\varphi)|^2 = 2\left|\cos\left(\frac{k}{2}d\cos\theta\right)\right|^2. \tag{4.24}$$

As discussed in Section 4.2.3, this can be recognized as the free-space array gain pattern, and in Fig. 4.6 this gain is illustrated as a function of direction θ for some different element separations d. From the expression in (4.24), it is noted that

- The free-space array gain has its maximum value two, $|\mathrm{AF}(\theta,\varphi)|^2 = 2$, for angles θ that satisfy $d\cos\theta = 0, \pm\lambda, \pm 2\lambda, \ldots$. For such angles, the received power is twice as high, or 3 dB higher, as compared to the case with a single element using the same total transmit power since the signals received combine constructively. From (4.4), $d\cos\theta$ can be recognized as the propagation path length difference between the two antennas, and hence constructive addition occurs when the path length difference equals a multiple of the wavelength;
- The free-space array gain attains its minimum value, $|\mathrm{AF}(\theta,\varphi)|^2 = 0$ for angles such that $d\cos\theta = \pm\frac{\lambda}{2}, \pm\frac{3\lambda}{2}, \ldots$ This means that when the propagation path length difference is an odd multiple of half the wavelength, the signals from the two antennas will cancel each other. For the case with $d = 0.7\lambda$, there will be zeros for $\theta \approx 90^\circ \pm 45^\circ$.

Thus, the signals from the two elements can add constructively in certain directions but also cancel out in other directions as stated in the introduction. However, the total gain in a certain direction depends not only on the array gain but also on the gain of the elements in the array in the same direction according to (4.23). In Fig. 4.7, gain patterns are illustrated, both for a single element of the array, and for a two-element array consisting of two such elements. The element pattern defined in [6] with 65-degree half-power beamwidth and 8 dBi gain is used, and the separation

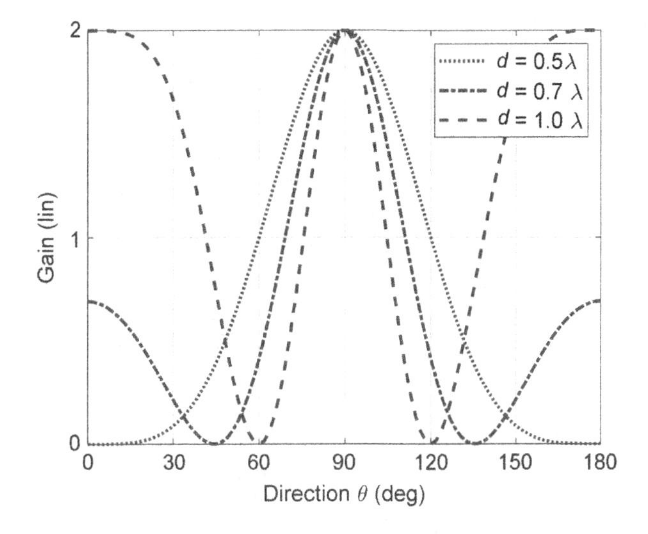

FIGURE 4.6

Gain on linear scale of using a two-element array relative, a single element for the case with isotropic elements.

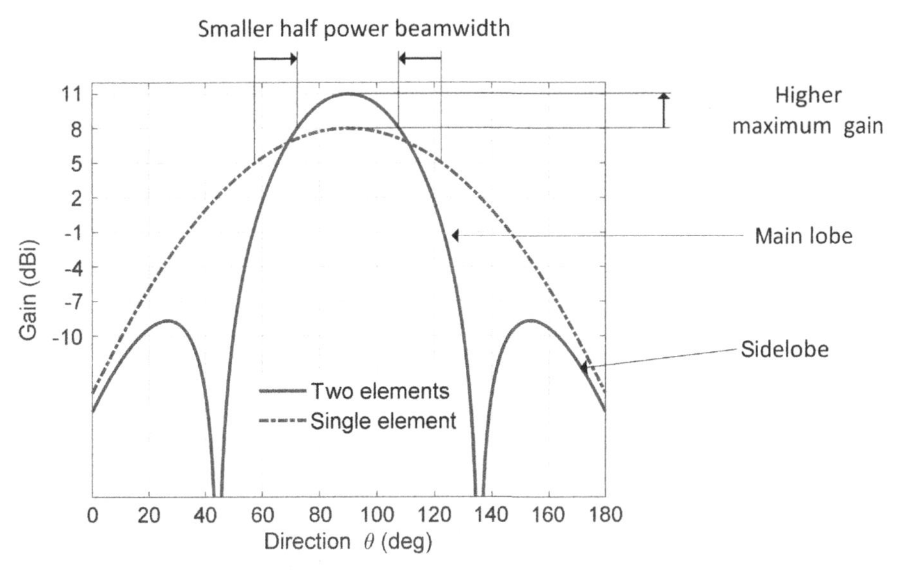

FIGURE 4.7

Gain as function of direction θ for $\varphi = 0°$ for a single element and an array with two such elements separated by 0.7λ. The element gain model in [6] is used and the elements are oriented so that the maximum gain occurs for $\theta = 90°$, $\varphi = 0°$.

between the element is $d = 0.7\lambda$. The impact of the element separation is elaborated on further in Section 4.4.2.1.

As can be seen, the maximum gain, which occurs for $\theta = 90^{\circ}$, is 3 dB higher compared to a single element aligned to have maximum gain in this direction.

Also, the half-power beamwidth of the main lobe, which is the range of angles for which the gain is within 3 dB of its maximum value, is smaller for the array as compared to the single-element case. Thus, the higher maximum gain comes at the cost of a narrower main lobe. This stems from the fact that the gain pattern is a product of the array gain and the element gain pattern. The gain pattern then typically has a beamwidth which is at least as narrow as the narrowest one of the two, in this case the array gain. Finally, for the chosen element separation $d = 0.7\lambda$, the array factor is zero for $\theta = 90^{\circ} \pm 45^{\circ}$ and the same holds for the total gain and this explains why there in addition to the main lobe also are sidelobes.

To complement the gain patterns given in Figs. 4.6 and 4.7, which illustrated only the gain as a function of zenith angle θ, the dependency on both the zenith angle and the azimuth angle φ is illustrated on a logarithmic scale using spherical coordinates in Fig. 4.8. For comparison, the case with a two-element horizontal array is also included. The difference as compared to the vertical array is the position of the second element which corresponds to replacing (4.3) with $\boldsymbol{d}_2 = \begin{bmatrix} 0 & d & 0 \end{bmatrix}^{\mathrm{T}}$. The same spacing $d = 0.7\lambda$ is used for both the horizontal and vertical arrays.

Both the vertical and horizontal arrays will have 3 dB higher maximum gain right along the x-axis as compared to the single-element case. The difference between the two patterns is that for the

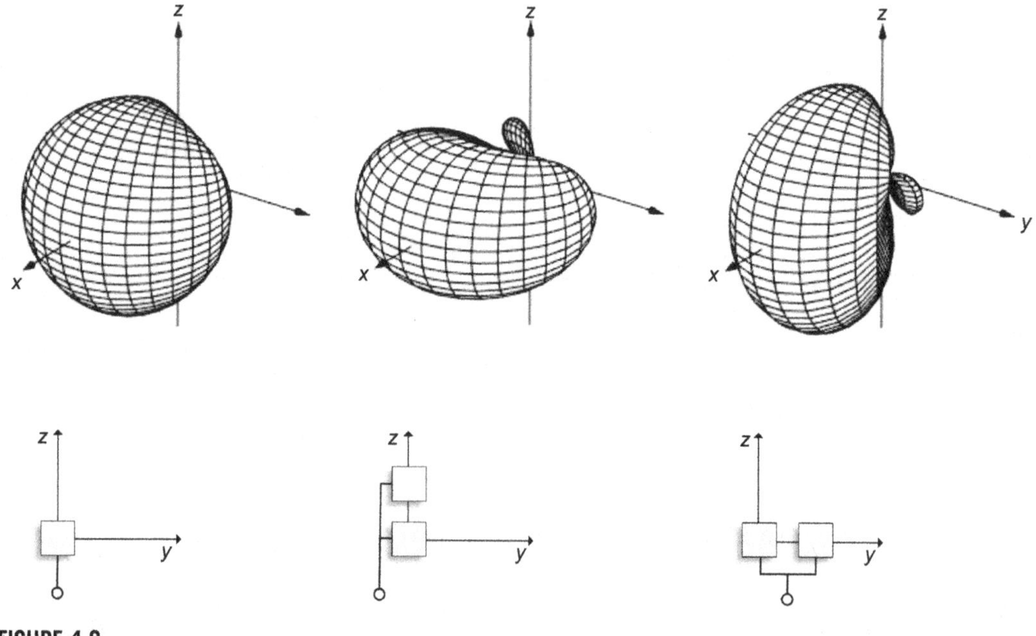

FIGURE 4.8

Illustration of gain in different directions for a single element (left), a vertical array with two elements (center) and a horizontal array with two elements (right). The element gain model in [6] is used in all three cases.

vertical array, the gain depends on the angle to the z-axis whereas it depends on the angle to y-axis for the horizontal array. The vertical array has a smaller main lobe width in the plane $\varphi = 0^\circ$ whereas the horizontal array has a smaller main lobe width in the horizontal plane $\theta = 90^\circ$.

To sum up, properties of the gain, for a two-element vertical array, have been considered. It was demonstrated that the gain can be increased with 3 dB while the main lobe width in the vertical plane was reduced. The same applies for a horizontal array except that the main lobe width in the horizontal plane is reduced.

4.2.5 EXTENSION TO NARROWBAND SIGNALS

The derivation of the gain is now extended also to narrowband signals. This essentially means that the assumption (4.18) is not used. Instead the key approximation done is

$$x(t - \tau_2) \approx x(t - \tau_1). \tag{4.25}$$

Such an approximation will be accurate for the case that the propagation delay difference $\Delta\tau = \tau_2 - \tau_1$ is small. Small in this context needs to be related to how fast the (baseband) signal $x(t)$ changes with time and as long as $B\Delta\tau \ll 1$, where B is the bandwidth of the signal, then (4.25) is a commonly used approximation in the field of array signal processing [7] (see also Section 3.6.2.5).

Eq. (4.25) may at first appear a bit unexpected since the propagation delay difference should give rise to a phase shift that is not visible in the equation. The reason for this is that the phase shift caused by the propagation delay is part of the channel coefficient in (4.15).

By using the relations (4.16), (4.17), and (4.25) in (4.15), the received signal can be expressed as

$$y(t) = \alpha e^{-j2\pi f_c \tau_1} \underbrace{\mathrm{AF}(\theta, \varphi) g(\theta, \varphi)}_{g_{\mathrm{AA}}(\theta,\varphi)} x(t - \tau_1) \tag{4.26}$$

From a receiver perspective, just as in the case with a single-frequency sinusoid, the signal appears as being transmitted from an antenna with amplitude pattern $g_{\mathrm{AA}}(\theta, \varphi)$, which coincides with the definition in (4.22).

This means that the results derived in the previous subsections are valid not only for sinusoidal signals but to a wider class of signals. It should though be remembered that the delay difference depends on the element separation d (see (4.16) and (4.17)), and this means that (4.25) becomes inaccurate when the element separation and/or the bandwidth of the signal become too large. In that case, the signal can be partitioned into sub-bands, for example, sets of adjacent subcarriers in the case with OFDM as described in Chapter 5 with bandwidths small enough for the approximation to hold.

4.3 UNIFORM LINEAR ARRAYS WITH MORE THAN TWO ELEMENTS

The case with two transmitting elements was considered in Section 4.2. Attention is now shifted to the case with $N > 2$ transmitting elements at arbitrary locations in the same free-space propagation scenario. The major difference as compared to the case with two elements is that the received signal is a sum of N versions of the transmitted signal.

To determine the gain as a function of direction for the array, a so-called *array response vector* will be defined. The array response vector includes the impact of the N elements' amplitude patterns and propagation delay differences, all depending on the direction of the receiver. By introducing a weight vector to represent the signal versions transmitted from all the elements, the complex amplitude for transmission with the array can be expressed as a product of the two defined vectors. The array response vector can also under certain assumptions be interpreted as a normalized free-space multiple-input-single-output (MISO) channel for transmission of a narrowband signal, and later in Section 5.3.2, the MIMO OFDM channel model will be formulated based on it.

The special but common case with a ULA is considered. For a ULA, the array is uniform in the sense that the separation between any two adjacent elements is the same and further all elements have the same amplitude pattern. The same signal is transmitted from all the elements, and at broadside, in the direction perpendicular to the array, the (power) gain as compared to transmission from a single element is N, stemming from coherent addition of contributions from all the N elements while keeping the total transmitted power the same as for the single-element case. However, as will be shown, the main lobe beamwidth scales approximately with the inverse of the length of the array, that is, as $\lambda/(dN)$, for an element separation d and wavelength λ. As a side note, it is observed that there are also nulls in the radiation pattern, indicating the possibility, not only to increase the gain in certain directions but also to reduce the gain in other directions.

4.3.1 ASSUMPTIONS

A free-space propagation scenario is considered, and the assumptions done are essentially the same as for the case with two elements described in Section 4.2.1, with the following differences (see also Fig. 4.9):

- The array has N identical elements and the positions of the elements are described by the vectors

$$\boldsymbol{d}_n = \begin{bmatrix} d_{x,n} & d_{y,n} & d_{z,n} \end{bmatrix}^{\mathrm{T}}, n = 1, \cdots, N, \tag{4.27}$$

relative to a common reference point in origin $\begin{bmatrix} 0 & 0 & 0 \end{bmatrix}^{\mathrm{T}}$.

- The propagation path between element n and a receiving antenna is denoted $\boldsymbol{r}_n$ and the distance to the receiver is large so that all propagation paths appear parallel

$$\boldsymbol{r}_n = R_n \hat{\boldsymbol{r}}, \; n = 1, \cdots, N. \tag{4.28}$$

Here, $\hat{\boldsymbol{r}}$ defined in (4.2) points in the direction from the reference point to the receiver and R_n is the length of the corresponding propagation path. Considering (4.4), the path length can be expressed as

$$R_n = R - \hat{\boldsymbol{r}}^{\mathrm{T}} \boldsymbol{d}_n, \tag{4.29}$$

where R is the path length from the reference point for the element positions and $\hat{\boldsymbol{r}}^{\mathrm{T}} \boldsymbol{d}_n$ is the projection of $\boldsymbol{d}_n$ onto $\hat{\boldsymbol{r}}$.

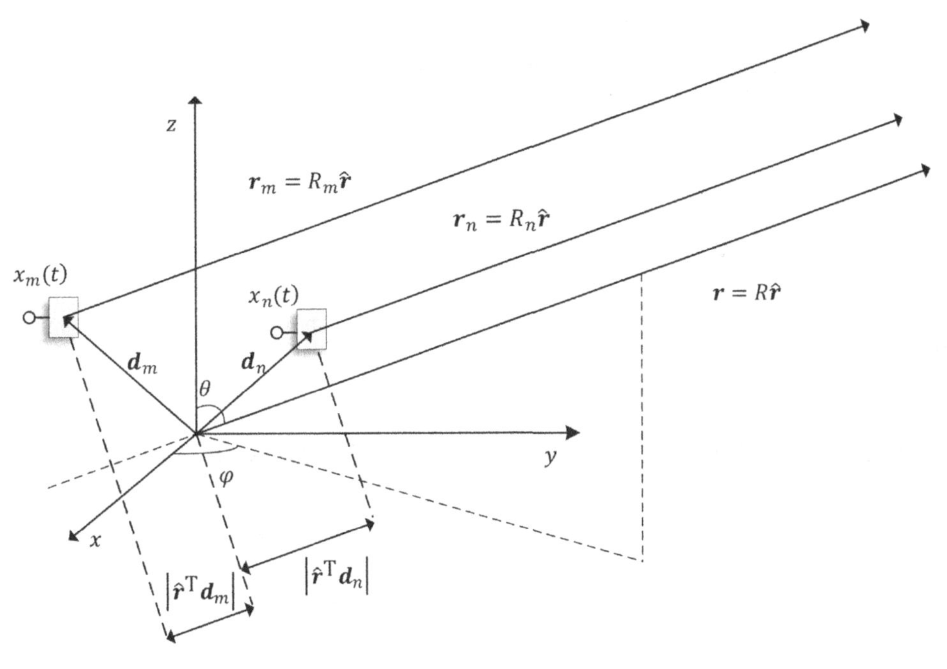

FIGURE 4.9

The distance to the receiver is large so that two paths $\boldsymbol{r}_n$ and $\boldsymbol{r}_m$ to the receiver (not shown in the figure) from two elements with positions $\boldsymbol{d}_n$ and $\boldsymbol{d}_m$ appear parallel.

- The signal transmitted from element n is taken as

$$x_n(t) = w_n x(t), \ n = 1, \cdots, N. \tag{4.30}$$

  For the case that the same signal is transmitted from all the elements the weights are chosen as

$$w_n = 1/\sqrt{N}, \ n = 1, \cdots, N, \tag{4.31}$$

  to allow a fair comparison between different array sizes that use the same amount of total power.
- The signal $x(t)$ represents a narrowband signal with bandwidth small enough so that

$$x(t - \tau_n) \approx x(t - \tau), \ n = 1, \cdots, N, \tag{4.32}$$

  where τ and τ_n represent the propagation delay from the reference point and element n, respectively, to the receiving antenna,

$$\tau = \frac{R}{c}, \tag{4.33}$$

$$\tau_n = \frac{R_n}{c}. \tag{4.34}$$

The approximation in (4.32) holds obviously for a single-frequency sinusoid as defined in (4.10), and is also, as discussed in Section 4.2.5, applicable for narrowband signals with bandwidth B such that the propagation delay differences, which can be related to the element positions using (4.29),

$$\Delta\tau_n = \tau_n - \tau = -\frac{\boldsymbol{r}^{\mathrm{T}}\boldsymbol{d}_n}{c}, \; n = 1, \cdots, N, \tag{4.35}$$

satisfy $B\Delta\tau_n \ll 1$. Recall from the discussion in Section 4.2.5 that the approximation may become inaccurate if the array or the bandwidth becomes too large.

This set of assumptions is next used to define an array response vector and a beamforming weight vector that leads to a compact formulation of the received signal.

4.3.2 ARRAY RESPONSE VECTOR AND BEAMFORMING WEIGHT VECTOR

For the case with two elements, it was found in Section 4.2.3 that the received signal appears to be transmitted from a single antenna with (complex) amplitude pattern determined as a product of an array factor and the element amplitude pattern as in (4.22) and this holds also for the case with N elements. The amplitude pattern can however also be expressed as a product, between an *array response vector* and a *beamforming weight vector*, and next these two vectors are defined, discussed and related to the gain.

The starting point is to generalize (4.15) to the case with N signals. By using the narrowband assumption in (4.32) for the relative propagation delays in (4.35), it follows that the generalized expression (4.15) can be written as

$$y(t) = \sum_{n=1}^{N} \alpha e^{-j2\pi f_c \tau} e^{-j2\pi f_c \Delta\tau_n} w_n g(\theta, \varphi) x(t - \tau). \tag{4.36}$$

The relation between the transmitted and received signal is illustrated in Fig. 4.10. As can be seen, a beamformer generates weighted versions of the signal $x(t)$ for the different antennas as described by (4.30). The weights $\{w_n\}$ are collected in a vector with N elements, a *beamforming weight vector*

$$\boldsymbol{w} = \begin{bmatrix} w_1 & w_2 & \cdots & w_N \end{bmatrix}^{\mathrm{T}}. \tag{4.37}$$

For now, it is assumed that the same signal is transmitted from all the elements, and the corresponding weights w_n are defined in (4.31). Other choices of (complex-valued) weights will be discussed in Section 4.4.

As shown in Fig. 4.10, there is also effectively a MISO channel between the N transmitting elements and the receiver. All the elements share the same path loss and a common propagation delay as given by (4.33). However, since the elements are at different positions the corresponding received signals experience different propagation delays, which depends on the direction, and this part of the channel may be represented by the array response vector. The elements' complex amplitude patterns depend also on the direction and may, in the general case, differ between the elements. For these reasons they are, together with the phase shift caused by the propagation delay difference relative to the common propagation delay, collected into the array response vector.

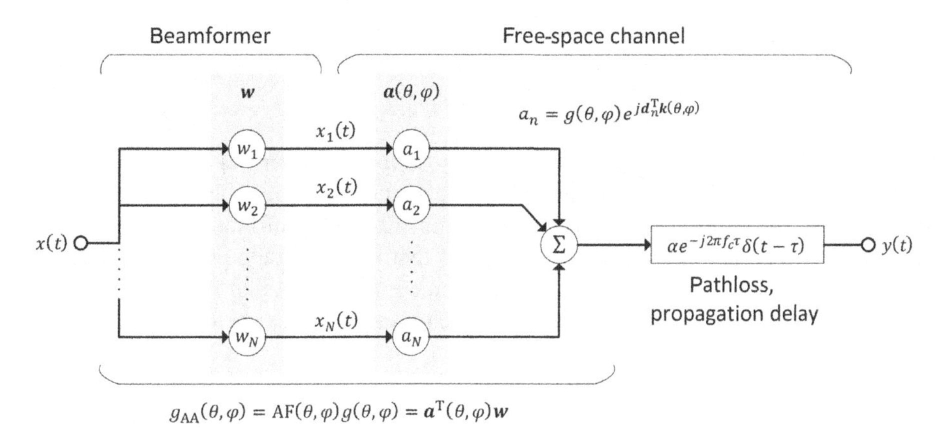

FIGURE 4.10

Relation between transmitted signal $x(t)$ and received signal $y(t)$.

The array response vector can therefore be seen as a normalized free-space MISO channel between the transmitted signals $\{x_n(t), n = 1, \cdots, N\}$ and the signal received by the isotropic polarization-aligned antenna $y(t)$. The normalization implies that it neither includes the impact of the propagation delay τ between the reference point of the array and the receiver, nor the path loss as represented with α.

Since all the elements are assumed to have the same complex amplitude pattern, $g(\theta, \varphi)$, the *array response vector* becomes

$$\boldsymbol{a}(\theta,\varphi) = g(\theta,\varphi)\left[\, e^{j\boldsymbol{d}_1^{\mathrm{T}}\boldsymbol{k}(\theta,\varphi)} \quad e^{j\boldsymbol{d}_2^{\mathrm{T}}\boldsymbol{k}(\theta,\varphi)} \quad \ldots \quad e^{j\boldsymbol{d}_N^{\mathrm{T}}\boldsymbol{k}(\theta,\varphi)} \,\right]^{\mathrm{T}}, \tag{4.38}$$

where the *wave vector* $\boldsymbol{k}(\theta, \varphi)$ is a scaled version of $\hat{\boldsymbol{r}}$ defined in (4.2),

$$\boldsymbol{k}(\theta,\varphi) = \frac{2\pi}{\lambda}\hat{\boldsymbol{r}} = \frac{2\pi}{\lambda}\left[\,\sin\theta\cos\varphi \quad \sin\theta\sin\varphi \quad \cos\theta\,\right]^{\mathrm{T}}. \tag{4.39}$$

By combining (4.35), (4.20), and (4.39) in sequence, the phase shift caused by the delay difference is related to the phase shifts of the array response vector in (4.38) as follows

$$e^{-j2\pi f_c\Delta\tau_n} = e^{j2\pi f_c\frac{\hat{\boldsymbol{r}}^{\mathrm{T}}\boldsymbol{d}_n}{c}} = e^{j\frac{2\pi}{\lambda}\hat{\boldsymbol{r}}^{\mathrm{T}}\boldsymbol{d}_n} = e^{j\boldsymbol{k}^{\mathrm{T}}(\theta,\varphi)\boldsymbol{d}_n} = e^{j\boldsymbol{d}_n^{\mathrm{T}}\boldsymbol{k}(\theta,\varphi)}, \tag{4.40}$$

for $n = 1, \cdots, N$. The received signal in (4.36) can then be written on the form

$$y(t) = \alpha e^{-j2\pi f_c\tau} g_{\mathrm{AA}}(\theta,\varphi)x(t-\tau), \tag{4.41}$$

where the amplitude pattern $g_{\mathrm{AA}}(\theta, \varphi)$ can be expressed as a product of the beamforming weight vector and the array response vector

$$g_{\mathrm{AA}}(\theta,\varphi) = \boldsymbol{a}^{\mathrm{T}}(\theta,\varphi)\boldsymbol{w}. \tag{4.42}$$

The combination of the beamformer and the array response vector thus appears as an antenna with amplitude pattern $g_{\mathrm{AA}}(\theta, \varphi)$. This is illustrated in Fig. 4.10 together with the relations to the beamformer and the array response vector. The corresponding gain pattern in terms of received power relative to an isotropic lossless antenna can be expressed by combining (4.23) and (4.42) as

$$G_{\mathrm{AA}}(\theta,\varphi) = |g_{\mathrm{AA}}(\theta,\varphi)|^2 = |\boldsymbol{a}^{\mathrm{T}}(\theta,\varphi)\boldsymbol{w}|^2. \tag{4.43}$$

This gain pattern is sometimes also referred to as *beam gain pattern* or *beam pattern*. Given an array response vector as a function of direction for a particular array in terms of element pattern and positions, such a beam pattern can thus be determined for a beamformer using (4.43).

Previously, in (4.22) and (4.26), it was also found that the amplitude pattern in (4.42) could be expressed as a product of an array factor and the element pattern. As all elements are assumed to be the same, this holds also for the case with N elements and the array factor can be written as

$$\mathrm{AF}(\theta,\varphi) = \sum_{n=1}^{N} w_n e^{j\boldsymbol{d}_n^{\mathrm{T}}\boldsymbol{k}(\theta,\varphi)}, \tag{4.44}$$

so that (4.22), repeated here for convenience, can be used to also express the amplitude pattern as

$$g_{\mathrm{AA}}(\theta,\varphi) = \mathrm{AF}(\theta,\varphi)g(\theta,\varphi). \tag{4.45}$$

Thus, for the case that all elements of the array are the same, the amplitude pattern can be expressed on two different but equivalent forms as given by (4.42) and (4.45). For the case that the elements are significantly different, (4.45) does not apply. In that case the array response vector needs to be modified and (4.42) can be used to determine the amplitude pattern.

In the next section some properties of the gain pattern $G_{\mathrm{AA}}(\theta,\varphi)$ are illustrated for the special but common case of ULAs.

4.3.3 UNIFORM LINEAR ARRAYS

A ULA is a special but very common antenna array used in practice. Here the array is uniform in the sense that the separation between any two adjacent elements is the same and all elements have the same amplitude pattern. Both vertical and horizontal array arrangements are possible (see Fig. 4.11). Vertical arrangements, also referred to as column antennas, of dual-polarized element pairs introduced in Section 4.5.2, are by far the most common structure for classical base stations antennas.

For analysis purposes the following characteristics are assumed:

- The positions of the elements $\boldsymbol{d}_n$ are taken along the positive z-axis for a vertical array and along the positive y-axis for a horizontal array

$$\boldsymbol{d}_n = \begin{cases} (n-1)\begin{bmatrix} 0 & 0 & d \end{bmatrix}^{\mathrm{T}} & \text{vertical array (along } z\text{-axis)} \\ (n-1)\begin{bmatrix} 0 & d & 0 \end{bmatrix}^{\mathrm{T}} & \text{horizontal array (along } y\text{-axis)} \end{cases}. \tag{4.46}$$

- All elements have the same complex amplitude pattern $g(\theta,\varphi)$ oriented so that the maximum (magnitude) occurs in the direction of the x-axis, $\theta = 90°$, $\varphi = 0°$.

The assumption that all elements have the same pattern is an approximation. In antenna arrays, the separation between adjacent elements is typically rather small, less than a wavelength as discussed in Section 4.4.2.1, and the elements will interact with each other. This interaction is referred

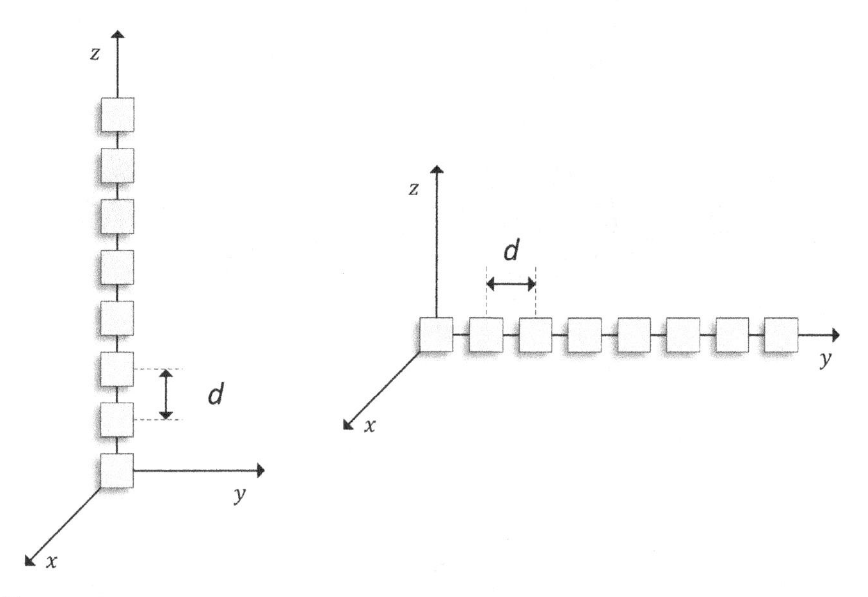

FIGURE 4.11

Examples with eight elements: vertical (left) and horizontal (right) uniform linear arrays.

to as mutual coupling [1,3] (see also Section 12.3.7). One consequence of this is that the gain patterns for the elements are different when they are used in an array as compared to when they are used in isolation. A common approach in analysis, design, and evaluations is therefore to incorporate the impact of mutual coupling by using a so-called embedded pattern rather than the pattern for the element in isolation, and in line with this all elements are assumed to have the same pattern.

4.3.3.1 Basic gain patterns

To explore some basic properties of a ULA, the gain as a function of the direction (θ, φ) will be considered for the case that the same signal is transmitted from all antennas. The approach taken is to first use the expression for the (complex) amplitude pattern in (4.42) or (4.45) and then determine the gain as the squared magnitude of it as given by (4.43). The array response vector for the ULA can be determined by combining (4.38) and (4.46) and the beamforming weight vector is chosen as

$$w = \begin{bmatrix} w_1 & w_2 & \cdots & w_N \end{bmatrix}^{\mathrm{T}} = \frac{1}{\sqrt{N}} \underbrace{\begin{bmatrix} 1 & 1 & \cdots & 1 \end{bmatrix}}_{N \text{ ones}}^{\mathrm{T}}, \tag{4.47}$$

to represent the case that the same signal is transmitted from all antennas as defined in (4.31).

The gain patterns for vertical and horizontal arrays with different number of elements are illustrated using a logarithmic scale in Figs. 4.12 and 4.13, respectively. In both cases, the element amplitude pattern defined in [6] with 65-degree half-power beamwidth and 8 dBi gain is used, and the separation between the elements is $d = 0.7\lambda$. Similar to the case with two elements, increasing the number of elements in either the vertical or horizontal dimension reduces the width of the main

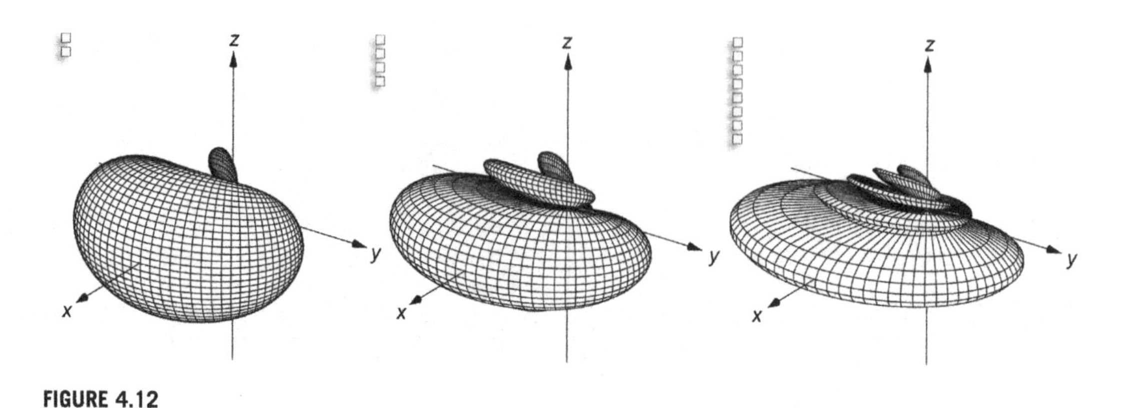

FIGURE 4.12

Illustration of gain pattern for vertical arrays with two, four, and eight elements.

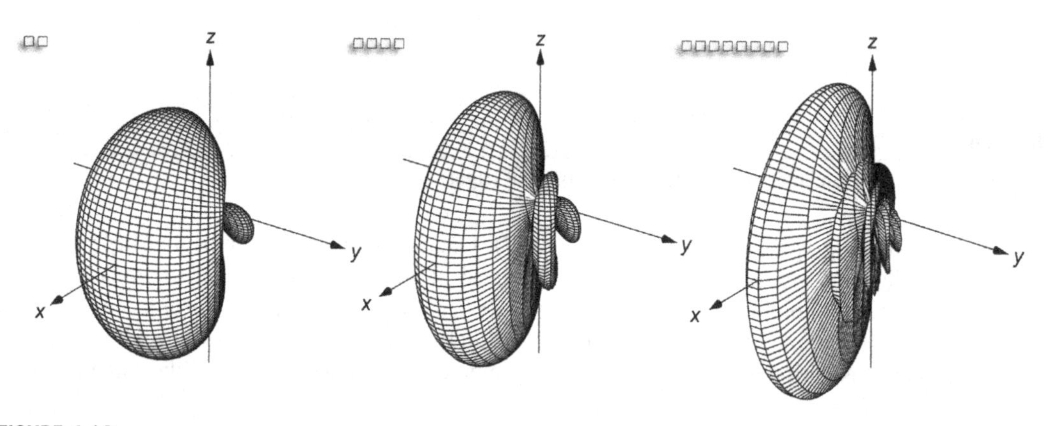

FIGURE 4.13

Illustration of gain pattern for horizontal arrays with two, four, and eight elements.

lobe in the corresponding dimension. Additionally, the gain patterns for the horizontal arrays are rotated versions of the corresponding vertical arrays and for this reason, only the case with vertical arrays is examined more closely in the following discussion.

To complement the three-dimensional illustrations of the gain patterns, Fig. 4.14 shows the gain pattern as a function of the zenith angle θ in the vertical plane $\varphi = 0^\circ$ for different vertical arrays.

As can be seen from Fig. 4.14, there is a peak as well as several nulls in the gain pattern. The width of the main lobe decreases and the number of nulls increases as the number of elements increases. To analyze this in somewhat more detail, the array factor is determined. Substitution of (4.47) and (4.46) into (4.44) results in

$$|\mathrm{AF}(\theta,\varphi)|^2 = \left|\sum_{n=0}^{N-1} \frac{1}{\sqrt{N}} e^{jkd\cos\theta}\right|^2 = \frac{1}{N}\left|\frac{\sin\left(N\frac{kd\cos\theta}{2}\right)}{\sin\left(\frac{kd\cos\theta}{2}\right)}\right|^2 \tag{4.48}$$

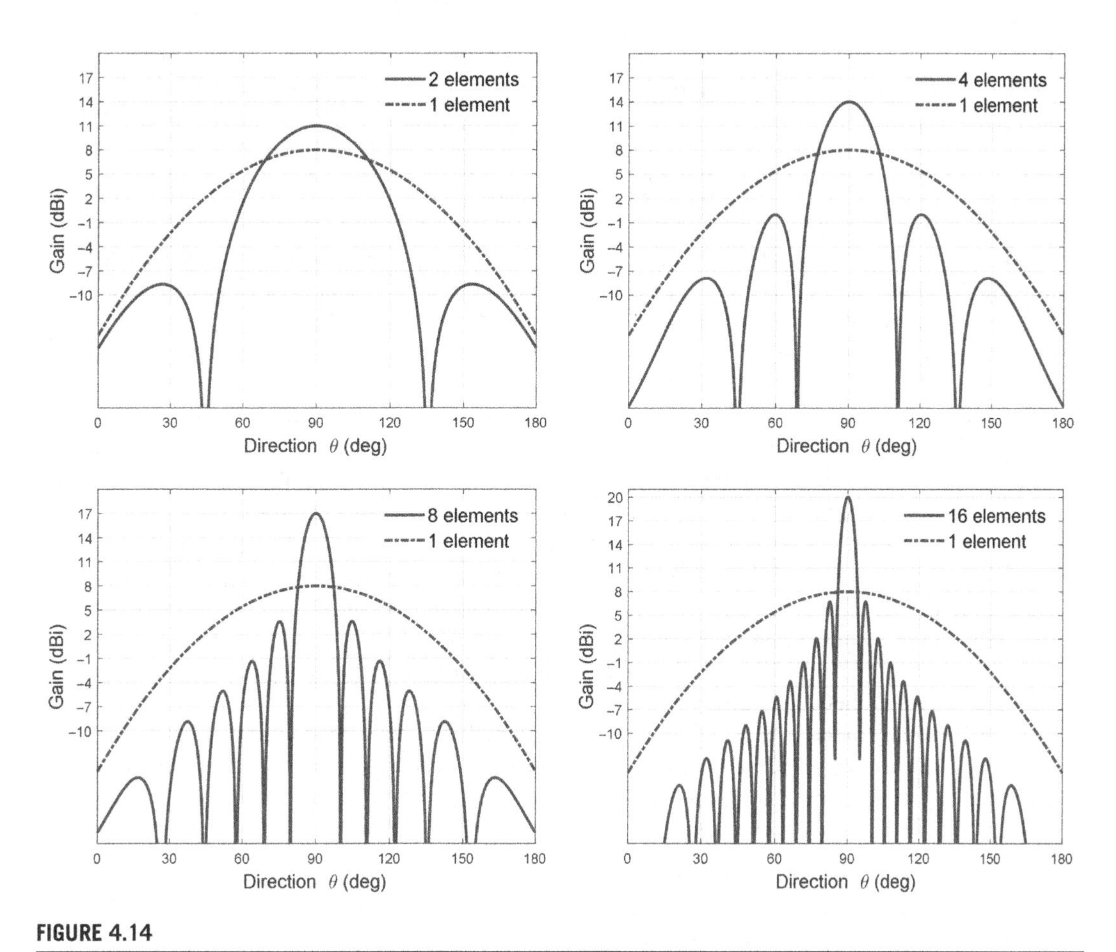

FIGURE 4.14

Gain pattern as a function of θ for $\varphi = 0^\circ$ for vertical arrays with different number of elements.

where $k = 2\pi/\lambda$. The last equality follows from the fact that the sum is a geometric series (see also [1] for a detailed derivation). The observations done for the case with two elements in Section 4.2.4 are now generalized to the case with more elements:

- The maximum gain for an array with N elements is N times higher compared to a single element. Maximum gain occurs where all the elements' contributions are in phase, and this together with the element orientation explains the peak at broadside, $\theta = 90^\circ$. At broadside, coherent addition of N terms, each with an amplitude $1/\sqrt{N}$, results in a power gain of N in addition to the element gain. As can be seen in Fig. 4.14, for two elements, the gain as compared to a single element is 3 dB, for four elements it is 6 dB and in the general case, it is $10\log_{10} N$ dB.

- There are nulls in the gain pattern. A null occurs when all the contributions sum destructively so that the array factor becomes zero. From the expression for the array factor in (4.48), the first nulls occur when

$$N\frac{kd\cos\theta}{2} = \pm\pi \rightarrow d\cos\theta = \pm\frac{\lambda}{N}. \tag{4.49}$$

  Here, $d\cos\theta$ can be recognized as the path length difference between two adjacent elements (see also Section 4.2.3), and it can be shown that there are nulls whenever this path length difference is a multiple of λ/N different from a multiple of the wavelength (see also [1]). This explains why the number of nulls increases with increasing number of elements and furthermore shows that it depends on the element separation d.
- The width of the main lobe decreases with increasing number of elements. More specifically, the width scales approximately with $\lambda/(dN)$ since the angular separation between the first nulls scales in this way for small $\lambda/(dN)$. This can be seen by first solving (4.49) to establish that the first nulls appear at angles $\pm\cos^{-1}(\lambda/(dN))$ and then use a Taylor expansion, $\cos^{-1}(\lambda/(dN)) \approx \pi/2 - \lambda/(dN)$. The reader is referred to [1] for more details as well as more exact expressions. It should though be remembered that the exact half-power main lobe width depends not only on the array factor but also on the element pattern, whereas the locations of the nulls do not. Although the nulls do not coincide with the directions corresponding to the half-power main lobe width, the scaling for the latter is nevertheless similar to the scaling of the width as measured between nulls.
- Thus, for fixed element separation, d, and wavelength, λ, the main lobe width scales with $1/N$. With reference to Figs. 4.12 and 4.13, it can be seen that vertical stacking of elements into a vertical array reduces the main lobe width in the vertical plane, whereas horizontal stacking of elements into a horizontal array reduces the main lobe width in the horizontal plane.

In summary, an array with N elements can increase the gain relative to a single element with N, or $10\log_{10}N$ dB on logarithmic scale. At the same time, the half-power beamwidth of the main lobe at broadside, which represents the range of angles where the gain is high, becomes narrower for a fixed element separation and scales approximately as $\lambda/(dN)$. Finally, it was also observed that there are nulls in the radiation pattern, which indicates that it is possible to not only improve the received signal power in a direction, it is also possible to reduce the signal power in other directions.

4.4 BEAMFORMING

So far, the same signal has been transmitted from all the elements and it has been demonstrated that by increasing the number of elements in the array the gain increases at broadside, at zenith angle $\theta = 90^\circ$ and azimuth angle $\varphi = 0^\circ$. At the same time, it was also seen that the width of the main lobe decreases when increasing the number of antennas in the array. In a cellular deployment, it is of interest to increase the signal strength not only at broadside but to terminals that may be at an arbitrary angle within a certain range. Beamforming allows adjusting the direction with maximum gain, that is, the direction of the main lobe, and this adjustment can be done dynamically without changing the mechanical orientation of the array.

More specifically the direction is adjusted by changing the transmission delay of the signal copies transmitted from different elements so the contributions from the different elements arrive at the same time at the receiver and therefore combine constructively. Equivalently, for a narrowband signal this can be accomplished by applying a corresponding element–specific phase shift, and this is what the beamforming weight vector represents. Thus, the beamformer adjusts the phase of the transmitted signals so that they add in phase at a receiver in a certain direction. This means that the gain as compared to a single element, the free-space array gain, will be equal to N, the number of elements, not only at broadside but in any desired direction. Some illustrations of this will be given for a ULA, followed by a short discussion on element separation and so-called grating lobes.

The technique, to steer a beam in a certain direction by adjusting the phase of the transmitted signals to maximize the gain to a receiver in a certain direction, leads to what in the present book is referred to as classical beamforming. In Chapter 6, multiple antenna techniques including more advanced beamforming techniques are considered in the context of 5G radio network deployments. However, already here, the opportunities beyond classical beamforming are illustrated. By choosing the beamforming weights more flexibly, including also adjusting the amplitude, it is possible to suppress sidelobes, place nulls in certain directions and increase the width of the main lobe.

4.4.1 CLASSICAL BEAMFORMING

So far in the present chapter, the focus has been on the case that the same signal is transmitted from all the elements of the array. This is now revisited so that copies of the same signal but with different time delays are transmitted from the elements. For this purpose, the transmission time delay of the signal from antenna n is denoted $\tau_{\mathrm{t},n}$ where the subscript t refers to transmission. The intention is to choose the delays so that the signals add constructively at a receiver in direction θ_0 and φ_0.

Furthermore, in the definition of the array response vector $\boldsymbol{a}(\theta, \varphi)$, which includes the impact of the propagation delay differences as function of direction, it was found (see (4.40)), that the relative propagation delay $\Delta\tau_n(\theta_0, \varphi_0)$ for element n satisfies

$$2\pi f_c \Delta\tau_n(\theta_0, \varphi_0) = -\boldsymbol{d}_n^{\mathrm{T}}\boldsymbol{k}(\theta_0, \varphi_0). \tag{4.50}$$

Thus, the propagation delay difference is related to path length difference which in turn is related to the position $\boldsymbol{d}_n$ [see (4.27)], relative to a reference point as well as the direction of the receiver in terms of the wave vector $\boldsymbol{k}(\theta_0, \varphi_0)$ defined in (4.39).

In Fig. 4.15, the signal paths are illustrated. In the figure, also the delay between the array's reference point used to define the element positions, and the receiving antenna, τ, is included. The total delay for the contribution from element n is

$$\tau_{\mathrm{tot},n} = \tau_{\mathrm{t},n} + \Delta\tau_n(\theta_0, \varphi_0) + \tau. \tag{4.51}$$

To make the contributions from the individual elements add constructively, the delays are chosen so that all the contributions have the same total delay. For this reason, they are chosen as

$$\tau_{\mathrm{t},n} = -\Delta\tau_n(\theta_0, \varphi_0) \rightarrow \tau_{\mathrm{tot},n} = \tau \text{ for all } n = 1, \cdots, N. \tag{4.52}$$

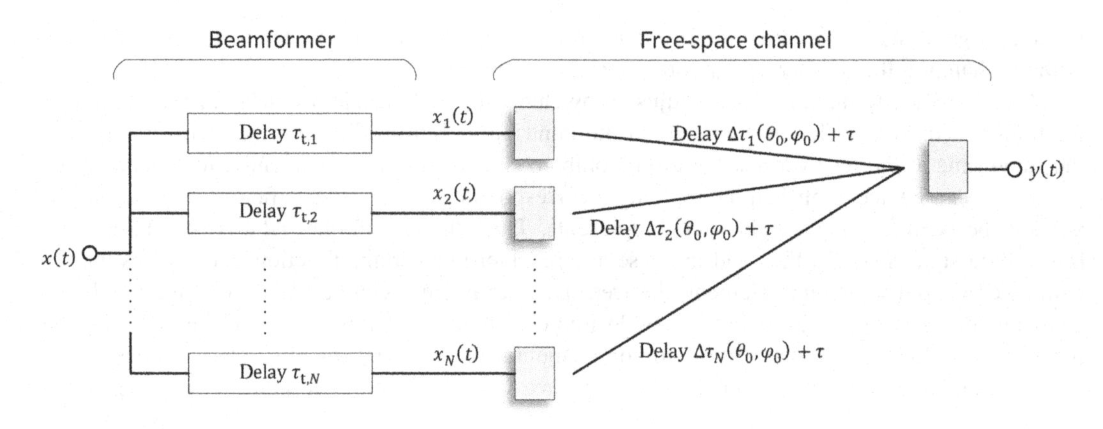

FIGURE 4.15

Copies of the same signal with different total delays reach the receiving antenna in the direction θ_0 and φ_0.

In the example of Fig. 4.15, where an element further down the array with increasing index n has a longer propagation delay, the transmission delays should be chosen as $\tau_{t,1} > \tau_{t,2} > \cdots > \tau_{t,N}$. Since the signal, $x(t)$ according to (4.8), is a complex baseband equivalent to a narrowband radio frequency signal $x_{RF}(t)$, a delay of the radio frequency signal corresponds to a phase shift of the baseband signal (see also Section 4.2.5),

$$x_{RF}\left(t-\tau_{t,n}\right) \leftrightarrow e^{-j2\pi f_c \tau_{t,n}} x(t). \tag{4.53}$$

This means that the baseband signal transmitted from element n is to be taken as

$$x_n(t) = w_n x(t),\ n = 1,\ldots,N, \tag{4.54}$$

where w_n is the complex weight associated with element n. To implement the time delay, the weight w_n is chosen as

$$w_n = \frac{1}{\sqrt{N}} e^{-j2\pi f_c \tau_{t,n}} = \frac{1}{\sqrt{N}} e^{j2\pi f_c \Delta\tau_n(\theta_0,\varphi_0)} = \frac{1}{\sqrt{N}} e^{-j\boldsymbol{d}_n^{\mathrm{T}}\boldsymbol{k}(\theta_0,\varphi_0)}, \tag{4.55}$$

where (4.52) and (4.50) have been used and the factor $1/\sqrt{N}$ represents the fact that the total power is the same independent on how many elements are used. The weights of all the N elements are then collected into a beamforming weight vector, denoted $\boldsymbol{w}(\theta_0, \varphi_0)$ to stress its dependency on the steered direction (θ_0, φ_0),

$$\boldsymbol{w}(\theta_0, \varphi_0) = \frac{1}{\sqrt{N}} \left[e^{-j\boldsymbol{d}_1^{\mathrm{T}}\boldsymbol{k}(\theta_0,\varphi_0)} \quad e^{-j\boldsymbol{d}_2^{\mathrm{T}}\boldsymbol{k}(\theta_0,\varphi_0)} \quad \ldots \quad e^{-j\boldsymbol{d}_N^{\mathrm{T}}\boldsymbol{k}(\theta_0,\varphi_0)} \right]^{\mathrm{T}}. \tag{4.56}$$

The derivations of the gain done in Section 4.3.2 did not make any specific assumptions on the beamforming weight vector and therefore (4.43) still applies with the array response vector as defined in (4.38). Using the expressions for the array response vector and the beamforming weights, the gain in the steered direction (θ_0, φ_0) becomes

$$G_{\mathrm{AA}}(\theta_0, \varphi_0) = \left|\boldsymbol{a}^{\mathrm{T}}(\theta_0, \varphi_0)\boldsymbol{w}(\theta_0, \varphi_0)\right|^2 = \left|\sum_{n=1}^{N} \frac{1}{\sqrt{N}}\right|^2 G(\theta_0, \varphi_0) = NG(\theta_0, \varphi_0), \tag{4.57}$$

where $G(\theta, \varphi) = |g(\theta, \varphi)|^2$ is the elements' gain pattern. Thus, the delays, or equivalently the phase shifts, make all the contributions sum up constructively. Considering (4.45), it furthermore follows that the (free-space) array gain, as given by the array factor indeed is as expected,

$$\left|\mathrm{AF}(\theta_0, \varphi_0)\right|^2 = N. \tag{4.58}$$

The Cauchy–Schwartz inequality can be used to establish an upper bound on the gain for any choice of weights $\boldsymbol{w}$ subject to the total power constraint $\|\boldsymbol{w}\|^2 = 1$

$$\|\boldsymbol{w}\|^2 = 1, \|\boldsymbol{a}(\theta_0, \varphi_0)\|^2 = NG(\theta_0, \varphi_0)$$
$$\rightarrow G_{\mathrm{AA}}(\theta_0, \varphi_0) = \left|\boldsymbol{a}^{\mathrm{T}}(\theta_0, \varphi_0)\boldsymbol{w}\right|^2 \le NG(\theta_0, \varphi_0)$$

Since the upper bound is achieved, it can be concluded that the choice of beamforming weights (4.56) does indeed maximize the gain in the direction θ_0 and φ_0. This comes from the fact that the conjugates of the beamforming weights, $\boldsymbol{w}^{\mathrm{c}}(\theta_0, \varphi_0)$, is a scaled version of the array response vector $\boldsymbol{a}(\theta_0, \varphi_0)$.

To sum up, the direction where the array gain pattern is maximized can be changed by adjusting the delays, or equivalently the phase shifts represented by the elements in the beamforming weight vector.

4.4.2 UNIFORM LINEAR ARRAYS

For the case with a ULA and the classical beamforming weights given by (4.56), it is possible to formulate an expression for the absolute value squared of the array factor similar to (4.48). For the case with a vertical array, the difference is that $\cos\theta$ is to be replaced by $\cos\theta - \cos\theta_0$, and the properties of the gain patterns discussed in Section 4.2.4 do therefore essentially carry over (see further [1–3]). One difference is though that the main lobe width also increases as the steering direction changes from broadside, $\theta_0 = 90^\circ$ and this will be seen in the examples below. Another difference is that the maximum gain, which includes both the array gain part and the element gain part, will not necessarily occur at the steered direction θ_0. The reason for this is that the gain in a certain direction depends not only on the array factor but also on the element gain in that direction.

A specific example, with a vertical ULA as defined in Section 4.3.3, is considered. The array consists of $N = 8$ elements separated $d = 0.5\lambda$ with the element pattern defined in [6]. Four different steering angles θ_0 are considered and the gain pattern as a function of zenith angle θ for $\varphi = 0^\circ$ is shown in Fig. 4.16.

As can be seen, it is indeed possible to change the direction of the main lobe by adjusting the transmission timing of signal copies transmitted from different elements. It may also be observed that due to the element gain pattern, the direction in which the antenna gain of the array has its maximum does not fully coincide with the direction for which the array gain is maximized. Another impact of the element gain is that the maximum gain decreases with increasing steering

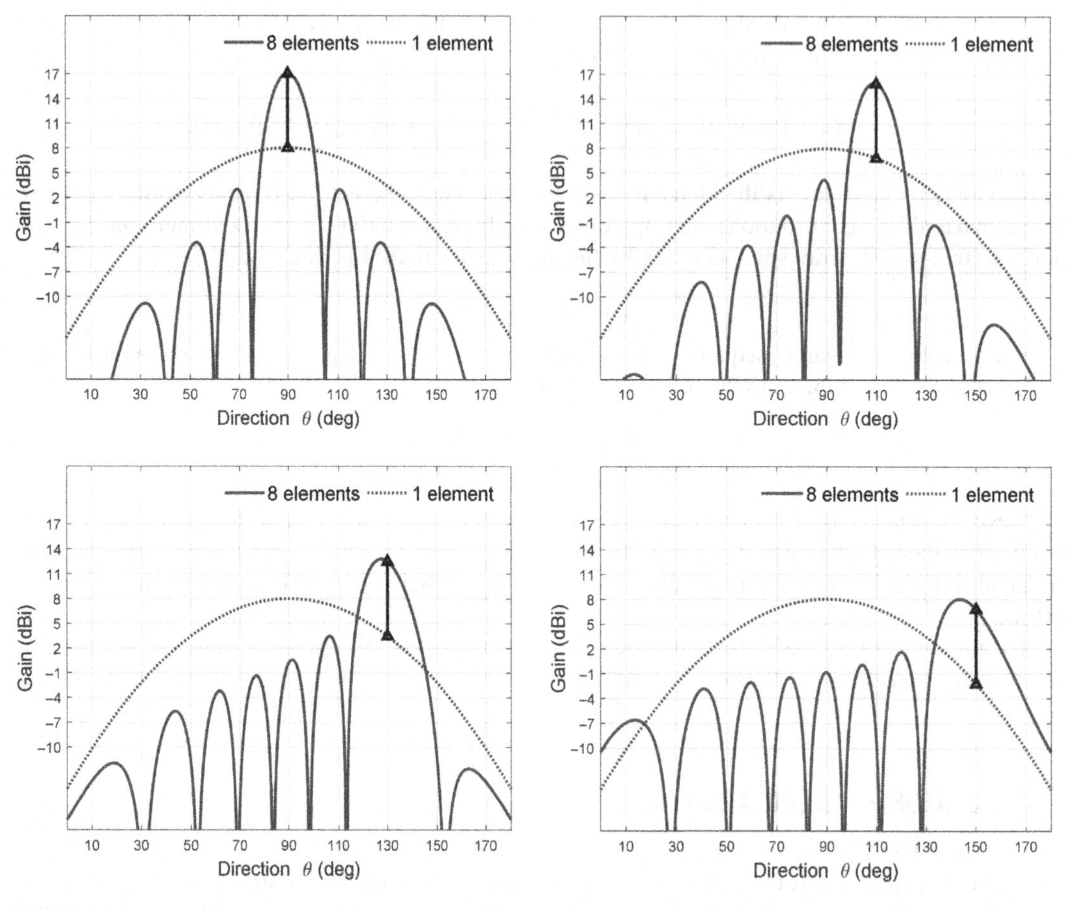

FIGURE 4.16

Gain pattern as a function of θ for $\varphi_0 = 0^{\circ}$ and four different steering directions $\theta_0 = 90^{\circ}$ (top left), $\theta_0 = 110^{\circ}$ (top right), $\theta_0 = 130^{\circ}$ (bottom left) and $\theta_0 = 150^{\circ}$ (bottom right) using an vertical array of eight elements with gain model according to [6] separated by $d = 0.5\lambda$. The black arrows represent the $10\log_{10}8 \approx 9$ dB where the array gain is maximized.

direction. However, the array gain is $10\log_{10}8 \approx 9$ dB for all the steered directions. Finally, the width of the main lobe increases as the angle between broadside and the steering direction increases.

In fact, since the gain in a given direction is the product of the corresponding element gain and the array gain, the envelope over all steered beams will have the same shape as the element pattern, and in the general case with N elements be $10\log_{10}N$ dB higher. This is illustrated in Fig. 4.17 where gain patterns for 20 different steering directions are shown.

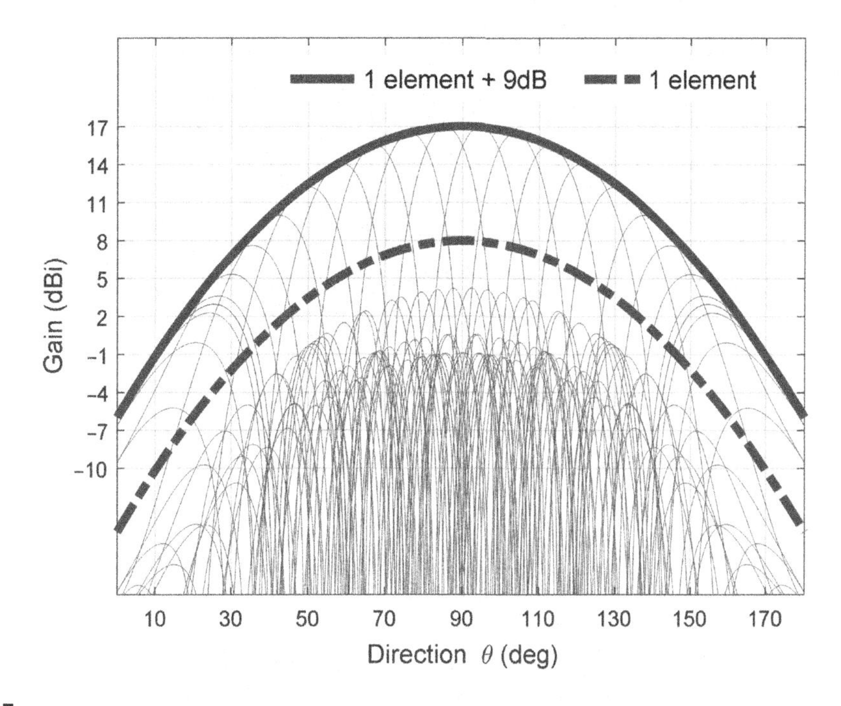

FIGURE 4.17

Gain pattern as a function of θ for $\varphi = 0^{\circ}$ and different steering directions θ_0 between 0° and 180° (thin lines) with a vertical array of $N = 8$ elements with gain model according to [6] and element separation $d = 0.5\lambda$. For comparison, the gain for a single element and the single element plus $10\log_{10}8 \approx 9$ dB are also shown.

4.4.2.1 On element separation

In the examples so far in the chapter, rather small element spacing has been considered, in the order of $0.5\lambda - 1.0\lambda$, since such are commonly used in practice. A detailed treatment of the impact of antenna separation would include accounting for the mutual coupling between the elements in the array and is beyond the scope of this book, even though some aspects are discussed in Section 12.3.7. However, common to many practical designs is that the maximum gain of the embedded pattern, which is different compared to the pattern of the element in isolation, can be increased when the spacing is increased. This appeals to intuition since a larger spacing would lead to a larger area per elements, which in turn allows collecting more incident power. Given a certain physical area, there is thus a choice between different partitioning in terms of number of elements and the area per element which corresponds to different array gains and element gains.

When designing an array, there are not only requirements on gain for signals transmitted to a single-intended user within a cell (for which it will become clear in Section 6.1.1.2 that the gain patterns may look substantially different as compared to the ones presented so far). There are also requirements to transmit signals with wider main lobes, for example intended to multiple users in a cell. Requirements here are not only related to gain and width of the main lobe but also related to

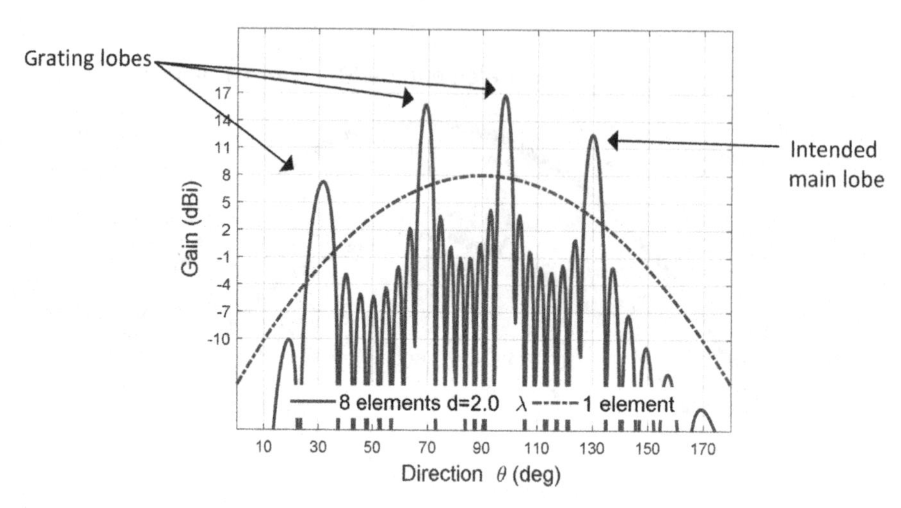

FIGURE 4.18

Gain pattern as a function of θ for vertical ULA with $N = 8$ elements separated by $d = 2\lambda$. The direction of the main lobe is steered to $\theta_0 = 130^\circ$.

sidelobes. Balancing different requirements with cost, for example, in terms of antenna size and complexity, typically leads to antenna separations in the above-mentioned range with a bias toward $0.5\lambda - 0.7\lambda$.

As a side note, it is well known that increasing the element separation above 0.5λ leads to aliasing in the sense that so-called grating lobes as defined in [8] can occur (see also [3]). A rather extreme example of this is given in Fig. 4.18 for an element separation $d = 2\lambda$ and the main lobe steered to $\theta_0 = 130^\circ$. The corresponding case for 0.5λ is given in Fig. 4.16, where only the intended main lobe is present.

A grating lobe occurs in a direction other than intended direction in which the signals transmitted from all the elements combine constructively. This means that there is an ambiguity in the sense that the maximum array gain can be obtained in more than one direction. Whether ambiguity is a problem or not depends on the application, and if the application is to position a terminal by receiving signals in the uplink, it is indeed a problem since the direction cannot be uniquely determined. However, it is still possible to demodulate transmitted data.

4.4.3 BEAMFORMING BEYOND BASIC BEAM STEERING

In Section 4.4.1, basic beam steering was introduced, and it is illustrated in Section 4.4.2 that the direction of the main lobe could be steered to maximize the (free-space) array gain in a particular direction. More specifically, the beamforming weights are a function of the intended pointing direction which can be specified in terms of zenith and azimuth angles θ_0 and φ_0. In the general case, the beamforming weights may be chosen more freely to generate quite arbitrary gain patterns. This is the topic of Chapter 6, in which multiple antenna techniques including beamforming techniques suitable for the multipath radio channels (cf. Section 6.1.1.2) are outlined. As will be seen there,

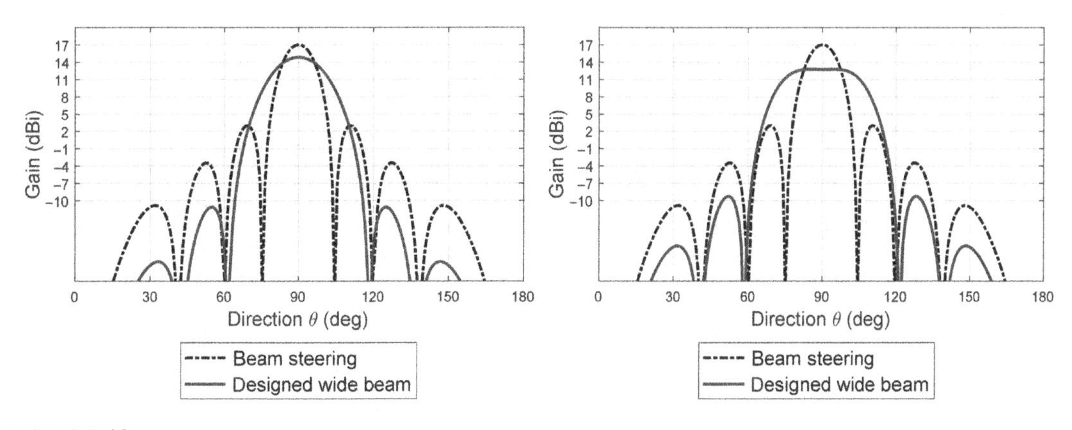

FIGURE 4.19

Examples gain patterns for two designed wide beams compared to the gain pattern with beam steering to $\theta_0 = 90^\circ$ for the case with vertical ULA with $N = 8$ elements.

the gain patterns can have quite arbitrary shapes and different from the ones illustrated so far. Thus, in general, it is possible to select beamforming weights to obtain gain patterns with certain desired properties, for example, to suppress sidelobe levels, widening the main lobe or have nulls in specific directions. There is a vast literature available on the topic, and [1,3,9,10] can serve as starting points for further exploration. Here, the ambition is not to go through in detail how to generate beamforming weights to meet arbitrary gain patterns, but merely to make the reader aware of the fact that the properties of the gain pattern for basic beam steering as illustrated in Sections 4.2.4, 4.3.3.1, and 4.4.2 depend on how the beamforming weights are determined.

In Section 4.3.3.1, it was seen that the main lobe width for basic beam steering with a ULA decreases with increasing length of the array. However, if the beamforming weights are chosen differently, this needs not to be the case. For example, by choosing weights such that only one of the elements is used for transmission, then the main lobe width will not change. At the same time, there will not be any gain from using the array. In Fig. 4.19, gain patterns for two different beams designed with the Fourier transform method in [3] are shown for the same case as considered in Fig. 4.16, which is a vertical ULA with $N = 8$ elements separated by $d = 0.5\lambda$. The width of the main lobe is wider, but at the same time, the maximum gain is lower as compared to the beam steered to broadside $\theta_0 = 90^\circ$. In practice, there is often a constraint on maximum power per element, and for this reason other methods are used.

Another example includes creating nulls in the gain pattern in specific directions. For the case that the main lobe is steered to $\theta_0 = 90^\circ$, two directions with nulls are given by (4.49). To create nulls in L specific directions $\{(\theta_l, \varphi_l), l = 1, \ldots, L\}$, the beamforming weights need to satisfy

$$a^{\mathrm{T}}(\theta_l, \varphi_l)w = 0,\ l = 1, \ldots, L. \tag{4.59}$$

One interpretation of this is that weight vector, w, is constrained to lie in the subspace defined by the constraints above, and one solution is to project the beamforming weight vector $w(\theta_0, \varphi_0)$ defined in (4.56) onto this subspace. In practice, more sophisticated approaches are used to strike a balance between high gain in a desired direction and low but not necessarily zero gain in directions

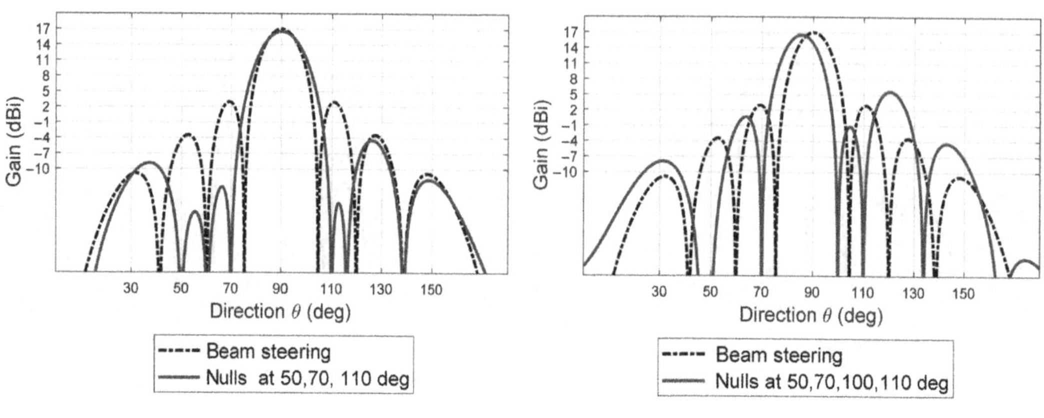

FIGURE 4.20

Example of gain patterns with null steering compared to the gain pattern with beam steering to $\theta_0 = 90^\circ$ for the case with a vertical ULA with $N = 8$ elements.

to be nulled. For this reason, the details on the solution to (4.59) are not further elaborated here. In Fig. 4.20 gain patterns illustrating nulls being positioned in different directions are shown for the same setup as in Fig. 4.19.

To sum up, it is possible to create rather arbitrary gain patterns, and properties such as gain, and main lobe widths depend on the chosen beam weights. Typically imposing additional constraints in terms of nulls and beamwidth will lead to lower gain in the steered direction as compared to classical beamforming.

4.5 DUAL-POLARIZED UNIFORM PLANAR ARRAYS

The dual-polarized UPA is probably the most common array configuration being considered for advanced 4G and 5G systems. A UPA, also referred to as uniform rectangular array, offers the beamforming capabilities of both a vertical and a horizontal ULA simultaneously.

The UPA is a special case of the general array introduced in Section 4.2.5 with element positions on a uniform grid. For the basic beam steering technique introduced in Section 4.4.1 which aims at maximizing the gain in a certain direction, the array factor can be written as a product of an array factor for a vertical array and array factor for a horizontal array. The gain as compared to using a single element is therefore given by the total number of elements and at broadside, the vertical and horizontal width of the main lobe scales with the inverse of the size of the antenna in the corresponding dimension.

A dual-polarized UPA is a UPA of dual-polarized element pairs, more specifically two elements with different, preferably orthogonal, polarizations. The difference as compared to the case with co-polarized elements is that the beamforming vector generates weighted version of the signal to be transmitted to elements with different polarization. This allows not only changing the gain pattern but also controlling the polarization of the transmitted signals, for example to mitigate polarization mismatch (see also Section 3.4.2).

4.5.1 UNIFORM PLANAR ARRAY

A UPA consists of a number of elements arranged on a uniform grid with N_v rows and N_h columns. Such an array will be referred to as an $N_v \times N_h$ UPA and it contains in total $N = N_v N_h$ elements. The spacing between adjacent elements is d_v in the vertical domain and d_h in the horizontal domain and the spacing in the two domains need not be the same for the array to be referred to as a UPA. In Fig. 4.21 an example with an 8×4 UPA is depicted.

It may be noted that a ULA, as defined in Section 4.3.3, is a special case of a UPA. More specifically, a $1 \times N$ UPA corresponds to a horizontal ULA and an $N \times 1$ UPA is a vertical ULA.

The following assumptions are done:

- The $N = N_v N_h$ elements are ordered (see also Fig. 4.21) such that their locations in Cartesian coordinates are given by the vectors

$$
\begin{gathered}
\boldsymbol{d}_{1+m+nN_v} = \begin{bmatrix} 0 & nd_h & md_v \end{bmatrix}, \\
m = 0, \cdots, N_v\text{-}1, \\
n = 0, \cdots, N_h\text{-}1.
\end{gathered}
$$

- All elements have the same complex amplitude pattern $g(\theta, \varphi)$ oriented so that the maximum (magnitude) occurs in the direction of the x-axis, $\theta = 90^\circ, \varphi = 0^\circ$.

The assumption on the element pattern is further discussed in Section 4.3.3 and with reference to Fig. 4.21, the ordering of elements is when looking toward the array from the intended coverage area.

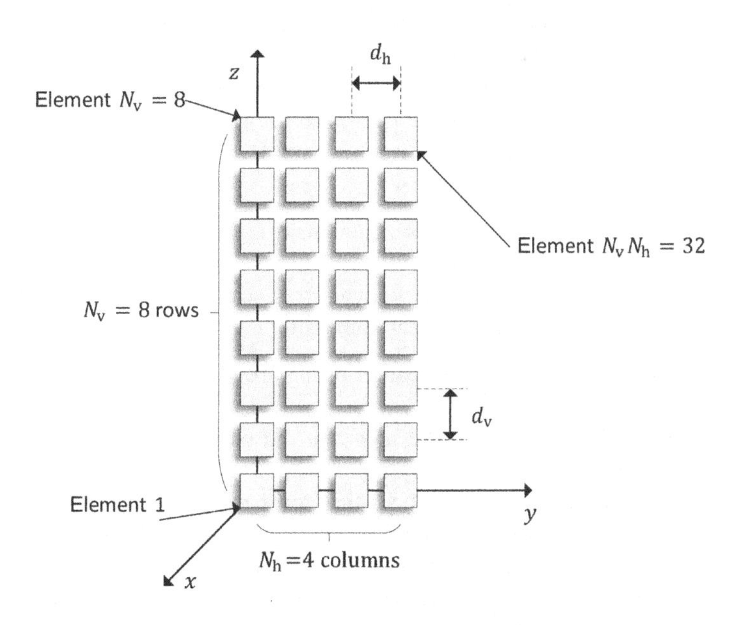

FIGURE 4.21

An 8×4 UPA with $N_v = 8$ rows and $N_h = 4$ columns.

4.5.1.1 Basic gain pattern

The UPA is a special case of an array with $N = N_v N_h$ elements, and under the assumptions listed in Section 4.3.1, the array response vector can be determined from (4.38) based on the element positions in (4.60) as

$$[a(\theta, \varphi)]_{1+m+nN_v} = g(\theta, \varphi) e^{jkmd_v \cos\theta} e^{jknd_h \sin\theta \sin\varphi}, \tag{4.61}$$

for $m = 0, \cdots, N_v - 1, n = 0, \cdots, N_h - 1$. From Section 4.4.1, it then follows that to maximize the gain in a direction θ_0 and φ_0, the weights should according to (4.56) be selected as

$$\left[w(\theta_0, \varphi_0)\right]_{1+m+nN_v} = \frac{1}{\sqrt{N_v}} e^{-jkmd_v \cos\theta_0} \frac{1}{\sqrt{N_h}} e^{-jknd_h \sin\theta_0 \sin\varphi_0}, \tag{4.62}$$

for $m = 0, \cdots, N_v - 1, n = 0, \cdots N_h - 1$. With the expression of the array response vector in (4.61) and the chosen beamforming weight vector (4.62), the gain pattern can be written as

$$G_{AA}(\theta, \varphi) = \left|a^T(\theta, \varphi) w(\theta_0, \varphi_0)\right|^2 = \left|\mathrm{AF}_v(\theta, \varphi)\right|^2 \left|\mathrm{AF}_h(\theta, \varphi)\right|^2 G(\theta, \varphi), \tag{4.63}$$

with

$$\mathrm{AF}_v(\theta, \varphi) = \frac{1}{\sqrt{N_v}} \sum_{m=0}^{N_v - 1} e^{jkd_v m(\cos\theta - \cos\theta_0)}, \tag{4.64}$$

$$\mathrm{AF}_h(\theta, \varphi) = \frac{1}{\sqrt{N_h}} \sum_{n=0}^{N_h - 1} e^{jkd_h n\left(\sin\theta \sin\varphi - \sin\theta_0 \sin\varphi_0\right)}. \tag{4.65}$$

Here, $\mathrm{AF}_v(\theta, \varphi)$ is the array factor for a vertical ULA with N_v elements separated a distance d_v and $\mathrm{AF}_h(\theta, \varphi)$ is the array factor for a horizontal ULA with N_h elements separated a distance d_h. The array factor can be factored into two array factors as in (4.63) as long as the beamforming weights are separable so that the weight for the element on row m of column n, $w_{m,n}$, can be written as $w_m w_n$. The beamforming weights in (4.62) are separable, and in that case, the UPA can be viewed in two ways:

- A vertical array where each element is a horizontal array so that $\left|\mathrm{AF}_v(\theta, \varphi)\right|^2$ is the array gain pattern and $\left|\mathrm{AF}_h(\theta, \varphi)\right|^2 G(\theta, \varphi)$ is the element gain pattern.
- A horizontal array where each element is a vertical array so that $\left|\mathrm{AF}_h(\theta, \varphi)\right|^2$ is the array gain pattern and $\left|\mathrm{AF}_v(\theta, \varphi)\right|^2 G(\theta, \varphi)$ is the element gain pattern.

For more details as well as further simplification of the array factors the reader is referred to for example [1].

From inspection of (4.63)–(4.65) it can be seen that the gain as compared to a single element obtained in the steered direction θ_0 and φ_0 is indeed equal to the total number of elements N. Furthermore, for ULAs it is seen in Section 4.3.3 that for a main lobe at broadside, $\theta_0 = 90^\circ$ and $\varphi_0 = 0^\circ$, the width of the main lobe scales approximately as $\lambda/(Nd)$. As shown to the left in the example in Fig. 4.22, the width of the main lobe is small in the vertical domain for an $N = 8$ vertical array whereas the width of the main lobe is small in the horizontal domain for a horizontal array. The right part of Fig. 4.22 shows the gain pattern for an 8×8 UPA and the width of the main lobe in this case is reduced in both the horizontal and vertical domains.

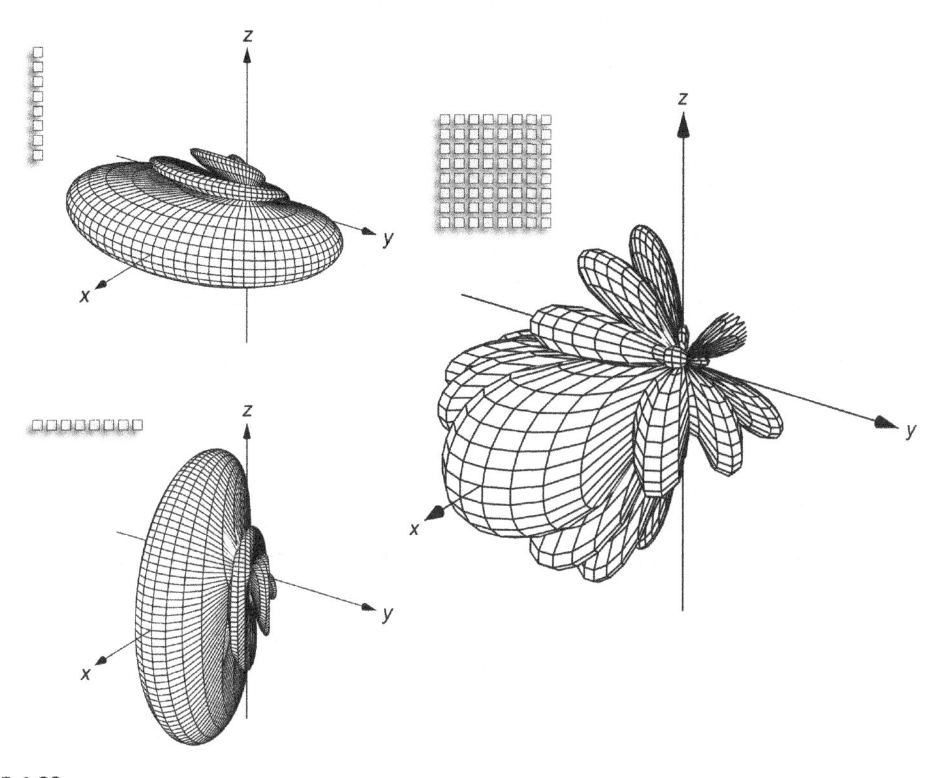

FIGURE 4.22

Illustration of gain pattern for an 8 × 8 UPA as well as vertical and horizontal ULAs with eight elements. The element is modeled according to [6], an element separation of $d = 0.5\lambda$ is assumed and the main lobe steered to toward broadside $\theta_0 = 90^\circ$ and $\varphi = 0^\circ$.

With a similar reasoning to that of Section 4.3.3, it can be argued that the width of the main lobe decreases with the array size as $\lambda/(d_h N_h)$ in the horizontal domain and $\lambda/(d_v N_v)$ in the vertical domain for steering toward broadside. To illustrate this in more detail, the gain patterns, both as a function of the azimuth angle φ in the horizontal plane $\theta = 90^\circ$ and as a function of the zenith angle θ in the vertical plane $\varphi = 0^\circ$, are shown in Fig. 4.23, for three different antennas arrays in addition to a single element. It may be noted that at broadside:

- Both a vertical array (8 × 1) and a horizontal array (1 × 8) offer $10\log_{10}8 \approx 9$ dB gain as compared to the single element.
- For the horizontal array (1 × 8) the main lobe is narrow in the horizontal plane, but as wide as the single element in the vertical plane.
- For the vertical array (8 × 1), the opposite is true in the sense that the main lobe is narrow in the vertical plane and wide in the horizontal plane.
- The 8 × 8 UPA offers another 9 dB over any of the two linear arrays, and the main lobe widths are narrow in both the vertical and horizontal planes.

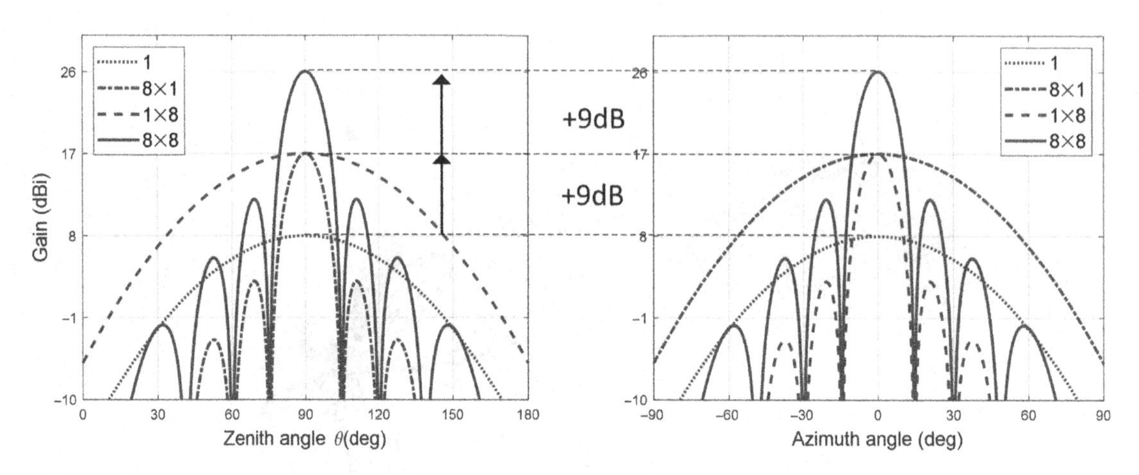

FIGURE 4.23

Gain pattern as a function of zenith angle θ for $\varphi = 0°$ (left) and as function of azimuth angle φ for $\theta = 90^{\circ}$ (right) for a single element (1), a vertical eight-element array (8×1), a horizontal eight-element array (1×8) and an 8×8 UPA (8×8). In all cases the element gain model in [6] is used, the element separation is $d = 0.5\lambda$ and the main lobe is steered to broadside.

To sum up, for basic beam steering with a UPA, the array factor can be factored into an array factor for a horizontal array and an array factor for a vertical array. The gain as compared to a single element is still given by the total number of elements. For a main lobe steered to broadside, the main lobe widths in the horizontal and vertical planes will however decrease inversely with increasing antenna size in the corresponding dimension.

4.5.2 DUAL-POLARIZED ARRAYS

Dual-polarized antennas have been used for a long time in radio access networks, both at the terminal side as well as on the base station side. On the base station side, multiple antennas were of interest initially to mitigate fading dips by diversity as well as losses due to polarization mismatch between transmitter and receiver antennas. More recently a key driver is to support spatial multiplexing as described in Section 6.3. The benefits of dual-polarized antennas can be leveraged also with antenna arrays, by using arrays of dual-polarized element pairs.

So far in the chapter, all the elements of the array have been assumed to have the same polarization, and to create a dual-polarized array, each element, such as a dipole, patch or Vivaldi element, is paired with an identical element with an as orthogonal polarization as possible (see also [3]). Often such dual-polarized element pairs are implemented in an area-efficient way. From the discussion in Section 4.4.2.1, with an assumed element spacing of around $0.5 - 1\lambda$, an element pair can conceptually be envisioned to be implemented within a square with such a side length. Most often an element pair with $\pm 45^{\circ}$ linear polarization is used, for example with a crossed dipole pair or a patch with two feed ports.

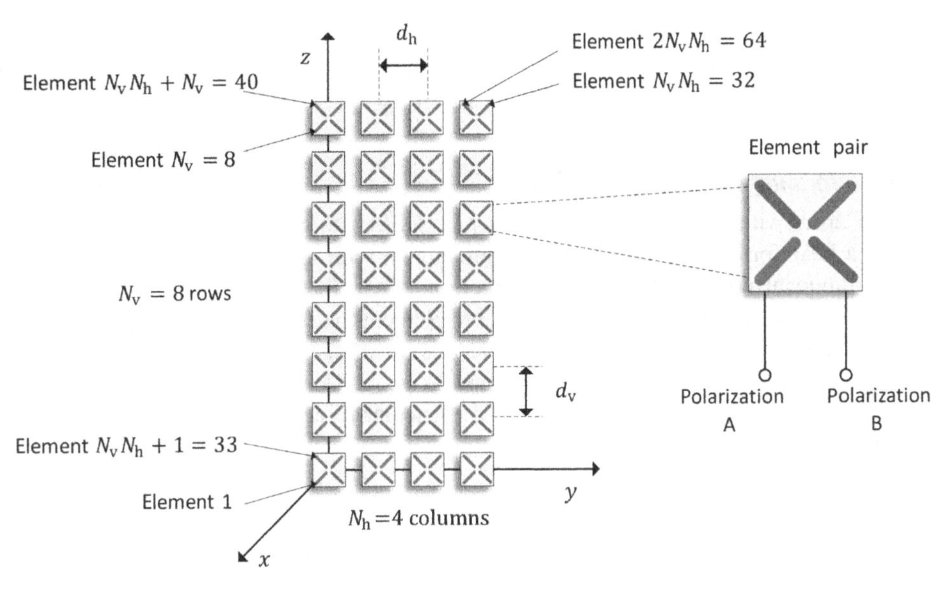

FIGURE 4.24

Example with uniform planar array of dual-polarized element pairs (left) and a single dual-polarized element pair (right).

In Fig. 4.24, a dual-polarized element pair is illustrated in addition to a UPA of dual-polarized element pairs. Each pair consists of two elements and has two ports, or feeds, one for polarization A and one for polarization B, that can be used for transmission and reception of signals implying that an $N_v \times N_h$ dual-polarized UPA has in total $N = 2N_vN_h$ elements.

Unless otherwise mentioned, the reader can assume that the N elements are ordered such that the first $N/2$ elements have a first polarization A and the last $N/2$ elements have a second polarization B. Moreover, the following can also be assumed:

- The $N = 2N_vN_h$ elements in the array are ordered (see Fig. 4.24) such that the locations of them in Cartesian coordinates are given by the vectors

$$\begin{aligned} &\boldsymbol{d}_{1+m+nN_v+pN_vN_h} = \begin{bmatrix} 0 & nd_h & md_v \end{bmatrix}, \\ &m = 0, \ldots, N_v - 1, n = 0, \ldots, N_h - 1, p = 0, 1 \end{aligned} \tag{4.66}$$

  where p enumerates the two polarizations.
- All elements have the same complex amplitude pattern $g(\theta, \varphi)$ oriented so that the maximum (magnitude) occurs in the direction of the x-axis, $\theta = 90^\circ, \varphi = 0^\circ$.
- The elements' polarizations are defined as

$$\hat{\psi}_p(\theta, \varphi) = \begin{cases} \hat{\psi}_A(\theta, \varphi) & p = 1, \ldots, N_vN_h \\ \hat{\psi}_B(\theta, \varphi) & p = N_vN_h + 1, \ldots, 2N_vN_h \end{cases}, \tag{4.67}$$

  where $\hat{\psi}_A(\theta, \varphi)$ and $\hat{\psi}_B(\theta, \varphi)$ represent the two polarizations (see also Section 3.3.3) which in the general case depend on direction, and not necessarily are orthogonal for all directions.

Assumptions on the element pattern are further discussed in Section 4.3.3. Furthermore, recall from Sections 3.3.3 and 3.3.4 that the product $g(\theta, \varphi)\hat{\psi}_p(\varphi, \varphi)$ characterizes the transmitted wave in the far-field in terms of the two orthogonal $\hat{\theta}$ and $\hat{\varphi}$ components.

4.5.2.1 Beamforming including polarization

The use of an array with dual-polarized element pairs allows to control the polarization. This can be done with a beamformer that transmits the same signal with appropriate weights from two elements with orthogonal polarizations (see also Section 3.2.3). In the case with dual-polarized element pairs, such a beamformer could operate on element pairs effectively making each dual-polarized element pair appear as single element. This is illustrated in Fig. 4.25.

In the example, the elements pairs are assumed to be $\pm 45^{\circ}$ linearly polarized. More specifically, it is assumed the polarizations in (4.67) in a particular direction, such as right in front of the element pairs, are given by

$$\hat{\varphi}_{\mathrm{A}} = \frac{1}{\sqrt{2}}\hat{\theta} + \frac{1}{\sqrt{2}}\hat{\varphi}, \; \hat{\varphi}_{\mathrm{B}} = \frac{1}{\sqrt{2}}\hat{\theta} - \frac{1}{\sqrt{2}}\hat{\varphi}. \tag{4.68}$$

Transmitting the same signal from both elements (left part of the figure) makes the element pair appear as to have vertical linear polarization whereas the other weighting (right part of the figure) makes the pair appear as an element with horizontal polarization. To be a bit more specific, consider a single dual-polarized element pair, that is, a UPA with $N_{\mathrm{v}} = N_{\mathrm{h}} = 1$, and assume similar to the case in Section 4.2, that the same signal is transmitted from the two elements with weights w_1 and w_2. Then the element pair will appear as an antenna with polarization $\hat{\psi}_{\mathrm{t}} = w_1\hat{\psi}_{\mathrm{A}} + w_2\hat{\psi}_{\mathrm{B}}$ and this allows to minimize the loss due to polarization misalignment between transmitter and receiver antennas by appropriately adjusting w_1 and w_2 (see also Section 3.4.2). So far in the chapter, the polarizations have been assumed to be aligned (see (4.14)), and with dual-polarized element pairs, this can thus be achieved. Furthermore, it can also be combined with a UPA of arbitrary size, including a ULA, which is illustrated in Fig. 4.26.

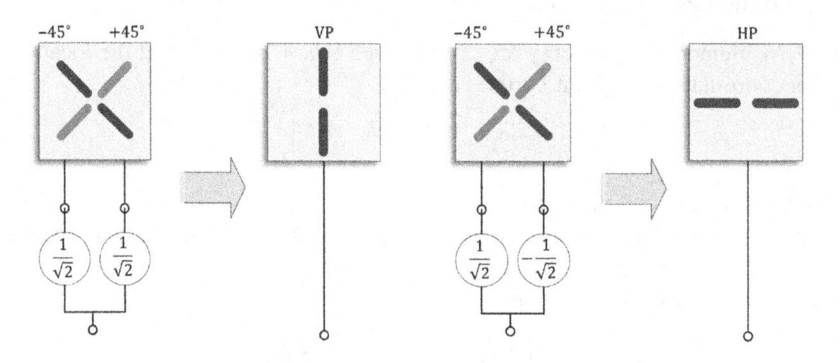

FIGURE 4.25

Beamforming with a $\pm 45^{\circ}$ linearly polarized element pair can appear as an element with for example vertical polarization (left) or horizontal polarization (right)

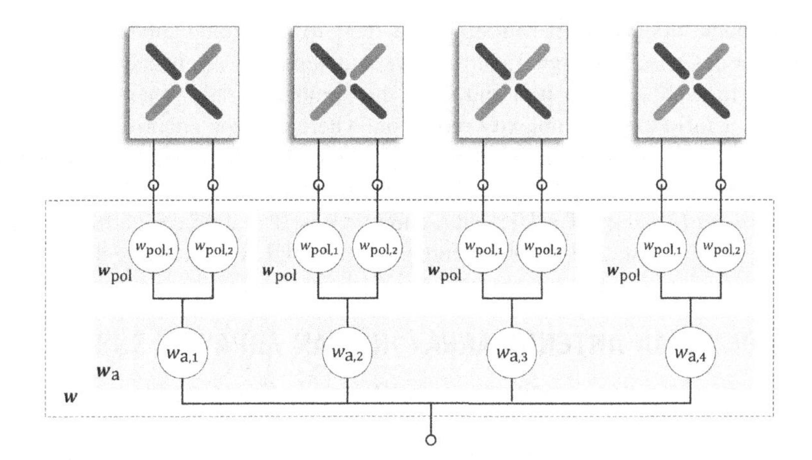

FIGURE 4.26

Example of beamforming with a dual polarized array where $\boldsymbol{w}_a$ is selected to steer a beam in a direction θ_0 and φ_0, and $\boldsymbol{w}_{pol}$ is selected to control the polarization.

In Fig. 4.26, there are two beamformers. A first beamformer, $\boldsymbol{w}_a$, is selected to steer a beam in a certain direction (θ_0, φ_0) for the UPA of element pairs. The second beamformer, $\boldsymbol{w}_{pol}$, is then selected to control the polarization. It should be noted that in the general case, the N coefficients of a beamforming weight vector $\boldsymbol{w}$ can be freely chosen, and that the order of the beamformers can be exchanged. The purpose with the example is merely to demonstrate the possibilities offered by a dual-polarized array. Additional possibilities, more specifically, use of dual-polarized antennas for spatial multiplexing will be discussed in Section 6.3.

4.6 ARRAYS OF SUBARRAYS

To transmit a signal with beamforming, copies of the signal weighted according to the beamforming weight vector need to be generated for all the elements. For classical beamforming, this results in a narrower main lobe as compared to using a single element and allows the direction of the main lobe to be steered. If, however, the set of needed steering directions fall within a limited range of angles, it is not necessary to apply individual adjustments for all the elements. The array can be partitioned into *subarrays*, where a subarray is here defined for a dual-polarized antenna array as a subset of dual-polarized elements pairs connected to two ports, one for each polarization. To change the direction, a copy is then needed only for every subarray port. As will be seen, the range of angles, where the main lobe can be directed without significant gain drop, typically decreases by increasing the size of each subarray. For a given total array size, this corresponds to reducing the number of subarrays and thereby reducing the complexity in the sense that the number of signals that needs to be independently weighted is lower (see also Section 12.3.3).

In the present book, a subarray partitioning will refer to an implementation at radio frequency through appropriate design of the signal paths between a subarray's port and its elements. A radio chain is connected to each subarray port and since the number of radio chains needed can then be reduced, it enables a substantial complexity reduction. Therefore, the subarray size is an important design parameter for practical advanced antenna system (AAS) deployments and will be dealt with in Chapters 13 and 14 in general and in Section 13.2.2 in particular.

In what follows, partitioning of a UPA into subarrays is first defined, and after that gain as a function of angle will be illustrated for partitioning a vertical ULA into different subarray sizes.

4.6.1 PARTITIONING AN ANTENNA ARRAY INTO AN ARRAY OF SUBARRAYS

It is assumed that the array to be partitioned is a UPA with K_v rows and K_h columns of in total $K = K_v K_h$ dual-polarized element pairs as defined in Section 4.5.2. The element pairs of such a UPA can be partitioned into non-overlapping rectangular subsets of element pairs, referred to as subarrays similar to [8]. All subarrays are, to ease the exposition, assumed to have the same size with M_v rows and M_h columns. Due to this the array can also be seen as an array of subarrays (AoSA), with N_v rows and N_h columns. Two examples of partitioning an 8×8 UPA of dual-polarized element pairs are shown in Fig. 4.27.

Due to the assumption that subarrays have the same size, it follows that the dimensions of the subarray (M_v and M_h), the AoSA (N_v and N_h), and the underlying UPA (K_v and K_h) are related as follows

$$K_v = N_v M_v, K_h = N_h M_h.$$

In the following, the notation

$$(M_v \times M_h)_{SA}(N_v \times N_h)$$

will be used as shorthand for a UPA with N_v rows and N_h columns of subarrays with M_v rows and M_h columns.

Since the subarray is a UPA with dual-polarized element pairs, it is in here defined to have a pair of ports, one for each polarization. This is different from [8] where a subarray has one port.

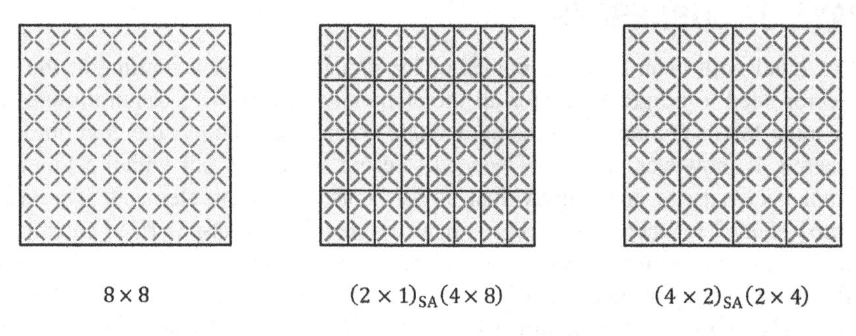

FIGURE 4.27

An 8×8 UPA with 64 dual-polarized element pairs (a) can, for example, be partitioned into a 4×8 array of subarrays with 2×1 subarrays (b) or into a 2×4 array of subarrays with 4×2 subarrays (c).

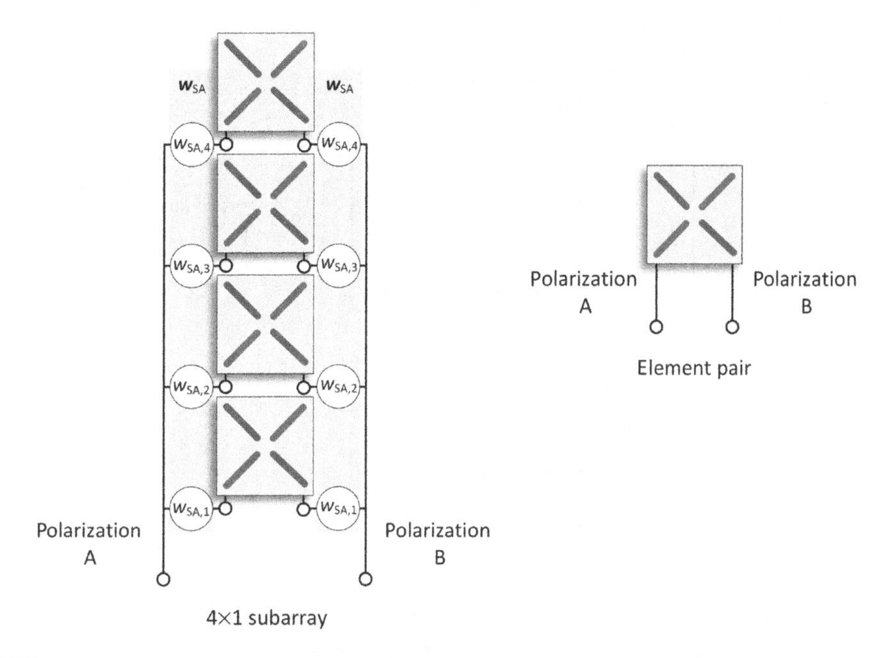

FIGURE 4.28

Example of a 4 × 1 subarray (left) consisting of four dual-polarized element pairs (right).

More specifically, an element pair has one element with polarization A and one element with polarization B. In line with this, the subarray has two ports, one for polarization A and another for polarization B. The elements with polarization A are connected to the subarray port with polarization A and the elements with polarization B are connected to the subarray port with polarization B.

Furthermore, a radio chain is connected to a subarray port and the signal transmitted from an element will depend also on the phase characteristics of the signal path between the subarray port and the element. If all signal paths are such that exactly the same signal is transmitted from the elements of the subarray, the gain will increase at broadside (see Sections 4.3.2 and 4.5.1.1). At the same time, it is also possible to design other characteristics for example by selection of signal paths' lengths', or equivalently phase shifts (see Section 4.4.1) to adjust the direction of the main lobe, similar to electrical tilt in conventional base station antennas. In any case, this means that the signal paths' characteristics of the subarray can be represented by a classical beamformer, $\boldsymbol{w}_{\mathrm{SA}}$, which in what follows, is assumed to be the same for both polarizations and also fixed in the sense that it is not dynamically changed to steer beams to different users in different time slots. The functionality of a subarray is illustrated with an example in Fig. 4.28.

From a radio perspective, the array of subarrays appears as a UPA of dual-polarized element pairs, where in this case each element pair is a subarray. This means that the radio needs to have in total $2N_{\mathrm{v}}N_{\mathrm{h}}$ radio chains, one per subarray port and weighted versions of the signal are generated based on the beamforming weight vector $\boldsymbol{w}_{\mathrm{AoSA}}$ which can be dynamically changed. This is illustrated in Fig. 4.29.

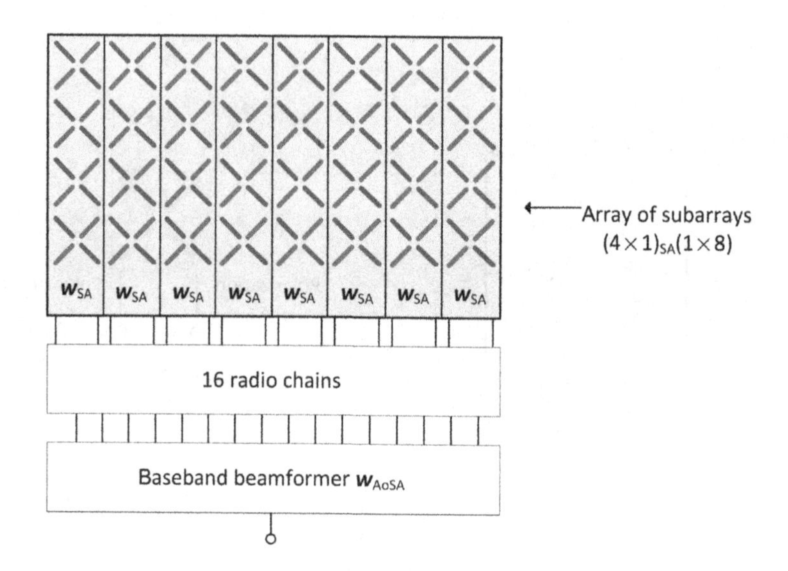

FIGURE 4.29

Example with beamforming at baseband seeing a 1×8 UPA of dual-polarized subarrays, converted to radio frequency using 16 radio chains and connected to a 4×8 UPA partitioned into a 1×8 array of subarrays with 4×1 subarrays.

The total beamforming weight vector as seen from the array of dual-polarized element pairs is thus the result of both the subarray beamformer $\boldsymbol{w}_{\mathrm{SA}}$ and the beamformer realized by the radio $\boldsymbol{w}_{\mathrm{AoSA}}$. Whereas the radio chains allow to dynamically adapt the beamformer $\boldsymbol{w}_{\mathrm{AoSA}}$, the subarray beamformer $\boldsymbol{w}_{\mathrm{SA}}$ is implemented at radio frequency close to the antenna elements and regarded as either fixed or changed on a slow basis for electrical down-tilt functionality. Thus, effectively, the set of beamforming weights that can be used for the array of element pairs is constrained.

In previous sections, it has been established that the total gain is a function of the elements' gain and the array gain and that the array gain for classical beamforming is given by the number of elements. In the present case, the subarray can be seen as an element, and when an AoSA is used with classical beamforming, the achievable gain pattern will have the same shape as the subarray pattern, similar to Fig. 4.17. Since the subarray beamformer is taken as a classical beamformer, with main lobe width decreasing with increasing subarray size, it follows for a fixed array size that the range of angles with high total gain decreases as the subarray size increases, or equivalently, as the number of radio chains is reduced.

Finally, it should also be noted that the concept of subarrays as described in here can be generalized, for example, to allow dynamic updating of the subarray beamformer weights but in a constrained manner such as constant over frequency in case of OFDM-based transmission introduced in Chapter 5. Such a generalization can be viewed as offering the possibility to dynamically change the intended coverage area within which the AoSA beamformer may operate completely flexible over also frequency. A detailed description of this is however beyond the scope of this book.

4.6.2 GAIN PATTERNS

To illustrate the impact of the subarray partitioning introduced in the previous Section 4.6.1, classic beamforming is considered. Only one of the polarizations is examined, and the receiver antenna is assumed to be aligned with respect to the chosen polarization. From the radio's perspective the AoSA then appears as an $N_v \times N_h$ UPA where each element is a subarray which in turn appears as an $M_v \times M_h$ UPA. The gain as a function of direction can therefore be expressed using (4.63) as

$$G_{AoSA}(\theta, \varphi) = |AF_v(\theta, \varphi)|^2 |AF_h(\theta, \varphi)|^2 G_{SA}(\theta, \varphi), \tag{4.69}$$

$$G_{SA}(\theta, \varphi) = |AF_{v,SA}(\theta, \varphi)|^2 |AF_{h,SA}(\theta, \varphi)|^2 G(\theta, \varphi). \tag{4.70}$$

The array factors for the AoSA, $AF_v(\theta, \varphi)$, and $AF_h(\theta, \varphi)$ are given by (4.64) and (4.65) with element spacing $M_v d_v$ and $M_h d_h$, respectively, and they depend on the dynamically controlled steering direction θ_0 and φ_0. The same expressions also apply for the subarray's array factors $AF_{v,SA}(\theta, \varphi)$ and $AF_{h,SA}(\theta, \varphi)$, but with element spacing d_v and d_v, respectively, and corresponding fixed or slowly varying steering directions θ_{SA} and φ_{SA} for the subarray.

In any steered direction θ_0 and φ_0, the array factors for the AoSA will achieve their maximum values, so that (4.69) becomes

$$G_{AoSA}(\theta_0, \varphi_0) = N_v N_h G_{SA}(\theta_0, \varphi_0), \tag{4.71}$$

and for the special case that the steered direction coincides with the fixed subarray steering direction, $\theta_0 = \theta_{SA}$ and $\varphi_0 = \varphi_{SA}$, the gain in (4.71) becomes using (4.70)

$$G_{AOSA}(\theta_{SA}, \varphi_{SA}) = N_v N_h M_v M_h G(\theta_{SA}, \varphi_{SA}). \tag{4.72}$$

Thus, in the direction of the subarrays' steered direction, the gain over using a single element is given by the total number of element pairs in the partitioned array $K = N_v N_h M_v M_h$. This is the same as the maximum gain obtained if the subarray constraint would be lifted so that there was one radio chain per element. Thus, the subarray partitioning does not reduce the maximum gain.

From (4.71) it further follows that the shape of the achievable gain pattern, which is the envelope of all steered beams, is given by the subarray gain pattern, which in turn depends on the size of the array in the corresponding dimension (see also Section 4.5.1.1). Taking the case when the subarray is a vertical ULA, the following is noted:

- The taller the subarray, the smaller is its vertical half-power beamwidth and the smaller is the range of elevation angles for which a gain within say 3 dB of the maximum gain can be achieved. At the same time, for a fixed total array size, the taller the subarray, the lower is the number of subarrays and the lower is the number of radio chains needed.
- The shorter the subarray, the larger is its vertical half-power beamwidth and the larger is the range of elevation angles for which a gain within say 3 dB of the maximum gain can be achieved. At the same time, for a fixed total array size, the shorter the subarray, the higher is the number of subarrays and the higher is the number of radio chains needed.

This is next further explored for an 8×1 vertical UPA. Such a UPA can be partitioned in four different ways with subarray sizes 1×1, 2×1, 4×1, and 8×1 with corresponding AoSA sizes 8×1, 4×1, 2×1, and 1×1 where the case $(8 \times 1)_{SA}(1 \times 1)$ corresponds to the case with no

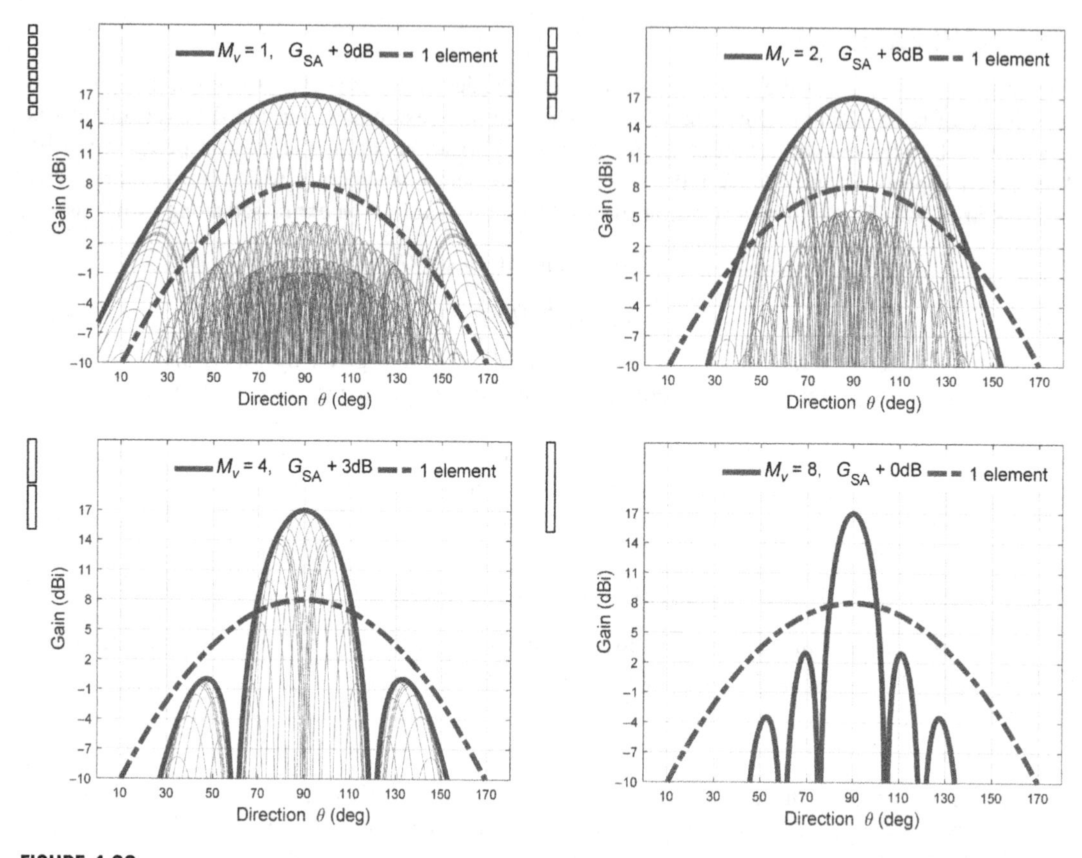

FIGURE 4.30

Gain pattern as a function of θ for different steering angles θ_0 between 0° and 180° (thin lines) by using a vertical array of $K_v = 8$ elements with gain model according to [6], element separation $d = 0.5\lambda$, and different subarray sizes M_v with $\theta_{SA} = 90^\circ$. For comparison, the gain for a single element and the gain for a single subarray plus $10\log_{10} N_v/M_v$ are also shown.

dynamic beamforming. Gain pattern as a function of the angle θ for $\varphi = 0^\circ$ is plotted for different steering angles θ_0 and for comparison the envelope of the steered main lobes, as given by (4.71), is plotted as well in Fig. 4.30.

Another perspective is given in Fig. 4.31, showing the gain pattern for the individual subarrays to the left, and further the envelopes for the different subarray sizes when used with beamforming to the right. Such an envelope, or achievable gain thus includes both gain of the subarrays' fixed beamformer as well as the gain of dynamic beam steering.

To sum up, the gain for beamforming using a UPA partitioned into AoSA has been considered, and it was illustrated that the range of angles for which high gain can be offered, for example to match an intended coverage area, depends on the size of the subarrays. If the range of angles is small, it is possible to use large subarrays, which, as further discussed in Section 12.3.3, offer cost-efficiency benefits.

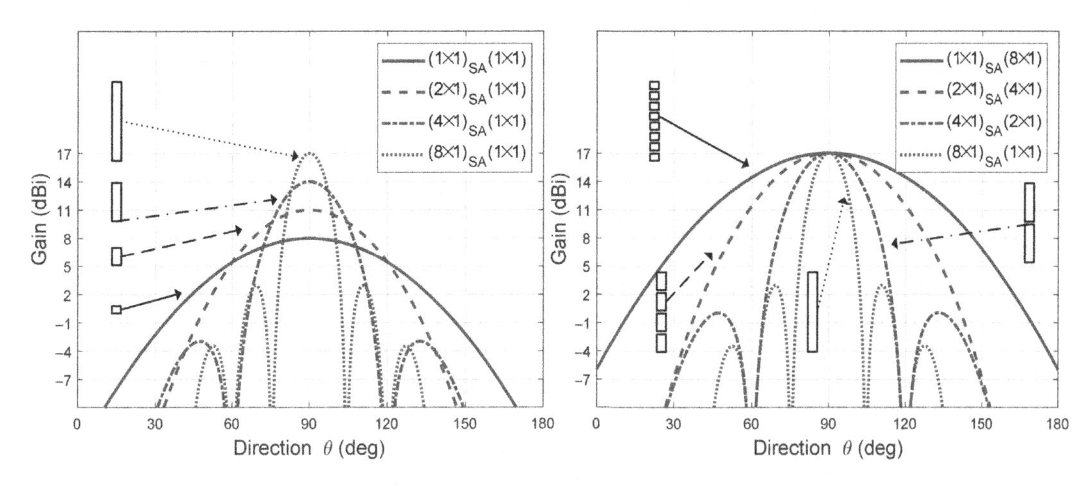

FIGURE 4.31

Illustration of gain pattern for a single subarray (left) and the envelope of all beams when beam steering is done using the array of subarrays (right).

Finally, the selection of a suitable antenna configuration needs to consider both the signal gain and interference, and this topic is addressed in Sections 13.4 and 14.3, where it is demonstrated that a suitable array partitioning depends on the deployment. Typically, such partitioning is done in the vertical domain since the range of vertical angles as seen from the base station is rather small. In this case, the partitioning allows benefiting from the antenna area without requiring an excessive number of radio chains.

4.7 SUMMARY

As stated in the introduction, the intention with this chapter is to explain the basic concept of classical beamforming. Similar to a flashlight beam, signal power can be directed in a desired direction and this has been illustrated in the form gain patterns for uniform linear and planar arrays.

A summary of the chapter is as follows.

- Classic beamforming is described in Section 4.4.1.
  - It can maximize the gain pattern in a steerable direction under free-space propagation conditions.
  - With an array of N elements, the gain in terms of received power compared to using a single element is N in the steered direction for the case of free-space propagation.
  - The power gain as compared to a single isotropic antenna in any steered direction is a product of N and the elements' gain in the steered direction.
  - For a ULA, the width of the main lobe steered to broadside is approximately proportional to $\lambda/(dN)$ in the dimension of the array's orientation, where λ is the wavelength and d is the element separation.

- For a UPA, the width of the main lobe steered to broadside in the vertical and horizontal domain is proportional to main lobe widths of the corresponding vertical and horizontal linear arrays.
- Gain and width of the main lobe depend not only on the array but also on the beamformer as illustrated in Section 4.4.3.

- Array configurations
  - A uniform linear array (ULA) is defined in Section 4.3.3.
  - A uniform planar array (UPA) is defined in Section 4.5.1.
  - A UPA of dual-polarized element pairs is defined in Section 4.5.2. This array configuration allows the beamformer to control also the polarization as illustrated in Section 4.5.2.1.
- The array response vector $\boldsymbol{a}(\theta, \varphi)$ is defined in Section 4.3.2.
  - It includes the elements' amplitude pattern and the phase shift due to propagation delay differences as a function of direction for the elements of the array.
  - It can be seen as a free-space channel between the transmitting elements of the array and an isotropic receiving antenna as function of direction.
  - It will be used in Section 5.3.2 to formulate a MIMO channel model.
  - The gain in a certain direction can be expressed as a product between the corresponding array response vector and the beamforming weight vector.
- Arrays of subarrays
  - An array can be partitioned into subarrays with fixed beamformers as described in Section 4.6.1.
  - Such a partitioning allows a tradeoff between the number of radio chains and the range of angles for which high gain can be achieved. This concept is used when identifying suitable AAS configurations for different deployment scenarios in Chapters 13 and 14.

In Chapters 5 and 6, the discussion will be extended beyond classical beamforming, free-space propagation and multiple antennas only at the transmitter side. More advanced multiple antenna techniques will be considered for OFDM-based transmission in multipath propagation scenarios with multiple antennas also at the receiver side.

REFERENCES

[1] C.A. Balanis, Antenna Theory: Analysis and Design, 3rd ed., John Wiley & Sons, 2005.
[2] D.K. Cheng, Field and Wave Electromagnetics, 2nd ed., Pearson Higher Education, 1989.
[3] R.J. Mailloux, Phased Array Antenna Handbook, 2nd ed., Artech House, 2005.
[4] D. Tse, P. Viswanath, Fundamentals of Wireless Communication, Cambridge University Press, 2005.
[5] J.G. Proakis, Digital Communications, 4th ed., McGraw-Hill, 2001.
[6] 3GPP TR 38.901, V15.1.0. (2019-09) Study on channel model for frequencies from 0.5 to 100GHz.
[7] H. Krim, M. Viberg, Two decades of array signal processing: The parametric approach, IEEE Signal Processing Magazine 13 (4) (April 1996).
[8] IEEE standard for definitions of terms for antennas, IEEE Std 145-2013, 2013.
[9] B.D. Van Veen, K.M. Buckley, Beamforming: a versatile approach to spatial filtering, IEEE ASSP Mag. 5 (2) (1988).
[10] L.C. Godara, Application of antenna arrays to mobile communications, Part II: Beam-forming and direction-of-arrival considerations, Proc. IEEE 85 (8) (1997).

CHAPTER

OFDM-BASED MIMO SYSTEMS

5

The main purpose of this chapter is to derive and describe a *multiple-input multiple-output* (MIMO) channel model for orthogonal frequency division multiplexing (OFDM)–based transmission and associated properties such as spatial correlation. This will be key when describing multi-antenna techniques in Chapter 6 and to understand the rationale for the 3GPP standard specifications for long-term evolution (LTE) and new radio (NR) introduced in Chapter 8 and Chapter 9.

5.1 INTRODUCTION

It will be shown in subsequent sections that for a MIMO OFDM-based system, the relation between transmitted and received signals *for a given subcarrier and at a certain point in time* can be written as the following simple matrix relation:

$$\boldsymbol{y} = \boldsymbol{H}\boldsymbol{x} + \boldsymbol{e}, \tag{5.1}$$

where the received signal, the transmitted signal, and additive impairments are represented by the vectors $\boldsymbol{y}$, $\boldsymbol{x}$, and $\boldsymbol{e}$, respectively. The impairments stem from thermal noise, distortions, and interference. The complex-valued matrix $\boldsymbol{H}$ represents the impact of the radio channel for the given subcarrier and time instance, and as multiple signals can be transmitted as well as received, the channel is referred to as a MIMO channel. The dimensions of the vectors and the matrix correspond to the number of transmit and receive antennas.

As will be explained in the following, using OFDM allows modeling the channel between a specific transmit-receive antenna pair, for each subcarrier, as a *complex-valued scalar*. This is a property that also makes OFDM attractive for wideband MIMO communications as it simplifies the receiver compared to other schemes such as spread spectrum-based transmission as used in 3G.

The model (5.1) will be formulated based on the so-called array response vectors introduced in Section 4.3.2 and the radio propagation characteristics discussed in Chapter 3. More specifically, the MIMO channel matrix will be expressed in terms of propagation path attributes such as angles and delays. Moreover, long-term channel properties by means of spatial correlation matrices will be introduced. In addition, whereas Chapter 3 and Chapter 4 considered mainly narrowband signals such as a sinusoid, a wideband OFDM signal will be considered in this chapter.

5.1.1 ORTHOGONAL FREQUENCY DIVISION MULTIPLEXING

The air interface waveform of LTE and NR, like many other modern digital communication standards, is based on *orthogonal frequency division multiplexing* (OFDM). OFDM is highly spectrally

Advanced Antenna Systems for 5G Network Deployments. DOI: https://doi.org/10.1016/B978-0-12-820046-9.00005-8

efficient and allows high data rate transmission with low receiver complexity even in a dispersive radio channel.

OFDM is a multicarrier modulation scheme that utilizes many narrowband so-called subcarriers that are densely packed in frequency while still being mutually orthogonal. The composite signal has wide bandwidth and facilitates an efficient and flexible sharing of resources in the frequency domain. Hence, rather than transmitting a single stream of modulated symbols (explained below) with short symbol duration, multiple modulated symbols are transmitted in parallel on different subcarriers with much longer symbol duration. Using a longer symbol duration makes the transmission more robust against channel delay spread, that is, each subcarrier experiences a non-frequency selective (i.e., flat) channel.

The orthogonality of OFDM transmissions comes from the specific frequency domain structure and the choice of subcarrier spacing equal to the inverse OFDM symbol duration (defined below). In particular, the receiver needs to correlate over a complete OFDM symbol time interval to preserve orthogonality. To maintain the orthogonality between subcarriers and counteract intersymbol interference (ISI) also in radio channels with time dispersion, a cyclic prefix (CP) that extends the symbol duration slightly is inserted.

There is a rich literature describing OFDM and its applications; see, for example, [1–3], and the references therein for further details.

5.1.2 OUTLINE

Section 5.2 starts by describing the case of OFDM transmission using a single transmit antenna and a single receive antenna, referred to as single-input single-output (SISO). The description will include OFDM modulation and demodulation as well as CP insertion that plays a key role when it comes to maintaining the orthogonality between so-called resource elements (REs) in radio channels with time dispersion. Moreover, some examples are given for how different signals, for example, intended for different users, can be multiplexed in time, in frequency, and even spatially. The section is concluded with a description of discrete Fourier transform (DFT)-precoded OFDM, which is used for uplink transmission in LTE and can be used also in uplink for NR.

Section 5.3 then builds on Section 5.2 and addresses the MIMO OFDM case. The basic MIMO model for a single subcarrier is introduced and the MIMO channel is expressed using the array response vector introduced in Section 4.3.2. For both single- and dual-polarized antenna arrays, the spatial structure of the MIMO channel can then be revealed by expressing it as a weighted sum of outer products. Understanding this channel structure is fundamental to MIMO OFDM performance and algorithm design as will be seen in later chapters.

Finally, the exposition of the spatial structure is concluded by introducing various spatial correlation matrices and discussing their properties. The point to be made is that even time-averaged metrics of the channel matrix that utilizes second-order statistics, such as correlation, contain spatial information. Hence, there is an underlying long-term property of the MIMO channel that can be exploited by the MIMO transmission scheme and precoding.

Having exposed some details of OFDM and MIMO, Section 5.4 returns to the radiation pattern (introduced in Section 3.3.4) and relates it to a two-element channel. This gives some fundamental basics that help understanding the meaning of the term *beam*, which is addressed in Section 5.5, both for the case of transmission and reception.

In Section 5.5 the terms *beam* and *beamforming* are discussed. As the use of multi-antenna technologies has changed over the years, a generalization and clarification of the traditional definition of a beam are motivated.

Finally, the chapter is concluded in Section 5.6.

5.2 SINGLE-ANTENNA OFDM TRANSMISSION AND RECEPTION

The basic concept of OFDM is as the abbreviation suggests, to use several parallel so-called subcarriers on different frequencies. This is illustrated in Fig. 5.1.

For example, with a 20 MHz channel bandwidth, an LTE carrier uses 1200 subcarriers with a subcarrier spacing of $\Delta_f = 15$ kHz. An NR carrier can use up to 3300 subcarriers with a configurable subcarrier spacing, $\Delta_f = 15 \cdot 2^{\mu}$ kHz, $\mu = 0, 1, 2, 3$, for data and bandwidths up to 100 MHz (see Section 9.3). There will be some empty subcarriers at the band edges to allow for means, such as filtering, to reduce the power outside the channel bandwidth where another carrier of the same or a different operator may be in use.

At the top of Fig. 5.2, a simplified illustration of transmission and reception in an OFDM-based system is given. In the transmitter, a block of data bits is encoded by the channel encoder (using a Turbo code in LTE and low-density parity check codes in NR) and further modulated to generate a block of quadrature amplitude modulation (QAM) symbols where each QAM symbol may contain 2–8 encoded bits depending on the used modulation order. See Table 8.1 for the different modulation constellations used in LTE. The block of QAM symbols is then mapped to the different subcarriers of an OFDM modulator according to the resources assigned for transmission.

In practice, an OFDM modulator is implemented using an inverse fast Fourier transform (IFFT) to obtain a sequence of time domain samples, where the full sequence is referred to as one *OFDM symbol*. To maintain orthogonality between QAM symbols transmitted on different subcarriers in radio channels with non-zero time dispersion, a so-called *cyclic prefix* (CP) is added where the last part of the OFDM symbol is copied and added at the beginning of the OFDM symbol, thereby

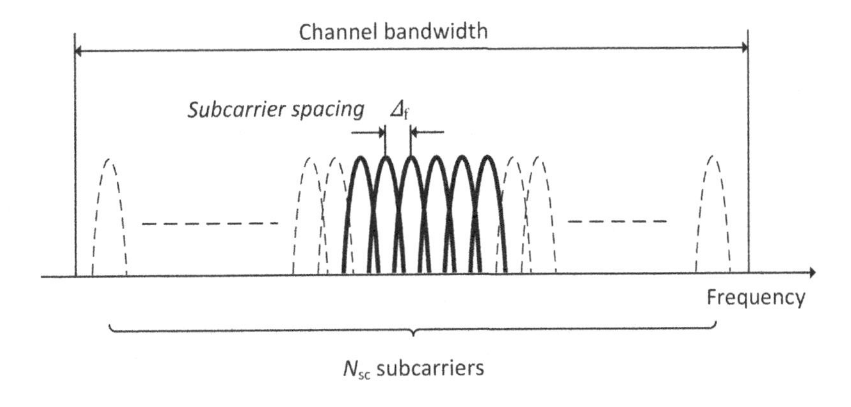

FIGURE 5.1

OFDM using N_{sc} subcarriers within an allocated channel bandwidth.

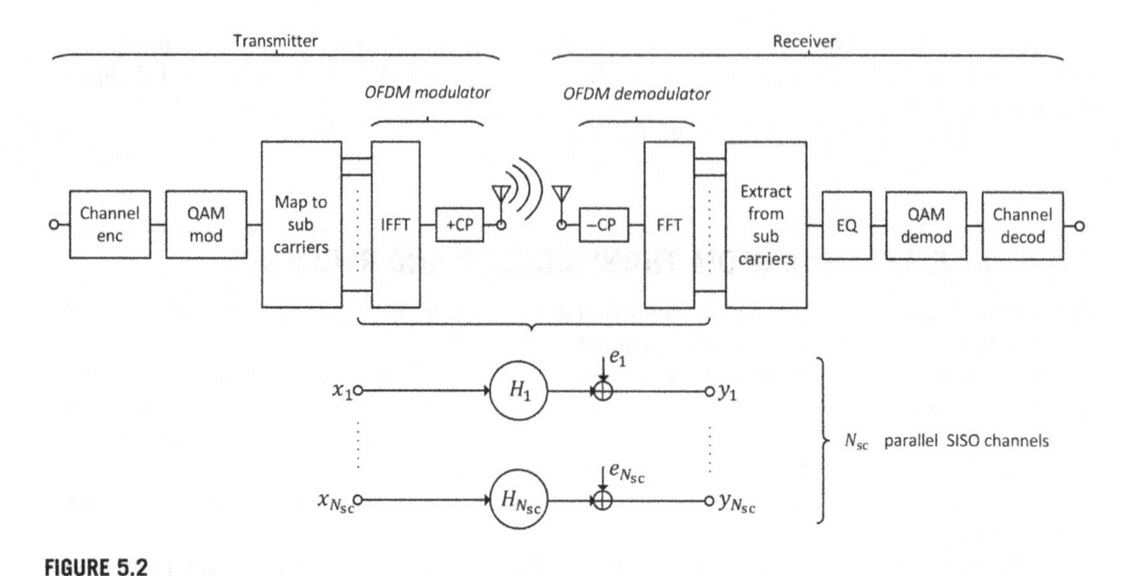

FIGURE 5.2

Single-antenna transmission and single-antenna reception of data using N_{sc} OFDM subcarriers (top) can be viewed as communicating over N_{sc} parallel and non-interfering channels (bottom).

making the OFDM symbol duration longer in time. The length of the CP should be larger than the time dispersion of the channel (see Section 3.6) in order to maintain subcarrier orthogonality in the receiver. However, adding a CP also reduces the spectral efficiency since it implies additional overhead, and this is the price for maintaining subcarrier orthogonality. The procedure of performing the IFFT and adding the CP is done by an *OFDM modulator.*

Note the difference between the *QAM symbols*, which are the stream of symbols that are input to the IFFT, and the single *OFDM symbol*, which is the resulting output from the OFDM modulator. The sequence of OFDM symbols is then fed to a radio chain that will generate an analog high-power signal at radio frequency for the transmission from the antenna. This radio chain processing includes digital-to-analog conversion, filtering, upconversion to radio frequency and amplification (see also Fig. 11.2).

In the receiver, a radio chain performs downconversion, filtering, and sampling of the antenna signal to generate a sequence of received samples (see also Fig. 11.4). An *OFDM demodulator* will remove samples corresponding to the CP and perform a fast Fourier transform (FFT) on a block of samples to recover the symbols transmitted on the different subcarriers. The received QAM symbols are then extracted from the output of the OFDM demodulator and after equalization, that is, compensation for the impact of the radio channel, a QAM demodulator generates soft bits that are used by the channel decoder to decode the transmitted block of data bits. A soft representation of a transmitted binary bit is a real-valued number, which represents a probability that the transmitted bit was either 0 or 1. The channel decoder takes soft bits jointly into account before deciding on the most probable transmitted sequence of binary bits in the block.

OFDM modulation and demodulation as well as the role of the CP will be described in more detail in the next sections. If the CP is chosen larger than the time dispersion in the radio channel, then the demodulator output on subcarrier k in a certain OFDM symbol, y_k, can for this SISO case be expressed as

$$y_k = H_k x_k + e_k, \tag{5.2}$$

where H_k is a complex scalar channel coefficient, x_k is the transmitted modulated symbol such as a data-carrying QAM symbol, and e_k represents impairments.

The channel coefficient H_k is determined by the channel transfer function defined in Section 3.6.2.4 evaluated at the frequency of the subcarrier and can be recognized as a Fourier transform of the channel impulse response (see also Section 5.2.3). The additive impairments stem from thermal noise, distortions, as well as intra- and intercell interference but they do not include any contributions from the other subcarriers or OFDM symbols. In particular, it may be noted from (5.2), that there is no interference from any other QAM symbols, and consequently equalization becomes simply a matter of compensating for the phase of the channel coefficient and taking the channel coefficient magnitude as well as the power of the additive impairment factor into account when estimating the soft bit values.

Hence, OFDM-based SISO transmission allows modeling the system as a set of multiple, ideally independent, single-carrier SISO links. Each modulated symbol x_k is carried over a scalar channel H_k and received without interference from the simultaneous transmissions on the other subcarriers.

5.2.1 ORTHOGONAL FREQUENCY DIVISION MULTIPLEXING MODULATION AND DEMODULATION

The OFDM modulation and demodulation are now described in more detail. As mentioned in the previous section, OFDM transmission is block-based in the sense that time is divided into symbol intervals, known as OFDM symbols, and in each OFDM symbol, multiple data-carrying QAM symbols are transmitted in parallel using the allocated subcarriers. For the single transmit and receive antenna case, the continuous-time baseband signal $x(t)$ to be transmitted can, in line with [4], be expressed as

$$x(t) = \sum_{k=1}^{N_{sc}} x_k e^{j2\pi\Delta_f kt}, \quad -T_{cp} \le t \le T, \tag{5.3}$$

where x_k is the QAM symbol of subcarrier k, Δ_f is the subcarrier spacing, N_{sc} is the number of subcarriers, T_{cp} is the length of the CP, and T is the duration of the OFDM symbol excluding the CP. In practice, the signal $x(t)$ is generated by using an IFFT of the QAM symbol sequence $\{x_k\}_{k=1}^{N_{sc}}$, as illustrated in Fig. 5.2, followed by appropriate upsampling, digital-to-analog conversion, and appropriate filtering.

To ensure that subcarriers become orthogonal, the subcarrier spacing Δ_f can be selected (shown below) as

$$\Delta_f = \frac{1}{T}, \tag{5.4}$$

and since the signal in (5.3) then is periodic with period T, the CP is a cyclic extension in the sense that the last part of the signal T is copied and used as the CP, inserted in the beginning. Hence, the OFDM symbol length becomes $T_s = T + T_{cp}$ after CP insertion.

Orthogonality means that the receiver can recover the symbol transmitted on a specific subcarrier by correlating the received signal with the corresponding complex exponential for the frequency over the symbol interval. In detail, assume that the received baseband signal before the OFDM demodulator is

$$y(t) = x(t) + n(t), \tag{5.5}$$

where $x(t)$ is the transmitted OFDM symbol in (5.3) and $n(t)$ represents additive impairments including thermal noise. To recover the symbol x_k transmitted on subcarrier k the OFDM demodulator correlates $y(t)$ with the complex exponential of the same frequency,

$$y_k = \frac{1}{T}\int_0^T y(t)e^{-j2\pi\Delta_f kt}dt, \tag{5.6}$$

where it may be noted that the receiver discards the CP which is the first part of the symbol with duration T_{cp}. By combining (5.3), (5.4), and (5.5), and using the identity

$$\int_0^T e^{j2\pi(l-k)\frac{t}{T}}dt = \begin{cases} T \text{ if } l = k \\ 0 \text{ if } l \neq k \end{cases},$$

it follows that the subcarriers are indeed orthogonal and (5.6) can be re-written as

$$\begin{aligned} y_k = \frac{1}{T}\int_0^T (x(t) + n(t))e^{-j2\pi\Delta_f kt}dt &= \frac{1}{T}\int_0^T \left(\sum_{l=1}^{N_{\text{sc}}} x_l e^{j2\pi l\frac{t}{T}} + n(t)\right)e^{-j2\pi k\frac{t}{T}}dt = \frac{1}{T}\sum_{l=1}^{N_{\text{sc}}} x_l \int_0^T e^{j2\pi(l-k)\frac{t}{T}}dt \\ &+ \frac{1}{T}\int_0^T n(t)e^{-j2\pi k\frac{t}{T}}dt = x_k + e_k, \end{aligned} \tag{5.7}$$

where e_k represents the contribution from the noise term $n(t)$ at the output of the correlator. Hence, there are no contributions from the other subcarriers after the correlation and this stems from the selection of subcarrier spacing in (5.4).

In a practical receiver, an OFDM demodulator is implemented in the digital domain after filtering and analog-to-digital conversion, and the integration is replaced by a summation, which can be implemented with an FFT as illustrated in Fig. 5.2.

5.2.2 CYCLIC PREFIX AND TIME DISPERSIVE CHANNEL

In practice, multiple consecutive OFDM symbols are transmitted and received over a radio channel with multipath propagation. In this case, the received signal before the OFDM demodulator is represented as

$$y(t) = \sum_n \gamma_n x(t - \tau_n) + n(t), \tag{5.8}$$

where expressions for the path coefficients $\{\gamma_n\}$ and delays $\{\tau_n\}$ can be found in Section 3.6 (although these parameters are not important for the following discussion).

In the previous section, the transmitter generated an OFDM symbol with duration $T_s = T + T_{\text{cp}}$. Due to the time dispersion introduced by the multipath propagation in the radio channel, each received symbol will have a duration $T_s + \Delta_\tau$ where the time dispersion, Δ_τ, is the difference in delay between the first and the last path,

$$\Delta_\tau = \max_n \tau_n - \min_n \tau_n. \tag{5.9}$$

This is illustrated in Fig. 5.3.

However, the OFDM demodulator introduced in the previous section discarded the first part of the symbol. This means that as long as the time dispersion is smaller than the CP,

$$\Delta_\tau \leq T_{cp}, \tag{5.10}$$

then it is possible for the receiver to position the "window" corresponding to the OFDM symbol duration T, so that there is no contribution from the preceding OFDM symbol. The CP thus acts as a guard that protects against interference from previously transmitted symbols. This is also illustrated in Fig. 5.3.

From an ISI perspective, interference between OFDM symbols could be equally well handled by having an *empty* guard period between every OFDM symbol. However, as described in [2] the CP also effectively maintains the orthogonality between the subcarriers within an OFDM symbol and is thus the method used in LTE and NR.

5.2.3 FREQUENCY DOMAIN MODEL AND EQUALIZATION

A CP with duration larger than the time dispersion in a radio channel with multipath propagation enables the receiver to recover the symbols transmitted on different subcarriers in different OFDM symbols without any mutual interference. The channel coefficient describes how the radio channel impacts the symbol transmitted on a subcarrier, as will be seen next.

Due to the propagation delays, between the transmitter and the receiver, the receiver needs to align its window accordingly so that it starts at

$$\tau_{\min} \triangleq \min_n \tau_n. \tag{5.11}$$

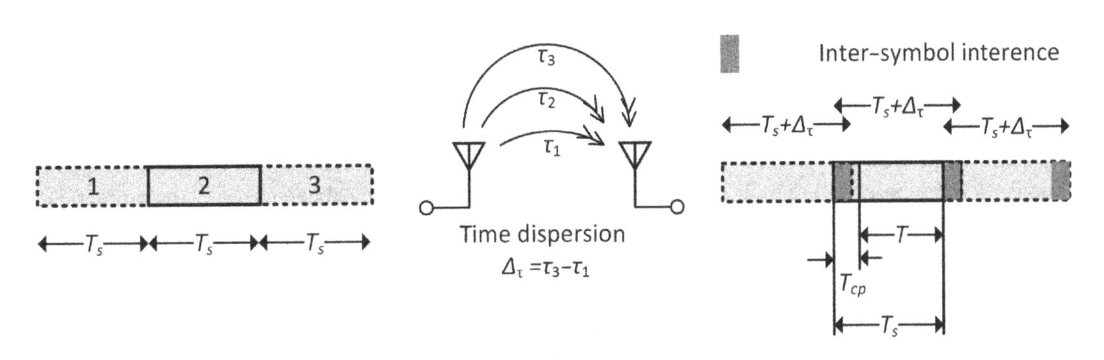

FIGURE 5.3

Due to time dispersion of the radio channel, the symbols appear longer at the receiver. If the time dispersion is smaller than the length of the cyclic prefix, interference between OFDM symbols can be avoided after OFDM demodulation.

The demodulator output in (5.6) for subcarrier k is then taken as

$$y_k = \frac{1}{T}\int_{\tau_{\min}}^{\tau_{\min}+T} y(t)e^{-j2\pi k\Delta_f(t-\tau_{\min})}dt. \tag{5.12}$$

Next, for notational simplicity it is assumed that the path and delays are defined from a receiver perspective so that the first path corresponds to zero delay,

$$\tau_{\min} = 0. \tag{5.13}$$

By using the expression for the received signal in (5.8) for the case with multipath propagation in the expression for the output of the OFDM demodulator in (5.12), and using the expression for the transmitted signal $x(t)$ from (5.3) with the assumptions on receiver alignment in (5.13) and time dispersion less than CP length (5.10), it follows that the subcarriers are still orthogonal and the demodulator output (5.14) can be shown to be

$$y_k = H_k x_k + e_k, \tag{5.14}$$

where e_k stems from the impairments and the channel coefficient experienced by subcarrier k is given by

$$H_k = \sum_n \gamma_n e^{-j2\pi k\Delta_f \tau_n}, \tag{5.15}$$

Before discussing the properties of the channel coefficient H_k, the first observation is that despite the time dispersion in the radio channel, the orthogonality between the subcarriers is maintained in the sense that the output for subcarrier k, y_k in (5.14), contains no contributions from symbols transmitted on any other subcarriers x_l for $l \neq k$. As mentioned above in the introduction in Section 5.2, this is the reason why equalization becomes simple and straightforward for OFDM with CP also in time dispersive radio channels.

To detect x_k from y_k in (5.14) implies that the channel H_k first needs to be estimated and this is done by transmitting a few known modulated symbols. This is referred to as inserting demodulation reference signal (DM-RS) among the transmitted symbols x_k on certain subcarriers where data symbols are not transmitted. By using these known transmitted signals, typically QAM symbols, the receiver can with appropriate averaging and interpolation between these reference symbol carrying subcarriers generate estimates of the channel H_k for all subcarriers and for all OFDM symbols. The placement of reference symbols is illustrated in Fig. 5.4 and the possible DM-RS configurations available in NR are given in Section 9.3.1.2.

The channel coefficient for subcarrier k in (5.15) can be recognized as the channel transfer function defined in Section 3.6.2.4, or equivalently the Fourier transform of the channel impulse response for the baseband signal, evaluated for the frequency of the subcarrier, $k\Delta_f$. In a practical digital implementation, the channel coefficients are given by a discrete-time Fourier transform rather than a continuous-time Fourier transform, and the impact of filtering is included. In both cases, the channel coefficient is referred to as the frequency domain channel.

Another important aspect related to orthogonality is the need for synchronization. In light of (5.12), the receiver needs to align the integration window and know the delay of the first tap, $\tau_{\min}$, at least to the extent that a decent amount of the signal power is within the assumed CP window. However, there is also a need to align with respect to frequency, and in practice there is often a frequency mismatch between the receiver and the transmitter. In fact, if the frequency offset is as

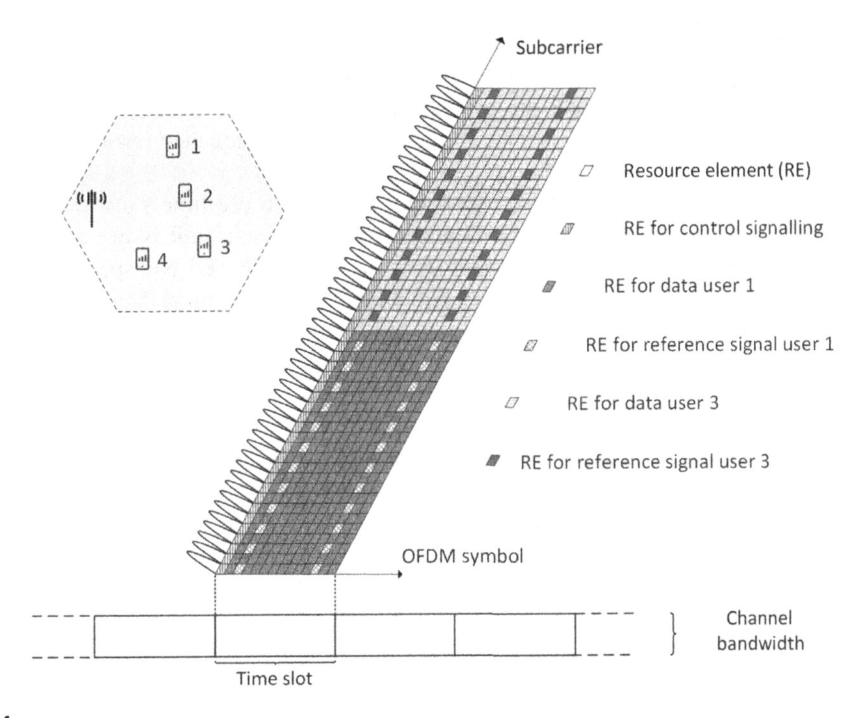

FIGURE 5.4

In each time slot, different signals can be multiplexed onto different resource elements, where a resource element refers to a single subcarrier in a single OFDM symbol. In this figure only a single layer is illustrated.

large as the subcarrier spacing, then due to the orthogonality, no signal power would be received. For this purpose, in standards such as LTE and NR, synchronization signals are transmitted from the base stations and additional reference signals are used not only for channel estimation and measurements, but also for fine-tuning the time and frequency synchronization. This fine-tuning is critical to be able to receive higher-order, highly spectral efficient QAM constellations as they are more sensitive to misalignment. In NR, the *tracking RS* was introduced for this purpose (see Section 9.3.1.1.1).

It should be noted that the orthogonality is lost, which may degrade demodulation performance, if there are channel variations during the OFDM symbol time, for example, caused by fast user equipment (UE) movement. In that case, the basic relation (5.13) needs to be modified for correct modeling, for example, by including intercarrier interference contributions in the impairments term e_k.

5.2.4 MULTIPLEXING ON A TIME–FREQUENCY GRID

So far in this chapter, transmission of data bits to a single receiver has been illustrated, and it was observed that reference signals are typically needed for time and frequency synchronization as well as channel estimation at the receiving end.

Additionally, in a 5G system, the available channel bandwidth needs to be shared to serve multiple users, both for uplink and downlink transmissions. Since the symbols transmitted on different subcarriers in different OFDM symbols are orthogonal, it is possible to multiplex transmissions of different data blocks in an orthogonal manner, utilizing time, and frequency resources, that is, subcarriers and OFDM symbols.

More specifically, the basic concept for user multiplexing is to use time slots and in each such time slot transmit multiple OFDM symbols. The smallest addressable unit is one subcarrier times one OFDM symbol, denoted *a resource element* (RE) in 3GPP LTE and NR specifications. A RE contains one modulation symbol and each such symbol carries one or more bits, depending on the used QAM constellation. Each such symbol carries either encoded data, control information, or a predefined reference symbol known to the receiver, which can be used for synchronization or channel estimation. An example of this is given in Fig. 5.4. The OFDM symbols and the subcarriers in a slot can thus be viewed as a time–frequency grid of REs. In addition, in multi-antenna transmission, *spatial multiplexing* can be used, where this third layer dimension is added on top of the time–frequency grid. In this case, a RE contains one modulation symbol per subcarrier, OFDM symbol, and MIMO layer. Note that while different time/frequency REs within a layer are orthogonal, different layers of a RE are not orthogonal by design. Hence, it is up to the transmitter to ensure that the interference between layers is minimized at the receiver, for example, by using multi-antenna precoding techniques, and also at the receiver side, multiple antennas can be used to separate the layers.

Examples and details on how the time–frequency grid is used will be given in Sections 8.1 and 9.2 for LTE and NR, respectively. In short, a given transmission will use a set of REs, that is, subcarriers, OFDM symbols and layers, for the transmission. The narrowest possible transmission in frequency of a data channel is by using 12 subcarriers in both LTE and NR (ignoring the so-called narrowband feature introduced in LTE Release 13). The number of symbols for a transmission varies between 2 and 14 and the number of layers transmitted from the network to the same user is between one and eight. Hence, the OFDM-based transmission allows for easy multiplexing of users in the time–frequency grid. Moreover, 5G systems typically share and assign these resources to different users very rapidly in the sense that control information is transmitted in the downlink frequently, at least once per millisecond. This enables that the time–frequency resources can be reallocated between different users quickly.

Returning to the example in Fig. 5.4, which applies for the downlink, a cell with four users is considered, and in the depicted time slot, data are transmitted to two of them (user 1 and user 3) using a single layer and multiplexed in frequency (users are assigned different subcarriers). Reference signals are embedded in the time–frequency grid for both users so that the receivers can estimate the channels. Furthermore, in the first OFDM symbol of the slot, control signaling, including both control information and associated reference signals, is transmitted. In the example, all four UEs, for example, mobile phones, monitor and decode control signaling in each slot. UE 1 and 3 have in this example detected, by decoding the control channel, that data have been scheduled for them. The control information contains the necessary information for the scheduling assignment such as modulation, coding rate, and the set of used REs for the data and reference signals. The UE then performs channel estimation and demodulate and decode the data. UE 2 and 4, on the other hand, have in this example not detected a scheduling assignment when decoding the control channel in this slot; therefore these UEs can remain idle until the next occasion of a control channel transmission (e.g., in the beginning of next slot).

5.2.5 LOW PAPR OFDM—DFT-PRECODED OFDM

The *peak-to-average power ratio* (PAPR) of a signal waveform is an important metric as a small value implies that the power amplifier used to transmit signals can operate more efficiently and thus save battery in the UE (see also Section 11.3.1.1). The PAPR is rather high in OFDM due to the linear combination of many (pseudo-random) QAM symbols in the IFFT operation (5.3). Hence, due to the central limit theorem, some output OFDM symbols will have very large PAPR. To mitigate this, a DFT precoding of the sequence of QAM symbols can be used. This is known *as DFT-precoded* OFDM, *DFT-spread* OFDM, or sometimes as *transform-precoded* OFDM.

Since the input to the DFT is a sequence with rather low amplitude variation due to the use of QAM symbols, the input sequence has rather low PAPR. Using a transform followed by an inverse transform, that is, the DFT and IFFT serially, the output from the OFDM modulator has similar PAPR properties as the input to the DFT precoder. Hence, it is ensured that the output of the OFDM modulator also has a low PAPR property provided that the output of the DFT to the input of the IFFT is mapped properly to preserve PAPR, for example, to contiguous subcarriers. Note that it is possible to reduce the PAPR further by using an even lower PAPR input sequence compared to QAM; hence, NR supports the use of $\pi/2$-BPSK modulation for uplink transmissions. In Fig. 5.5 the steps of the transmitter chain are illustrated.

For base stations, the impact of PAPR is slightly less critical, since more advanced power amplifier linearization techniques can be used to reduce the impact of a high PAPR waveform, whereas for UE such advanced linearization techniques are not feasible due to complexity and power consumption. Hence, DFT-precoded OFDM transmission is used in LTE and can be configured for NR but for UE transmission only. To improve the power efficiency and cope with PAPR for the downlink transmission, PAPR reduction techniques such as clipping and filtering are often used (see further [5]).

For DFT-precoded OFDM-based transmissions, there is an important difference compared to non-DFT-precoded OFDM in that no additional signals (such as reference signals) can be added outside the allocated subcarriers for data symbols at the input of the IFFT, if the low PAPR property should be maintained. Hence it is not possible to multiplex demodulation reference signals in the same OFDM symbol as the data, as that would destroy the low PAPR property of the DFT-precoded OFDM waveform. In LTE and NR, the DM-RS and associated uplink data

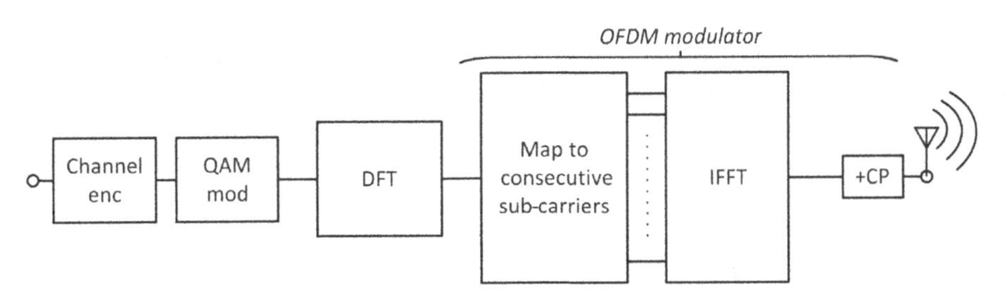

FIGURE 5.5

Illustration of DFT-precoded OFDM, where a block of QAM symbols are first DFT-precoded and mapped to consecutive subcarriers, before the "normal" OFDM modulation as in Fig. 5.2 applies.

(physical uplink shared channel, PUSCH) with DFT precoding enabled are time multiplexed in different OFDM symbols (see Section 8.3.1.6 for LTE).

It should also be noted that PAPR can be defined as power-to-average power ratio, but sometimes the term peak-to-average power ratio is used. In the former case, the statistical distribution of the power-to-average power ratio is generated and the PAPR is thus a statistical distribution function. The latter definition of PAPR is often taken as a percentile of the PAPR distribution, for example, the PAPR value is exceeded in 0.1% of the time. This percentile is often referred to as the peak-to-average power ratio.

5.3 MULTI-ANTENNA OFDM TRANSMISSION AND RECEPTION

The use of OFDM and MIMO fits very well together, since OFDM-based transmissions using CP ideally allow modeling each subcarrier as an independent MIMO system. This then circumvents the complexity of MIMO-based spread spectrum receivers as in 3G systems, since as long as the CP is longer than the time dispersion of the channel, an OFDM MIMO receiver can in principle operate per subcarrier, independent of the adjacent subcarriers.

In this section, the single-antenna model for OFDM (5.13) is extended to the MIMO case that involves multiple antennas at both the transmitter and receiver side. At the top of Fig. 5.6, a simplified illustration of transmission and reception in a MIMO OFDM-based system is given.

The difference compared to the case with a single transmitter is that the block of modulated QAM symbols (known in 3GPP as a codeword of encoded and then modulated bits) is further split into several layers by, what is known in 3GPP as, the codeword to layer mapping. For example, for spatial multiplexing with two layers, half of the QAM symbols are mapped to the first layer and the remaining half are mapped to the second layer. Each layer is then mapped to the multiple antennas by means of MIMO precoding. This means that the input of the OFDM modulator for a given transmit antenna may in the general case contain contributions, that is, a linear combination, of modulated symbols from all the layers.

Note that the figure exemplifies the structure of information-carrying channels such as the data channel in 3GPP standards such as LTE and NR. For those kinds of channels, streams of QAM-modulated symbols are referred to as layers. But there are other transmitted signals using at least parts of the transceiver chain but where the symbol stream is not referred to as a layer. Reference signals are a primary example of the latter, where the symbol stream constitutes some known sequences and thereafter is mapped to the antennas via a precoding step but where that symbol stream is not called a layer. Note also that the term layer is defined to correspond to a symbol stream at a very particular point in the transceiver chain. Thus "layer" is a more specific term encompassed by the more general term symbol stream. Although the development to follow is focusing on the structure of the data channel as an example, it should be understood that the resulting received data models are applicable for more general use as well.

The MIMO precoding operation can be described by a matrix where each column represents how each layer is mapped to the antennas, for example, a beamforming weight vector, as introduced in Section 4.3.2, for the corresponding layer. This is further described in Section 6.3, where it is referred to as spatial multiplexing and where the cases with spatial multiplexing to a single

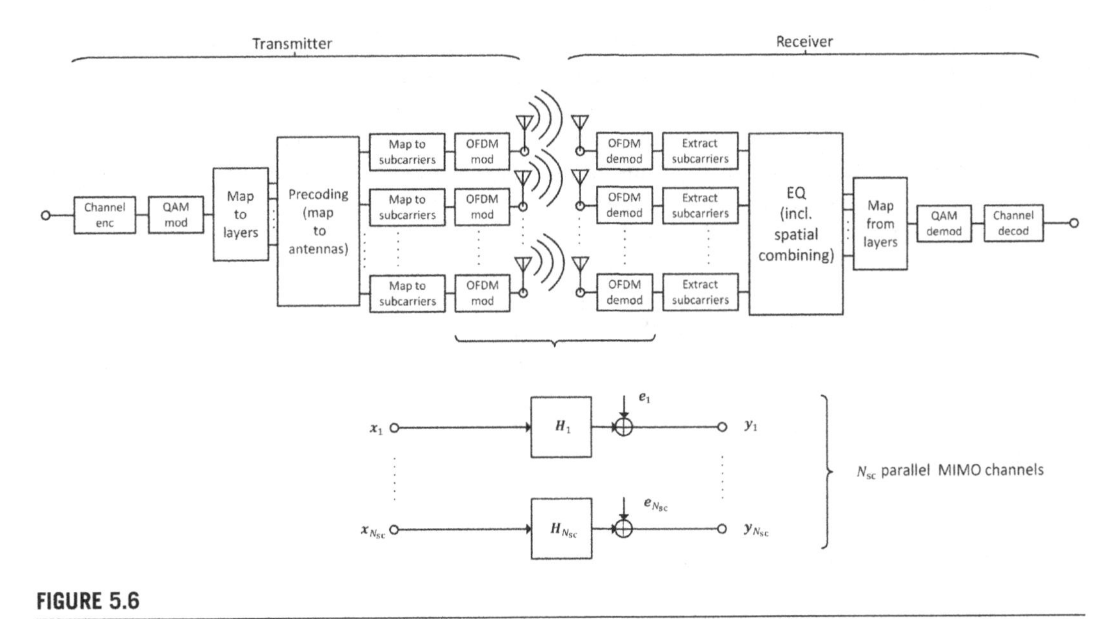

FIGURE 5.6

MIMO OFDM transmission and reception with N_{sc} subcarriers and multiple antennas at the transmitter and multiple antennas at the receiver (top) can equivalently be viewed as N_{sc} parallel MIMO channels, each having multiple layers, for each subcarrier and OFDM symbol.

user and to multiple users are both discussed. In the case with multiple users, there are multiple blocks of data that are separately encoded, modulated, mapped to corresponding layers, and pre-coded so that the input symbols to the OFDM modulator of each antenna are a sum of beamformed symbols intended for multiple users.

At the receiver side, the signal from each receive antenna is fed to an OFDM demodulator. Then for actual subcarriers used for the transmission, the outputs from all antennas are fed to an equalizer. The task for the equalizer is to not only compensate for the channel variations but also to combine the signals from the different receive antennas, referred to as spatial combining, as well as to separate the different MIMO layers. The layers are then combined into a sequence of equalized symbols from which the QAM demodulator generates soft (nonbinary) representations of the transmitted bits that are used by a channel decoder to decode the transmitted block of data bits.

In Section 6.3, spatial filtering on the receive side, which can be thought of as receiver beamforming that also mitigates interference between the layers, is further discussed. It may be noted, in light of the illustration in Fig. 5.6, that more advanced receivers, performing joint equalization and demodulation, are certainly possible. The orthogonality property between the subcarriers is in any case advantageous since it allows parallel processing of the different subcarriers. More specifically, under the assumption of a long enough CP and an appropriately synchronized receiver, the outputs from the OFDM modulator for subcarrier k can be expressed as

$$\boldsymbol{y}_k = \boldsymbol{H}_k \boldsymbol{x}_k + \boldsymbol{e}_k. \tag{5.16}$$

This is a generalization of the corresponding single-antenna case in (5.2). If N_t is the number of transmit antennas and N_r is the number of receiver antennas, then $\boldsymbol{y}_k$ is a vector with N_r elements representing the received symbols, $\boldsymbol{x}_k$ is a vector with N_t elements representing the transmitted symbols, $\boldsymbol{H}_k$ is a complex-valued matrix with N_r rows and N_t columns representing all the N_rN_t channel coefficients between all the receive and transmit antennas, and $\boldsymbol{e}_k$ is vector with N_r elements representing additive impairments such as thermal noise.

Thus, an interpretation of (5.16) is that the combination of MIMO and OFDM with N_{sc} subcarriers can be seen as N_{sc} parallel MIMO channels. This is also illustrated at the bottom of Fig. 5.6.

It may be noted that the formulation above in (5.16) assumes that there is one OFDM modulator for each transmitter antenna and one OFDM demodulator for each receiver antenna. This is illustrative but not a requirement on a practical implementation that may use lower number of modulators and demodulators. The possibility to do so will depend on the selected multiple antenna transmission techniques and other factors such as the number of different transmissions that need to be multiplexed in the same OFDM symbol.

5.3.1 FREQUENCY DOMAIN MODEL

Due to superposition principle described in Section 3.2.4, the signal received by each receiver antenna will be a sum of contributions from transmissions from all N_t transmit antennas. The channels coefficients experienced by the N_t different transmitters to the receiver (i.e., for each *transmit-receive antenna pair*) will, however, be different. One reason for this is that the antenna patterns for each transmitter can be different, but even if all transmit antennas are identical, there will in general be a difference. This is because the transmit antennas are at different locations in space and therefore the propagation path lengths, or equivalently propagation delays, are slightly different for the different transmit-receive antenna pairs, and this in turn leads to different channel coefficients as a result of the superposition of the multiple paths (see also Figs. 3.19 and 3.20). However, for typically used antenna configurations, the separation between the different antennas is small, both at the transmitter side and at the receiver side. Therefore, the time dispersion is essentially the same for all the antenna pairs and a sufficiently long CP will avoid interference between different OFDM symbols.

To formalize this in mathematics, it is noted that the OFDM demodulator is linear, and it is then possible to generalize (5.14) so that the output of the OFDM demodulator for subcarrier k and receiver antenna p, $y_{k,p}$, becomes

$$y_{k,p} = \sum_{q=1}^{N_t} H_{k,p,q} x_{k,q} + e_{k,p}, \quad p = 1, \ldots, N_r. \tag{5.17}$$

In this expression, $x_{k,q}$ is the input on subcarrier k to the OFDM modulator associated with transmit antenna q, which can be a sum of weighted QAM symbols of different layers as in the example of Fig. 5.6 and further outlined in Section 5.3.1.1. The term $e_{k,p}$ represents additive noise and impairments and the scalar complex channel coefficient from transmit antenna q to receive antenna p is denoted $H_{k,p,q}$. The channel coefficients are samples of the time- and space-dependent transfer function introduced in Section 3.6.2.5, and in Section 5.3.2 this will be further pursued to explore the spatial structure of the MIMO channel.

Next, introduce

$$y_k = \begin{bmatrix} y_{k,1} & y_{k,2} & \cdots & y_{k,N_r} \end{bmatrix}^T, \tag{5.18}$$

$$x_k = \begin{bmatrix} x_{k,1} & x_{k,2} & \cdots & x_{k,N_t} \end{bmatrix}^T, \tag{5.19}$$

$$e_k = \begin{bmatrix} e_{k,1} & e_{k,2} & \cdots & e_{k,N_r} \end{bmatrix}^T, \tag{5.20}$$

and let the MIMO channel matrix containing all the channels between transmit and receive antenna pairs be defined as

$$H_k = \begin{bmatrix} H_{k,1,1} & H_{k,1,2} & \cdots & H_{k,1,N_t} \\ H_{k,2,1} & H_{k,2,2} & & \vdots \\ \vdots & & \ddots & \\ H_{k,N_r,1} & \cdots & & H_{k,N_r,N_t} \end{bmatrix}. \tag{5.21}$$

With these definitions, the set of equations in (5.17) can equivalently be represented with Eq. (5.16). An illustration is given in Fig. 5.7 for the basic case with $N_t = N_r = 2$.

Before exploring the spatial structure of the MIMO channel matrices $\{H_k\}$, precoding in the transmitter and spatial combining in the receiver are next introduced.

5.3.1.1 Precoding and spatial combining

With the basic MIMO model in place, the functionality of the precoding operation at the transmitter side as well as part of the equalization, referred to as spatial combining, in the receiver as depicted in Fig. 5.6 can be outlined. This functionality is also referred to as transmitter beamforming and receiver beamforming respectively and is further described in detail in Section 6.1. One reason for briefly introducing the functionality already here is to facilitate the discussion on what the term beam means in Section 5.5.

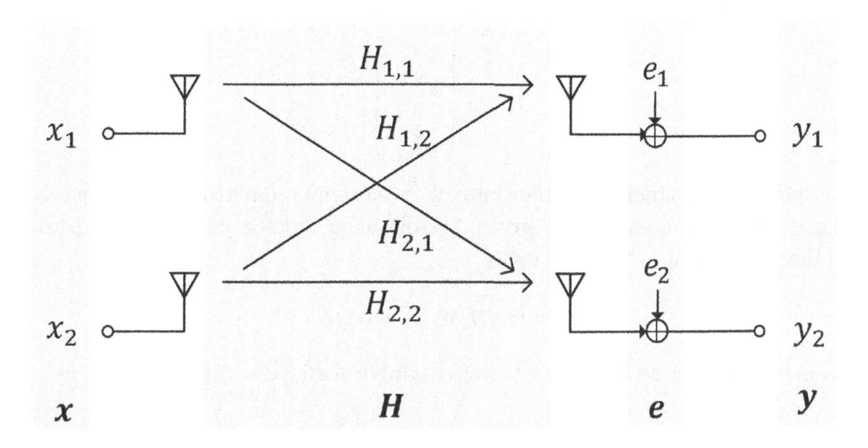

FIGURE 5.7

Illustration of the MIMO channel relation $y = Hx + e$ for a single subcarrier for the case with two receive and two transmit antennas.

For the case with L spatially multiplexed layers (or symbol streams), the precoding operation can for a specific subcarrier k be described with an $N_t \times L$ matrix $\boldsymbol{W}_{t,k}$ defined as

$$\boldsymbol{W}_{t,k} \triangleq \begin{bmatrix} \boldsymbol{w}_{t,k,1} & \boldsymbol{w}_{t,k,2} & \cdots & \boldsymbol{w}_{t,k,L} \end{bmatrix}, \tag{5.22}$$

where the mth column of $\boldsymbol{W}_{t,k}$ is the weight vector for the mth layer. The input symbols to the IFFT in the transmitter can then for subcarrier k be written as

$$\boldsymbol{x}_k = \boldsymbol{W}_{t,k}\boldsymbol{s}_k, \tag{5.23}$$

where the vector $\boldsymbol{s}_k$ represents the QAM symbols for the different layers to be transmitted on subcarrier k,

$$\boldsymbol{s}_k = \begin{bmatrix} s_{k,1} & s_{k,2} & \cdots & s_{k,L} \end{bmatrix}^{\mathrm{T}}. \tag{5.24}$$

This is illustrated in Fig. 5.8, where to ease the notation, the subscript k has been omitted.

With reference to Fig. 5.6, the equalizer will process the output from the OFDM modulators of all the N_r receive antennas. Due to the orthogonality between the subcarriers, the equalizer can process each subcarrier separately. For the case with L layers, a linear equalizer will estimate the QAM symbols on the different layers by forming linear combinations of the signals received by the different antennas. This means that the output of the spatial combiner for a subcarrier k is given by

$$\boldsymbol{z}_k = \boldsymbol{W}_{r,k}\boldsymbol{y}_k, \tag{5.25}$$

where $\boldsymbol{W}_{r,k}$ is an $L \times N_r$ matrix with spatial combining weight vectors, also referred to as receiver beamforming weights or spatial combining weights. The vector

$$\boldsymbol{z}_k = \begin{bmatrix} z_{k,1} & z_{k,2} & \cdots & z_{k,L} \end{bmatrix}^{\mathrm{T}}, \tag{5.26}$$

is the result of the spatial combining and may generally represent various quantities depending on what kind of signal information that needs to be extracted. In the present case of receiving QAM-modulated transmitted symbols, $\boldsymbol{z}_k$ represents estimates of those that are then fed to the QAM demodulator to generate soft bits for channel decoding. The rows of the combining weight matrix $\boldsymbol{W}_{r,k}$ can be recognized as combining weights for the different layers, that is,

$$\boldsymbol{W}_{r,k} \triangleq \begin{bmatrix} \boldsymbol{w}_{r,k,1}^{\mathrm{T}} \\ \boldsymbol{w}_{r,k,2}^{\mathrm{T}} \\ \vdots \\ \boldsymbol{w}_{r,k,L}^{\mathrm{T}} \end{bmatrix}, \tag{5.27}$$

and in Fig. 5.9 the receiver combining is illustrated with subscript k omitted for notational convenience.

Combining (5.23), (5.16), and (5.25) gives the following relation between the spatial combiner output $\boldsymbol{z}_k$ and the transmitted symbol vector $\boldsymbol{s}_k$

$$\boldsymbol{z}_k = \boldsymbol{W}_{r,k}\boldsymbol{H}_k\boldsymbol{W}_{t,k}\boldsymbol{s}_k + \boldsymbol{W}_{r,k}\boldsymbol{e}_k \tag{5.28}$$

To shed some more light on this, the equalized symbol for layer i can be written as

$$z_{k,i} = \boldsymbol{w}_{r,k,i}^{\mathrm{T}}\boldsymbol{H}_k\boldsymbol{w}_{t,k,i}s_{k,i} + \underbrace{\sum_{j=1,j\neq i}^{L} \boldsymbol{w}_{r,k,i}^{\mathrm{T}}\boldsymbol{H}_k\boldsymbol{w}_{t,k,j}s_{k,j}}_{\text{inter-layer interference}} + \underbrace{\boldsymbol{w}_{r,k,i}^{\mathrm{T}}\boldsymbol{e}_k}_{\text{noise}} \tag{5.29}$$

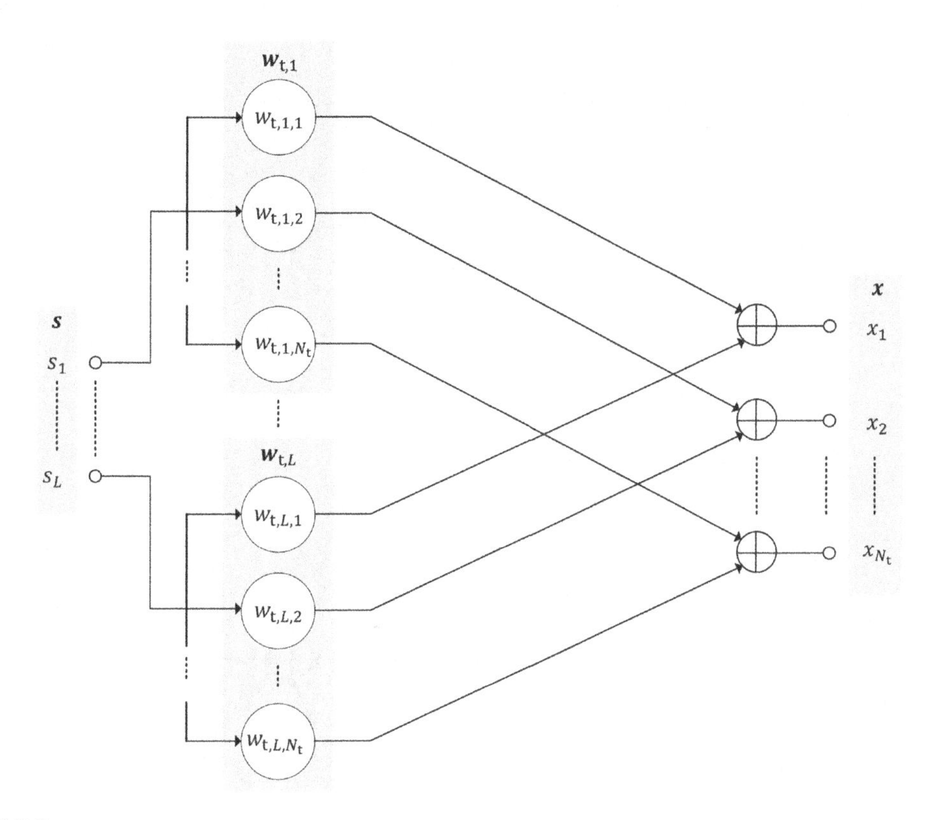

FIGURE 5.8

Illustration of precoding of L symbols into N_t antennas using precoding matrix $W_t \triangleq \begin{bmatrix} w_{t,l} & w_{t,2} & \cdots & w_{t,L} \end{bmatrix}$ for a specific subcarrier k.

This expression shows that the transmitted symbol is impacted by interference from the other transmitted layers. However, this interference can be minimized by selecting the transmit precoder and the receive combiner accordingly, with respect to the channel realization. As mentioned above, the selection of precoder and combiner is addressed further in Sections 6.3.2 and 6.3.3.

5.3.2 SPATIAL STRUCTURE OF THE MIMO CHANNEL

In Section 3.6, a radio channel model based on a directional propagation channel was described. This model for multipath propagation includes not only path delays and powers but also angles of departures and arrivals at the transmitter and receiver side, respectively, as well as polarization properties. The radio channel was then extended in Section 3.6.2 to include also the impact of the antennas at both the transmitter and receiver side, and based on that model a compact formulation using the array response vectors introduced in Section 4.3.2 will here be given for the MIMO channel matrices $\{\boldsymbol{H}_k\}$. In fact, whereas Chapter 4 only considered a single propagation path in a free-space scenario with multiple antennas at the transmitter side, the present formulation includes also

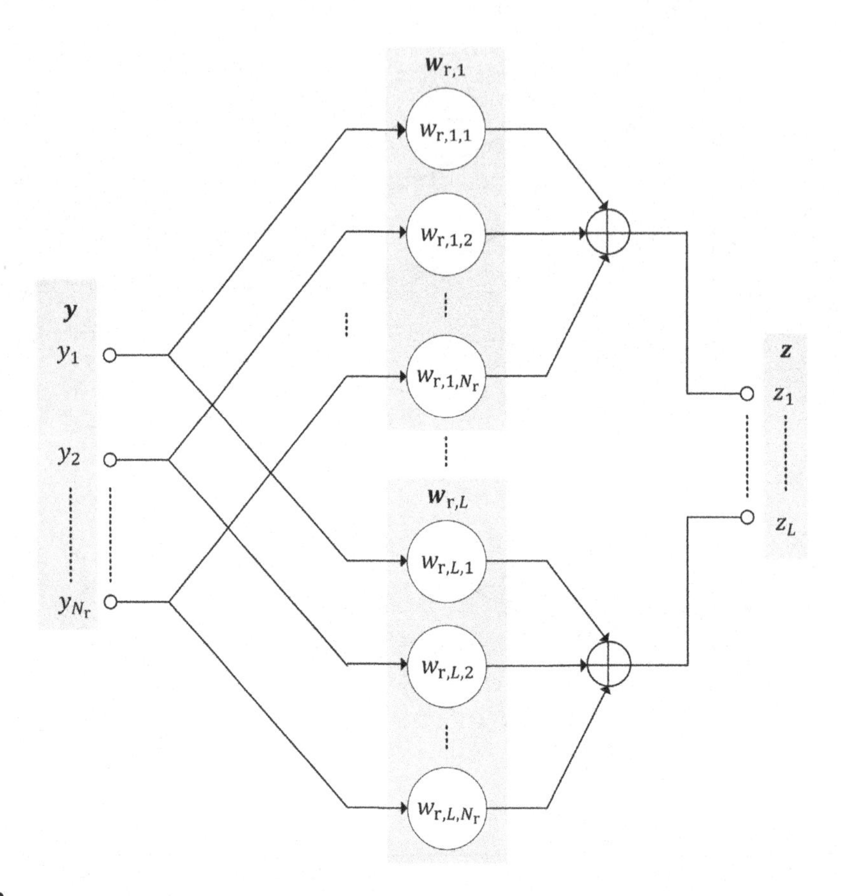

FIGURE 5.9

Illustration of receiver combining N_r antennas into L symbols using combining matrix $\boldsymbol{W}_r \triangleq \begin{bmatrix} \boldsymbol{w}_{r,1} & \boldsymbol{w}_{r,2} & \ldots & \boldsymbol{w}_{r,L} \end{bmatrix}^T$ for a specific subcarrier k.

multiple propagation paths, multiple receiver antennas, and furthermore multiple subcarriers since the expressions are intended for OFDM-based MIMO systems. The analysis and implications of the model are left for Chapter 6 since this is preferably done in the context of different transmissions and reception techniques.

As mentioned above in Section 5.3.1, the channel coefficients for different transmit and receive antenna pairs are not the same, and this comes mainly from the fact that the antennas have different spatial positions. For the purpose of modeling, let the vectors $\{\boldsymbol{d}_{r,p}, p = 1, \ldots, N_r\}$ and $\{\boldsymbol{d}_{t,q}, q = 1, \ldots, N_t\}$ represent the spatial positions of the antennas at the receiver and transmitter, respectively. The positions are relative to local origins for which the propagation paths are defined, that is, the set of vectors $\{\boldsymbol{d}_{r,p}\}$ is relative to one local origin and the set of vectors $\{\boldsymbol{d}_{t,q}\}$ is relative to (potentially) another local origin.

Furthermore, to simplify the description, it is assumed that all transmit antennas have the same complex amplitude pattern $g_t(\theta, \varphi)$ and similarly that all receive antennas have the same complex amplitude pattern $g_r(\theta, \varphi)$. However, to be able to address the common case with arrays with dual-polarized antenna pairs, it is not assumed that all antennas have the same polarization. In what follows, $\underline{\widehat{\psi}}_{r,p}(\theta, \varphi)$ denotes the receive polarization of antenna p at the receiver as a function of the direction (θ, φ) and $\underline{\widehat{\psi}}_{t,q}(\theta, \varphi)$ is the polarization of transmit antenna q,

$$\underline{\widehat{\psi}}_{r,p}(\theta, \varphi) = \begin{bmatrix} \psi_{r,p,\theta} \\ \psi_{r,p,\varphi} \end{bmatrix}, \ \underline{\widehat{\psi}}_{t,q}(\theta, \varphi) = \begin{bmatrix} \psi_{t,q,\theta} \\ \psi_{t,q,\varphi} \end{bmatrix}. \quad (5.30)$$

The antenna radiation pattern in terms of the complex amplitude pattern and polarization was introduced in Section 3.3.4, and an example of positions and polarizations for a uniform planar array with dual-polarized element pairs can be found in Section 4.5.2.

By combing the channel impulse response model from (3.76) in Section 3.6.2.6 with the expression of the subcarrier channel coefficient H_k in (5.15), the channel coefficient of receive antenna q from transmit antenna p for subcarrier k can be expressed as

$$H_{k,p,q} = \sum_n \gamma_{n,p,q} e^{-j2\pi k \Delta_f \tau_n}, \quad (5.31)$$

where

$$\gamma_{n,p,q} = \alpha_n g_r(\theta_{r,n}, \varphi_{r,n}) g_t(\theta_{t,n}, \varphi_{t,n}) \chi_{n,p,q} e^{j(\boldsymbol{d}_{r,p} + \boldsymbol{v}_r t)^T \boldsymbol{k}_{r,n}} e^{j\boldsymbol{d}_{t,q}^T \boldsymbol{k}_{t,n}}, \quad (5.32)$$

and

$$\chi_{n,p,q} = \underline{\widehat{\psi}}_{r,p}^T(\theta_{r,n}, \varphi_{r,n}) \underline{\Psi} n \underline{\widehat{\psi}}_{t,q}(\theta_{t,n}, \varphi_{t,n}), \quad (5.33)$$

since in expression (5.32), $\boldsymbol{d}^T\boldsymbol{k} = \boldsymbol{d} \cdot \boldsymbol{k}$ is the scalar product between two real-valued vectors. The channel coefficients in (5.31) can be recognized as samples of the continuous-time- and space-dependent transfer function defined by (3.75) in Section 3.6.2.5, and the vectors $\boldsymbol{d}_{r,p}$ and $\boldsymbol{d}_{t,q}$ represent the spatial locations where the transfer function is sampled. Also, it may be noted that the amplitude $\gamma_{n,p,q}$ defined in (5.32) is a function of time, and this is further discussed below. Also, similar to the SISO case in Section 5.2.3, and mainly to simplify the notation, the path delays have been redefined from a receiver perspective so that the delay of the first path is zero see (5.13). Moreover, the parameters for path n are defined in Sections 3.6.1 and 3.6.2 and the reader is referred to that section for more details, illustrations, and explanations. However, for the purpose of expressing the MIMO channel using the array response vectors, the meaning of the variables in (5.32) and (5.33) are for convenience repeated here.

- The direction of departure at the transmitter is specified by zenith and azimuth angles $\theta_{t,n}$ and $\varphi_{t,n}$ or equivalently by the unit vector$\widehat{\boldsymbol{r}}_{t,n}$ pointing outwards from the transmitting antenna in the direction of the transmitted wave. The directions are defined in the transmitting antenna array's coordinate system. If the transmitting array is rotated, then so is the coordinate system of the antenna (see also [6]). The wave vector $\boldsymbol{k}_{t,n}$ in (5.32) is defined as $\boldsymbol{k}_{t,n} = (2\pi f_c/c)\widehat{\boldsymbol{r}}_{t,n}$, where f_c is the carrier frequency and c is the speed of light;
- The direction of arrival at the receiver is specified by zenith and azimuth angles $\theta_{r,n}$ and $\varphi_{r,n}$ or equivalently by the unit vector $\widehat{\boldsymbol{r}}_{r,n}$, pointing outwards in a local coordinate system of the receiving array. The wave vector $\boldsymbol{k}_{r,n}$ in (5.31) is defined as $\boldsymbol{k}_{r,n} = (2\pi f_c/c)\widehat{\boldsymbol{r}}_{r,n}$ and points in the direction opposite to the propagating wave;

- The complex amplitude α_n includes phase shifts and attenuation along propagation path n such that $p_n = |\alpha_n|^2$ is the power of the path;
- The term $\chi_{n,p,q}$ defined in (5.33) includes the impact of the transmitter and receiver antenna polarizations as well as the relative attenuations and phase shifts between the different transverse field components through the 2×2 polarization scattering matrix $\underline{\boldsymbol{\Psi}}_n$ for path n defined in Section 3.6.1;
- The receiver antennas are moving with a velocity described by the vector $\boldsymbol{v}_\mathrm{r}$. Hence, the positions of the receiver antennas depend on time t and the location of receiver antenna p, and is given by $\boldsymbol{d}_{\mathrm{r},p} + \boldsymbol{v}_\mathrm{r} t$. As mentioned above, this means that the $\gamma_{n,p,q}$ defined in (5.32) is a function of time, and it should be stressed that for the model to be valid, the speed is assumed to be sufficiently low so that channel variations during an OFDM symbol can be neglected.

Next, the complex amplitude of path n as defined in (5.32) is further examined. As described in Section 3.6.2.3, the factor

$$e^{j(\boldsymbol{d}_{\mathrm{r},p}+\boldsymbol{v}_\mathrm{r} t)^\mathrm{T}\boldsymbol{k}_{\mathrm{r},n}} e^{j\boldsymbol{d}_{\mathrm{t},q}^\mathrm{T}\boldsymbol{k}_{\mathrm{t},n}}, \tag{5.34}$$

represents a phase shift stemming from differences in path lengths or equivalently propagation delays for the different antennas under the assumption that the paths to the different antennas appear parallel. This is illustrated in Fig. 5.10.

Furthermore, the array response vector defined in Section 4.3.2 includes the impact of the antenna amplitude pattern as well as the phase shift due to the propagation delay difference. More specifically, with

$$\boldsymbol{a}_\mathrm{t}(\theta_{\mathrm{t},n}, \varphi_{\mathrm{t},n}) = g_\mathrm{t}(\theta_{\mathrm{t},n}, \varphi_{\mathrm{t},n}) \begin{bmatrix} e^{j\boldsymbol{d}_{\mathrm{t},1}^\mathrm{T}\boldsymbol{k}_{\mathrm{t},n}} & e^{j\boldsymbol{d}_{\mathrm{t},2}^\mathrm{T}\boldsymbol{k}_{\mathrm{t},n}} & \dots & e^{j\boldsymbol{d}_{\mathrm{t},N_\mathrm{t}}^\mathrm{T}\boldsymbol{k}_{\mathrm{t},n}} \end{bmatrix}^\mathrm{T}, \tag{5.35}$$

$$\boldsymbol{a}_\mathrm{r}(\theta_{\mathrm{r},n}, \varphi_{\mathrm{r},n}) = g_\mathrm{r}(\theta_{\mathrm{r},n}, \varphi_{\mathrm{r},n}) \begin{bmatrix} e^{j\boldsymbol{d}_{\mathrm{r},1}^\mathrm{T}\boldsymbol{k}_{\mathrm{r},n}} & e^{j\boldsymbol{d}_{\mathrm{r},2}^\mathrm{T}\boldsymbol{k}_{\mathrm{r},n}} & \dots & e^{j\boldsymbol{d}_{\mathrm{r},N_\mathrm{r}}^\mathrm{T}\boldsymbol{k}_{\mathrm{r},n}} \end{bmatrix}^\mathrm{T}, \tag{5.36}$$

Eq. (5.32) can be rewritten as

$$\gamma_{n,p,q} = \alpha_n [\boldsymbol{a}_\mathrm{r}(\theta_{\mathrm{r},n}, \varphi_{\mathrm{r},n})]_p [\boldsymbol{a}_\mathrm{t}(\theta_{\mathrm{t},n}, \varphi_{\mathrm{t},n})]_q \chi_{n,p,q} e^{j\boldsymbol{v}_\mathrm{r}^\mathrm{T}\boldsymbol{k}_{\mathrm{r},n} t}, \tag{5.37}$$

where $[\boldsymbol{a}_\mathrm{r}(\theta, \varphi)]_p$ is element p of the array response vector for the receiving array and $[\boldsymbol{a}_\mathrm{t}(\theta, \varphi)]_q$ is element q of the array response vector for the transmitting array. Furthermore, by defining

$$\boldsymbol{\chi}_n \triangleq \begin{bmatrix} \chi_{n,1,1} & \chi_{n,1,2} & \cdots & \chi_{n,1,N_\mathrm{t}} \\ \chi_{n,2,1} & \chi_{n,2,2} & & \vdots \\ \vdots & & \ddots & \\ \chi_{n,N_\mathrm{r},1} & \cdots & & \chi_{n,N_\mathrm{r},N_\mathrm{t}} \end{bmatrix} \tag{5.38}$$

and the Doppler frequency for path n, $f_{\mathrm{D},n}$, such that

$$2\pi f_{\mathrm{D},n} = \boldsymbol{v}_\mathrm{r}^\mathrm{T}\boldsymbol{k}_{\mathrm{r},n}, \tag{5.39}$$

the MIMO channel matrix $\boldsymbol{H}_k$ in (5.21) containing all the $N_\mathrm{r} N_\mathrm{t}$ channel coefficients for subcarrier k can be expressed as

$$\boldsymbol{H}_k = \sum_n \alpha_n e^{j2\pi f_{\mathrm{D},n} t} e^{-j2\pi k \Delta_f \tau_n} \left(\boldsymbol{a}_\mathrm{r}(\theta_{\mathrm{r},n}, \varphi_{\mathrm{r},n}) \boldsymbol{a}_\mathrm{t}^\mathrm{T}(\theta_{\mathrm{t},n}, \varphi_{\mathrm{t},n})\right) \odot \boldsymbol{\chi}_n \tag{5.40}$$

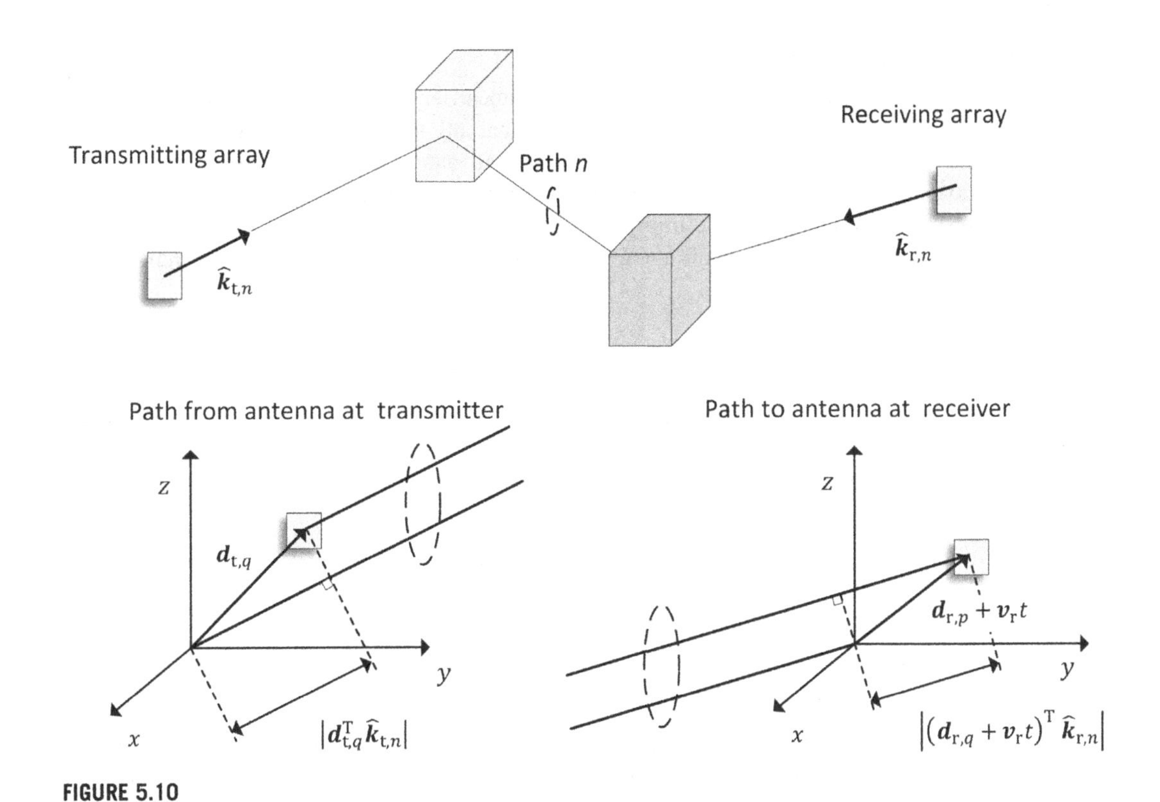

FIGURE 5.10

For each propagation path (top), the path lengths are slightly different at the transmitter (bottom left) and at the receiver (bottom right) depending on the antenna positions.

where $\odot$ denotes the Hadamard product (see Appendix 1) use appropriate size of hadamard product, not superbig.

Note that (5.40) is derived under a narrowband assumption (see also Sections 3.6.2.5 and 4.2.5), and that the array response vectors strictly speaking also are a function of the frequency or equivalently the wavelength. This means that if the bandwidth of the baseband signal $x(t)$ in (5.2) is large, so that the wavelength varies over the bandwidth, the model expression may lose its accuracy. In such cases, the bandwidth can be divided into parts small enough for the involved approximations to be valid (see also [6]).

Furthermore, the array response vector can be generalized to include the case where the elements of the array have different amplitude patterns and in that case the formulation above can be seen as a vector matrix formulation of the fast fading model in [6] for the case with OFDM modulation and demodulation. However, in practice, regular antenna structures, such as uniform linear or planar arrays introduced in Chapter 4 are often used, and therefore such special but common cases are used also in the following. Even though such array structures are idealized, they serve as good models for well-designed antennas and are often used in design and analysis, as they allow tractable analysis and give valuable insights into antenna behavior.

5.3.2.1 Single-polarized arrays

The expression for the MIMO channel for a certain subcarrier may be simplified for antenna arrangements using a single polarization. This means that all the antennas at the receiver are assumed to have the same polarization, $\underline{\widehat{\psi}}_{\mathrm{r}}(\theta, \varphi)$ and that all antennas at the transmitter have the same polarization, $\underline{\widehat{\psi}}_{\mathrm{t}}(\theta, \varphi)$ although the transmitter and receiver antennas may have different polarization. This restriction does, however, imply that the polarization coefficient $\chi_{n,p,q}$ in (5.33) is equal for all pairs of receive and transmit antennas for path n and therefore the antenna indices can be dropped and $\chi_{n,p,q}$ simplifies as

$$\chi_n = \underline{\widehat{\psi}}_{\mathrm{r}}^{\mathrm{T}}(\theta_{\mathrm{r},n}, \varphi_{\mathrm{r},n}) \underline{\boldsymbol{\Psi}}_n \underline{\widehat{\psi}}_{\mathrm{t}}(\theta_{\mathrm{t},n}, \varphi_{\mathrm{t},n}).$$

This in turn leads to that all elements of $\boldsymbol{\chi}_n$ defined in (5.38) become equal and that the expression for the channel in (5.40) can be written as the following weighted sum of outer products of the transmit-receive array response vectors:

$$\boldsymbol{H}_k = \sum_n \lambda_{n,k}(t) \chi_n \boldsymbol{a}_{\mathrm{r}}(\theta_{\mathrm{r},n}, \varphi_{\mathrm{r},n}) \boldsymbol{a}_{\mathrm{t}}^{\mathrm{T}}(\theta_{\mathrm{t},n}, \varphi_{\mathrm{t},n}), \tag{5.41}$$

where the new variable $\lambda_{n,k}(t)$ is introduced for simplicity, and defined as

$$\lambda_{n,k}(t) \triangleq \alpha_n e^{j2\pi f_{\mathrm{D},n} t} e^{-j2\pi k \Delta_{\mathrm{f}} \tau_n}. \tag{5.42}$$

To gain some insight into the channel expression (5.41), the case with a single path and a single receive antenna, $N_{\mathrm{r}} = 1$ is now further considered. In this case, the channel matrix $\boldsymbol{H}_k$ becomes a row vector, $\boldsymbol{h}_k^{\mathrm{T}}$, and the received signal y_k is a scalar as is also the additive noise e_k. Hence, for the case with single path and a single receive antenna, $N_{\mathrm{r}} = 1$, the received signal can be expressed as

$$y_k = \boldsymbol{h}_k^{\mathrm{T}} \boldsymbol{x}_k + e_k, \tag{5.43}$$

with

$$\boldsymbol{h}_k = \lambda_{1,k}(t) \chi_1 g_{\mathrm{r}}(\theta_{\mathrm{r}}, \varphi_{\mathrm{r}}) \boldsymbol{a}_{\mathrm{t}}(\theta_{\mathrm{t}}, \varphi_{\mathrm{t}}), \tag{5.44}$$

where $\theta_{\mathrm{r}} \triangleq \theta_{\mathrm{r},1}$, $\varphi_{\mathrm{r}} \triangleq \varphi_{\mathrm{r},1}$, $\theta_{\mathrm{t}} \triangleq \theta_{\mathrm{t},1}$, and $\varphi_{\mathrm{t}} \triangleq \varphi_{\mathrm{t},1}$ are introduced for notational simplicity.

The channel $\boldsymbol{h}_k$ is thus, to within a complex scaling factor, given by the array response vector, and the case with a single propagation path appears at the transmitter side as free-space propagation (see also Section 4.3.2). In this case, the only difference between the channels from different transmit antennas to the receive antenna is the propagation delay and this delay is indeed captured by phases of the elements of the array response vector. This property holds for all OFDM subcarriers as least if the bandwidth is not too large, even though the scalar $\lambda_{1,k}$ varies with subcarrier index when $\tau_1 \neq 0$.

When there are multiple propagation paths, there is more than one term in the sum over paths, so angular spread acts as a superposition of transmit array response vectors in different directions. Furthermore, when there are multiple receive antennas, the outer product expression in (5.41) is a generalization of the single path single receive antenna case in (5.44), to a superposition of outer products of array response vectors with different transmit and receive directions.

5.3.2.1.1 Free-space single path case

Another important special case of the single path case of (5.41) is the case with free-space propagation and stationary transmit and receive antennas at the respective origins. The free-space channel was introduced in Section 3.6.2.6, substantially used in Chapter 4 and will be related to in Chapter 6.

For the free-space case, the channel $\boldsymbol{H}_k$ in (5.41) is next further simplified. First, it is assumed that the single path has zero delay, $\tau_1 = 0$, see also discussion concerning (5.11)–(5.13). Furthermore, since the antennas are stationary the Doppler frequency is zero, $f_{\mathrm{D},n} = 0$. Then, (5.42) simplifies according to

$$\lambda_{1,k}(t) = \alpha_n \triangleq \alpha. \tag{5.45}$$

Moreover, as $\underline{\boldsymbol{\Psi}}_1$ is a 2×2 identity matrix for the free-space channel, χ_1 becomes

$$\chi_1 = \widehat{\underline{\psi}}_{\mathrm{r}}^{\mathrm{T}}(\theta_{\mathrm{r},1}, \varphi_{\mathrm{r},1}) \underline{\boldsymbol{\Psi}} \widehat{\underline{\psi}}_{\mathrm{t}}(\theta_{\mathrm{t},1}, \varphi_{\mathrm{t},1}) = \psi_{\mathrm{r},\theta}\psi_{\mathrm{t},\theta} + \psi_{\mathrm{r},\varphi}\psi_{\mathrm{t},\varphi} \tag{5.46}$$

where $\widehat{\underline{\psi}}_{\mathrm{t}}(\theta_{\mathrm{t},1}, \varphi_{\mathrm{t},1}) = \left[\psi_{\mathrm{t},\theta} \quad \psi_{\mathrm{t},\varphi}\right]^{\mathrm{T}}$ and $\widehat{\underline{\psi}}_{\mathrm{r}}(\theta_{\mathrm{r},1}, \varphi_{\mathrm{r},1}) = \left[\psi_{\mathrm{r},\theta} \quad \psi_{\mathrm{r},\varphi}\right]^{\mathrm{T}}$. Then, by using (5.45) and (5.46), $\boldsymbol{H}_k$ in (5.41) can be written as

$$\boldsymbol{H}_k = \alpha\left(\psi_{\mathrm{r},\theta}\psi_{\mathrm{t},\theta} + \psi_{\mathrm{r},\varphi}\psi_{\mathrm{t},\varphi}\right)\boldsymbol{a}_{\mathrm{r}}(\theta_{\mathrm{r}}, \varphi_{\mathrm{r}})\boldsymbol{a}_{\mathrm{t}}^{\mathrm{T}}(\theta_{\mathrm{t}}, \varphi_{\mathrm{t}}). \tag{5.47}$$

A major difference as compared to Chapter 4 is that the propagation delay has been omitted here and that a single polarization aligned isotropic antenna was considered in Chapter 4, that is, $\chi_1 = 1$ and $\boldsymbol{a}_{\mathrm{r}}(\theta_r, \varphi_r) = 1$.

5.3.2.2 Dual-polarized arrays

A dual-polarized uniform linear array was defined in Section 4.5.2, and for such an array the first $N/2$ antennas have one polarization and the last $N/2$ antennas have another polarization. Furthermore, each position in such an array is shared between two antennas with different polarizations. For the case with dual-polarized arrays at both transmitter and receiver side, the polarization of the antennas becomes

$$\widehat{\underline{\psi}}_{\mathrm{r},p}(\theta, \varphi) = \begin{cases} \widehat{\underline{\psi}}_{\mathrm{r,A}}(\theta, \varphi) p = 1, \ldots, N_r/2 \\ \widehat{\underline{\psi}}_{\mathrm{r,B}}(\theta, \varphi) p = N_r/2 + 1, \ldots, N_{\mathrm{r}}, \end{cases} \tag{5.48}$$

and

$$\widehat{\underline{\psi}}_{\mathrm{t},q}(\theta, \varphi) = \begin{cases} \widehat{\underline{\psi}}_{\mathrm{t,A}}(\theta, \varphi) q = 1, \ldots, N_t/2 \\ \widehat{\underline{\psi}}_{\mathrm{t,B}}(\theta, \varphi) q = N_t/2 + 1, \ldots, N_{\mathrm{t}}. \end{cases} \tag{5.49}$$

In this case, the polarization coefficient in (5.33) can take on four different values and these four different values are collected into a 2×2 matrix for each path n as

$$\chi_n \triangleq \begin{bmatrix} \hat{\underline{\psi}}_{\mathrm{r,A}}^{\mathrm{T}}(\theta_{\mathrm{r},n},\varphi_{\mathrm{r},n})\underline{\Psi}_n\hat{\underline{\psi}}_{\mathrm{t,A}}(\theta_{\mathrm{t},n},\varphi_{\mathrm{t},n}) & \hat{\underline{\psi}}_{\mathrm{r,A}}^{\mathrm{T}}(\theta_{\mathrm{r},n},\varphi_{\mathrm{r},n})\underline{\Psi}_n\hat{\underline{\psi}}_{\mathrm{t,B}}(\theta_{\mathrm{t},n},\varphi_{\mathrm{t},n}) \\ \hat{\underline{\psi}}_{\mathrm{r,B}}^{\mathrm{T}}(\theta_{\mathrm{r},n},\varphi_{\mathrm{r},n})\underline{\Psi}_n\hat{\underline{\psi}}_{\mathrm{t,A}}(\theta_{\mathrm{t},n},\varphi_{\mathrm{t},n}) & \hat{\underline{\psi}}_{\mathrm{r,B}}^{\mathrm{T}}(\theta_{\mathrm{r},n},\varphi_{\mathrm{r},n})\underline{\Psi}_n\hat{\underline{\psi}}_{\mathrm{t,B}}(\theta_{\mathrm{t},n},\varphi_{\mathrm{t},n}) \end{bmatrix}. \tag{5.50}$$

In this case, the matrix χ_n defined in (5.38) gets a block structure, and it can be shown that the MIMO channel in (5.40) can be expressed as

$$\boldsymbol{H}_k = \sum_n \lambda_{n,k}(t)\chi_n \otimes \left(\boldsymbol{a}_{\mathrm{r}/2}(\theta_{\mathrm{r},n},\varphi_{\mathrm{r},n})\boldsymbol{a}_{\mathrm{t}/2}^{\mathrm{T}}(\theta_{\mathrm{t},n},\varphi_{\mathrm{t},n})\right), \tag{5.51}$$

where $\boldsymbol{a}_{\mathrm{r}/2}(\theta,\varphi)$ is the array response vector of size $\mathrm{N_r}/2 \times 1$ for the positions of the receiver antenna pairs and $\boldsymbol{a}_{\mathrm{t}/2}(\theta,\varphi)$ is the array response vector of size $\mathrm{N_t}/2 \times 1$ for the transmit antenna pairs. Recall from the definition of a UPA with dual-polarized element pairs in Section 4.5.2 that there are two elements for each position; thus each element location occurs twice at transmit and at receive sides explaining why the half-size transmit/receive array response vector is repeated twice to cover all transmit/receive antennas. Furthermore, the expression in (5.51) with the Kronecker product allows for partitioning the channel matrix into four different parts,

$$\boldsymbol{H}_k = \begin{bmatrix} \boldsymbol{H}_{k,1,1} & \boldsymbol{H}_{k,1,2} \\ \boldsymbol{H}_{k,2,1} & \boldsymbol{H}_{k,2,2} \end{bmatrix} \tag{5.52}$$

where the four blocks are formed as

$$\boldsymbol{H}_{k,i,j} = \sum_n \lambda_{n,k}(t)\left[\chi_n\right]_{i,j}\boldsymbol{a}_{\mathrm{r}/2}(\theta_{\mathrm{r},n},\varphi_{\mathrm{r},n})\boldsymbol{a}_{\mathrm{t}/2}^{\mathrm{T}}(\theta_{\mathrm{t},n},\varphi_{\mathrm{t},n}). \tag{5.53}$$

A comparison with (5.41) reveals that all the four parts of (5.45) can be seen as MIMO channels for the case with single-polarized arrays at both the receiver and transmitter. In fact, spatial properties in terms of directions, or array response vectors, are the same for the four different parts, but the linear combinations are different since the elements of χ_n are different.

In the context of beamforming, the fact that the spatial properties of different parts of (5.45) are similar suggests that similar or even the same beamforming weights can be used, like in the example in Section 4.5.2.1, where it was shown that a dual-polarized array allows control of the polarization as well. However, there are also opportunities for spatial multiplexing utilizing the polarization. To see this, the special case with free-space propagation in Section 5.3.2.1.1 is revisited for the case that the receiver has a single dual-polarized antenna pair and furthermore that the polarizations of the receiver and transmitter antennas are orthogonal and aligned so that χ_1 becomes a scaled identity matrix

$$\chi_1 \sim \begin{bmatrix} 1 & 0 \\ 0 & 1 \end{bmatrix}.$$

In this case

$$\boldsymbol{H}_k \sim \begin{bmatrix} \boldsymbol{a}_{\mathrm{t}/2}^{\mathrm{T}}(\theta_{\mathrm{t},1},\varphi_{\mathrm{t},1}) & 0 \\ 0 & \boldsymbol{a}_{\mathrm{t}/2}^{\mathrm{T}}(\theta_{\mathrm{t},1},\varphi_{\mathrm{t},1}) \end{bmatrix}. \tag{5.54}$$

The meaning of the zeros in the channel matrix (5.47) is that there are two parallel and noninteracting channels, one for each polarization, and within each such a channel, classical beamforming introduced in Section 4.4.1 can be used.

The implications of the structure revealed by (5.51) and (5.52) are discussed further in Chapter 6, when it comes to how to design efficient codebooks for MIMO precoding.

5.3.3 SPATIAL CORRELATION MATRICES

As was discussed in Chapter 3, due to the superposition of the propagation paths, the magnitude and phase of the elements of the channel matrices will vary as the UE moves. For small movements, in the order of a few wavelengths, significant channel variations may occur, and such variations are sometimes referred to as fast fading variations (see Section 3.5.3.1 and Figs. 3.19 and 3.21). Note that the wavelength at 2 GHz carrier frequency is 15 cm while at 28 GHz it is 1 cm. For such small-scale movements, some channel defining parameters such as pathloss, delay spread and angular directions at both the transmitter and receiver side remain essentially constant. These parameters are thus referred to as large-scale parameters (see also Section 3.6.3.5 on how such parameters are used to define stochastic geometric channel models).

The spatial channel correlation properties introduced in Section 3.6.2.8 can be interpreted as how similar the channel coefficients are *on average* in local vicinity of a current location, for example, for different closely spaced antennas. In fact, the cross-correlations are determined by averaging out the fast fading variations and what then remains is an expression for the spatial channel correlation in terms of the channel's large-scale properties. Hence, the channel has a spatial structure even when the fast fading variations have been averaged out in the correlation metrics. In practice such large-scale properties can be estimated by averaging in time and frequency and be used by transmission and reception schemes as discussed further in Chapter 6.

To illustrate this structure, the so-called channel correlation matrices containing all the different correlations between antenna pairs at the receiver and transmitter side are here obtained by using the expectation operator E{} with respect to time, or equivalently position, as

$$\boldsymbol{R}_{\mathrm{r}} = \mathrm{E}\{\boldsymbol{H}_k \boldsymbol{H}_k^*\} \tag{5.55}$$

and

$$\boldsymbol{R}_{\mathrm{t}} = \mathrm{E}\{\boldsymbol{H}_k^{\mathrm{T}} (\boldsymbol{H}_k^{\mathrm{T}})^*\}. \tag{5.56}$$

Here, $\boldsymbol{R}_{\mathrm{r}}$ is the receive correlation matrix of size $N_{\mathrm{r}} \times N_{\mathrm{r}}$ and it can be written as a sum of the correlation matrices for all individual transmit antennas. Similarly, the transmit correlation matrix $\boldsymbol{R}_{\mathrm{t}}$ is of size $N_{\mathrm{t}} \times N_{\mathrm{t}}$ and can be expressed as a sum of the correlation matrices for all receiver antennas. As will be seen below, the correlation matrices are the same for all subcarriers, at least as long as the bandwidth is not too large so that the frequency dependency of the array response vectors can be neglected (see also Section 3.6.2.5). In practice, this means that the correlation matrices can be estimated by averaging both in time and in frequency over subcarriers.

For the case with single-polarized antennas, the receive spatial correlation matrix $\boldsymbol{R}_{\mathrm{r}}$ is determined by inserting (5.41) and (5.37) into (5.55) and applying the expectation operator with respect to time. The cross terms then vanish under the assumption of different directions leading to different Doppler shifts (cf. (5.39)), $\mathrm{E}\left\{e^{j2\pi(f_{\mathrm{D},n}-f_{\mathrm{D},m})t}\right\} = 0$ for $f_{\mathrm{D},n} \neq f_{\mathrm{D},m}$. This then leads to the receive spatial correlation matrix

$$R_r = \sum_n p_{r,n} a_r(\theta_{r,n}, \varphi_{r,n}) a_r^*(\theta_{r,n}, \varphi_{r,n}) \quad (5.57)$$

where relative contribution from each path n is scaled as

$$p_{r,n} = |\alpha_n|^2 |\chi_n|^2 \| a_t(\theta_{t,n}, \varphi_{t,n}) \|^2. \quad (5.58)$$

As can be seen, the receive covariance matrix is the same for all subcarriers since the covariance matrix does not depend on the subcarrier index k as $|\lambda_{n,k}(t)| = |\alpha_n|$, and the correlation matrix may with appropriate assumptions also be defined as an expectation with respect to (baseband) frequency. Alternatively, the elements of the receive correlation matrix can be expressed using samples of the spatial correlation function at the receiver side defined in (3.83) in Section 3.6.8.2.

In any case, the transmitter side impacts the receiver side correlation matrix only via a scalar term, and therefore does not affect the structure of the receive correlation matrix. The large-scale parameters including the directions, relations between polarizations and amplitudes including the impact of pathloss as well as the path delay are regarded as constant for movements in the vicinity of the receiver and are not affected by the expectation operator.

In a similar manner, the following expression for transmit spatial correlation matrix can be obtained as

$$R_t = \sum_n p_{t,n} a_t(\theta_{t,n}, \varphi_{t,n}) a_t^*(\theta_{t,n}, \varphi_{t,n}) \quad (5.59)$$

where

$$p_{t,n} = |\alpha_n|^2 |\chi_n|^2 \| a_r(\theta_{r,n}, \varphi_{r,n}) \|^2.$$

Similar to the case with the receive correlation matrix, the elements of the transmit correlation matrix can be expressed using samples of the spatial correlation function defined in (3.82) in Section 3.6.8.2.

Moreover, both the receive and transmit correlation matrices can alternatively be expressed more compactly with matrices. For example, the receive correlation matrix can be written as

$$R_r = A_r P_r A_r^* \quad (5.60)$$

where A_r contains the receive array response vectors for different paths as columns and where P_r is a diagonal matrix with the factors $p_{r,n}$ on the diagonal. Using linear algebra, this means that the columns of the channel matrix H_k fall into the space spanned by the columns of the matrix A_r, that is, the receive array response vectors. Also, the expression (5.60) resembles an eigenvalue decomposition (EVD) of the correlation matrix in case the array response vectors in A_r are close to orthogonal. In case the number of antennas is large, then the columns can be approximated as close to orthogonal, and the expression is close to an EVD. Also, if some of the array response vector directions are very close to each other (much smaller separation than the main lobe width given by the array response vector), then multiple array response vectors can be assumed to be lumped together. Note that this discussion is not strict in a mathematical sense but should be understood as a way to visualize the underlying properties of the correlation matrix.

The main purpose of defining and writing out expressions for the correlation matrices is that the properties of them will be used in Chapters 6 and 13. It can be observed from the above analysis that the array response vectors dominate and highly influence the spatial characteristics of the channel model. Furthermore, some observations are next done.

- As mentioned above, the elements of the correlation matrices can be expressed in terms of the spatial correlation functions introduced in Section 3.6.2.8 and illustrated in Fig. 3.28. As can be seen, elements corresponding to large antenna spacing tend to have lower magnitude relative to elements corresponding to smaller spacing. Furthermore, for a fixed element separation, the magnitude tends to decrease with increasing angle spread. The reason is that larger antenna spacing leads to faster phase shifts between the elements of the array response vectors $\boldsymbol{a}_{\mathrm{r}}(\theta_{\mathrm{r},n}, \varphi_{\mathrm{r},n})$ creating more randomness when the array response vectors of different directions are summed together over the different paths n. The receive correlation matrix has therefore a tendency to be diagonally heavy for large antenna spacings and with increasing number of paths;
- The off-diagonal elements are, however, complex-valued, and the phase of these elements depends on the directions through the array response vectors. To see this, consider the special case with a single path. In this case the receive correlation matrix will have the form

$$\boldsymbol{R}_{\mathrm{r}} \sim \boldsymbol{a}_{\mathrm{r}}(\theta, \varphi)\boldsymbol{a}_{\mathrm{r}}^{*}(\theta, \varphi). \tag{5.61}$$

  This means that the channel has a single dimensional subspace and hence all columns of the channel matrix $\boldsymbol{H}_k$ (for the case $\boldsymbol{R}_{\mathrm{r}}$ is the receive correlation matrix) are proportional to this single array response vector $\boldsymbol{a}_{\mathrm{r}}(\theta, \varphi)$. For the case with a vertical uniform linear array as defined in Section 4.3.3 it can be shown that

$$[\boldsymbol{R}_{\mathrm{r}}]_{m_1,m_2} \sim \left|g_{\mathrm{r}}(\theta, \varphi)\right|^2 e^{j\frac{2\pi}{\lambda}(m_1-m_2)d_{\mathrm{v}}\cos\theta} \tag{5.62}$$

  where d_{v} is the element separation. Thus the phase of the off-diagonal elements is determined by the angles of arrival. Alternatively phrased, there is direction information available in the phases of the off-diagonal elements;
- Often, a complex Gaussian process is used to model the MIMO channel elements, and the real-valued amplitude of the elements then follows a Rayleigh distribution, or in case there is a nonzero mean a Rice distribution (cf. Section 3.5.3.1). This is not the same thing as saying that all the channels are uncorrelated; there is, depending on the large-scale parameters, in the general case correlation, or structure between the elements. If the distributions of paths are such that the covariance matrix has a small number of dominant eigenvalues relative to the total number of antennas, then all realizations are well approximated as a linear combination of eigenvectors associated with the dominant eigenvectors (see also the EVD interpretation of $\boldsymbol{R}_{\mathrm{r}} = \boldsymbol{A}_{\mathrm{r}}\boldsymbol{P}_{\mathrm{r}}\boldsymbol{A}_{\mathrm{r}}^{*}$ above). When this EVD approximation holds, and the number of large diagonal values of $\boldsymbol{P}_{\mathrm{r}}$ is small, then fewer parameters are needed to describe the long-term properties of the channel and hence the possible channel realizations are fewer compared to the case where elements of $\boldsymbol{P}_{\mathrm{r}}$ have similar magnitude;
- The correlation matrices change according to the large-scale parameters of the propagation; hence they vary slowly in time. They are furthermore approximately the same for all subcarriers. As will be seen in Chapter 6 this can be exploited in the design of robust beamforming schemes.

The above exercise assumed the single-polarized antennas, and the corresponding correlation matrices can be derived also for dual-polarized UPAs. This is, however, not pursued in here.

5.4 RADIATION PATTERN INTERPRETED AS AN EFFECTIVE TWO-ELEMENT CHANNEL

The radiation pattern of an antenna is as described in Section 3.3.4 per definition equal to a 2×1 complex-valued Jones vector $\underline{g}(\theta,\varphi)$ that captures both gain and polarization properties of the radiation for the transmit direction. It is, however, somewhat more difficult to understand how to interpret a radiation pattern in the receive direction. To shed some light on this issue, it is instructive to describe how the radiation pattern can be seen as obtained from a hypothetical measurement setup. Using such an approach, it will be demonstrated that a radiation pattern in receive direction has a direct correspondence to a so-called receiving pattern. Toward this end, consider such measurements in a free-space channel for an antenna array with for simplicity identical elements all having the radiation pattern $\underline{g}(\theta,\varphi)$.

5.4.1 MEASURING RADIATION PATTERN IN TRANSMIT DIRECTION

To measure the radiation pattern of an antenna array in the transmit direction, assume that a dual-polarized antenna pair is used as measurement receiver to measure the received signal transmitted from the antenna array of interest as illustrated in Fig. 5.11. A weight vector $\boldsymbol{w}$ is used to adjust phase and amplitude for each antenna element of a single symbol stream transmitted to the measurement receiver. This is a special case of (5.16) combined with (5.23) and the model of the received signal for one of the antenna elements in the antenna pair can thus be written as

$$y = \boldsymbol{h}^{\mathrm{T}} \boldsymbol{w} s \tag{5.63}$$

where s is a symbol from the symbol stream, the noise term has been neglected, and $\boldsymbol{h}^{\mathrm{T}}$ denotes the single-row channel matrix. The weight vector and symbol s are here power normalized so that $\|\boldsymbol{w}s\|^2 = 1$.

The array response vector of the considered antenna array is assumed to be

$$\boldsymbol{a}(\theta,\varphi) = g(\theta,\varphi)\tilde{\boldsymbol{a}}(\theta,\varphi) \tag{5.64}$$

where the antenna element amplitude gain pattern $g(\theta,\varphi)$ has been factored out leaving $\tilde{\boldsymbol{a}}(\theta,\varphi)$ to denote the array response vector for a corresponding array of lossless isotropic elements. The channel matrix is in this case $1 \times N$ and can from (5.47) combined with (5.64) be written as

$$\boldsymbol{h}^{\mathrm{T}} = \alpha g_{\mathrm{r}}(\theta_{\mathrm{r}},\varphi_{\mathrm{r}})\left(\psi_{\mathrm{r},\theta}\psi_{\mathrm{t},\theta} + \psi_{\mathrm{r},\varphi}\psi_{\mathrm{t},\varphi}\right) g(\theta_{\mathrm{t}},\varphi_{\mathrm{t}})\tilde{\boldsymbol{a}}^{\mathrm{T}}(\theta_{\mathrm{t}},\varphi_{\mathrm{t}}) \tag{5.65}$$

Substituting for the channel in (5.65) into (5.63), gives the received signal

$$y = \alpha g_{\mathrm{r}}(\theta_{\mathrm{r}},\varphi_{\mathrm{r}})\left(\psi_{\mathrm{r},\theta}\psi_{\mathrm{t},\theta} + \psi_{\mathrm{r},\varphi}\psi_{\mathrm{t},\varphi}\right) g(\theta_{\mathrm{t}},\varphi_{\mathrm{t}})\tilde{\boldsymbol{a}}^{\mathrm{T}}(\theta_{\mathrm{t}},\varphi_{\mathrm{t}})\boldsymbol{w}s \tag{5.66}$$

The two antennas 1 and 2 in the receiving measurement antenna pair are assumed lossless and isotropic with orthogonal polarizations. From (3.12) and subsequent expressions it is understood that such an antenna pair can be modeled using radiation patterns that in receive direction corresponds to

$$g_{\mathrm{r},1}(\theta_{\mathrm{r}},\varphi_{\mathrm{r}}) = g_{\mathrm{r}}(\theta_{\mathrm{r}},\varphi_{\mathrm{r}})\psi_{\mathrm{r},\theta}\hat{\boldsymbol{\theta}}_{\mathrm{r}} = \hat{\boldsymbol{\theta}}_{\mathrm{r}}$$

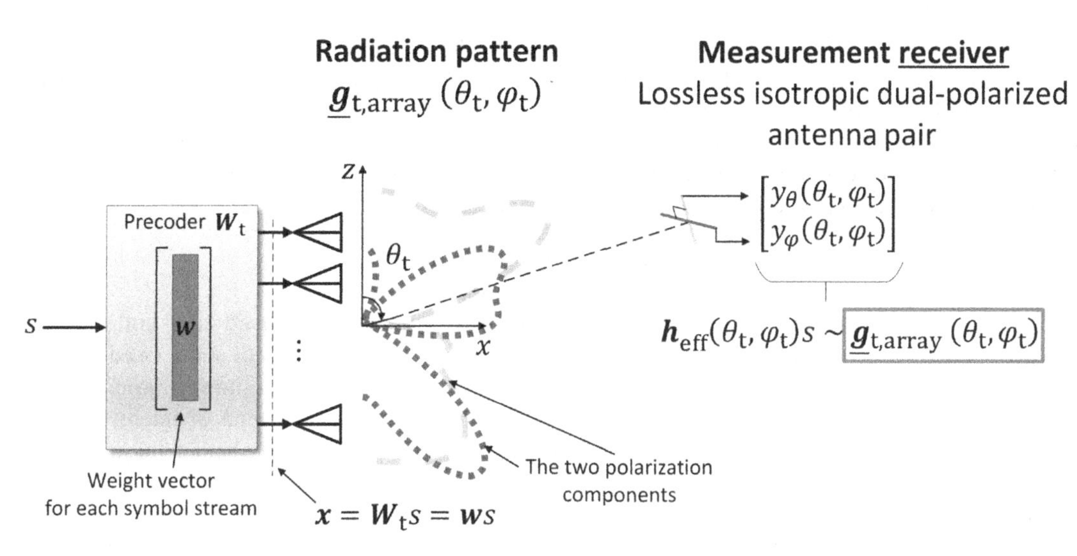

FIGURE 5.11

Measurement setup for determining the radiation pattern of an antenna/antenna array. The radiation pattern turns out to be proportional to the 2 × 1 effective channel between the antenna array and a measurement receiver.

$$g_{\mathrm{r},2}(\theta_{\mathrm{r}},\varphi_{\mathrm{r}}) = g_{\mathrm{r}}(\theta_{\mathrm{r}},\varphi_{\mathrm{r}})\psi_{\mathrm{r},\varphi}\widehat{\boldsymbol{\varphi}}_{\mathrm{r}} = \widehat{\boldsymbol{\varphi}}_{\mathrm{r}} \tag{5.67}$$

which after combining with (5.66) produces two corresponding received signals

$$\begin{aligned} \boldsymbol{y}(\theta_{\mathrm{t}},\varphi_{\mathrm{t}}) \triangleq \begin{bmatrix} y_\theta(\theta_{\mathrm{t}},\varphi_{\mathrm{t}}) \\ y_\varphi(\theta_{\mathrm{t}},\varphi_{\mathrm{t}}) \end{bmatrix} &= \alpha \begin{bmatrix} \psi_{\mathrm{t},\theta} g(\theta_{\mathrm{t}},\varphi_{\mathrm{t}})\tilde{\boldsymbol{a}}^{\mathrm{T}}(\theta_{\mathrm{t}},\varphi_{\mathrm{t}})\boldsymbol{w} \\ \psi_{\mathrm{t},\varphi} g(\theta_{\mathrm{t}},\varphi_{\mathrm{t}})\tilde{\boldsymbol{a}}^{\mathrm{T}}(\theta_{\mathrm{t}},\varphi_{\mathrm{t}})\boldsymbol{w} \end{bmatrix} s \\ &= \alpha \underline{\boldsymbol{g}}(\theta_{\mathrm{t}},\varphi_{\mathrm{t}})\tilde{\boldsymbol{a}}^{\mathrm{T}}(\theta_{\mathrm{t}},\varphi_{\mathrm{t}})\boldsymbol{w}s = \boldsymbol{h}_{\mathrm{eff}}(\theta_{\mathrm{t}},\varphi_{\mathrm{t}})s \end{aligned}$$

where

$$\underline{\boldsymbol{g}}(\theta_{\mathrm{t}},\varphi_{\mathrm{t}}) = g(\theta_{\mathrm{t}},\varphi_{\mathrm{t}})\begin{bmatrix} \psi_{\mathrm{t},\theta} & \psi_{\mathrm{t},\varphi} \end{bmatrix}^{\mathrm{T}}$$

is the Jones representation of the radiation pattern of an element in the antenna array. Here, the angular dependence of the received signals $y_\theta(\theta_{\mathrm{t}},\varphi_{\mathrm{t}})$, $y_\varphi(\theta_{\mathrm{t}},\varphi_{\mathrm{t}})$ is made explicit and

$$\boldsymbol{h}_{\mathrm{eff}}(\theta_{\mathrm{t}},\varphi_{\mathrm{t}}) \triangleq \alpha \underline{\boldsymbol{g}}(\theta_{\mathrm{t}},\varphi_{\mathrm{t}})\tilde{\boldsymbol{a}}^{\mathrm{T}}(\theta_{\mathrm{t}},\varphi_{\mathrm{t}})\boldsymbol{w} \tag{5.68}$$

is the effective 2 × 1 channel between the transmitted symbol s and the outputs in $\boldsymbol{y}$ of the measurement receiver, that is,

$$\boldsymbol{y}(\theta_{\mathrm{t}},\varphi_{\mathrm{t}}) = \boldsymbol{h}_{\mathrm{eff}}(\theta_{\mathrm{t}},\varphi_{\mathrm{t}})s \tag{5.69}$$

Consider now the special case that the transmitting antenna array only has $N = 1$ element. The factor $\tilde{\boldsymbol{a}}^{\mathrm{T}}(\theta_{\mathrm{t}},\varphi_{\mathrm{t}})\boldsymbol{w}$ in (5.68), which corresponds to the array factor in (4.44), then becomes 1 and thus

$$\boldsymbol{h}_{\mathrm{eff}}(\theta_{\mathrm{t}},\varphi_{\mathrm{t}}) = \alpha \underline{\boldsymbol{g}}(\theta_{\mathrm{t}},\varphi_{\mathrm{t}})$$

It is seen that the effective channel is proportional to the Jones vector form $\underline{g}(\theta_t,\varphi_t)$ of the element radiation pattern $g(\theta_t,\varphi_t)$. The channel $\tilde{h}(\theta_t,\varphi_t)$ can be estimated from the received signals and the path loss α can be determined by repeating the measurement but replacing the transmitting antenna with an antenna with known characteristics, for example, a lossless isotropic antenna. Thus, this measurement procedure results in a point on the 2×1 radiation pattern of the single transmitting antenna obtained from the effective channel as

$$\underline{g}(\theta_t,\varphi_t) = \frac{\boldsymbol{h}_{\mathrm{eff}}(\theta_t,\varphi_t)}{\alpha} \tag{5.70}$$

Imagine now a measurement where the measurement receiver is moved and collects the received signals and thus the channel $\boldsymbol{h}_{\mathrm{eff}}(\theta_t,\varphi_t)$ for all points on a sphere with some fixed radius. Clearly, such a measurement procedure using the measurement antenna pair moved around on a sphere in a free-space channel indeed gives the radiation pattern of an antenna of interest in the transmit direction from the effective channel. Note that this is regardless of whether the transmitting antenna is an antenna array or not. The same experiment can therefore be repeated but for an antenna array with $N > 1$ elements and would hence provide the corresponding radiation pattern for an array of antennas. The factor $\tilde{\boldsymbol{a}}^{\mathrm{T}}(\theta_t,\varphi_t)\boldsymbol{w}$ is now no longer equal to one and the 2×1 radiation pattern is understood from (5.70) and (5.68) to become

$$\underline{g}_{\mathrm{t,array}}(\theta_t,\varphi_t) = \frac{\boldsymbol{h}_{\mathrm{eff}}(\theta_t,\varphi_t)}{\alpha} = \underline{g}(\theta_t,\varphi_t)\tilde{\boldsymbol{a}}^{\mathrm{T}}(\theta_t,\varphi_t)\boldsymbol{w} \tag{5.71}$$

in the transmit direction of the array with the weight vector $\boldsymbol{w}$.

Although the analysis was here performed assuming identical elements in the antenna array for notational simplicity, the measurement procedure providing the radiation pattern as an effective channel holds for more general antenna arrays as well.

5.4.2 MEASURING RECEIVING PATTERN IN RECEIVE DIRECTION

The measurement procedure in the previous section gives the radiation pattern in the transmit direction. This provides a strategy to describe what a radiation pattern in the receive direction could correspond to. Consider the same antennas and procedure as above but now instead let the dual-polarized measurement antenna pair be transmitting and the antenna array be receiving. Furthermore, let the signals from the antenna array be linearly combined using the same weight vector $\boldsymbol{w}$ as used for transmitting from the array as shown in Fig. 5.12.

Consider the expression for a spatially combined received signal in (5.28). For a single transmitting antenna, the precoder $\boldsymbol{W}_t$ is 1×1 and can be set equal to 1. Based on this, the signal after the spatial combiner $\boldsymbol{w}$ can be modeled as

$$\begin{aligned} z &= \boldsymbol{w}^{\mathrm{T}}\boldsymbol{H}\boldsymbol{s} = \boldsymbol{w}^{\mathrm{T}}\begin{bmatrix}\boldsymbol{h}_1 & \boldsymbol{h}_2\end{bmatrix}\boldsymbol{s} \\ &= \boldsymbol{w}^{\mathrm{T}}\boldsymbol{h}_1 s_1 + \boldsymbol{w}^{\mathrm{T}}\boldsymbol{h}_2 s_2 = \begin{bmatrix}\boldsymbol{w}^{\mathrm{T}}\boldsymbol{h}_1 & \boldsymbol{w}^{\mathrm{T}}\boldsymbol{h}_2\end{bmatrix}\begin{bmatrix} s_\theta \\ s_\varphi \end{bmatrix} \\ &= \tilde{\boldsymbol{h}}_{\mathrm{eff}}^{\mathrm{T}}(\theta_r,\varphi_r)\boldsymbol{s} \end{aligned} \tag{5.72}$$

where $\tilde{\boldsymbol{h}}_{\text{eff}}^{\text{T}}(\theta_{\text{r}},\varphi_{\text{r}})$ is a 1×2 effective channel and $\boldsymbol{s}$ contains the transmitted symbol from each antenna in the measurement antenna pair. The noise term is again neglected. The channel between one of the transmitting antennas in the antenna pair and the antenna array, $\boldsymbol{h}_m$, is in this case $N\times 1$ and can from (5.47) combined with (5.64) be obtained as

$$\boldsymbol{h}_m = \alpha\tilde{\boldsymbol{a}}(\theta_{\text{r}},\varphi_{\text{r}})g(\theta_{\text{r}},\varphi_{\text{r}})\left(\psi_{\text{r},\theta}\psi_{\text{t},m,\theta} + \psi_{\text{r},\varphi}\psi_{\text{t},m,\varphi}\right)g_{\text{t},m}(\theta_{\text{t}},\varphi_{\text{t}})$$

As seen, this is similar to the channel in (5.65) but transposed, which is natural since reciprocity holds if the receiver and transmitter parts attached to the antenna array are properly aligned, which is the tacit assumption here. See Sections 3.4.1, 6.5, and 12.3.6 for further information about reciprocity and alignment requirements. The effective channel coefficient corresponding to the transmitted symbol s_θ $(m=1)$ or s_φ $(m=2)$ can now be written as

$$\boldsymbol{w}^{\text{T}}\boldsymbol{h}_m = \alpha\boldsymbol{w}^{\text{T}}\tilde{\boldsymbol{a}}(\theta_{\text{r}},\varphi_{\text{r}})g(\theta_{\text{r}},\varphi_{\text{r}})\left(\psi_{\text{r},\theta}\psi_{\text{t},m,\theta} + \psi_{\text{r},\varphi}\psi_{\text{t},m,\varphi}\right)g_{\text{t},m}(\theta_{\text{t}},\varphi_{\text{t}})$$

The same dual-polarized measurement antenna pair is now used but in transmit direction with the resulting radiation patterns

$$g_{\text{t},1}(\theta_{\text{r}},\varphi_{\text{r}}) = \widehat{\theta}_{\text{t}}$$
$$g_{\text{t},2}(\theta_{\text{r}},\varphi_{\text{r}}) = \widehat{-\varphi}_{\text{t}}$$

and thus $\psi_{\text{t},k,\theta}=1$, $\psi_{\text{t},k,\varphi}=-1$ where the sign change compared with (5.67) is due to using the antenna in the opposite direction as shown in (3.26). The effective channel can then be written as

$$\tilde{\boldsymbol{h}}_{\text{eff}}^{\text{T}}(\theta_{\text{r}},\varphi_{\text{r}}) = \left[\boldsymbol{w}^{\text{T}}\boldsymbol{h}_1 \quad \boldsymbol{w}^{\text{T}}\boldsymbol{h}_2\right] = \alpha\boldsymbol{w}^{\text{T}}\tilde{\boldsymbol{a}}(\theta_{\text{r}},\varphi_{\text{r}})g_{\text{r}}(\theta_{\text{r}},\varphi_{\text{r}})\left[\psi_{\text{r},\theta} \quad -\psi_{\text{r},\varphi}\right]$$

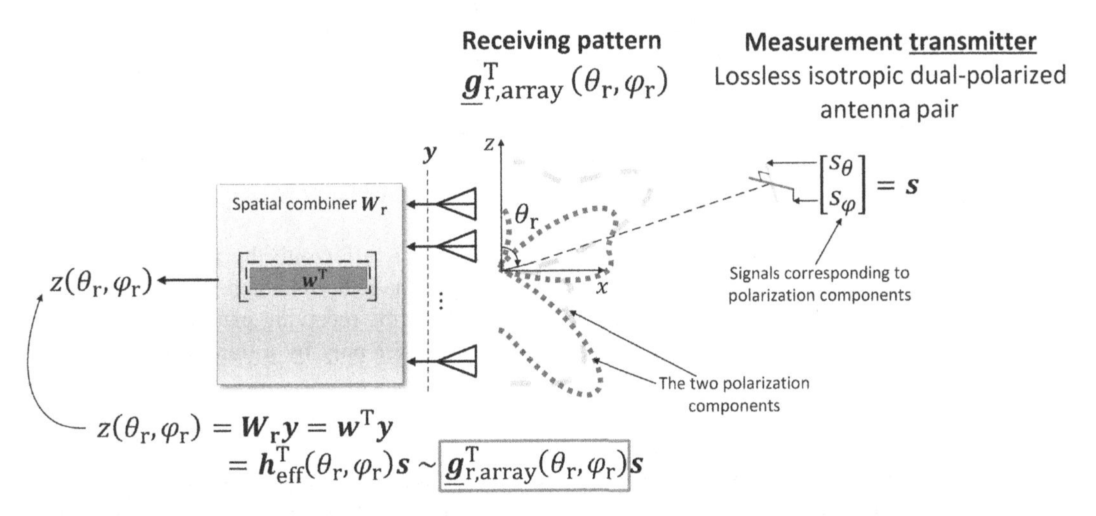

FIGURE 5.12

Measurement setup for determining the receiving pattern of an antenna/antenna array. The receiving pattern turns out to be proportional to the 1×2 effective channel between the antenna array and a measurement transmitter.

$$= \alpha \boldsymbol{w}^{\mathrm{T}} \tilde{\boldsymbol{a}}(\theta_{\mathrm{r}}, \varphi_{\mathrm{r}}) \underline{\boldsymbol{g}}^{\mathrm{T}}(\theta_{\mathrm{r}}, \varphi_{\mathrm{r}})$$

where $\underline{\boldsymbol{g}}(\theta_{\mathrm{r}}, \varphi_{\mathrm{r}})$ is the transmit radiation pattern of a single element in the array. Note how here the negative sign in front of $\psi_{\mathrm{r},\varphi}$ results in obtaining the Jones vector of the element radiation pattern $\underline{\boldsymbol{g}}(\theta_{\mathrm{r}}, \varphi_{\mathrm{r}})$ for the transmit direction via the sign change described in Section 3.3.3. Similarly as for the radiation pattern case in Section 5.4.1, the receiving pattern can now be obtained by removing the scaling factor α

$$\underline{\boldsymbol{g}}_{\mathrm{r,array}}^{\mathrm{T}}(\theta_{\mathrm{r}}, \varphi_{\mathrm{r}}) = \frac{\tilde{\boldsymbol{h}}_{\mathrm{eff}}^{\mathrm{T}}(\theta_{\mathrm{r}}, \varphi_{\mathrm{r}})}{\alpha} = \boldsymbol{w}^{\mathrm{T}} \tilde{\boldsymbol{a}}(\theta_{\mathrm{r}}, \varphi_{\mathrm{r}}) \underline{\boldsymbol{g}}^{\mathrm{T}}(\theta_{\mathrm{r}}, \varphi_{\mathrm{r}}) \tag{5.73}$$

This is what is referred to as the *receiving pattern* of the array for the spatial filter $\boldsymbol{w}$.

5.4.3 SUMMARY OF CONCEPT FOR INTERPRETING RADIATION/RECEIVING PATTERN

Comparing (5.71) and (5.73) it is seen that the transmitted and receiving patterns are the same when both directions use the same weight vector, which also follows from the previously mentioned reciprocity principle. Thus a radiation or receiving pattern for an antenna array and a weight vector can be obtained by measuring signals corresponding to sweeping the surface of a sphere with a dual-polarized lossless isotropic antenna pair with orthogonal polarizations and estimating the corresponding two-element effective channel. This measurement can be conducted for both transmitting from and receiving to the antenna array leading to a radiation or receiving pattern, respectively.

For transmit direction the radiation pattern is proportional to the 2×1 effective channel $\boldsymbol{h}_{\mathrm{eff}}(\theta_{\mathrm{t}}, \varphi_{\mathrm{t}})$ between the transmitted signal, that is mapped to the antenna array via a weight vector, and the received signals at the two outputs of the measurement antenna pair as a function of direction to said antenna pair.

Similarly, for receive the receiving pattern is proportional to the 1×2 channel $\tilde{\boldsymbol{h}}_{\mathrm{eff}}^{\mathrm{T}}(\theta_{\mathrm{r}}, \varphi_{\mathrm{r}})$ between the transmitting measurement antenna pair and the received signal obtained after spatial combining with a weight vector of the received signals at the antenna array as a function of direction to said antenna pair

The channels for transmit $\boldsymbol{h}_{\mathrm{eff}}(\theta_{\mathrm{t}}, \varphi_{\mathrm{t}})$ and receive $\tilde{\boldsymbol{h}}_{\mathrm{eff}}^{\mathrm{T}}(\theta_{\mathrm{r}}, \varphi_{\mathrm{r}})$ differ only by a transpose if the same weight vector is used in both cases, which is in line with the reciprocal nature of the propagation channel (cf. Section 6.4.2). This motivates also referring to the receiving pattern as a radiation pattern, with the understanding that they are equivalent and differ only by a transpose, i.e., corresponding to the direction of the signal.

The effective channel, and thus the radiation pattern, for transmit describes the phase and amplitude impact of the effective channel on the received signal for each antenna in the dual-polarized antenna pair. For receive, the effective channel, and thus the receive pattern, describes how signals transmitted from the dual-polarized antenna pair are combined to form the received spatially filtered signal on the antenna array side. In essence, the receiving pattern thus describes how an incoming radio wave, as a function of its direction, is attenuated or amplified and how the wave's polarization properties are combined to form the spatially filtered received signal.

5.5 WHAT IS A BEAM?

The term "beam" is a seemingly innocent and at a first glanze intuitively easy-to-understand term that has been used throughout the history of multi-antenna technologies. In this section, it is explained how this term has become open to different interpretations as the multi-antenna technologies have become more complex and new use cases have come to existence, and what possible alternative concepts are useful as complements or in some cases replacements of the traditional "beam" concept.

During the mid-1900s, the main focus areas for multi-antenna technologies were radar and satellite communication, which are typically characterized by free-space propagation, single narrowband signals, and a single stream of symbols. The typical beamforming application was to maximize the signal strength in a specific physical direction, which is referred to as classical beamforming in this book (cf. Sections 4.4.1 and 6.1.1.1). The power radiated in this direction was naturally referred to as a beam as it resembles the characteristics of other known beams, for example, a beam of light. Hence, there was a high correspondence between the technical term beam and the intuitive picture of a beam. The term beam was also incorporated in the technical terminology, here described by an IEEE definition from 1969 [7]:

> "beam (of an antenna): The major lobe of the radiation pattern of an antenna"

which is stated even in recent versions of [7]. The term worked well for the conditions in which it was adopted. In a modern context, several factors, for example, propagation environment, transmission bandwidth, number of symbol streams, etc., do, however, impact the beam concept and this leads to a need to clarify and generalize the definition. As will be shown later, all the factors above will have a significant impact for all regular mobile communication systems and hence the beam concept will need to be adapted or complemented with respect to these aspects. It should, however, be noted that the use of a dominating main lobe is still relevant in modern mobile communications systems, in particular for example in analog beamforming for mm-wave advanced antenna system (AAS) using the NR standard (see Section 9.3.6).

In the 1980s, multi-antenna techniques were studied also for multipath propagation environments. In such environments, there are typically many, rather than one, desired directions of communication. There are also fast fading phenomena to consider. In a paper from 1988 [8], Van Veen and Buckley defined beamforming inspired by the IEEE beam definition above, which is here referred to as "pencil beams":

> The term beamforming derives from the fact that early spatial filters were designed to form pencil beams in order to receive a signal radiating from a specific location and attenuate signals from other locations. "Forming beams" seems to indicate radiation of energy; however, beamforming is applicable to either radiation or reception of energy.

> When the spatial sampling is discrete, the processor that performs the spatial filtering is termed a beamformer.

This book takes *beamforming* to mean *spatial filtering conducted in a spatially discrete manner*, as in [8]. This is valid both for transmission and reception. In the case of beamforming, the spatially discrete spatial filtering is performed by an antenna array which, as a function of direction, can provide different gain and phase adjustments, as well as affecting polarization components differently. This is especially evident from the discussion on radiation and receiving patterns in Section 5.4.3. For reception, spatial filtering can be thought of as using spatial sampling of the signal, in the same way as temporal filtering uses time sampling of the signal.

Beamforming defined as spatial filtering does not put any limitations on the resulting radiation/receiving pattern, which is referred to as generalized beamforming in this book (see Section 6.1.1.2). If beamforming is taken as the process that forms a beam, it follows that the term beam should include the whole result of that process and not only a part of it. In other words, it should include the entire radiation/receiving pattern and not only the main lobe as in the IEEE definition.

Frequency-selective digital beamforming, for example, as commonly used in mid-band (frequency range FR1) AAS applications, can optimize the radiation pattern to adapt to complex multipath radio propagation environments, often resulting in no dominating strong lobe (see Section 6.1.1.2). It is therefore less meaningful to use the IEEE definition of beam as being the strongest lobe. There might be several lobes as well as one or several nulls that are of importance, which would be neglected if the focus would be on the main lobe, creating misunderstandings of what beamforming is and what possibilities it can provide.

There is even the special case with two equally strong dominating lobes, where that definition of beam is not applicable, since the beam was defined as the (one) strongest lobe. One possibility would have been to say that there are two beams in this case; however, in order to make beams countable in a deterministic way, the number of beams should not depend on the radio propagation environment or the beamforming algorithms used to decide the beamforming.

This book will therefore, *when the context does not indicate otherwise*, use a generalization of the IEEE definition, with the aim that the term *beam* should include the whole radiation/receiving pattern, and not only the main lobe. "A beam" is hence typically no longer "a beam" in a classical sense that it was used earlier, even though in several cases, the analogy to the ordinary usage of the word beam still provides support for the intuition.

In addition to this, modern systems for mobile communications transmit and receive multiple parallel symbol streams (see Section 5.3) over structured time/frequency, and even (spreading) code, resources (see Section 5.2.4). A beam is therefore considered to have a one-to-one correspondence with a symbol stream and operate over a specific set of time/frequency/code resources.

In summary, the main intention is that the meaning of the term beam is:

- extended from corresponding to the main lobe only, to the whole radiation/receiving pattern to include all the effects of beamforming in a multipath environment and
- clarified by stating the context of a beam to be per symbol stream over a time/frequency/code resource.

Based on this historical background and the identified needs in modern mobile communication systems, the meaning of the term "beam" can now be stated as below.

This book uses the term "beam" in the following way, unless the context indicates otherwise:

- *A beam in the transmit direction corresponds to the radiation pattern, over a time/frequency/code resource, obtained from the total linear processing of a symbol stream that is statistically independent of all other transmitted symbol streams;*

- *A beam in the receive direction corresponds to the receive pattern, over a time/frequency/code resource, corresponding to the total linear processing of the signal(s) received from an antenna, producing a spatially filtered symbol stream.*

The above also means that, for a given antenna array, a "beam" may correspond to a weight vector; a column vector in the precoder matrix of (5.22) and a spatial combining row vector of the spatial combining matrix in (5.27). The precoder and spatial combining matrix here represent the total spatial linear processing in transmit and receive directions, respectively.

Note that a radiation/receiving pattern can be thought of as an effective two-element channel as a function of direction in a measurement setup as discussed in Section 5.4. As also shown therein, for the same beamformer in transmit as in receive they are equivalent, which follows from reciprocity.

One example of what this book means by "beam" is in the context of the LTE and NR standards. In those standards, for the data channels (PDSCH / PUSCH) and assuming linear transmitter/receiver structures, a symbol stream corresponds to a *layer*, and the resource is a time/frequency resource corresponding to a group of *resource blocks*. This gives *one beam per layer* transmitted/received for a certain group of resource blocks.

An illustration of the term beam for transmit is given in Fig. 5.13 and likewise for receive in Fig. 5.14. More thorough and precise explanations on how to arrive at the above resulting beam descriptions are given by the following subsections where also some of the mathematical expressions in the figures will be elaborated upon.

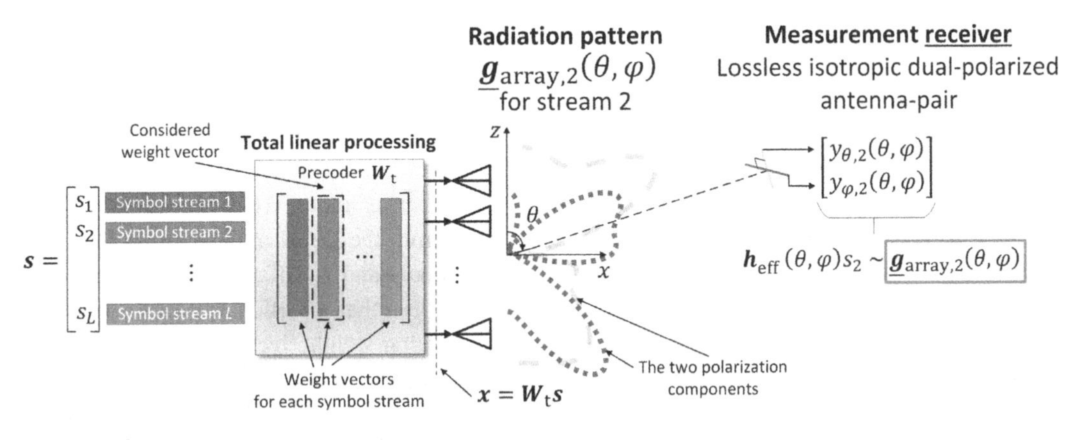

FIGURE 5.13

Illustration of meaning of "beam" in transmit direction for symbol stream 2. Recall from Section 5.4.1 that a radiation pattern can be viewed as an effective channel as a function of direction to a measurement receiver.

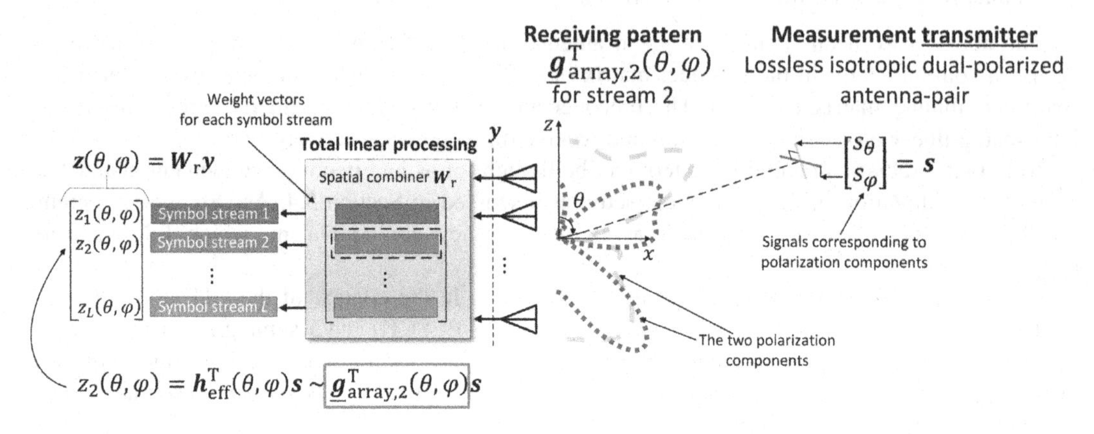

FIGURE 5.14

Illustration of meaning of "beam" in receive direction for symbol stream 2. Recall from Section 5.4.2 that a receiving pattern can be viewed as an effective channel as a function of direction from a measurement transmitter.

5.5.1 A BEAM IN THE TRANSMIT DIRECTION

The present section will step by step discuss the key components in the phrase describing the term beam in the transmit direction to give a deeper understanding. For this reason, consider a transmitter structured as in Fig. 5.13, which can be thought of as a condensed view of the transmitter depicted in Fig. 5.6. From (5.23), the transmitted $N \times 1$ signal is

$$\boldsymbol{x} = \boldsymbol{W}_{\mathrm{t}}\boldsymbol{s} = \begin{bmatrix} \boldsymbol{w}_{\mathrm{t},1} & \boldsymbol{w}_{\mathrm{t},2} & \cdots & \boldsymbol{w}_{\mathrm{t},L} \end{bmatrix} \begin{bmatrix} s_1 \\ s_2 \\ \vdots \\ s_L \end{bmatrix} = \sum_{l=1}^{L} \boldsymbol{w}_{\mathrm{t},l} s_l \tag{5.74}$$

on a particular subcarrier and OFDM symbol. Here, the subcarrier index k has been dropped and the $L \times 1$ vector $\boldsymbol{s}$ contains the modulation symbols output from the QAM modulators in Fig. 5.6. Each element in $\boldsymbol{s}$ corresponds to a separate stream of symbols and L different streams, in some cases called *layers*, are transmitted at once. From (5.74) it is seen that a particular symbol stream s_l is mapped to the antenna elements via the corresponding column weight vector $\boldsymbol{w}_{\mathrm{t},l}$. Note that this vector *completely* describes the mapping, that is, the symbol stream s_l does not affect the transmitted signals from the antenna array in any other way than via the vector $\boldsymbol{w}_{\mathrm{t},l}$ and the different symbol streams are considered to be statistically unrelated/independent. The latter requirement ensures that there is no information from symbol stream s_l in any of the other $L-1$ symbol streams, thus avoiding diluting what a symbol stream conveys.

5.5.1.1 A beam corresponds to a radiation pattern

Showing that the radiation pattern of the antenna array is a superposition of radiation patterns resulting from each transmitted symbol streams is at the heart of attaching a new meaning to the term beam. To see this, recall that the weight vector $\boldsymbol{w}_{t,l}$ adjusts the phase and amplitude of the symbol s_l for each antenna element. If it is assumed that this is the only transmitted symbol stream, then as described in Section 5.4.1 such a transmission leads to a corresponding radiation pattern. In particular, from (5.69) it is seen that the transmission of such an individual symbol stream in isolation can be viewed as being conducted over a 2×1 effective channel $\boldsymbol{h}_{\text{eff}}(\theta_t, \varphi_t)$ to the receive side, where a lossless isotropic dual-polarized antenna pair in direction (θ, φ) act as a hypothetical measurement receiver in obtaining the radiation pattern. This interpretation of radiation pattern can be used in the present context where the transmission of the lth symbol stream is represented by $\boldsymbol{w}_{t,l}s_l$. Assuming an antenna array with N different antenna elements all having the same radiation pattern $\underline{g}(\theta,\varphi)$, and normalization of the precoder $\boldsymbol{W}_t$ and symbols in $\boldsymbol{s}$ such that $\|\boldsymbol{x}\|^2 = 1$, then the resulting radiation pattern for the array for the transmission of a single symbol stream can from (5.71) be written as

$$\underline{g}_{\text{array},l}(\theta,\varphi) = \frac{\boldsymbol{h}_{\text{eff},l}(\theta,\varphi)}{\alpha} = \underline{g}(\theta,\varphi)\tilde{\boldsymbol{a}}^{\text{T}}(\theta,\varphi)\boldsymbol{w}_{t,l}$$

Similarly, when all the symbol streams are transmitted simultaneously, the transmitted signal is a superposition of the transmitted signals of each symbol stream. It follows that also the total radiation pattern of the antenna array is a superposition of the radiation patterns for each symbol stream and thus the radiation pattern becomes

$$\underline{g}_{\text{array}}(\theta,\varphi) = \frac{1}{\alpha}\sum_{l=1}^{L} \boldsymbol{h}_{\text{eff},l}(\theta,\varphi)s_{t,l}$$

$$= \sum_{l=1}^{L} \underbrace{\underline{g}(\theta,\varphi)\tilde{\boldsymbol{a}}^{\text{T}}(\theta,\varphi)\boldsymbol{w}_{t,l}}_{\sim \underline{g}_{\text{array},l}(\theta,\varphi)} s_{t,l} = \underline{g}(\theta,\varphi)\tilde{\boldsymbol{a}}^{\text{T}}(\theta,\varphi)\boldsymbol{W}_t\boldsymbol{s}_t$$

In other words, each symbol stream corresponds to a radiation pattern proportional to $\underline{g}_{\text{array},l}(\theta,\varphi)$ and the radiation pattern of the array with multiple symbol streams is a superposition of those radiation patterns for the individual symbol streams. That the processing from the symbol streams to the transmitted signals is linear is key here to uphold the superposition principle and it is this that allows each symbol stream contributing to the radiation pattern to be considered individually without being affected by the transmissions of all the other symbol streams.

Clearly, there is a one-to-one correspondence between a symbol stream s_l, the weight vector $\boldsymbol{w}_{t,l}$, and the radiation pattern $\sim\underline{g}_{\text{array},l}(\theta,\varphi)$ over all directions (θ,φ), as also evident from Fig. 5.13. This motivates letting the term beam correspond to any of these quantities. However, in accordance with the aim of generalizing the IEEE definition, the term beam is primarily said to correspond to a radiation pattern. This also provides meaning to the term when the antenna array just consists of a single-antenna element for which the weight vector is reduced to just a trivial scalar.

Although not treated in the book, nonlinear transmitter (and receiver) structures may make it challenging to at all describe a beam in terms of a radiation (or receiving) pattern since the radiation pattern is no longer a superposition of radiation patterns for individual symbol streams.

5.5.1.2 Possibilities to refer to equivalent quantities as beams

Transmission of L symbol streams may now be said to correspond to the transmission of L different beams, which in turn corresponds to L different weight vectors. It therefore makes sense to refer to the weight vector $\boldsymbol{w}_{\mathrm{t},l}$ as a "beam (weight) vector," or even just "beam," the latter with the tacit understanding that the application of that beam vector *leads to* a corresponding beam (radiation pattern). With this logic and implied understanding, even a particular symbol stream could be referred to as a "beam," although the reader should perhaps in the latter case be discouraged from doing so to avoid confusion.

5.5.1.3 Radiation pattern over a resource

So far, the discussions have ignored that the weight vector $\boldsymbol{w}_{\mathrm{t},l}$ may change over frequency and time for what is considered the *same* symbol stream. In fact, such frequency-selective precoding and spatial combining are common as evident from Sections 6.1.1.2.3, 9.3.3, and 13.2.3. The one-to-one correspondence between symbol stream, weight vector, and radiation pattern then only holds on a resource over which the weight vector is constant. In OFDM-based systems, such a resource is typically confined to some set of OFDM symbols and subcarriers, that is, essentially a time/frequency resource. Thus it makes sense that a "beam" is taken to be with respect to some resource over which the weight vector is constant, which in turn usually means the radiation pattern is essentially constant, unless the relative bandwidth of the resource is large (the radiation pattern varies with frequency but only slowly if relative bandwidth is small). A resource may also correspond to the code domain, for example, for a certain set of spreading codes in a code division multiple access (CDMA) system or for modulation schemes utilizing other forms of basis functions than the DFT-based basis functions of OFDM.

5.5.1.4 Total linear processing and end-to-end beams

In the transmitter, there may be many stages corresponding to the same symbol stream. For example, consider an overall precoding matrix $\boldsymbol{W}_{\mathrm{t}}$ implemented as a product of two precoders given the two-stage precoding operation

$$\boldsymbol{x} = \boldsymbol{W}_{\mathrm{t}}\boldsymbol{s} = \boldsymbol{W}_{\mathrm{t},1}\boldsymbol{W}_{\mathrm{t},2}\boldsymbol{s} = \boldsymbol{W}_{\mathrm{t},1}\tilde{\boldsymbol{x}}$$

where $\tilde{\boldsymbol{x}} \triangleq \boldsymbol{W}_{\mathrm{t},2}\boldsymbol{s}$ represents partially precoded symbol streams that act as input to the precoder $\boldsymbol{W}_{\mathrm{t},1}$. Unless the context says otherwise, a beam is to be considered end-to-end in the sense of capturing the overall weight vector $\boldsymbol{w}_{\mathrm{t},l}$ in the lth column of the overall precoder matrix $\boldsymbol{W}_{\mathrm{t}}$ corresponding to a symbol stream s_l. In essence, the precoder matrix $\boldsymbol{W}_{\mathrm{t}}$ captures the total linear processing performed by the transmitter and the end-to-end is with respect to parts within the transmitter.

However, the precoder $\boldsymbol{W}_{\mathrm{t},1}$ could also be said to contain beam weight vectors and thus form beams, but this time based on symbol streams $\tilde{\boldsymbol{x}}$ that have in turn been precoded. In addition, the precoding operation $\boldsymbol{W}_{\mathrm{t},2}\boldsymbol{s}$ could be said to lead to beams corresponding to the columns of the precoder $\boldsymbol{W}_{\mathrm{t},2}$. However, this time the beam is in a vector space since the output of the precoding is not directly fed to the antenna array so the correspondence to a radiation pattern is not clear. All these kinds of "beams" are clearly not end-to-end since there is either an earlier or later linear transformation applied not captured by the corresponding weight vector. Nevertheless, such usage of "beam" is common (see, e.g., the advanced CSI feedback schemes described in Sections 6.4.1.2.1 and 9.3.5.5).

In the former case of considering weight vectors in $\boldsymbol{W}_{t,1}$, this makes the input symbol streams in $\tilde{x}$ statistically dependent while in the latter case of weight vectors in $\boldsymbol{W}_{t,2}$ the connection to a radiation pattern is not clear. If it is apparent from the context, it is nevertheless reasonable to refer to them as beams.

It is also common to refer to the half-part of a weight vector applied to antenna elements of the same polarization in a dual-polarized antenna array to be a beam instead of using that term to refer to the whole weight vector. This is related to the convenient usage of array response vectors with half as many elements as the number of elements in the array (see, e.g., (5.54)). Since the relative phase and amplitude of orthogonal polarizations in the common case of a multipath channel tend to vary quickly over frequency and time and such fast fading behavior is not well-captured by radiation pattern–based analysis (cf. Sections 6.1.1.2.1 and 6.1.1.2.2), it could make sense to consider the notion of a beam to be within a subset of the antennas corresponding to the same polarization, even though that does not correspond to the radiation pattern of the antenna array.

All these alternative uses of the term beam are common and are not prohibited by the updated terminology used in this book. But as previously mentioned if there is no such context given, then it makes sense to interpret "beam" to mean an end-to-end beam. An end-to-end beam implies that the most complete linear transformation is assumed, that is, that the symbol stream is picked as early as possible in the transmitter chain. In other words, a (end-to-end) beam shall be the result of the total spatial linear processing in the transmitter. In Fig. 5.6 this corresponds to letting the symbol stream correspond to a layer taken directly after the output of the QAM modulation. Although not treated in the book, nonlinear transmitter (and receiver) structures may make it challenging to at all describe a beam in terms of a radiation (or receiving) pattern since the radiation pattern is no longer a superposition of radiation patterns for individual symbol streams.

5.5.1.5 A symbol stream could be a layer

In general, any, on the relevant time/frequency/code resource, transmitted single symbol stream that can be described to be fully mapped onto the antenna array using a beam weight vector could play the role of a symbol stream in the above description. For example, a known reference signal could be said to be transmitted in a beam (cf. CSI-RS and Class B CSI feedback in LTE, described in Section 8.3.2.3), even though the symbol stream of a reference signal is not referred to as a layer.

5.5.1.6 Summary

All this reasoning and mathematical derivations can now be condensed to a single sentence describing what this book means with a "beam" in the transmission direction, unless context indicates otherwise:

- *A beam in the transmit direction corresponds to the radiation pattern, over a time/frequency/code resource, obtained from the total linear processing of a symbol stream which is statistically independent of all other transmitted symbol streams.*

Keep in mind that the beam here is end-to-end and that an antenna array can also be considered an antenna. Note also that for the single-antenna element case, it is only the element itself that forms the radiation pattern. As previously mentioned, an illustration of this for a particular example is given in Fig. 5.13.

5.5.2 A BEAM IN THE RECEIVE DIRECTION

Since a beam in transmit direction is associated with a radiation pattern, it is reasonable to make a similar association also for a beam in the receive direction. Although a radiation pattern describes properties of the electric field when an antenna transmits, there is an equivalent receiving pattern describing how the electric field of an incoming radio wave is attenuated/amplified and how the polarization components are combined to form the received signal. This now plays the same role as radiation pattern does for the beam term in the transmit direction and much of the reasoning from the previous section therefore applies also in this case and will hence not be repeated.

To understand how the receiving pattern applies to the term beam, recall from Section 5.4 how a radiation/receiving pattern can be interpreted as a two-element effective channel between the antenna (array) of interest and a hypothetical lossless isotropic dual-polarized measurement antenna pair. This interpretation becomes useful now when attempting to describe why it is reasonable to associate a beam with a receiving pattern in the receive direction.

Fig. 5.14 shows an antenna array in the receive direction where spatial combining is used to produce a vector of symbol streams. Similar to as in (5.25), consider the use of an $L \times N$ spatial combining matrix $\boldsymbol{W}_{\mathrm{r}}$ applied to the received $N \times 1$ signal $\boldsymbol{y}$ from an antenna array of N identical elements producing the $L \times 1$ spatially combined vector

$$\boldsymbol{z}(\theta,\varphi) \triangleq \begin{bmatrix} z_1(\theta,\varphi) \\ z_2(\theta,\varphi) \\ \vdots \\ z_L(\theta,\varphi) \end{bmatrix} = \boldsymbol{W}_{\mathrm{r}}\boldsymbol{y} \triangleq \begin{bmatrix} \boldsymbol{w}_{\mathrm{r},1}^{\mathrm{T}} \\ \boldsymbol{w}_{\mathrm{r},2}^{\mathrm{T}} \\ \vdots \\ \boldsymbol{w}_{\mathrm{r,L}}^{\mathrm{T}} \end{bmatrix} \boldsymbol{y}$$

with obvious notation. Each row of the spatial filter matrix $\boldsymbol{W}_{\mathrm{r}}$ is seen to contain a corresponding weight vector $\boldsymbol{w}_{\mathrm{r},l}^{\mathrm{T}}$ that is used to combine the signals from the antenna array resulting in the signal or symbol stream $z_l(\theta,\varphi) = \boldsymbol{w}_{\mathrm{r},l}^{\mathrm{T}}\boldsymbol{y}$. If now the received vector $\boldsymbol{y}$ is conceptually thought to arise from a hypothetical measurement antenna pair in direction (θ,φ) transmitting from the other side of the link over a free-space channel, then similarly as in (5.72), the channel describing the relation between the transmitted signal and the spatially filtered symbol stream $z_l(\theta,\varphi)$ is a 1×2 effective channel $\tilde{\boldsymbol{h}}_{\mathrm{eff}}^{\mathrm{T}}(\theta,\varphi)$. From (5.73), it is seen that this effective channel is proportional to the receiving pattern

$$\underline{\boldsymbol{g}}_{\mathrm{r,array}}^{\mathrm{T}}(\theta,\varphi) = \tilde{\boldsymbol{h}}_{\mathrm{eff}}^{\mathrm{T}}(\theta,\varphi)/\alpha = \boldsymbol{w}_{\mathrm{r},l}^{\mathrm{T}}\tilde{\boldsymbol{a}}(\theta,\varphi)\underline{\boldsymbol{g}}^{\mathrm{T}}(\theta,\varphi)$$

In other words, the spatial combiner vector $\boldsymbol{w}_{\mathrm{r},l}^{\mathrm{T}}$ corresponds to a receiving pattern and there are L different such receiving patterns corresponding to the L rows of the spatial combiner matrix $\boldsymbol{W}_{\mathrm{r}}$. In line with the meaning of "beam" in the transmit direction, the application of each spatial combiner weight vector leads to a receiving pattern and thus may be taken to correspond to a beam. The spatial combining may be decomposed into several stages, similarly to the precoding in the transmit direction, and the combining in these individual stages may also be referred to as a beam. However, the term beam also in the receive direction refers without further context to the spatial filtering that includes all linear transformations of the received signals on a particular resource, that is, includes the total linear processing in the receiver producing that symbol stream. Unless context indicates otherwise, this book thus takes "beam" in receive direction to mean:

- *A beam in the receive direction corresponds to the receive pattern, over a time/frequency/code resource, corresponding to the total linear processing of the signal(s) received from an antenna, producing a spatially filtered symbol stream.*

As previously mentioned, an illustration of this for a particular example is given in Fig. 5.14.

5.6 SUMMARY

This chapter has shown the principles of OFDM-based transmission and reception and their characteristics. OFDM has become the preferred choice of waveform for several standardized telecommunication technologies, including LTE and NR. The use of OFDM with a CP has several advantages:

- allows wideband communication over frequency-selective channels without introducing too high complexity on the receiver side;
- simplifies user multiplexing due to its inherent orthogonal property also in the receiver;
- can be realized using the FFT and can thus utilize existing efficient implementations of the FFT, and
- allows for scalability to MIMO systems with a very large number of antennas and simultaneously to very wide channel bandwidths with a bounded complexity increase.

On the drawback side, the use of OFDM requires power amplifier linearization since it has a high PAPR and the system must ensure a high level of synchronization between receiver and transmitter. However, technology advances have progressed over the last decades to handle these difficulties.

In this chapter, it is also shown how the MIMO channel can be described in frequency domain, using direction, polarization, and frequency variables, thereby revealing the spatial structure of the channel. The chapter also includes the definition of spatial correlation matrices and how they can be expressed in terms of the so-called large-scale parameters of the channel. This background is needed to obtain the deeper understanding of MIMO algorithms and performance analysis introduced later in this book.

The chapter ends with an enlightening discussion around the definitions of the widely used terms "beam" and "beamforming," which are also central concepts in this book.

REFERENCES

[1] D. Tse, P. Viswanath, Fundamentals of Wireless Communication, Cambridge University Press, 2005.

[2] T. Hwang, C. Yang, G. Wu, S. Li, G.Y. Li, OFDM and its wireless applications, a survey, IEEE Trans. Veh. Technol. 58 (4) (2009).

[3] E. Dahlman, S. Parkvall, J. Sköld, 4G LTE/LTE-Advanced for Mobile Broadband, Academic Press, 2011.

[4] 3GPP TR 38.211, V15.5.0 (2019-03) NR; Physical channels and modulation.
[5] S.H. Han, J.H. Lee, An overview of peak-to-average power ratio reduction techniques for multicarrier transmission, IEEE Wirel. Commun. 12 (2) (2005).
[6] 3GPP TR 38.901, V15.1.0 (2019-09), Study on channel model for frequencies from 0.5 to 100 GHz.
[7] IEEE standard for definitions of terms for antennas, IEEE Std 145–1969, March 1969.
[8] B.D. Van Veen, K.M. Buckley, Beamforming: a versatile approach to spatial filtering, IEEE ASSP Mag. 5 (2) (1988).

CHAPTER

MULTI-ANTENNA TECHNOLOGIES

6

The use of multi-antenna techniques for wireless communication has by now been an active area of research for more than three decades. An almost endless plethora of schemes have been developed and analyzed. This chapter aims to explain some of the most important and basic techniques and most importantly provide the intuition needed to have a good understanding.

The chapter makes no attempt to be a complete exposé of all the possible schemes and variants. Although the multi-antenna area involves a lot of potentially advanced mathematics for a truly full understanding, this chapter deliberately attempts to limit the use of mathematics, although some mathematics is unavoidable. To facilitate a fuller understanding, it is recommended to first read about wave propagation and channel modeling in Chapter 3, antenna arrays in Chapter 4, and orthogonal frequency division multiplexing (OFDM)−based transmission and models in Chapter 5. This chapter also lays the foundation for understanding the multi-antenna concepts for long-term evolution (LTE) and new radio (NR) standards described in Chapters 8 and 9, as well as the performance evaluations in Chapter 13, and Section 14.3.

Topics that will be covered include different types of beamforming and how beamforming achieves benefits, nullforming to fight interference, cell shaping, spatial multiplexing for single-user multiple-input multiple-output (SU-MIMO) and multi-user MIMO (MU-MIMO), acquisition of channel state information (CSI) for TX beamforming/precoding, and radio alignment errors and their impact on the multi-antenna performance. Coordination of transmission/reception between different cells introducing the topic of coordinated multipoint (CoMP) is also covered. The chapter ends with a discussion on some of the misunderstandings and challenges when applying theories from the field of massive MIMO to real-life commercial networks and why the conclusions from theories sometimes are difficult to apply to commercial networks.

The present chapter makes very limited use of the term "subarray." The discussions are instead performed in terms of "antennas." This is deliberate since a subarray is just one of many ways of creating an antenna. The techniques covered herein can therefore obviously be applied on top of subarrays, in fact that is what is typically being done. Furthermore, discussions are generally tacitly assuming "typical" antenna arrays designs with antenna spacings on the order of 0.5−0.7 wavelengths and nonextreme antenna sizes, unless explicitly stated otherwise.

The techniques in this chapter all view the number of antennas as simply a design parameter. They are thus applicable both to the cases of few or massively many antennas — there is nothing special with the use of a large number of antennas in that sense, except that the potential performance improves. This is not surprising considering that the techniques are all well-known and have a long history that indeed predates the use of the term massive MIMO.

Note that this book in Section 1.4 defines massive MIMO to mean systems equipped with and exploiting massively many dynamically adaptable antennas. The definition intentionally attempts to avoid mixing in requirements on technology concepts such as MU-MIMO or time division duplex

Advanced Antenna Systems for 5G Network Deployments. DOI: https://doi.org/10.1016/B978-0-12-820046-9.00006-X

(TDD) with reciprocity into the massive MIMO definition, even though in the literature these techniques are often taken as mandatory for a massive MIMO system.

6.1 BASIC DYNAMIC CHANNEL-DEPENDENT BEAMFORMING CONCEPTS

The basic idea behind beamforming is to amplify transmitted/received signals more in some directions than in others, that is, introduce a spatially non-uniform gain. A main goal is usually to achieve a high gain in the direction of some device of interest so as to improve the link quality.

Spatially shaping the gain is typically achieved using antenna arrays that adapt the phase and/or amplitude of signals on the different antennas to achieve constructive addition of signals in some directions, corresponding to high gain, and destructive addition in other directions, corresponding to low gain. Direction is here defined relative to the location of the antenna array. This antenna array could for example correspond to an advanced antenna system AAS or to a user equipment (UE) having with multiple antennas. Direction is commonly described in spherical coordinates using zenith and azimuth angles.

The adaptation of phases/amplitudes at an antenna array is often based on knowing the channel properties at the device where the beamforming is applied. Channel knowledge is commonly referred to as *channel state information* (CSI), and beamforming that utilizes CSI may be referred to as *channel-dependent beamforming.* Knowledge of direction to a device to beamform toward is an example of an intuitively easy to understand kind of channel knowledge but as will be discussed in this chapter many other and richer forms of channel knowledge are possible. Beamforming can alternatively be channel-independent; the use of a static beam shape to cover a whole cell constitutes a primary example. This category includes classical single static beam systems such as single-column sector antennas but it can also include newer beamsweeping schemes as part of NR (cf. Section 9.3.6 on beam-based operation) where the beam as such is not static but where instead a static/predetermined sequence of beams is used for cell-defining signals to give a static shape of a cell via the envelope of all constituent beam shapes.

The spatially non-uniform gain distribution can be described in a graph with gain as a function of direction as shown in Section 4.3.3.1. Note that the gain here corresponds to a power ratio relative to the use of some spatially isotropic gain distribution (i.e., the same gain in all directions). If a peak in the graph is sufficiently narrow, it starts to visually resemble a beam, hence motivating the terms "beam" or "beamforming." The mentioned graph is for a particular signal referred to as a beam gain pattern or beam shape (see Section 4.3.2). It could also formally be referred to as an antenna gain pattern but that is a confusing terminology if the beam is the result of the use of an antenna array as it can be confused with the antenna gain pattern of each individual antenna in the array. In common cases (but far from all) the beam shape has a pronounced strongest direction (main lobe) so that it becomes meaningful to speak in terms of beamforming direction. Since the energy conservation principle applies, a stronger gain in some direction necessarily means lower gain in others. Thus beamforming is about reallocation of gain/power over directions to better fit a certain purpose.

A common application is to utilize beamforming on the base station side to provide a strong signal in the direction of the UE of interest, so-called UE-specific beamforming. Beamforming can be used when transmitting as well as when receiving. Although it can be used on any side of the communication link, the possibilities for more aggressive beamforming using a large number of antennas are far greater on the base station side than on the UE side. Constraints on physical size of the device are one obvious reason, among several others, why beamforming on UE side is often more limited.

Beamforming strives to reshape the spatial power distribution of signals. This can be exploited to ensure that a larger fraction of the signal power is in the direction of the UE of interest. This leads to improved signal strength, which via better signal-to-interference-and-noise ratio (SINR) in the end results in better performance. Fig. 6.1 compares a base station that has a completely isotropic beam shape, meaning directions are equally amplified, with a base station where most of the power has been focused toward a UE of interest. If the example corresponds to transmit beamforming, the total transmit power with these two beam shapes is the same, while the more narrow beam ensures that a larger part of the transmitted power becomes useful as it reaches the UE.

The beam shape can be static or vary more or less dynamically. The use of a static beam shape in elevation is common in cellular networks. For example, a single-column antenna whose beam shape is fixed and with a narrow main lobe in the vertical domain and a wide sector covering beam in the horizontal domain is often used to serve all the UEs in a cell. To obtain good beamforming gain with a fixed beam system, it is important that all the devices that are intended to be served by the fixed beam are located in a direction that is close to a strong peak of the beam; otherwise, beamforming gain quickly drops. With above-rooftop deployments and large cells this is often approximately fulfilled in the vertical domain for a relatively large portion of the UEs in a cell; as seen from the base station many UEs tend to be close to the horizon so the vertical *UE angular spread* is small. The horizontal spatial UE distribution is, however, typically entirely different with often a wider and rather uniform UE angular spread over the sector. This means that in horizontal direction beamforming gain is lost with a fixed beam approach, either since the fixed beam must be made wide to cover all UEs or because a narrow beam misses most of the UEs.

Dynamic beamforming avoids the drawback of the fixed beam approach by providing the possibility to tailor the beam shape to a device at each moment in time. This allows the beam to be narrow and still track movement of the intended device. For base station side beamforming it allows each UE in the cell to enjoy a specific beam shape with a strong peak instead of being forced to share the same beam over many geographically dispersed UEs. The difference between dynamic and static beamforming is illustrated in Fig. 6.2.

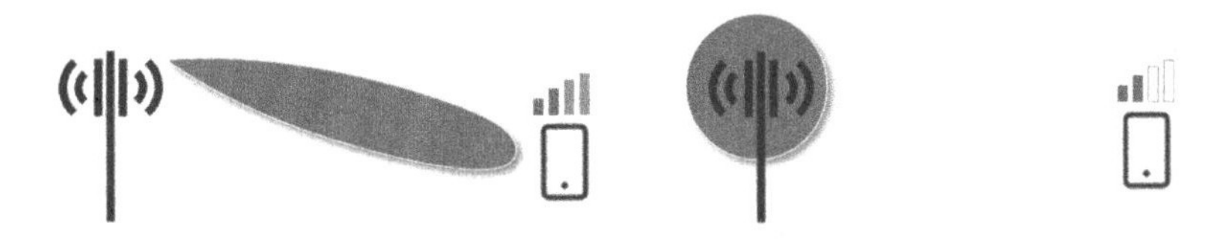

FIGURE 6.1

Beamforming focusing the gain in the direction of UE of interest (left), thereby improving signal strength over the use of an isotropic static beam (right). This signal-to-noise ratio (SNR) is seen to be improved with beamforming as more signal bars are active.

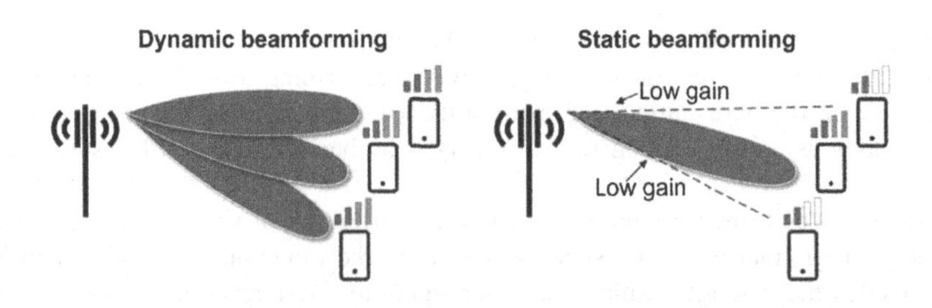

FIGURE 6.2

Dynamic, or UE–specific, beamforming allows each UE to get a high-gain beam of its own (left) in contrast to static beamforming where a single beam is shared among many UEs and thus often not well matched to the UE angular distribution (right).

Since each UE may with dynamic beamforming be served by a beam(s) of their own, this is commonly referred to as *UE–specific beamforming*. Beamforming may also be performed on the UE side. Receive beamforming is particularly common since both UEs and base stations have for a long time had at least two receive antennas and having at least two antennas is a prerequisite for dynamic beamforming.

Dynamic beamforming allows the UE to track the channel properties both in time and frequency. Since the environment surrounding a UE often constitutes a rich scattering environment, the channel angular spread may be large, which leads to fast fading, thereby increasing the need for dynamic beamforming that updates the beamformer both over frequency and time to track the fast fading. The impact of channel angular spread will be further discussed in Section 6.1.1.

Another motivation to be able to alter beam shape also over frequency is that the set of scheduled UEs in LTE/NR may not only vary from one slot to another but also between different frequency resources within a slot. The use of such time- and frequency-domain multiplexing of UEs makes it particularly important to support adaptation of beam shape both in time as well as over frequency.

6.1.1 THE TWO CATEGORIES OF CLASSICAL AND GENERALIZED BEAMFORMING

It is useful to classify beamforming into the two categories of *classical beamforming* and *generalized beamforming*. Classical beamforming is easy to understand and agrees well with intuition. The beam shape has a well-defined and often a relatively narrow dominating main lobe that can be pointed into a desired physical direction, for example, from a base station toward a UE. Classical beamforming is in the focus of traditional antenna theory and is considered in Chapter 4, and is introduced in Section 4.4.1. In free-space or line-of-sight (LOS) propagation, classical beamforming is, or is near, optimal in sense of maximizing signal strength achieving maximum gain. Such channels have (essentially) a single propagation path, that is, zero *channel angular spread*, and no or little multi-path fading; thus a beam with a single direction matches the channel very well. In contrast, generalized beamforming includes totally arbitrary beam shapes with potentially no obvious connection to propagation path directions in the channel. With channel angular spread classical beamforming is no longer optimal and leads at least in theory with ideal channel knowledge to a SNR loss as it is unable to match the fast fading properties of the channel.

6.1.1.1 Classical beamforming—optimal only for free space

An example of a classical beam shape shown in linear scale is depicted in Fig. 6.3. There is a single propagation path in the channel and its direction is marked with the dashed line. The beam shape is optimal, that is, maximizes signal strength, for this zero angular spread scenario and hence it can as expected be seen how the peak of the beam points *exactly* along the direction of that single propagation path.

Note that a two-dimensional (2D) uniform planar array (UPA), cf. Section 4.5.1, with four rows, eight columns, and lossless isotropic single-polarized antennas, was used to generate the beam. Antenna spacing was 0.5 wavelength both horizontally and vertically. The distance from the

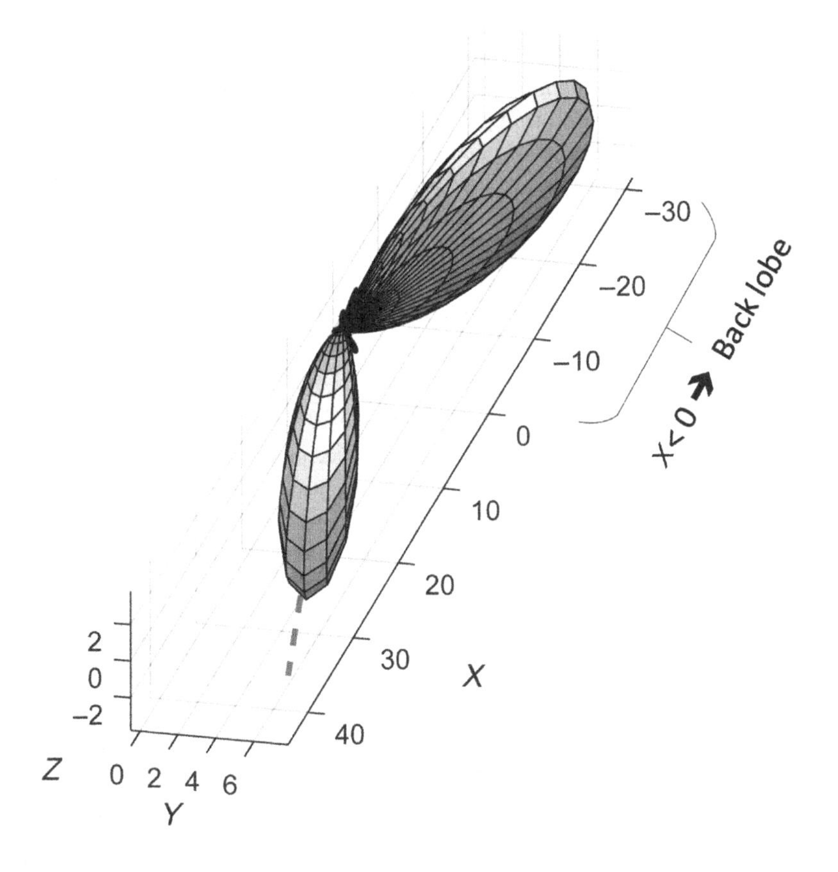

FIGURE 6.3

Beam gain pattern in linear scale for classical beamforming, which has a well-defined single pointing direction in the form of a clear main lobe and small side lobes within intended coverage area ($x < 0$ corresponds to back lobes normally suppressed by the antenna gain pattern of antennas in the array).

origin of the coordinate system, $(x, y, z) = (0, 0, 0)$, to a surface point represents the power gain in that direction relative to the power gain of a single lossless isotropic antenna, that is, the gain in power in said direction due to use of the more advanced antenna array. Although somewhat difficult to read out due to the viewing angle, the peak of the beam is at a distance of 32 corresponding to the maximum power gain of a 32-antenna array. Since lossless isotropic antennas are used, the figure also corresponds to the array gain (array factor) as a function of direction as previosuly described in Section 4.2.3. The use of isotropic antenna elements in this example also means there is a strong back-lobe pointing toward negative x-axis. This is an artifact of the isotropic antenna element assumption and would normally be suppressed by the common use of antenna elements with a much more directive antenna pattern. Further note that all the assumptions described in this paragraph are common to all the figures of beam shapes in this section and will therefore not be repeated. This includes an assumption of constant total transmit power if the beamformer is viewed as a transmit beamformer.

6.1.1.2 Generalized beamforming—optimal for all channel conditions

As the propagation environment becomes richer with multiple propagation paths, channel angular spread increases and classical beamforming is no longer optimal. Generalized beamforming with much more arbitrary beam shapes now becomes highly relevant. There is now a distribution of directions, each of different channel gain, that lead signals from/to the beamforming device (e.g., base station) to/from the device (e.g., UE) on the other side of the link. To match such a diverse multi-path channel characteristic, it turns out that an optimal beam needs to have a rather similar shape as the distribution of channel directions. To exemplify with transmit beamforming, this makes sense considering that only energy emitted in directions along the propagation paths reaches the receiver.

6.1.1.2.1 Limited angular spread case and the finite resolution problem

Fig. 6.4 illustrates a multi-path scenario with three distinct propagation directions. Three lobes of the beam are seen to point roughly in those three propagation directions confirming the intuition that an optimal beam shape should match the distribution of channel directions. In contrast, the use of classical beamforming in this case would not be optimal and lead to a lower SNR as it is unable to match multiple propagation directions at once.

Unlike the previous single propagation path case in Fig. 6.3, the lobe peaks do not exactly match the propagation directions. This is because the array has a finite resolution, and so is unable to completely resolve individual channel directions and in particular unable to resolve directions that are very close to each other.

The resolution of an antenna array can in this context be thought of as being proportional to the lobe width of a beam matched to a single channel direction in isolation as in the free-space case in Fig. 6.3; roughly speaking, the antenna array "sees the world" with (some factor of) a lobe width resolution. Compare with how a radar lets a beam scan over all directions and how it essentially samples the world a beamwidth at a time. As described in Section 4.5.1.1, lobe widths generally decrease with increasing antenna array size. Hence, the resolution of an antenna array generally increases with size so if the number of antennas is increased, eventually even directions very close to each other can be resolved to a high degree.

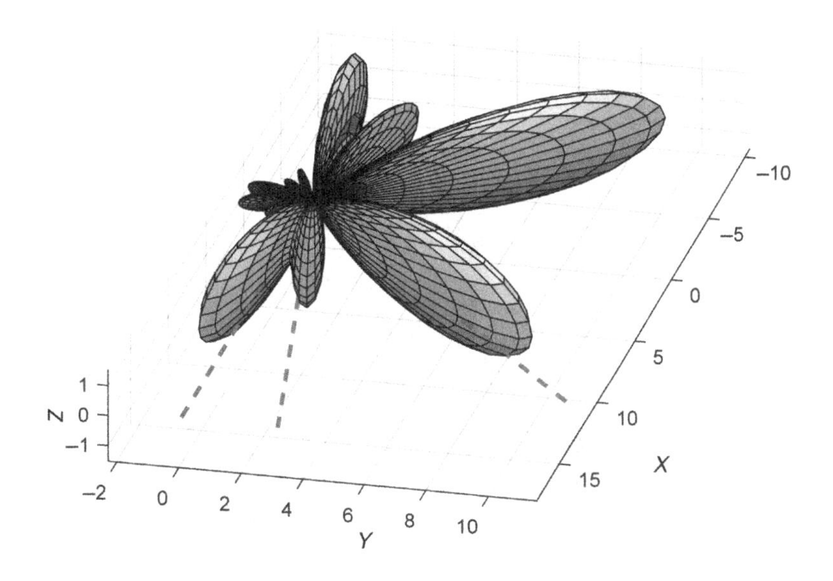

FIGURE 6.4

Generalized beamforming optimal for the illustrated three propagation path situation. Unlike classical beamforming there is no longer a single well-defined pointing direction.

The optimal beam in Fig. 6.4 is proportional to a superposition of the optimal beam for each channel direction taken in isolation. This means that Fig. 6.4 can be viewed as a sum of three beam shapes like in Fig. 6.3 but tailored for each channel direction. The superposition is, however, not conducted in the power domain as captured by beam shapes but rather in the complex-valued amplitude domain where also phase and not only power matters. Thus it is not as simple as just "adding three pictures."

If two channel directions come within the resolution limit, they, *both* for these directions, and other directions in the vicinity of the resolution granularity, contribute strongly in determining the gain in those directions. What happens is that the two channel directions lead in isolation to lobes that are partially overlapping in the superposition step. It is this that makes the beam peaks be offset relative to the corresponding channel directions. How exactly the peaks are offset, or what the resulting beam shape in the overlap region between two closely spaced channel directions looks like, is very arbitrary and cannot really be predicted based on studying beam shapes alone. This since a beam shape analysis only gives information about spatial power distribution but not about phase and hence it becomes impossible to carry out the superposition based on beam shape pictures. It is common practice in analysis to ignore phase and only draw diagrams of the power part of the beam pattern, that is, purely focus on the beam shape, even though formally the term antenna/beam pattern could be interpreted to be a complex-valued metric and hence include phase.

The missing information about phase can potentially have a very large impact since it can lead to complete cancellation or complete constructive addition of signals. To further exemplify this

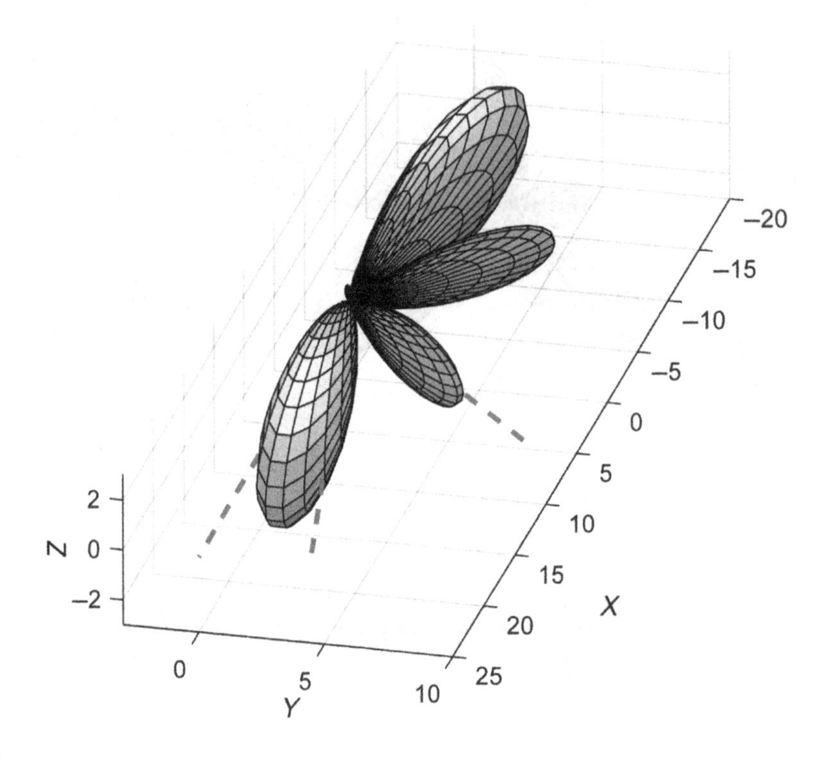

FIGURE 6.5

Generalized beamforming optimal for the illustrated three propagation path situation with propagation directions identical to the directions in Fig. 6.4 but where the corresponding radio waves have different phases leading to very different beam shape close to the two left-most channel directions.

issue, consider the beam shape in Fig. 6.5 which shows the optimal beam for a channel with *identical* directions as in the previous Fig. 6.4. The lobe pointing toward the isolated channel direction looks similar in both figures but the beam shape in the vicinity of the two channel directions that are not fully resolved looks entirely different—in Fig. 6.4 it corresponds to two distinctly different peaks while in Fig. 6.5 the two channel directions are covered by a single lobe. The *only* difference between the two pictures is the assumption on phases of the radio waves traveling along the different channel directions, creating entirely different results. The heavy reliance on phase can be seen as a fast fading effect. Multi-path channels thus constitute a challenge for the use of beam shapes or even more general complex-valued antenna/beam patterns as an analysis tool since directions of propagation paths no longer necessarily have a direct correspondence in the beam shape. Because channels in practice have multi-path characteristics and not free-space characteristics, care needs to be exercised to avoid drawing too strong conclusions based on antenna/beam patterns alone.

As evident from this section, an analysis based on physical directions may support some deeper insights when the radio environment allows for such interpretations, but it cannot capture all relevant phenomena. Thus, however attractive for building intuitive mental models, it must be applied with some caution. The following properties are not well-captured and may cause confusion:

- Realistic multi-path channels have angular spread and with angular spread follows fast fading effects in the beam shape that are highly phase-dependent;
- The resulting beam may have an arbitrary shape with a weak relation to physical directions;
- The combined effect of the channel and the beam shape is what matters and is not obvious in case of multi-path channels.

Where the effects above dominate, it must be understood that an analysis solely based on physical directions when studying beam shapes is inadequate. Despite these shortcomings, beam shapes and antenna/beam patterns remain an important analysis tool, particularly because channels in practice often have relatively small angular spread that can be modeled with a large number of propagation paths so that the impact of phase realizations of individual paths average out and drastic cancellation effects are replaced with more modest impact on loss of gain. The important thing is to understand the shortcomings of physical direction-based analysis and be able to combine it with another tool that, albeit more abstract and lacking a physical and intuitively appealing interpretation, has the benefit of properly describing fast fading effects. Such a tool will be touched upon in the next section without going into details and then a more fundamental understanding is only possible as part of the mathematical treatment of the topic in Sections 6.1.2 and 6.1.4.

6.1.1.2.2 Extreme angular spread making beam shape analysis completely irrelevant

Channel angular spread may vary depending on the scenario. Perhaps the most extreme case is to assume that the propagation directions are uniformly distributed in all directions. The beam shape can then look totally arbitrary with multiple lobes without any of them dominating and in general no well-defined characteristics. Fig. 6.6 shows an example of this where 36 different propagation

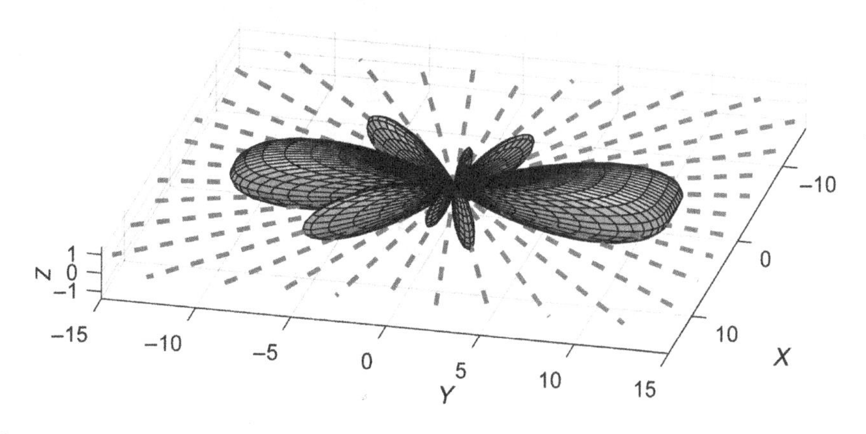

FIGURE 6.6

Channel directions uniformly distributed in the horizontal plane leading to a rather arbitrary (generalized) beam shape totally dependent on the phase realizations of each individual radio wave corresponding to each direction.

directions spaced 10 degrees apart in the horizontal plane lead to a rather arbitrary looking beam pattern. Needless to say, just by changing the phases of the radio waves of these directions may lead to a completely different pattern. The beam shape would have looked even more chaotic if the propagation directions would have been uniformly distributed around the sphere instead of all confined in the horizontal plane.

This clearly demonstrates that if the channel angular spread becomes sufficiently large, performing analysis in terms of intuitively pleasing physical directions and beam shapes becomes totally meaningless and the beamforming is better understood as corresponding to a direction in an abstract linear algebraic complex-valued vector space. This means analyzing in terms of the complex-valued coefficients in the beam weight vector and the vector/matrix describing the channel between the antenna array and the device on the other side of the link, making no attempt to make interpretations using physical directions. The fundamental issue with doing analysis purely based on beam shapes or more general antenna patterns (which may in principle include both phase and polarization) is that such an analysis is unable to capture the impact of a channel with angular spread and it is only the combination of the channel and the beamforming operation that in the end is interesting to study since both are needed to determine the properties of signals.

As angular spread and number of propagation paths increase, the channel coefficients of different antennas in the array tend to decorrelate and with a very large angular spread the correlation

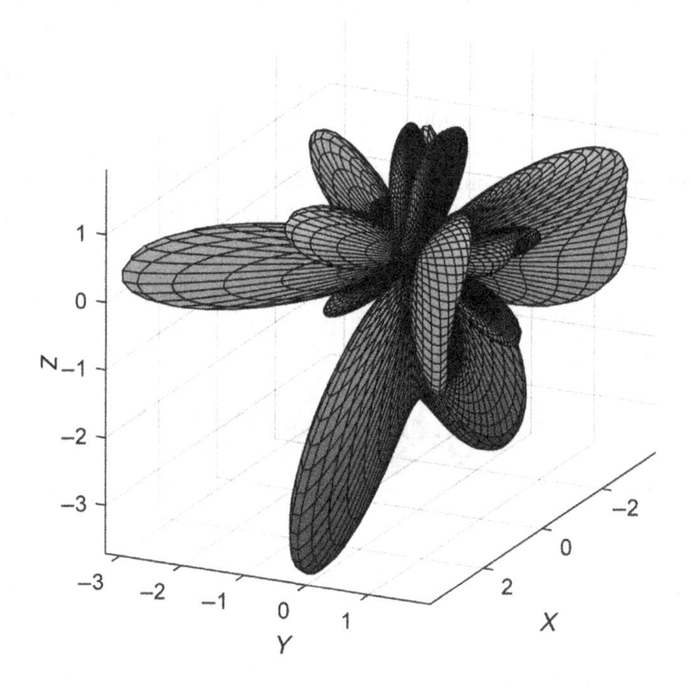

FIGURE 6.7

Optimal beam shape for one particular IID Rayleigh fading channel realization leading to a totally chaotic beam shape with no practical relevance for predicting any properties at all of wireless signals.

(cf. Section 3.6.2.8) may be so small that the channel coefficients may be reasonably modeled as spatially uncorrelated complex Gaussian or identically and independently distributed (IID) Rayleigh fading. This idealization has absolutely no meaningful interpretation as a beam shape and completely different beam shapes are obtained for each channel realization; in essence everything has been reduced to a fast fading effect and as has been previously discussed in this section, beam shapes and patterns are not really useful for capturing fast fading effects. Fig. 6.7 illustrates the chaotic yet optimal beam shape matched to one particular realization of an IID Rayleigh fading channel.

6.1.1.2.3 Generalized beamforming tracking fast fading over time and frequency

For generalized beamforming to be optimal it needs to change over both time and frequency to track the fast fading. As explained in Section 3.5.3.1, fast fading is a small-scale property of the channel and arises from the multi-path as several different propagation paths with different phase offsets are summed up to form a received signal. The fast fading changes the channel realization both over time and frequency, causing the distribution of channel directions to change over time and frequency as well. Contrast this to the zero channel angular spread scenario where there is no fast fading so only large-scale properties of the channel affect the beam shape.

6.1.1.2.4 Terminology related to generalized beamforming

To explicitly stress the use of potentially arbitrarily looking beam shapes, the term *generalized beamforming* is used in this book. Generalized beamforming encompasses classical beamforming as a special case. When the channel angular spread happens to be zero the use of classical beamforming is optimal; otherwise, it is not. Note that such beamforming may arise due to other reasons than the properties of the propagation channel of an individual link. For example, as will be discussed in Section 6.1.8, intentionally reducing beam gain in certain directions by means of null-forming to avoid interference may also create rather arbitrary beam shapes.

A popular term for generalized beamforming when used on the transmit side is *precoding*. For receive beamforming an alternative term is *spatial combining*. Regardless of transmit or receive, beamforming can also be referred to as *spatial filtering*.

The following will take a more detailed look at beamforming and its classical and generalized variants to give a deeper understanding. Toward this end, some basic mathematical models are useful.

6.1.2 SIMPLE NARROWBAND MODELS

A beamformer[1] can be represented by an $N \times 1$ complex-valued weight vector

$$\boldsymbol{w} = \begin{bmatrix} w_1 & w_2 & \cdots & w_N \end{bmatrix}^{\mathrm{T}} \tag{6.1}$$

where $w_k^{(\mathrm{a})}$ and $w_k^{(\mathrm{p})} \in [0, 2\pi)$ in beamformer coefficient $w_k \triangleq w_k^{(\mathrm{a})} \exp\left(j w_k^{(\mathrm{p})}\right)$ are the amplitude and phase applied to the kth antenna element, respectively. It is also referred to as a beam weight vector since it is used to weigh the signals on the different antennas. The beamformer spatially filters the

[1]Transmit beamforming is often referred to as precoding and the corresponding term for "beamformer" would then be "precoder." The beamforming terminology is somewhat confusing. This issue is further discussed in Section 6.1.5.

channel between the two sides of the communication link. For simplicity, the initial focus will be on narrowband signals, for example, a single subcarrier in an OFDM system.

Consider a communication link between a device A (e.g., a base station) with N_A antennas and a device B (e.g., a UE) with a single antenna and where beamforming is used in device A. For beamforming in the transmit direction, the complex baseband data model introduced in Section 5.3.2 specializes for the received signal at device B into

$$y_B = \boldsymbol{h}_{A\to B}^T \boldsymbol{x}_A + e_B = \boldsymbol{h}_{A\to B}^T \boldsymbol{w}_{A,t} s_A + e_B \tag{6.2}$$

where

$$\boldsymbol{h}_{A\to B}^T \triangleq \left[h_1^{A\to B} \quad h_2^{A\to B} \quad \cdots \quad h_{N_A}^{A\to B} \right]$$

is the $1 \times N$ channel vector from device A to B, $\boldsymbol{x}_A$ is the transmitted $N_A \times 1$ vector over the N_A antennas, $\boldsymbol{w}_{A,t}$ is the transmit beamformer, and s_A is the transmitted information-carrying symbol from device A. Noise, interference, and modeling errors are represented by the noise term e_B. Since the channel here has multiple inputs and only one output, it is referred to as a *multiple-input single-output* (MISO) channel. The whole setup is illustrated in Fig. 6.8. Note how each beamformer coefficient multiplies with the channel coefficient of the corresponding antenna.

For receive beamforming in the reverse link the corresponding expression for signal z_A at device A after processed by receive beamforming is

$$z_A = \boldsymbol{w}_{A,r}^T \boldsymbol{y}_A = \boldsymbol{w}_{A,r}^T \boldsymbol{h}_{B\to A} s_B + \boldsymbol{w}_{A,r}^T \boldsymbol{e}_A \tag{6.3}$$

where $\boldsymbol{w}_{A,r}$ is the $N \times 1$ receive beamformer, $\boldsymbol{h}_{B\to A}$ is the $N_A \times 1$ channel vector from device B to A, s_B is the transmitted information-carrying symbol from a single device A antenna, and $\boldsymbol{e}_A$ is an $N_A \times 1$ vector of noise. The channel may be described as a *single-input multi-output* (SIMO) channel. An illustration is provided in Fig. 6.9.

Without loss of generality and for notational simplicity and brevity, the focus is henceforth on transmit beamforming and consequently the subscripts/superscripts will be dropped.

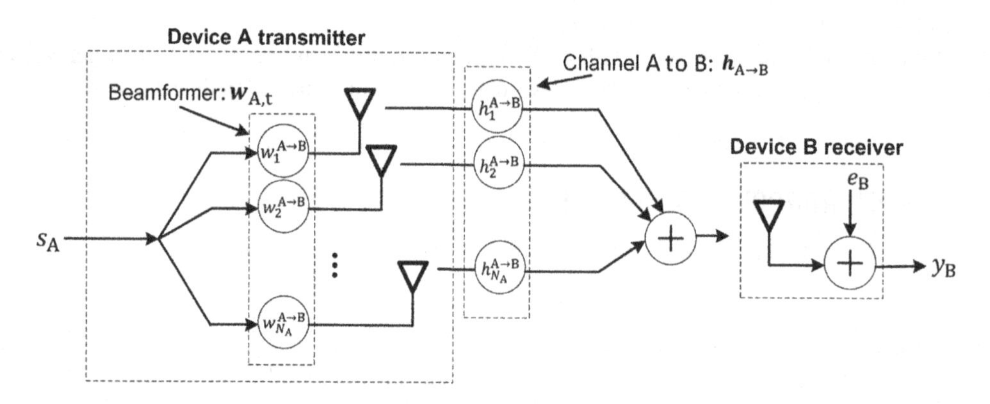

FIGURE 6.8

Received data model for *transmit* beamforming on a link between a multi-antenna device A and a single-antenna device B.

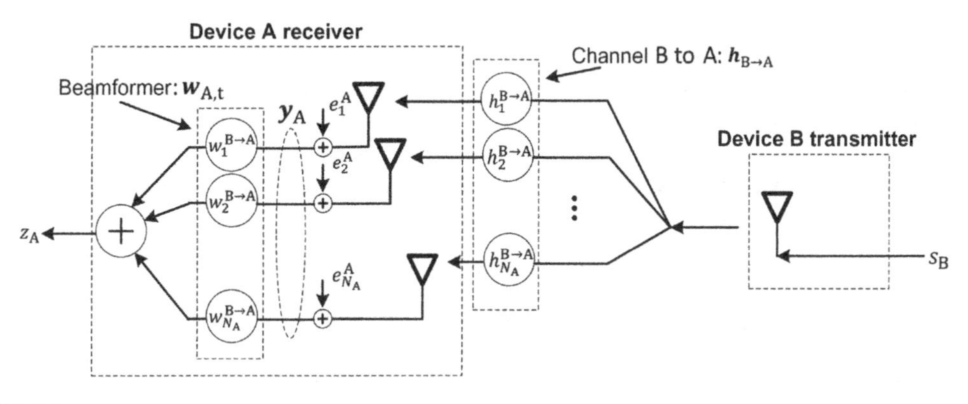

FIGURE 6.9

Received data model for receive beamforming on a link between a multi-antenna device A and a single-antenna device B.

6.1.3 CLASSICAL BEAMFORMING FOR PURE LINE OF SIGHT

This section derives a simple classical beamformer. Various other classical beamformers are possible. What all these beamformers have in common is a single dominating peak in the beam pattern, thus making it meaningful to think in terms of a well-defined pointing direction.

As discussed in Section 6.1.1, the properties of the channel can vary greatly depending on the scenario. In the extreme case of free-space propagation (or pure LOS) the angular spread of the channel is zero so there is only a single propagation ray in the direction of the UE and the channel vector is thus equal to a single *array response vector* $\boldsymbol{a}(\theta, \varphi)$ as explained in Section 5.3.2.1. This results in a channel vector

$$\boldsymbol{h} \triangleq \begin{bmatrix} h_1 & h_2 & \cdots & h_N \end{bmatrix}^{\mathrm{T}} = \boldsymbol{a}(\theta,\varphi) = g(\theta,\varphi)\begin{bmatrix} \exp\left(j\boldsymbol{r}_1^{\mathrm{T}}\boldsymbol{k}\right) & \exp\left(j\boldsymbol{r}_2^{\mathrm{T}}\boldsymbol{k}\right) & \cdots & \exp\left(j\boldsymbol{r}_N^{\mathrm{T}}\boldsymbol{k}\right) \end{bmatrix}^{\mathrm{T}} \quad (6.4)$$

where

$$\boldsymbol{r}_n = \begin{bmatrix} r_x^{(n)} & r_y^{(n)} & r_z^{(n)} \end{bmatrix}^{\mathrm{T}}$$

is the position in three-dimensional (3D) space of the nth antenna in the array and the so-called *wave vector*

$$\boldsymbol{k} = \frac{2\pi}{\lambda}\begin{bmatrix} \cos(\varphi)\sin(\theta) & \sin(\varphi)\sin(\theta) & \cos(\theta) \end{bmatrix}^{\mathrm{T}}$$

points in the direction from the device A performing the beamforming (e.g., base station) to device B (e.g., UE). See also Section 4.3.2 for a description of these entities. Note that for simplicity, it is here assumed that all antennas have the same pattern $g(\theta, \varphi)$, and that the path loss is 0 dB and the transmitter and receiver antennas are polarization aligned (i.e., the true channel is proportional to $\boldsymbol{h}$ and thus only differs by a scaling factor not affecting beam shape).

From (6.4) it is seen that the channel coefficient $h_n = g(\theta, \varphi)\exp\left(j\boldsymbol{r}_n^{\mathrm{T}}\boldsymbol{k}\right)$ in channel vector $\boldsymbol{h}$ rotates the signal in the complex plane of the nth antenna before being added to the signals from the other antennas. This rotation may lead to constructive or destructive addition, depending on the relative phase, or equivalently delay or path length, between signals of the different antennas.

To maximize the received power at the UE, all N channel coefficients need to be de-rotated by choosing the beamformer as

$$\boldsymbol{w} \sim \boldsymbol{a}^{\mathrm{c}}(\theta, \varphi) = g(\theta, \varphi)\left[\exp\left(-j\boldsymbol{r}_1^{\mathrm{T}}\boldsymbol{k}\right) \quad \exp\left(-j\boldsymbol{r}_2^{\mathrm{T}}\boldsymbol{k}\right) \quad \cdots \quad \exp\left(-j\boldsymbol{r}_N^{\mathrm{T}}\boldsymbol{k}\right)\ \right]^{\mathrm{T}} = \boldsymbol{h}^{\mathrm{c}} \tag{6.5}$$

This aligns the phases of the signals for all the antennas before their summation on the receive side in Fig. 6.8.

For transmit beamforming, the transmission power is limited by the capabilities of the device. Usually, equal size power amplifiers (PA) are used on all the antennas. Power cannot be reallocated from one PA to another. The present beamformer has the same power on each antenna since the beamformer coefficients all have the same amplitude. It is therefore straightforward to exploit all the totally available power P of the N antennas of the device by scaling the beamformer vector according to

$$\boldsymbol{w}_{\mathrm{opt}} = \kappa \boldsymbol{h}^{\mathrm{c}} = \frac{\sqrt{P}}{\|\boldsymbol{h}\|}\boldsymbol{h}^{\mathrm{c}} \tag{6.6}$$

where κ denotes some scaling factor set so as to fulfill some power constraint, in this case taking the value $\kappa = \sqrt{P}/\|\boldsymbol{h}\|$. It can be shown that the beamformer in (6.6) is optimal in the sense of maximizing the received power at device B. In fact, it corresponds to the matched filter of the channel, which is known to be optimal and leads to the downlink received signal

$$y = \sqrt{P}\|\boldsymbol{h}\| s + e$$

In other words, the complex conjugate of the array response vector maximizes the received power in the direction of the UE, thereby also, compared with an isotropic beam shape, reducing the transmission power in the other directions so that the total radiated power stays the same according to energy conservation principles. For receive beamforming, the reasoning is very similar except that no power normalization is needed.

Note that other power scaling schemes are also possible, but it is beyond the scope of this book to elaborate further on this for the general case. For brevity the scaling scheme will hence in the following be represented by the κ parameter without further details.

To illustrate the spatial power distribution of this kind of beamformer, consider the example of an antenna array with its elements arranged in $M_{\mathrm{v}} = 4$ rows and $M_{\mathrm{h}} = 8$ columns with 0.5 wavelength elements spacing horizontally as well as vertically. The resulting beam pattern was shown in Fig. 6.3 and is seen to have a distinct main lobe and small and regular side lobes. It very much looks like a "beam" —the layman's likely mental model of what a beam should look like and therefore falls under the category of classical beamforming. Such beamforming may not only arise in the particular scenario of free-space propagation, but it may also be the result of a conscious choice to use beamformers that resemble the classical beamforming shape with a clear main lobe, even though the propagation channel has multi-path components in other directions than the main lobe.

6.1.4 GENERALIZED BEAMFORMING FOR A MULTI-PATH CHANNEL

A channel consisting of a single propagation path is a special case. Usually the propagation environment introduces at least some degree of multi-path, scattering the signals on objects before reaching the receiver. This means there are now multiple directions from the beamforming device A that can convey power to the device B. As evident from Section 3.6.2.8, the resulting angular spread leads to that the correlation between channel coefficients corresponding to different antennas is lowered and thus the channel vector $\boldsymbol{h}$ becomes more random and unstructured than in free-space propagation. In the extreme limit of a totally isotropic angular spread around the base station, the correlation between the channel coefficients of different antennas may be very low. Such low correlation scenarios are often modeled by assuming the channel vector $\boldsymbol{h}$ is IID complex Gaussian, implying zero correlation among the channel coefficients in $\boldsymbol{h}$. The popular use of such an idealized spatially uncorrelated model will later in Section 6.7.4.2 be shown to lead to several misleading conclusions.

In scenarios with very low spatial correlation, both the amplitudes and phases of the different channel coefficients are completely, or largely, unrelated, which is the opposite of the single propagation ray pure LOS scenario. As previously mentioned in Section 6.1.1, the optimal beam that maximizes received power in presence of channel angular spread may then look *very* different from the pure LOS propagation case with its corresponding classical beamforming.

Under a sum power constraint P over all the transmit antennas, it can be shown [1] that the transmit beamformer that maximizes the received power is also in this case

$$\boldsymbol{w}_{\text{opt}} = \kappa \boldsymbol{h}^{\text{c}} = \frac{\sqrt{P}}{\|\boldsymbol{h}\|} \boldsymbol{h}^{\text{c}} \tag{6.7}$$

that is, a spatial filter matched to the channel that ensures all antenna signals add constructively at device B by aligning their phases. However, since the channel coefficients in $\boldsymbol{h}$ due to the angular spread are no longer of equal amplitude, the beamformer now has a corresponding amplitude weighting. The technique is referred to as *conjugate beamforming* or *maximum ratio transmission* (MRT).

In reality, each PA has a maximum output power rating, which implies that unused power of one PA cannot be transferred to another PA so instead a per PA maximum power constraint should be used when deriving transmit beamforming weights. The optimum beamformer would then have the same phases as $\boldsymbol{w}_{\text{opt}}$ but where the amplitudes of the beamformer coefficients are all equal corresponding to the power per PA. Again, the situation for receive beamforming is completely analogous except that there is no need to deal with the power constraint aspect. Such receive beamforming is called *maximum ratio combining* (MRC) and is described in, for example [2].

Fig. 6.6 provides an example of an optimal MRT/MRC beam shape for a multi-path radio channel, assuming the same antenna setup as in the classical beamforming case in Fig. 6.3. As mentioned in Section 6.1.1, such a beam constitutes generalized beamforming. It is observed that there are now several strong directions and which of these directions can be considered main lobe and which directions are side lobes can no longer clearly be distinguished. This beamformer optimally allocates power along all the propagation paths of the channel. Despite its optimality, in traditional antenna theory this would be characterized as an extremely ill-designed beam shape! This is because the target for traditional antenna theory is typically to obtain a distinct main lobe, while side lobes and grating lobes are considered harmful.

As described in Section 3.5.3.3, multi-path not only leads to angular spread it also makes the channel frequency-selective. The degree of frequency variation depends on the time dispersion of the channel, which is mainly affected by differences in propagation lengths between different propagation paths. A frequency-selective channel results in that the optimal beamformer for generalized beamforming becomes frequency-selective as well. This contrasts with classical beamforming in which the same beam shape can be used over a wide frequency range.

6.1.4.1 Revisiting the hand-waving beam shape analysis now using mathematics

Section 6.1.1 tried hard to avoid mathematics and instead utilized hand-waving to explain rather intricate phenomena concerning beam shapes in conjunction with multi-path channels. This subsection will use mathematics as explanation tool instead.

As shown in Section 5.3.2, the array response vectors play a key role in forming the channel. In the case of a multiple antenna base station and a single-antenna UE, the channel vector is a sum of weighted array response vectors where each array response vector corresponds to a propagation path. This is a generalization of the single-path model used in (6.4) and can in fact be seen as a superposition of multiple single-path contributions. In other words, and similar to as in Section 5.3.2.1 but for the MISO channel case, the channel vector $\boldsymbol{h}$ in (6.4) now generalizes in the case of L paths to

$$\boldsymbol{h} = \sum_{l=1}^{L} \lambda_l \boldsymbol{a}(\theta_l, \varphi_l) \tag{6.8}$$

where λ_l is complex-valued and captures the phase and amplitude of the l:th path. From (6.7) it is seen that the optimal MRT beam weight vector $\boldsymbol{w}_{\text{opt}}$ is proportional to the complex conjugate of the channel vector $\boldsymbol{h}$, that is,

$$\boldsymbol{w}_{\text{opt}} \sim \sum_{l=1}^{L} \lambda_l^{\text{c}} \boldsymbol{a}^{\text{c}}(\theta_l, \varphi_l) \tag{6.9}$$

Drawing beam shapes as in Figs. 6.3–6.7 can conceptually be thought of as moving around a single-antenna receiver on the entire sphere and measuring receive power in a noise-free free-space channel $\boldsymbol{h} = \boldsymbol{a}(\theta, \varphi)$. The resulting received signal $y(\theta, \varphi)$ in looking direction given by (θ, φ) can from (6.2) and (6.9) be written as

$$y(\theta, \varphi) = \boldsymbol{h}^{\text{T}} \boldsymbol{w}_{\text{opt}} \sim \boldsymbol{a}^{\text{T}}(\theta, \varphi) \sum_{l=1}^{L} \lambda_l^{\text{c}} \boldsymbol{a}^{\text{c}}(\theta_l, \varphi_l) = \sum_{l=1}^{L} \lambda_l^{\text{c}} \left(\boldsymbol{a}^{*}(\theta, \varphi) \boldsymbol{a}(\theta_l, \varphi_l)\right)^{\text{c}} = \sum_{l=1}^{L} \lambda_l^{\text{c}} \alpha_l(\theta, \varphi) \tag{6.10}$$

where transmitted symbol s has been set to identity, superscripts and subscripts have been dropped and $\alpha_l(\theta, \varphi) \triangleq \left(\boldsymbol{a}^{*}(\theta, \varphi) \boldsymbol{a}(\theta_l, \varphi_l)\right)^{\text{c}}$ is the conjugate of the complex-valued scalar product between the array response vectors of the "looking direction" and the direction of the l:th propagation path and can be viewed as a measure on the similarity between the array response vector of the looking direction and the array response vector of the l:th path. As will be evident in the following, it also constitutes the beam shape of a classical beam pointing in the direction of the l:th path and thus has a certain non-zero beamwidth corresponding to the, in Section 6.1.1.2.1, discussed resolution of the array.

Comparing (6.9) with (6.5), classical beamforming is seen to correspond to $L = 1$ channel propagation path. Thus $\alpha_l(\theta, \varphi)$ can be interpreted as the complex-valued amplitude beam pattern of an optimal classical beam along the lth propagation path, for example, as in Fig. 6.3 which shows the beam shape in the power domain based on $|y(\theta, \varphi)|^2$.

Similarly, the multiple propagation path case can from (6.10) be viewed as the superposition of multiple complex-valued amplitude patterns $\lambda_l^c \alpha_l(\theta, \varphi)$, each with a beam optimized for the corresponding propagation path. This simple model was in fact used to produce the beam shape illustration in Figs. 6.4 and 6.5 based on a channel

$$\boldsymbol{h} = \lambda_1 \boldsymbol{a}(0, 0) + \lambda_2 \boldsymbol{a}(0, 10) + \lambda_3 \boldsymbol{a}(0, 40)$$

where angles are measured in degrees, $\{\lambda_l\}$ is set to unit norm and assumed statistically independent and with 360-degree uniformly distributed phase. The only difference between the two figures are the values of the phases of $\{\lambda_l\}$.

If for a certain looking direction (θ, φ) one of the $\alpha_l(\theta, \varphi)$ terms in (6.10) dominates, then the gain in that direction becomes essentially equal to that of a classical beam targeting the corresponding propagation direction in isolation. If only one term in (6.10) remains for each possible looking direction, the phases of the propagation paths become irrelevant and the resulting beam gain pattern around the looking direction may be obtained by a superposition of beam shapes ignoring phase. On the other hand, if there is more than one term contributing significantly, as, for example, when the looking direction is close to two or more propagation directions, then the phase value as part of the propagation gain factor λ_l^c becomes crucial in determining the power gain $|y(\theta, \varphi)|^2$ and as earlier discussed in Section 6.1.1.2 intuition cannot predict the result of this fast fading effect.

6.1.5 THE CONFUSING BEAMFORMING TERMINOLOGY

Since generalized beamforming is often referred to as just "beamforming," the wording often leads people astray when their mental model of a beam clashes with such arbitrary looking beam patterns and is instead totally aligned with classical beamforming. In our experience this results in many misunderstandings. Therefore when we wish to emphasize the use of such kind of beamforming, while staying agnostic to whether it is transmit or receive beamforming, we instead explicitly refer to it as *generalized beamforming* although the two terms "generalized beamforming" and "beamforming" have formally the same definition of including all possible beam shapes, including classical ones. On the other hand, if the context is confined to generalized beamforming in the transmit direction, then we prefer to use the term *precoding* since this abstract sounding term leaves little room for deceiving mental pictures. This term is also used in 3GPP specifications and standardization work, as also evident from Chapters 8 and 9, which are devoted to the details of 3GPP technologies.

6.1.6 CLASSICAL VERSUS GENERALIZED BEAMFORMING

Loosely speaking, the transmit channel correlation matrix (see Section 5.3.3)

$$\mathbf{R}_{hh} \triangleq \mathrm{E}\left[(\boldsymbol{h}^{\mathrm{T}})^{*}\boldsymbol{h}^{\mathrm{T}}\right] = \mathrm{E}\left[\boldsymbol{h}^{\mathrm{c}}\boldsymbol{h}^{\mathrm{T}}\right] = \mathrm{E}[\boldsymbol{h}\boldsymbol{h}^{*}]^{\mathrm{c}}$$

captures the physical directions of the channel and their relative strengths as seen from the transmitter. The fast fading is, however, averaged away via the expectation operator E[·]. Assuming the common IID complex Gaussian channel model makes the correlation matrix proportional to an identity matrix. Other channel models, however, typically result in much better models in line with reality and then the corresponding correlation matrix is usually far from an identity. The channel correlation matrix represents large-scale channel properties, while the actual channel realization $\boldsymbol{h}$ is instantaneous and captures everything for the particular considered time and frequency, including the fast fading.

Channel correlation is slowly changing over frequency and time, while the opposite is true for the fast fading. In the special case of zero channel angular spread, that is, a single propagation path, the channel correlation matrix is rank 1 and contains the same information as the channel realization.

Since classical beamforming targets adapting transmission/reception to a single physical propagation direction, it can be said to target the long-term/wideband spatial correlation or large-scale properties of the channel in the case of zero channel angular spread. It is in fact optimal for this case.

When there is multi-path propagation leading to non-zero channel angular spread, the channel coefficients as shown in Section 3.6.2.8 start to decorrelate and classical beamforming is no longer optimal. Only generalized beamforming can be made optimal by appropriately shaping the beam in accordance with the spatial distribution of channel gain over direction. Under typical channel conditions, the short-term/frequency-selective channel properties of the fast fading also need to be exploited to create a time- and frequency-varying beamformer that achieves optimality. As the channel angular spread reduces, generalized beamforming in the form of MRT becomes more and more similar to classical beamforming.

Classical beamforming has the advantage of being simple and in most cases robust, even for higher device speeds. For example, assume the beam direction at the base station side is updated every 10 ms and a UE is moving with a speed of 100 km/h only 10 m from a base station. The direction to the UE then only changes 1.6 degrees between each beam direction update, which is not an issue, unless the beamwidth is much narrower than 1.6 degrees. Note that this represents an extreme case and normally the angular speed is much smaller than this.

UEs moving around a corner can, however, present a problem even for classical beamforming if the channel conditions change very abruptly. This may be the case for high-frequency applications such as mm-wave where the diffraction is very limited so the border to a radio shadow behind a building becomes very sharp. The particular challenges and issues with mm-wave are addressed in Chapter 7. Concepts and Solutions for High-Band Millimeter Wave.

For classical beamforming, very little CSI is needed to describe what beam to use, essentially only a single direction. In contrast, generalized beamforming needs a lot of information to describe its complex and arbitrary beam shape, a beam shape that moreover tends to vary quickly over frequency and time depending on propagation channel properties. Frequency variations increase with increasing multi-path time dispersion. Time variations increase with increasing mobility of the involved devices—moving even only a part of a wavelength could lead to a substantially different channel realization and a need for the beamformer to be updated at a high rate. These are fast fading effects that further increases the amount of CSI needed. The challenges in capturing fast fading channel properties have great ramifications on the system design as will be evident from subsequent sections.

6.1.6.1 Impact of angular spread on beamforming gain

As previously mentioned, classical beamforming is only optimal in maximizing SNR in free-space channels, while with multi-path channels generalized beamforming is required to guarantee a maximal SNR. To illustrate the performance difference between classical and generalized beamforming, a simplified example will now be analyzed.

Consider a transmitting base station antenna array configuration given by a co-polarized *uniform linear array* (ULA) consisting of N_t antennas with a lossless isotropic pattern and a receiving UE with one receive antenna (see Fig. 6.10). The noise impaired received data model for a single subcarrier in an OFDM system is given by (6.2). Recall from (6.8) that a narrowband multi-path channel can be written as a sum of array response vectors for the various paths. A single cluster of propagation paths is assumed with a large number of propagation paths with horizontal angles φ_l drawn according to a Gaussian distribution with a mean $m_\varphi \triangleq \mathrm{E}[\varphi]$ corresponding to the direction of the cluster and a standard deviation equal to the angular spread of the channel. The angle θ_l and angular spread in the vertical domain are assumed to be zero. The propagation paths all have equal strength ($|\lambda_l| = \text{constant}$ in (6.8)) and 360-degree uniformly distributed phase.

An estimate

$$\hat{\boldsymbol{h}} = \boldsymbol{h} + \tilde{\boldsymbol{e}}$$

of the true channel $\boldsymbol{h}$ is assumed to be known at the transmitting base station. The noise term $\tilde{\boldsymbol{e}}$ is here assumed to be spatially white and complex Gaussian with a variance such that the channel estimation SNR is SNR_h. This noise may model various impairments of the channel knowledge at the base station, for example, channel estimation errors.

For MRT, the beamformer/precoder is from (6.7) given as $\boldsymbol{w}_{\mathrm{opt}} = \sqrt{P}/\|\boldsymbol{h}\|\boldsymbol{h}^{\mathrm{c}}$, while for classical beamforming the base station selects a beam from a fixed grid of $2N_t$ different beams spanning all directions of the front half-plane of the array. The beam that maximizes the SNR given a channel equal to the channel estimate $\hat{\boldsymbol{h}}$ is selected. The beamforming method is referred to as discrete Fourier transform (DFT)-based grid of beam (GoB), which will be further discussed in Section 6.3.3.4.

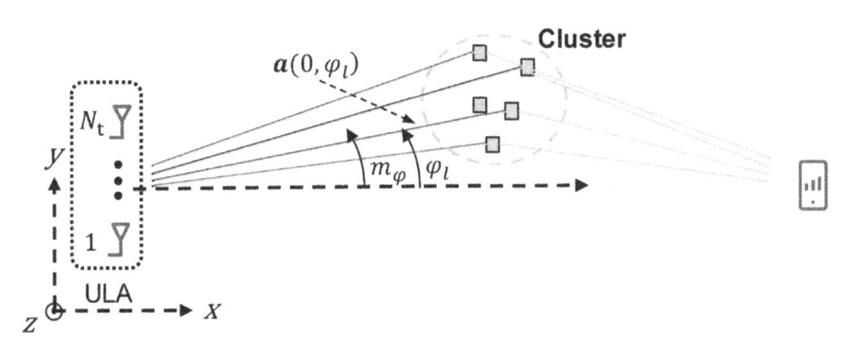

FIGURE 6.10

A base station ULA transmitting to a UE over a channel with a cluster of propagation paths.

Fig. 6.11 illustrates the beam shapes associated with some different number of antennas for the broad-side beam in the grid. The width of the main lobe is seen to decrease with increasing number of antennas (see also Section 4.3.3.1).

The average SNR of the received signal y in (6.2) is computed, where the average is not only taken over noise but also taken over many realizations of the propagation path phases in λ_l to create many different fast fading realizations of the channel. Fig. 6.12 shows the resulting average SNR as a function of the angular spread for GoB and MRT with different numbers of transmit antenna elements N_t, essentially infinite channel estimation SNR_h, and mean angle of departure $m_\varphi = 0$. MRT is seen to be insensitive to angular spread and reaches the maximum theoretical SNR gain of N_t (linear scale), whereas GoB performance clearly degrades with increased angular spread.

For classical beamforming such as GoB, a rule of thumb is that the received signal strength grows in accordance with the gain given by the beam shape as long as the (half power) main lobe width is sufficiently much larger than the angular spread of the channel, while for a main lobe width that is smaller than the angular spread, the beamforming gain starts to saturate. Thus the loss for GoB versus MRT for a fixed angular spread increases with increasing number of antennas as the lobe width is inversely proportional to the number of antennas (see Section 4.3.3.1). According to this kind of reasoning, the relation between angular spread and main lobe width is the key parameter in determining how effective classical beamforming is relative to the optimal MRT.

The right part of Fig. 6.12 shows the loss relative MRT as a function of the angular spread relative to the main lobe width. It is seen that the loss is roughly invariant to the number of antennas and when angular spread is significantly smaller than the main lobe width the loss is small. This thus confirms that at least for this simplified single-cluster multi-path scenario, the rule of thumb makes sense. So if the angular spread and beamwidth are known, then the loss is also known.

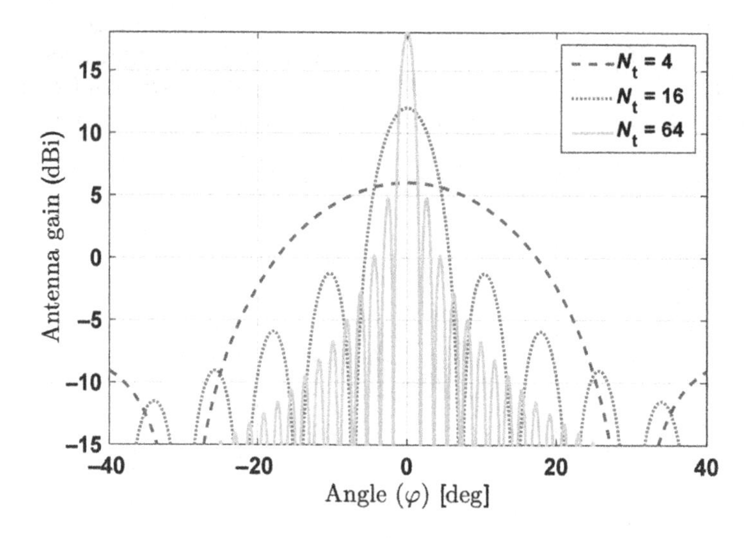

FIGURE 6.11

Beam gain pattern associated with the GoB weight vector of all ones pointing in broad side for $N_t = 4, 16, 64$ antennas.

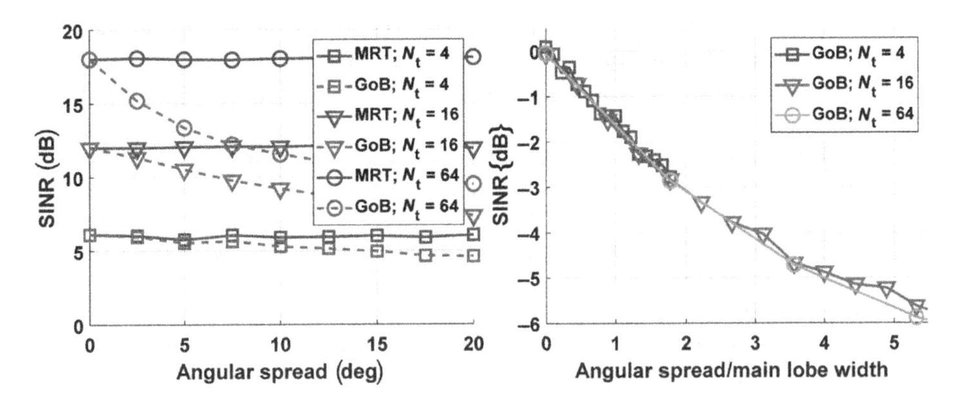

FIGURE 6.12

Illustration of impact of angular spread for performance of GoB and MRT for different number of transmit antennas, $N_t = 4, 16, 64$.

Fig. 6.13 shows SINR as a function of angular spread for GoB and MRT with $N_t = 16$ transmit antennas and different levels of channel estimation error SNR_h. Here $SNR_h = 100$ dB corresponds to ideal channel estimation with no errors. It is seen that MRT is generally more sensitive to channel estimation errors than GoB, and with estimation errors GoB might outperform MRT at low angular spread. This example illustrates that although generalized beamforming is optimal

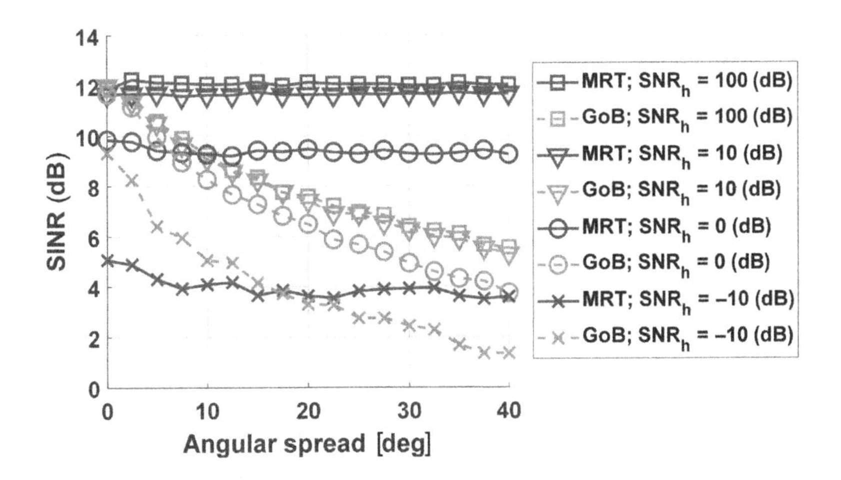

FIGURE 6.13

Illustration of impact of angular spread and channel estimation errors for performance of GoB and MRT for $N_t = 16$ antennas.

and better than classical beamforming when the channel knowledge is perfect and angular spread is large it is more sensitive to impairments than the latter. Hence, for UEs with poor coverage, classical beamforming such as GoB can yield better received signal strength compared with generalized beamforming such as MRT, whereas for UEs with good coverage, MRT may excel.

It needs to be emphasized that this was a simple single-cluster example to illustrate a principle. More realistic channel models (see Section 3.6.3.2) consist of multiple clusters with different delays, resulting in frequency-selective channels. Hence, for generalized beamforming such as MRT to work optimally, frequency-selective precoding is needed where the precoder for each sub-band (a group of consecutive subcarriers) is given by the corresponding channel. For classical beamforming, there is typically little benefit from making it frequency-selective as it exploits large-scale channel properties that by nature have a limited frequency impact on the channel, as long as the bandwidth relative to the carrier frequency is sufficiently small, which is often the case. The presence of multiple clusters tends to soften the negative impact of large angular spread on classical beamforming as there is a likelihood that the main lobe is able to resolve the strongest cluster from the other clusters, thus reducing the effective angular spread to that of a single cluster.

Also, here a single-dimension ULA was considered, while most AAS uses planar arrays consisting of both horizontal and vertical dimensions. In that case, the principles illustrated here can generally be applied per dimension. Note, though, that for most scenarios, the vertical angular spread is substantially less than the horizontal angular spread. See Section 3.5.3.3 for further details and typical numbers on angular spread for different channel models.

6.1.7 BEAMFORMING AND IMPACT ON INTER-CELL INTERFERENCE

Beamforming clearly increases the received power. But what does it do to the interference? This section focuses on the inter-cell interference case and assumes the beamforming is only designed to match the channel of the device of interest. The use of beamformers that are *actively* designed to suppress inter-cell interference is discussed in Section 6.6.2, while intra-cell interference is treated as part of describing spatial multiplexing techniques in Section 6.3.

In a typical cellular network there are many base stations and each base station intends to cover a particular area. Neighboring cells tend to interfere with each other, since the same time/frequency resource is reused between cells. In downlink, transmission from a neighboring base station may interfere with UEs close to the cell border. The interference can be modeled to be part of the $\boldsymbol{e}_{\mathrm{B}}$ noise term in (6.2) where device A and B now play the roles of a base station and a UE, respectively. In uplink, transmissions from UEs in neighboring cells may interfere with the reception at the base station and can be modeled to be part of the $\boldsymbol{e}_{\mathrm{A}}$ noise term in (6.3).

To get good performance the signal of interest should be strong relative to noise and interference, that is, the SINR should be high. SINR is a metric defined to be the power ratio between signal of interest and interference plus noise. Beamforming increases the received power of the signal of interest and that helps to increase SINR.

The impact of beamforming on the interference level is, however, plagued by misconceptions. A commonly made claim is that transmit beamforming in general reduces interference, presumably because beamforming focuses the energy in the direction of the intended receiver, thereby emitting less power in unwanted directions that only contribute to interference. This is, however, not

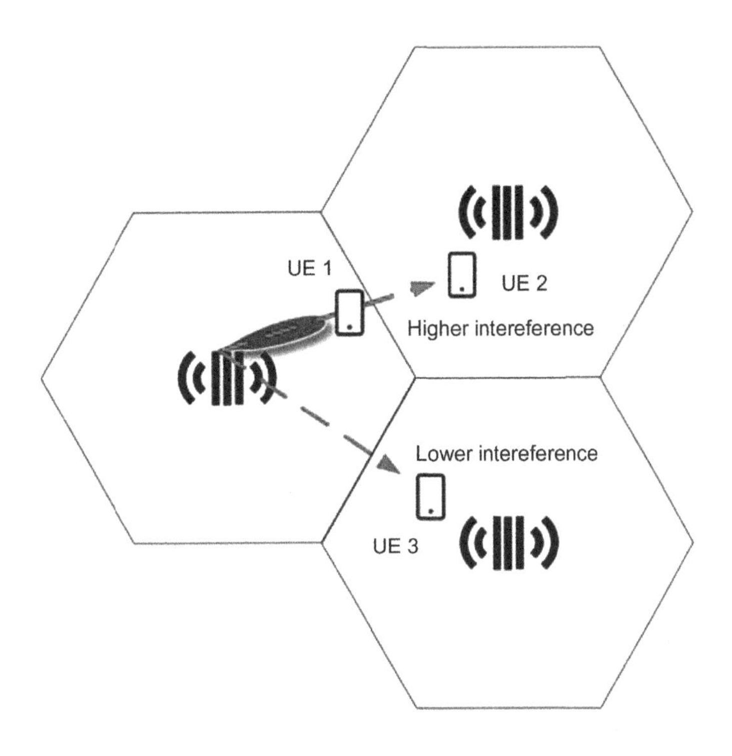

FIGURE 6.14

Beamforming can increase or decrease interference depending on the situation.

necessarily true and can easily be shown by a counter example. Consider the situation in Fig. 6.14 with pure LOS and a base station beamforming to UE 1, while that beam is also directly hitting UE 2 that belongs to another cell, thus acting as interference to that UE. UE 1 gets higher reception power (compared with broader beam transmissions) but the interfering power to UE 2 is also increased. Thus beamforming *increases* the interference! On the other hand, it is seen how UE 3 is missed by the main lobe of the beam intended for UE 1 and thus interferences level for UE 3 *decreases*.

The impact on interference level heavily depends on the situation, in particular, as is evident from the examples in Fig. 6.14, it is for typical propagation channels very much affected by the spatial UE distribution. Interference level sometimes increases and sometimes decreases. That said, in practice for cellular networks, the interference level often stays the same or only mildly increases due to increased levels of transmit beamforming from the base stations. For example, in the classical homogenous deployment of hexagonal macro cells with spatially uniformly distributed UEs, increased levels of horizontal UE specific transmit beamforming (i.e., more narrow beams with higher gain) from the base stations do not significantly affect the average inter-cell interference level seen at the UEs and moreover in several cases have even very little impact on the cumulative density function of the inter-cell interference level.

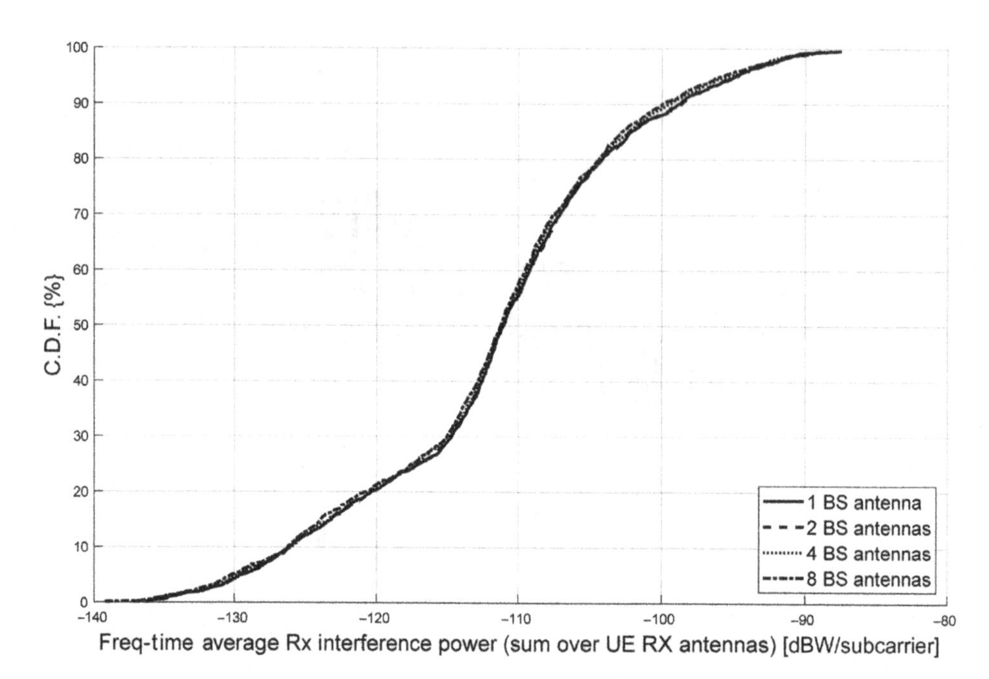

FIGURE 6.15

Average (over time and frequency for each UE) inter-cell interference power distribution relatively agnostic to different degrees of horizontal downlink beamforming from base station (BS) to UE with uniform UE distribution (3GPP 3D Urban Macro Scenario [4]).

The essentially constant average inter-cell interference level can be intuitively understood by considering the average power distribution from an interfering base station, that is, essentially the power envelope of the used data beams. A spatially uniform UE distribution leads approximately to that the angular UE distribution is uniform as well and that hence the data beams are used similarly often over the coverage area of the cell. Consequently, the power envelope of used data beams follows the antenna pattern to a large degree and hence essentially remains unchanged as the number of antennas performing beamforming increases. This is illustrated in Fig. 6.15. The classical hexagonal macro deployment case hence provides an example of when beamforming increases the received power without much impact on the interference level, thus providing an SINR increase similar to the increase in reception power.

It can be noted that the same hexagonal macro example but studied for transmit beamforming in the vertical domain exhibits a totally different behavior. For this case the inter-cell interference level increases with increased level of beamforming, which is explained by the fact that the spatial UE distribution in the vertical domain is not uniformly covering the antenna pattern. Rather, the beamformer tends to be directing an increasingly stronger beam close to the horizon, where many UEs in the serving cell are located (unless the cell is very small), creating more interference to other cells.

6.1.8 NULLFORMING—EXPLICIT INTER-CELL INTERFERENCE MITIGATION

Although beamforming as such does not in general reduce interference levels, it can be designed to explicitly mitigate interference. By *intentionally* designing the beam shape to have nulls or lower gain in directions where victim transceivers are, the harmful interfering signals can be filtered out. These techniques are often in general referred to as nullforming, so the term "null" should not necessarily be taken literally. Nullforming to mitigate inter-cell interference is illustrated in Fig. 6.16. Nullforming is also commonly used to reduce intra-cell interference in MU-MIMO as will be discussed in Section 6.3.

Forming a peak in the beam shape is relatively insensitive to errors in the channel knowledge, even with rather large errors the direction of the peak tends to stay more or less the same, while some gain may be lost. Nullforming is on the other hand very sensitive to errors. This difference in quality-level needs of CSI can be understood by observing that true nulls are deep and narrow with very large gain slopes as evident from Fig. 6.17. Hence, adding just a small error that moves the null slightly may result in a very much larger beam gain than in the minima of the null. In contrast, optimal peaks tend to be broader, so an error that moves the peak does not alter the gain much. Various regularization methods may be used to broaden the nulls and reduce the sensitivity, but that comes at the cost of loss in performance compared with the performance potential of a true (albeit in practice unachievable) null.

On the receive side nullforming is very common and performed via techniques such as *interference rejection combining* (IRC), which will be explained in more depth in Section 6.3.2. For the transmit side, the situation is more challenging since acquiring the accurate CSI needed to form the beam shape with its nulls is non-trivial, as discussed in more detail in Section 6.4.

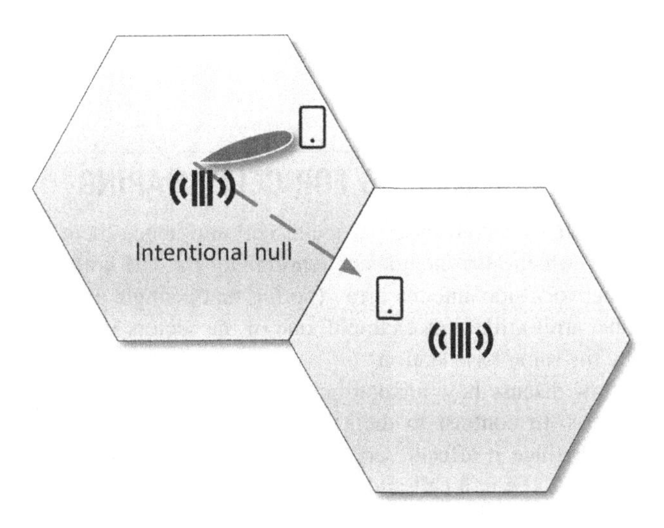

FIGURE 6.16

Beamforming designed with an intentional null toward a victim UE in a neighboring cell.

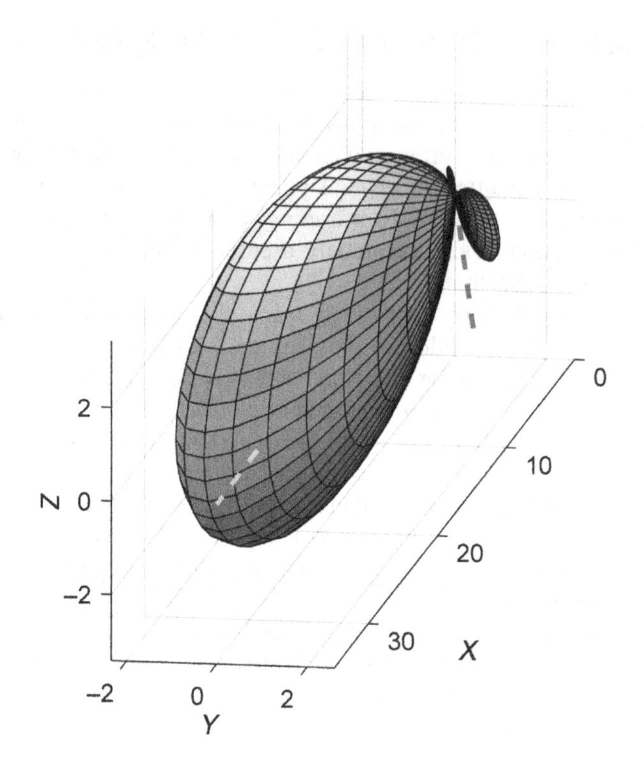

FIGURE 6.17

Nulls are narrow while peaks are relatively broad. Red dashed line marks the direction of the intentional null while green dashed line marks the direction of a UE of interest in a free-space channel model. Nullforming is seen to slightly offset the peak of beam relative to the UE direction.

6.2 SEMI-STATIC BEAM CONCEPTS FOR CELL SHAPING

In cellular networks, each UE is served by a particular cell and hence there is a need for a UE-to-cell association procedure, which also includes synchronizing the UE with the cell. Usually a cell corresponds to a single-network side antenna array (noting that a single antenna can be viewed as a special case of an antenna array), like, for example, one of the sectors in a site. All the sectors of a site are often handled by the same base station.

The sections that follow discuss how adaptation of cell shape via beamforming can be used to obtain performance benefits. In contrast to the UE−specific beamforming, the cell shape update rate is usually much slower since it affects very basic and fundamental procedures in the network that potentially affects all the UEs in a cell and neighboring cells. Thus cell shaping can be viewed as a semi-static beam concept.

Note that merely dynamically switching what antenna array a UE is served by for its data communication can be implemented as a pure physical layer procedure and could hence be completely dynamic. But the family of such schemes does not change the basic broadcast signals that define a cell and are hence not included in the concept of cell shaping.

6.2.1 BRIEF BACKGROUND ON ORDINARY USER EQUIPMENT TO CELL ASSOCIATION

Cell selection is one of the first steps for a UE to get initial access to the network. Determining a UE-to-cell association is also part of handover, which may be triggered when a UE moves enough to cross a cell border. This in turn creates a need to switch serving cell.

The conventional way for cell association is to select the serving cell as the strongest cell the UE receives signals from. This serves to maximize the coverage of each cell while also mitigating inter-cell interference since each individual interfering cell is then guaranteed to be weaker than the signal of interest from the serving cell.

To perform UE-to-cell association, so-called cell-defining signals are transmitted from the network side. Each cell transmits at least one unique such signal and a UE then measures on this signal and considers the strongest cell-defining signal as the strongest cell. This alone does, however, not ensure the cell to be the strongest also for the data transmission. But if the envelope of the beams corresponding to the measured cell-defining signals of a cell matches the envelope of possible beams for data transmission for that cell, then measurements on the former give a good indication of which cell is the strongest one also for data transmissions.

In the common case that there is a one-to-one mapping between cell-defining signals and antenna arrays, the cell association procedure can alternatively be viewed as selecting the strongest antenna array. The in general abstract concept of a cell has a very clear physical interpretation in this case. The selected cell becomes the serving cell for the UE, which means it is handling all the communication with the UE, downlink as well as uplink.

Cell-defining signals include synchronization signals and various other broadcast channels and messages containing information needed to access the cell. Taking NR as an example, such signals include synchronization signal block (SSB)/master information block and system information block 1 carried on physical downlink shared channel (see Section 9.3.2). Earlier wireless standards have more intrusive cell-defining signals, for example, cell-specific reference signal (CRS) in LTE, than NR that on purpose from the start was designed to have a looser coupling between different signal types for increased flexibility.

Before proceeding, the notion of strongest cell deserves some further elaboration. Note that just because a UE has been associated with the strongest cell-defining signal does not necessarily mean that all the signals from that cell for the actual *data* transmissions are the strongest. Only if the beam shapes of the cell-defining signal and the data transmission are the same is this guaranteed. Traditionally, this has been the case, or at least nearly so, as the degree of dynamic beamforming has been limited in commercial networks. This applies for the earlier and in practice most common forms of LTE.

With the introduction of AAS with more aggressive beamforming the difference in beam shape between cell-defining cell and data transmission is potentially growing large. For example, a cell-defining signal may use a fixed beam that covers an entire 120-degree sector horizontally and is very narrow vertically while the UE–specific data beams are dynamic and narrow both horizontally and vertically. With such divergence in beam shapes there is a risk that the selected cell is not the strongest for the data transmissions and hence performance may suffer compared with if indeed the cell would have been selected based on the strongest data beam. In NR, this discrepancy can be reduced somewhat since NR allows for the use of multiple beams for transmitting the same cell-defining channel. For example, multiple SSBs with different beams that together jointly define the cell may be used.

The set of geographical locations in which a particular cell is selected by a UE (i.e., is the strongest in the typical case) represents that cell's coverage area [4], or more precisely cell volume since geographical positions also have a height. On the border of the coverage areas between two or more cells, the so-called cell edge, the SINR levels are often very low due to inter-cell interference since the serving cell and interfering cell are similar in strength.

The shape of the coverage area naturally depends strongly on the channel properties and also on the beam shape used for the cell-defining signals. If for a moment it is pretended that the channel strength may be modeled as a uniformly decreasing function of distance, and the beam shape is isotropic, then the strongest cell is the cell with the corresponding antenna array (here assuming a one-to-one mapping between cell and antenna array) closest to the UE. This leads to the familiar situation with cell coverage areas that are hexagonal. In practice, the channel properties are far from this ideal situation. Objects in the environment interact with the propagating signals with a large impact on the cell shape. Shadow fading and spatially varying LOS/nonline-of-sight (NLOS) states also destroy a strict relation between distance and channel strength.

6.2.2 CELL SHAPING AS A MEANS FOR ADAPTING THE CELL COVERAGE

The cell shape depends on the beam shape of the cell-defining signals. This provides the possibility for the network to tailor the cell shape for some specific purpose by changing the beam shape and slowly update the cell shape instead of striving for fixed cell areas with potentially hexagonal structure. The beam shape of a cell-defining signal is controlled by the use of a beamforming vector (6.1) that maps the signal onto the antennas, just as in the dynamic beamforming case. By measuring various parameters in the network, suitable beam shapes can be determined. The technique of letting the system adapt and update the beam shape of cell-defining signals is called *cell shaping*. Cell shaping can be used together with dynamic and UE–specific beamforming to further improve performance or it can be used in a stand-alone fashion.

Cell shaping can be used to solve various problems. For example, the cell border with its interference impairments can be moved to avoid being in the middle of a location where typically there are many simultaneously connected UEs, known as a hotspot. An example of such a hotspot is a train station. Using cell shaping to avoid a cell border within a hotspot is illustrated in Fig. 6.18. Cell shaping also offers the possibility for load balancing among cells by moving UEs from one cell to another, thus altering the uptake areas of cells.

Furthermore, cell shaping can reduce interference in unwanted directions. Such a reduction is especially important in transmit directions that negatively impact particularly many UEs in neighboring cells. The possibility for a controlled interference reduction by cell shaping is in contrast to what is possible with downlink UE–specific beamforming (without explicit nullforming) that typically has only a moderate or little impact on inter-cell interference. Even if cell shaping is used together with UE–specific beamforming, and hence the inter-cell interference comes from

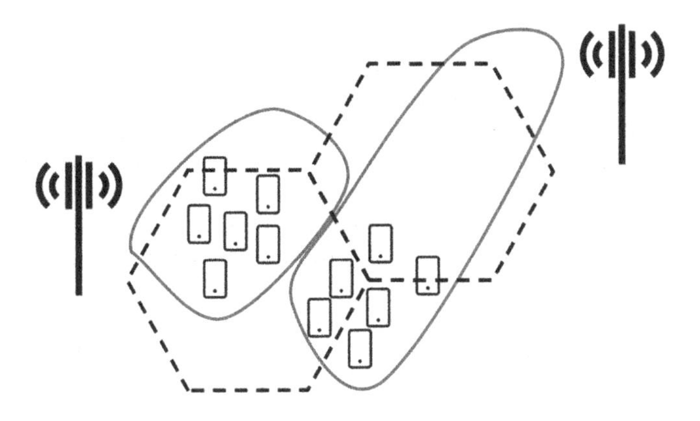

FIGURE 6.18

Cell shaping moving cell borders out of the hotspot area.

transmissions of the latter, interference levels can change substantially by cell shaping. To see why, recall that the UE-to-cell association controlled by cell shaping affects the UE angular distribution as seen from each base station and consequently, as already mentioned in Section 6.1.7, impacts spatial distribution of transmission power from each interfering base station. This obviously strongly influences the characteristics of the UE−specific interfering beamforming transmissions.

Cell-defining signals are shared by many UEs, that is, they constitute broadcast beams. They also affect mobility handling between base stations and the state of already connected UEs. It is hence risky to change them quickly. Cell shaping is therefore confined to relatively slow, semi-static, adaptation of beam shape to capture long-term variations in spatial UE distributions and traffic load. The actual shape can be determined based on various parameters related to load and channel properties.

Cell shaping can be seen as a form of coordination between cells to control antenna array to UE association and reduce inter-cell interference via slow beam adaptation of broadcast signals. Thus it is a form of *coordinated multipoint* (CoMP). CoMP will be further described in Section 6.6.2.

6.3 SPATIAL MULTIPLEXING

A transmit beamformer takes an information-carrying symbol stream and maps it to the antenna elements of the array by multiplying it with a beamforming vector (precoder vector). The result is a transmission along some "direction" or more specifically transmission using some specific spatial signature as generated by the corresponding beam weight vector (6.1). By using different spatial signatures for different information-carrying symbol streams multiple such streams can be multiplexed on the same time/frequency resource to increase the total data rate, so-called spatial multiplexing. The symbol streams are commonly referred to as layers. The multiplexed symbol streams can all go to the same device or to different devices. The case when all the layers belong to the same transmitting/receiving device is referred to as single-user MIMO (SU-MIMO) while the other case involving spatial multiplexing of multiple transmitting and/or receiving devices is called multi-user MIMO (MU-MIMO). An illustration of the two kinds of spatial multiplexing is given in Fig. 6.19.

As will be shown in the following, to be able to efficiently multiplex multiple layers, multiple antennas are needed both on the transmit and receive sides or else the layers will interfere too much with each other since the spatial signatures for the different layers cannot be effectively separated by the receiving device. In other words, the total channel matrix describing the channels between all transmitting and receiving devices involved in the multiplexing needs to be of MIMO type. Note that the antennas on transmit or receive side for the MIMO channel matrix may for the MU-MIMO case span multiple devices. Hence, it is not necessarily so that an individual device must have multiple antennas for efficient spatial multiplexing to be possible.

For MU-MIMO it is easy to intuitively understand how spatial multiplexing is achieved. The UEs are often well separated in terms of angle of departure as seen from the base station, so pointing a narrow beam pattern toward each UE results in a strong signal for the UE of interest while only incurring weak interference on the other co-scheduled (victim) UEs. Each layer in Fig. 6.19 thus corresponds to a beam with a corresponding pointing direction as given by a well-defined main lobe.

Note that the discussion in Section 6.1.7 on the often small impact of transmit beamforming on interference levels specifically considered *inter*-cell interference and plain beamforming without any coordination between different beam transmissions. Hence, this reasoning is not applicable to the present MU-MIMO context where typically a scheduler attempts to only co-schedule UEs with sufficiently spatially distinct channel properties, thereby introducing a form of beam coordination.

In contrast to the MU-MIMO case, spatial multiplexing for SU-MIMO is, however, comparatively more difficult to intuitively understand, since then all the receiving antennas are effectively placed in the same direction as seen from the transmitting device. MU-MIMO to/from multiple devices very close to each other is similarly challenging to understand. Spatial multiplexing here relies on that the propagation channel typically offers several different propagation paths. Roughly speaking and for a moment ignoring polarization, each layer with a corresponding spatial signature can then be thought of as targeting a particular propagation path. The transmitted layers can then be separated out on the receive side by means of receive beamforming (spatial filtering) if those layer-carrying paths have different arrival angles. This is illustrated in Fig. 6.20. As intuition

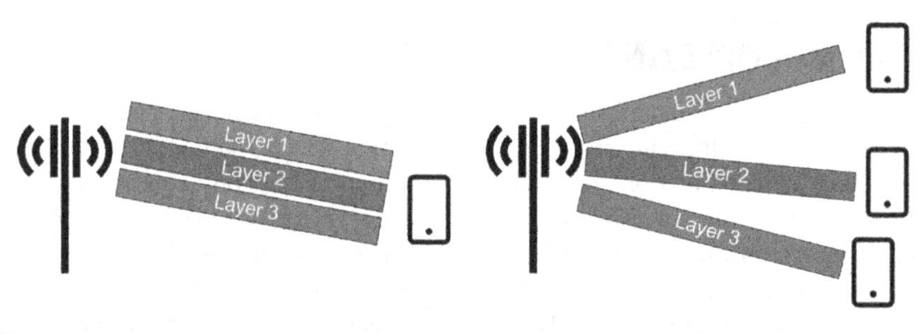

FIGURE 6.19

Spatial multiplexing of three different data streams (layers) to/from either a single UE (SU-MIMO) or to/from multiple UEs (MU-MIMO) on the *same* time/frequency resource.

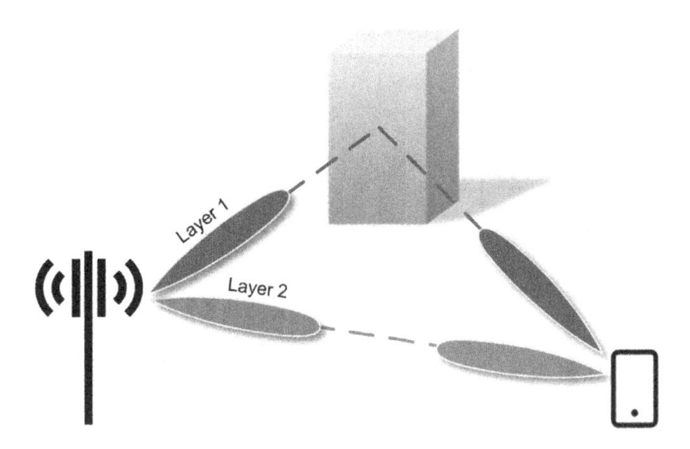

FIGURE 6.20

Spatial multiplexing of two layers for SU-MIMO relying on multi-path propagation to be able to separate the multiplexed layers (similarly for closely spaced UEs in MU-MIMO).

suggests, the number of layers that can be efficiently multiplexed, also commonly referred to as the *channel rank*, is indeed upper bounded by the number of sufficiently different propagation paths in the channel.

Note that this mental model holds reasonably well as long as the main propagation paths are well separated compared with the width of the main lobe. However, for sufficiently large angular spread or antenna separation leading to low spatial correlation of the channel, the mental model breaks down and fast fading effects such as in Section 6.1.1.2.1 start to dominate. The proper way of understanding this is then, just as in the Section 6.1.4.1 described case of generalized beamforming in channels with multi-path, that the beamforming and propagation paths take place in an abstract linear algebraic vector space instead of the more tangible and intuitive physical 3D space.

In channel conditions close to pure LOS, only a single propagation path exists, the channel rank therefore seems to be one and efficient transmissions would hence be restricted to use only one layer. This explains why you commonly hear claims that SU-MIMO does not work in LOS scenarios. Such statements are, however, somewhat misleading since the typical case in practice is to have dual-polarized antenna setups at both the base station and UE sides (see Fig. 6.21 which shows dual-polarized antenna pairs with a +45 and −45-degree polarization). Two layers can then be multiplexed, one on each polarization, and since signals with orthogonal polarizations are well-isolated from each other in a close to pure LOS setting, there is little inter-layer interference and SINR levels are thus high. The ability for antennas to transmit/receive with different polarization properties is described in Section 3.3.3 and also in Section 4.5.2.1.

Lack of cross-talk between layers can intuitively be understood from the figure as follows. In free-space a −45 degree antenna on the UE only hears signals transmitted from the −45 degree antennas on the base station and correspondingly for the +45 degree antennas, thus providing two

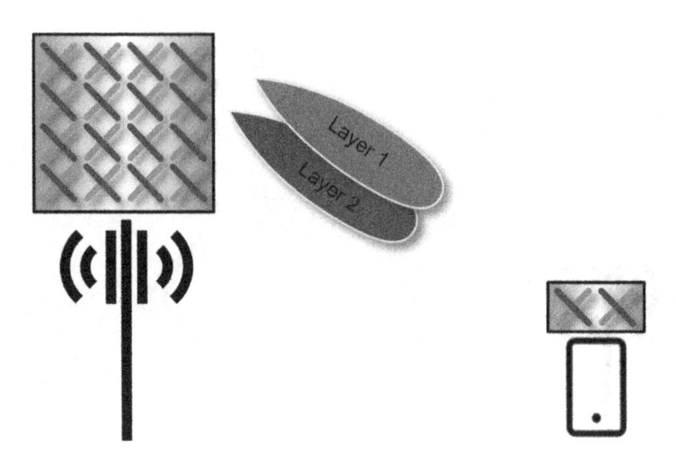

FIGURE 6.21

LOS link with a base station with 4 × 4 dual-polarized antenna pairs and a UE with two dual-polarized antenna pairs. Dual-layer MIMO is supported since each polarization can carry a layer of its own.

independent signal pipes from transmitter to receiver upon which two different layers can be multiplexed without disturbing each other.

In practice, there are often NLOS propagation components as well creating some cross-talk between the polarizations. In pure NLOS situations the cross-talk indeed increases but is still limited by the typical levels of cross-polar discrimination (XPD) (cf. Section 3.5.3.5) encountered in realistic channels. Consequently, despite practical impairments dual-layer transmission exploiting polarization nevertheless represents a very attractive SU-MIMO technique.

6.3.1 RECEIVED DATA MODEL

Multilayer MIMO transmission is conveniently described in a mathematical form. As explained in Section 5.3.1, assuming N_r receive antennas, N_t transmit antennas and r layers transmitted over an $N_r \times N_t$ channel matrix $\boldsymbol{H}$, the received $N_r \times 1$ signal vector $\boldsymbol{y}$ can be written as

$$\boldsymbol{y} = \begin{bmatrix} \boldsymbol{h}_1^{\mathrm{T}} \\ \boldsymbol{h}_2^{\mathrm{T}} \\ \vdots \\ \boldsymbol{h}_{N_r}^{\mathrm{T}} \end{bmatrix} \boldsymbol{x} + \boldsymbol{e} = \begin{bmatrix} \boldsymbol{h}_1^{\mathrm{T}} \\ \boldsymbol{h}_2^{\mathrm{T}} \\ \vdots \\ \boldsymbol{h}_{N_r}^{\mathrm{T}} \end{bmatrix} \left(\sum_{k=1}^{r} \boldsymbol{w}_{t,k} s_k \right) + \boldsymbol{e} = \boldsymbol{H}\boldsymbol{W}_t \boldsymbol{s} + \boldsymbol{e} = \boldsymbol{H}_e \boldsymbol{s} + \boldsymbol{e} \tag{6.11}$$

where $\boldsymbol{x}$ is the $N_t \times 1$ transmitted vector. Furthermore, $\boldsymbol{w}_{t,k}$ and s_k are the precoding/beamforming vector and information-carrying symbol for the kth layer, respectively, and $\boldsymbol{e}$ is a noise term. To get a more compact expression, the precoding vectors are collected into a precoding/beamforming matrix $\boldsymbol{W}_t$, whose kth column is $\boldsymbol{w}_{t,k}$, and an $N_r \times r$ *effective channel* matrix $\boldsymbol{H}_e \triangleq \boldsymbol{H}\boldsymbol{W}_t$ is formed by the channel matrix and the precoding matrix. It is seen that (6.11) is a direct and straightforward generalization of the single-layer single-receive antenna model in (6.2) where the received signals are a superposition of the layer contributions $\boldsymbol{w}_{t,k} s_k$ and the channel $\boldsymbol{H}$ now contains multiple rows $\boldsymbol{h}_n^{\mathrm{T}}$ corresponding to multiple antennas on the receive side.

Henceforth assume that number of layers r is neither larger than the number of transmit antennas N_t nor the number of receive antennas N_r. It turns out this is a necessary, but not sufficient, condition for the spatial signature of each layer to be separable on the receive side. Loosely speaking, this can be thought of as that the maximum number of simultaneously sufficiently different transmit "directions" to put layers on is limited by the minimum of the number of transmit antennas and number of receive antennas N_r.

6.3.2 SPATIAL FILTERING ON RECEIVE SIDE TO EXTRACT LAYERS

The receiver seeks an estimate $\hat{x}$ of the transmitted layers in $\boldsymbol{x}$ by employing an $r \times N_r$ receive beamforming matrix, referred to as spatial combiner $\boldsymbol{W}_r$, to spatially filter the received vector according to

$$\hat{x} = \begin{bmatrix} \boldsymbol{w}_{r,1}^T \\ \boldsymbol{w}_{r,2}^T \\ \vdots \\ \boldsymbol{w}_{r,r}^T \end{bmatrix} \boldsymbol{y} = \boldsymbol{W}_r \boldsymbol{y} = \boldsymbol{W}_r(\boldsymbol{H}_e \boldsymbol{s} + \boldsymbol{e}) = \boldsymbol{W}_r \boldsymbol{H}_e \boldsymbol{s} + \boldsymbol{W}_r \boldsymbol{e} \tag{6.12}$$

There are many different designs of the spatial combiner $\boldsymbol{W}_r$. Perhaps the easiest to understand is to select it to in a sense invert the effective channel matrix $\boldsymbol{H}_e$ as that would ideally separate all the transmitted layers and remove any cross-layer interference, thereby completely undoing the effects of the effective channel. Assume that the number of layers is not larger than the number of receive antennas, $r \leq N_r$. This condition $r \leq N_t$ ensures almost surely that it is possible to invert the channel matrix. The $N_r \times r$ effective channel matrix $\boldsymbol{H}_e$ may not be square and thus a pseudo-inverse has to be employed in the general case (a "normal" matrix inverse can only be employed on square matrices). A pseudo-inverse may be obtained as

$$\boldsymbol{W}_r = \boldsymbol{W}_{r,zf} = \left(\boldsymbol{H}_e^* \boldsymbol{H}_e\right)^{-1} \boldsymbol{H}_e^*$$

which is easily verified to completely cancel the effects of the channel on the signal of interest since

$$\boldsymbol{W}_{r,zf} \boldsymbol{H}_e = \left(\boldsymbol{H}_e^* \boldsymbol{H}_e\right)^{-1} \boldsymbol{H}_e^* \boldsymbol{H}_e = \boldsymbol{I} \rightarrow \hat{s} = \boldsymbol{s} + \boldsymbol{W}_{r,zf} \boldsymbol{e} \tag{6.13}$$

This is a well-known technique [5–7] and is commonly referred to as *zero-forcing* since it removes all inter-layer interference, that is, forces it to zero. Thus the interference from the $r-1$ other layers is completely removed for each layer of interest. Typically, zero-forcing also simultaneously provides receive beamforming SNR gain, although there is a clear trade-off between beamforming gain and inter-layer suppression and the zero-forcing solution priorities the latter at the possible expense of the former. Zero-forcing is one way of implementing nullforming and in this case the nullforming is used to mitigate intra-cell interference between spatially multiplexed layers.

Depending on the effective channel properties, the effective channel matrix $\boldsymbol{H}_e$ may be more or less easy to invert. In theory, $\boldsymbol{H}_e$ is only invertible if all the r columns are linearly independent, that is, the matrix $\boldsymbol{H}_e$ is of (mathematical) rank r. Assuming this is the case, the rank is, however, upper-bounded by

$$r = \text{rank}(\boldsymbol{H}_e) = \text{rank}(\boldsymbol{H}\boldsymbol{W}_t) \leq \text{rank}(\boldsymbol{H}) \leq \min(N_r, N_t) \tag{6.14}$$

The number of layers r, the so-called transmission rank, must therefore be less or equal to the number of antennas on both sides of the link and this explains the previously introduced conditions of $r \leq N_r$ and $r \leq N_t$. It follows from the former that an N_r antenna receiver is able to receive at most N_r different signals at the same time in an interference-free fashion. In other words, an N_r antenna receiver can at most completely suppress $N_r - 1$ interfering signals (by means of linear spatial filtering).

The strict mathematical concept of channel matrix rank is useful for obtaining an upper bound on the number of layers but it does not reveal what number of layers give good performance. In fact, the optimal transmission rank depends on detailed properties of the channel related to how similar, or more precisely linearly dependent, the columns of the effective channel $\boldsymbol{H}_e$ are. The kth column represents the spatial receive signature of the kth layer. The closer the effective channel is to having linear dependence among its columns, the harder it becomes for the zero-forcing to keep the receive beamforming SNR gain high while removing all interference. The effective channel is then close to being singular and the inversion procedure roughly corresponds to dividing the received signal with a number that tends to zero. The power in the spatially filtered noise term $\boldsymbol{W}_{r,zf}\boldsymbol{e}$ will therefore increase a lot, thus reducing SNR. This is called noise enhancement and is a well-known problem with zero-forcing type of receivers.

In practice spatial receive filters with a better trade-off between interference mitigation and signal SNR than what is offered by strict zero-forcing is used. A simple fix to the zero-forcing beamforming matrix is to introduce some regularization that reduces the aggressiveness of the matrix inversion

$$\boldsymbol{W}_r = \left(\boldsymbol{H}_e^*\boldsymbol{H}_e + \lambda \boldsymbol{I}\right)^{-1}\boldsymbol{H}_e^* \tag{6.15}$$

where $\lambda \geq 0$ is a tunable regularization factor. Using $\lambda = 0$ leads to the pure zero-forcing solution while the other extreme of $\lambda \rightarrow \infty$ results in the matched filter solution [2]

$$\boldsymbol{W}_r = \boldsymbol{H}_e^* \tag{6.16}$$

that performs the opposite trade-off of maximizing the SNR of each layer while completely ignoring the inter-layer interference. This technique is often referred to as MRC.

As long as the regularization factor is small in comparison to the term $\boldsymbol{H}_e^*\boldsymbol{H}_e$ (more precisely, small compared with the term's singular values), this book will continue to refer to (6.15) as zero-forcing since zero-forcing in practice always includes some element of regularization.

To strike an optimal balance between improving SNR and fighting inter-layer interference so-called minimum mean square error (MMSE) filters [5] are commonly used. Such filters can be viewed as a generalization of the zero-forcing family of filters to take the properties of the noise term into account according to

$$\boldsymbol{W}_{r,mmse} = \left(\boldsymbol{H}_e^*\boldsymbol{R}_{ee}^{-1}\boldsymbol{H}_e + \boldsymbol{I}\right)^{-1}\boldsymbol{H}_e^*\boldsymbol{R}_{ee}^{-1} \tag{6.17}$$

where $\boldsymbol{R}_{ee} \triangleq \mathrm{E}[\boldsymbol{e}\boldsymbol{e}^*]$ is the spatial covariance matrix of the noise term. This matrix captures not only the noise level but also the spatial distribution of the noise. Since the noise term includes not only thermal noise but also interference, the noise term may have a strong bias toward some particular spatial directions. The interference may, for example, consist of inter-cell interference from neighboring cells. The MMSE beamforming matrix attenuates directions where the noise is strong while amplifying directions where the signals of interest appear. The net result is a maximization of the SINR for each layer.

Note that the term direction can in this context often be interpreted to correspond to a physical direction for ease of understanding, even though in the general case it is only a direction in an abstract vector space; the direction becomes more and more abstract with increasing angular spread of arrival and in the limit of totally uniformly distributed arrival angles over 360 degrees there is no longer any connection between such an abstract direction and any physical direction.

6.3.3 PRECODING ON THE TRANSMIT SIDE

The precoding matrix $\boldsymbol{W}_{\mathrm{t}}$ is used for adapting the transmit signal properties so that\ the effective channel $\boldsymbol{H}_{\mathrm{e}}$ is improved. Receive power may be increased by means of beamforming and interference between layers may be suppressed already on the transmit side. To be able to direct the transmission toward intended receivers, channel-dependent precoding must be employed, which naturally requires CSI to be available on the transmitter side. Ways to acquire CSI will later be discussed in detail in Section 6.4.

Most receive side beamforming schemes have direct counterparts on the transmit side as will be evident from the following.

6.3.3.1 Maximum ratio transmission

MRT, or conjugate beamforming, corresponds to MRC on the receive side and hence the precoding matrix

$$\boldsymbol{W}_{\mathrm{t,mrt}} = \kappa \boldsymbol{H}^{*} = \kappa \begin{bmatrix} \boldsymbol{h}_1^{\mathrm{c}} & \boldsymbol{h}_2^{\mathrm{c}} & \cdots & \boldsymbol{h}_{N_r}^{\mathrm{c}} \end{bmatrix}$$

is a direct generalization of the matched filter approach in (6.16) to a MIMO channel and multiple layers. Recall that κ represents some power scaling method, for example, $\kappa = \sqrt{P/\|\boldsymbol{H}\|_{\mathrm{F}}}$ to ensure a fixed sum power P transmitted over the array. The number of layers is here chosen to be the same as the number of receive antennas N_{r}, that is, the maximum possible. To use fewer layers different approaches could be conceived. The most straightforward and obvious method would be to form the precoder targeting the r strongest receive antennas.

The MRT precoder is typically good at achieving beamforming gain even for channels with large angular spread, assuming the precoder is updated both in time and frequency to track the fast fading. But it does not make any attempts at deliberately mitigating inter-layer interference. For arrays with sufficiently many antennas and under some rather general channel conditions, the effective channel becomes essentially orthogonal since the off-diagonal elements $\boldsymbol{h}_m^{\mathrm{T}}\boldsymbol{h}_n^{\mathrm{c}}, m \neq n$ become very small *in comparison to* the on-diagonal elements $\boldsymbol{h}_m^{\mathrm{T}}\boldsymbol{h}_m^{\mathrm{c}} = \|\boldsymbol{h}_m\|^2$ $\boldsymbol{H}_{\mathrm{e}}$ and thus

$$\boldsymbol{H}_{\mathrm{e}} = \boldsymbol{H}\boldsymbol{W}_{\mathrm{t,mrt}} \approx \kappa \mathrm{diag}\left(\|\boldsymbol{h}_1\|^2, \|\boldsymbol{h}_2\|^2, \cdots, \|\boldsymbol{h}_{N_r}\|^2 \right)$$

In this situation, inter-layer interference does not limit the SINR increase of higher beamforming gain.

Channel conditions ensuring that off-diagonal elements become relatively small include, for example, when the channel is modeled as IID complex Gaussian (spatially uncorrelated Rayleigh fading), corresponding to a very large angular spread. Another example is the other extreme of pure LOS with zero angular spread with single-antenna UEs well separated in the angular domain as seen from the base station side in downlink. This makes the victim UEs be hit by progressively smaller amount of interference relative to the strength of peak of beam as the number of antennas increase.

6.3.3.2 Precoding based on singular value decomposition

In singular value decomposition (SVD)–based precoding [8] the channel $\boldsymbol{H}$ is factorized into three matrices using SVD

$$\boldsymbol{H} = \boldsymbol{U}\boldsymbol{\Sigma}\boldsymbol{V}^* = \begin{bmatrix} \boldsymbol{u}_1 & \boldsymbol{u}_2 & \cdots & \boldsymbol{u}_{r_H} \end{bmatrix} \mathrm{diag}\begin{pmatrix} \sigma_1 & \sigma_2 & \cdots & \sigma_{r_H} \end{pmatrix} \begin{bmatrix} \boldsymbol{v}_1^* \\ \boldsymbol{v}_2^* \\ \vdots \\ \boldsymbol{v}_{r_H}^* \end{bmatrix} = \sum_{k=1}^{r_H} \sigma_k \boldsymbol{u}_k \boldsymbol{v}_k^* \tag{6.18}$$

where $\boldsymbol{U}$ is $N_r \times r_H$ containing r_H different orthogonal *left singular vectors* $\{\boldsymbol{u}_k\}$, $\boldsymbol{\Sigma}$ is diagonal $r_H \times r_H$ carrying the singular values $\sigma_k = [\boldsymbol{\Sigma}]_{kk}$, and $\boldsymbol{V}$ is $N_t \times r_H$ containing the r_H different orthogonal *right singular vectors* $\{\boldsymbol{v}_k\}$. Here, r_H is the rank of the channel matrix in the true linear algebra sense, not to be confused by the commonly used and not strictly defined term "channel rank" that loosely refers to have many layers r that can be efficiently transmitted over the channel.

Note that the channel is in (6.18) represented as a weighted sum of outer products of the left and right singular vectors, where the singular values are the weightings. The form of the expression is similar to the formula for the channel realization in terms of transmitter- and receiver-side array response vectors (see Section 5.3.2). Indeed, the left singular vector $\boldsymbol{u}_k$ and right singular vector $\boldsymbol{v}_k^*$ can be thought of as describing the kth receiver-side and transmitter-side channel directions, respectively, in an abstract vector space. Thus an outer-product term $\sigma_k \boldsymbol{u}_k \boldsymbol{v}_k^*$ represents an *abstract* propagation path in the vector space and the corresponding singular value σ_k describes the strength of that path. The distribution of singular values effectively determines how many significant abstract propagation paths there are. If the *physical* propagation paths are sufficiently well-separated and co-polarized antennas are used, then the singular vectors can be shown to be well-approximated by the array response vectors. A similar but somewhat more complicated analogy can be made for the dual-polarized antenna case.

The SVD-based precoder for *full rank transmission* of $r = r_H$ layers is now obtained as

$$\boldsymbol{W}_{\mathrm{t,svd}} = \kappa \boldsymbol{V} \tag{6.19}$$

This orthogonalizes the channel on the transmit side since

$$\boldsymbol{H}_{\mathrm{e}} = \boldsymbol{H}\boldsymbol{W}_{\mathrm{t,svd}} = \kappa \boldsymbol{U}\boldsymbol{\Sigma}\boldsymbol{V}^*\boldsymbol{V} = \kappa \boldsymbol{U}\boldsymbol{\Sigma}$$

and the receiver can therefore easily separate the layers by spatial filtering using

$$\boldsymbol{W}_{\mathrm{r}} = \boldsymbol{U}^* \rightarrow \hat{\boldsymbol{s}} = \boldsymbol{W}_{\mathrm{r}}\boldsymbol{y} = \boldsymbol{W}_{\mathrm{r}}(\boldsymbol{H}_{\mathrm{e}}\boldsymbol{s} + \boldsymbol{e}) = \kappa \boldsymbol{\Sigma}\boldsymbol{s} + \boldsymbol{U}^*\boldsymbol{e} \tag{6.20}$$

It is seen how the layers no longer interfere with each other since $\boldsymbol{\Sigma}$ is diagonal. If a lower transmission rank $r < r_H$ is used, the first r columns of $\boldsymbol{V}$ are used since they correspond to the right singular vectors with the strongest singular values. The mathematical end result is similar to the data model in (6.13) with a completely orthogonalized channel, except that there is no noise enhancement since the columns of $\boldsymbol{U}$ are orthonormal.

Note that a receiver is in practice free to do something better than using $\boldsymbol{U}^*$ as a filter, for example, taking also the spatial interference properties in $\boldsymbol{e}$ into account to obtain an optimal trade-off between inter-layer and inter-cell interference by using the MMSE filter in (6.17).

The transmission occurs along the right singular vectors of the channel. There is in general no exact intuitive interpretation on what physical directions these correspond to; in essence, the directions are in an abstract linear algebraic vector space as also previously mentioned in Section 6.1.1. However, if propagation paths are well separated in angle on the transmit and receive sides, they roughly correspond to transmitting along the r strongest such paths, similar to what is illustrated in Fig. 6.20.

6.3.3.3 Zero-forcing precoding

The technique of zero-forcing can also be applied on the transmit side to prioritize the removal of inter-layer interference [9,10], thus performing nullforming. The precoder for full rank transmission $r = N_\mathrm{r}$ is then given by

$$\boldsymbol{W}_\mathrm{t} = \boldsymbol{W}_\mathrm{t,zf} = \kappa \boldsymbol{H}^*(\boldsymbol{H}\boldsymbol{H}^* + \lambda \boldsymbol{I})^{-1} \tag{6.21}$$

which for the regularizing factor $\lambda = 0$ can be easily verified to invert the channel

$$\boldsymbol{H}\boldsymbol{W}_\mathrm{t,zf} = \kappa \boldsymbol{H}\boldsymbol{H}^*(\boldsymbol{H}\boldsymbol{H}^*)^{-1} = \kappa \boldsymbol{I} \rightarrow \hat{\boldsymbol{x}} = \kappa \boldsymbol{x} + \boldsymbol{e} \tag{6.22}$$

It is here assumed that $N_\mathrm{r} \leq N_\mathrm{t}$ so as to ensure that $\boldsymbol{H}\boldsymbol{H}^*$ is invertible and hence all inter-layer interference can be canceled. There are different ways to handle the case of transmitting with fewer than the maximum of $r = N_\mathrm{r} \leq N_\mathrm{t}$ layers. One option is to form a new channel matrix out of the r strongest rows of $\boldsymbol{H}$, corresponding to the r strongest receive antennas, and then based on the new channel matrix compute the zero-forcing precoder similarly as above.

Performing inter-layer interference mitigation on the transmit side makes the job of the receiver easier since the layers are already well separated when they arrive at the receiver. This is important especially for MU-MIMO. With MU-MIMO with multiple separate receivers, an individual receiver does not have complete information, that is, it only sees a subset of the receive antennas of the MIMO channel and has therefore limited possibilities to suppress interference.

6.3.3.4 Fundamentals of precoding based on discrete Fourier transforms

The array response vectors are often highly structured for common antenna array setups. For example, for ULAs the mth element has position $\boldsymbol{r}_m = \boldsymbol{r}_0 + m\boldsymbol{\Delta}_r$ and the phase difference between two consecutive antenna elements in the array response vector is from (6.4) thus seen to be a constant $\Delta_\phi \triangleq \boldsymbol{\Delta}_r^\mathrm{T}\boldsymbol{k}$, thereby leading to that (6.4) specializes into an array response vector of the form

$$\mathbf{a}(\theta, \varphi) = g(\theta, \varphi)\exp(j\phi_0)\begin{bmatrix} 1 & \exp(j1\Delta_\phi) & \cdots & \exp(j(N-1)\Delta_\phi) \end{bmatrix}^\mathrm{T} \tag{6.23}$$

where $\phi_0 = \boldsymbol{r}_0^\mathrm{T}\boldsymbol{k}$ and the antenna pattern $g(\theta, \varphi)$ has been assumed to be the same for all antennas. The elements of the array response vector are seen to have a linearly increasing phase, which is like the linear phase increase of the multiplicative factors in the DFT. Such an array response vector is therefore sometimes referred to as having DFT structure.

Recall from (6.8) that the multi-path channel $\boldsymbol{h}^{\mathrm{T}}$ in the case of a single-receive antenna and single-polarized antennas can be written as a sum of array response vectors.

$$\boldsymbol{h}^{\mathrm{T}} = \sum_{l=1}^{L} \lambda_l \boldsymbol{a}^{\mathrm{T}}(\theta_l, \varphi_l)$$

If the receive side now also has multiple antennas, the channel consists of multiple rows with a structure similar to $\boldsymbol{h}^{\mathrm{T}}$ and there is an array response vector that describes how the channel coefficients change from one receive antenna (or row) to another for a particular propagation path l. In other words, each propagation path contributes to the channel matrix with a term

$$\lambda_l \boldsymbol{a}_r(\theta_l^{\mathrm{r}}, \varphi_l^{\mathrm{r}}) \boldsymbol{a}_{\mathrm{t}}^{\mathrm{T}}(\theta_l^{\mathrm{t}}, \varphi_l^{\mathrm{t}})$$

consisting of an outer product of the receive array response vector $\boldsymbol{a}_r(\theta_l^{\mathrm{r}}, \varphi_l^{\mathrm{r}})$ and the transmit array response vector $\boldsymbol{a}_{\mathrm{t}}^{\mathrm{T}}(\theta_l^{\mathrm{t}}, \varphi_l^{\mathrm{t}})$, thus leading to the channel matrix expression

$$\boldsymbol{H} = \sum_{l=1}^{L} \lambda_l \boldsymbol{a}_{\mathrm{r}}(\theta_l^{\mathrm{r}}, \varphi_l^{\mathrm{r}}) \boldsymbol{a}_{\mathrm{t}}^{\mathrm{T}}(\theta_l^{\mathrm{t}}, \varphi_l^{\mathrm{t}}) \tag{6.24}$$

The directions on the receive $(\theta_l^{\mathrm{r}}, \varphi_l^{\mathrm{r}})$ and transmit side $(\theta_l^{\mathrm{t}}, \varphi_l^{\mathrm{t}})$ are now seen to be different, which supports modeling that a propagation path may have a totally different direction of arrival on the receive side than direction of departure on the transmit side. Further details on the structure of the channel matrix and the key role the array response vectors in general play, including for the case of dual-polarized antennas, are given in Section 5.3.2.2.

The channel is according to above basically a linear combination of outer products of such DFT-structured array response vectors, one outer-product term for each propagation path. Consequently, the transmit side of the channel can be thought of as consisting of a linear combination of array response vectors corresponding to the set of *angles of departures*. From Section 6.1.3, it was discussed how the complex conjugate of an array response vector results in a beamforming vector that is matched to the particular channel direction of a single array response vector. Although a linear combination of array response vectors cannot be exactly described by a single array response vector, if the channel angular spread is not too large, then the linear combination still has much in common with the key property of a single array response vector of having a trend of linearly increasing phase over the array. In other words, the channel on the transmit side has approximately a DFT structure as well, buried in more or less fast fading variations. It is then not surprising that a DFT-based precoder often constitutes a rather good match to the channel providing a large portion of the maximum beamforming gain.

This DFT structure of the channel reflects the physical propagation directions on the transmit side of the channel. As previously mentioned in Section 3.6.2.8, propagation directions are a large-scale channel property and as such are captured by the spatial channel correlation characteristic, in this case the transmitter side-channel correlation matrix

$$\mathbf{R}_{HH} \triangleq \mathrm{E}[\boldsymbol{H}^* \boldsymbol{H}]$$

is the most relevant. Recall that a row of $\boldsymbol{H}$ contains the channel coefficients $\boldsymbol{h}_n^{\mathrm{T}}$ in (6.11) that determine how a transmission vector $\boldsymbol{x}$ is affected by the channel to produce $\boldsymbol{h}_n^{\mathrm{T}}\boldsymbol{x}$ at the nth receiver side antenna. The transmit correlation matrix describes the average properties of a row of the channel matrix $\boldsymbol{H}$

This can be viewed as corresponding to an average channel direction in a vector space and if the angular spread is sufficiently small there is a direct linkage between the average direction in the vector space and the physical channel direction, as will be evident in the following.

The various forms of channel correlation matrices are described in detail in Section 5.3.3 and are in this particular case shown to be

$$\mathbf{R}_{HH} = \sum_{l=1}^{L} \lambda_l^{\mathrm{c}} \left\| \boldsymbol{a}_{\mathrm{r}}(\theta_l^{\mathrm{r}}, \varphi_l^{\mathrm{r}}) \right\|^2 \boldsymbol{a}_{\mathrm{t}}^{\mathrm{c}}(\theta_l^{\mathrm{t}}, \varphi_l^{\mathrm{t}}) \boldsymbol{a}_{\mathrm{t}}^{\mathrm{T}}(\theta_l^{\mathrm{t}}, \varphi_l^{\mathrm{t}}) \tag{6.25}$$

If all the transmit propagation path directions $(\theta_l^{\mathrm{t}}, \varphi_l^{\mathrm{t}})$ are close to each other, the correlation matrix can be approximated with a single direction, for instance the first term's direction, that is,

$$\mathbf{R}_{HH} \sim \boldsymbol{a}_{\mathrm{t}}^{\mathrm{c}}(\theta_1^{\mathrm{t}}, \varphi_1^{\mathrm{t}}) \boldsymbol{a}_{\mathrm{t}}^{\mathrm{T}}(\theta_1^{\mathrm{t}}, \varphi_1^{\mathrm{t}})$$

This says that the average channel direction on the transmit side is $\boldsymbol{a}_{\mathrm{t}}(\theta_1^{\mathrm{t}}, \varphi_1^{\mathrm{t}})$ in an N_{t} dimensional complex-valued vector space and is $(\theta_1^{\mathrm{t}}, \varphi_1^{\mathrm{t}})$ in the physical space, providing the previously mentioned linkage between the two domains. The correlation is high; in fact it is the maximum possible. To see this, observe that the amplitude of each element in the correlation matrix, $\left|[\mathbf{R}_{HH}]_{k,l}\right|$ is equal to a constant c since different elements in $\boldsymbol{a}_{\mathrm{t}}(\theta_1^{\mathrm{t}}, \varphi_1^{\mathrm{t}})$ only differ via the constant amplitude exponential as given by $\exp(j(n-1)\Delta_{\phi}), n = 1, 2, \cdots, N_{\mathrm{t}}$, in (6.23). The so-called normalized correlation coefficient

$$\rho_{k,l} \triangleq \frac{[\mathbf{R}_{HH}]_{k,l}}{\sqrt{[\mathbf{R}_{HH}]_{k,k}}\sqrt{[\mathbf{R}_{HH}]_{l,l}}}, \left|\rho\right| \leq 1$$

is thus at its maximum amplitude of 1 since $\left|\rho\right| = c/(\sqrt{c}\sqrt{c}) = 1$. It is therefore common to refer to the small channel angular spread case by saying that the (amplitude of the normalized) correlation is close to 1.

The transmit channel correlation matrix in (6.25) is seen to be a linear combination of outer products of each path's transmit array response vector with itself highlighting how important role the array response vectors play also in determining correlation, and thus, large-scale properties. The receive correlation matrix is similarly structured and is found in Section 5.3.3.

DFT-based precoding can due to its relation to physical channel directions be said to target the correlation properties of the channel. Recall that those properties are of long-term and wideband nature, so the same DFT precoder can be used over a large bandwidth without much performance loss.

When the channel angular spread increases, the channel coefficients decorrelate more and more, that is, the normalized correlation coefficient $\left|\rho_{k,l}\right|$ decreases and the DFT structure approximation of the channel worsens. The latter can be understood from the fact that a linear combination of DFT vectors (transmit array responses $\boldsymbol{a}_{\mathrm{t}}^{\mathrm{T}}(\theta_l^{\mathrm{t}}, \varphi_l^{\mathrm{t}})$) as in (6.24) is far from being a DFT vector unless the combined DFT vectors are very similar (corresponding to similar directions $(\theta_l^{\mathrm{t}}, \varphi_l^{\mathrm{t}})$).

For large channel angular spread, the conventional and popular DFT-based precoding using only a single DFT vector as in classical beamforming is not effective since it no longer matches the channel structure. Generalized beamforming schemes such as MRT and SVD where the precoder is updated to track the fast fading both over time and frequency are then more appropriate. Another option to efficiently handle this case that will be explored in Section 6.4.1.2.1 is to form the precoder as a linear combination of multiple DFT vectors as will be illustrated in Fig. 6.36. Such a linear combination then matches the structure of the channel and has therefore better possibilities to provide performance.

6.3.3.5 DFT – based precoder codebook structures for one-dimensional arrays

In DFT-based precoding, the columns, or parts of columns, of the precoder matrix are DFT-based. Recall that each column constitutes the beamforming weight vector for a particular layer. Returning to the ULA example, each column in a suitable precoder is DFT-structured, meaning that for a rank r transmission

$$W_{\mathrm{t,1d-dft}} = \begin{bmatrix} w_{\mathrm{1d-dft},1} & w_{\mathrm{1d-dft},2} & \cdots & w_{\mathrm{1d-dft},r} \end{bmatrix}$$

where $\{w_{\mathrm{1d-dft},k}\}$ are all obeying a one-dimensional (1D) DFT structure. In practice it is common that the precoder columns are taken as a column subset of an oversampled DFT matrix Q defined as

$$[Q]_{1+m,1+n} = \exp\left(j\frac{2\pi mn}{O_s N_t}\right), m = 0, \cdots, N_t - 1, n = 0, \cdots, N_t O_s - 1 \tag{6.26}$$

where O_s is the oversampling factor. The column vectors in Q span all directions and can be thought of as forming a grid of beams (GoB). Fig. 6.22 illustrates the GoB beams obtained when there is no beam oversampling. It is seen that a peak of a beam corresponds to nulls of the other

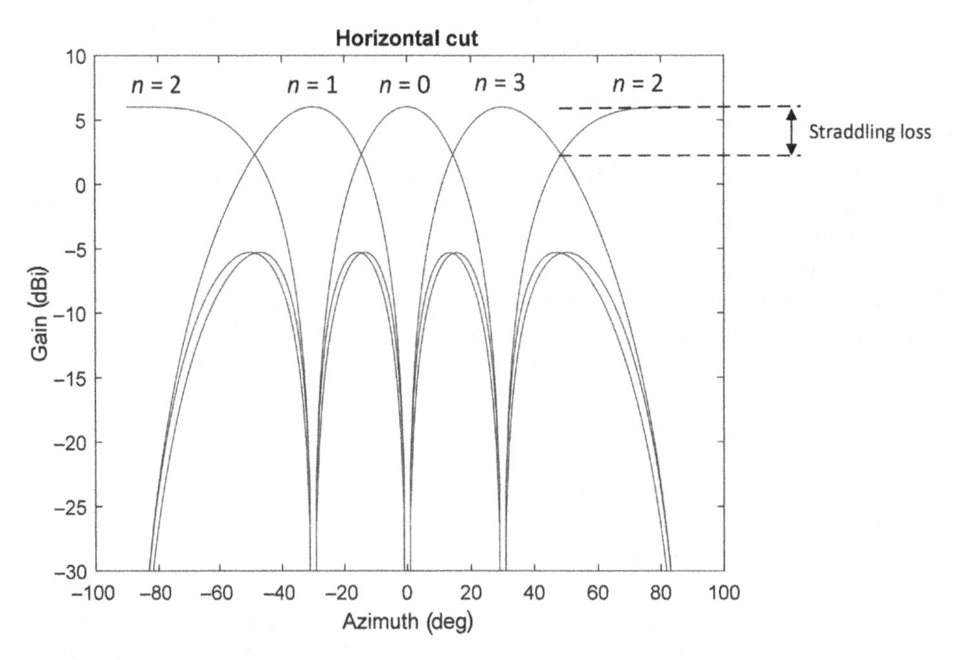

FIGURE 6.22

Beam shapes of DFT-based beamforming for a horizontally oriented uniform linear array with $N_t = 4$ antennas and oversampling factor $O_s = 1$.

beams; hence the beams are orthogonal. This is not surprising as the DFT matrix $\boldsymbol{Q}$ has orthogonal columns when the oversampling factor $O_s = 1$. The column subset of $\boldsymbol{Q}$ that forms the precoder is usually selected so that all the beamforming vectors are orthogonal to reduce inter-layer interference.

GoB is one example of a so-called codebook-based technique, in which the precoder is selected from a finite and countable set of precoders. Codebook-based techniques will be revisited as part of discussions in Section 6.4.1 on closed-loop techniques aiding the selection of precoders.

Fig. 6.22 shows that there is quite significant loss of beamforming gain in directions between the beams, around 4 dB of so-called *straddling loss* can be observed. To mitigate this problem, it is possible to insert extra beams in-between the orthogonal beams by increasing the beam oversampling factor. A good rule of thumb is to use an oversampling factor value of at least 2 so that the beamforming gain does not decrease too much between two angularly consecutive beams. All beams are then no longer orthogonal. Orthogonality is maintained only between beams spaced O_s beam indices apart.

For antenna arrays with dual cross-polarized antennas it is common to do DFT-based GoB separately for each polarization. Since the channel directions are the same in both polarizations, albeit differently linearly combined, it makes sense to transmit using the same DFT beam in both polarizations. The fast fading of orthogonal polarizations tends to be very different as explained in Section 3.5.3.5. It could therefore be useful on top of the DFT beams to adjust the phase between the two polarizations to account for the different fast fading leading to a relative phase difference between the channel polarizations. With these design considerations in mind, the precoder structure becomes

$$\boldsymbol{W}_{\mathrm{t,dft}} = \begin{bmatrix} \boldsymbol{w}_{\mathrm{1d-dft},1} & \boldsymbol{w}_{\mathrm{1d-dft},2} & \cdots & \boldsymbol{w}_{\mathrm{1d-dft},r} \\ e^{j\phi_1}\boldsymbol{w}_{\mathrm{1d-dft},1} & e^{j\phi_2}\boldsymbol{w}_{\mathrm{1d-dft},2} & & e^{j\phi_r}\boldsymbol{w}_{\mathrm{1d-dft},r} \end{bmatrix} \quad (6.27)$$

where $\{\phi_k\}$ are the inter-polarization phase offsets and the antenna elements are assumed to be ordered so that the first $N_t/2$ rows of the matrix correspond to the first polarization and the remaining rows correspond to the second polarization. An illustration of a column of this structure for the case of eight antennas is given in Fig. 6.23. Since the fast fading of orthogonal polarizations tends to be very different, it would be beneficial to update the inter-polarization phase offsets at a similar rate as the fast fading alters the channel. Fast fading properties in general change the channel over both time and frequency and interpolarization phase offsets thus need to be updated both in frequency and time to track the fast fading.

6.3.3.6 DFT – based precoder codebook structures extended to two-dimensional antenna arrays

The above two DFT-based precoder structures assumed linear arrays, that is, the antenna array is 1D. With the introduction of AAS, 2D arrays are receiving considerable interest. Such arrays support UE–specific beamforming in both the horizontal and vertical domains. Arrays with the antenna elements arranged in a rectangular lattice are particularly popular. A GoB and DFT-based structure is still highly relevant. However, the beam vectors need to be altered to take the 2D nature of the array into account. This is usually achieved by forming the beam vectors out of two 1D DFT vectors, one DFT vector along the rows of the antenna array to perform vertical beamforming and the other DFT vector along the columns to support horizontal beamforming. Such a combination is

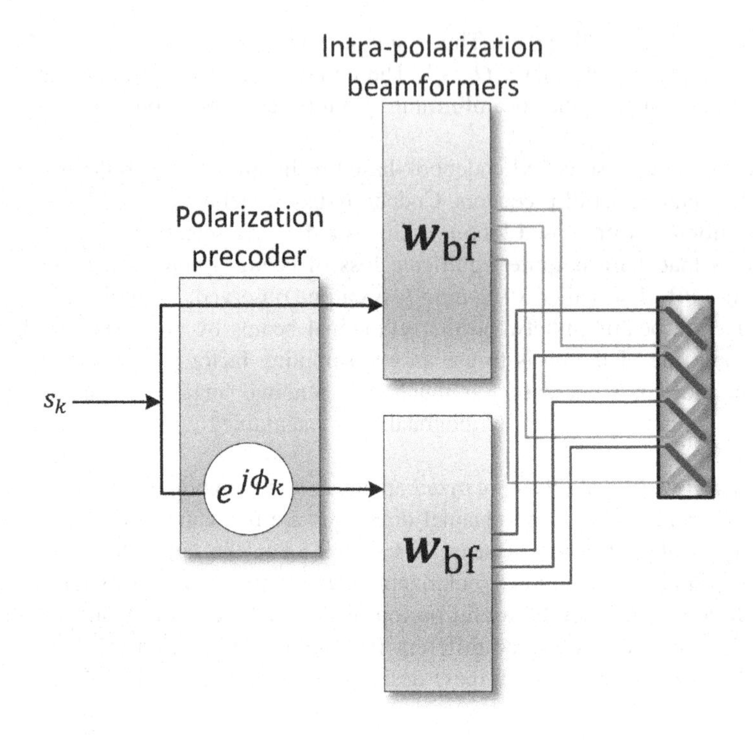

FIGURE 6.23

A single column corresponding to layer k in the DFT-based precoder structure for dual-polarized antenna arrays. The vector w_{bf} is in the 1D case equal to $w_{1d-dft,k}$.

effectively corresponding to a 2D DFT vector and can be written as the Kronecker product of two 1D DFT vectors as

$$w_{2d-dft} = w_{1d-dft,h} \otimes w_{1d-dft,v}$$

and used to form the per-polarization beamforming in a precoder matrix according to

$$W_{t,2d-dft} = \begin{bmatrix} w_{2d-dft,1} & w_{2d-dft,2} & \cdots & w_{2d-dft,r} \\ e^{j\phi_1} w_{2d-dft,1} & e^{j\phi_2} w_{2d-dft,2} & & e^{j\phi_r} w_{2d-dft,r} \end{bmatrix} \tag{6.28}$$

The underlying 1D DFT vectors are in turn taken from the columns of some possibly over-sampled 1D DFT matrix as in the 1D case in (6.26).

6.3.4 DOES SPATIAL MULTIPLEXING ALWAYS PROVIDE GAIN?

Spatial multiplexing does not always provide gain. Sometimes performance is maximized with only one layer, that is, rank-1 transmission corresponding to no spatial multiplexing. Sometimes a low number of layers are preferred and sometimes the use of many layers is optimal. The best choice of transmission rank r depends on several factors including channel and noise properties as well as transmission power and receiver algorithms.

To give some intuitive understanding on the fundamental reasons behind spatial multiplexing gain, consider the special case when the symbol estimate formula in (6.13) reduces to

$$\hat{s} = s + e \tag{6.29}$$

where the symbols are power normalized $\mathrm{E}\left[\|s\|^2\right] = P/r$, and the noise term is taken to be complex Gaussian with $\mathrm{E}[ee^*] = \sigma^2 I$. The model represents a set of r independent and parallel additive white Gaussian noise (AWGN) channels as illustrated in Fig. 6.24 and may arise, for example, in pure LOS conditions with a receiver with two antennas of orthogonal polarizations and a transmitter with multiple dual-polarized antenna elements creating a 2×2 identity matrix as effective channel matrix. In this case, the optimal receive side beamforming matrix is also an identity matrix, leading to (6.29) with $r = 2$.

But pure LOS is not the only case when a set of parallel AWGN channels is a reasonable model. As seen in Sections 6.3.2 and 6.3.3, receive and transmit spatial filtering can be used to mitigate cross-layer interference approximately creating a set of parallel AWGN channels. Admittedly, channel fading may lead to variations in time and frequency of the variance of the noise term e, which is not part of this simple model and the simplified analysis to come. Nevertheless, the

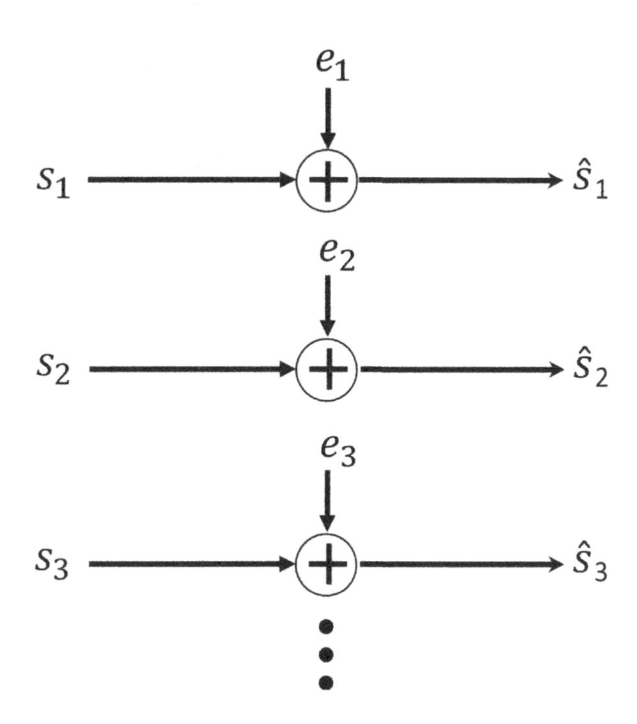

FIGURE 6.24

A parallel set of AWGN channels to model the symbol estimate formula in (6.29).

analysis captures the basics behind spatial multiplexing and the conclusions and explanations remain valid even with more accurate models.

The data rate possible to transmit without errors, or more precisely the information theoretic channel capacity, for a single-input single-output (SISO) AWGN channel with transmit power P and noise variance σ^2 is given by [11]

$$C_{\text{SISO}} = \log_2(1+\rho) \text{ (bits/channel use)} \tag{6.30}$$

where $\rho = P/\sigma^2$ is the SISO channel SNR. The channel capacity for (6.29) is obtained as the sum of the individual AWGN channel capacities. Thus,

$$C = \sum_{k=1}^{r} \log_2(1+\rho_k) = \sum_{k=1}^{r} \log_2\left(1+\frac{P}{r\sigma^2}\right) = r\log_2\left(1+\frac{P}{r\sigma^2}\right) = r\log_2\left(1+\frac{\rho}{r}\right) \quad \text{(bits/channel use)}$$

It is seen that increasing the rank increases the number of parallel "data pipes" but on the other hand the data pipes share a constant total transmission power P so the SNR is reduced per data pipe to ρ/r and data rate per data pipe thus decreases. To understand what wins of these two conflicting trends and why, consider Fig. 6.25 where the two cases of single and dual layers are compared. The single-layer case results in the capacity curve from (6.30). It is seen that the high SNR value ρ does not fully translate to a data rate increase since the slope of the curve flattens for high SNR values, that is, single layer operates in the saturation region of the curve. When the number of layers is increased to two, the SNR per layer is halved since power is shared. But at $\rho/2$ the SISO AWGN capacity curve is much steeper, so this SNR value gives a better payoff in terms of data rate. And now there are two SISO channel capacities to sum up giving a net gain over the single-layer case. As more and more layers are used the SNR for each layer goes down making the layer operate more and more on the linear part of the capacity curve, continuing to increase the total capacity. Increasing the number of layers, however, gives diminishing returns since from $\rho \to 0$, it follows that $\log_2(1+\rho) \to \rho$ which is the completely

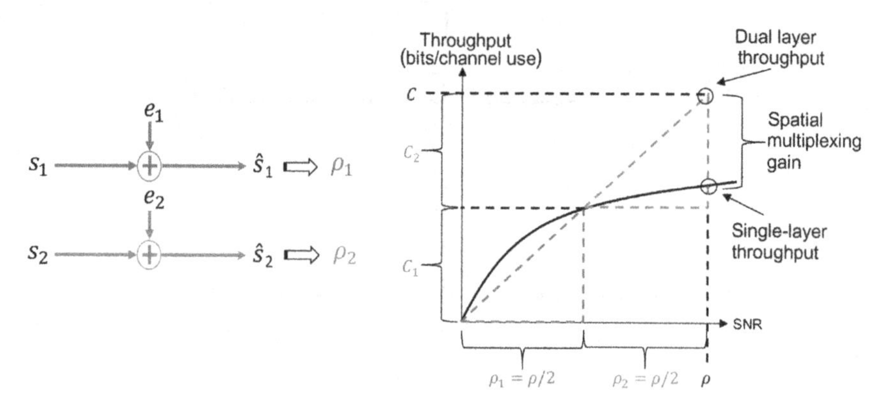

FIGURE 6.25

Example with dual layers to show how spatial multiplexing brings gain by operating at the steeper parts of the throughput curve.

linear part of the capacity curve. The increase in throughput due to more layers is thus completely canceled by the loss in data rate per layer.

Note that in this idealized example, there is never any loss by increasing the number of layers. This is, however, not true in practice, where impairments such as inter-layer interference and channel estimation errors reduce the number of layers that should be used. The type of transmission scheme also matters. The simpler MRT has a tendency of preferring more layers than zero-forcing, although zero-forcing typically performs better as later discussed in Section 6.7.2.

The idealized analysis herein makes it clear that spatial multiplexing is a way of sharing "too high" SNR levels among more data pipes to in total getting higher throughput than if fewer data pipes would be used. As such spatial multiplexing requires sufficiently high SNR to show significant gain.

In practice, layer SNR is actually an SINR, that is, the noise term contains contributions from residual inter-layer and inter-cell interference as well that remains after the receive and transmit spatial filtering. The inter-cell interference can on the cell edge be very strong. This impairment heavily contributes to reducing the number of layers. Roughly speaking, the number of layers decreases with distance from the base station as path loss and inter-cell interference increases. This is illustrated in Fig. 6.26. It is not uncommon that only single-layer transmission is performed on cell edge. Hence, spatial multiplexing can be said to be a technique that is particularly suitable for cell center, and/or isolated cells. For SU-MIMO lower network loads are also beneficial since that reduces inter-cell interference, while for MU-MIMO network load needs to be high in order to at least fulfill the minimum requirement for MU-MIMO of two simultaneously active UEs present in the same transmission instant to potentially spatially multiplex. MU-MIMO is therefore to be considered as a capacity enhancement feature for highly loaded cells.

The inter-layer interference is often substantial leading to much reduced SINR levels. Unless idealized zero-forcing is assumed, such interference is heavily influenced by the properties of the channel matrix. If it is close to being singular, that is, very uneven distribution of singular values creating an ill-conditioned channel, interference will be higher than if the opposite is true.

It is important to keep in mind that even such a rank-deficient channel can support many layers if the SNR is sufficiently high—rank deficiency just increases the requirements on the SNR level as evident from the capacity curve analysis and in view of that rank deficiency acts as lowering the per layer SINR values. But according to (6.13), it does not for sufficiently good receivers remove the possibility for those SINR values to grow with, for example, increased transmit power.

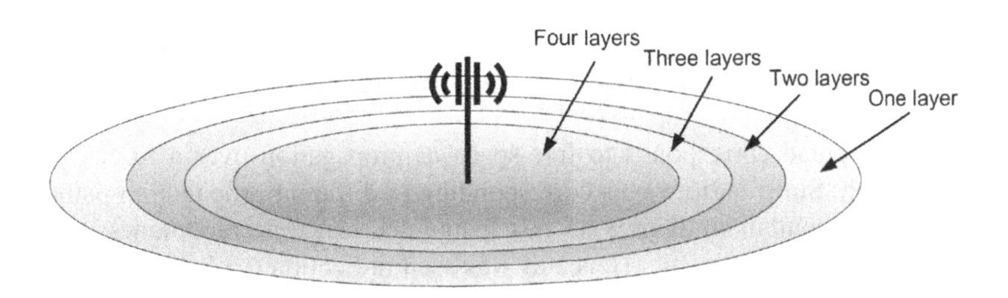

FIGURE 6.26

Number of usable layers tends to decrease with increasing distance to transmitter.

6.3.5 CHANNEL PROPERTIES MATTER!

As alluded to in the previous section, the properties of the combined channel over which all spatially multiplexed layers are transmitted have great impact on how many layers can be efficiently supported. The effective channel rank can be low or high depending on the circumstances. It is important to keep in mind that the effective channel rank is not the same thing as the linear algebraic rank of the channel matrix. The latter is well-defined and given by the number of non-zero singular values $\text{rank}(\boldsymbol{H}) \leq \min(N_r, N_t)$ of the combined channel $\boldsymbol{H}$ while the former is loosely defined as the number of layers r that leads to the maximum total data rate. From (6.14) it is seen that $r \leq \text{rank}(\boldsymbol{H}) \leq \min(N_r, N_t)$. To better understand how the values of the actual channel matrix may further restrict the effective channel rank r, the following sections will consider the two extreme cases of

1. zero angular spread—pure LOS conditions
2. maximum angular spread—uniformly distributed over all directions (modeled as spatially uncorrelated Rayleigh fading)

for SU-MIMO and MU-MIMO, respectively. For simplicity, co-polarized arrays on both sides of the link will be assumed in the mathematical expressions so as to ignore the polarization aspect. This simplifies the expressions and facilitates focusing on building intuition for the problem.

6.3.5.1 SU-MIMO with zero angular spread

In this scenario, only $L = 1$ propagation path exists. Assuming single-polarized antenna setups at both transmitter and receiver, the channel matrix can from (5.41) in Section 5.3.2.1 be written on the form

$$\boldsymbol{H} \sim \sum_{l=1}^{L} \alpha_k \boldsymbol{a}_r(\theta_{r,l}, \varphi_{r,l}) \left(\boldsymbol{a}_t(\theta_{t,l}, \varphi_{t,l})\right)^T \qquad (6.31)$$

$$\sim \boldsymbol{a}_r(\theta_{r,los}, \varphi_{r,los}) \left(\boldsymbol{a}_t(\theta_{t,los}, \varphi_{t,los})\right)^T$$

The channel matrix is seen to consist of only a single outer-product term and hence the channel matrix is not invertible and can therefore be described as *ill-conditioned*. This channel thus only supports one layer at a time, corresponding to the single propagation path. In other words, the effective channel rank is 1 and equal to the linear algebraic rank. When dual-polarized antennas are used on both sides of the link, the above analysis holds for the channel within each polarization and since there are two orthogonal polarizations, the effective, as well as linear algebraic, channel rank is 2.

The zero angular spread corresponds to free-space or propagation over a LOS channel with a single propagation path. Since radio waves corresponding to different propagation paths in this scenario are never linearly combined, there is no fast fading and hence no randomness in the channel; the channel is completely deterministic. Hence as was seen in Section 6.3.3.4 the spatial correlation is maximally high (i.e., the normalized correlation coefficient between any pair of channel coefficients in $\boldsymbol{H}$ is one). If the channel angular spread is instead non-zero but still small, then the propagation paths are similar direction-wise and the sum of the corresponding outer products in (6.31) can hence be approximated by a single outer-product term, that is, a single propagation path. From this it is concluded that a small channel angular spread leads to high correlation.

Although a small angular spread (or high correlation) limits the effective channel rank, it does have the benefit of providing stable long-term and wideband channel properties that are easy to track. For example, even if a UE moves at a high speed, the angular speed of the UE as seen from the base station is typically rather low and hence beamforming weights only need to be updated relatively seldom compared with techniques that rely on the fast fading properties of the channel to provide gains.

6.3.5.2 SU-MIMO with maximum angular spread

With maximum angular spread, the channel matrix is the sum of infinitely many spatially uniformly distributed propagation paths around the sphere.

$$\boldsymbol{H} \sim \sum_{k=1}^{\infty} \alpha_k \boldsymbol{a}_r\left(\varphi_{\mathrm{r},k}, \theta_{\mathrm{r},k}\right)\left(\boldsymbol{a}_{\mathrm{t}}\left(\varphi_{\mathrm{t},k}, \theta_{\mathrm{t},k}\right)\right)^{\mathrm{T}}$$

Like in the zero-spread extreme case, the above can be thought of as the channel within a polarization for dual-polarized antenna setups. Since there are many substantially different propagation paths, and each such path can be thought of as potentially carrying a layer, it is conceivable that the channel supports the transmission of many layers.

Assuming typical antenna designs where inter-element distances are not abnormally small (e.g., at least half a wavelength), there are many outer-product terms that are, due to the large angular spread, very different—the array response vectors have phase slopes over the arrays that differ substantially and they are weighted by arbitrarily phased α_k factors. Each channel coefficient in the channel matrix $\boldsymbol{H}$ is therefore the result of the summation of complex-valued numbers with very different phases making the channel coefficient appear random. These phases differ for different antennas in the array so one channel coefficient may look very different from another. In other words, a large channel angular spread leads to low correlation, sometimes even reaching the lowest possible correlation (normalized or not) of zero.

The maximum angular spread case may therefore be modeled by spatially uncorrelated channel coefficients using the common IID complex Gaussian channel model. This is an idealization as even though the spatial correlation may typically be small with a maximum angular spread assumption, it does not necessarily have to be equal to zero.

The two extreme examples of zero and maximum angular spread with their corresponding correlation levels together explain why spatial correlation tends to increase with decreasing channel angular spread.

Due to the low or zero spatial correlation, the channel matrix $\boldsymbol{H}$ is almost surely of full rank. The linear algebraic rank is therefore $\mathrm{rank}(\boldsymbol{H}) = \min(N_{\mathrm{r}}, N_{\mathrm{t}})$. The *effective* channel rank is a much more useful concept, which, however, depends on the particular channel realizations that the fading produces for the data block to be sent and also on the SNR level. A distribution of effective channel rank is obtained and the mean effective channel rank of that distribution increases with increasing SNR; increasing the SNR means a larger portion of the set of all channel realizations support at least a given number of layers. In general, and as expected from the availability of many distinct layer-carrying propagation paths, the maximum angular spread case is a scenario very friendly toward spatial multiplexing and it becomes even more friendly when combined with dual-polarized antenna setups.

In practice, the angular spread is somewhere in-between the minimum of zero and the maximum case. Often with above-rooftop deployments, the channel angular spread is rather small even in NLOS situations. As explained in Table 3.3, it tends to be particularly small in the vertical domain, although also the horizontal domain is the angular spread typically relatively small and spatial correlation thus high. The layer-carrying propagation paths become more similar in direction, making it progressively harder for the receiver to separate them, that is, the effective channel becomes harder to "invert" on the receive side as it comes closer to being ill-conditioned. This reduces the effective channel rank, pushing it closer and closer to the zero-spread extreme as the array response vectors in the outer-product terms become more and more similar.

The above argumentation can alternatively be turned around—a decreasing angular spread, or increasing spatial correlation, means a higher SNR level is required to compensate for the more difficult receiver side-channel inversion to keep the effective channel rank the same as for a higher angular spread or lower spatial correlation.

The findings for SU-MIMO can now be summarized as

- lower channel angular spread ➔ higher channel correlation;
- lower channel angular spread ➔ fewer layers;
- higher channel angular spread ➔ lower channel correlation;
- higher channel angular spread ➔ more layers;
- any number of layers possible with typical channels if SNR can be made sufficiently high and impairments ignored;
- dual-polarized antenna setups great for SU-MIMO.

6.3.5.3 MU-MIMO with zero angular spread

With zero channel angular spread, using MU-MIMO to co-schedule multiple devices on the same time/frequency resource requires the devices to be sufficiently well separated in the angular domain. This is in line with intuition that beamforming in the direction of each device only works if the beam directions are sufficiently different so that there is little overlap between the beams belonging to different UEs. In this case the spatial correlation of the channel of each individual multiplexed link is high and beamforming vectors therefore have a direct physical interpretation in the angular domain.

Classical beamforming using, for example, GoB works reasonably well in this scenario. GoB works particularly well if the devices are fortunately placed so that an orthogonal set of GoB beams points toward the devices. The orthogonality then completely avoids introducing any inter-layer interference. This would correspond to that the victim (i.e., interfered) device locations coincide with nulls in the beam patterns of aggressor transmissions. Techniques for explicitly removing inter-layer interference such as zero-forcing also work well, even though interference levels are quite low in cases of angularly well-separated UEs.

The potential for a large effective channel rank is high. It is, however, important that the scheduler is smart and properly groups devices for coscheduling that are spatially compatible, that is, that have sufficiently large angular separation. For example, performing MU-MIMO on all the UEs in a hotspot of UEs located in close proximity of each other, resulting in that they all essentially have the same direction relative to the base station, would be ill-advised. Those UEs should instead be multiplexed in some other dimension like time or frequency.

The maximum effective rank is 1 with a single-polarized antenna setup and two with the dual-polarized counterpart when all the devices have essentially identical directions, just as for SU-MIMO.

Reducing the device separation creates more overlapping beams and thus higher inter-layer interference when classical beamforming such as DFT-based techniques are used. If zero-forcing kind of beamforming is used, the similarity in array response vectors makes the channel matrix closer to singular and thus it becomes harder to separate the layers by inverting the channel without punishing the beamforming gain. Regardless of beamforming technique, the SINR levels of the layers are reduced for a fixed number of layers r when device separation shrinks, thereby also reducing spatial multiplexing gains.

6.3.5.4 MU-MIMO with maximum angular spread

Maximum angular spread scenarios are often modeled using spatially uncorrelated Rayleigh fading (IID complex Gaussian distributed) for the channels of the individual links as well as for the combined channel from all the multiplexed links. In contrast to zero angular spread scenarios, such maximum angular spread scenarios make it hard to make intuitive physical interpretations—each device is effectively "everywhere" as seen from the other side of the link. Instead, just as for SU-MIMO, the beamforming can only fully be understood as being performed in an abstract vector space over a combined channel that almost surely is full rank in a linear algebraic sense.

Classical beamforming–based techniques do not work well at all in this case. Their strongly restrictive structure is unable to provide a good match to the highly random nature of decorrelated channels that are far from having DFT characteristics and moreover are dominated by fast fading effects. To mitigate the inter-layer interference, it becomes critical to employ techniques such as zero-forcing or MMSE spatial filtering. Coscheduling of devices that are essentially in the same physical direction is then no problem; in fact it is not more problematic than if the devices are well separated. This is a fact that comes as a surprise to many that tend to forget that separation of layers now exploits the small-scale fast fading properties of the channel and not the much more intuitive large-scale properties such as propagation directions. A main drawback of the necessity to exploit fast fading properties is the need to update the spatial filtering in both time and frequency with usually a fine granularity and the corresponding challenging requirements on availability of accurate CSI.

For closely spaced devices, the situation is rather similar to the SU-MIMO case with the exception that the processing at each device typically does not have access to all the information that the other devices have access to. Hence, the maximum angular spread channel deserves to be classified as highly MU-MIMO friendly, at least if the need for accurate CSI is ignored. It also reduces the need for clever coscheduling grouping of devices in the scheduler.

In practice channel angular spread is obviously also for MU-MIMO somewhere in-between the two considered extremes. Angular spread is seldom so large so that physical directions almost completely lose their meaning. On the other hand, even in LOS conditions, there is usually some angular spread and thus several UEs can be co-scheduled even if they are essentially in the same location, if interference mitigation techniques exploiting fast fading channel properties are exploited, for example, zero-forcing, and SNR levels are sufficiently high.

Zero forcing–like techniques can, however, not perform miracles. The increasingly ill-conditioned channel as UEs move closer to each other makes even advanced techniques more sensitive to noise enhancement and channel estimation errors. Interference avoidance on the transmit side (e.g., zero-forcing precoding) does indeed manage to separate the layers but at the expense of increasingly uneven Tx power distribution over the radio chains. This effectively reduces the transmit power that can be used for the communication since power cannot be borrowed across PAs.

LOS or low angular spread for the channels of individual links (high spatial correlation) is good for robustness due to less fast fading effects but on the other hand makes clever scheduling more important and makes it harder to co-schedule multiple closely located devices.

6.3.5.5 Performance impact of angular spread on MU-MIMO

The simple example on impact of angular spread in Section 6.1.6.1 is now revisited but extended to cover the case of MU-MIMO with two different single-antenna UEs receiving one layer each on the same time–frequency resource as illustrated in Fig. 6.27. UE 2 now has a channel with a single propagation path cluster at a fixed mean angle $m_{\varphi}^{(2)} = 15$ while the mean cluster angle $m_{\varphi}^{(1)}$ for UE 1 will be varied from 0 to 15 degrees to see what happens when the two UEs are located increasingly close to each other.

The received signals y_1 and y_2 for UE 1 and 2, respectively, can from (6.11) be written as

$$y_1 = \boldsymbol{h}_1^{\mathrm{T}}\left(\boldsymbol{w}_{\mathrm{t},1}s_1 + \boldsymbol{w}_{\mathrm{t},2}s_2\right) + e_1 = \boldsymbol{h}_1^{\mathrm{T}}\boldsymbol{w}_{\mathrm{t},1}s_1 + i_1$$

$$y_2 = \boldsymbol{h}_2^{\mathrm{T}}\left(\boldsymbol{w}_{\mathrm{t},2}s_2 + \boldsymbol{w}_{\mathrm{t},1}s_1\right) + e_2 = \boldsymbol{h}_2^{\mathrm{T}}\boldsymbol{w}_{\mathrm{t},2}s_2 + i_2$$

where the noise plus interference terms i_1 and i_2 have obvious definitions. The receive SINR for the layer of UE k is given by

$$\mathrm{SINR}_k \triangleq \frac{\left|\boldsymbol{h}_k^{\mathrm{T}}\boldsymbol{w}_{\mathrm{t},k}\right|^2 \mathrm{E}[|s_k|^2]}{\mathrm{E}[|i_k|^2]}$$

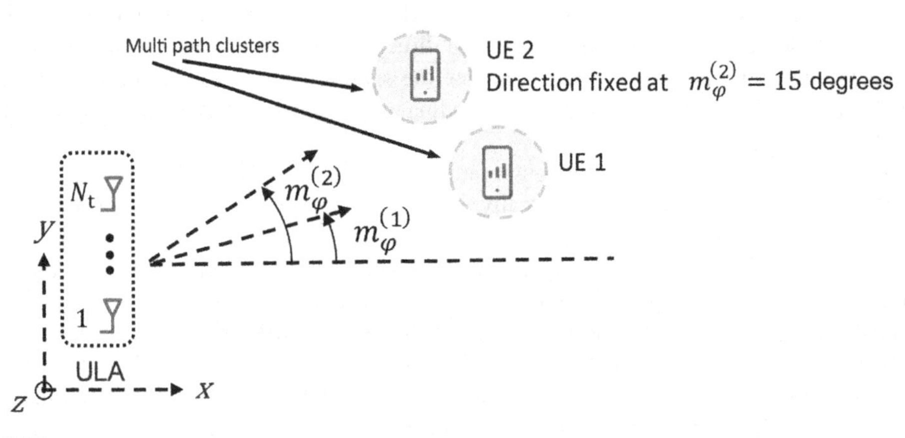

FIGURE 6.27

MU-MIMO from a transmitting base station to two single-antenna UEs getting one layer each.

Two different methods for determining the precoder vectors $w_{t,k}$ will be compared. The first method is regularized zero-forcing as in (6.21). The second method is DFT-based precoder vectors implementing GoB as in the original example in Section 6.1.6.1 with the precoder selected for a UE so as to maximize its received signal strength. Both methods assume the same total transmit power of the N_T base station antennas and that each layer receives half that power.

The SINR based on zero-forcing or GoB for the received layer for UE 1 as a function of the channel angular spread is provided in Fig. 6.28. Different angular separations $15 - m_\varphi^{(1)}$ of the two users for $N_T = 8$ and $N_T = 64$ transmit antennas are evaluated, and ideal channel estimation is assumed. As expected, zero-forcing outperforms GoB. GoB is sensitive to angular spread and does not take inter-layer interference into consideration. Hence, the performance of GoB drops drastically when the angular spread increases or when the two UEs are closely positioned, while with low angular spread and well-separated UEs GoB works rather well.

One can also observe some strange effects for zero-forcing at low angular spread and closely positioned users (see the $m_\varphi^{(1)} = 14, 15$ curves). The reason is that when channel spread goes to zero, the two channels h_1^T and h_2^T become more and more similar, making it increasingly difficult for the zero-forcer to invert the overall channel to suppress the interference.

It is also interesting to see that with enough channel spread, users can be separated even though they are overlapping. It should, however, be pointed out that in this simple example, channel realizations are independent, but in practice it is likely that the channels for overlapping users may be correlated.

Comparing results with $N_T = 64$ antennas and $N_T = 8$ antennas, increasing the number of antennas gives more degrees of freedom (better resolution of the spatial filter); hence better performance can be achieved for closely spaced users with less channel spread.

6.3.5.6 Beam shapes may say little about received signals for multi-path channels

Nullforming was illustrated previously in Section 6.1.8 by studying the resulting beam shape. The important combined effect of the beam shape and the channel that are necessary to determine properties of the received signals has, however, so far been ignored. Taking the example of transmit precoding, recall that for multi-path channels the beam shape of the precoder may not say much

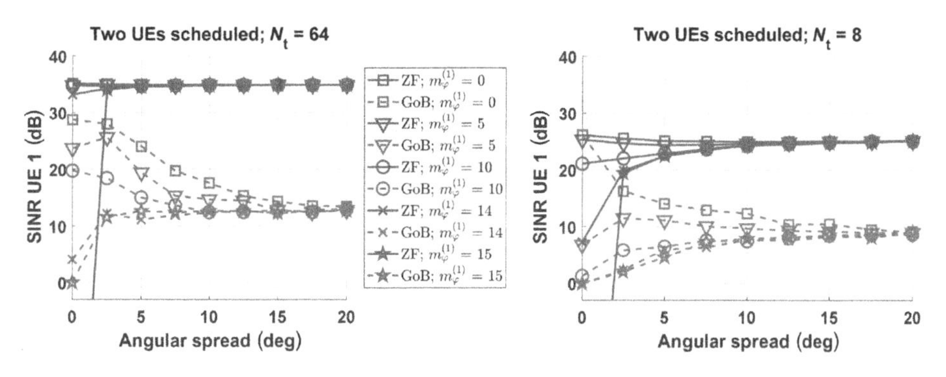

FIGURE 6.28

MU-MIMO with two UEs. SINR for UE 1 for different UE separation as a function of channel angular spread.

about how the electrical field strength of the resulting electromagnetic waves varies as a function of location in the environment. In other words, the impact on the receive side of using transmit beamforming is hard to predict based on the transmit side beam shape. To shed some light on the issue, this section will illustrate how the signal power on the receive side varies as a function of receiver location.

Consider an example with a base station using a UPA with four rows and eight columns of co-polarized antennas that spatially multiplexes four single-antenna UEs on the same time/frequency resource according to Fig. 6.29. The base station knows the channel perfectly (see Section 6.4 for how it may acquire such knowledge). Zero-forcing as in (6.22) is used without regularization, meaning that all inter-layer interference is canceled. The precoder vector for UE 1 thus makes sure that the corresponding layer is heard by UE 1, while the other three UEs receive exactly zero power from that layer. The other three layers operate correspondingly. All UEs are positioned on $z=0$, that is, ground level and a channel with a LOS path and some non-LOS paths is assumed. The exact channel characteristics are not important; it suffices to know there is some angular spread.

The receive power of the signal intended for a certain UE varies over receiver location. Fig. 6.30 shows how the received power is distributed over different geographical locations for the layer intended for UE 1. The base station is here located along the y-axis at negative x and positive z and transmits toward positive x. The z coordinate of a point on the displayed surface is proportional to the received power level and is measured in dB. Thus peaks represent high-power levels, while strongly negative values represent nulls, that is, zero power.

It is seen how three nulls are created at the locations of the three victim UEs 2, 3, and 4, and thus they receive zero power. The signal level at the UE 1 position is non-zero corresponding to a good signal level.

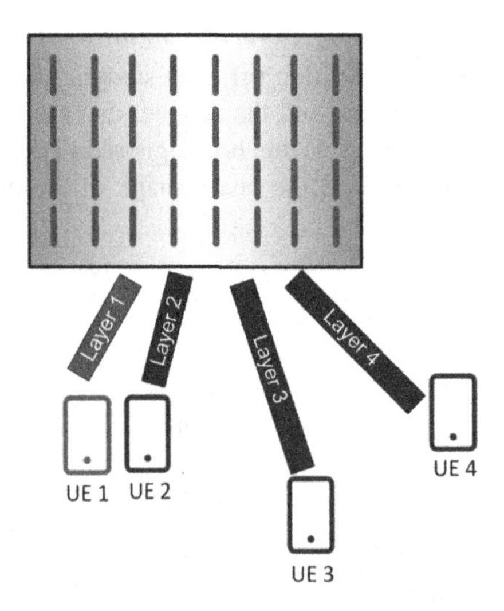

FIGURE 6.29

Base station transmitting four layers to four different single-antenna UEs using zero-forcing, one layer to each UE.

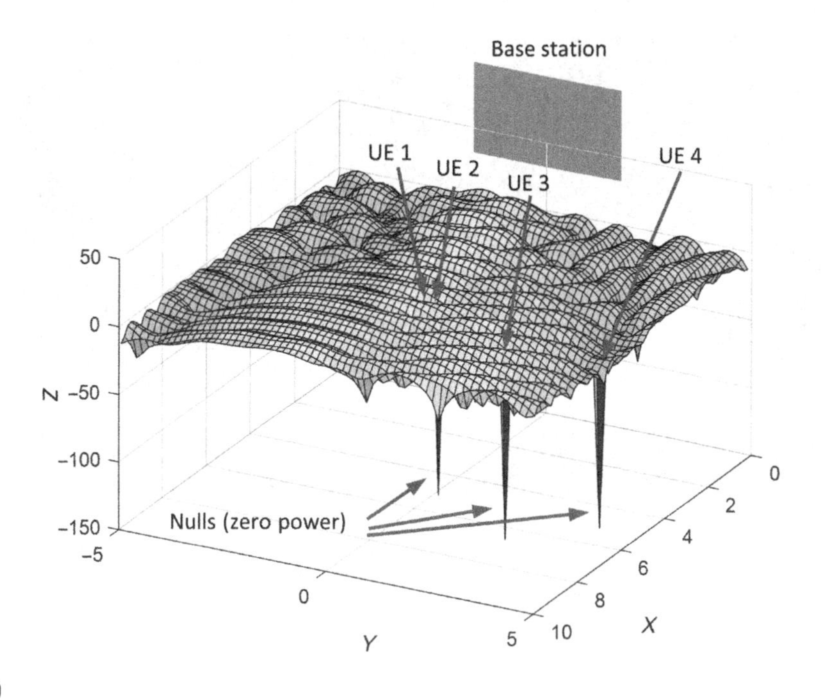

FIGURE 6.30

Power in dB for different receiver positions for the layer intended for UE 1 in a *multi-path* channel scenario. The layer is nullformed by means of zero-forcing to create zero power at the exact locations of victim UEs 2, 3, and 4 while giving decent power level to UE 1.

Note how the power level varies with (x, y) location and is only zero at the three victim UE locations. The x- and y-axis are in units of wavelength, which makes it clear that the signal power varies rapidly over the plane and can be totally different some fraction of a wavelength away. This is a fast fading effect and in fact, the time variations of the power level that fast fading leads to can be studied by moving around over the x-y plane. The nulls are seen to be deep for a very small area, maybe 10% of a wavelength. At a 2 GHz carrier frequency, this would correspond to merely 1.5 cm. Just next to a null is the signal power strong again illustrating that the nullforming ability may quickly degrade with increasing mobility, causing the precoder to become outdated once it is applied. In practice, there are methods to partly mitigate this impairment by broadening the nulls.

Fig. 6.31 gives an alternative view of the power distribution, this time in linear scale to show more of the characteristics of the power peaks rather than the nulls. The beam shape of the UE 1 precoder vector is also shown and the LOS direction from the base station to each UE is illustrated. The beam shape is seen to have a clear null in the direction of UE 4 in line with the resulting received power level of zero. Nullforming is also performed for UE 2 and 3, but despite that the beam shape has no nulls in those UE directions. In fact, those directions are close to peaks of the beam shape! This is due to multi-path and fast fading effects when propagation paths with different phases are combined and is yet another example where caution has to be exercised in not

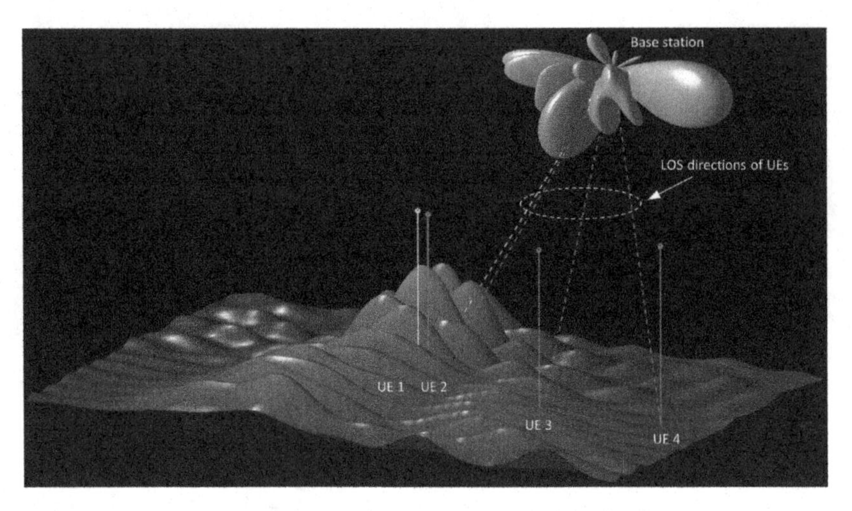

FIGURE 6.31

Beam shape and resulting signal power for different receiver positions for the layer intended for UE 1 in a *multi-path* channel scenario. The layer is nullformed by means of zero-forcing to create zero power at the exact locations of victim UEs 2, 3, and 4.

overinterpreting the implications of the beam shape and an analysis should be complemented with a mathematical analysis based on abstract complex-valued signals in a vector space.

UE 1 and UE 2 are very close to each other but that does not prevent UE 2 to be nulled and have zero power while UE 1 has a good, albeit not maximum signal strength. Both UEs lie in essentially the same direction with similar power gain on the beam shape. The beam shape, however, does not reveal any information about the phase, which could make the difference between constructive or destructive addition of signals. Since there is multi-path, two points in entirely different directions on the beam shape could transmit power in two different paths that add constructively at UE 1 location and destructively at UE 2 location. It is therefore not strange that a null on the receive side may be achieved without any visible null in the beam shape of the transmitter.

Multi-path channels are clearly risky to study only based on analyzing beam shapes. Free-space channels, however, give rise to much more intuitive results. Fig. 6.32 show exactly the same setup as previously but with no multi-path by assuming a free-space channel like in a pure LOS situation. All three victim UEs (UE 2, 3, and 4) now have a null in their receiver location and a corresponding null in the beam shape in their respective direction. The nullforming thus fulfills its goal of removing interference and works in a manner as intuitively expected.

The nullforming is unfortunately achieved at the expense of the received power of UE 1. The power level has now been strongly reduced to be almost zero and the power level in general over the area is much lower than in Fig. 6.31. This is a consequence of the free-space channel assumption, which makes receiver side signals be determined by the properties of the transmitter side beam shape and since an array has a limited lobe width it becomes impossible to maintain a good power level for UE 1 while insisting on zero power for a UE in almost the same direction (UE 2).

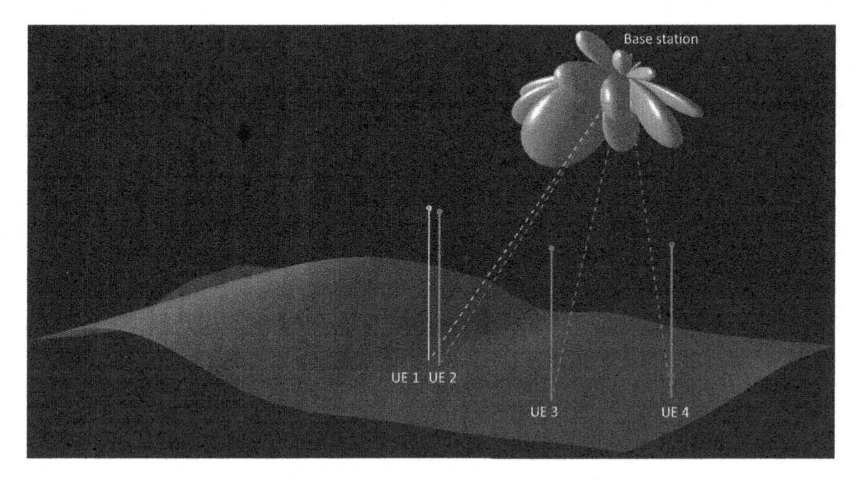

FIGURE 6.32

Beam shape and resulting received signal power distribution for the layer intended for UE 1 in a *free-space* channel scenario. The layer is nullformed by means of zero-forcing to create zero power at the exact locations of victim UEs 2, 3, and 4.

The multi-path creating fast fading effects thus helps in separating layers to closely space UEs as also seen in the example of Section 6.3.5.5.

6.3.5.7 Concluding remarks on multiple user multiple-input multiple-output

The findings for MU-MIMO can be summarized as

- lower channel angular spread ➔ more angular separation among multiplexed devices needed for a given number of layers;
- higher channel angular spread ➔ less angular separation among multiplexed devices needed for a given number of layers;
- any number of layers possible with typical channels if SNR can be made sufficiently high and impairments neglected, and;
- MU-MIMO can be, and typically is, combined with SU-MIMO depending on scheduling decisions.

6.4 CHANNEL STATE INFORMATION FOR TX PRECODING

The discussed beamforming techniques are all dependent on knowing properties of the channel to be able to determine a spatial filter that is well matched to the channel. In essence, channel knowledge is needed to know where to beamform. The issue of how to obtain this so-called CSI has, however, so far been tacitly ignored.

For receiver side spatial filtering the situation is conceptually rather straightforward if the receiver is able to receive signals from all directions at once from all the antennas. This is the case in a fully digital receiver implementation having radio chains for each antenna, providing the baseband processing with a digitally sampled signal for each antenna. Things, however, get considerably more complicated if such an architecture is not available as in, for example, beamforming implemented by analog phase shifters commonly used in mm-wave systems. This chapter discusses transmit and receive techniques that in some cases may require systems with capabilities equivalent to the fully digital case, in line with that commercial multi-antenna systems in the lower FR1 frequency range so far tend to more or less be of this type. The additional complications due to the system not being able to see all channel directions at once are left to Section 7.4 where the implications of a more restrictive implementation for mmW are discussed.

The transmitter is usually sending some form of reference signals. These signals, also referred to as training pilots, are known by the receiver and can hence be used by the receiver to accurately estimate the channel realizations over frequency. The channel estimates may be somewhat noisy but nevertheless constitute a very good basis for forming efficient spatial combining filters for receive side beamforming gain and interference suppression via nullforming.

Achieving efficient transmitter side spatial filtering is, however, considerably more challenging. Such precoding needs to obtain CSI from somewhere. There are fundamentally two different ways to do this. The first approach is to determine the CSI on the receiver side by channel measurements, which after possibly further processing, are fed back over a digital feedback channel to the transmitter. This is called closed-loop CSI feedback and is illustrated in Fig. 6.33.

The second approach is instead to do channel measurements on the transmit side and exploit the fact that some or all channel properties in forward and reverse link are more or less the same, that is, rely on reciprocity. This is shown in Fig. 6.34. Various combinations of these two fundamental CSI acquisition approaches are obviously also possible.

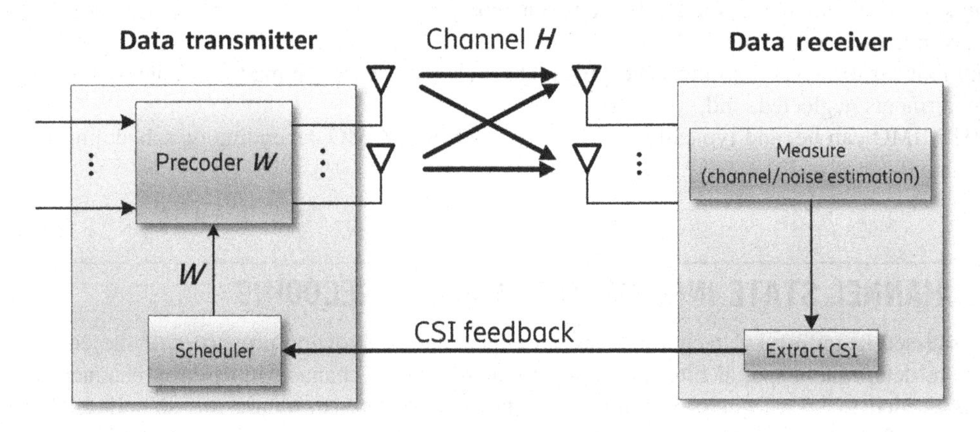

FIGURE 6.33

CSI acquisition from closed-loop CSI feedback.

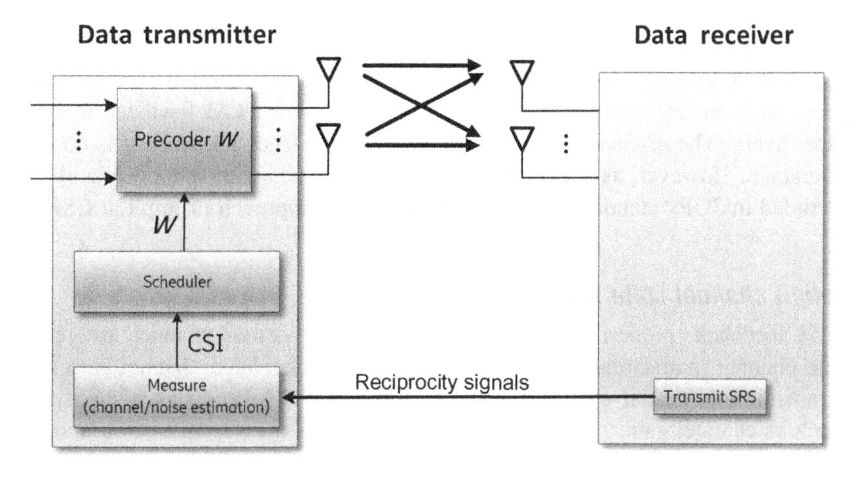

FIGURE 6.34

CSI acquisition based on reciprocity.

6.4.1 CLOSED-LOOP CHANNEL STATE INFORMATION FEEDBACK

In closed-loop CSI feedback, the receiver is usually estimating the channel based on known reference signals sent from the transmitter. Based on these channel estimates, and possibly other estimates for noise and interference, the receiver determines some kind of processed CSI.

In multi-antenna systems, the channel has MIMO characteristics and thus several reference signals are needed to estimate all the individual channel coefficients between each transmit-to-receive antenna pair. Assuming the MIMO channel matrix is $N_r \times N_t$ in size, N_t different *reference signals* (RS), one for each transmit antenna, are needed to efficiently estimate the entire MIMO channel matrix. Designs usually strive for making the reference signals orthogonal to avoid that one reference signal disturbs another. Orthogonality can be achieved by placing different reference signals on different time/frequency resources. Different spreading codes may also be used. Regardless of which, maintaining orthogonality consumes resources proportional to the number of transmit antennas, resources that could instead have been used for the actual data transmission. This overhead can become significant for large antenna arrays.

To limit the RS overhead, it is possible to transmit using nonorthogonal, or pseudo-orthogonal, RS that reuses the same resource for multiple RS. The different reference signals are then distinguished by their different symbol sequences and those sequences have low, but not zero, cross-correlation properties over sufficiently long averaging intervals. This suppresses interference between RS in the channel estimation process but does not mitigate it to the same degree as orthogonal RS and is for this reason often deemed to be inadequate in practice.

The processing of CSI usually targets compressing the CSI into as few bits as possible to keep the overhead low on the digital feedback channel whose purpose is to convey the processed CSI to the transmitter. The challenge here is to minimize the feedback overhead while still keeping as much useful channel information as possible in the CSI to ensure that efficient precoders can be determined on the transmitter side.

There is almost an endless amount of different CSI feedback schemes. To structure this vast field, it is common to broadly classify them into so-called explicit CSI feedback and implicit CSI feedback, respectively. The discussion starts with explicit CSI feedback since it is conceptually the easiest to understand. However, as will be discussed, it has issues with testability and has therefore so far been avoided in 3GPP standards in favor of the second approach of implicit CSI feedback.

6.4.1.1 Explicit channel state information feedback

In explicit CSI feedback, properties *explicitly* related to the *actual* channel are fed back. This includes whole channel matrix realizations, like $\boldsymbol{H}$ in (6.11), or related channel parameters thereof such as the transmit side spatial correlation matrix $\boldsymbol{R}_{\boldsymbol{H}}^{(t)} \triangleq \mathrm{E}[\boldsymbol{H}^{*}\boldsymbol{H}]$. Feedback of nonspatial channel parameters such as channel gain, received power, or noise plus interference covariance also belong to this category.

Having access to the entire channel matrix provides the possibility to directly use any of the beamforming/precoding methods outlined in Section 6.3.3. Whole channel realization feedback is the most direct and obvious approach. The scalar-valued channel coefficients for each $\boldsymbol{H}$ can each be quantized and sent over the CSI feedback channel. The channel changes with the fast fading over both frequency and time, so this naive method tends to lead to excessive feedback overhead. It is possible to remedy that to some extent with more advanced compression schemes where the structures of typical channels are exploited for more efficient compression. For example, channel taps may be sparsely located in the time domain. Doppler spectrum may also be sparse and the sparseness may be exploited for improved compression.

Probably the most basic channel compression method is, however, to exploit spatial correlation properties. The transmit side spatial correlation matrix can be viewed as representing a form of average channel over the averaging region of choice. As such it can also be used as input to the previously mentioned techniques and could therefore make sense to feedback. For example, together with utilizing the eigenvector decomposition

$$\boldsymbol{R}_{\boldsymbol{HH}} = \mathrm{E}[\boldsymbol{H}^{*}\boldsymbol{H}] = \boldsymbol{V}\boldsymbol{\Lambda}\boldsymbol{V}^{*}, \tag{6.32}$$

it can be used for determining a precoder as

$$\boldsymbol{W}_{\mathrm{t}} = \kappa\boldsymbol{V}$$

As seen, this expression is similar to the SVD-based precoding and in fact can be viewed as a generalization of it to average channel knowledge. The eigenvectors $\boldsymbol{V}$ can thus be viewed as the "average" right singular vectors in the SVD precoding method. If the averaging region over time and frequency is sufficiently small in (6.32), then it specializes into the pure SVD precoding of (6.19).

The above kind of correlation-based CSI can also be used to *select* a precoder out of a codebook of precoders. A codebook is a finite and countable set of precoder matrices. The family of GoB techniques belongs to this category. The transmitter can upon receiving the channel correlation feedback search for a precoder that is well matched to the correlation matrix. For example, the precoding matrix used for transmission can be obtained via a criterion to maximize received power

$$W_t = \arg \max_{W \in C(r)} \left\| W^* R_H^{(t)} W \right\|_F$$

where $C(r)$ denotes the codebook of precoders for transmission rank r.

Compared with channel matrix feedback, feeding back spatial channel correlation metrics has the benefit of greatly reducing the feedback overhead. The correlation properties of the channel are related to large-scale channel parameters such as propagation directions, path loss, and shadow fading. These change relatively slowly with receiver position, so the correlation CSI does not need to be updated at a high rate to still work well for high mobility scenarios. The correlation properties are also stable over large frequency ranges, making it usually sufficient to only feedback one spatial correlation matrix over the entire bandwidth of a carrier. Basing the precoding on such long-term/wideband channel properties tends in practice to offer a robust performance baseline. The obvious drawback is that if the spatial correlation is low, there is not much useful information available for the precoder selection. Thus scenarios with large channel angular spread on the transmit side are hostile to spatial correlation feedback, resulting in low beamforming gain. But in many cases the channel angular spread is low, particularly for above-rooftop deployments, or due to LOS conditions, and spatial correlation–based precoding offers competitive beamforming gain.

However, CSI based on spatial correlation alone has issues with achieving good nullforming performance for MU-MIMO. The lack of accurate fast fading information in the CSI is detrimental to the nullforming performance. The averaging of channel realizations over the fast fading means that an originally low-rank channel for a link with only a few non-zero singular values (cf. channel realization expression in Section 6.3.3.2) becomes full rank with N_t singular values being non-zero. The CSI can be thought of as giving a "fuzzy" view of the channel directions. This fuzziness forces a nullforming algorithm to spend efforts on placing unnecessarily many nulls, thereby reducing performance. Note that to be precise, channel directions are here in an abstract vector space.

Channel realization or spatial correlation explicit CSI feedback has the advantage of being easy to conceptually work with since they directly mimic the channel, a channel that is anyway used in the algorithms for SU-MIMO and MU-MIMO. Channel realization feedback has the added advantage of both handling SU-MIMO as well as efficient nullforming with MU-MIMO. A potential drawback is that feedback bits are spent on representing some intermediate quantity in the form of a channel property instead of spending this overhead on directly representing efficient precoders.

6.4.1.1.1 Difficulties in practice—testing

Explicit CSI feedback may sound attractive and simple. However, it exhibits a major drawback when attempting to apply it for systems based on multivendor standards such as 3GPP.

To see why, note that such standards have specifications written to ensure compatibility between both ends of the link, for example, between base stations and UEs, even if these are developed by different vendors. Specifications on the physical layer are typically written to define the behavior of the equipment and in a way to allow as much implementation freedom as possible given the intended observable behavior. To enforce that the equipment indeed behaves correctly, performance requirements and corresponding tests are developed. A feature not carefully tested cannot in practice be expected to work according to the specifications. Hence, testing of equipment is crucial in a multivendor ecosystem.

Since explicit CSI feedback directly deals with properties of the channel, the problem from a specification point of view becomes to define what is a channel? Is it just the propagation channel?

Does it include antenna properties? Are transmit and receive filters included as well? The fundamental issue is that the channel that, for example, a UE sees is not an observable quantity from outside since parts of the channel are hidden inside the transmitter/receiver. And if it is not observable it is hard to devise a test that ensures that the UE reports the explicit CSI correctly—the question arises what to compare the CSI report with? This is a challenging problem to solve and is a major reason why the 3GPP standards both for LTE and NR have stayed away from explicit CSI feedback.

6.4.1.2 Implicit channel state information feedback

In contrast to its explicit counterpart, implicit CSI feedback does not strive for representing properties of the channel. It instead entails feeding back recommendations on transmission properties, sometimes referred to as a *transmission hypothesis*, and the hypothetical consequence of using these recommendations. Transmission property recommendations usually correspond to what precoder to use for the different frequency resources. Note that this means that also transmission rank is recommended. The so-called channel quality indicators (CQIs) say something about the quality of the effective channel formed by the recommended precoder and the radio channel and constitute the consequence part. CQIs are often roughly interpreted as a form of effective SINR value for one or more layers hypothetically transmitted over the mentioned effective channel, but their formal definition is more involved than that (cf. Section 5.2.2.1 of [12]). An illustration of typical implicit CSI feedback is given in Fig. 6.35.

In essence, implicit CSI feedback is about providing transmission recommendations and consequences thereof as opposed to directly feeding back channel properties. Good transmit recommendations typically need to match the properties of the channel. The consequence part is obviously also a function of the radio channel. Information about the radio channel is in this way implicitly obtained. The distinction between the two feedback approaches is important from a testability perspective in a multivendor ecosystem. As previously mentioned, explicit CSI feedback is difficult to devise tests for but implicit CSI feedback with its transmit recommendations is considerably easier. Such tests collect implicit CSI reports from a receiver device, apply those recommendations on the transmitter side and measure the resulting performance. A requirement on minimum acceptable performance then ensures that the precoder recommendations are sufficiently good, meeting the intentions of the physical layer part of specifications.

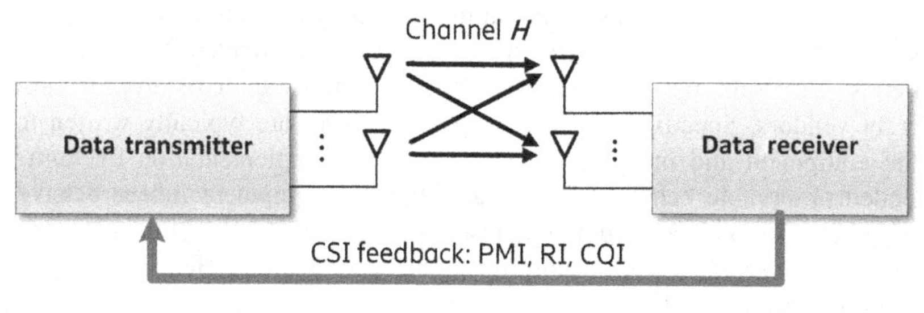

FIGURE 6.35

The three components of typical implicit CSI feedback.

All the previously described transmitter side precoder algorithms in Section 6.3.3 could be used to determine a precoder to feedback. Hence, similar such algorithms can take place in the data receiver with the target to maximize the performance for a *hypothetical* data transmission.

In implicit CSI feedback, the receiver of data transmissions usually first estimates some channel properties (e.g., channel matrix realizations) and then based on those properties it determines a precoder recommendation that intends to cover a certain frequency region. The frequency region can be of a very large bandwidth spanning an entire carrier, so-called wideband precoding, or much more narrowband leading to frequency-selective precoding. The latter has the potential for better performance via better matching of the precoder with the fast fading properties over frequency, but it also has the drawback of substantially increasing the feedback overhead.

The transmitter of the data can choose to slavishly follow the precoder recommendation from the data receiver. But it can alternatively just treat it as one of many information sources used as input to precoder determination, including interpreting it as some form of channel property.

The precoders are typically selected from a precoder codebook, which is just another name for a finite and countable set of precoder matrices. This leads to so-called codebook-based precoding where the feedback signaling for a precoder consists of sending an index pointing out a precoder matrix element in the codebook. This index is usually further divided into two parts—a *rank indicator* (RI) specifying the number of layers and a *precoder matrix indicator* (PMI) that points to one of the precoder matrices in the codebook for the number of layers given by RI. An example of this typical kind of CSI feedback is illustrated in Fig. 6.35.

Different feedback reporting modes are available in NR and LTE. Some modes only provide a single PMI for the entire carrier bandwidth, so-called *wideband precoding*, while other modes provide different PMIs for different sub-bands supporting *frequency-selective precoding*. In contrast, RI tends to always be reported in a wideband fashion. Much effort has been spent in 3GPP standardization to design efficient codebook structures that minimize feedback overhead while maximizing performance.

Strictly speaking, it is the CSI feedback that is codebook based. The actual precoders used in the transmission may or may not be codebook based, regardless of the characteristic of the CSI feedback. If the transmitter follows the precoder recommendations, then the precoding on the transmit side is codebook based as well, while if some other quantity which is essentially not finite and countable is part of determining the precoder, then so-called noncodebook-based precoding is obtained.

Note that GoB is per definition representing a codebook. It is hence perhaps not surprising that the most common codebook structures in practice are largely designed using GoB with DFT beam vectors at its core. DFT-based GoB combined with polarization precoding similar to that in (6.27) and (6.28) form the recipe for high performance as it is well matched to the structure of commonly encountered propagation channels. The GoB part takes care of the wideband/long-term correlation properties of the channel and can hence be constant over the bandwidth of an entire carrier, while the polarization precoding deals with short-term/frequency-selective channel properties. Taking the precoder expressions in (6.27) and (6.28) and slightly generalizing them by allowing different beam vectors on different polarizations of the same layer leads to a structure of this family of codebooks as

$$W_t = \begin{bmatrix} w_{\text{dft},1}^{(a)} & w_{\text{dft},2}^{(a)} & \cdots & w_{\text{dft},r}^{(a)} \\ e^{j\phi_1} w_{\text{dft},1}^{(b)} & e^{j\phi_2} w_{\text{dft},2}^{(b)} & & e^{j\phi_r} w_{\text{dft},r}^{(b)} \end{bmatrix}$$

where the DFT-based beams vectors can be either 1D or 2D depending on the application. In 3GPP it is popular to rewrite the expression as a product of two precoders $W_t = W_1 W_2$. The first precoder is block-diagonal

$$W_1 = \begin{bmatrix} B & 0 \\ 0 & B \end{bmatrix}$$

with the $N_t/2 \times K$ block B containing DFT beam vectors as columns. The second precoder can be written as

$$W_2 = \begin{bmatrix} e_{k_{a,1}} & e_{k_{a,2}} & \cdots & e_{k_{a,r}} \\ e^{j\phi_1} e_{k_{b,1}} & e^{j\phi_2} e_{k_{b,2}} & \cdots & e^{j\phi_r} e_{k_{b,r}} \end{bmatrix}$$

where $\{e_{k_{a,l}}\}_{l=1}^{r}$ are selection vectors for polarization "a" with element $k_{a,l}$ equal to one and the rest are zero. The selection vectors for polarization "b" are similarly defined. In this precoder product structure, W_2 selects beamforming vectors out of W_1 for each polarization and adjusts the phase between the polarizations. Wideband and long-term CSI properties are handled by W_1 while W_2 deals with short-term and frequency-selective CSI properties. Thus the needed update rate in time and frequency differs between W_1 and W_2. For W_1 it can be low while for W_2 it needs to be high.

By varying the selection vectors in W_2 over the bandwidth, different beamforming vectors can be selected on different sub-bands allowing some degree of tracking the fast fading properties of the channel. Even though such "beam jittering" is supported both by LTE and NR standards, it typically does not bring much gain to jitter among DFT vectors. This since the DFT vectors are not a good match to the *fast fading part* of a channel which starts to spatially decorrelate due to increasing transmitter side angular spread. As previously mentioned in Section 6.3.3.4, the use of a single DFT vector at a time is indeed a good match to the large-scale or correlation properties of the channel but that part is already well-catered for by wideband beamforming so hence *additional* gain due to jittering is small.

6.4.1.2.1 Advanced channel-state information feedback

The so far described product precoder scheme works really well for SU-MIMO. In SU-MIMO, good performance is achieved mainly via maintaining a high beamforming gain. Nullforming is not well-supported since that requires much more accurate CSI, which better tracks the fast fading properties of the channel. But that does not matter much for SU-MIMO since the receiver anyway has the ability to do powerful inter-layer interference mitigation for the layers that are intended for it. This since for those layers typically corresponding reference signals are known to the receiver and can be used to obtain accurate channel estimates that are suitable for forming efficient interference mitigating spatial filters.

Efficient nullforming is, however, very important for MU-MIMO to limit the interference between layers belonging to different users. Hence, although MU-MIMO using the presently considered CSI feedback with GoB kind of beams is not well-equipped to handle MU-MIMO, just like full transmit side spatial channel correlation CSI, it may provide some gains in some scenarios but often the gains are far from impressive.

To obtain better performance, the above product precoder structure can be further generalized by employing more flexible structures inside $\boldsymbol{W}_1$ and $\boldsymbol{W}_2$. The fundamental principles are still the same; minimize feedback overhead by letting $\boldsymbol{W}_1$ match with channel correlation properties and the other part $\boldsymbol{W}_2$ match with fast fading properties. By using $\boldsymbol{W}_1$ to essentially focus the transmit power along on average the strongest directions of the channel, the much more overhead consuming $\boldsymbol{W}_2$ part can operate over a new effective channel $\boldsymbol{H}_e = \boldsymbol{H}\boldsymbol{W}_1$ with fewer inputs. The size of $\boldsymbol{W}_2$ consequently shrinks. Since a small precoder matrix in general can be signaled using fewer bits, signaling overhead for $\boldsymbol{W}_2$ that has to track the fast fading is reduced.

One of these alternative product precoder structures that is particularly relevant due to its availability in LTE and NR is to maintain a DFT-based $\boldsymbol{W}_1$ structure but substantially relax the structure of $\boldsymbol{W}_2$ to extend beyond its previously described limited beam selection. By letting $\boldsymbol{W}_1$ perform aggressive dimensionality reduction to only a few DFT beams, the size of $\boldsymbol{W}_2$ becomes so small that it becomes feasible to essentially separately quantize each coefficient in $\boldsymbol{W}_2$ both in phase and amplitude. This advanced CSI feedback scheme is illustrated in Fig. 6.36.

With such a structure, the DFT vectors in $\boldsymbol{W}_1$ target multiple channel directions on the transmit side and $\boldsymbol{W}_2$ forms a flexible linear combination among those directions. The beams in $\boldsymbol{W}_1$ should roughly capture the most import channel directions so an appropriate value of number of beams K may depend on the transmitter side-channel angular spread.

Note that the linear combining performed by $\boldsymbol{W}_2$ means an overall precoder $\boldsymbol{W}$ is created, whose columns do not follow the DFT structure of a linearly increasing phase slope within a polarization—a linear combination of DFT vectors is in general not a DFT vector. The overall precoder

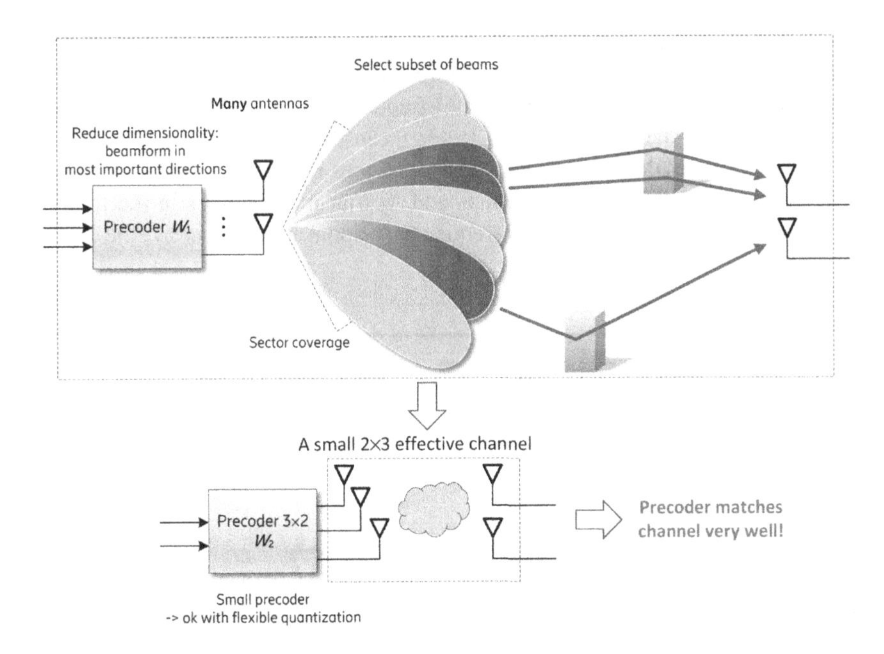

FIGURE 6.36

Principles of advanced implicit CSI feedback.

can because of this track the fast fading much better than what is possible with the beam jittering approach, which maintains a pure DFT structure of the precoder columns. It turns out that the tracking of fast fading is sufficiently good to allow even nullforming and hence MU-MIMO to work well (see the results in Sections 13.1.4 and 13.6.2).

This more advanced CSI feedback scheme extends the use of implicit CSI feedback beyond SU-MIMO to the more challenging MU-MIMO case. It is therefore not surprising that such advanced CSI feedback schemes are supported by recent LTE and NR standards as explained in Sections 8.3.2.4 and 9.3.5.5, respectively.

6.4.2 RECIPROCITY-BASED CHANNEL STATE INFORMATION

Closed-loop CSI feedback is not the only option for acquiring CSI to the data transmitting side. Measurements in the reverse link can be conducted to estimate channel properties and these can then be exploited for precoding in the forward link. This relies on that the properties are sufficiently similar for the reverse and forward link channels.

The propagation channel is the same regardless of the link direction, given a certain time instant and frequency. This is called reciprocity (with respect to link direction). If the propagation channel frequency response between a transmit antenna n and receive antenna m at time t in the forward link is denoted $h^{(\mathrm{f})}_{m,n}(f;t)$, then it holds for the reverse-link channel frequency response

$$h^{(\mathrm{r})}_{n,m}(f;t) = h^{(\mathrm{f})}_{m,n}(f;t)$$

where it should be noted that the subscripts m, n are swapped when switching link direction to take into account that what is a transmitter (receiver) in forward link becomes a receiver (transmitter) in the reverse link. For an OFDM system the channel frequency response corresponds to the channel as a function of subcarrier. The frequency parameter f can thus alternatively be taken to represent the subcarrier index in an OFDM symbol. Fig. 6.37 illustrates reciprocity for a given antenna pair.

For a MIMO system with N_r receive antennas and N_t transmit antennas in the forward link, the corresponding MIMO propagation channel frequency response is given by the $N_\mathrm{r} \times N_\mathrm{t}$ matrix $\boldsymbol{H}^{(\mathrm{f})}(f;t)$ with coefficients

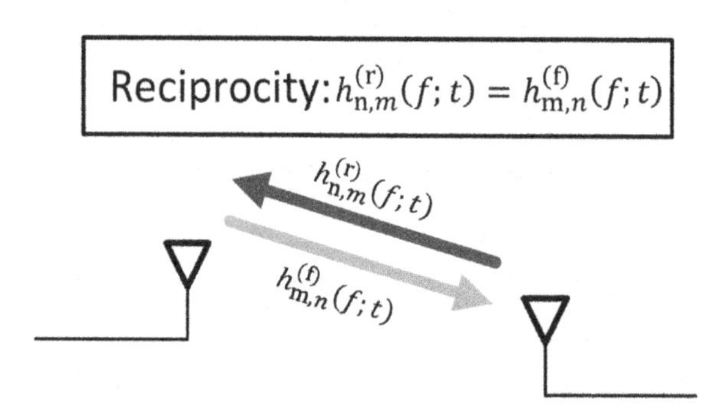

FIGURE 6.37

Propagation channel is reciprocal given the same time instant and frequency in forward and reverse link.

$$\left[\boldsymbol{H}^{(\mathrm{f})}(f;t)\right]_{m,n} = h^{(\mathrm{f})}_{m,n}(f;t)$$

It follows that the corresponding received data model in line with (6.11) becomes

$$\boldsymbol{y}^{(\mathrm{f})} = \boldsymbol{H}^{(\mathrm{f})}\boldsymbol{W}_{\mathrm{t}}^{(\mathrm{f})}\boldsymbol{s}^{(\mathrm{f})} + \boldsymbol{e}^{(\mathrm{f})}$$

where for notational convenience the frequency and time parameters have been dropped. In the reverse link, the former channel *inputs* in forward link now become *outputs* and the former *outputs* now become *inputs*. The MIMO propagation channel frequency response matrix in the reverse link is thus the transpose of its forward link counterpart, that is,

$$\boldsymbol{H}^{(\mathrm{r})}(f;t) = \left(\boldsymbol{H}^{(\mathrm{f})}(f;t)\right)^{\mathrm{T}}$$

and the reverse link received data model hence becomes

$$\boldsymbol{y}^{(\mathrm{r})} = \boldsymbol{H}^{(\mathrm{r})}\boldsymbol{W}_{\mathrm{t}}^{(\mathrm{r})}\boldsymbol{s}^{(\mathrm{r})} + \boldsymbol{e}^{(\mathrm{r})} = \left(\boldsymbol{H}^{(\mathrm{f})}\right)^{\mathrm{T}}\boldsymbol{W}_{\mathrm{t}}^{(\mathrm{r})}\boldsymbol{s}^{(\mathrm{r})} + \boldsymbol{e}^{(\mathrm{r})}$$

Conversely, the received data model in the forward link can be written in terms of the reverse-link channel as

$$\boldsymbol{y}^{(\mathrm{f})} = \boldsymbol{H}^{(\mathrm{f})}\boldsymbol{W}_{\mathrm{t}}^{(\mathrm{f})}\boldsymbol{s}^{(\mathrm{f})} + \boldsymbol{e}^{(\mathrm{f})} = \left(\boldsymbol{H}^{(\mathrm{r})}\right)^{\mathrm{T}}\boldsymbol{W}_{\mathrm{t}}^{(\mathrm{f})}\boldsymbol{s}^{(\mathrm{f})} + \boldsymbol{e}^{(\mathrm{f})}$$

The fact that only a transpose differs between forward- and reverse-link channels is illustrated in Fig. 6.38. CSI for the transmitter can now be obtained by estimating the channel in reverse link and using it to determine suitable precoders in forward link. Since the CSI is in the form of channel properties, reciprocity-based CSI acquisition has many similarities to explicit CSI feedback. In fact, the former may be viewed as a form of analog CSI feedback, in contrast to the digital CSI feedback of the latter.

In TDD systems, the forward and reverse links are using the same carrier frequency. Reciprocity capturing also fast fading channel aspects can then be exploited, as long as the channel estimate in reverse link is taken sufficiently close in time to when the resulting precoder is used in the forward link so that the forward- and reverse-link channels are similar, that is,

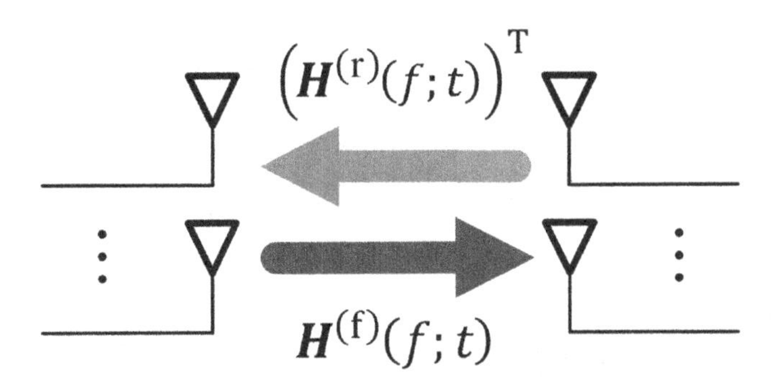

FIGURE 6.38

Relation between forward- and reverse-link MIMO propagation channels.

$$H^{(f)}(f;t_1) \approx \left(H^{(r)}(f;t_0)\right)^{T}$$

with t_0 and t_1 representing the time instants for channel estimation and precoder use, respectively. To properly capture fast fading channel effects, the time difference $t_1 - t_0$ should be substantially smaller than the channel coherence time (see Section 3.5.3.4 for a discussion on coherence time). This condition is typically fulfilled for low mobility situations for relevant 5G frequency bands in the important 2–6 GHz range, although that depends on how often reference signals for reverse-link measurements (cf. Section 9.3.1.5.1) are transmitted. At least in theory, walking speed users can hence benefit from detailed CSI obtained from fast fading reciprocity supporting both beamforming and nullforming.

Frequency division duplex (FDD) systems employ significantly different carrier frequencies in forward and reverse links. The forward- and reverse-link channels are no longer similar since the fast fading is strongly frequency-selective and hence the channel realizations become very different. Compared with TDD, this limits the possibilities for reciprocity. However, large-scale parameters of the channel such as angles of departure/arrival, angular spread, path loss, and shadowing fading are much more stable across frequency as explained in Section 3.5. Reciprocity with respect to large-scale parameters can thus be assumed for typical duplex frequency offsets between forward- and reverse-link carriers. This makes it possible to, for example, do reverse-link measurements to determine strong physical directions toward the receiver of interest and then in forward link form a precoder with beams that points in said direction.

Note that from Section 5.3.3 that the large-scale properties determine the spatial channel correlation matrix. Alternative precoding schemes that work also for FDD therefore can be based on the latter. Fast fading effects have then been averaged away leaving only the typically over frequency (and time) slowly varying large-scale characteristics in the spatial correlation matrix.

A necessary condition for a reciprocity-based scheme to work well that directly uses the spatial channel correlation matrix to form precoders is that the duplex frequency offset between forward- and reverse-link carriers must be small in comparison to the carrier frequency of the forward link. This ensures the wavelength and hence the arrays response vectors stay approximately constant over the relevant bandwidth. Such array response vectors result in that the directional information in the spatial channel correlation matrix stays similar between forward- and reverse-link frequencies as well, which is understood by recalling from Section 5.3.3 that the spatial channel correlation matrix can be expressed in terms of array response vectors. In case the duplex frequency offset is too large, methods for approximately transforming spatial correlation matrices from one frequency to another can be used [13].

Precoding based on large-scale reciprocity can work well for SU-MIMO. But for MU-MIMO, which relies heavily on high-performance nullforming, such large-scale reciprocity-based techniques are not enough. For efficient MU-MIMO the CSI needs to at least be complemented with closed-loop CSI feedback, for example, the advanced Type II feedback found in LTE and NR (see Sections 8.3.2.4 and 9.3.5.5, respectively).

6.4.2.1 Reference signals for reciprocity

As previously mentioned, exploiting fast fading reciprocity implies the need to frequently update the CSI both in time and over frequency. Preferably the whole system bandwidth should be covered by up to date CSI to give the transmitter maximum flexibility in how to schedule. Such scheduling flexibility is important in modern packet–based systems where the traffic may be bursty and difficult to predict.

Since data signals in reverse link is not guaranteed to cover the entire system bandwidth extra reverse-link transmissions of reference signals that can be used for channel estimation are needed, that is, beyond demodulation reference signals (see Section 9.3.1.2 for demodulation RS in NR) that are transmitted with the data transmission. In LTE and NR these extra signals are in uplink referred to as sounding reference signals (SRSs). With SRS, reverse-link CSI acquisition can largely be decoupled from reverse-link data transmission and scheduling.

Since accurate CSI needs to be available over the whole or large parts of the system bandwidth and the transmitter power is limited, the power per frequency unit therefore easily becomes low. For example, in LTE and NR, the total maximum allowed UE transmission power is for normal handhelds often fixed to 23 dBm, which is much lower than typical macro base station power levels. Carrier bandwidths can be many tens of MHz or more. Transmission power is thus easily spread thin as shown in Fig. 6.39 and coverage for fast fading reciprocity CSI acquisition may therefore be challenging.

Potential remedies to boost the power per frequency include to transmit SRS on a smaller bandwidth at a time and sweep the SRS transmission over different frequency resources as time progresses. This, however, means it takes longer time to cover the system bandwidth, so CSI may easily get outdated when it is unable to track the fast fading. Hence, regardless of whether wideband SRS or scanning SRS is used, a large system bandwidth presents a challenge for fast fading reciprocity.

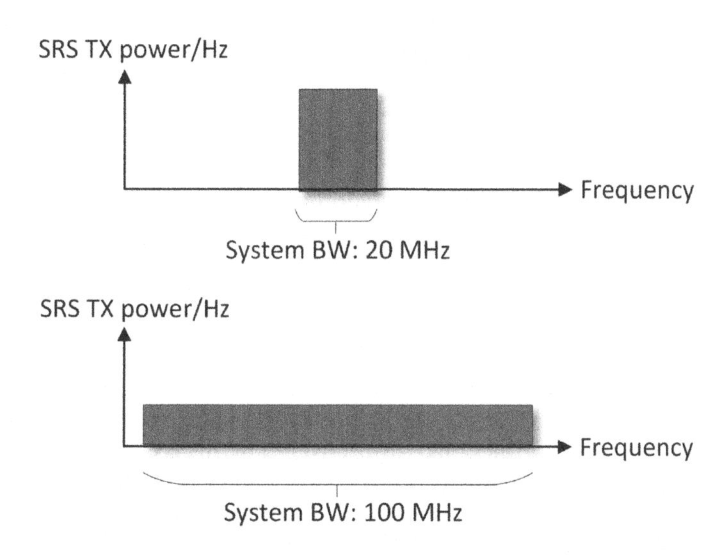

FIGURE 6.39

The same total UE Tx power (e.g., 23 dBm) is spread thinner and thinner per Hz as system bandwidth increases.

6.5 RADIO ALIGNMENT

One important aspect tacitly ignored in the discussions so far is that in reality the propagation channel is not the whole part of the channel matrix. Impairments in the analog parts of a device such as transmit and receive filters as well as antenna responses also contribute to the channel matrix, introducing so-called *radio alignment* errors. Alignment errors can substantially reduce or even completely ruin the performance. Most multi-antenna schemes therefore require the level of alignment errors to be limited.

In fact, alignment errors can be viewed as lowering the quality of the CSI since the channel matrix undergoes an unknown or possibly a channel structure destroying distortion. Different multi-antenna schemes exhibit different resilience to CSI impairments. Classical beamforming is relatively robust with respect to alignment errors while generalized beamforming conducting nullforming is as can be understood from the discussion in Section 6.1.8 much more sensitive.

Explicit procedures are often necessary for reducing alignment errors. These procedures often go under the somewhat misleading name of *antenna calibration*, even though more than just the antennas are calibrated. Implementation aspects of the radio alignment topic are treated in Section 12.3.6.

The radio alignment topic can be divided into

- transmitter alignment;
- receiver alignment;
- transmitter relative receiver alignment.

where achieving the latter is crucial for schemes relying on reciprocity. The sections to follow will first introduce a simple model for how the channel is affected by distortions due to alignment errors. The discussion proceeds with addressing separate alignment of transmitter and receiver and thereafter considers alignment of transmitter relative receiver.

6.5.1 RADIO MISALIGNMENT MODEL

Radio misalignment introduces phase and amplitude offsets that modify the undisturbed channel and effectively create a new *misaligned channel*.

The frequency response between the nth transmit antenna and the mth receive antenna of the misaligned channel can be modeled as

$$h_{m,n}(f;t) = \delta_m^{(r)}(f)\tilde{h}_{m,n}(f;t)\delta_n^{(t)}(f)$$

where $h_{m,n}(f;t)$ is the misaligned channel and $\tilde{h}_{m,n}(f;t)$ is the nominal channel in absence of any distortions $\delta_m^{(r)}(f)$ and $\delta_n^{(t)}(f)$ that represent the frequency response functions of the receiver and transmitter side misalignments of the link, respectively. In matrix form this is conveniently expressed as a matrix product

$$\boldsymbol{H} = \text{diag}\left(\delta_1^{(r)}(f), \delta_2^{(r)}(f), \cdots\right)\tilde{\boldsymbol{H}}\text{diag}\left(\delta_1^{(t)}(f), \delta_2^{(t)}(f), \cdots\right) = \boldsymbol{\Delta}^{(r)}\tilde{\boldsymbol{H}}\boldsymbol{\Delta}^{(t)}$$

Note that the misaligned channel $\boldsymbol{H}$ takes the role of the channel matrix in all the previous expressions in this chapter. Fig. 6.40 shows how misalignment may give another beam shape than the intended one. This is due to that the misalignment matrix $\boldsymbol{\Delta}^{(\cdot)}$ can be thought of as lumped together with the beamforming weight vector, effectively forming a new distorted weight vector over the undistorted channel matrix $\tilde{\boldsymbol{H}}$. Classical beam shapes like in the figure tend to be rather insensitive to misalignment errors and are hence in general not a good choice for verifying that requirements on misalignment are met for much more demanding applications such as generalized beamforming with nullforming.

6.5.2 SEPARATE ALIGNMENT REQUIREMENTS FOR TRANSMITTER AND RECEIVER

Many transmissions as well as reception schemes require the radio misalignment direction to be limited between antennas. This section focuses on the case of radio alignment considerations between antennas in the same link direction, that is, between different antennas when transmitting or between different antennas when receiving.

The problem with radio misalignment is that the phase and amplitude modifications in $\boldsymbol{\Delta}^{(r)}$ and $\boldsymbol{\Delta}^{(t)}$ impair the often strong channel structure stemming from the propagation part of the channel, creating a new channel misaligned channel with less pronounced, or even destroyed, properties. Many schemes exploit the channel structure for achieving good performance but if those properties are not sufficiently found in the misaligned channel, then performance will suffer.

For example, recall that the channel is typically a linear combination of DFT-based array response vectors. Phase misalignment, however, changes the relative phases between the antennas. If these phase changes are sufficiently large, the DFT structure with linearly increasing phase over the antennas may effectively be destroyed. This presents a major problem for DFT-based precoding, which is unable to match the non-DFT-like structure of the misaligned channel if the misalignment $\boldsymbol{\Delta}^{(t)}$ on the transmitter side is large.

Receiver algorithms that do their processing or part of their processing using components based on classical beamforming or explicitly exploit the channel structure may also suffer similarly if the misalignment $\boldsymbol{\Delta}^{(r)}$ between antennas when receiving is too large. Direction of arrival estimation is another receiver processing scheme that requires receiver misalignments to be sufficiently small.

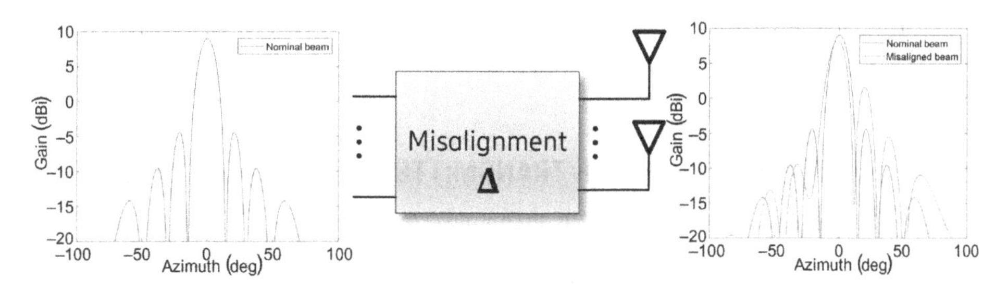

FIGURE 6.40

Misalignment alters the beam shape. The misalignment consists of phase errors that are IID over eight antennas and uniformly distributed between 0 and 100 degrees. The blue curves represent the nominal beam shape without misalignment, while the red curve shows how the beam shape is altered by the misalignment.

The radio misalignments may vary over frequency. For example, the common case of a timing misalignment between antennas translates to a linearly increasing phase over frequency. Such misalignments effectively decrease the coherence bandwidth of the channel. This means precoders and spatial combiners may need to be updated with a finer granularity in frequency. Estimation algorithms often rely on channel coherence across a frequency interval to obtain enough data for good estimates. For example, the IRC receiver algorithm in (6.17) needs an estimate of the interference covariance matrix $\boldsymbol{R}_{ee}$ and then averaging across frequency is usually needed to get a good estimate. But receiver side time misalignments decrease the coherence bandwidth of the interference, which means that either the averaging over frequency has to be reduced, at the expense of more noisy estimates, or the averaging level is kept but more of the fast fading properties are averaged away. Regardless of which, performance may suffer.

Requirements on the phase and amplitude misalignments tolerated between antennas depend on the scheme under consideration. Antenna calibration procedures running continuously that explicitly attempt to compensate for the misalignments introduced by the hardware often have to be introduced to meet the requirements. This is beyond possible efforts to calibrate the radio parts in factory as part of producing the hardware.

As long as schemes only exploit CSI obtained by measurements in the same link direction of the scheme, there is no need to calibrate transmitters relative receivers. Consider, for example, closed-loop CSI feedback from a UE to a base station. The UE is measuring on *transmitted* reference signals from the base station and the CSI is therefore completely based on signals originating from the base station transmitters. Hence, if the resulting CSI at the base station is used for precoding in the downlink, then only antenna calibration among the base station transmitters is needed. For uplink reception usually all used CSI is obtained from uplink measurements and hence only receiver antenna calibration is needed.

It could be argued that calibration in case of the above example of closed-loop CSI schemes is not necessary, the UE is after all capturing the effect of the misalignment when estimating the channel based on the transmitted reference signals and the feedback could convey that misaligned channel to the base station and hence no harm done. This is true, but only if the closed-loop feedback is very flexible and does not attempt to exploit structure in the channel. The argument is thus of little practical relevance since exploiting channel structure is useful for minimizing the feedback overhead and is therefore almost always desirable. The CSI feedback modes of contemporary standards such as LTE and NR indeed all rely on exploiting channel structure (see Sections 8.3.2.2 and 9.3.5.3, respectively, for LTE and NR codebook design).

6.5.3 MODEL FOR MISALIGNMENT OF TRANSMITTER RELATIVE RECEIVER

Although the propagation part of the channel is guaranteed to be reciprocal, there are no such guarantees for the other parts of the channel formed by the radio front end. The transmitter and receiver misalignment of the same device may thus be different. The misaligned channel in forward link could be very different from the misaligned channel in the reverse link, even though the propagation channel is completely reciprocal (like in TDD).

To formulate this mathematically, consider a link between a device A and a device B and assume the frequency is the same in both link directions. The transmitter and receiver misalignments at device A are represented by $\mathbf{\Delta}^{(\mathrm{t,A})}$ and $\mathbf{\Delta}^{(\mathrm{r,A})}$, respectively, and corresponding notation for device B. The misaligned channel from A to B (forward link) is then

$$\boldsymbol{H}^{(\mathrm{A}\rightarrow\mathrm{B})} = \mathbf{\Delta}^{(\mathrm{r,B})}\tilde{\boldsymbol{H}}^{(\mathrm{A}\rightarrow\mathrm{B})}\mathbf{\Delta}^{(\mathrm{t,A})}$$

while from the misaligned channel in the opposite link direction from B to A (reverse link) it is seen that

$$\left(\boldsymbol{H}^{(\mathrm{B}\rightarrow\mathrm{A})}\right)^{\mathrm{T}} = \left(\mathbf{\Delta}^{(\mathrm{r,A})}\tilde{\boldsymbol{H}}^{(\mathrm{B}\rightarrow\mathrm{A})}\mathbf{\Delta}^{(\mathrm{t,B})}\right)^{\mathrm{T}} = \mathbf{\Delta}^{(\mathrm{t,B})}\tilde{\boldsymbol{H}}^{(\mathrm{A}\rightarrow\mathrm{B})}\mathbf{\Delta}^{(\mathrm{r,A})}$$

It can be seen that despite that the forward- and reverse-link channels without misalignment are the same, the misaligned counterparts are in general not equal, unless the transmitter and receiver misalignments of device A are equal and the transmitter and receiver misalignments for device B are equal.

Parts of the transmit and receive radio chains in a device may be different and have different amplitude and phase characteristics. This means that the forward- and reverse-link channel matrices may be substantially different. In some cases, different antennas may even be used in forward- and reverse-link of the same device and that obviously ruins reciprocity and also makes the above simple model inadequate for describing the relation between forward- and reverse-link channels.

6.5.4 ALIGNMENT OF TRANSMITTER RELATIVE RECEIVER

For efficient use of reciprocity, it is important to ensure radio alignment of the transmitter and receiver parts in a device so that they have as similar phase and amplitude characteristics as possible. Such relative antenna calibration serves to compensate for the difference between forward- and reverse-link misalignment at the *same* device. If, for example, device A is considered, the goal is to make the resulting misalignment matrices for receive $\mathbf{\Delta}^{(\mathrm{r,A})}$ and transmit $\mathbf{\Delta}^{(\mathrm{t,A})}$ as similar as possible with each other. A corresponding goal holds for device B if relative calibration is performed at device B.

Without sufficiently tight requirements on transmitter/receiver relative alignment, a scheme that relies on measurements of CSI in reverse link to be used for transmissions in forward link may be seriously impaired. For example, a precoder determined based on reverse-link measurements may be totally wrong, the relative phase errors between transmit and receive may lead to that the transmissions are not at all focused along the intended directions.

It is imperative that relative misalignments between transmitter and receiver of a data transmitting device A are sufficiently small, while the other device B representing the data receiver side of the link may have less stringent requirements. The latter is usually the case for UEs in commercial networks. Assuming perfect alignment at transmitting device A leads to that the channel matrix in forward and the corresponding channel matrix based on reverse link channel are multiplied by different diagonal misalignment matrices from the left, that is,

$$\boldsymbol{H}^{(\mathrm{A}\rightarrow\mathrm{B})} = \mathbf{\Delta}^{(\mathrm{r,B})}\tilde{\boldsymbol{H}}^{(\mathrm{A}\rightarrow\mathrm{B})}$$

$$\left(\boldsymbol{H}^{(\mathrm{B}\rightarrow\mathrm{A})}\right)^{\mathrm{T}} = \mathbf{\Delta}^{(\mathrm{t,B})}\tilde{\boldsymbol{H}}^{(\mathrm{A}\rightarrow\mathrm{B})}$$

and the reciprocity is hence not perfect. The receiver side spatial filtering however also acts as a matrix multiplication from the left (cf. (6.12)), so the receiver can implicitly compensate for this difference if it deems that to be beneficial. It is, for example, within the realms of an MMSE spatial filter as in (6.17) to make such a determination. Lack of forward/reverse-link calibration on the receive side of the link is thus less important.

For TDD, targeting fast fading reciprocity is possible and in principle it is sufficient to only have tight alignment requirements of transmitter *relative* receiver. In other words, the two diagonal misalignment matrices $\boldsymbol{\Delta}^{(t,A)}$ and $\boldsymbol{\Delta}^{(r,A)}$ should be close to being equal up to a complex-valued scaling factor, while each still can be far from a scaled identity matrix and thus introduce large misalignments for the transmitter and the receiver seen separately. The amplitude and phase variations due to these impairments may in this case thus be substantial as long as these transmitter and receiver misalignment variations closely track each other.

Leaving large misalignments between the antennas of a transmitter, or between antennas of a receiver, is, however, in practice not an attractive option. This is because misalignments tend to randomize the channel structure in space as well as frequency as already discussed in Section 6.5.2. Thus in practice requirements on alignment need to be imposed both for transmitter relative receiver and transmitter and receiver separately.

For FDD it does not make sense even in principle to only have relative alignment requirements between transmitter and receiver. Since reciprocity-based beamforming in the FDD case invariably is of wideband nature utilizing spatial correlation or large-scale properties of the channel, the beamforming gain is hurt regardless of whether the transmitter and receiver misalignments are similar in phase and amplitude or not, if the misalignments within the transmitter/receiver are large.

6.6 COORDINATION IN THE FORM OF COORDINATED MULTIPOINT

Traditionally, different cells operate in a largely independent manner. Scheduling and beamforming are performed opportunistically for each cell without explicit regard to the consequences to other cells. There is no explicit coordination between the cells, except static configurations efforts such as deployment and cell planning. As a result, the inter-cell interference seen from other cells becomes to a large extent statistically independent of the processes in the cell of interest.[2]

It is, however, possible to coordinate different cells with each other. The possibilities are almost endless and are commonly in the industry referred to as *coordinated multipoint* (CoMP) [14]. A basic assumption in coordination is that the cells that are mutually coordinated must somehow exchange information so as to know how to coordinate. Such information exchange is commonly assumed to be conveyed over the backhaul but over-the-air signaling could also be possible. An example use of CoMP is illustrated in Fig. 6.41 where so-called coordinated scheduling is used to avoid interference between neighboring cells.

The use of the term "point" deserves a digression into a more detailed discussion as it is a commonly used term in 3GPP standardization contexts and it is related to the common confusion of

[2]This is a first-order approximation since measurements in a serving cell may capture interference from other cells and those measurements may in turn be used in the serving cell to affect its scheduling.

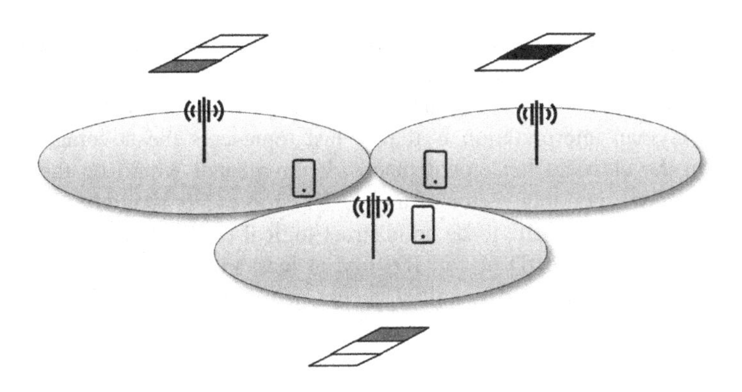

FIGURE 6.41

Example of CoMP in the form of coordinated scheduling where the three base stations coordinate their transmission in a slot so as to put the cell-edge UEs on different frequency resources to avoid inter-cell interference.

what a cell is. Thus being able to correctly use the term point becomes important. Roughly speaking, point refers to an antenna array at a certain geographical location, where the term antenna array includes the special case of a single antenna. A more precise definition is that

- A point is an antenna array at a certain geographical location intending to cover a certain geographical area.

Considering that the precise definition sounds very much like what a layman would view as a cell, it is not surprising that there are a lot of misunderstandings plaguing this topic.

6.6.1 RELATION BETWEEN CELL, SECTOR, POINT, AND BASE STATION

Cell, sector and base station are all commonly used terms in the field of cellular networks. Indeed, Section 2.2.1 in this book used those terms to explain cellular network basics. In that early chapter, the explanations tried to be very concrete and specific to keep it simple and relate all these terms to physical entities as opposed to logical entities. While discussing the point concept, this section will also revisit these other terms and show that it may not always be as simple as in Section 2.2.1, where the terms are introduced by means of a maximally simple yet common and relevant example.

To make the concept of a point more concrete, consider, for example, the sectors of a three-sector site. They correspond to three separate points since they have different coverage areas although their geographical location is essentially the same. Furthermore, sectors at different sites correspond to different and separate points. For a classical hexagonal macro deployment, each sector is therefore a separate point. Remote radio heads placed at different sites, but all connected to the same base station represent separate points. These examples are illustrated in Fig. 6.42.

The motivation behind introducing "point" as a new term is to fulfill the need to distinguish between physical and logical entities. This need becomes particularly strong when coordination is considered, which blurs the border between existing terms such as base station or cell. A base station can be connected to one or more antenna arrays. There is thus a potential one-to-many relation between a base station and associated point(s).

The term cell has an unfortunate dual meaning depending on context which is the source of endless confusion. The layman interpretation is that it just represents the coverage area of a single "antenna," that is, in the classical hexagonal macro deployment it would be the area of a single hexagon covered by a corresponding sector of a site. This represents a purely physical concept. In contrast, 3GPP standard defines "cell" to be an abstract logical entity, defined in its purest form on the physical layer to just be some ID (a cell ID) tied at least to the most basic broadcast signals, such as synchronization channels used in initial access or mobility. The ID is then, for example, used to determine various scrambling sequences, for example, for reference signals.

Often, the logical and physical concepts of a cell coincide. For example, in most deployments each sector/point is using a separate cell ID and all the transmissions/receptions in the coverage area of the sector are based on the corresponding cell ID. In fact, earlier 3GPP standards all the way up till and including LTE Release 10 were all in their foundation built on an assumption that everything is associated to the cell ID. In other words, there is a strong tie between the early channels in initial access and all other uses of the system; the mental model was that the coverage area of the initial access signals is the same as the coverage area of the data signals in connected mode— "all signals are transmitted/received the same way." This corresponds to a simple mental model, which is still deeply entrenched.

The traditional rigid cell concept can be seen as a restriction imposed on how the network is configured and deployed. To gain more flexibility the concept of a cell was considerably loosened starting with LTE Release 11, which allowed various UE−specific signals to not necessarily be functions of the synchronization channel's cell ID and hence not even transmitted from the same point as the synchronization channel. This led to the necessity to introduce the concept of quasi

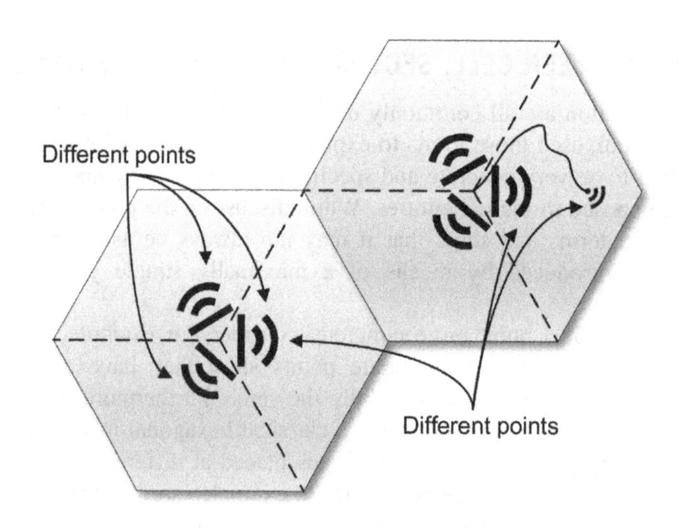

FIGURE 6.42

Illustration of the term "point."

colocation (QCL) in 3GPP (see Section 8.3.1.2). It is not a coincidence that this paradigm shift entered in LTE Release 11 as that was the first release that focused on the standardization of features targeting general use of CoMP. With NR the trend has continued, and the cell concept is loose and largely minimalistic already for the initial release as discussed in Section 9.1.

With coordination in the picture a rigid connection between how cell-defining signals and UE–specific signals are transmitted/received becomes limiting. A UE could, for example, acquire initial access from the same sync and broadcast signal transmitted over several points, while the UE–specific data in connected mode are received from only one of those points. This is referred to as the shared cell concept [15] and is illustrated in Fig. 6.43. The opposite case also exists where the initial access is obtained from cell-defining signals transmitted from a single point but where the UE–specific data transmission is performed jointly over a whole set of points in connected mode. The latter is referred to as joint transmission (CoMP) and the corresponding scheme for uplink is referred to as joint reception (CoMP).

6.6.2 BEAMFORMING FOR COORDINATED MULTIPOINT

CoMP not only includes techniques for coordinating the resource allocation and link adaption among different points. Beamforming can be used to intentionally reduce inter-cell interference by dynamically shaping the beam so that the main lobe points toward the UE of interest in the own cell, while nulls are in the directions of victim UEs in neighboring cells, as illustrated in Fig. 6.44. To design nulls CSI concerning the victim UEs and on what time/frequency resource they are scheduled is needed. This information can, for example, be conveyed via the backhaul of the points performing the coordination.

Contrast this to the uncoordinated case where there is no control over the resulting interference level, which may sometimes even increase with unfortunate UE distributions. Performance gains from employing coordinated beamforming (CBF) over noncoordinated schemes may thus be significant as evident from simulation results in Section 13.6.4.

Note that in uplink, nullforming toward interfering UEs served by other points is in principle automatically achieved by the use of conventional interference rejection receivers, which from the estimate of an interference covariance matrix get to understand how to suppress the received signals in certain directions. The coordination is here implicit and occurs over the air instead of

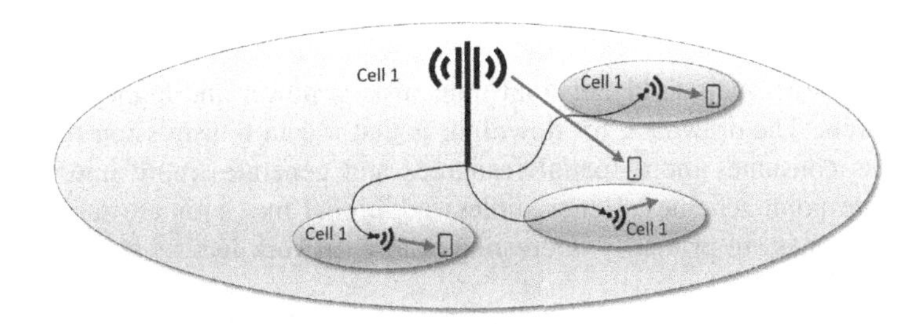

FIGURE 6.43

A so-called shared cell where multiple points all belong to the same cell (i.e., have the same cell ID as transmitted on initial access signals) despite data transmission being performed only from "closest" point.

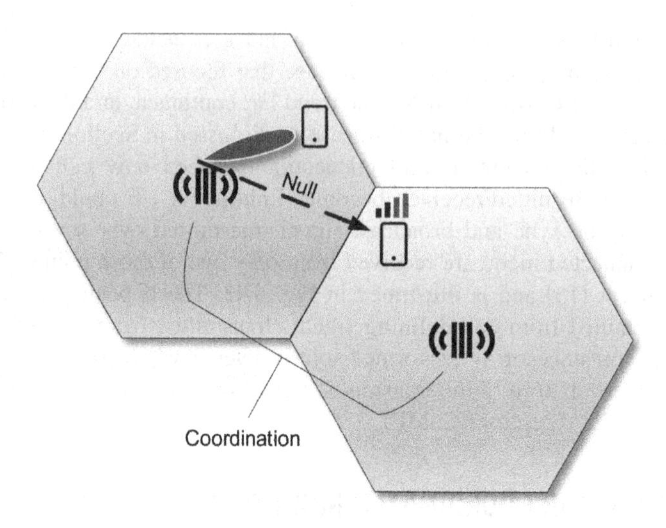

FIGURE 6.44

Coordinated beamforming (CBF) is used to form nulls toward victim UEs in neighboring cells.

explicit by the exchange of information between points over some backhaul. It is debatable whether this automatic and common way of utilizing IRC receivers should be classified as CoMP. Inter-cell nullforming techniques in downlink, however, needs intentional steps to function via appropriate CSI acquisition even in their over-the-air variants and hence fall under the framework of CoMP. Less ambitious CBF can also be performed that does not involve nullforming but rather just attempts to coordinate the main lobe directions simultaneously used from several points.

Another kind of coordination is to instead let the beam span multiple cells. For example, the three sectors of a site are usually connected to the same baseband and then joint processing of signals over all three sectors/cells may be performed. If there is sufficiently good inter-site communication, that is, sufficiently capable backhaul, the beamformer may involve multiple sites as well. This is referred to as joint (transmission/reception) CoMP and provides macro-diversity. Fig. 6.45 illustrates how joint transmission/reception involves multiple points for the same communication signal.

In downlink the benefits include higher total transmission power and higher beamforming gain, thus raising SNR levels. The drawback for downlink is that a data transmission now occupies multiple points and thus consumes more spatial resources and generates more inter-cell interference than traditional single point schemes. Hence, unless additional measures are taken to increase the spatial reuse, such schemes are primarily interesting at low network loads since their capacity tends to be low.

To improve the capacity of joint transmission, it is imperative that multiple UEs can simultaneously transmit on the same frequency/time resources from the set of points participating in the joint transmission. Multi-user spatial multiplexing is hence needed. Such multiplexing is similar to MU-MIMO in principle but where the antenna array, and hence the beam/precoder, can now be

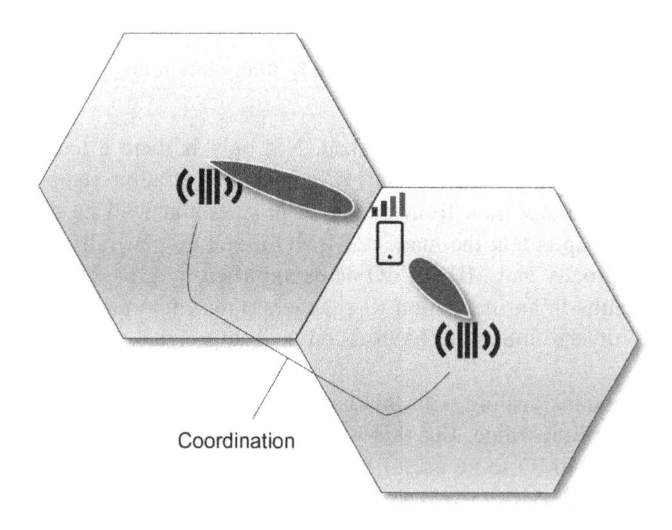

FIGURE 6.45

Joint transmission/reception CoMP where same UE is served from multiple points at once.

thought of spanning multiple points. This kind of scheme is referred to as multi-user joint transmission. In contrast to single-user joint transmission, the capacity is much higher if the quality of the CSI is sufficiently good so that nullforming can be conducted across the points. The challenges in practice are, however, large, especially to get a robust system, considering that the bulk of transmissions now heavily would rely on sensitive fast fading nullforming to work instead of naturally exploiting the spatial separation offered by transmissions from individual points.

Another coordination technique is *dynamic point selection*. In contrast to joint transmission/reception only one point at a time is involved in data transmission/reception to/from a UE. The used point can change dynamically from one slot to another, unlike in coordinated scheduling.

6.7 MASSIVE MIMO IN COMMERCIAL NETWORKS — PUTTING IT ALL TOGETHER

The previous sections described basic multi-antenna concepts and terminology. This is put to use in the present section for discussing massive MIMO techniques in the context of wide-area deployment of commercial cellular networks. Recall the massive MIMO definition in Section 1.4 of systems exploiting massively many dynamically adaptable antennas. Unfortunately, the massive MIMO field is plagued with many misunderstandings that lead to wrongful conclusions for common cellular deployments. This section aims at rectifying some of these misunderstandings. Covered topics include on a high level:

- how fast fading reciprocity is challenged by network trends;
- SU-MIMO versus MU-MIMO and number of layers;
- the pitfall of unfair SU-MIMO versus MU-MIMO comparisons;

- simpler signal processing or fewer antennas?;
- how close are commercial deployments to "infinitely many antennas"?;
- challenges with limited UE capabilities.

Several myths and misconceptions are debunked. Not only is there a tendency in the wireless industry to sometimes exaggerate benefits via artificial demonstration setups but there are also many aspects missed in the translation from theory to practice that will be discussed. A common theme that will be brought up is that the massive MIMO literature generally has a rather one-sided focus on fast fading reciprocity and MU-MIMO as exemplified by [16–25], which implicitly, and sometimes explicitly, results in an inclination to exaggerate the importance and feasibility of MU-MIMO. In fact, most of the literature focuses on and highlights the attractiveness of these techniques.

The subsequent discussions dealing with this issue might easily be interpreted to mean that we find these techniques have little value. But this would be a misunderstanding. We find fast fading reciprocity as well as MU-MIMO to be useful techniques in the toolbox of a massive MIMO system (see, for example, the MU-MIMO versus SU-MIMO results in Section 13.1.4). They are, however, not the *only* techniques and they are like anything else not without their drawbacks. One of the intentions of the subsections to follow is to give realistic expectations on these techniques and provide the reader with an understanding of why it can be challenging in commercial networks to achieve the benefits indicated by much of the massive MIMO literature.

6.7.1 FAST FADING RECIPROCITY CHALLENGED BY NETWORK TRENDS

The use of TDD to ensure fast fading reciprocity is often emphasized to be a key technology for massive MIMO [18,20]. The literature in the area, however, tends to neglect taking several important aspects into account that makes the use of fast fading reciprocity more challenging.

The core concept of estimating detailed channel realizations in reverse link to determine efficient precoders for forward link appears very appealing. In comparison with closed-loop CSI feedback, the reference signal overhead scales with the number of receiver side instead of transmitter side antennas. This constitutes a clear factor in favor of reciprocity in the case of a very large number of base station antennas and relatively few UE antennas.

In the 1990s during the early days of the field of multi-antennas for cellular communication, systems were relative narrowband, traffic was symmetric with respect to downlink/uplink and circuit-switched. UEs generally had only one antenna. These were conditions ideal for fast fading reciprocity. The narrowband assumption implied a high-power per frequency unit and likely full cell coverage of reciprocity-based channel estimation. Since the traffic was circuit-switched the UE was anyway regularly and predictably performing data transmissions over the full bandwidth with, due to high-power per frequency unit, full coverage. Thus there was no need to transmit extra signals for the sake of exploiting reciprocity such as SRS. Channel estimates obtained from data transmissions in uplink could be reused for downlink purpose as well. CSI acquisition via fast fading reciprocity for TDD came for free as long as tight radio alignment requirements were fulfilled.

Wireless standards have evolved and with modern standards such as LTE and NR many of the assumptions from the 1990s that heavily worked in favor of reciprocity no longer hold. The following subsections will discuss these aspects.

6.7.1.1 Wider bandwidths leading to decreasing fast fading reciprocity coverage

The ever-increasing data rate needs lead to increased bandwidth needs. Consequently, the amount of bandwidth used by cellular systems has been increasing for a long time. From the narrowband 2G GSM's carrier of 200 kHz, to 5 MHz for 3G WCDMA, 10–20 MHz for first 4G LTE release and now with 5G NR in practice up to 100 MHz for lower frequency ranges (FR1, i.e., below 7 GHz) and total system bandwidths surpassing 1 GHz for mm-wave bands. Large bandwidths pose a challenge for downlink transmission schemes exploiting fast fading reciprocity since as previously discussed in Section 6.4.2.1 and illustrated in Fig. 6.39, the UE transmit power is fixed and limited and hence spreads thinner and thinner with increasing bandwidth. The SINR of the SRS in the uplink thereby decreases, making it hard to maintain a sufficiently good channel estimate quality in all parts of many commonly encountered cells.

Ways to mitigate the SINR decrease for uplink channel estimation include sounding smaller portions of the bandwidth at a time. Such narrowband scanning, however, takes time and during that time the channel changes according to the fast fading. Scanning increases the SINR of the channel estimates at the expense of a more outdated channel estimate, so the end result is that despite this mitigation technique, a larger bandwidth still makes things worse for fast fading reciprocity. Note that obtaining the large-scale properties of the channel by measurements on narrowband signals does in contrast not suffer from any significant coverage issues since plenty of averaging over both time and frequency can be performed.

In practice there is often coverage for obtaining fast fading reciprocity CSI for only a portion of the cell area that is "close" to the base station, close in a radio signal strength sense. Outside that part of the cell other means of acquiring CSI have to be used. This is illustrated in Fig. 6.46 by an inner ring around the base station with fast fading reciprocity coverage.

CSI feedback or exploiting long-term/wideband reciprocity can be used in the outer ring since they have better coverage. With CSI feedback the channel estimation is, for example, done by a UE on downlink reference signals transmitted by a high-power base station as opposed to the base station doing that operation on low power transmissions from a UE. Furthermore, for long-term/wideband parameters such as spatial channel correlation or wideband GoB there is plenty of averaging and thus processing gain is high, which improves coverage. The CSI feedback channel can have coverage issues if the feedback report contains a large number of bits. But for the reporting types that essentially only capture large-scale wideband channel effects, such as Type I CSI feedback in NR as described in Section 9.3.5.3, the payload is rather small and coverage is therefore substantially better than the fast fading reciprocity coverage.

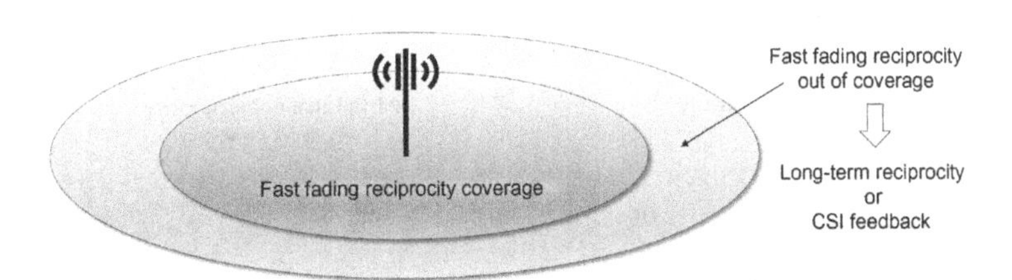

FIGURE 6.46

Fast fading reciprocity may not have full cell coverage.

To summarize, the coverage of fast fading reciprocity decreases as the trend of increased bandwidth continues. This is exacerbated by the related trend of increasing carrier frequency, which is driven by the quest for new spectrum at higher frequencies to facilitate extending bandwidth usage. Path loss increases with frequency, while operators have a strong interest in reusing existing sites even as carrier frequencies are increased. This makes systems in general more vulnerable to coverage issues. Fast fading reciprocity is indeed an interesting and important technique, but it is clearly not ubiquitous, and can as such not be seen as a *defining* technique for massive MIMO.

6.7.1.2 The partial channel reciprocity problem

The need for more capacity and higher data rates pushes wireless standards and vendors to equip UEs with more receive antennas to boost performance. LTE made two receive antenna UEs the norm from the very start and it now looks that for NR the typical UE will have four receive antennas, since it is specified to be mandatory in the lower FR1 frequency range according to 3GPP standards as discussed in Section 9.1.

The number of UE transmit antennas is, however, a different story. For LTE the norm is a single TX radio chain and a single-transmit antenna. The standard indeed supports antenna switching to offer the possibility to sound on all UE antennas without needing several separate radio TX chains. However, in practice and for various reasons, many UEs do not support such antenna switching. For NR the situation is not yet entirely clear and only time can tell where the market will end up. But judging from LTE experience it is not unconceivable that the situation will be similar with many UEs supporting less transmit antennas for uplink sounding than for downlink reception.

Having a UE where not all receive antennas are sounded in uplink creates a problem for reciprocity. Instead of obtaining CSI for the whole downlink channel matrix $\boldsymbol{H}$ only some of its rows are known. This is called the partial channel reciprocity problem and impairs the performance of reciprocity-based systems. Fig. 6.47 illustrates the issue. For LTE the typical case is to have only half channel reciprocity (2 receive, 1 transmit), while for NR the norm could in principle turn to an even worse problem of quarter channel reciprocity (1 transmit, 4 receive).

From a base station perspective, it appears only a subset of the UE antennas is active and the base station can hence not take those unknown UE receive antennas into account when determining the precoding for downlink transmission. If nullforming is performed, it can only mitigate inter-layer interference on the antennas for which there is sounding, while the remaining antennas would have considerably lower SINR levels. Consequently, the beamforming gain due to differences in fast fading may be substantially less on the unsounded antennas. The partial channel reciprocity

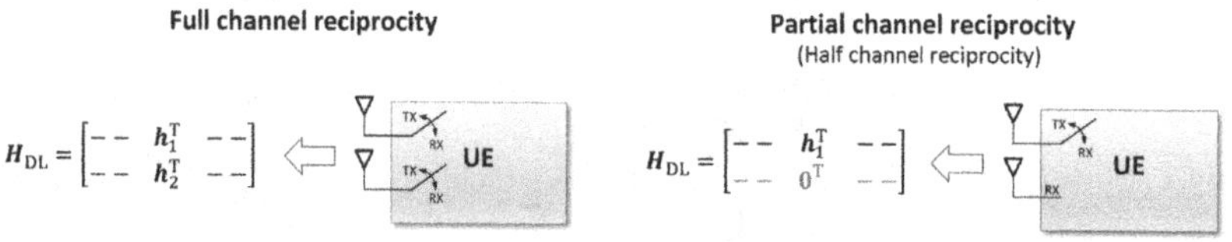

FIGURE 6.47

The partial channel reciprocity problem exemplified for a two-antenna UE.

problem is real and contributes to lowering the performance potential of fast fading reciprocity-based schemes in commercial networks.

In contrast to commercial networks, the massive MIMO literature in general, including, for example, [16–25], has an essentially total focus on the case of single-antenna UEs. This also means that the number of transmit and receive antennas are the same. Hence the issue of partial channel reciprocity is not taken into account in the massive MIMO literature, thus risking leading to overestimation of the gains of proposed techniques.

6.7.1.3 Carrier aggregation and how it limits reciprocity opportunities

The needs of ever-increasing bandwidth have traditionally been addressed in commercial standards by means of carrier aggregation. A standard in its initial release usually has a design target of a certain maximum bandwidth and designs a carrier around that. In a later release supporting further increase of bandwidth is achieved by aggregating multiple such previously designed carriers. This allows the single carrier design to be largely reused as a building block for so-called *carrier aggregation*. Unlike previous 3GPP standards, NR is a bit unusual in this sense since it even in its first release was designed to support carrier aggregation. The design choice is motivated by today's world where fragmented spectrum is prevalent and to offer a smoother evolution from LTE to NR.

The supported amount of bandwidth and number of carriers often differ between downlink and uplink. It is common that a UE supports the reception of several simultaneous carriers in downlink while in uplink it supports less, perhaps only one. An example is illustrated in Fig. 6.48. This presents yet another challenge for fast fading reciprocity since the UE may be unable to sound the entire bandwidth corresponding to downlink. Large parts of the downlink bandwidth may therefore lack CSI for at least the fast fading properties of the channel, making the concept of fast fading reciprocity more difficult to exploit. The carrier aggregation issue appears to not be considered in the literature.

In an attempt to improve fast fading reciprocity support for carrier aggregation, Release 14 of LTE [26] and NR from first release support a feature called SRS carrier-based switching with the target to offer a faster way of switching carrier in the uplink for the sole purpose of transmitting SRS in uplink carriers that correspond to all carriers in the downlink. The switching itself incurs an overhead and generally complicates the system design since there are switching transients as evident from the information element SRS-SwitchingTimeNR in [27]. The commercial UE support so far appears in practice to be limited.

Another option is to simply accept that fast fading reciprocity is not available over all parts of the downlink bandwidth and instead more semi-statically allocate different UEs on different carriers. The drawback of such an approach is that a scheduler restriction is incurred that, due to the highly dynamic and heterogeneous traffic common in mobile broadband, can quite significantly reduce the performance of the system.

6.7.2 DOWNLINK SU-MIMO VERSUS MU-MIMO AND NUMBER OF LAYERS

A prerequisite for MU-MIMO is that there are at least two, preferably more, UEs that want to have data in the same time unit. In other words, it is a capacity feature and the load in the cell/point should be high. At the same time the SINR levels should also be high as explained in

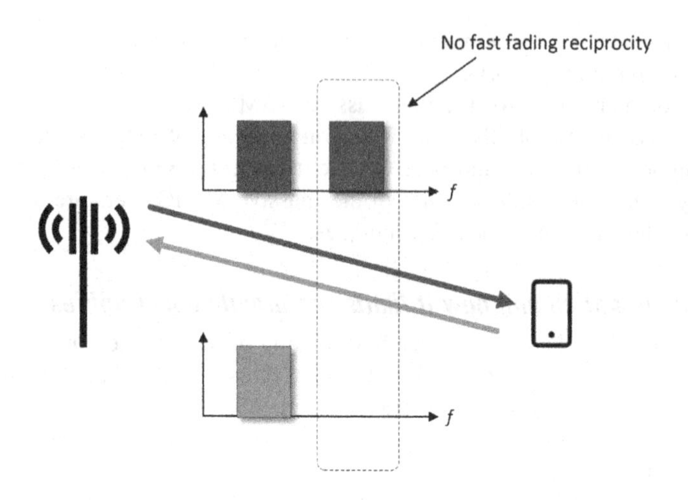

FIGURE 6.48

A UE supporting two downlink carriers but only one uplink carrier makes fast fading reciprocity difficult.

Section 6.3.4. These are somewhat contradictory requirements since the inter-cell interference increases with load, resulting in reduced SINR levels.

The ideal situation for MU-MIMO would therefore be a highly loaded *isolated* cell. This is, however, quite uncommon in real networks. Such a scenario is nevertheless popular to use in the wireless industry for demonstration purposes since it is a scenario showing huge MU-MIMO benefits with the support of a large number of layers. Full buffer traffic (cf. Section 6.7.4.1) is commonly generated where multiple UEs in the same cell want to have an endless amount of data in every time unit, sometimes combined with careful placement of the UEs. This artificially favors MU-MIMO. Fig. 6.49 compares such a demo case with more realistic scenarios. Needless to say, the demo case has little to do with real network traffic patterns for mobile broadband. Fixed wireless scenarios may, however, come closer with respect to traffic characteristics.

Another requirement for efficient MU-MIMO is the availability of accurate CSI that captures the fast fading properties of the channel. This allows the use of nullforming to mitigate inter-layer interference. Fig. 6.50 demonstrates the large negative performance impact of skipping nullforming and instead using the simpler MRT. It is seen that zero forcing–like schemes beat MRT with a considerable margin. Without fast fading–based nullforming, the MU-MIMO performance suffers greatly. MU-MIMO using classical beamforming such as GoB or based on long-term channel properties such as spatial channel correlation also show disappointing performance.

Obtaining accurate fast fading CSI is challenging since it is very detailed and hence the needed information transfer is large. There may thus not be full coverage of the cell area for fast fading CSI, regardless of whether reciprocity or feedback techniques are used.

The bursty and highly dynamic and random nature of the traffic in mobile broadband create additional challenges for the CSI acquisition. Very high load is needed to make MU-MIMO beneficial. But the reason behind the high load matters. A situation with only a moderate number of UEs where each UE is demanding a lot of traffic would be relatively easy to support. But in mobile

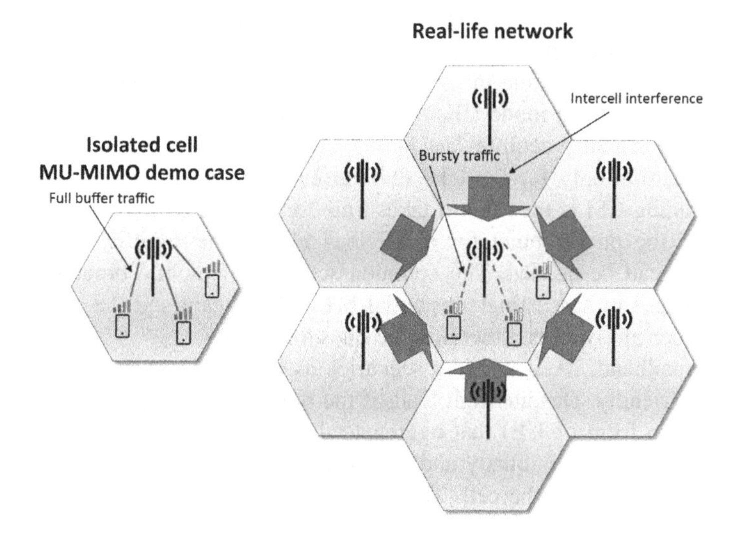

FIGURE 6.49

Contrasting the maximally MU-MIMO friendly isolated cell demo case with real-life networks facing inter-cell interference and bursty traffic that contributes to limiting MU-MIMO performance.

FIGURE 6.50

3GPP 3D UMi scenario with 32 antennas arranged as four rows and four columns of dual-polarized antenna pairs. MRT is seen to be considerably worse than nullforming based on fast fading channel properties as in zero-forcing.

broadband the load is instead typically high because there is a very large number of UEs in connected mode in the cell, while the amount of data per UE is not that large. The subset of active UEs wanting data in a particular time instant is small but varies in a highly dynamic fashion over the much larger set of connected mode UEs as illustrated in Fig. 6.51. Due to overhead reasons, it may be unfeasible to constantly acquire fast fading CSI for all the connected mode UEs. Instead on-demand CSI acquisition only for the UEs that currently want data needs to be employed. The challenge with fast fading CSI is then that it takes time for the CSI to arrive to the base station side and during that time the packet burst for a UE may already be significantly over and CSI for another UE would instead be needed. The common scenario of mobile broadband therefore often presents major challenges in harvesting substantial MU-MIMO gain and the usefulness of supporting a very large number of layers can therefore be questioned.

Unlike mobile broadband, fixed wireless scenarios as further described in Section 2.1.6.1 may be more MU-MIMO friendly. Despite high load, if the bulk of traffic go via rooftop customer premises equipment (i.e., a form of UE) that aggregate the traffic from each house it is conceivable that the overall traffic may be less bursty and the load *per* UE high while only a moderate number of UEs are in connected mode in the cell. The bursty nature of packet-based wireless communication in general is discussed in Section 2.2.2. The CSI acquisition under such aggregated circumstances becomes easier. An above-rooftop placement of the UEs likely also reduces channel angular spread benefiting MU-MIMO and perhaps even making MU-MIMO with classical beamforming competitive.

The requirements of at the same time high SINR, high load, and accurate fast fading CSI naturally limit the application of MU-MIMO. In contrast, SU-MIMO with UE−specific beamforming does not have these requirements and can provide substantial gains in essentially all cells. Although the need for a large number of layers has been greatly exaggerated and there have been tendencies of overhyping MU-MIMO in general, MU-MIMO appears to be a valuable capacity enhancement feature for some cells in mobile broadband and likely for a larger portion of cells in

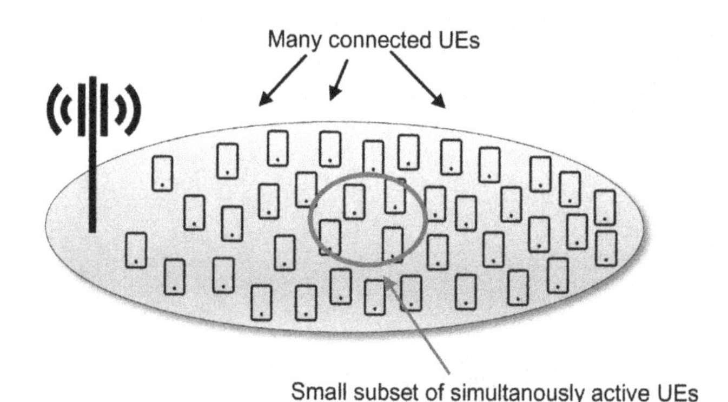

FIGURE 6.51

A highly loaded, that is, MU-MIMO friendly, cell often has many connected UEs but only a small and dynamically varying subset of UEs has a need for data to be transmitted at a given time.

fixed wireless. However, UE–specific SU-MIMO beamforming remains due to its ubiquitous benefits the cornerstone of massive MIMO performance and brings great performance advantages on its own. Thus UE–specific beamforming with SU-MIMO provides sufficiently large gains to on its own motivate the use of massive MIMO capable AAS.

6.7.3 UPLINK SU-MIMO VERSUS MU-MIMO AND NUMBER OF LAYERS

Channel-dependent beamforming and SU-MIMO spatial multiplexing in uplink UE transmissions are so far much less common than in downlink. The primary reason for that is likely the drawback of needing to equip the UE with multiple transmit radio chains. Increased cost, battery consumption, size, etc., have been deemed to outweigh the possible advantages in terms of higher throughput. The performance benefits of multiple layers in uplink can be questioned. Due to limited UE transmit power compared with base station transmit power, SINR levels are often much lower in uplink than in downlink, thus reducing spatial multiplexing gains. Alternatively, uplink power control shrinks the scheduling bandwidth to increase the SINR but spending bandwidth for transmitting multiple layers does not lead to total performance increase as the resources in the frequency domain are typically much more orthogonal than the interference-impaired layer resources in the spatial domain. UEs are often in rich scattering environments (unlike the base stations), so the channel angular spread is high, making it more difficult achieving high beamforming gains with techniques that avoid relying on fast fading CSI. Despite this, for mm-wave applications in the higher frequency range FR2, UEs may support some form of transmit beamforming as indicated in Section 7.4.

Since SU-MIMO spatial multiplexing may not be supported by the UEs in uplink, MU-MIMO becomes more favorable relatively speaking and often the only possibility for any spatial multiplexing at all. In uplink, accurate CSI is already present via the usually dense demodulation reference signals, so fast fading–based nullforming toward interferers in both own and neighboring cells is relatively easy both for FDD and TDD in case of fully digital receiver architectures, as also previously discussed in Section 6.4. This is unlike in the downlink where possibly outdated CSI first has to be acquired from feedback or reciprocity. Furthermore, the imbalance in link quality between downlink and uplink acts to increase the uplink resource utilization assuming same traffic load in downlink and uplink. Keep in mind that MU-MIMO thrives on high resource utilization. This speaks in favor of that the prospects of reaching good gains of using MU-MIMO may be better in uplink than in downlink.

There are, however, some factors that mitigate the benefits of MU-MIMO in uplink, including reducing the number of layers that need to be supported:

- SINR levels in uplink tend to be lower than in downlink due to the low UE transmit power. This reduces the number of layers needed for MU-MIMO;
- Furthermore, the uplink is power controlled to reduce near-far problems and to control inter-cell interference. The power control roughly strives for maintaining a certain power spectral density of the received signals at the base station by adjusting both the UE transmit power and its scheduling bandwidth. This means that UEs far away from the base station tend to transmit at maximum power and also use more narrow scheduling bandwidths. Consequently, scheduling bandwidths overall tend to be smaller relative to downlink and more UEs can therefore be

frequency multiplexed (FDM) over the system bandwidth. The need for additional multiplexing in the spatial domain using MU-MIMO may thus decrease because of the FDM that anyway takes place;

- There is usually much more traffic going in the downlink than in the uplink. Downlink to uplink traffic ratio for MBB is so far often somewhere around 10:1 as mentioned in Section 2.2.2. This greatly reduces the resource utilization in uplink, if the uplink and downlink provide the same amount of time/frequency resources. Since MU-MIMO relies on a high load level, the need for MU-MIMO may be smaller for FDD where the possibilities to shift resources from uplink to downlink are limited and thus the likelihood that free uplink resources are available is higher.

For TDD, the downlink/uplink slot radio can be used to adjust the relative amount of downlink and uplink resources. Theoretically, the resource ratio can then be set to correspond to the traffic ratio and hence the resource utilization could with such a strategy be high also in uplink even if the difference in link quality is ignored. In practice this is made difficult by the fact that TDD downlink/uplink patterns often need to be quite static (and can hence not dynamically track the traffic variations) and aligned among different operators in the same band to avoid devastating base station to base station interference. This is certainly the case for macro deployments, while smaller cells could possibly use dynamic TDD patterns. For the static TDD pattern case, agreements between competing operators in a region need to be formed, which naturally reduces the flexibility in the choice of TDD pattern. It is not unreasonable to expect that the rigidness in negotiation processes between operators creates a need to allocate uplink resources with a substantial margin, which obviously lowers uplink resource utilization and thereby reduces the number of layers needed for MU-MIMO also for TDD.

An extra uplink resource margin is likely also needed to mitigate the uplink coverage challenges of NR deployments. Full cell coverage is becoming increasingly difficult as carrier frequencies are increased while striving for reusing existing deployment grids.

To summarize, it is currently unclear whether the resource utilization in uplink will be substantially higher for TDD than for FDD so as to affect the dimensioning of the spatial multiplexing. It is also unclear whether uplink is more or less MU-MIMO friendly than downlink. As discussed, there are several conflicting effects that impact the dimensioning so it is hard to form a conclusion based on reasoning alone. Uplink MU-MIMO evaluation results in Section 13.7.3, however, indicate that the number of layers to reach good performance is rather moderate and does at least not show a sign of needing more layers than the moderate number needed in downlink (cf. Section 13.6.4.2).

6.7.4 THE PITFALL OF UNFAIR SU-MIMO VERSUS MU-MIMO COMPARISONS

The discussions in the previous sections highlight many of the challenges with MU-MIMO and indicate that much of the massive MIMO benefits are due to plain UE–specific beamforming for SU-MIMO. The discussions intend to set realistic expectations on the MU-MIMO benefits. The conclusions, however, contradict the common understanding, in the literature [18–20] as well as often in industry, that massive MIMO is almost synonymous with MU-MIMO due to the massive benefits expected from the latter.

What are some of the possible reasons for these contradicting conclusions? As is often the case, the answer is in the assumptions and the following discusses some of the key diverging assumptions and how they impact the conclusion on the relative merits of MU-MIMO over SU-MIMO.

6.7.4.1 Full buffer versus bursty packet–based traffic

It is not uncommon that studies in the literature assume full buffer traffic. In fact, the traffic assumption topic is typically not brought up as evident from, for example, some of the most influential overview papers [18–20] that attempt to summarize the research in the area. Full buffer is also a popular assumption when demonstrating new wireless technologies such as in test setups as already mentioned in Section 6.7.2. Full buffer traffic means every UE wants an infinite amount of data all the time. This is perfect for MU-MIMO since it only takes two active UEs to have ample MU-MIMO scheduling opportunities and with just a few more such UEs there is in addition a lot of multi-user diversity gains to benefit from.

In practice, and as discussed in Section 2.2.2, traffic is instead packet based and highly bursty with a plethora of silent periods in a session. The occurrence of small packets is not uncommon. The likelihood that there are at least two UEs to co-schedule over the same time/frequency resources is then much reduced. This greatly lowers MU-MIMO gain over SU-MIMO compared with the very MU-MIMO friendly full buffer assumption as can be seen from the simulation results in Table 6.1.

It can be argued that the negative impact of non-full buffer traffic patterns relative to full buffer can be circumvented by merely assuming higher network load and then the results should align with their full buffer counterparts. Higher load helps of course in boosting MU-MIMO gains relative to SU-MIMO, but it should also be kept in mind that non-full buffer traffic always runs the risk of overload that congests the system and leads to that the data queues keep expanding. It is desirable to ensure that the system never or very rarely enters this very unstable system operation point. Thus deployment practices usually aim at site densities that have a margin against the instability region. This is, for example, the reason why evaluations in 3GPP often measure the UE throughput performance at a 50% and not a 100% resource utilization. The former is still considered to be "high" load and would have some margin toward queue buildup, while the latter would represent a completely unstable and broken system. Because resource utilization must be significantly lower than the 100% of full buffer traffic, a non-full buffer penalty on MU-MIMO gains appears unavoidable.

Table 6.1 Example of that gain of MU-MIMO over SU-MIMO can drastically reduce when more realistic non-full buffer traffic model is used (3GPP 3D UMi scenario with a base station antenna setup of four rows, eight columns of dual-polarized antenna pairs).

4 × 8 Antenna UMi, MU Versus SU Gain	
Traffic model	**Average User Throughput Gain (%)**
Full buffer	32
Non-full buffer (at 50% resource utilization)	11

6.7.4.2 IID Rayleigh fading versus a proper three-dimensional channel model

Highly simplified channel modeling is often used in the massive MIMO literature [2,17–22]. In fact, the most commonly used assumption is that the channel is spatially uncorrelated and IID Raleigh fading [2,17,18,20]. The underlying assumption is a very large channel angular spread in both the horizontal and vertical directions. The very notion of channel direction loses significance and all directions become equally good. Co-scheduling possibilities for MU-MIMO are greatly enhanced as different UE channels are with high probability substantially different. The physical properties of the antenna setup are neglected, and it does not matter whether antennas are placed along horizontal or vertical dimensions.

In contrast, channels encountered in practice often have a rather limited angular spread as seen in Table 3.3 and Fig. 3.23. For LOS conditions it is particularly small but the spread may also be small on the base station side in NLOS in the common case of above-rooftop deployments. In Fig. 6.52, the commonly more realistic small angular spread is contrasted to the commonly used modeling assumption of IID Rayleigh fading corresponding to a very large angular spread. The channel spread moreover differs significantly between the horizontal and vertical dimensions. The spread in the horizontal direction is often larger than the spread along the vertical direction.

Also the UE angular spread differs greatly between the two dimensions as illustrated in Fig. 6.53. In the horizontal direction, UEs are distributed over a large angular interval, often covering the whole of a 120-degree sector. The vertical direction is in most cases entirely different with a small UE angular spread with a significant bias toward directions close to the horizon. With above-rooftop deployments, the vertical spread is especially small for anything but high-rise cities and in the latter case vertical spread can also be small unless the target is to cover indoor high-rise users with outdoor deployments. The existence of indoor systems in high rises, however, mitigates this need.

Neither channel nor UE angular spread is captured by the commonly employed highly idealistic assumption of spatially uncorrelated Rayleigh fading. The small vertical UE angular spread in combination with the small vertical channel angular spread generally lowers the performance gain of UE–specific beamforming compared with static cell-common beamforming. It especially lowers the performance potential for MU-MIMO in the vertical dimension since there are small possibilities to spatially separate such closely spaced UEs with highly correlated channels. Antennas added

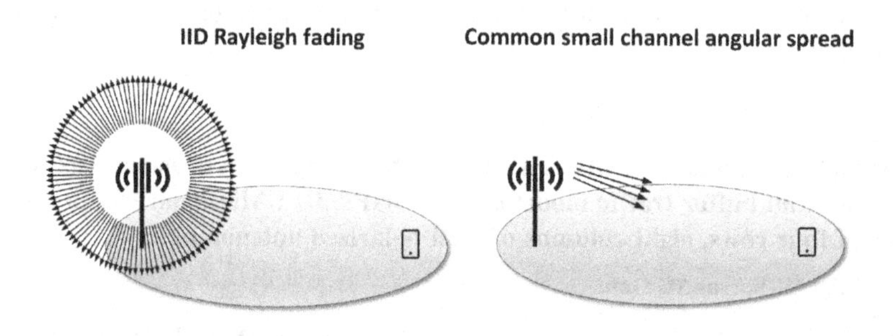

FIGURE 6.52

A common modeling assumption of IID Rayleigh fading corresponding to very large channel angular spread versus the in real-life common small spread.

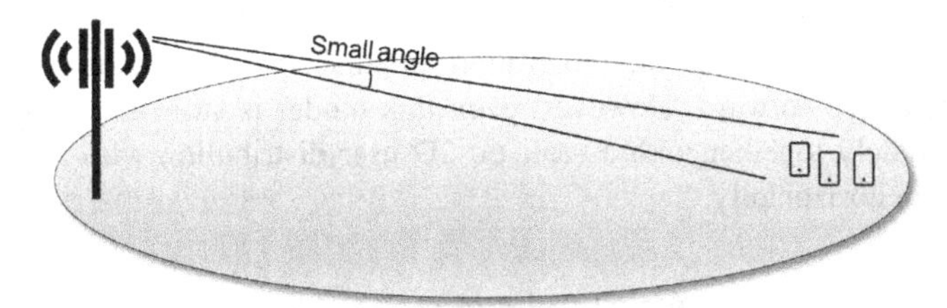

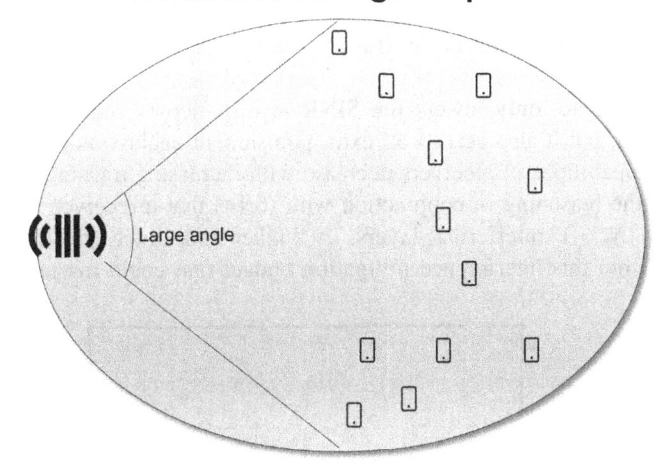

FIGURE 6.53

UE angular spread is usually much larger in horizontal than in vertical domain.

to the vertical dimension therefore contribute relatively little to the performance gain of UE–specific beamforming over static beamforming; it is almost as if the antennas in the vertical dimension should not be counted when comparing with results in the literature. This can explain much of the divergence in conclusions.

Take for example, at least in the initial phase of 5G, the popular AAS of 64 different "antennas" or more precisely 64 TRX. This 2D antenna array consists of the radio chains arranged in eight columns and four rows of dual-polarized antenna ports as illustrated in Fig. 6.54. Each dual-polarized antenna port pair is in turn connected to a 2V × 1H or 3V × 1H subarray of antenna elements giving in total an array of 128 or 192 antenna elements. The use of 64 TRX clearly qualifies as massive MIMO but since only eight columns are used the system acts more as if it only would have 8 TRX when it comes to comparing MU-MIMO with SU-MIMO performance (the dual-polarizations are beneficial for spatial multiplexing regardless of MU- or SU-MIMO). Since MU-MIMO gains generally increase with more TRX, or degrees of freedom, it is not surprising that the literature predicts much larger gains than encountered in practice.

To address the shortcomings of the spatially uncorrelated Rayleigh fading assumption, some recent literature [23] also mentions the other extreme of the pure free-space propagation channel. This resembles LOS conditions and very small channel angular spread situations. The model is overall probably more relevant than the much more popular spatially uncorrelated Rayleigh fading and thus represents a step forward. However, even this model is still misleading in a similar way unless used appropriately together with a realistic 2D user distribution with much less UE angular spread vertically than horizontally.

6.7.4.3 Single-cell versus interfering cells in a network

As already mentioned in Section 6.7.3, it is popular to demonstrate the merits of MU-MIMO with a large number of layers by employing a single isolated cell setup, often combined with careful placement of UEs and the use of full buffer traffic. The important impairment of inter-cell interference is thus neglected.

Inter-cell interference not only lowers the SINR levels, thereby reducing spatial multiplexing gains from MU-MIMO, but it also acts as an extra punishment against more layers since the interference suppression capabilities of receivers decrease with increasing transmission rank of the interference; recall from the reasoning in conjunction with (6.14) that a receiver with N_r antennas may completely suppress $N_r - 1$ interfering layers. A higher-rank inter-cell interference consumes degrees of freedom from this interference mitigation budget that could instead have been spent on

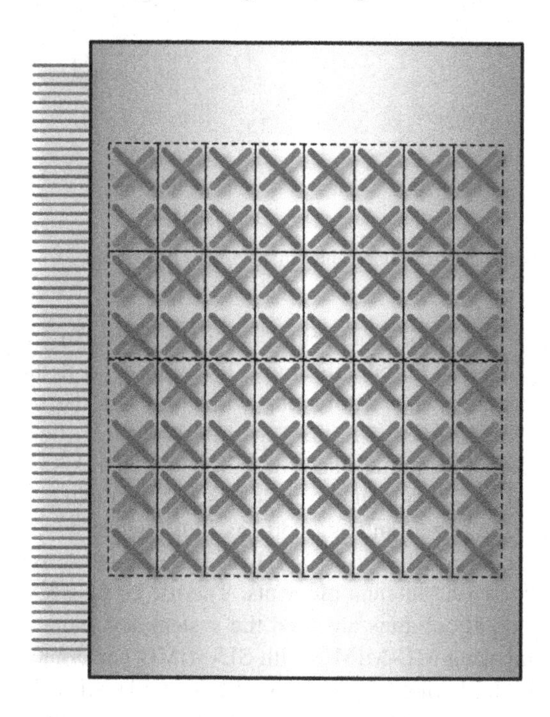

FIGURE 6.54

64 TRX AAS with 2V × 1H subarrays.

better mitigating interference between the layers intended for the UE of interest. Neglecting inter-cell interference by studying single isolated cell is therefore yet another factor that contributes to artificially increasing observed MU-MIMO performance relative performance of SU-MIMO.

6.7.4.4 One TX/RX user equipment versus multi-RX with few TX user equipment

It appears all massive MIMO literature assume the UEs have the same number of receive as transmit antennas. In fact, essentially all focus seems to be on the case of a single-receive and single-transmit antenna [2,17–25].

Assuming same number of receive and transmit antennas is perfect for reciprocity as the whole channel can then be estimated on the base station side. As previously mentioned in Section 6.7.1.2, there is a risk that commercial UEs will continue to have fewer transmit antennas than receive antennas even for NR. In such a situation, the resulting half or quarter channel reciprocity will significantly impair schemes like MU-MIMO that rely on fast fading reciprocity techniques. SU-MIMO will be less affected since it does not need to exploit so detailed CSI to start with and hence relative gain of MU-MIMO over SU-MIMO decreases.

The fact that commercial UEs have two or four receive antennas tilts the comparison in favor of SU-MIMO in other ways as well. With a single UE antenna assumption, the only way to exploit spatial multiplexing is to use MU-MIMO, while SU-MIMO becomes severely crippled due to the resulting restriction to a single layer. This obviously favors MU-MIMO over SU-MIMO, especially considering that because of dual-polarized antennas, two-layer SU-MIMO is very common and highly beneficial even in LOS conditions.

The presence of multiple UE receive antennas also provides UEs with inter-cell interference rejection capabilities. The rejection works better if the interference is of low rank but MU-MIMO increases the transmission rank and hence this punishes MU-MIMO over SU-MIMO and is yet another example of how assumptions in the literature may inadvertently raise the observed MU-MIMO gains compared with reality.

6.7.4.5 The low power amplifier utilization problem

It is common when comparing different transmission schemes to assume a fixed total power summed over all the radio chains [2,17–19,21]. This, however, neglects that power cannot be reallocated from one PA to another. Hence the appropriate assumption is to instead use a per PA (per "antenna") power constraint.

DFT-based precoding such as GoB uses the same amplitude on all the antennas and full power utilization is ensured regardless of whether a sum power or per antenna power constraint is used. However, schemes that use different power on different antennas will with the per antenna power constraint lose in total transmission power. Both MRT and zero forcing–like MU-MIMO suffer from uneven antenna power distribution and thus power loss. For MRT the issue is easily rectified by just keeping the phase while scaling up different antennas differently so they all run at full power. But such scaling would for zero-forcing impair the inter-layer interference suppression and is therefore less desirable. Thus MU-MIMO tends to suffer from a power loss compared with GoB like schemes. Fig. 6.55 compares the power distribution of GoB with schemes employing nullforming such as zero-forcing. In view of that GoB is quite suitable for SU-MIMO while inadequate for MU-MIMO, the conclusion is that using a realistic power constraint lowers the MU-MIMO gains. Furthermore, the power loss worsens the coverage of MU-MIMO.

The PA power is determined as the summation of power over all frequencies/subcarriers over the system bandwidth. This summation of power reduces to some extent the effect of power imbalances that vary over frequency. Nevertheless, power imbalance and resulting power inefficiency remain a significant drawback of nullforming kind of schemes.

Note that the power loss tends to increase, the more similar the channels of co-scheduled UEs are. The combined channel then becomes harder to invert, which is resulting in potentially large power variations (cf. how the dynamic range of inverting a small number may be very large). Hence, a LOS situation with two UEs very close to each other may result in a larger power loss than if the channel angular spread is larger and/or the UEs are more spatially separated.

6.7.4.6 Difficulties performing uplink time-domain filtering

Time-domain filtering of channel estimates is a well-known technique for improving the quality of the estimates. Such filtering is often proposed as a way of dealing with the coverage problems of SRS. In commercial networks this may, however, not be feasible. Time-domain filtering requires both the amplitude and phase of the SRS transmissions to be stable over longer periods of time. But in 3GPP standards, there are no strict requirements on coherence between transmissions in different slots and UE vendors have so far shown tendencies to resist the introduction of such. Technical reasons for this seem to include that it is difficult to maintain coherence when transmit power level is varying as the power level may determine which stages of the PAs may then be involved.

The end result of lack of phase and amplitude time-domain coherence in UEs is that the base station cannot be sure that the UE SRS transmissions are coherent over any significant time (across

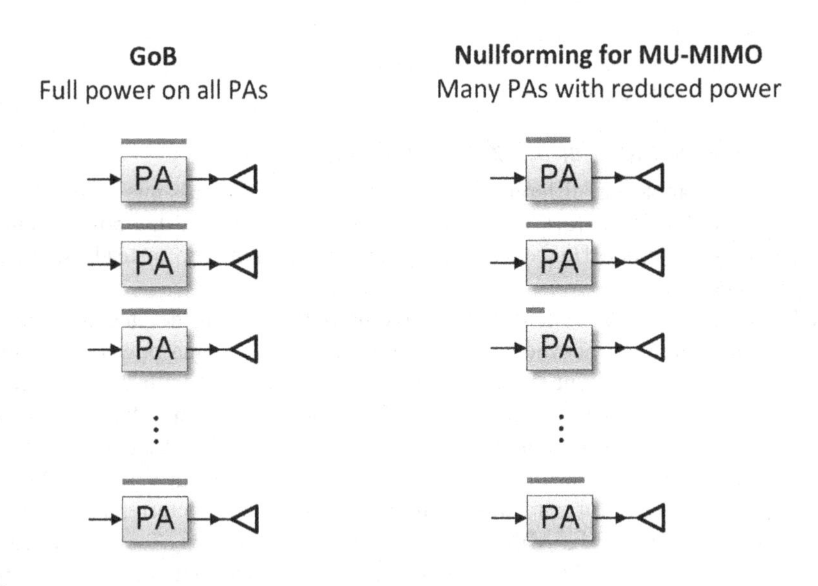

FIGURE 6.55

The power is uneven across the antennas when using nullforming for MU-MIMO leading to total power loss compared with GoB. The length of a red bar is proportional to the power of the corresponding PA.

slots in NR) and therefore does not dare perform time-domain channel filtering. Channel processing gain improvements due to time-domain filtering that some studies in the literature [25] assume may therefore be difficult to realize in practice.

6.7.5 SIMPLER SIGNAL PROCESSING OR FEWER ANTENNAS?

Due to the popular study of MRT and the focus on a very large number of antennas in the literature [17], it is sometimes easy to get the impression that nullforming is not needed for massive MIMO and that instead the simpler MRT technique can just as well be used. Theory predicts that as the number of antennas goes to infinity, so does the SINR and hence the performance keeps increasing even without nullforming. Such asymptotic reasoning, however, often does not consider how fast the performance increases as a function of the number of antennas (radio chains) in practical systems and therefore is hard to utilize for conclusions concerning commercial networks. The question to ask is how many antennas it takes for the asymptotic result to kick in and is the so obtained number *relevant* for the deployment at the hand.

With more realistic assumptions and with $N_t = 32$ transmit antennas, a number practical for the important FR1 frequency bands (7.125 GHz and less), Fig. 6.50 shows that nullforming by means of zero forcing–like schemes beats MRT with a considerable margin. Without fast fading–based nullforming, the MU-MIMO performance suffers greatly. MU-MIMO based on classical beamforming such as GoB or based on long-term channel properties such as covariance feedback also shows disappointing performance.

Thus if simpler signal processing that avoids fast fading–based nullforming is selected, then the number of antennas needs to be further increased to compensate for the performance loss. It is typically not an attractive trade-off to have to use more antennas rather than performing smarter signal processing in the baseband.

For higher bands such as mm-wave the number of antennas may be an order of magnitude higher, which may possibly be enough to make explicit nullforming redundant. The question of nullforming or not for those bands may, however, be determined mainly by another factor. Those bands typically offer much larger bandwidth and this makes it due to SRS coverage reasons (cf. Section 6.7.1.1) difficult to obtain fast fading CSI from reciprocity over the entire bandwidth. In addition, the channel varies very quickly over time since Doppler frequencies scale with the carrier frequency. Nullforming as well as MRT may therefore be challenging. The cells may in the mm-wave case also be smaller, which increases the LOS probability, decreases channel angular spread, and thus makes MU-MIMO using classical beamforming based on long-term/wideband CSI more competitive. On the other hand, it could be argued that coverage and not capacity is the main challenge at the often very bandwidth-broad mm-wave bands and therefore SU-MIMO beamforming is crucial, while MU-MIMO plays in general a less important role.

6.7.6 HOW CLOSE ARE COMMERCIAL DEPLOYMENTS TO "INFINITELY MANY ANTENNAS?"

As previously mentioned in Section 6.7.4.2 in conjunction with discussing the implications of overly simplistic channel model assumptions when comparing MU-MIMO with SU-MIMO, it is mainly the antennas in the horizontal dimension that act similar to the models in the literature. Thus the *effective* number of antennas in commercial deployments is likely much less than the nominal number of antennas. Hence, the common case of a 64 TRX system with eight columns is quite far away from behaving like a system with a very large number of antennas, perhaps more behaving like an eight-antenna system in the literature. It could in this sense be claimed to not be a massive MIMO system or it could be argued that deployment strategies should be changed to instead use very wide but short antenna arrays. A strong reason speaking against the latter is, however, that it is tempting to exploit the vertical dimension by just using taller subarrays since that avoids needing to increase the number of radio chains and is thus simpler while still often providing substantial gains (cf. Fig. 13.27). In that case the antenna area is already spent on the vertical dimension, so making the antenna wider would necessarily imply that the total antenna area increases, which may not be deemed acceptable from a deployment perspective.

6.7.7 CHALLENGES WITH LIMITED USER EQUIPMENT CAPABILITIES

Closed-loop CSI feedback schemes heavily rely on standardization support. The entire CSI acquisition procedure needs to be agreed in the standardization process and the UE vendors must thereafter choose to implement the corresponding functionality. Often, at least some parts of the more advanced CSI schemes are optional to implement. Indeed in practice far from all features are implemented and the ones that are may face delayed introductions, creating a heterogeneous UE population where many UEs lack a certain feature.

With CSI feedback the UE must for a massive MIMO system be capable of estimating a large number of channels between all transmit–receive antenna pairs. Ideally, there should be one downlink reference signals for each base station antenna but currently, the number of such reference signals, referred to as CSI-RS ports, is in both LTE and NR limited to at most 32 as further described in Section 9.3.1.1. The complexity of the channel estimation and the associated CSI determination may be considered high by the UE vendor, motivating delayed introduction or even completely skipping the feature. LTE is here in particular at disadvantage since it has a long history and thus has a huge installed base of legacy UEs lacking various features related to multi-antenna techniques. This in practice brings down the AAS network-level gains considerably for LTE FDD since large parts of the UE population in that case is limited to demodulation and measurements on at most four antenna ports (corresponding to the CRSs as described in Section 8.3.1.3). In contrast, LTE TDD is here in a better position when it comes to the UE support of needed reference signals. The newer NR standard starts from a clean sheet with all stakeholders now realizing the importance and potential of massive MIMO systems from the very start, so FDD and TDD will then from reference signals perspective be on an equal footing and there are reasons to believe the overall situation will be considerably better than in LTE.

Lack of UE feature support may make it hard for base station vendors to implement advanced features requiring detailed CSI. MU-MIMO is per definition a feature that requires the support from a large proportion of the UE population or the gain is lost. Hence, even if advanced CSI feedback schemes exist in the standard and in principle offer very competitive performance, lack of UE support may make widespread adoption difficult.

Historically, the need for UE support has been the Achilles heel of CSI feedback relative to reciprocity. The latter has more relied on proprietary features in the base station and hence has put less stringent requirements on the UE capabilities. Although the reliance on UE support continues to be strong for CSI feedback, reciprocity is now starting to face similar challenges. For example, if reciprocity is to avoid the dreaded partial channel reciprocity problem, the UE must support sounding on all the antennas that are used for reception in the downlink. Support of fast carrier switching for SRS might be another reciprocity feature not supported by all UEs.

6.8 SUMMARY

This chapter surveyed the most common and basic multi-antenna techniques for transmission and reception. There are many misunderstandings in the area and considerable effort was spent on discussing those. The chapter entails a vast body of material whose most important points can be summarized as below.

Beamforming

- A beam may be a beam or may have a completely arbitrary shape and may hence not look like a beam (Section 6.1.1);
- Classical beamforming has beams that look like a beam (Fig. 6.3);
- Beam shape of generalized beamforming may look totally arbitrary (Fig. 6.6);
- Understanding beamforming in terms of physical directions becomes more problematic as channel angular spread increases (Section 6.1.1);
- Beamforming can be dynamic, such as in UE–specific beamforming providing SINR improvement almost regardless of UE position (Section 6.1);
- Beamforming can be semi-static and cell-specific such as in cell shaping providing interference coordination and load balancing (Section 6.2).

Popular base station transmission and reception methods

- Transmission: precoding via DFT-based GoB (Section 6.3.3.4);
- Transmission: zero-forcing precoding to perform inter-layer nullforming (Section 6.3.3.3);
- Reception: MMSE spatial filtering, also called IRC (Section 6.3.2).

Popular CSI acquisition methods for determining base station transmit precoding

- Closed-loop CSI feedback containing recommended precoder(s) out of DFT-based GoB codebook and CQIs (Section 6.4.1.2);
- Measurements on uplink signals such as SRS (Section 6.4.2.1), relying on propagation channel reciprocity (Section 6.4.2), and small antenna misalignment (Section 6.5.4).

Spatial multiplexing

- Number of layers in spatial multiplexing tends to increase with SINR — high SINR desirable and fewer layers on cell edge than in cell center (Section 6.3.4);
- SU-MIMO and MU-MIMO are the two main and popular techniques (Section 6.3);
- Dual-layer SU-MIMO is highly beneficial, particularly for LOS due to use of dual-polarized antennas (Section 6.3);
- MU-MIMO is an important capacity feature but as such requires high load in addition to high SINR (Section 6.3.4);

- Nullforming via, for example, zero-forcing is important for limiting interference in MU-MIMO (Fig. 6.50).

Fast fading reciprocity
- Fast fading reciprocity challenged by network trends such as more bandwidth, more carriers, more UE RX antennas, and higher Doppler frequencies (Section 6.7.1).

MU-MIMO versus SU-MIMO
- Performance obtained from UE–specific beamforming for SU-MIMO sufficiently high to motivate massive MIMO capable AAS (Section 6.7.2);
- MU-MIMO gains over SU-MIMO vary depending on the situation but can be significant (Section 6.7);
- Risk for overestimation of MU-MIMO gains relative to SU-MIMO (Section 6.7.4);
- MU-MIMO gains are boosted by tendency in industry to focus on artificial demo cases (Fig. 6.49).

MU-MIMO gains reported in massive MIMO literature benefit from modeling assumptions not in line with practice (Section 6.7.4), including, for example:
- Use of full buffer instead of more realistic non-full buffer traffic (Section 6.7.4.1);
- Abundant use of simplistic channel models not capturing the difference in angular characteristic between horizontal and vertical domains. Realistic channel leads to substantially lower number of effective antennas— "64 antennas may act as 8 antennas" (Section 6.7.4.2);
- Strong focus on UEs with only a single receive and transmit antenna (Section 6.7.4.4).

Coordinated multipoint—CoMP
- Coordination of transmission/reception between multiple points (Section 6.6);
- A point is an antenna array at a certain geographical location intending to cover a certain geographical area (Section 6.6);
- A cell or base station can be mapped to one or more points (Section 6.6.1);
- Coordinated scheduling, CBF, and joint transmission/reception are important CoMP categories of schemes (Section 6.6.2).

Radio alignment errors
- The phase and amplitude characteristics of the radio circuitry need to be sufficiently aligned among the antennas to avoid distortion (Section 6.5);
- With insufficient alignment, signals may be impaired as effectively new and unwanted beam weights are formed (Section 6.5.1) or channel structure destroyed (Section 6.5.2);
- Alignment requirement may be stated for transmitter side, receiver side, or transmitter relative receiver side (Section 6.5);
- Nullforming sets particularly tough requirements on alignment (Section 6.5).

REFERENCES

[1] R.A. Monzingo, T.W. Miller, Introduction to Adaptive Arrays, Wiley, 1980.

[2] L. Kahn, Ratio squarer, Proc. IRE 42 (11) (1954) 1704.
[3] 3GPP TR 36.873 V12.7.0, Study on 3D channel model for LTE.
[4] V.H. Macdonald, Advanced mobile phone service: the cellular concept, Bell System Tech. J. 58 (1) (1979) 15–41.
[5] P. Balaban, J. Salz, Optimum diversity combining and equalization in digital data transmission with applications to cellular mobile radio. I. Theoretical considerations, IEEE Trans. Commun. 40 (5) (1992).
[6] J.H. Winters, J. Salz, R.D. Gitlin, The impact of antenna diversity on the capacity of wireless communication systems, IEEE Trans. Commun. 42 (234) (1994).
[7] D. Gore, R.W. Heath, A. Paulraj, On performance of the zero forcing receiver in presence of transmit correlation, in: Proceedings of IEEE International Symposium on Information Theory, June 2002.
[8] E. Telatar, Capacity of multi-antenna Gaussian channels, Bell Labs Technical Memorandum, October 1995.
[9] C.B. Peel, B.M. Hochwald, A.L. Swindlehurst, A vector-perturbation technique for near-capacity multi-antenna multiuser communication-part I: channel inversion and regularization, IEEE Trans. Commun. 43 (1) (2005). Feb.
[10] T. Haustein, C. von Helmolt, E. Jorswieck, V. Jungnickel, V. Pohl, Performance of MIMO systems with channel inversion, in: IEEE 55th Vehicular Technology Conference, Spring 2002.
[11] T.M. Cover, Elements of Information Theory, Wiley, 1991.
[12] 3GPP TS 38.214 V15.6.0, Physical layer procedures for data.
[13] K. Hugl, J. Laurila, E. Bonek, Downlink beamforming for frequency division duplex systems, in: Proceedings of IEEE Global Telecommunications Conference (GLOBECOM'99), vol. 4, 1999, pp. 2097–2101.
[14] 3GPP TR 36.814 V9.0.0, Further advancements for E-UTRA physical layer aspects (Release 9), March 2010.
[15] 3GPP R1–110649, Aspects on distributed RRUs with shared cell-ID for heterogeneous deployments, February 2011.
[16] T.L. Marzetta, How much training is required for multiuser MIMO?, in: 2006 Fortieth Asilomar Conference on Signals, Systems and Computers, Pacific Grove, California, USA, November 2006.
[17] T.L. Marzetta, Noncooperative cellular wireless with unlimited numbers of base station antennas, IEEE Trans. Wirel. Commun. 9 (11) (2010).
[18] T.L. Marzetta, Massive MIMO: an introduction, Bell Labs Tech. J. 20 (2015).
[19] F. Rusek, D. Persson, B.K. Lau, E.G. Larsson, T.L. Marzetta, O. Edfors, et al., Scaling up MIMO: opportunities and challenges with very large arrays, IEEE Signal. Process. Mag. 30 (1) (2013).
[20] E.G. Larsson, O. Edfors, F. Tufvesson, T.L. Marzetta, Massive MIMO for next generation wireless systems, IEEE Commun. Mag. 52 (2) (2014).
[21] L. Lu, G.Y. Li, A.L. Swindlehurst, A. Ashikhmin, R. Zhang, An overview of massive MIMO: benefits and challenges, IEEE J. Sel. Top. Signal. Process. 8 (5) (2014).
[22] J. Hoydis, S. ten Brink, M. Debbah, Massive MIMO in the UL/DL of cellular networks: how many antennas do we need? IEEE J. Sel. Areas Commun. 13 (2) (2013). Jan.
[23] E. Bjornson, E.G. Larsson, T.L. Marzetta, Massive MIMO: ten myths and one critical question, IEEE Commun. Mag. 54 (2) (2016).
[24] J. Flordelis, F. Rusek, F. Tufvesson, E.G. Larsson, O. Edfors, Massive MIMO performance—TDD versus FDD: what do measurements say? IEEE Trans. Wirel. Commun. 17 (4) (2018).
[25] K.T. Truong, R.W. Heath Jr, Effects of channel aging in massive MIMO systems, J. Commun. Netw. 15 (4) (2013).
[26] RP-160639, SRS carrier based switching for LTE, 3GPP TSG RAN Meeting #71.
[27] 3GPP TS 38.331 V15.6.0, Radio resource control (RRC) protocol specification (Release 15).

CHAPTER

CONCEPTS AND SOLUTIONS FOR HIGH-BAND MILLIMETER WAVE

7

High-band, often referred to as millimeter wave (mm-wave) or as frequency range 2 (FR2) by 3GPP is a new area of operation for 5G. The conditions for operation on high-band and the state of the hardware technology today leads to some specific solutions being applicable for high-band. The purpose of this chapter is to give a high-level overview of some key issues specific to high-band, particularly in the areas of antenna design, spectrum, the use of time-domain analog beamforming versus frequency-domain digital beamforming (DBF), beam management, and high-band use cases. High-level examples of solutions that can be applied in practice to achieve good performance are given. Further aspects at high-band such as 3GPP details and hardware implementation and network performance are handled in later chapters.

7.1 BACKGROUND

To meet the increasing requirements on capacity and network performance, new frequency bands are continually adopted. It has been recognized for some time that it is difficult to find larger chunks (of the order of hundreds of MHz) of spectrum on low- and mid-frequency bands, that is, within today's FR1 range (<7.125 GHz), that are also available globally. Large chunks of spectrum are however available in the mm-wave range in different regions, typically in bands higher than 24 GHz. In this spectrum, frequency bands with bandwidths in the range of 1 GHz or larger are now being made available for commercial use for 5G in different parts of the world. In these bands, very high data rates of the order of tens of Gbps can be supported with suitably designed systems.

In general, most of the antenna array and beamforming concepts described in earlier chapters are agnostic to frequency band, that is, they apply equally for high-band and for mid-band advanced antenna system (AAS) systems. However, in practice, the high frequencies and large bandwidths mean that the conditions for operation and the solution characteristics are somewhat different between mm-wave bands and lower bands. High-band solutions are new for new radio (NR), and the hardware technologies needed to support them, and their levels of maturity and cost are also different compared to mid-band solutions (see Chapter 12). Mid-band solutions have been deployed and developed for longer and are more mature. High-band solutions also have some specific challenges with implementation issues like phase noise as described in Section 11.3.5.

As a result, it has been necessary to develop high-band specific solutions. One important factor in the early development of high-band solutions has been the choice of beamforming implementation, that is, analog beamforming (ABF) versus digital beamforming (DBF) with the first high-band solutions tending to use analog beamforming.

Advanced Antenna Systems for 5G Network Deployments. DOI: https://doi.org/10.1016/B978-0-12-820046-9.00007-1

The intention of this chapter is to describe some of the rationale for high-band specific solutions and give a high-level overview of how such solutions can work and their consequences. Details of the hardware and implementation are covered in Chapter 12 and network performance is covered in Chapter 13.

The outline of the chapter is as follows. The essentials of the high-band characteristics are described in Section 7.2, particularly a brief summary of the propagation conditions and antenna array characteristics as described in Chapters 3 and 4, and their relation to high-band frequencies. The different realizations of beamforming, for example, analog versus digital, are described at high level in Section 7.3. In Section 7.4, solutions specified to facilitate high-band are outlined. This section describes beam management procedures for NR and how they can be used to solve the outlined issues especially for time-domain analog beamforming–based systems used at high-band. Section 7.5 summarizes the chapter.

7.2 ISSUES ON HIGH-FREQUENCY BANDS

In this section some specific characteristics related to higher frequency bands are summarized.

7.2.1 SPECTRUM

High-band spectrum is unique for NR and so not supported by long-term evolution (LTE). In 3GPP release 15, the spectrum for high-band is specified as FR2 and contains new bands in the range 24.25–52.6 GHz as shown in Fig. 7.1. These FR2 bands are designed for time division duplex (TDD) operation.

For the purposes of this chapter high-band, or synonymously mm-wave, will be referred to as being equivalent to bands in today's FR2 range above. In future, NR or 6G may include even higher frequency bands but impacts at those bands are not the focus of this book. The FR2 bands that are defined in 3GPP at the time of writing this book are shown in Fig. 7.2.

Notice that the bands are generally some GHz wide. NR carriers within these FR2 bands can have large bandwidth, with bandwidth for single carriers of up to 400 MHz, compared to maximum

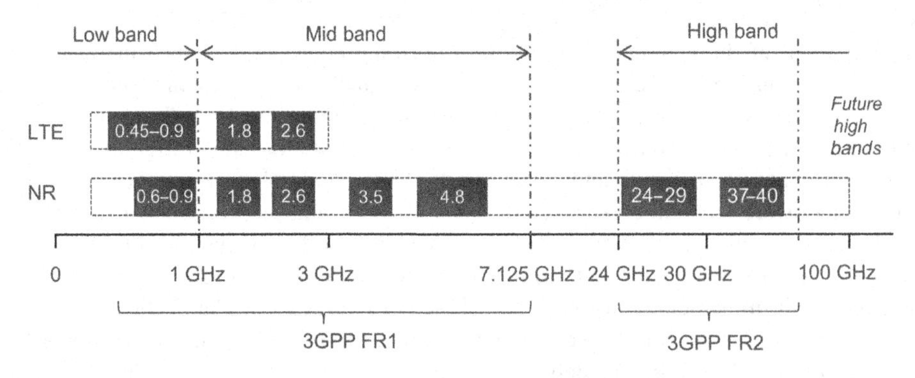

FIGURE 7.1

Frequency spectrum showing typical LTE and NR bands, and where high-band fits in the overall picture.

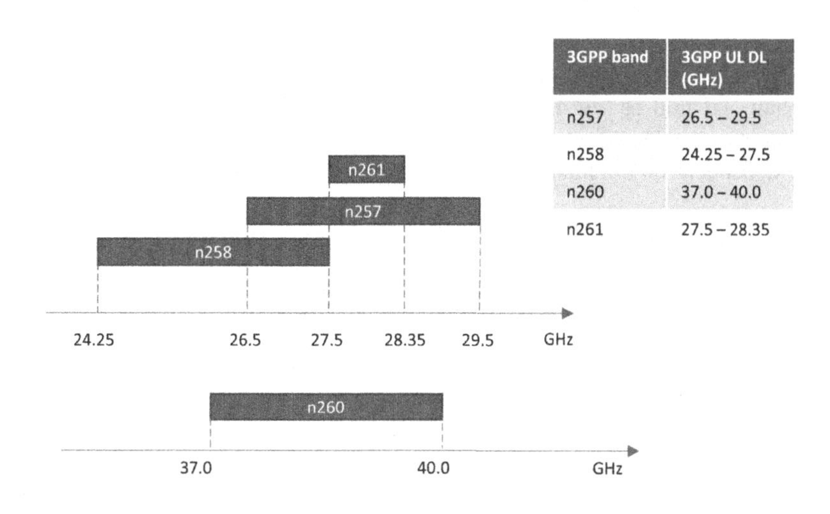

3GPP band	3GPP UL DL (GHz)
n257	26.5 - 29.5
n258	24.25 - 27.5
n260	37.0 - 40.0
n261	27.5 - 28.35

FIGURE 7.2

The first spectrum bands defined in FR2 in 3GPP.

of 100 MHz per carrier for FR1 bands (i.e., from 450 to 7.125 MHz) in 3GPP release 15. In addition to large maximum bandwidth per carrier, it is possible to aggregate carriers, with the maximum possible number of aggregated carriers allowed by 3GPP (at least from a physical layer point of view) equal to 16. This leads to a total maximum bandwidth of 16×400 MHz = 6.4 GHz.

However, such a maximum is not yet allowed by 3GPP RAN4 specifications today. At the time of writing this book 3GPP in release 15 supports FR2 combinations with aggregated downlink bandwidth of up to 1200 MHz for contiguous carriers (see Table 5.5A.1-1 in Ref. [1]) and a maximum frequency span of 1400 MHz for non-contiguous carriers, including any gaps (see Table 5.5A.2-2 in Ref. [1]), and so it can be expected that the first user equipment (UE) and radio access network (RAN) systems do not support more than this. Typical mm-wave aggregated bandwidths for the first NR systems will be of the order of 400 MHz, 800 MHz, or possibly 1.2 GHz, depending on operator spectrum allocations and commercial AAS and UE capabilities.

The large amount of spectrum possible to use means that the FR2 bands can support very high peak rates and throughput. Already in 2019, user throughput in field with early commercial 5G AAS systems and UEs reaches above 2 Gbps. Although not practical to realize today, the theoretical lower layer maximum throughput in FR2 is over 100 Gbps if assuming six layers, 256-QAM, and 16-component 400 MHz carriers. However, this is not supported by 3GPP release 15, for example, 256-QAM is not yet standardized for FR2, and the number of carriers that can be aggregated is limited. Still, end-user rates of more than 10 Gbps are feasible from a 3GPP standard point of view with release 15.

7.2.2 RADIO PROPAGATION AND ANTENNA CHARACTERISTICS

There is a general perception that coverage is lower at higher frequency bands. Sections 3.3.8 and 3.4.2 explain the physics behind this, and for the case of free-space propagation show that a main reason for the perceived difference is related to the concept of pathloss including the impact of the *effective antenna area* of an antenna, which becomes smaller with increasing frequency for a fixed antenna gain. In short, free space coverage can be good with large antenna arrays even at high band. For more general propagation, that is, through different materials, there is also a difference between high-band and mid or low bands, where the attenuation through some specific materials such as modern coated windows can be much greater at higher bands, although the amount of attenuation is highly dependent on the material (see Section 3.5.2 for more information).

7.2.2.1 Antenna characteristics and size

Performance, implementation, and deployment requirements can have an influence in determining the antenna size. Two important performance requirements are that:

1. Coverage should be enough for the planned deployment, that is, to satisfy a certain "cell" coverage area and a minimum cell-edge bit rate.
2. Latency for mobile broadband (MBB) and new 5G services should be low.

Coverage of a single node can be increased by increasing transmit power or by increasing the size of the antenna array to achieve high antenna gain or both. The highest gain beams for large (relative to λ) antennas with very many antenna elements and with a fixed antenna element spacing become increasingly narrow as the number of elements is increased (see Chapter 4, showing the impact of number of elements and element spacing on beamwidth and maximum beam gain).

If beams are very narrow, it can increase the total number of such beams to cover a given cell area, and for certain analog beamforming solutions as described in more detail in Section 7.4.3, this can increase the time needed to find the best high-gain beam which in turn can impact latency. Thus in a given AAS system solution, there may be a balance chosen between wanted maximum beam gain and total number of beams to manage for latency and overhead, which can set a practical limit on the minimum beamwidth in order to have full coverage. However, it should be noted that large antennas can also use wider beams with lower gain and so can act in a similar way to a smaller antenna, with an additional benefit of higher gain narrower beams when needed (see Section 7.4).

It should also be noted that very narrow beams may not be the most suitable for all deployments. In real deployments there is a variation in the angular spread of the radio channel (see Chapter 3). For scenarios where there is a large angular spread in the radio channel, for example, indoors with a lot of reflections, the generalized beams (see Chapter 6) that handle multipath from different directions are useful to exploit that spread. A narrow beam (i.e., a beam with a well-defined narrow lobe) that is much narrower than the channel angular spread may not be able to exploit all the multipath directions.

For illustration purposes the relative sizes and gains of mid-band and high-band antennas are now compared for two simplified cases, case 1 and case 2 below. First recall that Section 3.4.2 explains that in free space the signal power at the receiver (P_r) can be related to the signal power at the transmitter (P_t) using the Friis equation (Eq. 3.42). The simplified version of the equation (Eq. 3.43 in Section 3.4.2) where the transmit and receive antennas point toward each other and polarizations are aligned between antennas is:

$$P_r = P_t G_t G_r \left(\frac{\lambda}{4\pi r}\right)^2$$

G_t, and G_r are the maximum gains of the transmit and receive antenna, λ is the wavelength of the radio signal, and r is the distance between the transmit and receive antenna. Now consider the two cases:

Case 1: A mid-band and a high-band array both use the same fixed number of antenna elements and the same fixed transmit power P_t.

For a given antenna spacing in terms of λ (e.g., 0.5λ), and *a fixed number of elements*, K on both the transmit and receive antenna arrays, then both the maximum transmitter gain, G_t, of the transmit antenna array, and the maximum receiver gain, G_r, of the receive antenna array are constant. Hence the total gain in power (for this case when the two antennas are aligned toward each other) is given by:

$$G_t G_r \left(\frac{\lambda}{4\pi r}\right)^2$$

At a fixed distance r the gain is thus proportional to λ^2, that is, to $1/f^2$. In this case, given a fixed transmit power, the received power is lower on higher frequencies (f_{high}) compared to lower frequencies (f_{low}) by a factor $20\log_{10}(f_{\text{high}}/f_{\text{low}})$ in dB terms. Since the number of elements K is fixed the antenna array becomes smaller as the frequency increases as illustrated in Fig. 7.3. The lobe width is also fixed (assuming the antenna element pattern is the same for all antenna elements) and scales inversely proportional to K (see Section 4.3.3.1) for a horizontal uniform linear array, or with $\sqrt{K}$ in either horizontal or vertical dimension for a square uniform planar array (UPA). See Section 4.5.1 and Fig. 4.26 in that section showing how the width of the main lobe scales with elements in vertical and horizontal directions for a UPA.

In short, at high-band when using the *same number* of antenna elements as at mid-band, *and* with both antenna element spacing and antenna element size set to a fixed proportion relative to

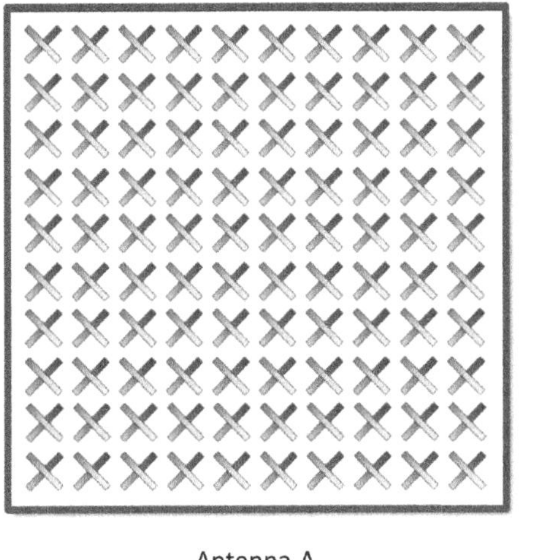

FIGURE 7.3

Two antennas A and B with the same element spacing (in terms of λ) and the same number of antenna elements, but operating on different frequencies $f_B = 10f_A$. Antenna B contains the same number of elements as antenna A but has 100 times smaller area.

wavelength, the antenna array is much smaller at high-band than mid-band. This leads to overall lower total gain as defined above, which in turn leads to worse coverage.

Case 2: The same antenna array size for mid-band and high-band with the same fixed transmit power P_t

In this case the *antenna array area* is kept constant for both the transmit and receive antenna and is the same for mid-band and high-band. With given antenna spacing in terms of λ (e.g., 0.5λ) the number of elements in the array, K, and hence each of G_t and G_r, scales up with $1/\lambda^2$, that is, with f^2 (assuming a UPA that is a square or rectangular array, see Section 4.5.1). Thus the total gain in received power

$$G_t G_r \left(\frac{\lambda}{4\pi r}\right)^2 \sim \frac{1}{\lambda^2}$$

is now proportional to f^2, that is, a *higher* total gain is possible on higher frequencies if *both* TX and RX antenna array areas are of the same size as at lower frequencies. In this case the high-band transmit and receive antennas would have better coverage in free space than the mid-band transmit and receive antennas. However, with increasing K, the lobe width (again inversely proportional to $\sqrt{K}$ for a square UPA) is much narrower on high-band.

In this case, as illustrated in Fig. 7.4, to have the same physical size for a 30 GHz high-band antenna as for a 3 GHz mid-band antenna where the latter wavelength is 10 times longer means a factor of 100 times more antenna elements are needed. Matching the size of a typical mid-band AAS with 128 antenna elements would then require 12800 elements at high-band.[1] Such an antenna would give very high-gain beams with about 100 times (or 20 dB) higher gain that are also around 10 times narrower (in both the horizontal and vertical dimension) than the mid-band beams. So, for example, if the mid-band width of the main lobe is in the order of 10 degrees wide in the horizontal domain, the high-band widths would be in the order of 1 degree. In this simplified example where

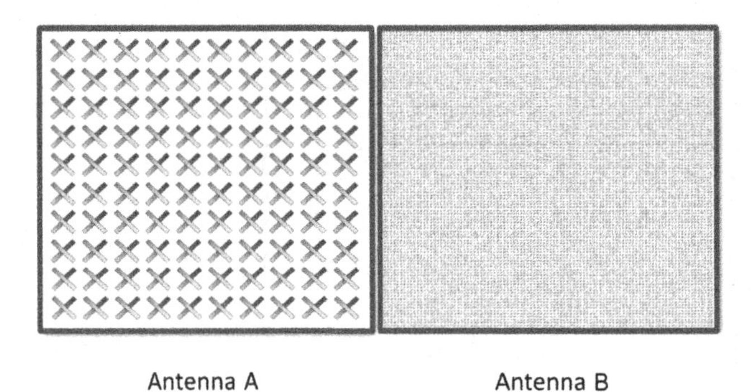

FIGURE 7.4

Two antennas A and B with the same element spacing (in terms of λ) and physical area but operating on different frequencies $f_B = 10f_A$. Antenna B contains 100 times more elements than antenna A.

[1]Assuming in this case for the sake of simplicity that all the antenna elements have equal element gain. mm-wave antenna elements could either have lower or higher element gain than mid-band elements.

transmit and receive antennas are the same size as each other and including the frequency-dependent component the two high-band antennas have a total gain that is 20 dB higher than the two mid-band antennas. It is possible to build such antenna arrays at high-band, but there are many practical building considerations when dealing with so many antenna elements (see Chapter 12).

The simplified analysis in these two cases also leads to the question of what antenna size increase gives exactly the right amount of gain to overcome the frequency-dependent pathloss difference? Assuming again that both arrays use the same total transmit power P_t. This occurs when the gain as below scales independently of λ:

$$G_t G_r \left(\frac{\lambda}{4\pi r}\right)^2$$

Thus in this case the product of G_t and G_r should scale with $1/\lambda^2$. This could be achieved for example if the gain of each of G_t and G_r scales with $1/\lambda$, that is, with f. So, for the example above where the high-band frequency of 30 GHz is 10 times the mid-band frequency of 3 GHz, 10 times more elements on both transmit and receive antenna are needed to overcome the frequency-dependent pathloss difference. Assuming the simplified case of 128 antennas on both transmit and receive antennas at mid-band, this case would lead to 1280 antennas being needed on both transmit and receive antennas for high-band, to cancel the free-space frequency-dependent pathloss, and would give antenna arrays that are 1/10th the size of the mid-band antenna arrays.

In reality, a UE is much smaller than a base station and has less space for antenna elements, so the case of both transmit and receive antennas being the same size as each other is unrealistic. However, since only the product of G_t and G_r needs to scale with $1/\lambda^2$ to cancel out the frequency-dependent pathloss, the number of elements needed to do so can be distributed differently between the transmit and receive antennas.

For example, if the receive antenna on the UE side has the same number of antenna elements for both mid-band and high-band (and in this example a constant gain G_r), then the base-station transmit side gain alone must scale with $1/\lambda^2$ to cancel out the frequency-dependent pathloss difference. Using the previous example, where the mid-band transmit antenna has 128 elements, and assume for simplicity that both the high-band and mid-band receive side have a single antenna element with the same gain, then the high-band antenna needs 100 times more elements, that is, 12,800 elements to cancel the frequency-dependent pathloss difference, and so will be the same size as the mid-band antenna. However, modifying this example to where the UE receive side has four times as many antennas elements in its array on high-band compared to the mid-band UE (and using fixed antenna element spacing such as 0.5λ), then the high-band base-station transmit side only needs 25 times as many antenna elements as the mid-band base station for the total gain to cancel out the free-space frequency-dependent pathloss. In this case the high-band antenna ends up at a quarter of the size of the mid-band antenna and has 3200 elements.

In practice today, AAS antenna arrays on high-band especially those intended for outdoor MBB applications generally have a somewhat larger number of antenna elements than on mid-band and so have larger antenna array gain but are still considerably smaller in physical size. Depending on the array sizes in both the AAS and the UE, the increase in the antenna array gain at high-band may still not be enough to compensate for the frequency-dependent pathloss.

In addition, the maximum output power of the AAS is generally much lower on high-band than mid-band (see Chapter 12). Given both antenna array size and maximum power allowed, the free-space coverage on high-band can be lower than on mid-band. On top of that there are other losses as mentioned in Section 7.2.2.2. In practice this can mean lower coverage for the same site density as mid-band. On the other hand, there are deployment advantages due to smaller size, that is, if the overall AAS size and weight are smaller, it may become easier to find sites where you can deploy, for example, on street poles, which can make densification easier.

7.2.2.2 Other propagation properties on high-band

Besides free-space propagation, there are other propagation-related issues on high-band. As discussed in Section 3.5.1, the higher frequency waves interact with materials such as walls, windows, and foliage, the degree of diffraction (i.e., bend around obstacles) is lower on high-band compared to mid-band. For outdoor non line-of-sight (NLOS) propagation, the frequency-dependent pathloss can increase from 20 log(f) dB, to a range of 20–23 log(f) dB, as shown in Section 3.5.1, Table 3.2. To overcome this would require even more gain and so even more antenna elements.

The radio waves at high-band frequencies also in general experience higher attenuation due to materials (for example, in walls and buildings) than at lower bands, with the amount of attenuation difference being very specific to the material type (see Section 3.5.2). Therefore outdoor-to-indoor coverage is generally easier to achieve on lower bands than on high-band.

7.2.3 USE CASES

Different deployment use cases can have different requirements on performance and cost, and there are numerous possible deployments for mm-wave solutions. To suit different types of deployment, the size and capability of individual AAS nodes can vary widely. The acceptable deployment density of AAS nodes can also vary significantly depending on the use case requirements and type of nodes available. For example, the requirements on throughput and reliability are one thing for a fully automated factory—requiring ultra high reliability and low latency to numerous devices including work stations, robots, and automated guided vehicles, and another for a typical MBB network handling typical consumer applications.

At the time of writing this book 5G mm-wave deployments in live networks are at an early stage but are mostly used for MBB deployments (using mm-wave AAS either on buildings or on street poles) or fixed wireless access (FWA). Due to the variety of deployments possible, different adaptations of solutions can be expected as the industry settles on the most preferred and cost-effective mm-wave solutions for different types of deployments both outdoor and indoor.

The high-band characteristics as described earlier have an impact on the deployment use cases where high-band solutions are most effective, and some example cases are described at high level below.

7.2.3.1 Mm-wave standalone solutions

Standalone mm-wave solutions here mean solutions where mm-wave is used alone, without the support of lower bands. Due to the propagation characteristics described above and in Chapter 3, scenarios that are very suitable for standalone mm-wave deployment are those with significant *line-of-sight (LOS) or near LOS propagation*, where the propagation path between the AAS either is relatively unobstructed or where there are unobstructed paths through reflections from nearby objects.

Since the potential data rates are very high, mm-wave is particularly useful for geographical open areas with very high traffic requirements, for example, in outdoor cases that can include stadiums, squares, parks, streets, and housing suburbs, and in indoor cases that can include airports, factories, shopping malls, concert venues, commercial buildings, etc.

For FWA-type deployments at high-band a mm-wave AAS communicates with a customer premise equipment (CPE) device that is, for example, mounted on the outside of or inside a building. The CPE devices are generally stationary and can be more capable (with higher maximum transmit power and higher antenna array gain) than standard smartphones and so the range of mm-wave can be extended significantly compared to smartphones.

Many deployments with *non line-of-sight (NLOS)* propagation, where the propagation path between the AAS and the device can be obstructed are also quite suitable for high-band, for example, the typical outdoor NLOS case mentioned in Section 7.2.2.2, and for some outdoor-to-indoor cases where signals can penetrate low-loss buildings quite well.

However, some NLOS scenarios can be more challenging, for example, scenarios that require ubiquitous coverage over large areas and where there are significant obstacles such as walls and buildings (especially high-loss buildings) between the user and the AAS that significantly increase the signal attenuation. For these scenarios, mid-band of the same site density can generally better serve those users in more challenging radio conditions, for example, users in higher loss buildings, shadowed from the radio signal, or far from the AAS, again a main reason being larger arrays at mid-band and that the mid-band signal propagates better through certain materials (see Chapter 3). This does not mean that mid-band will always be better for *all* users even in this type of scenario, as mid-band solutions with lower bandwidth may have lower peak rate than mm-wave for users in the best radio conditions and may even have lower peak rate for *some* indoor users in reasonably good radio conditions, for example, in those cases where mm-wave signals can propagate through walls in low-loss buildings and still provide high rates.

To improve mm-wave coverage in more challenging environments, site density can be increased to improve signal strength. Multiple transmission points can also be exploited to seamlessly switch a user between different transmission points. Note that full support of seamless switching of beams between transmission points was not completed in 3GPP release 15 (see Chapter 9) but is being further developed in release 16.

Another technology component to support a flexible deployment is *integrated access backhaul* (IAB). That can allow a mm-wave AAS to provide both access to end users and backhaul toward other access nodes over the radio interface. This removes the need for fixed backhaul (e.g., fiber) at every site. It should be noted that the backhaul does not have to be on the same band as the access to the end-user. For example, mm-wave can be used as backhaul to mid-band sites.

7.2.3.2 High-band in combination with lower bands

The benefits of mm-wave solutions can be substantially enhanced if they are used in combination with solutions on lower bands. See Section 13.7.6.2 for some performance results. The combination with lower bands (which are typically using larger antennas) can increase the coverage compared to a standalone mm-wave system as described below. The combination of bands also provides a boost effect to capacity, where the total capacity using high-band and low-band together is greater than the sum of the capacities if both bands were used individually.

One intuitive way to think about this is to consider cell-edge users in poor coverage. If they are handled at high-band alone (in the case that they can be handled at all), it is difficult to achieve high rates and they also occupy a lot of system resources. That is because for a cell-edge user in poor radio conditions it takes much more resources in terms of bandwidth and power to send the same number of bits compared to a cell center user in good radio conditions. The resources taken by these "difficult" users—of which there may not be so many—cannot then be used by other users. If such users are offloaded to a lower band, which has better coverage and can easily handle them, then a lot of high-band resources become freed up. These resources can then be used by the remaining high-band users, which are the users in better radio coverage. Users in better radio coverage can achieve much higher rates than users in poor radio coverage while using the same amount of resources, and so system wide a capacity boost can be achieved.

Mobile network operators usually have many frequency bands in FR1, where they can deploy LTE or NR. Deployment on FR1 bands is usually done on the existing site grid to ensure ubiquitous coverage in an area. mm-wave bands can then be used to provide extra high user performance or extra capacity wherever there is an area with higher throughput or capacity needs.

The use of both low/mid-band and high-band together can be done in different ways using, for example, either dual connectivity, or 3GPP release 15 carrier aggregation. In NR, carriers from FR1 and FR2 bands can be aggregated together as well as within each of FR1 and FR2.

Carrier aggregation with lower bands is useful for several reasons. It can improve downlink coverage and assist mobility and it is helpful in scenarios where the uplink becomes the limiting link for a mm-wave transmission. This is often the case (although not necessarily always, since the link budget per physical channel in a given configuration can depend on for example how much bandwidth is used, and the required minimum SINR of that channel, and the required minimum cell edge data rate of the channel) due to that the UE maximum transmit power at high-band is much lower than base station maximum transmit power, which can lead to an imbalance in the coverage of uplink and downlink physical channels. For example, a UE transmitting at 13.2 dBm (around 0.02 W) versus an AAS transmitting at 27.5 dBm (0.56 W) gives a difference of 14.3 dB between uplink and downlink transmit power alone. The UE maximum transmit power is also much lower than the base station maximum transmit power at mid-band, leading to an imbalance at mid-band too, which can be even larger due to the high power of a typical mid-band AAS. However, the uplink is more challenging at high-band than at mid-band for the same reasons described earlier, for example, the receiving AAS antenna array size is usually smaller than for mid-band and so may not compensate for the frequency-dependent pathloss, and some materials cause higher signal attenuation.

With carrier aggregation it is possible, for a given UE connection where the uplink is limiting, to reduce the uplink/downlink imbalance by, for example, placing the limiting uplink channels on a lower-frequency FR1 band that has larger coverage, while the downlink channels still run on the mm-wave FR2 band providing a high downlink throughput. One benefit of this is that the high downlink bit rates of mm-wave can be kept for longer when moving further (or into worse radio conditions) from the AAS. The uplink bit rate in poor radio conditions can also benefit and as transmission control protocol (TCP) ACKs for downlink data transmissions are sent on the uplink data channel, this also improves DL throughput in those cases where the uplink bit rate is limiting the DL throughput.

7.2.4 NEED FOR NEW MULTI-ANTENNA FUNCTIONALITY

As described above, the use of high-gain beams to increase coverage is important at mm-wave, and this results in that the beams can become narrow.

Wider beams, covering either the full area intended to be served by a cell, or a subset of that area are useful for the UEs to read common channels and to do cell selection and access the network. Wide beams can be created for these purposes even with arrays using many antenna elements (see Section 4.4.3). Wider beams have lower maximum gain than narrow beams; however, in many cases that can be enough for low rate control and broadcast information, and for data transmission to UEs in good radio conditions. For increased coverage and higher data rate possibilities, the network can switch to narrower higher gain beams, or in the most ideal case use generalized beams to adapt to the radio channel as described in Chapter 6.

Especially with time-domain analog beamforming systems, where one beam is transmitted at a time and the beams are usually predefined, then the use of *beam management* functionality becomes motivated. This is needed to handle both the cell-specific wider beams and the UE–specific wide or narrow beams. Time-domain analog beamforming and the relation to beam management are described in the next sections.

7.3 TIME-DOMAIN VERSUS FREQUENCY-DOMAIN BEAMFORMING

The purpose of this section is to describe at high-level different methods of beamforming, that is, time-domain and frequency-domain beamforming (Section 7.3.1), and to briefly describe their respective properties (Section 7.3.2) and their relation to analog beamforming and DBF.

On mid-band, DBF in the frequency domain has been the main industry preference for AAS systems since frequency-domain beamforming is very flexible and supports high performing solutions. In addition, the needed hardware technology was available for both UE and the network at feasible cost points when AAS was first introduced for LTE at mid-band and it also became a natural choice for mid-band NR.

The choice of time-domain analog beamforming as a beamforming method gained high interest during the early development of high-band solutions for NR. The reasons for this were mainly due to the maturity and cost of hardware technologies for high-band at that time. It was clear at the time that analog beamforming would become the main initial beamforming method for high-band and so the early standardization in 3GPP took measures to enable analog-based time-domain solutions. The implementation drivers and consequences for the use of one type versus the other are explored further in Chapter 12.

7.3.1 BEAMFORMING TYPE DESCRIPTIONS

Fig. 7.5 illustrates the different types of beamforming. From left to right, that is, in the downlink transmission direction, the baseband orthogonal frequency-division multiplexing (OFDM) signal is processed in the digital frequency domain, and then it is converted from frequency domain to the

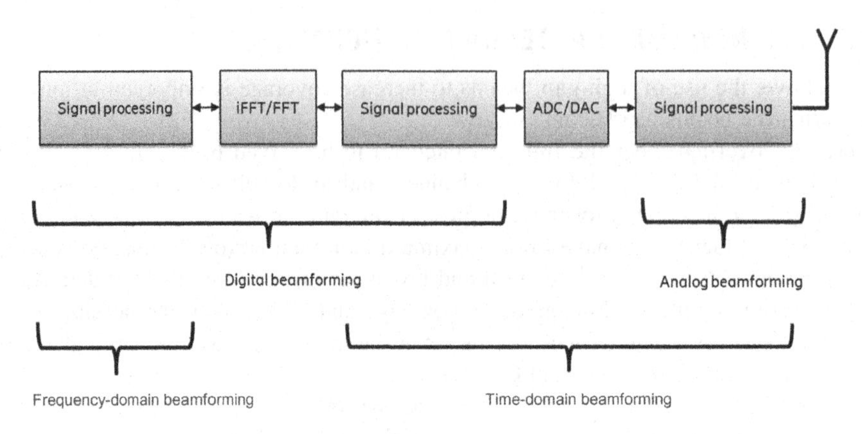

FIGURE 7.5

Simplified view of an OFDM system showing where analog and digital beamforming are performed relative to the ADC and DAC, and also where frequency-domain beamforming and time-domain beamforming are performed relative to the iDFT/DFT (which is typically performed using the iFFT/FFT algorithm as shown in the picture).

time domain using an inverse discrete Fourier transform (iDFT).[2] Then after some further (optional) signal processing in the time domain, the signal is converted from a digital time-domain signal to an analog time-domain signal via a digital-to-analog converter (DAC). Further signal processing takes place in the analog domain.

Analog beamforming refers to signal processing operations that are applied to the analog signal, that is, analog beamforming is applied after the DAC when transmitting, and before the ADC (analog-to-digital converter) when receiving. Beamforming of signals in an analog form is done in the time domain only.

DBF is signal processing operations that are applied to a signal represented in digital form. For signals in the digital domain, both time-domain and frequency-domain beamforming operations can be performed depending on where the operations are performed before or after the DFT/iDFT. Moreover, any combination of analog, digital, time, and frequency-domain beamforming operations can be performed on a transmitted or received signal to achieve the desired functionality. Fig. 7.5 only shows a simplified view. For AAS multiple antenna signal processing is used where the operations between the different paths are interdependent. For example, for the transmitter direction, at some point the transmitted signal needs to branch out to multiple antenna branches (see also Chapter 12).

7.3.2 ASPECTS OF FREQUENCY-DOMAIN VERSUS TIME-DOMAIN BEAMFORMING

Digital frequency-domain beamforming is very flexible. The signal is processed digitally, so there is freedom to create beams in different ways. For example, precoding (see Chapter 6), can be

[2]DFT/iDFT are almost always implemented through fast Fourier transform (FFT) and inverse fast Fourier transform (iFFT). Therefore DFT/iDFT and FFT/iFFT will be used interchangeably.

applied differently for different parts of the frequency band in order to follow the channel and/or interference characteristics across the frequency range. In the OFDM-based systems of NR and LTE, this essentially means that different physical resource blocks (PRBs) or groups of PRBs (known as precoding resource block groups) in the frequency domain can have different beam weights. This is shown in Fig. 7.6 for a single layer, but it also applies to multiple layers (see example in Fig. 7.7). These different beams can be used, for example, to separate users in frequency, or in layers, or to support multiple frequency selective beams for a single user in different parts of the carrier (see also Chapter 6).

There is no inherent difference in creating a single beam (with one set of beam weights applied to an array) using analog beamforming or DBF since the beam is determined by the beam weights and they can be applied to the array equivalently in either domain to generate the weighted signal replicas for each antenna. Frequency-domain beamforming, however, is quite suitable for the implementation of frequency selective beamforming, which as stated implies the application of different beam weights to different parts of the carrier.

Fig. 7.7 illustrates some capabilities of frequency-domain DBF with a simplified view for how a scheduler can use the possibilities in time, frequency, and layers to schedule several users that have data to transmit. Both single-user multiple-input, multiple-output (SU-MIMO) and multiple user MIMO (MU-MIMO) are shown. The spectrum can be used efficiently regardless of whether small or large packets are transmitted for each user.

Fig. 7.8 illustrates time-domain beamforming with a simplified view of (1) how a scheduler can use the possibilities in time, frequency, and layers to schedule two users with one layer and (2) two layers, respectively. Note that to have two layers requires two parallel analog beamformers to apply two sets of beam weights.

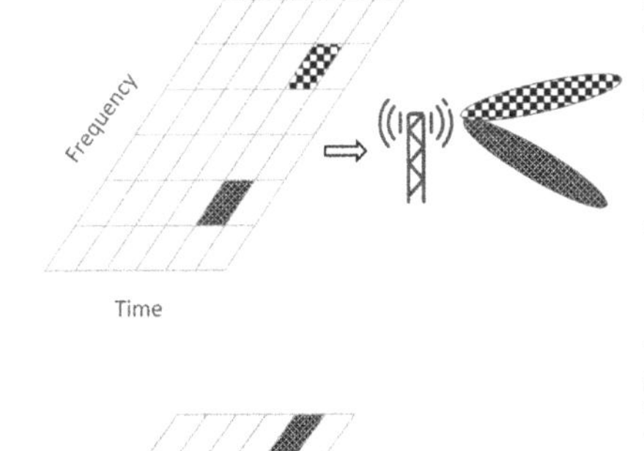

Digital Frequency Domain Beamforming
- Supports different weights (beams) for different parts of the frequency band, for example per PRB
- Supports frequency selective beamforming per user
- Supports frequency multiplexing of different users using different beams
    - Efficient for sending small packets to many users
- Supports different beams in the same frequency block at same time (layers) – not shown here
    - useful for SU-MIMO and MU-MIMO.

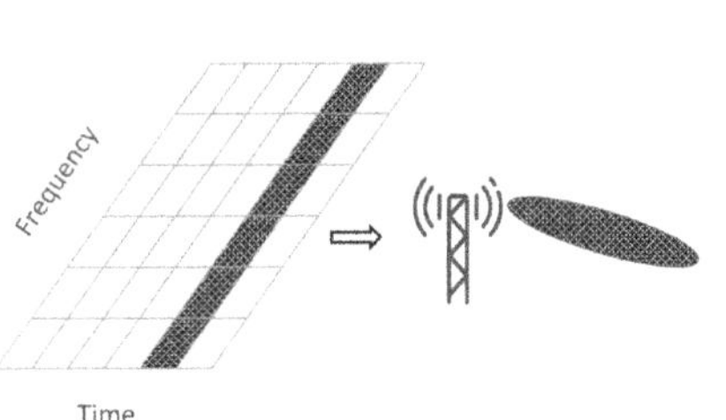

Time-Domain Beamforming - used with Analog beam forming
- Beam weights are applied in the time-domain
- One beam for one timeslot for one frequency carrier using one time-domain beamformer
    - High rate possible to single user
    - Less efficient use of spectrum when needing to send many small packets (that do not require the full carrier BW) to many users not covered by the same beam.
- More beams per timeslot require more parallel time-domain beamformers e.g. two parallel beamformers needed to
    - split the spectrum into two separate beams
    - support two layers for either SU-MIMO or MU-MIMO

FIGURE 7.6

Time-domain beamforming versus frequency-domain beamforming.

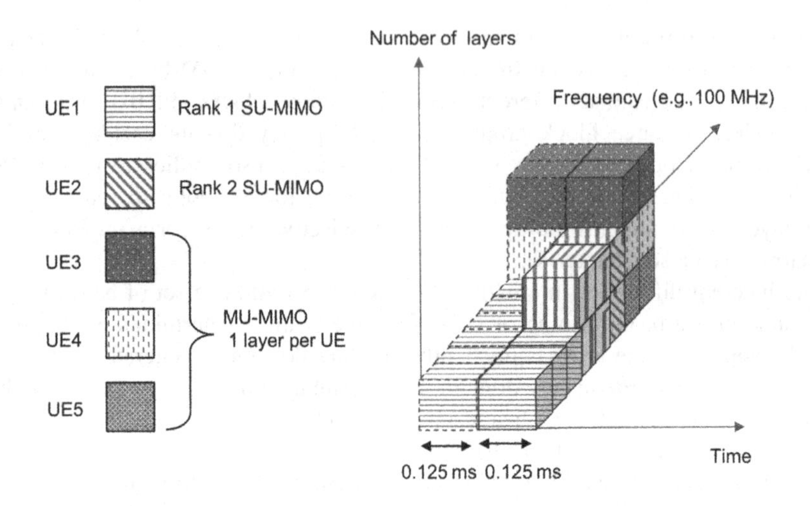

FIGURE 7.7

Scheduling in time, frequency, and layers to support SU-MIMO and MU-MIMO.

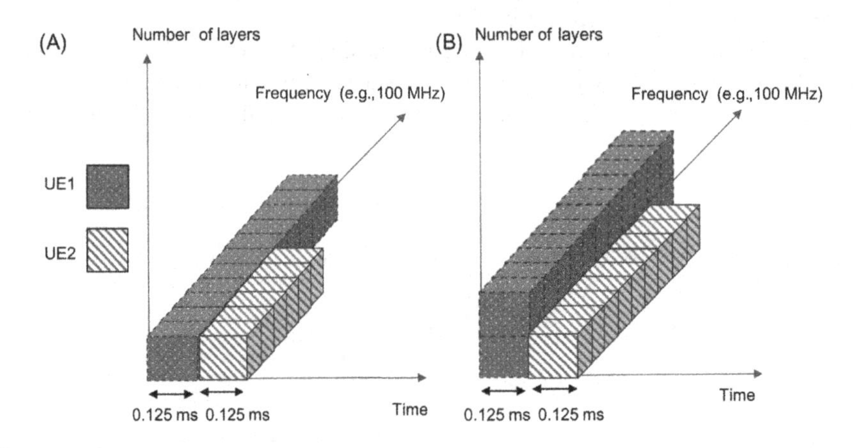

FIGURE 7.8

(A) Example of scheduling two users over 100 MHz, with a different beam for each user. (B) Scheduling with two simultaneous analog beams for UE1 supporting two-layer SU-MIMO for that user. In these examples, UE2 does not need the entire available bandwidth to send its data.

Assume a simple implementation of analog time-domain beamforming with one layer, where one beam at a time can be sent per carrier and compare to a frequency-domain beamforming implementation where many beams can be sent simultaneously per carrier. When multiple simultaneous users need to be served, this means that latency can increase with the analog implementation, as in the worst case only one user using one beam can be served at a time. However, one factor that can help latency at high-band is that users can be multiplexed in time with higher granularity than mid-band. The slot length of 0.125 ms (when using NR with 120 kHz subcarrier spacing at high-band) is eight times shorter than in LTE, or four times shorter than mid-band NR when using 30 kHz subcarrier spacing numerology, which is typical at mid-band.

A drawback with only being able to transmit in one beam a time with such large bandwidths is that users with very small packets to send may not need to use the full bandwidth in a timeslot, and so are allocated more spectrum than they need, which can reduce efficiency especially with a high load of small packets. However, there are also some possibilities to support more than one user at a time with analog beamforming—for example, by using more parallel analog beamformers (one per beam) to create more beams at a time, which can then be used to frequency multiplex users or create more layers, or alternatively by frequency multiplexing users in the same beam at the same time, that is, for those cases where all the users are in the coverage of the same beam. 3GPP has also made it possible to schedule in time using a smaller granularity than a slot (which consists of 14 symbols), for example, by using mini-slots (see Chapter 9) where less OFDM symbols are used to send the data.

7.4 PRINCIPLES OF BEAM MANAGEMENT

For high-band the use of beamforming is important to increase coverage and overcome frequency-dependent pathloss, and as mentioned for initial solutions time-domain beamforming solutions are common. To support handling of such solutions, beam management functionality has been introduced in 3GPP. This section outlines general principles for beam management.[3] The outlined procedures and principles are mostly[4] agnostic to the frequency band and so can be partly used in FR1 bands. As FR1 solutions typically use frequency-domain digital beamforming, many of the beam management procedures are not needed at FR1, for example, a fully digital receiver can receive beams from many directions at once. The beam management procedures are especially useful for systems using time-domain analog beamforming that can transmit or receive one or few beams at a time.

7.4.1 A BEAM-BASED APPROACH

As described in Chapter 6, beamforming can be made on both the transmitting side and the receiving side. One issue to address in the solution is how to find appropriate transmitter and receiver beams.

7.4.1.1 Beam pair

A key for beam management is the term *beam pair*, as illustrated in Fig. 7.9.[5] The network transmits a signal on one transmit beam and the UE receives that signal on a receive beam. These two beams are then called a downlink beam pair. Similarly, for the uplink the UE transmits using a transmit beam and the network receives on a receive beam. This is an uplink beam pair. For TDD,

[3]The procedures discussed in this chapter are aligned with what is adopted in the 3GPP standard. Other solutions than these are possible.

[4]RAN4 does not support beam correspondence requirements for FR1.

[5]Fig. 7.9 shows a case where the two beams point toward each other. In real-world environments where beams reflect and scatter off objects and buildings, the best performing beam pair may in some cases point in completely different directions.

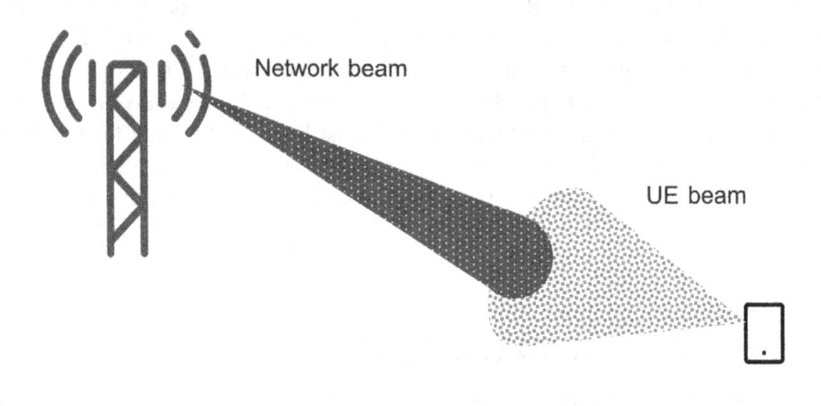

FIGURE 7.9

A beam pair where the UE beam corresponds to the network beam.

the channel is reciprocal in uplink and downlink. In this case, the uplink and downlink beam pairs are the same, that is, consist of the same transmit and receive beams, respectively. For this to be supported in practice on the network side the transmit beam and the receive beam on the network side must be the same or very similar, which requires some TX-RX calibration. Similarly, for the UE side, if the UE can determine the uplink transmit beam to use based on measurements on the downlink receive beam, then it is said that the UE supports beam correspondence (see Ref. [1] for the 3GPP definition). If beam correspondence does not hold, there is a need to determine uplink and downlink beam pairs separately. See also Section 9.3.6 on beam correspondence.

How to determine a beam pair using beam management procedures is described in the next section.

7.4.2 BEAM MANAGEMENT FUNCTIONALITY OVERVIEW

The purpose of the beam management functionality as specified in NR for 3GPP is to find and maintain the best beam pair between the network and the UE. This is especially useful (and needed) for a time-domain analog beamforming–based system where one or a few beams can be sent or received at a time. For a fully digital system (as is typically available at mid-band), the digital receiver can process the signal to "listen" in different directions simultaneously and detect the beam regardless of direction, and so much of the standardized functionality as listed below is not necessary.

The high-level issues to solve for using beam management are then for downlink:

- How the network finds the correct transmit beam to use for each UE (e.g., during initial access);
- How each UE knows which receive beam to use for a given network transmit beam (beam measurement and beam selection);

- How the network indicates a change in transmit beam so that the UE can change the receive beam accordingly (beam indication);
- How to recover if the quality suddenly drops, related to selecting a wrong beam or if the UE moves/rotates suddenly (beam recovery);
- If beamforming is used for access signals such as synchronization and system information, how should the access procedures work (beam discovery).

And, similarly for uplink:

- How the UE knows what beam it should transmit on, so that it can be sure that the network will use the corresponding receive beam (beam selection);
- How the network knows what beam it should use to receive when the UE is transmitting in a beam (beam selection).

A basic beam management procedure can be constructed where the network periodically transmits a reference signal in each of the predefined and static beams (i.e., beamformed reference signals). The UE is aware of the time and frequency locations of these reference signals and their periodicity. The UE can thus measure, for example, *reference signal received power* (RSRP) per beam and since the signal is periodic, the UE can for each network beam test different UE RX beams to try and improve the beam-pair link. This process can be rather slow but algorithms in the UE can converge toward a selection of a network beam and a good corresponding UE RX beam. Hence, the UE autonomously establishes a beam-pair link and stores the preferred RX beam in memory. The UE can also be triggered or configured to report the best beam or a set of top-N best beams. Note that the UE does not need to report the preferred UE RX beam for each reported network beam, this information is unnecessary for the network to know and so such feedback signaling to the network has not been standardized in 3GPP.

It may happen that the best beam pair to serve a user can change quickly. For example, this will normally happen with a grid-of-beams (GoB) type setup when the UE moves from the coverage area of one beam to another. The beam may often switch to an adjacent beam in the grid covering an adjacent part of the cell, but that is not guaranteed, it can also switch to a beam in some different part of the GoB pointing in a completely different direction, for example, due to reflection and scattering, moving the UE, etc.

Beam management is designed to handle these types of sudden events. The network also has the flexibility if necessary, to shorten the configured periodicity of the beam-specific reference signals and/or the reporting periodicity of these signals and hence make the system less sensitive to rapid changes in the most useful beam-pair link. However, the cost is increased reference signal overhead.

The beam management measurement framework is divided up into distinct procedures as follows. Details of the standardized procedures are described in Chapter 9. These procedures are used for downlink beam-pair establishment, but as mentioned earlier, if beam correspondence holds, then the same beam pair can be used for the uplink.

- *Procedure 1 (P1)*: Establishment of initial beam pair, that is, a coarse transmit and receive beam;
- *Procedure 2 (P2)*: Transmit beam refinement;
- *Procedure 3 (P3)*: Receive beam refinement;
- *Beam Recovery*.

Note that the terms P1, P2, and P3 are not used in the 3GPP specifications although the functionality to support these is specified.

7.4.2.1 Procedure 1 (P1): Establishment of a coarse transmit and receive beam

With this procedure a beam pair is set up between the AAS and the device.

Initially at setup of a connection between the AAS and the UE there is no knowledge of the beam pair(s) that should be used. Initial beam establishment procedures are used to determine such a coarse beam pair.

With an analog beamforming system, one beam at a time needs to be transmitted, where each beam contains a synchronization signal and the necessary system information for the UE to perform the random-access procedure. Moreover, if each of these synchronization beams can be uniquely identified and if their position in the radio frame is then given by this index, then the UE can obtain frame synchronization at the same time as a preferred beam is detected. This is the approach taken by NR specifications (see Chapter 9). An illustration of sweeping through different transmit beams is given in Fig. 7.10.

Furthermore, if the beam that the UE has detected is associated with the resource used for the random-access response from the UE to the network, then the transmission of this response automatically reveals which network beam the UE prefers. In addition, provided also that beam correspondence holds at the network, then the base station knows which receive beam to use when a certain random-access signal is transmitted from the UEs in the cell. Hence, with these few steps, an initial transmit and receive beam is established for the network.

Simultaneously, in the UE, the UE will upon detection of the synchronization signal, know which random-access resource to use for the response to the network, and if the UE supports beam correspondence it can use a similar beam (within the definition of beam correspondence in Ref. [1]) for transmission of the response as it used for reception. As a result, the UE has also

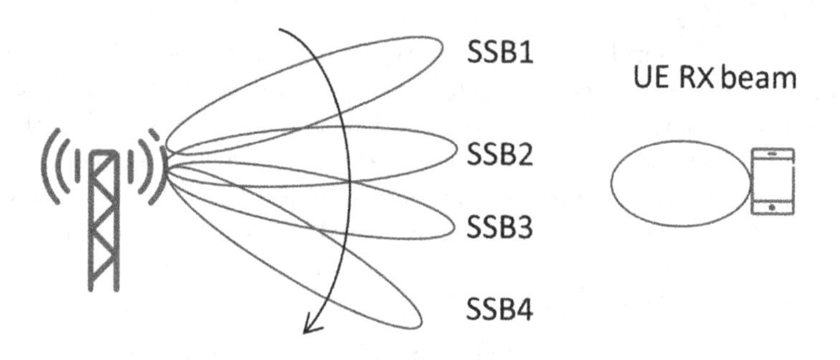

FIGURE 7.10

Beam sweeping of different transmit beams using different synchronization blocks (SSBs). The UE measures the SSB to determine best received SSB, and at the same time, determines a receive beam to use. Note there are max 64 SSB beams on FR2.

determined a receive and transmit beam and thus the network and UE initial beam-pair link is established and the P1 procedure is completed.

Note that the network does not need to know which receive beam (e.g., direction or beam shape or internal beam number) the UE determines in this procedure and vice versa. Hence, neither of the two receive beams in the beam pair needs to be known by the "other" side and the UE transmit beam is also agnostic to the network. It is only the label (e.g., synchronization signal identity) of the network transmit beam that defines a beam pair. This approach gives implementation freedom to both the UE and network side.

7.4.2.2 Procedure 2 (P2): Transmit beam refinement

Procedure P2 allows the network to refine the beams used in P1, for example, to use narrower, and thus higher gain beams compared to the beams used in P1 if needed. The measurement reference signals for P2 beam refinement [e.g., CSI—reference symbol (CSI-RS) reference signals as described in Chapter 9] can be configured with low overhead, for example, a-periodically triggered, which can be compared to the P1 procedure that typically uses periodic reference signals and may thus already generate quite some overhead depending on the number of beams used in P1. With P2, UE reports on the refined set of beams, and the network chooses the best beam to use

A "local" beam sweep in the sense that the sweep does not need to cover the whole cell can be used for P2, where only beam directions that are believed to reach the UE with sufficiently high power are used in the sweep. For example, the beam reporting based on the P1 procedure may already have found one or more suitable beam directions that then can be further probed with a P2 procedure (Fig. 7.11).

7.4.2.3 Procedure 3 (P3): Receive beam refinement

The beam pair may also change as the UE moves, so, for example, either the UE measures a different transmit beam that has a stronger signal strength than the beam in the current beam pair, or the UE receives the same transmit beam, but needs to adjust its own receive beam to have the best reception of that transmit beam.

The P1 process will also let the UE find a suitable receiver beam, so it would be possible to rely on this in practice, but the process can be slow (if the periodicity of the reference signals used

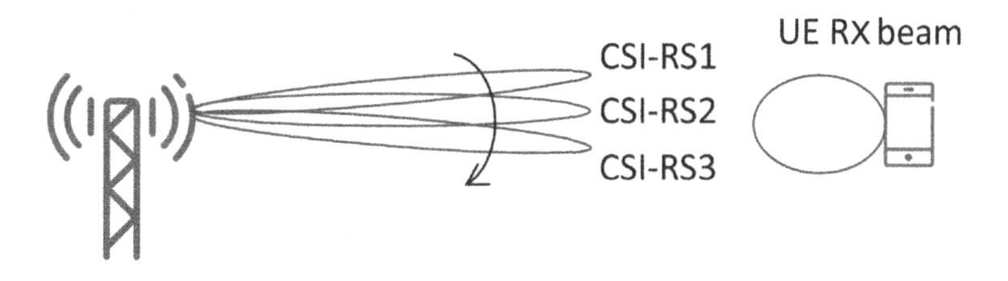

FIGURE 7.11

Beam adjustment/refinement of transmit beam using three CSI-RS configured for beam management measurements. The UE RX beam is kept fixed.

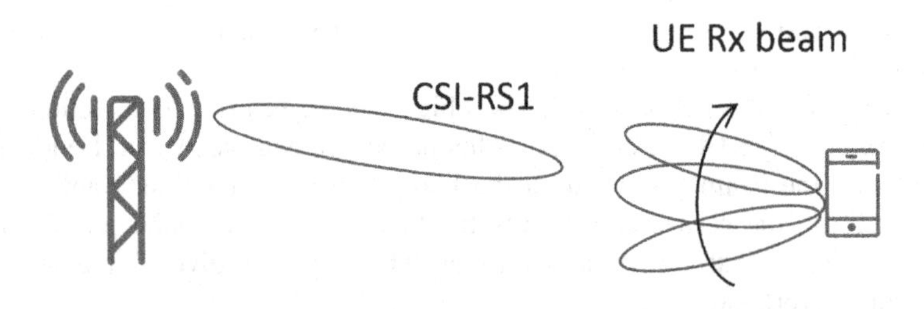

FIGURE 7.12

Network transmit beam kept fixed, while UE adjusts the UE receive beam.

for P1 is large), and there is a risk that beam failure is declared if the UE is unsuccessful in finding a receiver beam quickly.

Instead, the network can trigger a special procedure, P3, which is typically an aperiodic measurement procedure. In P3, the UE measures on a single network transmit beam, whose reference signals are repeatedly transmitted in multiple OFDM symbols in a row, so that UE can probe different receive beams in each symbol. The UE can then update the receive beam associated with the transmit beam internally, and so the beam pair is updated. As usual, the network does not need to know which receive beam that the UE used or selected, but the P3 procedure may be associated with a received signal power report from the UE, so the network can verify that the quality of the beam-pair link is maintained (Fig. 7.12).

7.4.2.4 Beam recovery

Due to the dynamics of the radio channel, it can happen that a beam pair is abruptly lost. For example, a beam or some beams are blocked when a UE moves into an area where the current beam pair or beam pairs are not useful for communications, or a blocking object moves in front of the UE or transmitter.

When this happens both the network and the UE need to quickly act; otherwise, the signal will be lost, and data transmission is interrupted (Fig. 7.13).

In short, a UE or device can first detect if a beam in a used beam pair is lost. This can be done by monitoring the RSRP of the beam measurement reference signals of the used beam pair, and if it drops below a certain level over a certain period, the UE may consider the beam pair as lost. The UE can then try to identify a new beam pair. This can be done by the UE measuring reference signals associated to a set of candidate beams for which measurements have been configured by the network. If one of the measured candidate beams is received strongly enough, then that new beam can be considered a beam that can be used for restoration of the connection. Note that the system has an opportunity to identify "likely" candidate beams for the UE to measure depending on the current beam.

Finally, once a viable candidate beam has been identified, the UE can initiate a beam recovery procedure. This is done by the UE performing a random access on a random access channel (RACH) occasion corresponding to a reference signal associated with the candidate beam. If there is no unique RACH occasion associated with the candidate beam (e.g., if all candidate beams use the same RACH occasion), then the beam recovery request just indicates to the network that a beam failure has happened, and the network side then can try to recover a new beam pair.

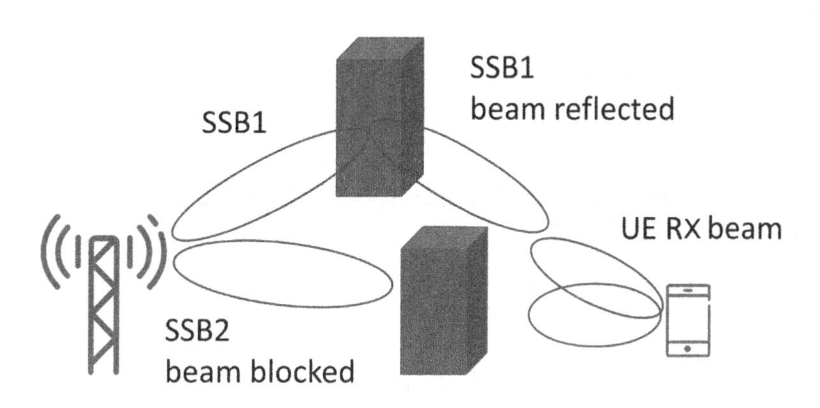

FIGURE 7.13

One beam pair is blocked, but another beam pair is possible to use.

What is important for recovery procedures from the network and the UE side is also the same as for the legacy supervision and recovery procedures, that is, to keep latency low and avoid interrupting transmissions. So, if a beam pair is lost, the faster a new beam pair is recovered the better.

7.4.2.5 Beam-based mobility and beam switching

The procedures described above outline methods for measurements and for finding a good beam pair to use. The network may then suddenly decide to switch to a new transmit beam, based on reported measurements. How then does the UE know which receive beam to use?

It is important to keep the implementation flexibility for beamforming; hence beam indication of a new beam can refer to a measurement that the UE has done previously. For example, the network transmits a set of measurement reference signals in a set of beams. These reference signals each have an index. The indication of a new transmit beam for data transmission can thus use the index of the measurement signal. It is then understood by the UE that the network will transmit data or control with the same beam as the indicated measurement signal (see Chapter 9 for details of the quasi colocated (QCL) framework to support this).

Since a UE can measure on a set of beams defined by the measurement signals, it is transparent for the UE whether these are transmitted from the same network node or across multiple, different network nodes. Hence, this allows for beam-based mobility, where the UE roams around in the network and reports based on beams, and can always be connected to the best beam, without knowing which network node is used for the transmission of the currently best beam. Beam-based mobility is supported in NR release 15 although only with beams transmitted from network nodes within the same cell (see Chapter 9 for details). Work is ongoing in release 16 to also allow beam-based mobility across different cells.

The above procedures describe downlink-based beam pair finding and recovery, but it may also be useful to do network receive beam adjustment based on uplink measurements on the network side. This is then based on network measurement on sounding reference signal (SRS) reference signals transmitted in the uplink by the UE (see Section 9.3.1 for details of SRS configuration for beam management). In this case the network sweeps through receive beams, while the SRS transmit beam from the UE is kept fixed in a repeated transmission, allowing the network to adjust the receive beam. Assuming beam correspondence holds on the network side, the network has the possibility to use the refined receive beam to also refine its own downlink transmit beam.

7.4.3 SIMPLIFIED BEAM HANDLING SOLUTION APPLIED TO AN ANALOG BEAMFORMING SYSTEM

In this section, some typical uses of wide beams and narrow beams are described, and a simplified example solution using fixed beams is described for an analog beamforming system.

7.4.3.1 Beamwidths and serving cell coverage

This section describes in a simplified way how the coverage area of a beam relates to the coverage area of a cell or sector and how to cover a cell with a set of beams.

A given cell *area* (or more accurately *volume*, as beams are directed in both an elevation and an azimuth direction) can be fully covered by a set of beams, for example, by many narrow beams or by fewer wide beams. Wide and narrow beams are illustrated in Fig. 7.14.

An analogy for wide beams versus narrow beams that is simple to visualize is to think of a common garden hose, where a water jet can be concentrated in one direction to achieve a high flow of water and distance in that direction. The water jet can also be adjusted to spread widely to cover a wider area at the cost of some distance. The same amount of power, or water flow, is distributed differently in space depending on the used approach.

There is a trade-off between beamwidth and beam gain for a given antenna array, that is, the narrower the beams, the higher the gain and vice versa. Thus to maximize gain, narrow high-gain

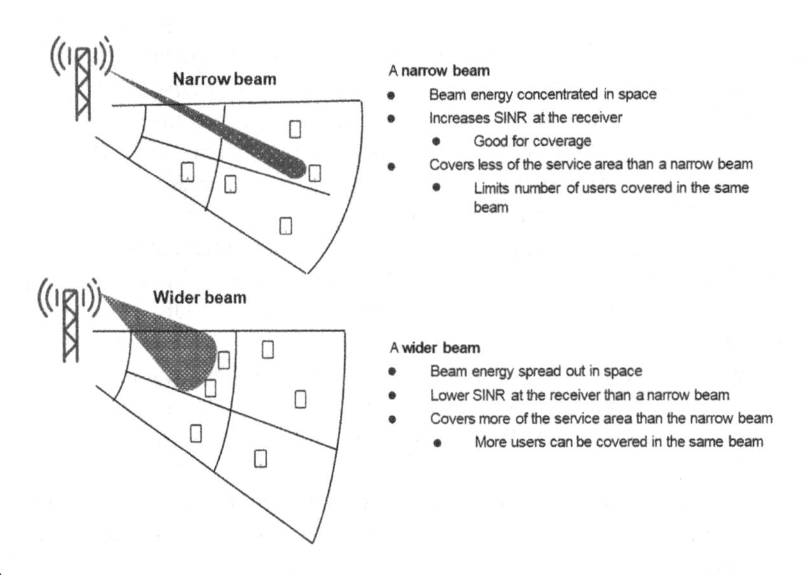

FIGURE 7.14

Narrow versus wide beams.

beams are appropriate. On the other hand, the larger the beamwidth, and the fewer the beams needed to cover a given area, the faster an appropriate *beam pair* can be found when that is needed.

Thus the fewer the beams on the network or the UE side, the simpler the beam management problem becomes. In the simplest case of the network transmitting on one wide beam that covers the entire cell, there is only one beam to transmit on, and if the UE only has a single wide beam available, then there is only one beam to receive on, and there is no beam management issue.

However, at high-band, multiple beams are required on the network side for coverage. Multiple beams can also be useful on the UE side. Typically, the UE side which has a simpler antenna structure will support fewer beams, and so wider beams than the network side.

As the gain is lower for a wider beam, the signal-to-interference-and-noise ratio (SINR) is generally lower and the achievable data rate is lower than when using a narrow beam, at a given UE location. However, signals of broadcast type that require full cell coverage, for example, control signals, are typically transmitted with low rates, and are designed to be robust, and so do not require a high SINR. Typically for mid-band NR systems one wide beam "the broadcast beam" covering the entire cell is enough for control information (though it is possible to have more than one if needed to increase coverage), and high-gain beams are used for UE–specific data. For high-band it is also beneficial to increase coverage of broadcast information using a larger set of more narrow high-gain beams, still it may be beneficial to limit the total amount of those beams to keep latency and overhead low, where overhead is caused by measurement reference signals (e.g., SSB) sent per beam.

Narrow beams are useful for UE-specific beamforming, that is, when the transmitter already knows the preferred direction to the UE. The increased beam gain of narrow beams can help increase received signal power that typically, depending on the radio conditions and interference situation, results in increased data rate at the UE for the data transmission. Measurement overhead can be controlled for UE-specific beams by only sending measurement reference signals (such as CSI-RS) when needed.

Beams can also be fixed, meaning that they have a fixed shape that is predefined, or dynamically computed where the shape can be arbitrary and is formed "on the fly" based on channel info. To summarize this discussion, some examples of usage of fixed beams and dynamically computed beams in a 5G system and their relation to wide and narrow beams are given below, while noting that this is not meant to be a general classification and so there can be exceptions.

Typical beam usage:

- A "wider" fixed beam or a set of fixed beams that covers an entire cell[6] or sector service area;
  - The beam directions are "cell-specific", for example, do not depend on the channel to any served UE;
  - The beamwidth is typically wider than the smallest width that is possible with the given AAS, and in the extreme case one beam covers the entire service area;
  - Each fixed beam can be defined by an associated beam measurement reference signal (e.g., SSB), which is periodically transmitted;

[6]For simplicity, the area covered by a single transmission point is described. In NR one cell can have multiple transmission points, each transmitting one or several beams; however, the concepts above still apply.

    - Note that 3GPP limits maximum number of SSBs to 8 for FR1 and 64 for FR2.
  - Can be used for broadcast signals and channels;
  - Can be used for UE-specific channels, for example, data and control;
  - Typically selected by received power per beam (e.g., as measured at the UE for DL);
  - The fixed beam can be used by a single UE or shared by multiple served users simultaneously;
  - When a UE using this beam needs to change to a different beam, it requires beam switching, where a new beam used for downlink is indicated to the UE;
  - Typically, fixed beams are not changed often, but different beam sets or "shapes" could be used in different scenarios and deployments or at different times. Changing the cell defining beam shapes is often called cell shaping[7]
- Fixed beams intended to be used for measurement or to transmit to one UE only;
  - The beamwidth may be narrow to increase beam gain toward a certain area of the cell;
  - The beam only needs to be transmitted if there is an ongoing transmission or measurement.
  - Beam measurements (e.g., via CSI-RS) can either be periodically triggered or a-periodically triggered to reduce overhead;
  - The predefined beam shapes populate a set (i.e., a codebook, such as a GoB) of beams that together cover the entire cell area;
  - A beam to use for transmission to a UE at any given time is chosen from the set based on measurements.
- Dynamically computed beams intended for a certain UE only;
  - The beam shape continually adapts to the radio channel between the AAS and the UE, so there is no need for explicit beam switching;
  - Generally used for UE-specific channels, for example, data and control;
  - The beam weights that determine the beam shape for a given time instant are created on the fly, by algorithms using actual channel measurements, so the beam may be narrow or wide or an arbitrary shape and is a result of whatever gives the best performance according to the used algorithm (see Chapter 6).

An AAS can be designed only using fixed beams as in the examples above, and this is a simple first solution for a high-band system. An AAS can also be designed using a combination of fixed beams for common channels and broadcast information and dynamically computed UE-specific beams, as typically used for reciprocity-based DBF systems. In the latter case the cell defining set of fixed beams fulfill the role of providing broadcast information where the location of the UE is unknown or when the transmitted information is intended for many UEs in the cell.

7.4.3.1.1 Example of an analog-based beamforming system solution using fixed beams

In an example outlined below, it is discussed how beam management with both wide and narrow fixed beams can be used for a high-band solution based on *time-domain analog beamforming*. In a

[7]In fact, cell shaping based on channel measurements can also be done using algorithms for example based on measurement of UE spatial characteristics or location; hence the cell defining beams can be dynamically shaped also. However, this is not common yet in real networks.

simplified case of an analog beamforming system with a single beamformer, it is only possible to transmit in one beam at a time and the amount of time it takes to sweep over all beams in the service area as described in the earlier beam management functionality scales up with the number of beams.

In this scenario there are plenty of choices in the design of a set of predefined beams. A simple and commonly used choice is to design a set of narrow beams that are placed side by side in azimuth and elevation to cover the entire cell or service area. As discussed in Chapter 6, this type of design leads to a so-called GoB (see Fig. 7.15).

One approach is to use such a grid of narrow beams for both cell level control beams and for UE-specific beams. The advantage of such a solution would be to give high coverage for control channel beams. However, if the link budget for coverage of control channels is not limiting relative to the wanted cell-edge data rate for the data channels, then such extra coverage for those channels may be unnecessary. The disadvantage of such a solution is the case when the best beam to use for a UE is unknown, for example, at initial access or in a recovery case. If only many narrow beams are defined, then both the latency to find a good beam pair and beam measurement overhead increase. Note also that the maximum number of SSB beams is limited to 64, so if more narrow beams are needed than 64, it is not always possible to use the same beams for data and control.

Another solution is to use a hierarchical approach, which is also illustrated in Fig. 7.15. In that solution a separate set of wide beams is specified to cover the whole cell area. Each wide beam covers the same area as several narrow beams. The entire cell area can be swept quickly when

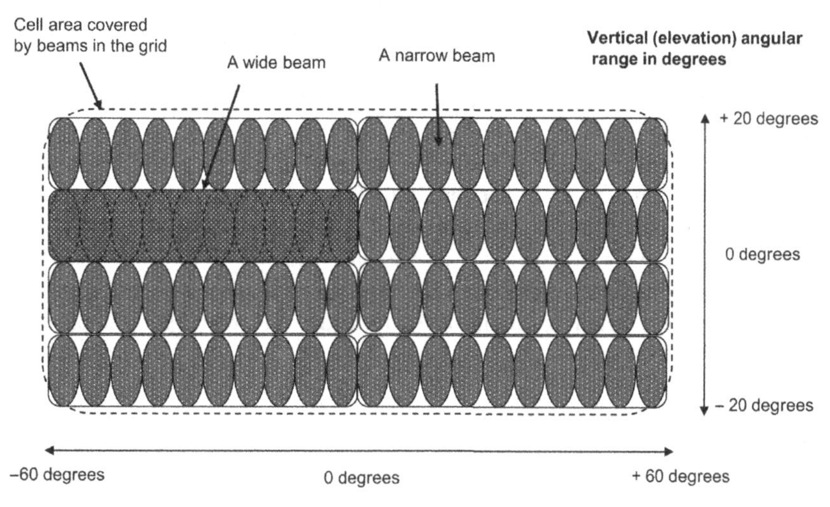

FIGURE 7.15

View of an antenna cell service area, showing horizontal and vertical coverage of the cell area using a group of fixed predefined beams. In this example the cell area is covered by either 80 narrow beams or 8 wide beams. Ten narrow beams cover the same area as one wide beam. Note that this is a beam view and not an array view. All beams can be created using the full antenna array.

needed, by first sweeping through all the wide beams and locating which wide beam covers a UE (e.g., the P1 procedure described earlier). Then, to find which narrow beam to use, all the narrow beams that correspond to the wide beam are swept (e.g., using the P2 procedure). The measurement signals used for the beam sweep of narrow beams can also be a-periodically triggered to a certain UE, since these measurements are only needed for that UE. This avoids the unnecessary reference signal overhead of transmitting periodic measurement signals in all available narrow beams.

In this solution, one of these GoB beams can be dynamically selected to be used for transmission towards the UE. Beam management will thus find a preferred GoB beam for the UE using beam management measurements and indicate to the UE the selection to be used for control and data channel transmissions. One approach could have been to number the beams from 1 to N and indicate to the UE which beam is used, for example, a codebook describing a GoB. However, as is described in Chapter 9, a more flexible approach was taken in 3GPP, where the number of beams and the beam shapes are not bound to a codebook but are related to reference signals and connected using the QCL framework that is used to specify how the different reference signals relate to each other. This allows maximal freedom in beam design and implementation on the network side and allows, for example, the possibility to optimize different sets of beams based on the deployment scenario. Similarly, the number and shape of UE beams are not specified but are up to each UE's implementation, giving similar flexibility on the UE side.

One design consideration is that when the beams are placed side by side in a set of beams as shown in Fig. 7.15, then there are some points of lower beamforming gain in between the beams. The loss in beamforming gain between two beams is known as *straddling loss*. To reduce straddling loss, more beams can be added to the grid, for example, with pointing directions in between the existing beams' directions. This, however, increases the total number of beams that need to be handled and managed by the system.

This hierarchical example solution thus provides high antenna gain, both from the transmitter and the receiver sides, and it also limits the latency and overhead by finding the beams for user communication in two steps and thereafter adjusting beams as needed. Further signaling details on such a solution using the 3GPP toolbox are given in Section 10.6. This solution is just one example that gives a picture of some ways to use beam management with a simple analog-based beamforming system. There are numerous other possible solution variants, for example, using different possible beam shapes. In addition, the performance of solutions like this can be improved significantly once it is possible to send and receive more beams simultaneously, for example, when DBF is applied at high-band.

7.5 SUMMARY

In the high-band frequency range, there are substantial amounts of spectrum available, allowing very high data rates to be supported of the order of many Gbps, and hence high-frequency bands are an important spectrum range to make use of to satisfy communication needs for 5G and beyond.

There are however several issues to be handled in order to make the best use of the high-band spectrum.

- Beamforming is necessary to achieve good coverage for many deployments because of frequency-dependent pathloss and the signal attenuation in materials at high frequencies;

- Due to technology maturity, the first solutions at high-band use time-domain analog beamforming where one or a few beams are transmitted or received at a time;
  - The transmitter and receiver have a reduced view of the total spatial domain at a given time and need to constantly align and keep track of the beams being used and change them as needed.
- Time-domain analog beamforming over the wide carrier bandwidths (e.g., 100 or 200 MHz) that are typical in FR2 is very suitable for high data rate transmissions but can lead to inefficient use of bandwidth when only a small number of bits need to be sent;
  - For example, when only single beam at a time is available, control information and small packets sent in a beam in a given time symbol block the entire carrier bandwidth for other users that also need to send at the same time but in a different beam.

Both proprietary and standardized solutions can be used to help deal with these issues, including using beam management support in 3GPP. Beam management is part of the NR toolbox that can be used to assist either analog beamforming or DBF solutions but is primarily aimed at analog beamforming solutions.

Time-domain analog beamforming is a useful and practical solution for today's high-band AAS solutions and is already deployed in operator's networks to support MBB and FWA. DBF techniques can also be used at high-band and will become more cost-effective as the technology matures. For example, frequency-domain DBF can help further with issues of inefficient use of bandwidth and remove the need for some beam management functionality, which can help increase capacity and reduce latency.

The high bands in FR2 are very suitable for certain deployment scenarios. These include outdoor relatively open areas such as stadiums, squares, parks, and streets. Propagation can also be excellent in indoor relatively open areas such as shopping malls, indoor open office landscape, and open area factories. High-band deployment is not limited to these scenarios as, for example, mm-wave penetration into low-loss buildings, or through low-loss internal walls can be good. However, penetration into more modern high-loss buildings is challenging.

For wide area coverage in environments with low site density or for scenarios where more outdoor-to-indoor coverage is required, mm-wave works best when used as a complement to lower bands. This can complement and offload traffic from lower band sites that can serve as the wide area coverage layer.

For NR, mm-wave can be used together with mid-band spectrum, for example, using carrier aggregation, which allows uplink and downlink to be handled on either the same or different bands. In such cases the different bands complement each other and increase network coverage and capacity compared to using the individual bands stand alone.

FWA scenarios with good propagation between the AAS and a typically stationary CPE are very suitable for high-band and long range is achievable as CPE devices can be more capable (e.g., with a larger antenna) than a standard smartphone.

For deployments with high density where there is a lack of fiber availability to all sites, IAB can be used to provide wireless backhaul between some of the sites. Due to the large amount of spectrum available on mm-wave and the high throughput achievable, this spectrum is very suitable for use with IAB. IAB with mm-wave backhaul can be used to provide backhaul for high-band radios or for radios on lower bands.

Due to the high-throughput and low-latency mm-wave solutions are now being investigated and trialed for use in highly automated factories, where they can support the control and monitoring of, for example, robots, production lines, and automated vehicles.

The first commercial MBB and fixed wireless NR solutions using mm-wave are already deployed today fulfilling the 5G promise of very high throughput and low latency, and they will continue to evolve performance in all aspects including supporting ultra-reliable low-latency communication (see Chapter 9). In general, the technology and techniques to use this spectrum will continue to evolve over the coming years as 5G use cases and applications develop and new deployment scenarios are introduced.

REFERENCE

[1] 3GPP TS 38.101-2 v15.6.0.

FURTHER READING

Ericsson white paper, Advanced antenna systems for 5G networks. <https://www.ericsson.com/en/white-papers/advanced-antenna-systems-for-5g-networks>, 2018.

F. Kronestedt, et al., The advantages of combining 5G NR with LTE, Ericsson Technology Review. <https://www.ericsson.com/en/reports-and-papers/ericsson-technology-review/articles/the-advantages-of-combining-5g-nr-with-lte>, 2018.

CHAPTER

3GPP PHYSICAL LAYER SOLUTIONS FOR LTE AND THE EVOLUTION TOWARD NR

8

The purpose of this chapter is to describe the parts of the 3GPP long-term evolution (LTE) specifications that are relevant for advanced antenna system (AAS). In Chapter 9, the corresponding new radio (NR) specifications are described. Also, in Chapter 6, the underlying concepts related to multi-antenna operation are introduced and motivated and this chapter describes how these concepts have been standardized and specified in *third generation partnership program* (3GPP) LTE specifications. It is recommended to be familiar with the key concepts in Chapter 6 to fully understand the motivation for multi-antenna standardization in 3GPP.

Moreover, to understand the motivation behind the concepts in the NR specifications, it is useful to get a grip on the fundamentals and shortcomings of LTE. This should motivate a reader interested in NR to also spend some time to understand LTE. Furthermore, central definitions such as antenna port and quasi colocation (QCL) are the same in LTE and NR and described more thoroughly in this LTE chapter.

The standardization work by 3GPP creates a multivendor solution by means of industry cooperation. This collaboration has created several successful telecommunication standards such as global system for mobile communications, wideband code division multiple access (WCDMA), LTE, and now NR. In the communication industry with global roaming of devices and many network components, there is no market without standards and interoperability. The key to the success has been open specifications and industry cooperation where all parties involved pool ideas to ensure the solution has cutting-edge technology that is practical to implement and can scale globally. A drawback with such cooperation is that compromises must always be made in the decision process which in some cases leads to an excess in the number of possible configurations. A single vendor telecommunication system would likely have done some parts slightly more efficient and streamlined, but the benefits such as crowdsourcing of ideas, resulting in global roaming technology and benefits of scale would have been lost.

The chapter is organized as follows: the basic principles of LTE are provided in Section 8.1 and the history and evolution of LTE from release 8 to 15 is given in Section 8.2. For the interested reader, a more detailed description of the LTE physical layer is provided in Section 8.3. A summary of LTE and "lessons learned" from this LTE evolution can be found in Section 8.4.

Finally, note that this section will by no means provide a complete description of 3GPP LTE physical layer specifications. The focus here is an in-depth explanation of the multi-antenna aspects, and less focus is given in this section to control signaling and other aspects such as connection establishment (random access procedures). For a broader presentation of LTE specifications, refer to Ref. [1]. There are also verticals evolving that build on LTE, for example, see the use of

Advanced Antenna Systems for 5G Network Deployments. DOI: https://doi.org/10.1016/B978-0-12-820046-9.00008-3

LTE for *Internet of things* (IoT) and *machine-type communications* as described in detail in Ref. [2]. Such technologies will not be discussed in this book.

8.1 LTE PHYSICAL LAYER—BASIC PRINCIPLES

The purpose of the standard is to define a common understanding of signals, procedures, and protocols exchanged between the network [for LTE physical layer specifications, the evolved Node B (eNB)] and the terminals (the user equipment, UEs) to allow for interoperability between network components from different vendors and also enable global roaming.

Furthermore, the core of the physical layer standard specification for data transmission and reception can be categorized into three main areas of functionality:

- Functionality to provide *synchronization* between eNB and UE, for tracking of the fading radio channel, including *channel analysis* where the UE estimates large-scale channel properties (see Section 3.5). This functionality is enabled by *synchronization and tracking reference signals (RSs)*;
- Definition of the *transmission schemes* for the data transmission and the associated *demodulation reference signals*;
- Functionality to provide *link adaptation* through *channel measurements*, the associated reporting of these measurements from UE to eNB, and the associated *measurement RSs*. Link adaptation is the procedure to adjust the modulation scheme and code rate of the transmission to the quality of the link. Link adaptation also includes a recommendation from the UE to the eNB of a multi-antenna transmit precoder, or a preferred transmit beam.

A brief overview of LTE is first presented before the three main areas of functionality as outlined above are discussed further.

In Fig. 8.1, an eNB and UE are communicating, and the different physical layer channels are illustrated, consisting of downlink (DL; eNB to UE) and uplink (UL; UE to eNB) transmissions. Note that eNB and UE are 3GPP terminology that corresponds to base station and wireless terminal in other sections of this book.

On the highest level, channels can be divided into control channels and data channels, where the purpose of the control channel is to assist the data channel reception, for example, carrying a data channel scheduling assignment. Also, channels can be of broadcasted type (to many UEs simultaneously or to a single, yet unknown UE) or of unicast type, to a single specific UE.

Broadcast channels cover the entire served area in DL. They are used by UE to find the network and perform synchronization [provided by the primary and secondary synchronization signals (PSS, SSS)] and to obtain the basic information necessary to access the network, known as broadcast information [carried by the physical broadcast channel (PBCH)]. Part of the control information is also broadcasted [common physical downlink control channel (PDCCH) or physical control format indicator channel (PCFICH)], while the other part is unicasted [dedicated PDCCH or physical hybrid ARQ indicator channel (PHICH)], intended for a single UE only, such as containing scheduling of data transmission or reception.

The data transmission is carried by one of multiple different transmission schemes and is using the unicast physical downlink shared channel (PDSCH) and physical uplink shared channel (PUSCH) for DL and UL, respectively. Finally, the physical random access channel (PRACH) is transmitted in UL for random access, e.g. in the initial access procedure and PUCCH is used for UL control signaling. In addition to these defined channels, there are RSs transmitted to support synchronization, mobility, and measurements.

These channels will be described further in detail in this section and are summarized in Table 8.2.

LTE uses orthogonal frequency division multiplexing (OFDM) (Section 5.1.1) for the DL transmissions and discrete Fourier transform (DFT)-spread OFDM (Section 5.1.5) for the UL. The basic LTE physical resource can thus be illustrated in a frequency domain representation by a time–frequency grid as illustrated in Fig. 8.1. The smallest physical resource in the OFDM based system is one *resource element* (RE) and if OFDM without DFT precoding is used, each RE contains one modulated symbol containing a variable number of bits depending on the link adaptation, that is, link signal-to-noise ratio. The used modulation constellations in LTE are shown in Table 8.1. Refer

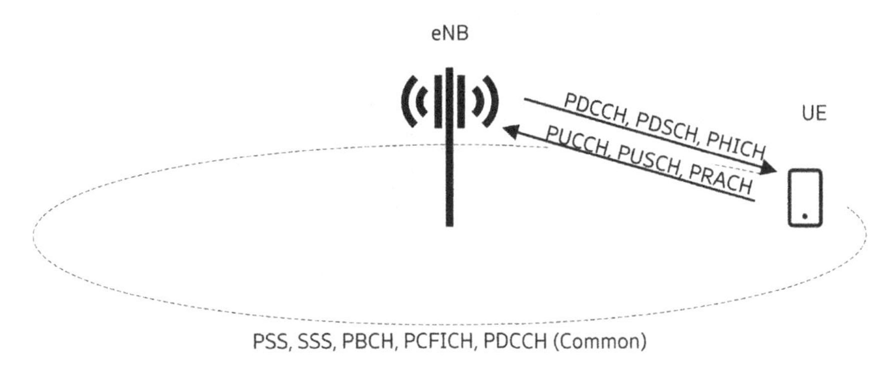

FIGURE 8.1

The downlink and uplink physical channels used for initial access, synchronization, broadcast, and unicast transmission in LTE. The unicast channels are only received by a single UE and can therefore be beamformed toward that UE. Other channels must have cell wide coverage as they may be received by many UEs simultanously.

Table 8.1 Modulation Constellations in LTE.

Modulation	Bits per Constellation Symbol
BPSK	1
QPSK	2
16-QAM	4
64-QAM	6
256-QAM	8

to Section 5.1 for an introduction to quadrature amplitude modulation (QAM) modulation. Each RE corresponds to one OFDM subcarrier during one OFDM symbol interval.

Furthermore, when scheduling data (see Section 2.1.1.2), that is, the PDSCH and PUSCH, the resource allocation is described in multiples of (physical) *resource blocks* (RBs) in frequency domain consisting of 12 contiguous subcarriers and a duration of 1-ms subframe in time domain (although later LTE versions introduced shorter scheduling durations to improve latency). The RB contains a control region in the beginning of the subframe (which can vary between 1 and 4 OFDM symbols) and the remainder of the subframe is a data region (see Fig. 8.2).

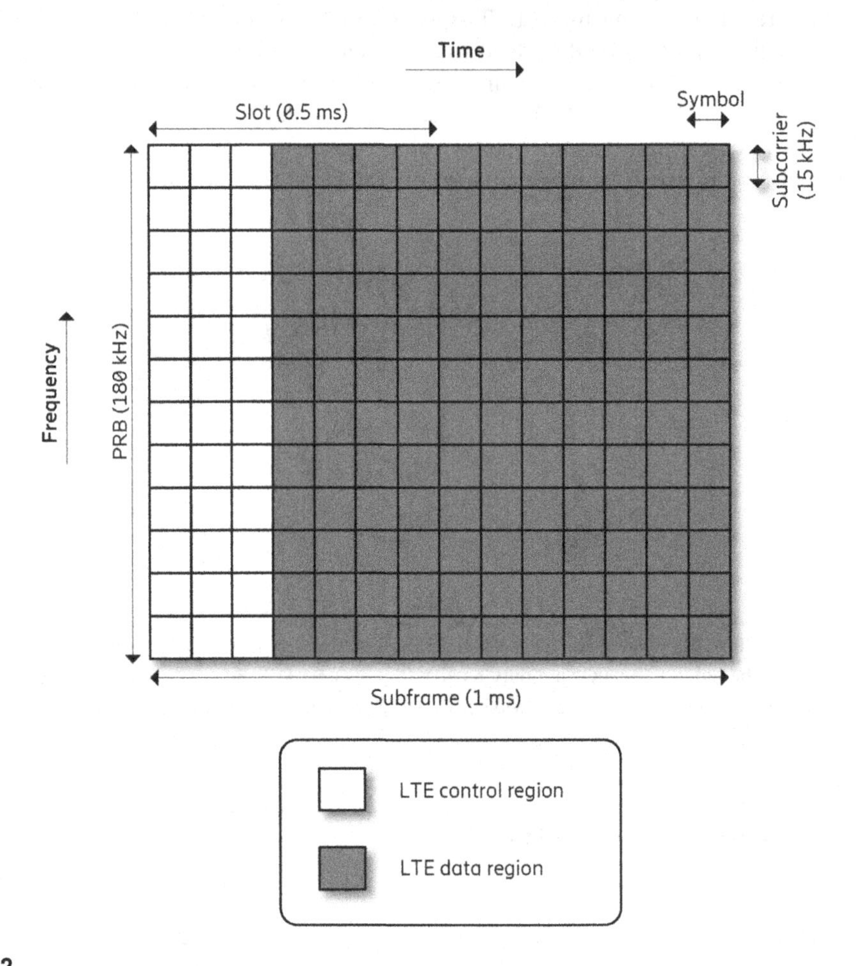

FIGURE 8.2

The LTE downlink physical resources showing an RB of 12 subcarriers and one LTE subframe (1 ms) of 14 OFDM symbols. It is divided into a control region and a data region (assuming FDD) where in this figure a control region length of three OFDM symbols is illustrated. LTE supports between one and four OFDM symbols for the control region.

Data can be scheduled contiguously (multiple adjacent RBs) or non-contiguously, to a given UE in the DL, and this allows for user multiplexing in frequency (see Fig. 8.3). The non-contiguous resource allocation is useful in case the channel has a higher signal-to-noise ratio (SNR) in some RBs than others (and the scheduler is SNR aware), then the transmission can use only these, so-called opportunistic frequency scheduling.

In the time domain, LTE DL subframes are organized into radio frames of 10 ms and each radio frame consisting of ten subframes of length 1 ms each (see Fig. 8.4).

The physical channels introduced in the initial release of LTE, release 8, are shown in Table 8.2 with a brief description of the content and use. The RS used to receive the channel is also given in the table.

For a UE to receive data in a subframe, the UE first decodes the PCFICH located in the first symbol of the subframe, to determine the length of the control region. This length can be

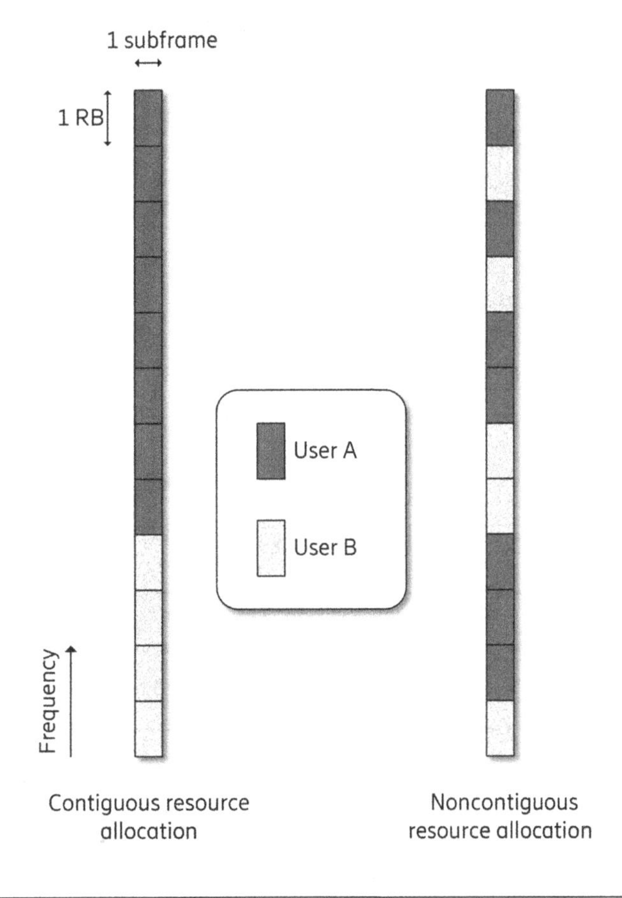

FIGURE 8.3

Resource allocation in basic LTE is performed in units of resource blocks in frequency and subframes in time. Resource allocations in the DL to a user can be both contiguous and non-contiguous.

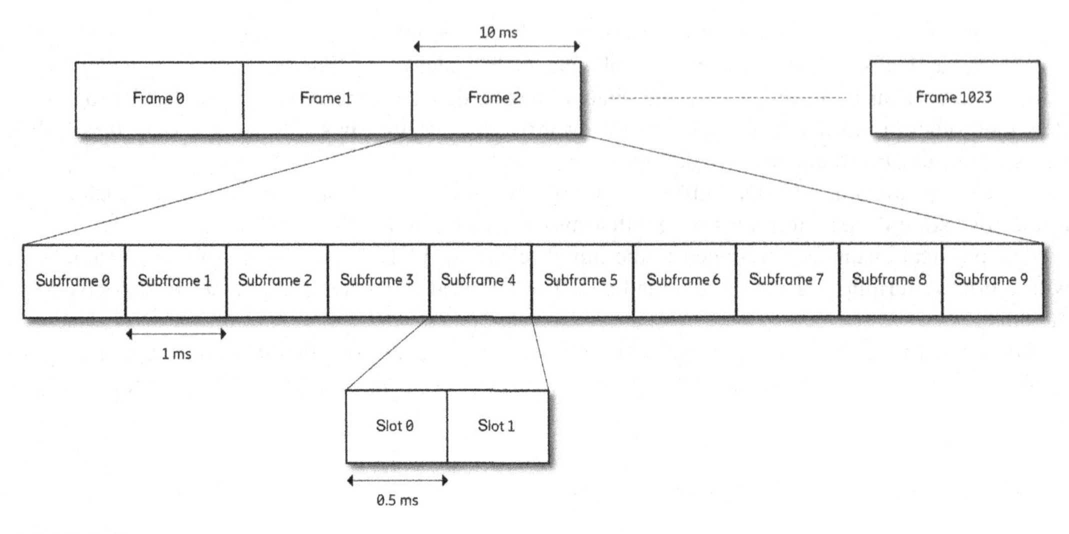

FIGURE 8.4

LTE time-domain frame, subframe, and slot structure.

Table 8.2 Physical Channels in LTE Release 8.

Physical Channel	Name of Physical Channel	Content	Demodulation Reference Signal
PBCH	Broadcast channel	Part of the system information to all UEs in the cell	CRS
PDCCH	Downlink control channel	DCI, for example, for scheduling PDSCH and PUSCH	CRS
PCFICH	Control format indicator channel	DCI to indicate the extent of the PDCCH in the same subframe	CRS
PHICH	HARQ indicator channel	DCI to indicate HARQ-ACK for PUSCH	CRS
PDSCH	Downlink shared channel	Downlink unicast data, paging, and part of the system information	CRS or DM-RS
PMCH	Multicast channel	Multicast data	MBSFN RS
PUSCH	Uplink shared channel	Scheduled uplink data transmission	DM-RS
PUCCH	Uplink control channel	UCI to indicate HARQ-ACK for PDSCH and for CSI feedback	DM-RS
PRACH	Random access channel	Nonscheduled uplink channel used, for example, in random access	N/A

dynamically changed by the network to 1, 2, or 3 OFDM symbols depending on the needed control channel (i.e., PDCCH) capacity (where four control channel symbols are only available for the smallest LTE system bandwidth of 1.4 MHz).

A PDCCH transmission contains downlink control information (DCI) to schedule the DL or the UL and is mapped to the REs in the control region, where these REs are distributed across the full system bandwidth to achieve diversity against fading and interference. The eNB can transmit multiple PDCCH simultaneously in nonoverlapping REs, and a UE decodes a set of PDCCHs in each subframe. This is known as blind decoding since the UE must decode multiple PDCCHs before it can determine whether any PDCCH included a DCI that was intended for the UE. The DCI may contain a scheduling assignment to receive a PDSCH in the data region of the same subframe or to transmit a PUSCH in UL in a later subframe. For frequency division duplex (FDD), the PUSCH is scheduled four subframes later than the reception of the PDCCH, for the UE to have time to prepare the UL transmission. See Fig. 8.5 for the timing for PDSCH and PUSCH (see also Section 8.3.3 on PDCCH transmission procedures and Section 8.3.4 on PDSCH transmission procedures). The acknowledgment (positive) (ACK) or negative ACK (NACK) (i.e., hybrid automatic repeat request (HARQ)-ACK) for the PDSCH reception is then reported to the eNB using uplink control information (UCI) mapped to either PUCCH or PUSCH, and for FDD, it is reported four subframes after the PDSCH (see Fig. 8.5). For time division duplex (TDD), the scheme gets more complicated as when determining the distance between PDSCH and ACK/NACK, then whether the ACK/NACK transmission subframe is a UL, a DL, or a mixed (special) subframe must be considered.

After scheduling and reception of a PUSCH at the eNB, the associated ACK or NACK (i.e., whether to perform a retransmission) may be conveyed to the UE, either by the PHICH or indirectly by a new PDCCH.

The UE also receives a broadcast channel, the PBCH that carries the master information block, that contains the information necessary to connect to the cell and to receive, for example, PDCCH. The PBCH is not scheduled by a PDCCH but is instead transmitted in a fixed location and format.

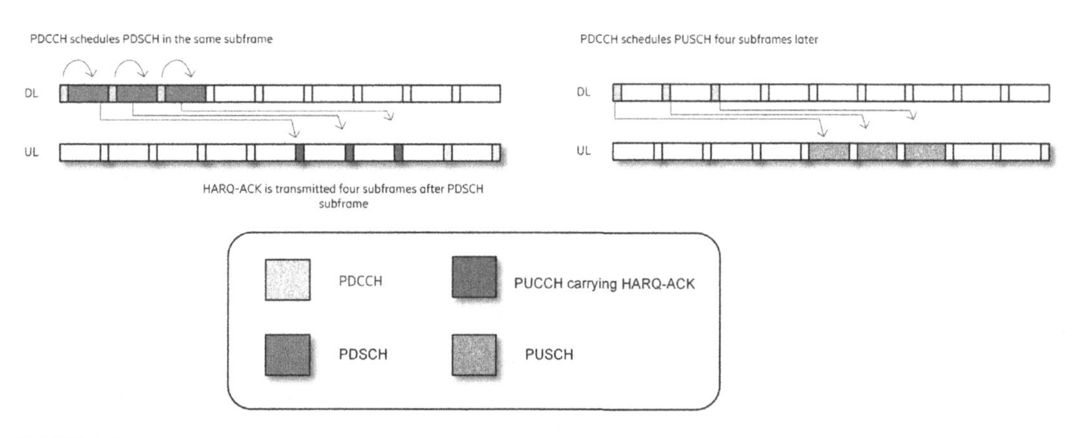

FIGURE 8.5

Scheduling timing for downlink (left) and uplink (right) for FDD.

In addition, the PMCH is a channel that optionally can be transmitted by the network, to support multicast services. Finally, PRACH is defined for UL transmissions to perform, for example, initial access to the network.

Now, how does a UE receive the DL transmissions with good quality? For coherent detection, the UE needs to know the channel over which the data has propagated so it can compensate for, for example, phase rotations induced by the channel. For this purpose, RSs are defined in specifications and transmitted by the eNB, and as these are known to the UE, the UE can then use these RSs to estimate the channel that they have propagated through. If the REs containing RS are dense enough, the UE can interpolate the channel for REs in between the RS and thus estimate the channel for each RE in between, i.e. for each transmitted modulated data symbol (e.g., quadrature phase shift keying, QPSK). The frequency and time separation between RSs in the time-frequency grid must be smaller than the coherence bandwidth and coherence time of the channel, respectively. Refer to Section 3.6 on multi-path propagation and channel coherence.

In LTE, there are two alternatives for RS that can be used for PDSCH demodulation, either *cell-specific RS* or *UE–specific RS*. The *cell-specific reference signals* (CRSs) transmitted in DL are shared by all the UEs served by the eNB (cell centric design), while the *UE–specific demodulation reference signal* (DM-RS) is dedicated for the demodulation of the transmission to or from one UE only (UE centric design). As this is an important differentiation, these two approaches will be further elaborated in the following paragraphs. The first LTE release relied heavily on the cell-specific RS, while a paradigm shift to UE–specific RSs occurred in LTE release 10. The UE–specific RSs were later used as a key principle of the design of NR (see Chapter 9).

8.1.1 KEY FUNCTIONALITIES

A first key functionality is *synchronization*, where the UE aligns the time–frequency grid of REs with the eNB transmitter; hence frequency offset estimation and time delay need to be estimated and compensated. In addition, the UE needs to find the frame and subframe boundaries as in Fig. 8.4 to be able to find, for example, the PDCCH so that the eNB can reach the UE with a scheduled message.

It is also beneficial for the UE to perform tracking of channel estimates over time, that is, across subframes, as this provides better performance compared to estimating the channel independently, on a per subframe basis. In addition, the receiver benefits from knowing the channels' large-scale properties, such as the Doppler shift and average delay (see further Section 3.5 on channel large-scale properties). If the UE knows these statistical properties of the channel prior to performing channel estimation, it can tune its channel estimation filter by using this as a priori information. This is important to ensure good reception performance, especially when higher-order modulation (e.g., 64- or 256-QAM) is used as these are more sensitive to channel estimation errors and transmitter-receiver misalignment.

For an LTE UE, this tracking and analysis functionality is always performed using measurements on the CRSs, irrespective of the configured transmission mode (TM). The CRS is present in every subframe and distributed across both the time and frequency, that is, with a density within the channel coherence limits (see Chapter 3) and is therefore a reliable, always present reference to base a channel analysis on.

Prior to LTE release 11, it was not specified that which RS the UE shall use to estimate the large-scale channel parameters; hence this was left open for UE implementation. However, since CRS is always present it is a reliable RS for this purpose and is thus typically used. In the later LTE releases and when TMs using DM-RS for demodulation was introduced, the specifications instead describe which RS the UE can use (or better use) for an estimate of these large-scale parameters. For this purpose, five such large-scale channel–fading parameters were defined in LTE release 11. As will be explained later, this was necessary to enable support for more advanced network deployments such as shared cells (see Section 6.6.1).

An LTE UE can be configured to operate according to 1 of 10 different *transmission modes* (TM) (see Section 8.3.4 for an overview of the TMs). If the UE supports LTE release 11 and when TM 9 or TM 10 is configured, not all large-scale channel–fading parameters should be derived from CRS, some should instead be derived from CSI-RS, and this is controlled by the introduction of the *quasi colocation* (QCL) feature in release 11. As will be seen in Section 9.3.1.6 for NR, a QCL framework is even more important and extensive, especially for high-band operation. Hence, understanding the basics of QCL principles is necessary for grasping the technical details of NR physical layer design (refer to Section 8.3.1.2 for a first introduction).

A stream of information bits is encoded by a Turbo encoder and then modulated (e.g., to QPSK, 16-QAM, 64-QAM, or 256-QAM) and mapped to the one or multiple multiple-input, multiple-output (MIMO) layers, thus supporting, for example, spatial multiplexing as described in Section 6.3. A *transmission scheme* for PDSCH or PUSCH describes this as the code word to layer mapping and the MIMO precoding. This mapping and precoding can be performed in various ways to achieve different purposes (e.g., maximize spectral efficiency or maximize diversity) (see Section 6.3.2.1). Hence, multiple such transmission schemes are specified in LTE. Each transmission scheme has a main design target, for example, providing maximal spectral efficiency, providing robustness by antenna diversity or providing robustness combined with increased spectral efficiency which is useful at higher UE velocities.

One scheme to maximize diversity in LTE is the well-known Alamouti diversity encoding [3] specified for LTE PDSCH and PDCCH. It can be described as a fixed MIMO linear dispersion block code [4], which can further be viewed as a generalization of the precoder concept in Section 6.3.3 to precoders spanning both the spatial and another domain such as frequency or time. Another scheme to achieve diversity that also supports spatial multiplexing is the large delay *cyclic delay diversity* [5] described as a set of precoders where one precoder from this set is cyclically selected and used per subcarrier, across the scheduled bandwidth.

When CRS is the RS used for demodulation of a certain transmission scheme, one unique CRS (from an available set of CRSs) is typically transmitted per eNB antenna. A typical case where CRS-based schemes are used is the conventional system with up to four radios and where the antenna consists of an antenna column with antenna elements of the same polarization. The mapping/precoding of a code word to these CRSs that is performed by the eNB, that is, the transmission scheme, must be known to the UE for proper PDSCH demodulation.

For transmission schemes where instead DM-RS is used for data demodulation, a set of DM-RS is used, and one unique DM-RS is associated with each PDSCH spatial layer. The same mapping of DM-RS as the data for the layer is used, which makes the de-mapping trivial and the used precoding by the eNB is then transparent to the UE. This is known as *noncodebook-based* transmission (see the concept described in Section 6.4.1.2). This also allows for the more advanced precoding techniques such as the singular value decomposition and zero-forcing precoding (see Sections 6.3.3.2 and 6.3.3.3, respectively).

Therefore the DM-RS-based transmission schemes give greater flexibility to the transmitter to optimize the transmission and improve the link performance. Still, the transmitter needs to acquire some information on how to select the precoder and there are two main approaches to this as discussed in Section 6.4. Either obtained by measurements by the UE in the DL see Section 6.4.1, where the UE chooses a preferred precoder and uses feedback reporting to the network known as channel state information (CSI) reporting. Alternatively, this information can be obtained by the transmitter (eNB) itself by utilizing channel reciprocity (see Section 6.4.2) and thus by UL measurements using *sounding reference signals* (SRS).

The first approach, using CSI feedback will now be discussed further, as a key functionality of LTE. Note that even in the case of reciprocity-based operation, some CSI feedback as in the following section may still be necessary as a complement as discussed in Section 6.4.2.

Channel state information (CSI) measurements and reporting from the UE are necessary as described in Section 6.4.1, and in LTE it can be configured to assist the eNB in determining the precoder. Implicit feedback is used in LTE (see Section 6.4.1.2) and the most basic CSI report contains the channel quality indicator (CQI), a precoding matrix indicator (PMI), and a rank indicator (RI).

CSI measurements can be done using the CRS for CRS-based transmission schemes, or using dedicated CSI measurements signals, the CSI-RS for the DM-RS-based transmission schemes.

Fig. 8.6 shows the CSI feedback for a *CRS-based transmission scheme* (i.e., LTE TM 4), where in this example, four CRS are transmitted (see the related description of the *antenna ports* in Section 8.3.1 to fully understand the notion of a CRS port as in Fig. 8.6).

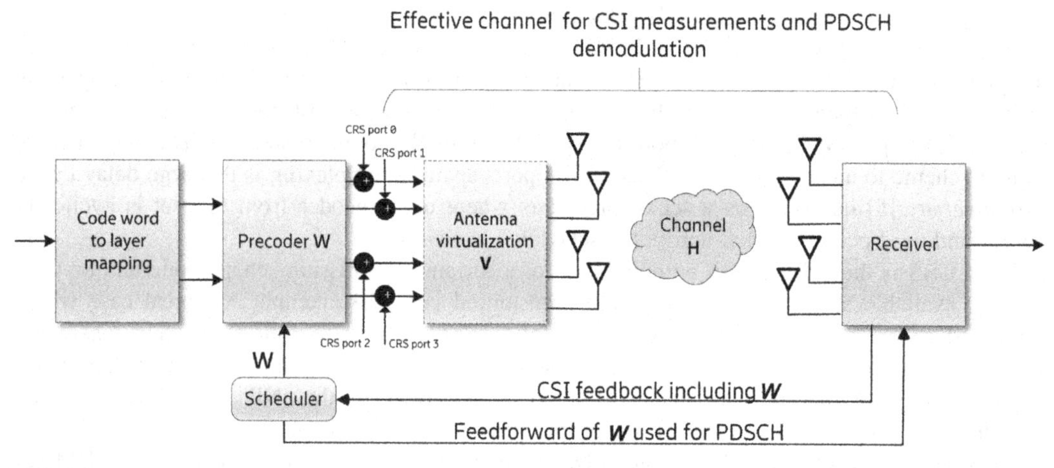

FIGURE 8.6

CSI feedback for a CRS-based transmission scheme, also illustrating the insertion point of CRS. The precoder is not part of the estimated combined channel ***HV***, and therefore the used precoder ***W*** must be indicated to the UE when scheduling PDSCH.

Antenna virtualization is used which maps the measurement signal (CRS in this case) to the four (or more) physical antennas. This virtualization step allows for cell shaping (cell-specific beamforming, see Section 6.2). Since the antenna virtualization is inserted *after* the measurement RSs (CRS in this case) in the direction of the signal flow, the used antenna virtualization matrix $\boldsymbol{V}$ is transparent to the UE as it becomes part of the combined channel which is the channel the UE observed by measuring on these CRSs. Hence, the UE cannot distinguish a change in $\boldsymbol{V}$ from a change in $\boldsymbol{H}$ as they are lumped together from a UE perspective. In the CRS-based scheme, the antenna virtualization $\boldsymbol{V}$ is the same for all UEs and typically does not change over time.

Antenna virtualization is necessary in the case where the eNB has more physical antennas than the maximum number of CRS that can be four. The input to the antenna virtualization can thus be described as a "virtual" antenna or antenna port. In the example shown in Fig. 8.6, there are however only four physical antennas illustrated and a one-to-one mapping can be used; one CRS can be transmitted from each of the four physical antennas of the eNB (i.e., the antenna virtualization matrix $\boldsymbol{V}$ is an identity matrix).

As described in Section 6.4.1, the UE performs channel estimation as well as interference estimation, in this case using the CRS. It selects the preferred precoder $\boldsymbol{W}$ from the specified precoder codebook and it selects the preferred rank. For the selected precoder and rank, the UE determines an associated CQI per code word (either one or two code words depending on the rank). It then feeds back the precoder, rank, and CQI information in the CSI report.

The receive data model from Section 6.3.1 is now used in the following description. When eNB decides to schedule a PDSCH transmission, the used precoder $\boldsymbol{W}$ from the codebook, which typically is obtained from the most recent CSI report, is indicated to the UE in the scheduling message. The UE then performs the CRS-based channel estimation again to obtain an estimate of the combined channel $\boldsymbol{HV}$ per subcarrier and then internally multiplies $\boldsymbol{HV}$ with the indicated precoder $\boldsymbol{W}$ from the eNB, to arrive at the effective channel $\boldsymbol{H}_\mathrm{e} = \boldsymbol{HVW}$, which is a matrix with the same number of columns as the number of layers, which describes the effective channel of each transmitted PDSCH layer.

Note that the CRS is used for *both* measurements and demodulation in this case. Also, since the used precoder $\boldsymbol{W}$ is not part of the combined channel that the UE estimates, it must explicitly be indicated to the UE in the scheduling (DCI) message.

Fig. 8.7 shows the CSI feedback for a *DM-RS-based transmission scheme* (e.g., LTE TM 9) where antenna virtualization is again used, which maps the measurement signal (CSI-RS in this case, which is a UE–specific RS) to the physical antennas. Since CSI-RS is configured per UE, different UE can use different CSI-RS and this allows for beamforming of the RSs, for example, UE–specific beamforming of CSI-RS, tailored for a specific UE. Hence, the antenna virtualization can when CSI-RS is used be dynamic and change, as opposed to the antenna virtualization in the CRS-based scheme. As antenna virtualization is inserted after the measurement RSs, the used $\boldsymbol{V}$ is also in this case transparent to the UE as it becomes part of the effective channel.

The antenna virtualization $\boldsymbol{V}$ can be wideband and fixed by antenna implementation as in the CRS-based case, or allow some functionality to provide for some adaptation of virtualization on a slow basis, hence to be semi-static over a long time. As mentioned, for DM-RS-based schemes it may vary dynamically, although still with some restriction on its time dynamics. The reason for such restriction is if the UE performs a channel measurement when a certain antenna virtualization

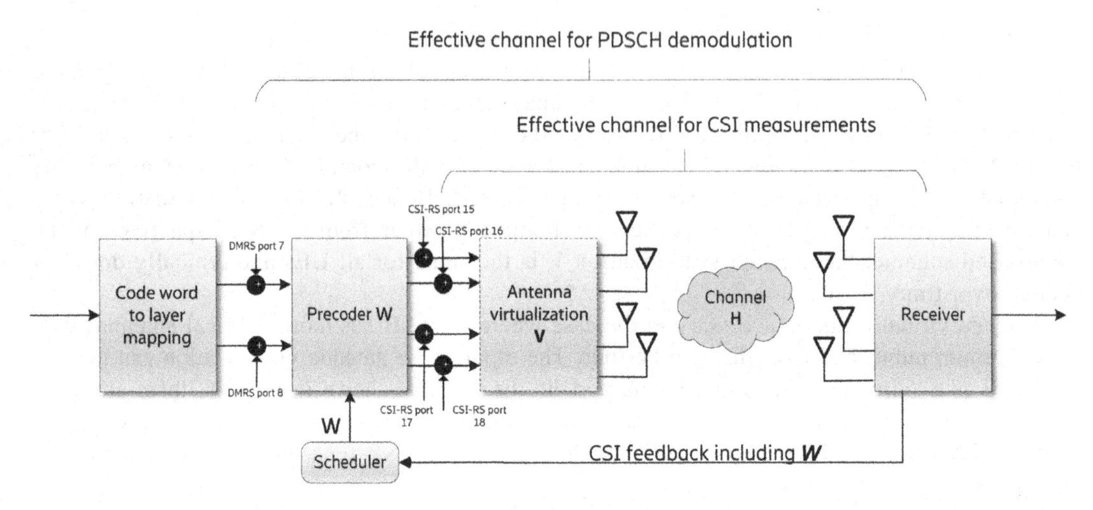

FIGURE 8.7

CSI feedback for a DM-RS-based transmission scheme illustrating the different insertion points of the DM-RS ports and CSI-RS ports. The precoder $\boldsymbol{W}$ is part of the effective channel for PDSCH demodulation; hence there is no need to indicate the choice of precoder to the UE and this gives full freedom to the network to select $\boldsymbol{W}$. Note that even in DM-RS-based transmission schemes in LTE, the CRS ports are always transmitted as well since they are cell-specific.

$\boldsymbol{V}$ is applied, then the same antenna virtualization $\boldsymbol{V}$ must be used in the eNB when transmitting the subsequent PDSCH. Otherwise, if a new virtualization $\boldsymbol{V}'$ is applied when PDSCH is transmitted, then the CSI report (especially the CQI and rank) and hence PDSCH link adaptation is valid for the wrong, that is, the previous virtualization $\boldsymbol{V}$.

To summarize, the CSI-RS is mapped to the virtual antennas, and the virtual antennas are further mapped to one or more physical antennas by a specification transparent mapping $\boldsymbol{V}$. Such virtual mapping could, for example, be a UE–specific beamforming of the CSI-RS in some direction. This is known as Class B operation in LTE CSI configuration. It could also be a static virtualization, typically using a one-to-one mapping between virtual antennas and physical antennas. This is referred to as Class A operation. In this case, as there is no UE–specific virtualization, typically all UEs in the cell share the same CSI-RS for their measurements of the DL channel. For Class A, the number of simultaneous CSI-RS is usually large since typically one measurement RS is needed per physical antenna. LTE supports up to 32 such simultaneous CSI-RS (referred to as 32 CSI-RS *antenna ports* in Section 8.3.1).

The CSI estimation principles for the DM-RS-based transmission schemes are the same as in the CRS-based schemes, using implicit CSI reporting, as defined in Section 6.4.1.2. The UE measures the channel by the CSI-RS over the antenna virtualization; it is thus measuring the combined radio channel $\boldsymbol{HV}$. It suggests a preferred precoder $\boldsymbol{W}$, preferred rank, and associated CQI computed based on the effective channel $\boldsymbol{H}_e = \boldsymbol{HVW}$ per subcarrier and feeds back the jointly determined precoder(s), rank and channel quality indicator(s) (CQI) to the eNB.

The eNB then schedules PDSCH and transmits one DM-RS per PDSCH layer and each DM-RS precoded in the same way as each associated PDSCH layer and the UE then estimates the effective channel $\boldsymbol{H}_e = \boldsymbol{HVW}$. The choice of $\boldsymbol{W}$ for PDSCH transmission is transparent to the receiver, giving full freedom to the eNB to modify. This is different compared to the CRS-based scheme where $\boldsymbol{W}$ is bound to a specified codebook. Note however that if eNB changes $\boldsymbol{W}$, or even changes the rank of the transmission, compared to what was indicated in the CSI report, then the CQI in the report is no longer valid and the eNB needs to recompute CQI to adjust the modulation and code rate of the transmission. This recomputation will then be an approximation since the exact effect on CQI when changing the precoder is not known to the eNB.

In addition, since one DM-RS is associated with each layer, the DM-RS overhead is in principle proportional to the number of MIMO layers. For CRS-based demodulation on the other hand, the number of CRSs is the same as the number of base station antennas (or virtual antennas in case of antenna virtualization); hence four CRSs are needed for a base station with four antennas even if only a single layer is transmitted, which gives an overhead drawback for low-rank transmissions.

8.2 LTE HISTORY AND EVOLUTION

In this section, the evolution in 3GPP of LTE multi-antenna features from the first to current release will be described, and in Fig. 8.8, this evolution is summarized in an illustration. Refer also to Section 6.3 for a general discussion to multi-antenna principles and Section 8.3.4 for a release-by-release introduction of LTE enhancements.

The first three releases of a radio access technology generation are the most important and contain the basic functionality. Although features added in later releases are well motivated and useful, they are most often optionally supported by the UE, so they have a higher threshold to be implemented and deployed. Also, they are in some cases functionality-wise hampered by the backward compatibility constraints. This is unfortunate, since very good technology components can be added in a later release but never realized in implementation. A striking example is the TM 10 in LTE, which has superior performance compared to TM 9 and "older" TMs, but it is not used in real networks due to lack of UE implementation support. However, the key TM 10 concepts made a strong "come back" in NR as the first NR release is functionality-wise based on LTE TM 10.

Multi-antenna-related features have been extensively studied in 3GPP over the years and every release includes enhancements in this area. In 3GPP, a release takes roughly a year to complete and is organized into time-limited projects denoted *work items* (WI). These are used to enhance a specific technical area or a feature and can be carried out by one or multiple 3GPP working groups. Sometimes, benefits of a proposed enhancement are less known, and it is too uncertain to start a related WI. It is then desirable to first carry out an initial feasibility study, resulting in a technical report. This is undertaken in the context of a study item (SI) and may include commercial as well as technical considerations. If the result of the SI is positive, it is likely to be followed by a WI, leading to new or an update to an existing technical specification.

There have been three occasions of multi-antenna-related SI for LTE (in release 9, 11, and 13, respectively). For NR, an SI took place in release 14 that preceded the NR release 15 WI that carried out the NR standardization. In addition to technical enhancements, there have been two SI on spatial channel modeling in release 12 and 14 that have relevance for AAS since they extended

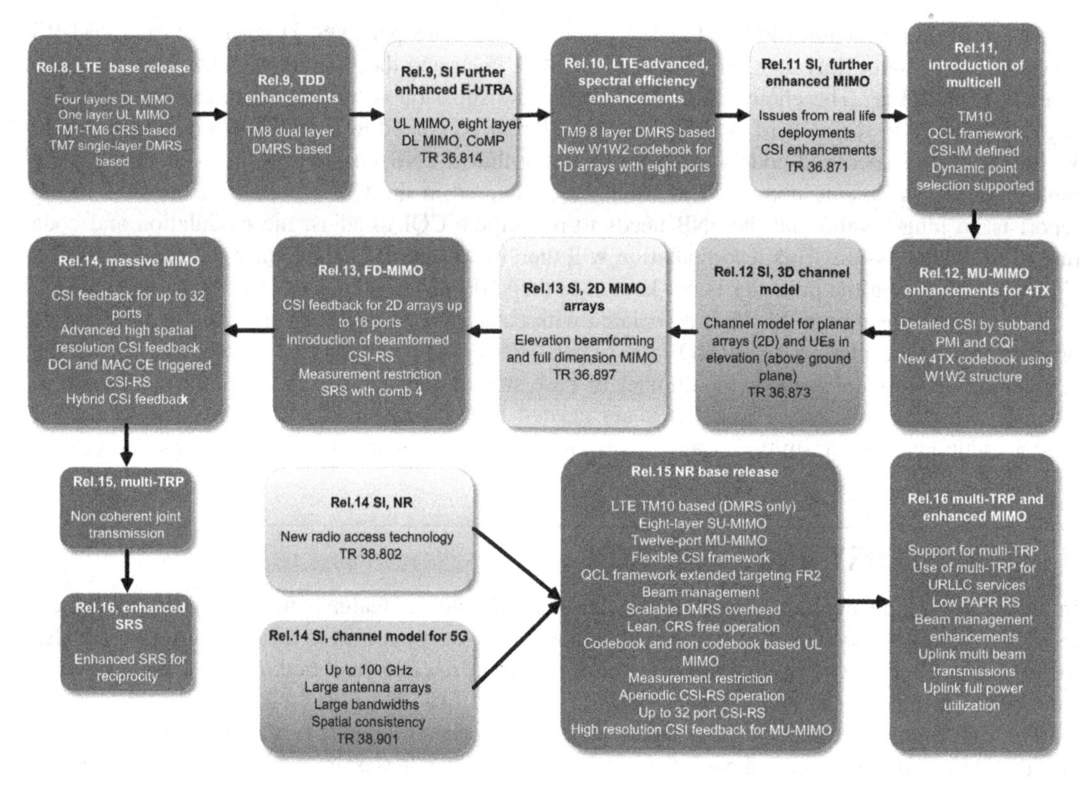

FIGURE 8.8

Overview of multi-antenna evolution in 3GPP, release by release for LTE and toward and including the introduction of NR. Blue box indicates an LTE work item, green box an NR work item, yellow box are channel model study items (SI), and peach-colored boxes represent LTE/NR SIs.

existing channel model to a three-dimensional (3D) channel model and further extended to the 5G channel models, respectively. See Section 3.6 for more in-depth discussion of channel modeling.

Release 8 of LTE specification was designed based on the requirements to support the operation of a single carrier component in a frequency band assigned for either FDD or TDD. The first release provided frequency flexibility supporting carrier bandwidths between 1.4 and 20 MHz multi-antenna transmission in DL was an integral part of release 8, supporting spatial multiplexing of up to four layers in the DL to the same UE, that is, single-user MIMO (SU-MIMO), and some basic multiple user MIMO (MU-MIMO) functionality. For the UL, release 8 did not contain any SU-MIMO transmission feature, although UL MU-MIMO is possible from release 8. Even in this first release, there was a large degree of configurability and hence a toolbox of features with seven different DL TMs, targeting different deployments and operating points.

LTE multi-antenna operation was then evolved during the following releases adding support for up to eight-layer reception in DL to a single-user, four-layer transmission in UL, support for measurements of eNB antenna arrays with up to 32 antenna ports (where antenna ports are defined in

Section 8.3.1) and in the later releases support for multipoint transmission of data using either *dynamic point selection* (DPS) or *noncoherent joint transmission* (NC-JT) which are flavors of coordinated multipoint transmission (see Section 6.6).

In release 10, IMT-Advanced requirements were defined by ITU, having a key feature of 1 GBit/s for low-mobility UEs, which then led 3GPP to conclude 3GPP requirements [6] for release 10 on a high-peak spectral efficiency of 30 bps/Hz for DL and 15 bps/Hz for UL.

The first LTE release did not fulfill the requirements for IMT-Advanced and enhancements of spectral efficiency was necessary. Therefore, the scope of release 10 included introduction of up to eight layer SU-MIMO in DL and four layers in UL. This release was preceded with a release 9 SI on physical layer enhancements in which multi-antenna aspects were included [7].

In release 11, yet another SI [8] concluded that MIMO operation could be further enhanced considering realistic deployment aspects such as noncolocated antennas at the network side (see Section 6.6). It was also observed that proposed methods to enhance the CSI feedback gave DL performance benefits at the expense of more feedback overhead. These findings were then explored in the WI of the following releases.

Prior to release 13, the CSI feedback framework supported only one-dimensional (1D) transmission steerability (i.e., in either vertical or in horizontal domain depending on how the antenna array is oriented relative to the horizon). In the scoping of release 13 enhancements of LTE, it was observed that the popularity of smartphone applications led to dramatic increase of wireless data traffic and MIMO was identified as one of the key technologies to address the increased capacity demands.

It was observed that to consider only the horizontal dimension to achieve the MIMO benefits is insufficient, especially in dense urban areas where most of the traffic demands arise, and where traffic distribution is both in vertical and horizontal dimensions. Therefore it was in release 13 decided to introduce standard support for CSI feedback for a two-dimensional (2D) planar antenna array to also exploit the vertical dimension. In parallel to the 3GPP working group (WG1), WG4 started an AAS SI in release 11, which turned into a WI in release 12. This work continued in release 13 where the first release of AAS specification was finalized (see Section 11.4). For details on how and when to utilize one or two dimensions, see the network performance analysis in Chapters 12 and 13.

At the time of writing, 3GPP has developed a large toolbox of features facilitating efficient multi-antenna operation for LTE. In Fig. 8.9, the 3GPP release toolbox for DL MIMO transmission for LTE is schematically described, where in a certain deployment and for a certain UE, the operator needs to choose which features should be enabled and which should remain disabled. A feature can be enabled only if the UE has reported capability to support the feature and if the network has implemented the feature. See a discussion on mandatory versus optional features in Section 10.1.3.

In Fig. 8.9, the 3GPP LTE feature tools have been grouped into which area of multi-antenna operation they mainly enhance. Note that the groups show enhancements over the baseline, for example, MU-MIMO can be used from release 8 with TM 7 and 8, while dedicated MU-MIMO enhancements were introduced in release 12 and 14.

In the following section, the tools specified in the LTE evolution from release 8 to release 15 are described in more detail. How and why these tools were developed and added to the 3GPP specifications will also be discussed.

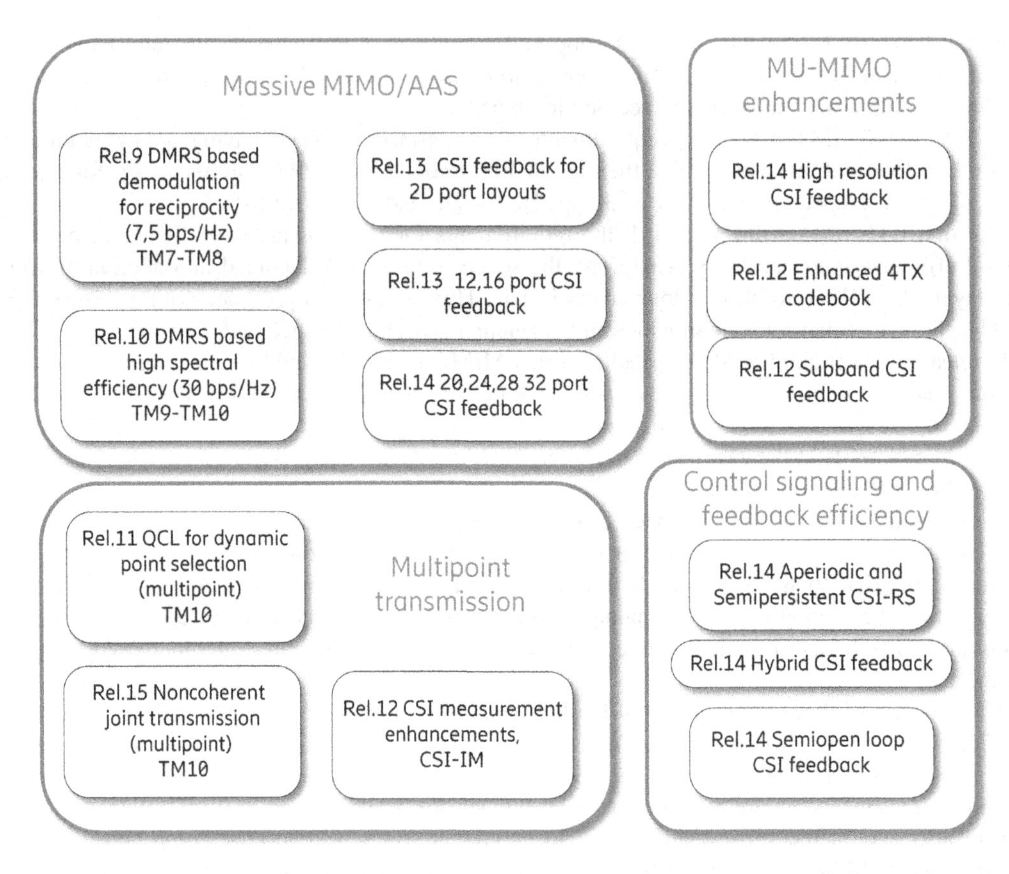

FIGURE 8.9

Overview of the multi-antenna toolbox provided by 3GPP standardization for LTE release 8–release 15. Note that since TM 9 and TM 10 support up to eight-layer SU-MIMO, the spectral efficiency per UE is as high as 30 bps/Hz, while TM 8 can only provide 7.5 bps/Hz per UE as there are at most two layers per UE. If MU-MIMO is used, then the sum spectral efficiency can be higher than the values indicated here.

8.3 LTE PHYSICAL LAYER SPECIFICATIONS FOR AAS

This section describes in more detail the specified functionality for the LTE physical layer where focus has been given to parts that relate to AAS.

8.3.1 ANTENNA PORTS, REFERENCE SIGNALS, AND QUASI COLOCATION

The channel that the UE estimates using transmitted RSs includes not only the propagation channel, but also the characteristics of the radio hardware such as the antenna elements, feeder cables, and power amplifiers (PA), on both the transmitter and the receiver side, plus the possible use of antenna virtualization as discussed in Section 8.1.1.

Hence, from the UE radio channel estimation perspective and in 3GPP physical layer specifications (as defined by 3GPP RAN1), it is undesirable and unnecessary to define "physical antenna," since what matters for demodulation is to have an accurate enough representation of the channel for which a channel (e.g., PDSCH) is transmitted over, the UE does not care how an RS has been mapped to one or more physical antennas. Specifying physical antennas would restrict the flexibility in applying the specifications to the real-world base station designs as such antennas need to consider many factors (such as multi-band arrays, combined LTE and 3G operation, etc.).

Therefore 3GPP has in the physical layer specifications instead introduced the notion of *antenna ports* as a logical concept. The 3GPP definition is [9]:

> An antenna port is defined such that the channel over which a symbol on the antenna port is conveyed can be inferred from the channel over which another symbol on the same antenna port is conveyed.

Such definition allows for assuming and specifying relations between different channels and RSs without explicitly mentioning channel estimation in specifications. The term antenna port can therefore be an identifier for a channel. A modulated symbol of one or more physical channels is transmitted "on" a certain antenna port, which means the UE can use the estimated channel using the RS symbols associated with that antenna port to demodulate the physical channel symbols mapped on the same antenna port.

Consequently, when a UE is said to measure the channel from an eNB antenna port, it measures the effective radio channel at the receiver using the transmitted RS(s) associated with that antenna port. Refer to Fig. 8.7 where the different CSI-RS mapped to the input of the antenna virtualization would in a 3GPP specification correspond to different CSI-RS antenna port numbers.

Some antenna ports are in LTE specifications numbered as in Table 8.3. In addition, there are RSs in LTE for special purposes such as port 4 for demodulating PMCH, port 6 used for positioning measurements, and ports 107–110 for the enhanced control channel [enhanced PDCCH (EPDCCH)] DM-RS in release 11.

Finally, note that for radio performance requirements and conformance testing (see Chapter 10), the term antenna reference point is used, which should not be confused with the logical concept of antenna port defined here.

Table 8.3 Region of Validity for Some LTE Antenna Ports in Downlink.

Port Number	Downlink RS	Validity Bandwidth	Validity Time Duration
0–3	CRS	Whole system bandwidth	Across all time
7–14	PDSCH DM-RS	One PRG (1, 2, or 3 RB depending on system bandwidth)	One subframe
15–46	CSI-RS	Whole system bandwidth	Across all time *without* measurement restriction enabled One subframe *with* measurement restriction enabled

8.3.1.1 Validity region of an antenna port

Another aspect of antenna ports that often leads to confusion is that, for example, port 7 can be used twice in two adjacent subframes but they are not actually the same antenna port since a port is only valid in a certain region (time and frequency). Hence, in this example, the UE is not allowed to use measurements of port 7 in one subframe to demodulate a PDSCH in the next subframe.

This is because antenna ports have a specified region of validity. If an antenna port with a given port number is used in two different validity regions, then the two estimated channels may be completely different since the eNB may have changed the precoder $\boldsymbol{W}$ between the two regions (e.g., subframes), and the UE shall therefore not interpolate these estimates across the regions. For the port 7 example, the validity region in time is one subframe, so cross-subframe interpolation is not allowed even though the number of the port is 7 in both subframes.

In addition, there is another source of confusion when it comes to how antenna ports are used in 3GPP specifications. A UE can be configured to measure on multiple CSI-RS resources simultaneously, where each such resource, for example, have four ports. The use of multiple CSI-RS resources is for example used in the CSI feedback mode of Class B, where each resource is transmitted in a different beam. The four antenna ports in the example will be numbered as 15, 16, 17, and 18 in all these CSI-RS resources, but this does not mean that port 15 in the first resource is the same as port 15 in the second CSI-RS resource and so on. Antenna ports belonging to different CSI-RS resources are completely independent even if the antenna port number is the same.

For PDSCH DM-RS, the *precoder resource group* (PRG), defined as 1, 2, or 3 RB wide in frequency, is introduced to define the validity region of the port in the frequency direction. The purpose is that the eNB may use a different precoder in each PRG to optimize the precoding gains (also known as sub-band precoding) and hence the effective channel is different for each PRG.

An overview of the region of validity for some DL antenna ports is given in Table 8.3. Note that the release 13 feature *measurement restriction* provides a means for eNB to control the validity region in time for CSI-RS, Measurement restriction configurations are discussed more in Section 8.3.4. Note also that the validity region of CRS is unlimited, the UE can average the estimates across time and frequency as it desires, leading to very good channel estimation performance on CRS.

8.3.1.2 Quasi co-location

The UE receiver algorithm typically performs some channel analysis prior to channel estimation in order to tune the channel estimator filters and to set the correct gain of the receiver front end to utilize the full dynamic range of the receiver. For example, it is useful for a channel estimation algorithm to know the delay spread of the channel (see Section 3.6.11), as this a priori information generally improves the estimation performance.

Delay spread can be measured using a wide bandwidth CSI-RS and used to demodulate a PDSCH even if it is only scheduled to be transmitted in a single physical resource block (PRB). Using the PDSCH DM-RS only, for delay spread estimation purpose would lead to a poor estimate in this example since the DM-RS is only a single RB wide. Hence, there is a need to specify an association between different RSs, such as the wide bandwidth CSI-RS and the variable bandwidth PDSCH DM-RS, to make clear which channel properties estimated from one RS may be exploited to improve the estimate of a channel from another RS.

The 3GPP LTE specification therefore introduces the concept of *quasi co-location* (QCL) between antenna ports of the DL transmissions, represented by five large-scale channel parameters: delay spread, Doppler spread, Doppler shift, average gain, and average delay. Hence, loosely speaking, the channel of two antenna ports has, if they are QCL, the same channel *property/properties*, although the *actual channel realization* is not the same. This strong relation of properties is useful information for the receiver in the UE.

When two antenna ports are QCL with respect to one or more of these five parameters, then the UE may estimate those large-scale parameters by a measurement of the first antenna port. When the UE, possibly at a later point in time, desires to perform channel estimation of the channel using the second port, then it can base the selection of the channel estimation filter (e.g., using the estimated delay spread) on these estimated large-scale parameters. Hence, the UE can prepare the filter in advance to receiving the second port since it already knows these large-scale parameters (i.e., the channel properties) of the channel for the second port.

Note also that the QCL relation works in both directions so the second and first ports could in principle be exchanged in the previous paragraph. However, it is common that one port is more suitable than the other to determine large-scale channel properties (see Section 3.5), for example, one port may have a higher time and frequency density or have a larger bandwidth and may be periodically transmitted. So, in practical use cases, QCL relations are one-directional and sometimes the term source and target RSs are used to denote the first and second RS, particularly when they are of different type (e.g., CSI-RS (source) and DM-RS (target), respectively).

As an example of a QCL relation, the LTE specification describes that the UE may assume that CRS ports 0–3 are mutually QCL with respect to delay spread, Doppler spread, Doppler shift, average gain, and average delay. For example, delay spread of CRS port 0 and 1 is assumed to be the same.

The consequence of such specified QCL relation is that the specifications do not guarantee the support of a network deployment where CRS port 0 and port 1 are transmitted from two largely separated physical locations. This is because that such deployment may imply that, for example, average gain and average delay of the two-channel measurements in the UE receiver are far from being the same; hence the specified QCL relation of at least "average gain" between CRS port 0 and 1 would be violated.

Therefore the network has incentives to ensure that two ports in this example have the same value of the large-scale parameters, which can be ensured by having them transmitted physically close to each other (i.e., two neighboring antenna elements) or at least the channels measured from these two ports at the receiver side must behave as they are transmitted from antennas physically close to each other in order for performance to be guaranteed. This is also the origin of the term "quasi co-located," which comes from the Latin word *quasi*, meaning "as it were."

8.3.1.3 Downlink cell-specific reference signals

The CRSs are fundamental in LTE from the first release as to provide a stable, dense, and always present DL RS which the UE uses for multiple and widely different purposes. These purposes are fine time and frequency synchronization, estimation of large-scale channel parameters, measurements for CSI feedback, as well as serving as demodulation reference for receiving the DL physical channels PBCH, PDSCH, PDCCH, PCFICH, and PHICH.

Since the CRS is used for all different purposes it is designed to be very dense in time and frequency to perform well also in the most demanding case of data channel demodulation at high UE speed where the channel coherence in time (see Chapter 3) can be very short. A CRS port of the CRS occupies every sixth subcarrier in frequency (i.e., 90 kHz distance between frequency samples) and in total two or four OFDM symbols within a subframe (see Fig. 8.10).

The CRS is not configured by dedicated signaling from the network to the UE, it is instead implicitly configured. This allows UEs even in IDLE mode to determine and use the CRS of any cell. A UE can therefore read the information in the PBCH to obtain the system information of any cell without any action by the eNB.

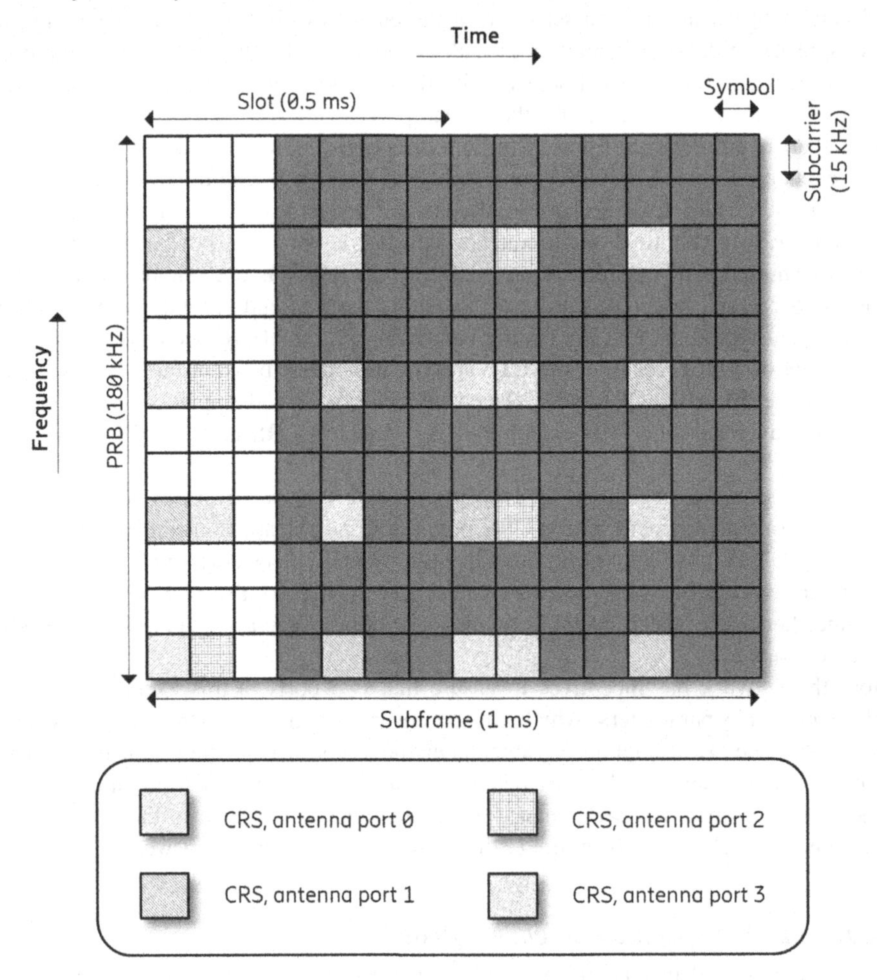

FIGURE 8.10

LTE CRS mapping to RE in subframe and RB assuming a control region of three symbols and a data region of 11 symbols. The figure shows four CRS ports in a normal subframe and with a subcarrier offset zero. Alternative CRS mappings is obtained by shifting the CRS pattern one or two subcarriers

The network configures the CRS transmission by the number of CRS ports (which can be a single port, two or four ports) and a subcarrier offset of one to three or one to six subcarriers. The available offsets depend on the number of configured CRS ports and which offset is used is linked to the physical cell ID (PCI) which the UE obtains from the PSS and SSS. The QPSK modulated symbol sequence used for CRS is obtained from a pseudo-noise (PN)-generated sequence for which initialization also depends on the PCI. In 3GPP specifications, the PN sequence is obtained by a specified pseudo-random sequence generator.

Hence, after the UE has synchronized to a cell by detecting the PSS and SSS, the PCI is obtained and also the CRS position and sequence become known to the UE. However, the number of ports the CRS is configured with is signaled in the PBCH message. To demodulate the PBCH, CRS is used, which leads to a chicken and egg problem as the number of CRS ports is needed to know the REs where the PBCH is mapped in the time–frequency grid. To resolve this paradox, the UE always assumes four ports for the PBCH mapping to REs, even though in reality fewer CRS ports may be transmitted by eNB.

CRS ports 2 and 3 are typically configured to be transmitted in case the eNB has capability to transmit (and UE to receive) more than two layers of PDSCH using a CRS-based transmission scheme. These ports have a lower time density compared to port 0 and 1 with only two OFDM symbols per subframe (see Fig. 8.10). The motivation was that higher-order MIMO transmission is only used by UEs with lower UE speed and thus two-channel time samples per subframe are enough, thereby saving some CRS overhead. Ports 2 and 3 are not used in mobility and reference signal received power measurements even if they are present, only ports 0 and 1 are used for this functionality.

Since the CRS is present in all DL subframes and covers the entire served cell, the UE can continuously track the channel, leading to excellent channel estimation performance and robustness for the CRS-based transmission schemes.

However, there is also a downside to this "always present" CRS. As LTE networks evolved in the subsequent LTE releases, the paradigm shifts toward using UE–specific RS for demodulation and dedicated CSI-RS for measurements instead of the "all-purpose" CRS took place. As there are legacy terminals connecting to the cell that use CRS and do not support the DM-RS-based TMs introduced in later releases of LTE, and since also IDLE UEs use CRS when measuring and connecting to a cell, the CRS must be always transmitted by the eNB in every subframe.

This "always on" CRS transmission leads to energy consumption by the base station even where there is currently no served UE. In addition, the CRS is an unnecessary overhead for UEs configured with the DM-RS and CSI-RS-based transmission schemes, since for such a UE, the CRS is only used for fine synchronization and for this purpose a dense RS in every subframe is unnecessary.

Moreover, the CRS from neighboring cells creates interference to PDSCH and PDCCH if the CRS has a subcarrier offset compared to the CRS of the serving cell, even if the interfering cell is not transmitting any data. On the other hand, if the same subcarrier shift is used, then there is another drawback related to CSI reporting. Since it is common for the UE to measure inter-cell interference as the residual power on the received serving cell CRS it means that the UE will measure high inter-cell interference even if the neighboring cell is not transmitting data, which gives incorrect CQI. These problems were one reason why dedicated interference measurements resources were introduced in TM 10 (see Section 8.3.4.3).

As mentioned earlier, the CRS ports 0–3 of a serving cell are QCL with respect to *delay spread, Doppler spread, Doppler shift, average gain, and average delay*, implying that these ports are all transmitted from the same eNB and with the same radiation pattern (in order to fulfill the QCL assumption on same average gain).

8.3.1.4 Downlink channel state information reference signals

The CSI-RS is introduced as a dedicated RS used for CSI measurement purposes and replaces the CRS for measurement purpose, at least for some TMs. When evolving LTE network deployments for heterogeneous networks and when introducing beamforming with advanced antenna arrays it was observed that the CRS-based measurement signals were too inflexible and restrictive. The LTE-Advanced requirements of a peak spectral efficiency of 30 bps/Hz implied a requirement of eight-layer transmission and extending the CRS-based transmission for this purpose (i.e., increasing the number of CRS ports to eight) was inefficient.

In addition, CRSs are shared by all UEs in the cell and hence, a CRS transmission cannot be channel-dependent, that is, it cannot be beamformed to a specific UE. It makes it difficult to fully utilize the antenna gains of AAS since the measurement RS will not benefit from beamforming gain. Moreover, to fully benefit from reciprocity-based operation, channel-dependent precoding/beamforming toward a specific UE was seen necessary *both* for the measurement signals and for the PDSCH transmission. When using TM 8 with reciprocity, the full benefit is not achieved since measurements are made on the CRS.

Therefore a UE specifically configured and *dedicated measurement signal*, the CSI-RS, was introduced in release 10 with 1, 2, 4, or 8 orthogonal antenna ports. In release 11, there was a need to be able to configure multiple CSI-RS to a UE simultaneously. This enables the UE to perform measurements and reporting of CSI for more than one transmission point for example. Therefore the concept of a *CSI-RS resource* was defined and a UE in TM 10 can have multiple such resources configured simultaneously and perform independent measurements on each of them.

Furthermore, in release 13 the definition of a CSI-RS resource was extended to have a larger number of antenna ports. The need for more ports was motivated by the introduction of 2D antenna port layouts. The 2D layouts allow precoding or beamforming in a given direction defined by both a vertical and a horizontal angle.

Note that the 2D antenna port layout (i.e., $N_1 \times N_2$ ports in first and second dimensions) is not part of the CSI-RS resource configuration, but instead it is part of the MIMO precoder codebook configuration for CSI feedback. The number of ports N_1 and N_2 in each dimension is configured by the network for the UE to use the correct codebook.

Although it is not specified that how CSI-RS ports are mapped to antennas (cf. the transparent antenna virtualization block in Fig. 8.7), the ports are numbered and the codebook is designed assuming a convention of mapping ports in a first dimension first, then a second dimension, and lastly across polarization. See Fig. 8.11 for the used convention.

The CSI-RS is UE specifically configured, that is, per dedicated signaling to each UE. However, the same CSI-RS resource may be configured to many or all UEs served by the eNB to save CSI-RS overhead, that is, they all measure on the same CSI-RS resource. In addition, the sequence used for CSI-RS is a PN sequence and the initialization is configured by dedicated radio resource control (RRC) signaling, hence it is not necessarily tied to the PCI (it is however possible to use the PCI as the initialization, if desired).

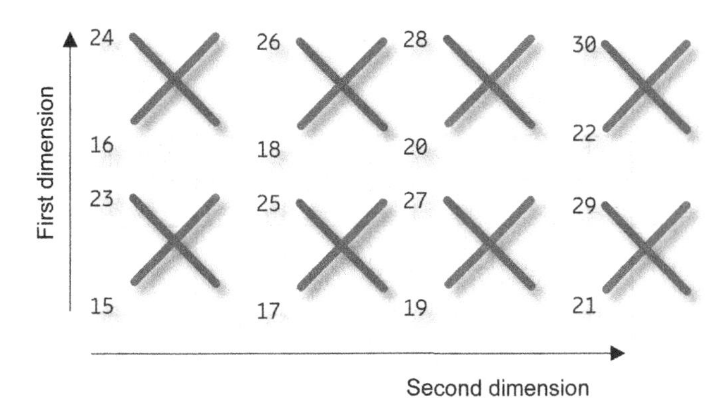

FIGURE 8.11

3GPP convention of CSI-RS port numbering and mapping to antennas for the 2 × 4 port layout. Numbering is across first dimension first (for one polarization) followed by second spatial dimension and lastly across polarizations. The UE is placed at the position of the reader, where port 15 is in the lower left corner. This example shows an antenna port layout with 2 and 4 ports in first and second dimensions, respectively, a total of 16 ports. Note that CSI-RS port numbering for a CSI-RS in LTE always begins with port 15.

Since CSI-RS is used only for measurements and not for demodulation or synchronization, the density is much lower than CRS; only 1 RE/port/PRB pair is used compared to 8 RE/port/PRB pair for CRS port 0. This density was seen enough during standardization since the CSI report is anyway quantized and there was not more performance benefit of having a larger density.

Instead, in release 14, even further reduced density for CSI-RS was introduced to reduce overhead allowing CSI-RS to be configured with 1/2 and 1/3 RE/port/PRB pair, but only for CSI-RS resources with 20 ports or more.

Fig. 8.12 shows a PRB pair and the total set of 40 REs that are available for a CSI-RS resource to be configured. As the first 1, 2, or 3 OFDM symbols in the subframe may carry control signaling, they are excluded from REs available for CSI-RS configuration. Also, symbols that may contain CRS are avoided. A CSI-RS resource is always allocated as symbol pairs and are thus located in OFDM symbol {6,7}, {10,11}, or {13,14}, respectively.

For example, if a resource with one or two ports is configured, only 2 REs out of the 40 available is used while if a CSI-RS resource of larger number of ports is required, then more REs will be used for the resource.

The CSI-RS, when configured, spans the whole system bandwidth and as can be seen in Table 8.3, the UE can utilize RS measurements from the whole bandwidth to estimate the channel of every subcarrier.

In specifications, the CSI-RS used to measure the channel is known as the *non-zero power* (NZP) CSI-RS. It is also possible to configure a CSI-RS that occupies the configured RE but the eNB does not transmit any energy in these RE, that is, they are empty. These REs are known as a *zero power* (ZP) CSI-RS resource, which has several use cases. For example, the NZP CSI-RS transmitted from cell A can be overlapped with a ZP CSI-RS from cell B. When the UE is measuring the channel using NZP CSI-RS, nothing is transmitted from cell B on these Res; hence there is

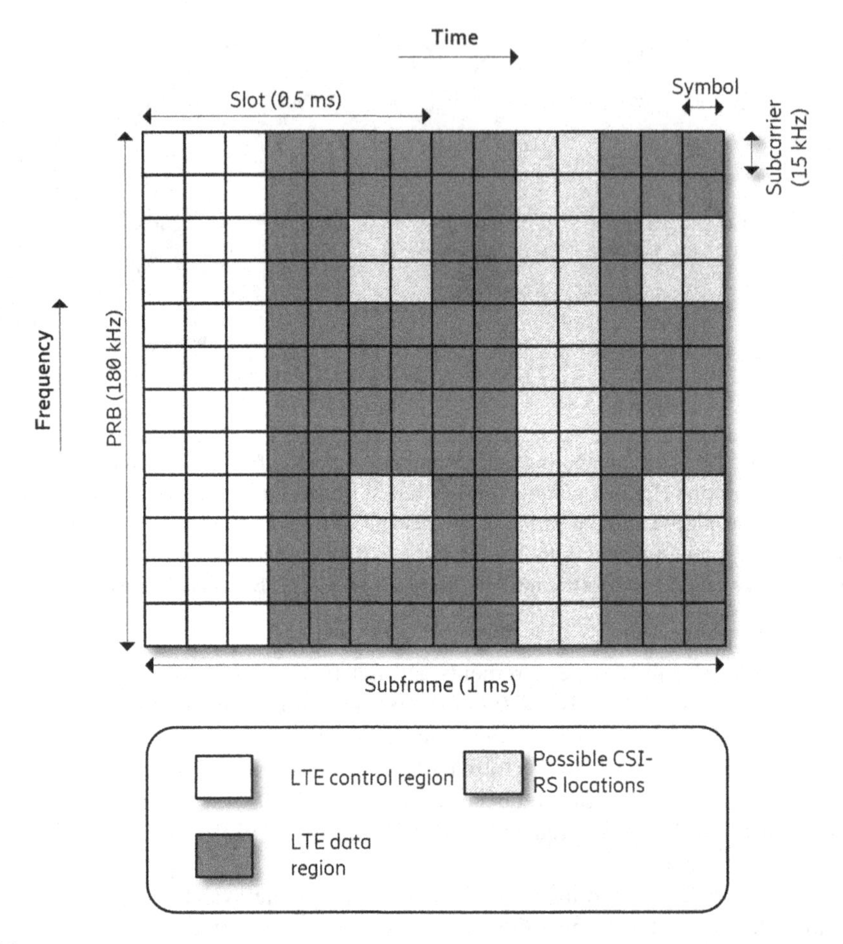

FIGURE 8.12

Possible REs where the LTE CSI-RS can be mapped in a subframe and one RB. For example, a 2 port CSI-RS use two of these 40 possible REs.When configured and present in a subframe, it is present in all RB in the system bandwidth.

no interference from cell B (provided that the propagation delay from these cells is comparable so the resources overlap). This improves the measurement performance of the channel from cell A. Hence, the ZP CSI-RS can be used *to protect* a configured NZP CSI-RS transmission in an adjacent cell.

In addition, the ZP CSI-RS is also used to define an *interference measurement resource* (CSI-IM). This provides the tool for making a "resource element hole" in the PDSCH transmission of the serving cell. The UE can measure the received power in this "hole" and thus it measures the level of interference from ongoing transmissions in neighbor cells without measuring the power received from the own cell (provided that a "hole" is not configured also for the interfering cell, which is another possibility, It all depends on which *transmission hypothesis* the network wants the UE to feed back).

Finally, a note on QCL for NZP CSI-RS. The baseline is that CRS and CSI-RS are QCL with respect to Doppler shift, Doppler spread, average delay, and delay spread; this is denoted as *QCL Type A* in LTE. However, in TM 10 (introduced in Section 8.3.4), the UE can alternatively be configured so the CRS and CSI-RS ports are only QCL with respect to the large-scale parameters Doppler spread and Doppler shift (see Section 3.5). This is known as *QCL Type B* in LTE and is useful if *dynamic point selection* (DPS) feature is deployed, in which case CRS is transmitted from either or both transmission points and one CSI-RS is transmitted from each transmission point respectively.

By configuring QCL Type B, it implies that the UE cannot estimate, for example, average gain from the CRS when receiving CSI-RS of the serving cell as CRS and CSI-RS are not QCL with respect to average gain. From the deployment aspect, this means that the CSI-RS may be transmitted from a non-colocated position compared to the position where CRS is transmitted, for example, a different eNB than the one transmitting CRS. It enables the possibility to acquire a CSI report from the UE assuming PDSCH transmission from a different eNB than the serving eNB.

8.3.1.5 Demodulation reference signals for physical downlink control channel

The PDSCH is the physical channel carrying the data payload to the UE. The DL RS for demodulation (DM-RS) is a reference signal that is associated to PDSCH. It is only present in the subframe and RB where the PDSCH is transmitted and it is only used by the UE that receives the PDSCH. Hence, it is a UE−specific RS and each PDSCH layer is associated with a DM-RS port as discussed earlier in Section 8.3.

The PDSCH DM-RS existed already in LTE release 8 but only for a single-layer PDSCH transmission (port 5). In release 9, it was extended to two ports (7 and 8) and later in release 10 a further extension of up to 8 ports (port 7−15) was specified, to support the IMT-Advanced requirement of eight-layer SU-MIMO.

The DM-RS overhead is small, when the number of transmitted PDSCH layers is one or two, and the DM-RS is self-contained within the scheduled PDSCH resources. Hence there is no transmission outside the PDSCH RBs that could cause interference or prevent scheduling of transmissions to other UEs, which is a principle that also was adopted for NR.

In Fig. 8.13, the DM-RS ports mapped to RE in a PRB pair is illustrated (for FDD or for a DL subframe in TDD). As can be seen, a DM-RS port is mapped to three subcarriers in the last two symbols of each slot. Four ports are mapped to 12 REs, hence the density is three times higher than for CSI-RS. Note that the REs used for DM-RS do not collide with CRS or CSI-RS which was a design requirement since CRS and sometimes CSI-RS will be transmitted in the same subframe as the PDSCH DM-RS.

If the PDSCH is scheduled on one or two layers, then port 7 or port 7 and 9 is used and the DM-RS uses 12 REs, while if more than two-layer PDSCH is scheduled, then 24 REs are used for DM-RS since more than two ports are needed for the demodulation. For rank 5 and higher, the overhead remains at 24 REs used for DM-RS and additional DM-RS ports are then obtained by the *code division multiplexing* (CDM) of DM-RS ports. The CDM is implemented by an *orthogonal cover code* (OCC) which in this case is applied across all four DM-RS RE in a subcarrier to effectively obtain up to four DM-RS ports per 12 RE. Since the OCC code length is across RE in

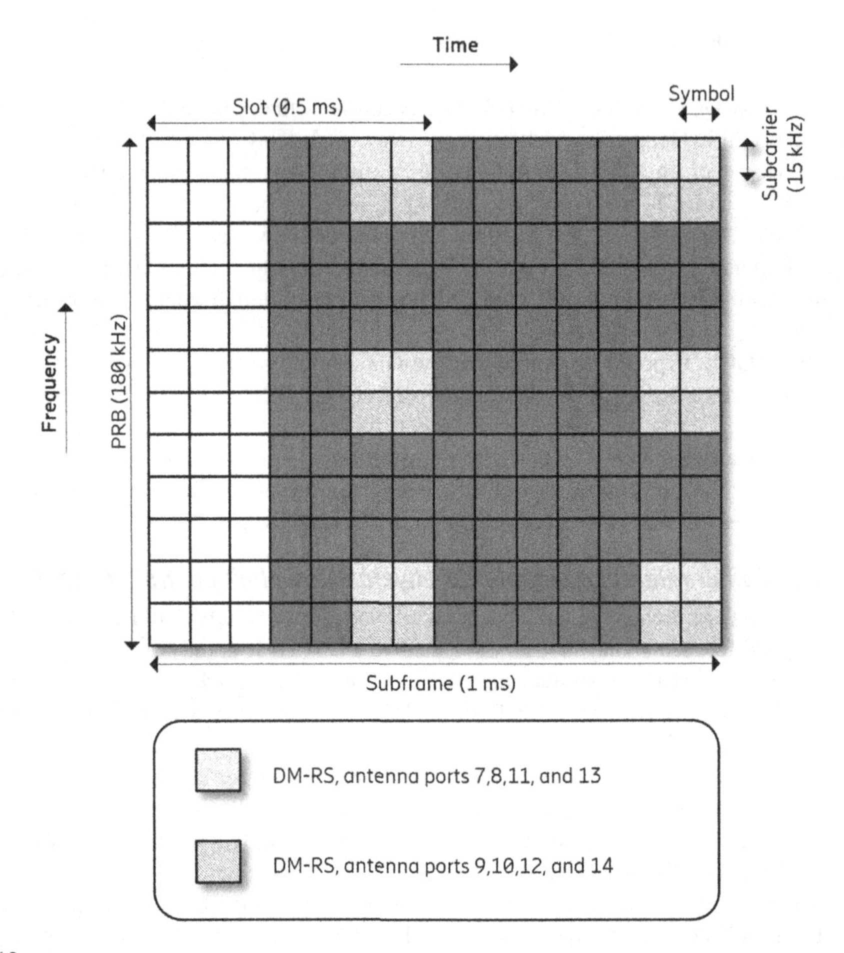

FIGURE 8.13

LTE DM-RS pattern for one subframe and one RB. Note that the CRS is not shown in this figure but will consume additional overhead as it is present in all subframes.

OFDM symbols of both slots, it means that such high-rank transmission is more sensitive to time variations in the channel as the channel should ideally be constant across all REs occupied by the OCC.

The sequence used for PDSCH DM-RS is a PN sequence and two sequence initialization seeds are configured by dedicated RRC signaling per UE. When scheduling the UE, the scheduling control message contains a $n_{SCID} = \{0,1\}$ parameter that selects either of the two PN sequences to be used to generate DM-RS of the scheduled PDSCH. If two DM-RS use different n_{SCID}, then these DM-RS are nonorthogonal.

When scheduling multi-user (MU) MIMO, that is, two or more UEs simultaneously in the same physical resources, then it is beneficial for channel estimation performance if orthogonal DM-RS ports are used. For example, one UE is scheduled on one layer with port 7 and the other UE on one layer with port 8. For both UEs, the same n_{SCID} value is indicated in the scheduling message to ensure orthogonality between DM-RS ports.

In LTE release 10, the use of MU-MIMO is restricted to two UEs if orthogonal ports are used, since if a single PDSCH layer is scheduled for a UE, then it is only possible to indicate port 7 or port 8. Hence, it is not possible to schedule, for example, three UEs with orthogonal ports since the standard does not support signaling of port 9 as a single-layer PDSCH to a third UE.

There is however a possibility to use nonorthogonal DM-RS ports and schedule additional UEs in MU-MIMO since the n_{SCID} parameter can be utilized. In this case, one UE uses port 7 with $n_{SCID} = 0$, while another UE uses port 7 with $n_{SCID} = 1$. Using nonorthogonal ports may have poor performance unless the multi-antenna precoder ensures that the spatial suppression of cross-interference between the transmission to the two UEs is good enough.

In LTE release 12, MU-MIMO enhancements were introduced and the signaling framework for DM-RS antenna ports in the scheduling DCI for TM 9 and 10 was extended so that up to four UEs could be scheduled in MU-MIMO also with orthogonal DM-RS ports. There were also possibilities to schedule, for example, one UE with two layers and two UEs with single layer, and still maintain orthogonal DM-RS ports between these UEs. This should be compared to TM 8 of release 9, where nonorthogonal ports must be used between UEs for the same example.

As a final note on DL DM-RS; in release 11, the *enhanced PDCCH* (EPDCCH) was introduced to be able to benefit from UE–specific beamforming also for the control channel in addition to the already supported feature for the data channel. Hence, EPDCCH uses DM-RS and the same RE mapping as PDSCH DM-RS, but to avoid confusion, these ports were numbered as 107–110 instead of ports 7–10. The reason for having four ports for EPDCCH is not to enable spatial multiplexing of control channel transmission; instead they are used to support up to four EPDCCH transmissions in one PRB pair with individual and orthogonal ports so that they can be transmitted to four different UEs simultaneously.

8.3.1.6 Demodulation reference signals for physical uplink shared channel

The DM-RS for PUSCH is similar in functionality as the PDSCH DM-RS introduced in release 10, where each layer has its own DM-RS port. However, the UL transmission is codebook based, where the eNB selects the precoding matrix the UE should use for the PUSCH transmission. Note that even if the UE has four physical antennas, if a single PUSCH layer is transmitted with a precoder across the four antennas, then a single DM-RS port is transmitted by the UE.

Since PUSCH in LTE is based on DFT-spread OFDM, which generates a waveform with low *peak-to-average power ratio* (PAPR), the PUSCH DM-RS is designed with a low PAPR property as well.

The set of orthogonal DM-RS ports is obtained by a time-domain cyclic shift (CS) of the DM-RS sequence and by using an OCC across the two OFDM symbols carrying the DM-RS in the subframe. Due to the properties of the DM-RS sequence, using carefully selected CS, orthogonality between the DM-RS ports is ensured provided that the different RSs are time-aligned at the receiver and that the channel is sufficiently nonfrequency selective.

In case of a very frequency selective fading channel (large channel delay spread), ports separated by CS will not be orthogonal at the receiver and there will thus be cross-port interference. As an alternative, port separation using the time-domain OCC can be used instead. Each scheduling

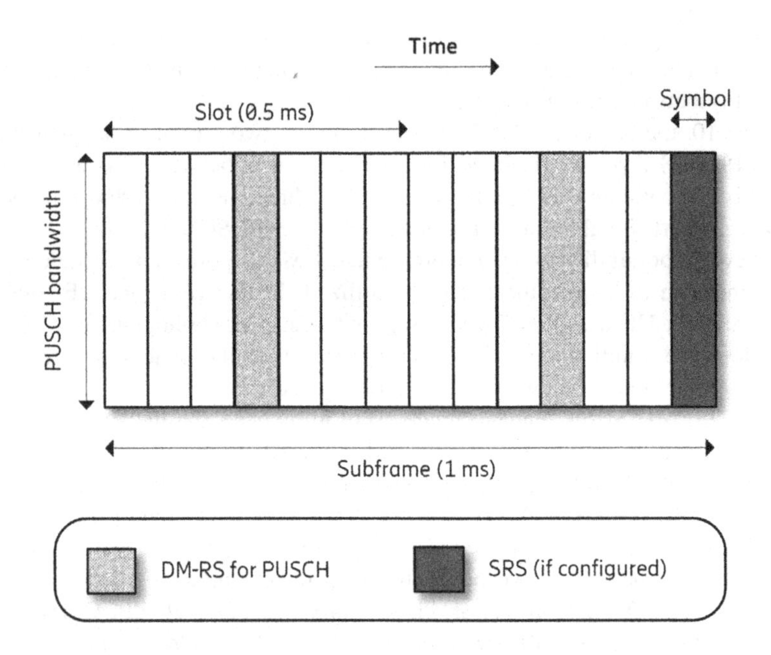

FIGURE 8.14

DM-RS and SRS positions in the subframe for the LTE uplink in release 8. In release 14, a comb-based mapping to subcarriers was introduced for DM-RS.

grant of the UL selects a CS and one out of the two available OCC codes, that is, a DM-RS port, for the PUSCH DM-RS.

The PUSCH DM-RS maps to the fourth and eleventh OFDM symbol in the subframe, and it maps to all subcarriers for the scheduled PUSCH (see Fig. 8.14).

In LTE release 14, the number of available DM-RS ports was increased further by introducing a comb-based DM-RS structure for PUSCH. This means that the RS of a DM-RS port is in this case mapped in an OFDM symbol to either odd or even subcarriers instead of every subcarrier. This theoretically doubles the number of PUSCH DM-RS ports, with the motivation of MU-MIMO with a larger number of simultaneously scheduled users in the UL. However, using a comb structure also increased the sensitivity to channel delay spread, like using CS, so the expected doubling of number of available ports is only valid for channels with very small delay spreads, for example, indoor deployments. To maintain the low PAPR property, the PUSCH and its associated comb-based DM-RS are not frequency multiplexed in the same OFDM symbol but are still time multiplexed as in Fig. 8.14.

8.3.1.7 Uplink sounding reference signals

A UL SRS transmission can be configured to assist in link adaptation for the UL (including determining a UL precoder) and for providing a UL channel estimate in reciprocity-based operation (see Section 6.4.2). The RS used for sounding the radio channel in the UL can be configured to be transmitted with a periodicity between 2 and 160 ms. The SRS is transmitted in the last symbol of the subframe in case of FDD (see Fig. 8.14). UL sounding is also used in reciprocity-based schemes, to provide the channel for the DL.

In LTE release 10, aperiodic SRS was introduced in addition to the periodic SRS, where a UL measurement could be triggered by a DCI on a need basis. In addition, for special subframes in TDD which has both a DL and a UL part, more than one SRS symbol can be configured in the subframe. This is because for TDD configurations with mostly DL subframes, the possibilities for UL transmission are few and there is a need for more SRS transmission opportunities.

When using SRS, there is a tradeoff between coverage and sounding bandwidth (see also Section 6.4.2.1). Since the transmit power of a UE is limited, typically to 23 dBm, one way to improve SRS coverage is to configure the UE to transmit the SRS over a smaller bandwidth. Given a constant transmit power, the *power spectral density* (in W/Hz) of the SRS transmission is then increased and thus SRS coverage improves. The drawback is that only a part of the bandwidth is sounded and then multiple such SRS transmissions hopping across different parts of the band are needed. The smallest SRS transmission bandwidth is 4 RBs. Since each of these multiple narrow SRS transmissions are in different subframes, the channel may have changed, i.e. faded between each transmission. When the whole bandwidth has been sounded, some parts may be outdated while other parts are more recent and still a valid representation of the channel.

The closer the UE is to the eNB, the wider SRS bandwidth can be configured as the path loss in the channel is lower. Ideally, the UE transmits SRS across the whole bandwidth in a single transmission.

The release 8 SRS transmission is mapped to either odd or even subcarriers (i.e., one of two available *combs*) to increase the SRS capacity. The SRS use the same type of low PAPR sequence as PUSCH DM-RS and different SRS resources are thus separated by a CS and the comb identifier. Separating using a comb simplifies the receiver compared to separating using CS as the different combs are strictly orthogonal.

In release 13, the support for receiving SRS from an even larger number of UEs simultaneously was introduced. Theoretically, three times more SRS resources were obtained per OFDM symbol. This extends the support for UL MU-MIMO or extends use of reciprocity-based operation. The subcarrier mapping configuration to include also every fourth subcarrier and more CS was introduced. Note however that even if the SRS capacity increases by comb 4 and with additional CS, using these enhancements increases the sensitivity to large channel delay spreads. A sparser sampling in frequency domain (i.e. increasing comb value) and/or using a larger number of CS with smaller shift separation between each CS may lead to that cross-interference between different SRS resources occur. Hence these enhancements typically are targeting small cell or indoor deployments where small delay spreads are more commonly encountered. In NR, the same SRS structure of using comb 2 and comb 4 and CS is used as in LTE release 13 (refer to Fig. 9.9).

UEs often have more receive branches (receive antennas and associated baseband processing) than transmit branches since transmitters require power-consuming PA. See, for example Fig. 8.15, where the UE has a single PA for transmission but two antennas for reception.

To support reciprocity-based transmission in the DL, there is a need to sound the channel in the UL from all (receive) antennas at the UE and this becomes an issue in the typical case shown in Fig. 8.15 as the second antenna cannot transmit anything. In this case, the eNB is only able to get the partial UL channel by using the SRS and this degrades the performance of reciprocity-based operation and thus the DL PDSCH transmissions.

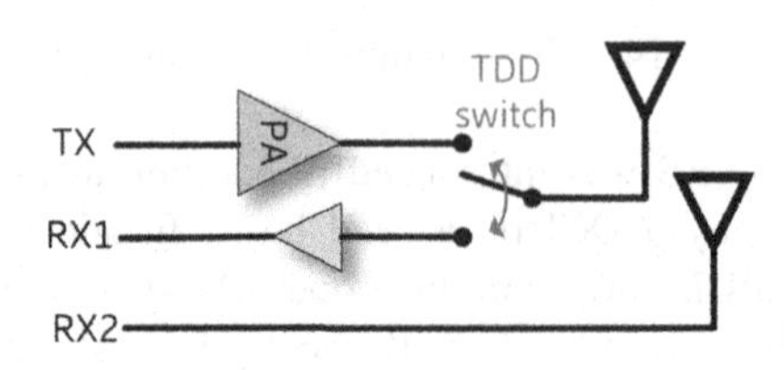

FIGURE 8.15

One transmit antenna and two receive antennas at the UE. The SRS can thus only be transmitted from one of the antennas and only partial channel sounding can be achieved. This UE thus supports 1T1R although it has two RX antennas.

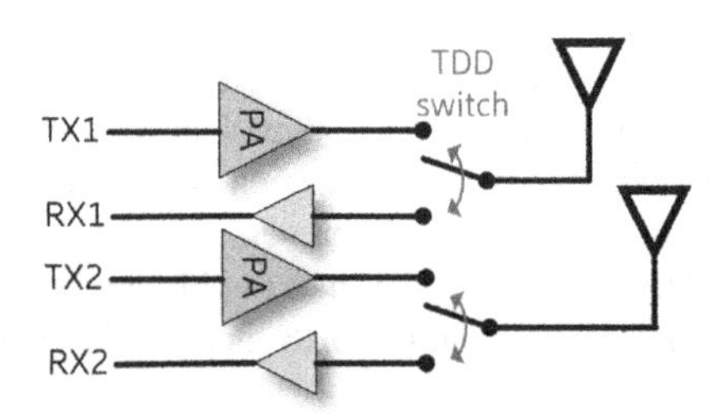

FIGURE 8.16

Full channel sounding with 2T2R, that is, both antennas can receive and transmit the SRS simultaneously.

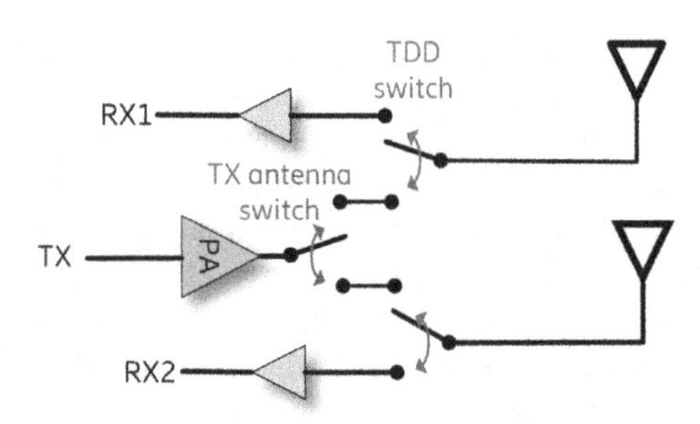

FIGURE 8.17

SRS antenna switching for 1T2R where the transmitter is dynamically switched to be connected to different antennas to allow full channel sounding.

It is thus desirable from reciprocity operation perspective (to improve DL performance) to support full channel sounding, where SRS can be transmitted from all antennas, as shown in Fig. 8.16.

In 3GPP, the used nomenclature for the case of Fig. 8.16 is 2T2R, which means the terminal has two transmit antenna and two receive antennas and thus supports full reciprocity. Another capability of a UE-supporting full reciprocity with four PA is 4T4R.

To support reciprocity-based operation also for terminals that do not have implemented the capability of full reciprocity channel sounding, antenna switching has been introduced for SRS. See Fig. 8.17 where the UE only has a single transmit PA. Here, UE capabilities that indicate that the UE support such switching are defined as 1T2R, 1T4R, or 2T4R. The scheme in Fig. 8.15 would imply that the UE needs to report 1T1R capability since it can only transmit SRS from one of the receive antennas as there is no antenna switching implemented.

The UE in the 1T2R case transmits SRS from the first antenna (corresponding to one of the two receive antennas) in a given subframe and then switches to the second antenna in a following subframe (assuming a single SRS symbol per subframe). Thereby the full UL channel can be sounded, although with at least 1-ms delay between the two measurements of the channels from the different antennas since this is the length of a subframe.

In TDD, the transmission of more than one SRS symbol in the same subframe is possible, and it is then possible to configure antenna switching within the same subframe, and in this case there is a guard (empty) symbol inserted between each SRS transmission to allow time for the UE to perform the antenna switching. The release 12 TDD supports 1 and 2 symbols used for SRS in TDD special subframes (known as UpPTS) and release 13 terminals support up to 6 SRS symbols in UpPTS.

In FDD, only a single symbol can be used for SRS per subframe although ongoing release 16 specification work supports UL subframes where all 14 OFDM symbols can be configured for SRS transmissions even from the same UE (a significant amount of SRS repetition in the same frequency band may thus be possible to configure for extended SRS coverage).

8.3.2 CSI REPORTING

8.3.2.1 Overview

The purpose of CSI feedback reporting from the UE to the eNB is to aid the scheduler in the DL link adaptation, that is, setting the modulation scheme and code rate for a scheduled PDSCH transmission. For the multi-antenna base station, the CSI can also include a *rank indicator* (RI), which indicates the preferred number of MIMO layers, and a preferred MIMO precoding matrix using a *precoding matrix indicator* (PMI) pointing to a matrix in a *codebook* of such matrices. In 3GPP, implicit CSI feedback is used (see a discussion of explicit vs implicit CSI feedback in Section 6.4.1).

The CSI reporting is computed under a hypothetical PDSCH transmission that can be wideband (the whole system bandwidth) or per sub-band. If sub-band CQI reporting is used, and the eNB decides to schedule more than one sub-band, then the eNB needs to perform additional calculations to combine these multiple per sub-band CQIs to obtain correct parameters, for example, code rate, for the link adaptation.

The bandwidth of a sub-band for CSI reporting depends on the used system bandwidth and ranges from 4 PRBs for bandwidths up to about 5 MHz up to eight PRBs for the largest bandwidths (where 20 MHz is the maximum bandwidth of an LTE carrier). This variable sub-band size keeps the CSI feedback overhead reasonably maintained even when the number of PRBs in the system is large.

Both periodic and aperiodic CSI reporting is supported, using PUCCH and PUSCH, respectively. For aperiodic CSI reporting over PUSCH, the report from the UE can be triggered by the eNB sending PDCCH with a UL scheduling grant. The PUSCH containing the aperiodic CSI report is transmitted k subframes after the subframe of the triggering PDCCH. The distance $k = 4$ for FDD and for TDD, k is variable but at least 4. For TDD, the variable k is to ensure the indicated subframe is a subframe valid for UL transmissions.

The purpose of the aperiodic and periodic CSI feedback is different. The periodic feedback is carried by PUCCH and is intended to give eNB less detailed channel information; hence feedback is coarse, and payload is low. The periodic feedback is received by eNB even if there is no data to transmit to the UE and it is thus important to keep the overhead low, for example, long periodicity.

If data arrives to eNB for DL transmission to a UE, then PDSCH can be scheduled where link adaptation is based on the periodic CSI, while at the same time, an aperiodic CSI report can be triggered. The aperiodic CSI report provides the eNB with more detailed MIMO channel information, to maximize the DL throughput. Since the payload of the aperiodic CSI is large, it is triggered "on demand" only. The drawback of aperiodic reporting is of course the need for a trigger, that is, it occupies a PDCCH resource and a larger UL overhead that steals resources from UL data transmissions. In addition, since the aperiodic report is more detailed, the UE needs more time to compute it and there is a specified delay of at least four subframes between the trigger and the report is transmitted. Hence, it is only beneficial to trigger such report if the data packet to deliver to the UE is so large that it is expected to take more than four subframes. Otherwise the packet is completely delivered before the aperiodic CSI report has been received, using the periodic CSI for link adaptation.

For the aperiodic CSI reporting over PUSCH, several CSI reporting modes are defined (see Table 8.4), and the eNB configures the UE with one of these modes by higher-layer signaling. The PMI can be wideband (a single PMI) or per sub-band (multiple PMIs) where the latter has higher CSI feedback overhead but provides more detailed information to the eNB. There is also a possibility to configure modes with "No PMI," which is used for the single antenna eNB, transmit diversity, and open-loop precoding transmission schemes (e.g., TM 3), respectively.

Moreover, "No PMI" can also be configured for DM-RS-based transmission schemes, when used for reciprocity-based operation since the eNB computes the precoder based on SRS measurements in this case and the UE need not feedback PMI.

The channel quality information (CQI) is a 4-bit value that indicates the highest modulation and code rate for a received transport block that meets a block error rate target of at most 10% (as estimated by the UE). It can be reported either as a single wideband value or a wideband value plus per sub-band values where the sub-band option has two flavors, UE selected (where the UE also

Table 8.4 CQI and PMI Feedback Modes for CSI Reporting on PUSCH.

	No PMI	Single PMI	Multiple PMI
Wideband CQI	Mode 1-0	Mode 1-1	Mode 1-2
Sub-band CQI, UE selected sub-bands	Mode 2-0	X	Mode 2-2
Sub-band CQI, higher layer–configured sub-bands	Mode 3-0	Mode 3-1	Mode 3-2

All modes were introduced in release 8 except Mode 3-2, which was introduced in release 12.

Table 8.5 CQI and PMI Feedback Modes for CSI Reporting on PUCCH.

	No PMI	Single PMI
Wideband CQI	Mode 1-0	Mode 1-1
Sub-band CQI, UE selected sub-bands	Mode 2-0	Mode 2-1

reports which sub-bands have highest CQI) or higher-layer configured, which report CQI for all sub-bands.

PUSCH CSI reporting Mode 3-2, which is sub-band CQI plus sub-band PMI, was introduced in release 12 to better support MU-MIMO as this gives a somewhat more detailed CSI report. The decision to introduce this mode 3-2 was made after the 3GPP RAN1 SI on enhanced MIMO where it was concluded that more detailed CSI feedback gives DL performance benefits at the cost of more CSI signaling overhead.

Depending on the configured TM to the UE (see Section 8.3.4), the RS used for CSI measurements is different (either CRS or CSI-RS based) and the subset of CSI reporting modes (see Table 8.4) that is available for configuration depends also on the configured TM. For example, the TMs that are using codebook feedback (TM 4, TM 6, and TM 8–TM 10) allow configuration of CSI reporting modes 1-2, 2-2, 3-1, and 3-2.

Periodic CSI reporting over PUCCH is also supported, but due to the limited maximum payload of 11 bits in PUCCH in LTE release 8, simultaneous PMI reporting or CQI reporting for all sub-bands cannot be obtained and as mentioned above only coarse CSI is provided in the report. The PUCCH periodicity is configured by the network and typically used values are 20 or 40 ms.

To reduce the PMI payload, codebook subsampling is used for PUCCH reporting, where the UE cannot use the full codebook for the PMI reporting, but can only select from a subset of the precoding matrices in the codebook.

Table 8.5 shows the four different CSI reporting modes available for PUCCH reporting, where PMI can be included or not in the report, and CQI can be either wideband or per a preferred sub-band that the UE selects and indicates.

8.3.2.2 Codebooks in LTE release 8–release 12

Precoding for spatial multiplexing is introduced in Section 6.3. The purpose of a MIMO precoding codebook is to define *a set* of precoders, and one example is discussed Section 6.3.3.4. The codebook is thus standardized and increasing the size of the set (by proper codebook design) should ideally give better performance at the expense of increased feedback overhead since the precoding matrix indicator (PMI) is needed to select one precoder from the set.

Hence, in design of codebooks in standardization, the challenge is to find the best possible set of precoding matrices given the overhead constraints, that is, given the codebook size. A model-based approach can be taken, where a priori assumptions on port layouts (physical distance between phase center of each antenna port in one or two dimensions) are made, such as assuming dual-polarized antenna pairs and some typical channel models.

Typically, the receive signal correlation between RSs transmitted from closely spaced antennas is high, while correlation is low between signals transmitted from different polarizations. Such knowledge is utilized in the codebook design and this model-based approach reduces the PMI feedback overhead.

A description of a precoder for a layer is represented by a $N_T \times 1$ vector of complex-valued elements with the same modulus, valid for either the full system bandwidth or per CSI reporting sub-band. When the reported rank is $r > 1$, a $N_T \times r$ matrix $\boldsymbol{W}$ instead of a vector represents the precoder, hence for r layers. Moreover, a finite number of such matrices $\boldsymbol{W}$ are defined per each rank r and number of ports N_T, and this superset of matrices $\boldsymbol{W}$ constitutes the entire LTE codebook.

If a CSI report is configured, the UE shall report a joint selection of preferred precoder(s) (PMI), rank indication (RI), and the largest CQI (per code word) that meets the 10% block error rate target.

In release 13, a *CSI-RS resource indicator* (CRI) was also added to the parameters in a CSI feedback report, which is present in the case a UE is configured to measure on more than one CSI-RS resource and select one of them for the CSI report. The motivation for introducing this is beamformed CSI-RS, for example, the eNB transmits four different CSI-RS resources in four different beam directions. The UE selects a preferred beam and feeds back the "normal" CSI report for a selected beam only (i.e., the CSI-RS resource) plus the CRI to indicate the selected resource. This is also known as Class B operation in LTE.

Sometimes, the network may want to restrict the precoders the UE can select from. Then there is a possibility for the eNB to restrict the UE to select only a subset of precoders within a codebook and/or a subset of the ranks. This is known as *codebook subset restriction* (CSR) and can be used if the network knows that it will never schedule a transmission in a certain direction (e.g., since this direction interferes heavily with users in an adjacent cell). Hence the precoders associated with this direction are restricted. Note that for simplicity, the use of CSR does not change the number of PMI and/or RI feedback bits, even if the number of "allowed" PMI and/or RI are significantly reduced.

In LTE, there are codebooks defined for $N_T = 2$ and 4 ports to be used with CRS-based measurements and $N_T = 2$, 4, 8, 12, 16, 20, 24, 28, and 32 ports with CSI-RS-based measurements. The antenna model used when designing the CSI-RS-based codebooks is uniformly spaced antenna ports in one or two dimensions, where the codebooks for 2D port layouts were introduced in release 13.

In the following, the basics of LTE codebooks will be presented, and general principles and key aspects will be discussed without digging into details. In the NR codebook description (Section 9.3.5), a more thorough walk through of the codebook design is given.

It is illustrative to first introduce codebooks with the LTE release 8 codebook for $N_T = 2$ CRS antenna ports, which for rank $r = 1$ are the following four vectors:

$$\boldsymbol{W} = \frac{1}{\sqrt{2}}\begin{bmatrix} 1 \\ 1 \end{bmatrix}, \frac{1}{\sqrt{2}}\begin{bmatrix} 1 \\ -1 \end{bmatrix}, \frac{1}{\sqrt{2}}\begin{bmatrix} 1 \\ j \end{bmatrix}, \frac{1}{\sqrt{2}}\begin{bmatrix} 1 \\ -j \end{bmatrix}$$

Hence, the PMI, which in this case requires two bits, indicates the best possible cophasing of the two antenna ports using a QPSK alphabet. In cophasing transmission, the eNB transmits the same PDSCH layer from both antenna ports but where the transmission from the second antenna port applies the phase shift x as was recommended by the UE. Coherent combining of the two transmitted layers is then achieved at the receiver antenna in the UE.

For $N_T = 2$ CRS antenna ports and $r = 2$, these are the three specified precoding matrices in the LTE codebook

$$W = \frac{1}{\sqrt{2}}\begin{bmatrix} 1 & 0 \\ 0 & 1 \end{bmatrix}, \frac{1}{2}\begin{bmatrix} 1 & 1 \\ 1 & -1 \end{bmatrix}, \frac{1}{2}\begin{bmatrix} 1 & 1 \\ j & -j \end{bmatrix}$$

The first matrix simply indicates to transmit one layer per CRS antenna port, while the two others transmit each layer across both CRS antenna ports with a cophasing. Note that the columns of the precoding matrix are by design orthogonal and that each matrix element has the same amplitude (constant modulus requirement), which are general principles used for codebook design in 3GPP. In addition, the precoding matrices are power normalized, so each layer uses an equal portion of the total transmitter power.

For the release 8 codebook design for $N_T = 4$ CRS antenna ports, one underlying model assumption was dual-polarized antenna pairs spaced some distance apart. However, in case the distance between the two antennas is large, then the spatial correlation between the fading variations experienced between antennas of the same polarization in different pairs is low (see discussion in Section 3.7). In the release 8 codebook design, the aim was to design a codebook suitable for both high and low spatial correlation and it therefore contains a mix of precoding matrices where some are motivated by high-correlation and dual-polarized antenna setups, while others are not bound by this underlying model but more "random."

Ultimately, a compact way to describe such a codebook was by using a Householder transformation, since the whole codebook for rank 1–4 could simply be described by 16 different 4×1 vectors $\boldsymbol{u}_n, n = 0, \ldots, 15$ and for each $\boldsymbol{u}_n$ a 4×4 matrix is obtained as

$$\boldsymbol{W}_n = \boldsymbol{I} - \frac{2\boldsymbol{u}_n\boldsymbol{u}_n^*}{\boldsymbol{u}_n^*\boldsymbol{u}_n},$$

where $\boldsymbol{I}$ is the identity matrix. The codebooks for $r = 1$, 2, and 3 are then obtained by specifying which r columns to select from each $\boldsymbol{W}_n$. Note that the use of the Householder transformation implies a certain restriction on the possible precoders compared to individually optimizing each matrix without such constraint, but ease of specification and simple description was prioritized. Alternative and improved four-port codebooks were specified in release 12 as will be discussed further below.

In LTE release 10, the 3GPP specifications for multi-antenna transmissions were expanded to target up to eight-layer MIMO transmission. A new codebook structure was then introduced for eight CSI-RS ports and the codebook design principle was different from release 8, since for eight CSI-RS ports, it was seen as likely that an array antenna with four linearly spaced dual-polarized antenna pair elements would be used in real deployments and hence high correlation between spatially separated antenna ports and low correlation between polarization separated antenna ports (with no spatial separation) was exploited in the eight-port codebook design. This new structure has then been used since then in both LTE evolution and in NR.

The precoder $\boldsymbol{W}$ structure for this codebook is referred to as the factorized precoder structure as discussed in more depth and motivated in Section 6.4.1.2 and is defined as a matrix multiplication of an inner and outer precoding matrix:

$$\boldsymbol{W} = \boldsymbol{W}_1\boldsymbol{W}_2$$

where the first, inner, $N_T \times \tilde{N}_T$ precoder matrix $\boldsymbol{W}_1$ is block diagonal with a matrix $\boldsymbol{B}$ repeated twice in the matrix diagonal; hence

$$\boldsymbol{W}_1 = \begin{bmatrix} \boldsymbol{B} & \boldsymbol{0} \\ \boldsymbol{0} & \boldsymbol{B} \end{bmatrix}$$

and where $\tilde{N}_T$ is a design parameter which is related to the number of remaining beams after the channel dimension reduction by $\boldsymbol{W}_1$.

The matrix $\boldsymbol{B}$ is an $N_T/2 \times N_b$ matrix that contains N_b DFT vectors and the purpose of $\boldsymbol{B}$ is hence to match the spatial channel from a uniformly and linearly spaced array of equally polarized antenna elements. The use of two blocks is thus to match the channel from each of the two different polarizations respectively. The channel from closely spaced antenna ports of the same polarization are likely to be well approximated with a DFT vector at least for modest angle spread, and thus it is expected that $\boldsymbol{W}_1$ is similar over the system bandwidth and over time. Hence, $\boldsymbol{W}_1$ can be wideband and updated rather infrequently as discussed more in Section 6.4.1.2. One could also see the different DFT vectors of the columns of $\boldsymbol{B}$ as different beam directions, an analogy that holds perfectly well under the "ideal" conditions of line of sight channel and half wavelength−spaced antenna ports per polarization.

The outer precoder $\boldsymbol{W}_2$ has then the purpose of performing beam selection and cophasing of the beams from the two polarizations, that is, combining the two diagonal $\boldsymbol{B}$ matrices. As correlation between different polarizations typically is small, the outer precoder needs to be selected per sub-band to follow the channel variations across frequency.

Another interpretation (see Fig. 6.23 in Section 6.4.1.2) is that $\boldsymbol{W}_1$ serves to create a new effective (dimension reduced) $N_R \times \tilde{N}_T$ channel matrix as measured by the UE, $\boldsymbol{H}_{eff} = \boldsymbol{H}\boldsymbol{W}_1$ for the outer $\tilde{N}_T \times r$ precoder $\boldsymbol{W}_2$ to be applied to. The inner precoder thus introduced virtual antennas (one port per beam polarization if that analogy can be used). Since the number of virtual antennas $\tilde{N}_T$ is much smaller than the number of antenna ports N_T, this provides a considerable dimension reduction and thus a smaller codebook (fewer elements) for the outer precoder. As the outer precoder typically is reported per sub-band, the smaller outer precoder codebook helps to keep the total PMI overhead down.

For rank 1 and rank 2 codebooks, a grid of beams of 32 DFT vectors are used and since the release 10 LTE codebook is designed for a 1D antenna port layout with $N_T = 8$ antenna ports, there are $N_T/2 = 4$ spatially separated ports in the linear array (per polarization) implying four spatially orthogonal DFT vectors. In the codebook design, if only these four orthogonal DFT vectors would have been chosen to represent the different precoders, then there would have been a large signal power loss if the UE is in a direction between these orthogonal DFT beam directions. To mitigate this power loss, spatial oversampling has been introduced, where additional beams (DFT vectors) are placed in between orthogonal beams. Since the rank 1 and rank 2 codebooks for four linear and spatially separated ports (of same polarization) have 32 DFT vectors, the oversampling factor is $O = 32/4 = 8$, for the eight-port codebook defined in LTE.

Another benefit of such a large degree of spatial oversampling is that it makes the codebook less sensitive to the physical distance between the four antenna ports compared to if, for example, only the four orthogonal beams would have been used. Even if a physical spacing larger than the half wavelength is implemented in the AAS, the codebook performance is still reasonably good thanks to this spatial oversampling property.

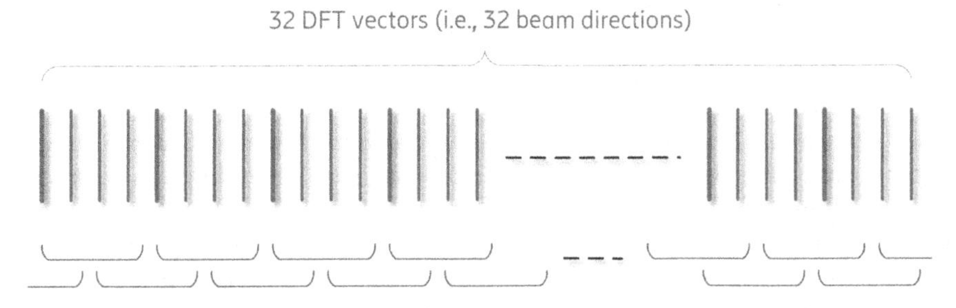

FIGURE 8.18

LTE codebook design for rank 1 and 2 where 32 DFT vectors are specified and the UE selects a subset of four adjacent DFT vectors for wideband feedback. The UE further selects one of the four, per sub-band.

From these 32 vectors, 16 different $\boldsymbol{B}$ matrices are defined in specifications where each contains $N_b = 4$ columns hence a sheaf of four DFT beam directions. The 16 matrices consist of subsets of the 32 DFT vectors as illustrated in Fig. 8.18.

The wideband PMI feedback, PMI_1, thus selects the inner precoder $\boldsymbol{W}_1$, that is, selects one of the 16 different $\boldsymbol{B}$ matrices and thus a sheaf of adjacent DFT beams, and the outer precoder $\boldsymbol{W}_2$ then performs selection of one column (i.e one beam) from each $\boldsymbol{B}$ so that effectively a dimension reduction from $N_T = 8$ to $\tilde{N}_T = 2$ is achieved, which simplifies the complexity of the following per sub-band cophasing.

The release 8 codebook for two ports is then re-used to perform the polarization cophasing. Hence, the outer precoder $\boldsymbol{W}_2$ performs both beam selection and cophasing. In general terms, the codebook for the outer precoder $\boldsymbol{W}_2$ can be written as

$$\boldsymbol{W}_2 = \begin{bmatrix} \boldsymbol{e}_{k_{\mathrm{a},1}} & \boldsymbol{e}_{k_{\mathrm{a},2}} & \cdots & \boldsymbol{e}_{k_{\mathrm{a},r}} \\ e^{j\phi_1}\boldsymbol{e}_{k_{\mathrm{b},1}} & e^{j\phi_2}\boldsymbol{e}_{k_{\mathrm{b},2}} & \cdots & e^{j\phi_r}\boldsymbol{e}_{k_{\mathrm{b},r}} \end{bmatrix}$$

where $\{\boldsymbol{e}_{k_{\mathrm{a},l}}\}_{l=1}^{r}$ are 4×1 selection vectors with a single vector element $k_{\mathrm{a},l}$ equal to one and the remaining elements equal to zero. The factors $e^{j\phi_i}$ are the polarization cophasing factors, per layer.

The rank 3 and 4 codebooks for eight ports are similar in principle although the oversampling factor is reduced, $O = 4$, and $N_b = 8$ DFT vectors are used in the matrix $\boldsymbol{B}$ since higher rank transmission requires selection of more spatially separated vectors. In addition, only four different $\boldsymbol{B}$ are defined.

For rank 5–8, there is no oversampling in matrix $\boldsymbol{B}$ design and there is no UE selection of cophasing or beam selection by $\boldsymbol{W}_2$ hence a fixed $\boldsymbol{W}_2$ is specified per rank. One out of four $\boldsymbol{W}_1$ matrices can be selected by the UE for rank 5–7 and for rank 8, the codebook consists of a single matrix $\boldsymbol{W}$. The reason is that the benefit of closed-loop precoding where UE selects beams and cophasing diminishes when the number of layers is above four for a system with eight transmit antennas and eight receive antennas. Hence, the PMI feedback overhead is wideband only and thus small for rank 5–8 feedback.

In release 12, an alternative four-port codebook was specified for the DM-RS-based transmission schemes. It uses the $\boldsymbol{W} = \boldsymbol{W}_1\boldsymbol{W}_2$ structure that can replace the Householder based codebook design from release 8. This codebook was specifically designed for two closely spaced cross-pole antennas, as well as for rank 1 transmission of four linearly spaced antennas with the same polarization, and thus has some advantage over the release 8 codebook for such deployments. It also reduces the CSI feedback overhead for large scheduling bandwidths since the release 8 codebook requires 4 bits per sub-band, while the new release 12 codebook has one wideband component $\boldsymbol{W}_1$ plus a low overhead cophasing component $\boldsymbol{W}_2$ per sub-band. A design difference compared to the eight-port codebook in release 10 is that this four-port codebook uses orthogonal, that is, nonadjacent beams in the block matrix $\boldsymbol{B}$. The motivation was that the beam width per DFT beam is doubled compared to the eight-port case and thus the use of adjacent beams in $\boldsymbol{B}$ is less motivated as they are rather similar. For rank 3 and 4, there was no change and the release 8 codebook is used but incorporated in the new codebook framework by using release 8 Householder codebook for $\boldsymbol{W}_2$ (allowing per sub-band reporting) and setting the identity matrix for $\boldsymbol{W}_1$.

8.3.2.3 Introducing two classes of channel state information feedback in long-term evolution release 13

In release 13, the LTE CSI framework was extended to 2D port layouts, and two classes of CSI feedback were defined, A and B, where Class A corresponded to the existing framework and Class B was a newly introduced framework that targeted beamformed CSI-RS.

The reason was that 3GPP concluded an SI on elevation beamforming and full-dimensional beamforming that followed the release 12 completion of a 3D channel model. The SI also included modeling of deployments with 2D antenna arrays. Prior to this SI, the used channel models assumed that all UEs are 1.5 m above ground, hence only distributed in horizontal as seen from the base station. The new 3D channel model allowed for UEs to be placed in buildings, indoor on different floors, which more realistically models an urban deployment.

MIMO in LTE had prior to release 13 been limited to adaptability of transmission in either the vertical or the horizontal domain and codebooks were designed for ID port layouts (uniform linear array of cross-pole antennas) that allowed beam steerability in one dimension only. Prior to release 13, there was no support for using horizontal and vertical beam steering simultaneously.

The release 13 SI report [10] concluded that significant throughput gains were achievable in realistic non-full-buffer evaluation with the joint vertical and horizontal beamforming, over the baseline, which was release 12–based MIMO features.

The concept of beamformed CSI-RS was introduced, motivated by the SI conclusion. One motivation was coverage, and another was that an envisioned product would create multiple vertical beams and in each of these beams there is a 1D horizontal port layout. The UE would then select a preferred vertical beam and use the "normal" CSI feedback assuming the codebook designed for a 1D port layout, as had been specified in release 10, within the selected beam. These principles laid the foundation for Class B type of operation.

Class A, also known as nonprecoded CSI-RS, is based on the principle of the same number of CSI-RS antenna ports as the number of antennas (although an "antenna" in this context may consist of multiple antenna elements in a subarray) (see Section 4.8 or the input to the antenna virtualization, according to below). The principle of Class A is thus that each CSI-RS port covers the whole serving cell (in contrast to Class B, see below).

The UE then computes the preferred precoder for a given rank, assuming transmission from these ports. Class A thus resembles what had been assumed for the codebook-based feedback in all

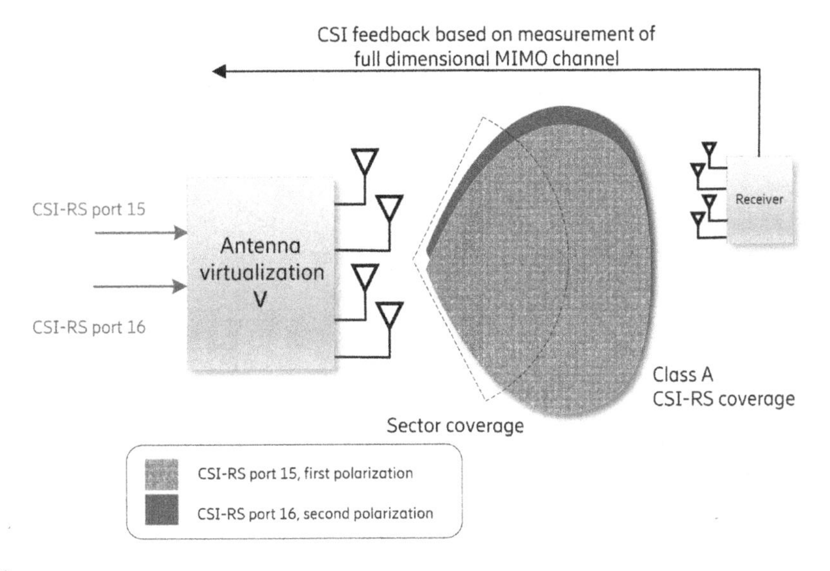

FIGURE 8.19

The Class A CSI feedback defined in release 13, also known as nonprecoded CSI-RS, in which each CSI-RS port has full sector coverage. The UE feeds back a precoder based on a full-dimensional channel since the UE can measure the full MIMO channel from all virtual antenna elements used by CSI-RS. The antenna virtualization is in this case static and same for all UEs served in the sector.

the previous LTE releases. In release 13, the introduced Class A operation also supports 2D CSI-RS antenna port layouts, and hence 3GPP introduced a corresponding codebook of 2D precoders (Fig. 8.19).

Class B is the beamformed CSI-RS where a transmitted CSI-RS does not need to have full sector coverage in vertical and/or horizontal. Hence, the eNB can be seen to perform a spatial prefiltering by the antenna virtualization, and the UE computes the preferred precoder on this spatially filtered channel. This operation requires either some knowledge at the eNB about a suitable spatial filter/beam (e.g., using reciprocity), or by providing multiple, different, beamformed CSI-RS resources (different virtualizations **V**) and let the UE select and feedback a preferred resource that is, beam (Fig. 8.20).

The Class A configuration on the other hand always uses a single CSI-RS resource in release 13 of 8, 12, or 16 ports and the codebook uses the $\mathbf{W}_1\mathbf{W}_2$ precoder structure as introduced in Section 8.3.2.2.

To support this new functionality, codebooks were specified in release 13 for closely spaced cross-poles in various 1D as well as 2D port layouts from rank 1 up to rank 8. The 2D port layout defines N_1 and N_2 antenna ports per first and second dimensions, respectively, where it is up to each deployment whether to use N_1 ports in horizontal or vertical dimensions and vice versa for N_2.

In a similar way as the 1D codebook introduced in release 10, the 2D codebook uses DFT beams with spatial oversampling but with beams pointing in both a vertical and a horizontal angle. The spatial oversampling factor O_1 and O_2 was defined as configuration parameters for the first

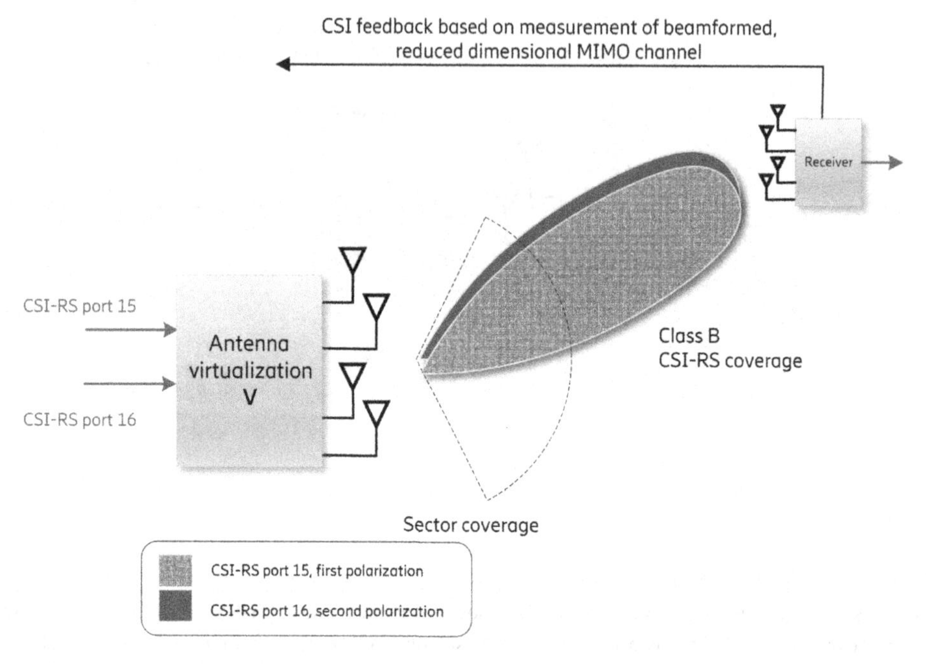

FIGURE 8.20

The Class B CSI feedback introduced in release 13, also known as beamformed CSI-RS, in which each CSI-RS resource is spatially filtered (e.g. beamformed) and has only partial sector coverage. The UE feeds back a precoder based on a low dimensional channel since the eNB has already reduced the number of dimensions by the spatial prefiltering. The antenna virtualization is in this case dynamic and can even be UE–specific.

and second dimensions, respectively. Note that the specification is agnostic to whether N_1 and N_2 correspond to vertical and horizontal dimensions of the antenna arrays, or vice versa, as this is up to the deployment and thus the physical rotation of the antenna array at installation.

The total number of beams for a given 2D codebook configuration in the two dimensions is $N_1O_1N_2O_2$ and where each of these four parameters are configurable. More beams imply more feedback overhead but potentially better performance.

The more detailed 2D codebook structure will be elaborated for the NR codebook in Section 9.3.5.3 since it is based on the same principles as LTE, using a grid of DFT beams in two dimensions, that is, vertical and horizontal.

For the LTE 2D codebook there is a great deal of configurability in the number of beams in each dimension (due to spatial oversampling), and different port layouts. Table 8.6 shows the supported configurations for the release 13 Class A codebooks where for instance (2,3) is interpreted as a 2D antenna array with 2 and 3 ports per polarization, respectively (thus in total $2 \times 3 \times 2 = 12$ ports), and where spatial oversampling can be configured to be either eight times in first dimension and four times in second dimension or eight times in both dimensions.

As discussed above, Class B codebooks are designed for beamformed CSI-RS, where a UE can be configured with multiple, $K \geq 1$ CSI-RS resources. In Class B operation, $\boldsymbol{W}_1$ is set to the identity

Table 8.6 Supported Codebook Configurations for LTE Release 13 Two-Dimensional Port Layouts.

(N_1,N_2)	(O_1,O_2)
(8,1)	(4,-), (8,-)
(2,2)	(4,4), (8,8)
(2,3)	(8,4), (8,8)
(3,2)	(8,4), (4,4)
(2,4)	(8,4), (8,8)
(4,2)	(8,4), (4,4)

N_1 *and* N_2 *are the number of ports in first and second dimensions, respectively (assuming same polarization), and* O_1 *and* O_2 *are the spatial oversampling factors per dimension. The two dimensions represents horisontal and vertical, respectively, or vice versa.*

matrix in the feedback. This is rather natural since eNB precodes the CSI-RS to mimic $\boldsymbol{W}_1$ in a transparent manner in Class B, as opposed to the codebook-based UE selection of $\boldsymbol{W}_1$ when using the Class A codebook.

In addition, two different configurations are possible for Class B operation, $K=1$ resource or $K>1$ resources, where each of resource $k=0,1,\ldots,K-1$ is configured with N_k ports where $N_k=\{1,2,4,8\}$.

The Class B, $K>1$ codebook uses the release 10 (i.e., 1D port layout) codebook for the selected CSI-RS resource. Up to $K=8$ CSI-RS can be configured, and a 1D port layout is used per CSI-RS resource. For this operation with $K>1$, the eNB beamforms each CSI-RS resource in a beam (e.g., vertical beam) and the UE selects a preferred beam on a wideband basis, that is, one of the K resources and feeds back its selection (by CRI, the CSI-RS resource indicator) together with CSI for the selected resource, including a PMI for the N_k ports (which then may represent the horizontal precoding).

Alternatively, if a single resource $K=1$ is configured with Class B operation, the intended use case targets UE–specific beamforming, potentially where channel reciprocity is utilized by the eNB to acquire a set of useful beam directions for the UE. Here, $\boldsymbol{W}_1$ equals the identity matrix (and thus not included in CSI feedback) while $\boldsymbol{W}_2$ is used to performing beam selection (out of a set of four dual-polarized beams) and cophasing of polarizations within the selected beam. For this cophasing reporting purpose, it is possible to use the release 10 codebooks, or a new release 13 codebook tailored for this feedback configuration is used.

Together with the introduction of beamformed CSI-RS in release 13, *measurement restriction* was introduced as a new feature. One use case with a beamformed CSI-RS is that a periodic CSI-RS resource is configured and the eNB then changes the beam direction at each instant of that CSI-RS transmission, so that multiple UEs at different directions can measure on the same configured CSI-RS resource. This reuse scheme saves overhead but also introduced a problem since the UE may apply filtering across multiple instances of the CSI-RS. If the eNB changes the beam at each instant, the averaging of the channel across these subframes does not make sense. This problem led to the introduction of measurement restriction configuration to the UE, where if configured, the UE shall only use a single instance of a periodic CSI-RS when requested to feed back a CSI report. By configuring measurement restriction, the problem of UE averaging over time is removed and the eNB knows which beam the UE has measured for each report. In release 13, measurement restriction for channel and/or interference measurements is supported.

In the following release, release 14, the work continued toward supporting massive MIMO, and Class A codebooks were further extended with 20, 24, 28, and 32 CSI-RS ports for both 1D and 2D port layouts.

8.3.2.4 Advanced channel state information feedback

Channel reciprocity-based operation is beneficial, and as seen in the performance evaluations in Section 13.6, and mainly beneficial for MU-MIMO scheduling. But it relies on many critical circumstances for it to work (see the discussion in Section 6.7). As an alternative to reciprocity based operation, one could rely on richer channel feedback using the CSI framework, in order to reach MU-MIMO performance similar to reciprocity-based operation. This feature thus enables very good MU-MIMO performance independently of whether FDD or TDD is deployed.

Hence, in release 14 the *advanced* CSI feedback reporting was introduced for 4–32 CSI-RS ports, with the aim to provide more detailed, high-resolution CSI feedback to the network. Only rank 1 and 2 feedback is supported in this advanced mode. The goal is to achieve comparable channel knowledge for an FDD system using CSI feedback as reciprocity-based SRS measurements can provide for TDD. Note that the advanced feedback is then also useful for TDD systems where channel reciprocity is not available, for example, due to the partial channel reciprocity problem discussed in Section 6.7.1.2. Using advanced feedback is more costly in feedback overhead and targets MU-MIMO performance gains although it should be noted in this context that reciprocity-based operation also consumes a lot of UL overhead since each UE needs to transmit SRS frequently.

Feeding back a single DFT beam direction as is used in feedback design of the previous codebooks for rank 1 and 2 is in many cases insufficient for MU-MIMO. The reason is that for MU-MIMO precoding, not only the strongest DFT beam direction of the channel matters, but also other directions since these other directions are relevant for interference toward co-scheduled UEs.

Hence, it was observed that a "richer" multi-antenna channel feedback could provide a rather large performance increase for MU-MIMO. This resulted in advanced CSI feedback using a 2D codebook.

The design principle of this codebook is illustrated in Fig. 8.21.

Two DFT beams ($\begin{bmatrix} \boldsymbol{b}_0 & \boldsymbol{b}_1 \end{bmatrix}$) are in a first step chosen from a group of beams (constituting an orthogonal 2D DFT basis of the channel space) to represent the two strongest directions of the spatial channel. The power of the second beam is scaled by $\sqrt{p_1}$ to represent the different propagation conditions, that is, path loss between the two beams.

These two steps will generate the outer, wideband matrix $\boldsymbol{W}_1$ followed by a per sub-band cophasing using $\boldsymbol{W}_2$. This co-phasing using $e^{j\alpha_1}$ of the second beam will also be carried out over the polarizations even if that is not illustrated in the figure.

For dual-layer transmission, that is, rank 2, individual beam co-phasing is used for the two layers. Note that the "normal" CSI feedback is then a special case of the advanced feedback where only a single beam is selected in the first step.

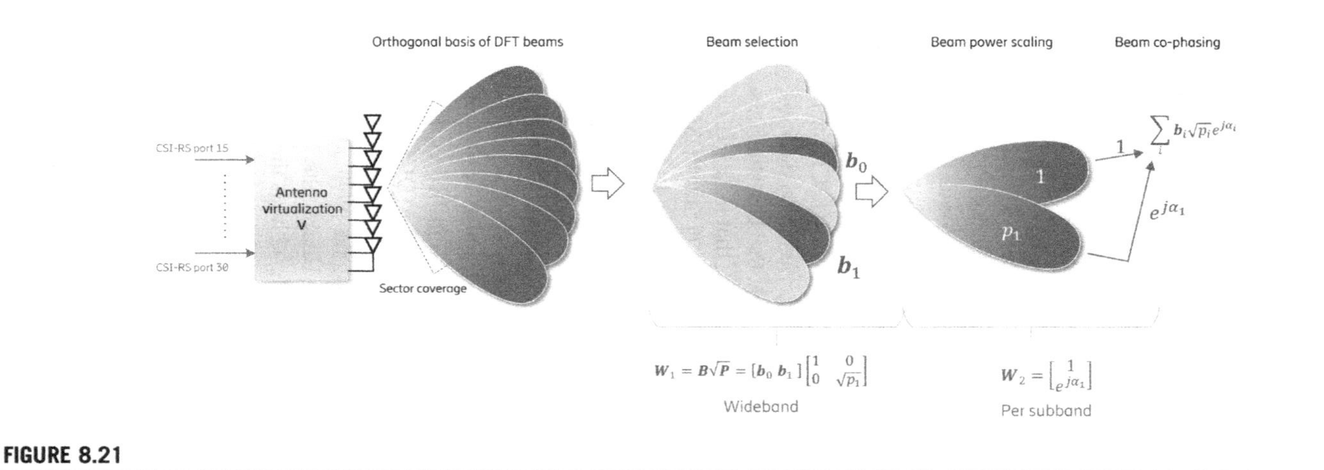

FIGURE 8.21

Codebook for advanced CSI feedback. The illustration only shows the cophasing of beams for one of the two polarizations (for simplicity).

8.3.3 PHYSICAL DOWNLINK CONTROL CHANNEL TRANSMISSION PROCEDURES

There are multiple TMs in LTE and the UE is simultaneously configured with one DL and one UL TM. The modes will be introduced in the next section. Prior to that, the control signaling is briefly introduced as it is important to understand how TM switching is performed without losing the connection to the UE and to briefly discuss how scheduling is performed.

DL control signaling in LTE and NR used to schedule UL and DL transmissions uses *downlink control information* (DCI), which is mapped onto the *physical downlink control channel* (PDCCH) in the control region of the subframe (see Fig. 8.2). For PDCCH, link adaptation is used where a PDCCH can consist of 1, 2, 4, or 8 *control channel elements* (CCEs). By selecting the number of CCEs (referred to as the aggregation level) for a PDCCH transmission the coverage can be controlled.

In each subframe, the UE shall decode a set of potential PDCCH transmissions, known as PDCCH candidates, for each aggregation level. The set of PDCCH candidates and their positions on the subframe is known as the PDCCH search space. Since there are multiple candidates, the network can choose which one to use for PDCCH transmission to a specific UE, and thereby avoid collision with another PDCCH transmitted simultaneously to another UE that is scheduled in the same subframe.

Note that the UE may also receive more than one PDCCH in the same subframe to schedule, for example, PDSCH and PUSCH simultaneously. Hence, in addition to blindly decoding a number of PDCCH candidates for each aggregation level, the UE shall also for each PDCCH candidate decode twice for two different information payloads, that is, a UL DCI payload and a DL DCI payload. These two DCIs are potentially scheduling PUSCH and PDSCH, respectively.

For DL scheduling, there is both a small DCI payload format and a large DCI payload format. The larger DCI payload has a variable size depending on the used TM, depending on which RRC configured features are enabled, such as carrier aggregation and SRS triggering. The smaller DCI payload is always the same across all LTE releases and irrespectively of any enabled features. The small payload has restrictions as spatial multiplexing cannot be scheduled from this smaller DCI format. Hence, it is always possible to reach the UE with a scheduling command with the smaller DCI payload, which is useful when the UE is reconfigured to a different TM as this reconfiguration takes a few milliseconds and it is ambiguous exactly when the UE has changed to the new TM. The smaller DCI is therefore often called the fallback DCI.

The fallback DCI for DL scheduling has the same payload size as the UL DCI for PUSCH scheduling, except when UL MIMO is configured. Hence, the UE can, after detecting the fallback DCI, further detect whether it is a UL or a DL DCI, without the need to blindly decode yet another DCI payload size.

When UL MIMO as introduced in release 10 is enabled for the UE, then a third payload size is used. UEs with UL MIMO enabled must therefore decode three different payload sizes, for DL, UL MIMO, and fallback, respectively.

Search spaces come in two types, the *common search space* (CSS) and the *UE–specific search space* (UESS). The CSS is used for broadcast messages, such as system information, paging, and random access response messages. Also, the fallback DCI is monitored by the UE in the CSS only. The large payload DCI is monitored in the UESS.

In release 11, the EPDCCH was introduced, allowing for beamforming of the control channel to the UE, but EPDCCH only supports the UESS; hence it can be configured as an alternative. The CSS must be broadcasted covering the whole cell and cannot benefit from UE–specific channel aware transmissions. For some of the messages scheduled from CSS, the preferred "beam" toward the UE is unknown (as in paging) or the message is a multicast message (such as system information).

8.3.4 PDCCH TRANSMISSION PROCEDURES AND DOWNLINK TRANSMISSION MODES

In LTE, there are 10 different *transmission modes* (TMs) and the first LTE release, release 8, contained the first seven, while the other three were added during LTE evolution as new features were introduced.

This section will describe these TMs. The reason for many modes in the initial release is mainly due to the use of CRS-based demodulation, which requires one mode for each transmission scheme as the actual mapping of bits to antennas must be clearly specified (e.g., single antenna, spatial diversity, spatial multiplexing, etc.). In hindsight, it may have been done differently and with fewer modes by allowing additional configurations on top of a TM, and this philosophy was also adopted in later LTE releases beyond release 11 where no new modes have been introduced.

In the first release of LTE, CRS was the all-purpose RS, used for fine synchronization, channel analysis, measurements, and demodulation. Hence, TM 1–6 all use CRS for demodulation, and to support different transmission schemes one mode was needed for each scheme. These are the release 8 TMs:

- **TM 1** and **TM 2** are single-layer schemes;
  - TM1 is configured when eNB only has a single CRS port, that is, a single transmit antenna;
  - TM2 is single-layer transmit diversity across two or four CRS ports;
- **TM 3** and **TM 4** are spatial multiplexing schemes used to increase spectral efficiency and are based on open-loop and closed-loop MIMO transmission, respectively;
  - TM 3 uses a predefined cycling of a precoding matrix selection from a codebook of matrices for a given rank, where cycling implies that one precoder from the codebook is selected pseudo-randomly for each RE and these precoders are cycled through within the PDSCH transmission to get spatial diversity for the PDSCH. The cycling pattern is given by the specifications and hence known to both eNB and UE. Supported transmission rank is 1, 2, 3, and 4;
  - TM 4 uses a single eNB selected precoder (wideband or per PRG) from a codebook of predefined matrices in the specifications. Hence, TM 4 is a codebook-based TM and supported transmission rank is 1, 2, 3, and 4;
- **TM 5** and **TM 6** are special configurations of TM 4 where only a single layer can be scheduled; hence transmission rank is restricted to 1. TM 5 has the possibility to signal in DCI a PDSCH power reduction of 3 dB. This is used for simultaneous transmission to another UE since then the power per UE is halved. TM 5 is therefore intended for MU-MIMO scheduling of two UEs with one layer each;

- **TM 7** is a single-layer transmission using DM-RS instead of CRS for demodulation. It can be used for UE–specific beamforming and reciprocity operation as the used precoder is transparent to the UE and not bound to be taken from a codebook. Hence, this is noncodebook-based operation since the codebook used for PDSCH transmission is not specified in the standards.

8.3.4.1 Release 9: Time division duplex enhancements and TM 8

In release 9, the scope of the enhancement was to improve the spectral efficiency for reciprocity-based operation, hence mainly targeting TDD. It was observed that most eNB have dual-polarized antennas and UEs have two RX antennas, yet only single-layer transmission using UE–specific RSs for demodulation was supported by release 8 (by TM 7).

Hence, dual-layer transmission was introduced in this release with new UE–specific RSs for demodulation, associated with a new TM, TM 8. As in TM 7, this is also a noncodebook-based TM and the precoder $\boldsymbol{W}$ is noncodebook based and hence derived by the eNB from UL measurements. During specification it was foreseen that even more layers with DM-RS would be introduced in release 10, so the release 9 DM-RS design for TM 8 was constructed to be forward compatible, that is, could be extended with more ports in a future release.

Note that no CSI feedback enhancements were made in this release, so even if reciprocity-based operation is used with TM 8, the CSI feedback is based on CRS measurements. Hence, if UE–specific beamforming is used for PDSCH, then the eNB needs to perform further calculations based on the CSI report to obtain a new CSI that better matches the link quality of a beamformed PDSCH. This problem was addressed in the next release by the introduction of dedicated measurement signals, the CSI-RS.

8.3.4.2 Release 10: Spectral efficiency enhancements and TM 9

Release 10, which was the LTE-Advanced release, took a big step in spectral efficiency by extending from four- to eight-layer transmission to the same UE. The UE–specific RS paradigm was used, and TM 9 was introduced, which, as TM 7 and TM 8, is a noncodebook-based TM. Also, CSI-RS was introduced to allow more flexibility in obtaining a measurement, specific for a certain UE (i.e., using channel-dependent beamforming of CSI-RS).

Due to these enhancements, the IMT-Advanced peak spectral efficiency target of 30 bps/Hz could be met. Furthermore, MU-MIMO operation was improved since the DM-RS is precoded in the same way as the associated data layer and MU-MIMO can therefore be supported with one layer per UE using orthogonal DM-RS or two layers per UE using different pseudo-orthogonal DM-RS sequences (see Section 8.3.1.5).

The DM-RS always have the same power as the PDSCH layer; hence there is no need to signal the power offset between DM-RS and PDSCH as was needed for the CRS-based MU-MIMO mode of TM 5. Precoding matrix codebooks were designed for CSI feedback with eight TX antennas (i.e., eight CSI-RS ports) at the eNB as was discussed in Section 8.3.2.2.

8.3.4.3 Release 11: Dynamic point selection and TM 10

In release 11, the scope of MIMO enhancements was targeting multipoint transmissions and improved CSI feedback. A definition of how the UE measures interference was specified and support for *dynamic point selection* (DPS), where the eNB that transmits data to the UE can dynamically change from one PDSCH transmission to the next, was introduced. Yet another enhancement to the noncodebook-based transmission was specified by a new mode, TM 10. It basically has all the features of TM 9, using the same codebooks for CSI feedback but additionally defined *CSI interference measurement resources* (CSI-IM) to precisely specify the REs the UE uses for interference estimation. Refer to the problem of CRS-based interference measurements in Section 8.3.1.3.

The CSI-IM gives network information on how the UE estimates SINR and thus CQI reporting. Up until TM 9, the interference measurement procedure for the UE was not specified. The introduction of CSI-IM was important as it improves network performance with more advanced deployments.

The concept of CSI processes was also introduced in release 11, where up to four CSI processes can be configured to the UE so that the network can get CSI feedback regarding transmission not only from the serving eNB but also for DPS transmissions from other eNB. Each process thus reflects CSI for a transmission from one eNB.

To fully support the DPS use case, the QCL framework was introduced, since the UE needs to perform synchronization and channel analysis for possible upcoming transmissions from not only one but any one in a set of multiple different eNB (see Section 8.3.1.2 for introduction of QCL).

8.3.4.4 Release 12: Further MU-MIMO enhancements

In release 12, a new precoding matrix codebook was introduced for rank 1 and 2 in 4 TX operation in TM 8, 9, and 10, that is, the TMs that use DM-RS for demodulation. At this time, 4TX eNB were gaining popularity in the market and there was a need to further optimize the performance by a new codebook that is better matched to actual products in the field. The new codebook has the same $\boldsymbol{W}_1\boldsymbol{W}_2$ structure assuming grid of beams as the 8 TX codebook that was introduced in release 10.

In addition, a new CSI feedback mode (denoted mode 3-2) was introduced for TM 4, 6, 8, 9, and 10, which provides precoding matrix index (PMI) selection as well as CQI per sub-band across the system bandwidth. This higher resolution feedback provided some benefit in performance, especially for MU-MIMO scheduling.

In release 12, a 3GPP SI completed a 3D channel model that allowed to continue exploiting multi-antenna transmission benefits in vertical domain. The previous channel model assumed that all UEs are on the "ground plane", while in reality they are distributed vertically, for example, on different floors in a building facing the eNB. The new 3GPP 3D UMi and UMa models distributed UEs between 1.5 and 22.5 meters height. Moreover, support for modeling of two-dimensional antenna arrays was added.

8.3.4.5 Release 13: Full-dimensional MIMO enhancements

In release 13, the term *full-dimensional MIMO* (FD-MIMO) was introduced, with the purpose of exploiting the vertical domain as well as the horizontal domain in the precoding and focusing of radiated energy. Hence, release 13 is the first release that aims at exploiting the benefits of massively large antenna arrays with many steerable antennas. TM 10 supports all features introduced for FD-MIMO, while TM 9 supports a subset of these new features. Hence, new CSI feedback precoding matrix codebooks were defined for 8, 12, and 16 TX antennas where both 1D and 2D array layouts were supported (assuming a model of equally spaced antennas). Hence, 2D precoding codebooks were defined as well.

There was also in release 13 scope an aim to reduce the UE complexity in measuring so many antenna ports for CSI measurement, by introducing precoded or beamformed CSI-RS, which also improves the coverage for each port. The principle is that the eNB may roughly know the preferred channel direction to the UE seen from the eNB (e.g., the vertical angle) and can then transmit one or multiple beams with CSI-RS to the UE (e.g., in different horizontal angles), and then the UE selects a preferred beam before reporting CSI for that selected beam. The number of CSI-RS ports per beam can then be kept small and since the eNB does part of the precoder down selection (e.g., by selecting vertical beam), the complexity is somewhat reduced for the UE while somewhat increased for the eNB. To support this in release 13, a Class A and Class B CSI feedback terminology as well as measurement restriction was introduced (see Section 8.3.2.3).

In addition, SRS was enhanced in release 13 for small cell operation (short delay spreads) and when using TM 3, 8, 9, and 10. The enhancement involved the use of a lower density SRS (a comb structure with every fourth subcarrier) allowing increased SRS capacity.

8.3.4.6 Release 14: Massive MIMO enhancements

In release 14, the number of CSI-RS ports was increased to 32 with associated precoding matrix codebooks for CSI feedback, supporting both 1D and 2D antenna arrays (modeled as equally spaced with dual-polarized antennas). The CSI-RS overhead with so many ports becomes large and therefore an optional reduced-density CSI-RS was introduced. For Class B operation, some further enhancements to reduce CSI-RS overhead were introduced, including aperiodic CSI-RS transmission triggering in any subframe and medium access controlled (MAC) control element (CE)–based signaling to disable and enable a periodic CSI-RS transmission. Using MAC CE is faster than the RRC control supported by earlier releases and uses the PDSCH to carry the control information. This MAC CE–controlled transmission was termed *semipersistent* CSI-RS transmission.

Release 14 also introduced advanced CSI feedback (see Section 8.3.2.4), developed mainly for enhancing MU-MIMO performance by enabling a richer, more detailed channel feedback at the expense of increased feedback overhead. A hybrid CSI feedback mode was also introduced, which allows for switching back and forth between Class A and Class B type of feedback, to save CSI-RS overhead. Another CSI feedback mode introduced in release 14 was the semiopen-loop CSI feedback mode, where only the long-term channel properties (precoder) are fed back to the network.

8.3.4.7 Release 15: Multi-point transmission enhancements

In release 15, support for *non-coherent joint transmission* (NC-JT) was introduced, where a UE can receive one PDSCH containing layers from two different cells simultaneously. This required some extensions of the QCL framework where PDSCH DM-RS can be split into two groups where each group has its own source RS for QCL purpose.

8.3.4.8 Release 16: Enhancement targeting reciprocity-based operation

In release 16, SRS enhancements were targeted, with aim to increase the SRS capacity to better support reciprocity-based MIMO operation for DL transmission. This is achieved by allocating additional SRS symbols in a subframe where a release 16 UE can use a configurable number of symbols for SRS, including all the symbols in a subframe, if desirable.

8.3.5 PUSCH TRANSMISSION PROCEDURES AND UPLINK TRANSMISSION MODES

For UL, release 8 supported only single-layer PUSCH transmission, but in release 10, two TMs (TM 1 and TM 2) were defined for PUSCH, the single-port TM 1 (single-layer data) and the closed-loop spatial multiplexing mode that allows for up to four-layer PUSCH transmission (TM 2). LTE uses DFT-precoded OFDM for PUSCH (see Chapter 5) to reduce the PAPR of the waveform, sometimes referred to as the "single carrier property."

In release 8, two UEs could not be paired for MU-MIMO in the UL if their scheduled bandwidths were partially overlapping (i.e., the two UEs had different scheduling bandwidths). This is because UL DM-RS sequences with different CS for PUSCH are only orthogonal if they have the same length. Thus only CS can be used to separate users.

In release 10, a length 2 OCC was introduced for DM-RS to separate two DM-RS ports by OCC in addition to CS (see Section 8.3.1.6). Hence, MU-MIMO multiplexing capacity using orthogonal DM-RS was doubled compared to previous release, and partially overlapped MU-MIMO scheduling was possible if UEs are assigned different OCCs. Since one UE may be power-limited scenario, it can only transmit on a few RBs, while another UE closer to the eNB can use more; the situation that two UEs with different scheduling bandwidths are scheduled simultaneously is not uncommon. Hence, support for partially overlapping frequency resources was motivated.

In release 14, the UL DM-RS was further enhanced with the aim to support UL MU-MIMO with even higher multiplexing capacity. This was enabled by a comb-based DM-RS, in which a DM-RS sequence for an antenna port is mapped to either odd or even subcarriers (instead of all subcarriers in the previous releases). Two UEs with DM-RS on different combs can thus be scheduled with independently assigned resource allocations without losing DM-RS orthogonality.

8.4 LTE SUMMARY

An LTE evolution stretching over eight releases and several SI has resulted in a rich multi-antenna toolbox containing a variety of features including those useful for AAS deployments. Codebook-based, noncodebook-based, and reciprocity-based MIMO are supported for both TDD and FDD. The evolution moved from a robust design with few transmitting antenna ports to a very large number of antenna ports allowing for both horizontal and vertical beamforming simultaneously.

Current LTE has a limitation to two MU-MIMO scheduled UEs if orthogonal DM-RS ports are to be used. Hence, LTE was not primarily designed for MU-MIMO with many layers although the basic functionality is supported and somewhat enhanced to four UEs in a later release. Hence, non-orthogonal DM-RS must be used for higher-order MU-MIMO.

There are strong dependencies on UE capability support for reaching optimum performance since most UEs cannot implement the full toolbox but only the basic features from early releases.

An LTE problem in its evolution is the always present CRS, which causes a large overhead for DM-RS-based TMs. For these DM-RS based TMs, the UE does not need CRS for PDSCH demodulation or measurements, only for fine synchronization. When used only for this purpose, the CRS density is unnecessary high.

In addition, the CRS transmitted in neighboring cells causes interference and is transmitted even when there is no traffic in a cell, and the CRS cannot be disabled since it is a common signal to the entire cell. Hence, the full potential of MIMO and AAS scalability to many steerable transmit antennas cannot fully be exploited with LTE due to the interference and overhead by presence of CRS.

8.4.1 LTE—WHAT IS IN THE PIPE?

It is likely that LTE and NR in carrier frequencies below 6 GHz will evolve side by side when it comes to AAS functionality. The reason is that the same radio hardware and baseband software are commonly shared between the two radio access technologies, and it makes sense that functionality is similar for the two.

As discussed, LTE has several limitations, and the most critical ones are related to the CRS. Over time, traffic will be carried over to NR networks and dynamic spectrum sharing between LTE and NR is likely to be used for a foreseeable time.

However, one obstacle for LTE MIMO evolution toward AAS is the lack of commercial availability of UEs that support an advanced MIMO feature (e.g., 32-port CSI-RS) as those are introduced in the later releases.

REFERENCES

[1] E. Dahlman,, S. Parkvall,, J. Sköld,, 4G LTE/LTE-Advanced for Mobile Broadband, Academic Press, 2013.
[2] O. Liberg,, et al., Cellular Internet of Things: Technologies, Standards, and Performance, Academic Press, 2019.
[3] S.M. Alamouti, A simple transmit diversity technique for wireless communications, IEEE J. Sel. Areas Commun. 16 (8) (1998) 1451–1458.
[4] B. Hassibi, B. Hochwald, Linear dispersion codes, in: Proceedings of 2001 IEEE International Symposium on Information Theory, 2001.
[5] G. Bauch, T. Abe, On the parameter choice for cyclic delay diversity based precoding with spatial multiplexing, in: Proceedings of IEEE GLOBECOM 2009.
[6] 3GPP TR 36.913, V8.0.1 (2009-03), Requirements for further advancements for evolved universal terrestrial radio access (E-UTRA).
[7] 3GPP TR 36.814, V2.0.1. (2010-03), Further advancements for E-UTRA physical layer aspects.
[8] 3GPP TR 36.871, V11.0.0. (2011-12), Downlink multiple input multiple output (MIMO) enhancement for LTE-Advanced.
[9] 3GPP TR 38.211, V15.5.0. (2019-03), NR; Physical channels and modulation.
[10] 3GPP TR 36.897, V1.0.1. (2015-06), Study on elevation beamforming/full-dimension (FD) MIMO for LTE.

CHAPTER

3GPP PHYSICAL LAYER SOLUTIONS FOR NR 9

This chapter describes the parts of the 3GPP new radio (NR) release 15 specifications related to advanced antenna system (AAS), that is, the content of the developed multi-antenna toolbox of features. It is recommended that the reader is familiar with the concepts and terminology of the basic long-term evolution (LTE) as introduced in Section 8.1, such as antenna ports and quasicolocation (QCL) since these are fundamental concepts also for NR. In addition, the reader should be familiar with OFDM–based modulation of Chapter 5, OFDM–Based MIMO, multi-antenna concepts in Chapter 6, and high-band aspects of Chapter 7.

This chapter will by no means provide a complete overview of 3GPP NR specifications. The focus of this book is an in-depth explanation of the multi-antenna aspects of lower layers. For a broader presentation of NR specifications, please refer to, for example, Ref. [1].

9.1 NR BACKGROUND AND REQUIREMENTS

The standardization of NR began in April 2016 and the first version of the release 15 NR specifications was completed in December 2017. Corrections and maintenance of these release 15 specifications continued toward the beginning of 2020. The physical layer specifications for NR can be found in Refs. [2–6].

The starting point of NR description in this chapter is to first reexamine LTE, as to highlight the main lessons learned from LTE multi-antenna standardization and the associated LTE network operation. These findings are summarized as follows:

- The presence of cell-specific reference signal (CRS) in every subframe in LTE and no possibility to disable it prohibit energy-efficient operation, create unnecessary interference, and make LTE more difficult to evolve (since CRS creates both a backward and a forward compatibility issue). A "leaner" carrier design for NR where presence of any "always-on" signal is minimized is desirable, which also enable forward compatibility for NR;
- The density of demodulation reference signals (DM-RS) in the time–frequency grid is for LTE designed for the worst case (for robustness) and is in many practical scenarios unnecessarily high, which creates a large RS overhead that limits the spectral efficiency. A configurable DM-RS density is thus desirable for NR to adapt depending on scenario;
- The channel-state information (CSI) feedback in LTE is tightly coupled to the transmission modes, which means that large standardization efforts are needed to introduce new network features as both new transmission modes and new CSI feedback schemes need to be designed. Decoupling of CSI framework and data transmission framework is thus desirable for NR;

Advanced Antenna Systems for 5G Network Deployments. DOI: https://doi.org/10.1016/B978-0-12-820046-9.00009-5

- Transmission modes in LTE are configured by higher-layer signaling, which means that fast switching between them is not possible. This is a drawback that has been observed in live LTE networks. Faster switching of the way data is transmitted and/or minimizing the number of specified transmission modes are desirable for NR;
- LTE does not scale well with an increased number of base station antennas since broadcast and common signals rely on CRS for demodulation and CRS is limited to four antenna ports. Better and more efficient utilization of massive multiple-input, multiple-output (MIMO) antenna arrays including inherent multiple user MIMO (MU-MIMO) support is thus desirable for NR;
- Using discrete Fourier transform (DFT)-precoded OFDM in the uplink (UL) limits the efficiency of UL MIMO operation, as the UL precoding matrix must be designed to maintain the single carrier property of the waveform. Hence, NR should be able to operate without DFT precoding for UL, thus making UL and downlink (DL) transmit waveforms equal.

Hence, these learnings were part of the foundation for the standardization work and some key and overall design principles for NR were expressed in Ref. [7] and can be summarized as follows:

- NR should have an inherently forward compatible design, making it easy to add features in coming releases;
- NR should provide ultra-lean transmission where transmissions are self-contained, for example, DM-RS used for demodulation are embedded in the scheduled physical downlink shared channel (PDSCH)/physical downlink control channel (PDCCH) resource elements;
- NR should have a scalable numerology with a subcarrier spacing (SCS) in multiples of 15 kHz, as 15 kHz is used in LTE;
- NR should have support for dynamic TDD where any slot can be decided to be a DL or UL slot "on the fly."
- NR should have support for massive MIMO with hundreds of antenna elements, beamforming, MU-MIMO, and reciprocity-based operation;
- NR should be possible to operate in both unlicensed spectrum and licensed spectrum.

Moreover, the requirements on NR release 15 include support for

- frequency bands up to 52.6 GHz carrier frequency including support for analog beamforming implementations;
- user equipment (UE) speeds in the range from stationary to high speed, up to 500 km/h; and
- MU-MIMO operation with a large number of co-scheduled UEs in both FDD and TDD.

These requirements are all differentiating from LTE. On the UE side, a significant difference from LTE is that reception using four receive antennas is mandatory for some frequency bands (typically bands above 2 GHz and below 7 GHz) in case the frequency band is used as a standalone band. The four receive antennas likely improve coverage and performance compared to LTE that use only two receive antennas as the baseline (although UE with four receive antennas exist in LTE as well).

Moreover, compared to the first LTE release, there are very few signals and channels that are tied to the cell ID in NR. The consequence of this, plus that the QCL framework is central and more elaborate in NR compared to LTE, is that the cell association becomes looser when a UE is connected. The NR UE connects to a cell at initial access, but after that, the UE may receive data from any of multiple transmission points within the cell, since the parameters that configure the transmission are controlled by dedicated signaling and not derived from cell ID as in LTE release 8. However, a complete decoupling of the UE from a cell in connected mode is not achieved in NR, only locally. When a UE moves a larger distance, inter-cell handover procedures are used, as in LTE. Nevertheless, NR allows for more flexible and creative network deployments such as advanced heterogenous deployments, then what is possible with LTE.

9.2 NR PHYSICAL LAYER—BASIC PRINCIPLES

While LTE targeted a single design, robust for any scenario, NR on the other hand introduces configurability that facilitates scenario-dependent optimization, where overhead and density of reference signals, the SCS, the bandwidth, etc., can be configured on a need basis. For example, in a less demanding scenario where UEs are stationary and with flat fading channels, the reference signals and control signaling overhead can be kept low. On the other hand, in a more demanding scenario, more densely configured reference signals with associated increased overhead need to be used to ensure good performance. In addition, the periodicity of mandatory (always-on) transmitted signals from the network has been minimized to every 20 ms in NR compared to mandatory transmitted CRS in every subframe (hence, less than every millisecond) for LTE.

Similar to the LTE overview in Section 8.2, a brief overview of NR physical resources and channels will now be given. Section 9.3 and forward will describe in more depth different reference signals, transmission schemes, and channel measurement functionalities.

NR uses OFDM with a cyclic prefix (CP) for DL and UL transmissions (see Chapter 5). For the UL, it is also possible to configure an alternative DFT-based OFDM transmission for UL coverage extension purpose. The use of non-DFT-precoded OFDM can enable low implementation complexity for wide bandwidth operations and supports MIMO features well, since no consideration is needed to maintain the single carrier property of the waveform. Therefore, the DFT-based configuration in UL supports a single physical uplink shared channel (PUSCH) layer transmission while the OFDM-based UL supports up to four layers.

NR supports a scalable numerology as a very wide range of spectrum needs to be supported. Therefore a configurable SCS of $2^{\mu} \cdot 15$ kHz ($\mu = 0, 1, \ldots, 4$) is used where the basic 15 kHz SCS is used in LTE. The CP is scaled by a factor of $2^{-\mu}$ from the CP length of 4.7 μs used in LTE, which reduces the robustness against, for example, channel delay spread. However, a scalable design allows support for a wide range of deployment scenarios and carrier frequencies instead of only supporting the worst case of delay spread as in the LTE design.

3GPP has so far specified operation in two *frequency ranges* (FR), denoted as FR1 and FR2, respectively.

- *FR1*: 450 MHz to 7.125 GHz, initially known as "sub-6"[1]
- *FR2*: 24.25–52.6 GHz, commonly known as "mm-wave"

In FR1 deployments, large coverage deployments can be targeted in which case also channel delay spreads can be large. In this case, the subcarrier spacings of 15 and 30 kHz are suitable. At

[1] In early phases of release 15 standardization, the range for FR1 was up to 6 GHz, hence the term "sub-6."

millimeter-wave frequencies, phase noise becomes problematic. Hence, FR2 deployments use a larger subcarrier spacing that reduces the sensitivity to phase noise. A subcarrier spacing of 60 or 120 kHz is used for data and 120 or 240 kHz for the SS/PBCH block (SSB) used for initial access in FR2. Larger subcarrier spacing is suitable in deployments where delay spread typically is smaller. The commonly used beamforming in FR2 where the transmit beamwidth is narrower than the sector width in angle typically changes the channel delay spread (see Section 3.6 on multi-path propagation).

The bandwidth of an NR carrier is UE−specific and thus configured by dedicated signaling to the UE as opposed to LTE where all UEs served by the cell assume the same system bandwidth (although the LTE evolution includes narrowband operation as well, cf. LTE-machine type communication and narrowband Internet of things, see Ref. [8]).

Hence, even if an NR carrier has 100 MHz bandwidth from the network perspective, some NR UEs may only utilize a 20 MHz part of this bandwidth. This allows for diverse UE implementations where, for example, a low-cost UE (e.g., a sensor equipment) may only need a small bandwidth and still connect to the same wide NR carrier as other UEs. Also, NR introduces the concept of a *bandwidth part* (BWP), which is the current active bandwidth for a UE. A UE can be simultaneously configured with up to four BWP, while only one is active at a given point in time. Dynamic BWP switching of active BWP is supported. An application of this is if a UE has low data rate demands, it can be configured a small bandwidth BWP (e.g., 20 MHz) to save UE battery, while if the data demand temporarily increases, the network can switch to a large BWP (e.g., 100 MHz) for that UE.

The NR time structure is divided into frames, subframes, and slots as in LTE although *a slot in NR is equivalent to a subframe in LTE and they have equal time duration for 15 kHz subcarrier spacing case*. A frame duration is 10 ms and consists of 10 subframes. Since LTE uses the same structure, NR and LTE can coexist and share the same bandwidth, known as spectrum sharing. An NR subframe consists of 2^{μ} slots, ($\mu = 0, 1, \ldots, 4$) where μ controls the SCS and where each slot contains 14 OFDM symbols. A slot is the unit for scheduling of PDSCH and PUSCH and as μ increases, the slot duration is reduced as (1, 0.5, 0.125 ms,...), thereby providing reduced latency.

NR additionally supports PDSCH and PUSCH transmissions that start and end at any OFDM symbol in the slot, and there could be multiple of these in a slot. This is sometimes named "mini-slot" transmission and it can be as small as a single OFDM symbol, facilitating very low latency for time-critical data which was one of the important requirements of NR.

A physical resource block (PRB) consists of 12 consecutive subcarriers in the frequency domain and a slot in time domain (see Fig. 9.1). An NR carrier is limited to at most 3300 subcarriers, or 275 PRB. The maximum carrier bandwidth in FR1 is 100 MHz, and the maximum carrier bandwidth in FR2 is 400 MHz which should be compared to the maximum LTE carrier bandwidth of 20 MHz.

A difference to LTE is that the control channel, the PDCCH, can be configured to be in any symbol of the slot and in multiple positions in the slot. A use case of this is the ultralow-latency operation, where it is important to transmit a scheduling assignment using PDCCH as soon as possible as the data is arriving from higher layers. The configuration of multiple such control regions in a slot, that is, more frequently than once per slot, reduces the scheduling latency to wait for the next available control region. However, support for receiving the control channels very frequently is not mandatory for UE to support and will primarily be supported by UEs supporting lower latency.

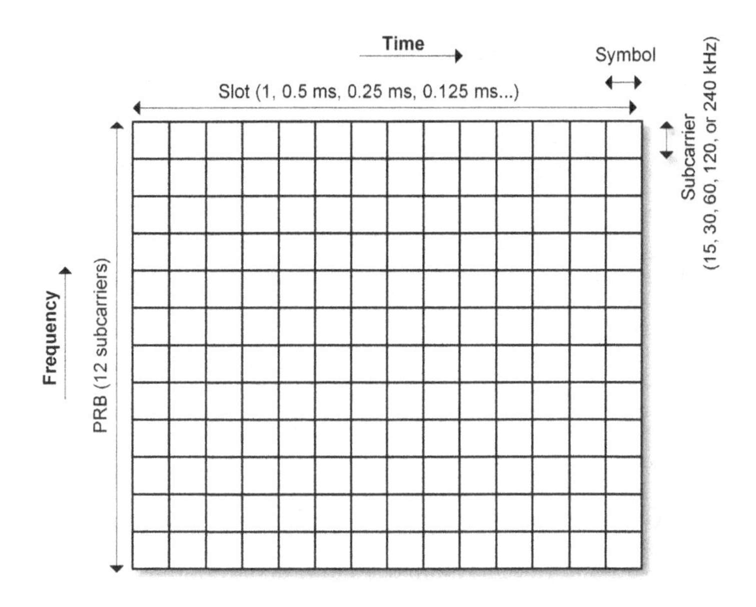

FIGURE 9.1

The NR downlink physical resource showing a PRB having 12 subcarriers by 14 OFDM symbols (one slot).

Considering only what is mandatory for UEs to support, then the NR control channel location in the slot becomes more like LTE, that is, using a control region in the beginning of the subframe (1–4 first OFDM symbols), while the remainder of the subframe is a data region, as was shown in Fig. 8.2 for LTE. An important difference with LTE is, however, that the control region does not have to span the whole bandwidth. Only a part of the bandwidth can be used and PDSCH can be mapped outside this control region, even in the same OFDM symbol as the PDCCH. This is useful for the large NR bandwidths, where it may be unnecessary to utilize the whole bandwidth for control channel transmissions. It is also forward compatible since a future feature can use all 14 symbols of a slot without colliding with the control region of legacy operations.

In Table 9.1, the physical channels in NR are listed and comparing with the LTE channels in Table 8.2, it can be observed that NR has fewer defined control channels.

NR standardization also addressed the coverage issue at high frequencies by enabling the possibility to beamform any channel (see the introduction of concepts related to high-frequency operation in Chapter 7). The procedures in NR that facilitate efficient beamforming of common channels are referred to as *beam management* procedures. In NR, beamforming is performed for dedicated, for common, as well as for broadcast and synchronization channels.

The NR physical layer specification is mostly FR-independent, but there are a few occasions where FR2 is mentioned explicitly. One example is when it comes to separation of UE capabilities, where a UE may support a feature related to beam-based operation for FR2 as mandatory and in FR1 as optional.

Beam management features are mainly targeting, but not exclusive to, operation in FR2, in case the UE supports them in FR1 they can be used there as well as the specifications does not

Table 9.1 Physical Channels in NR Release 15.

Physical Channel	Name	Content
PBCH	Physical broadcast Channel	Part of the system information to all UEs in the cell
PDCCH	Physical downlink control channel	Downlink control information (DCI), for example, for scheduling PDSCH and PUSCH
PDSCH	Physical downlink shared channel	Downlink unicast data, paging, and parts of the system information
PUSCH	Physical uplink shared channel	Uplink data
PUCCH	Physical uplink control channel	UCI to indicate HARQ-ACK for PDSCH and for CSI feedback
PRACH	Physical random access channel	Uplink channel used in random access

forbid this. The principles of beam management were laid out in Section 7.4 and how these were implemented in NR specifications will be further discussed in detail in Section 9.3.6.

To summarize beam-based operation in NR, the aim is to:

- support a beam-based initial access procedure;
- handle measurements to find preferred DL and UL beam candidates for data and control transmission;
- indicate a DL beam for the UE to assist the UE to select a corresponding receive beam when scheduling DL;
- indicate a UL beam to allow the gNB to select a corresponding receive beam when scheduling UL;
- support a beam recovery procedure in case the link is lost without involving higher-layer radio link failure procedure;
- manage mobility in a beam-based network.

The specified beam management framework also considers the PDCCH decoding time and the time it takes for a UE to perform beam switching. For example, the PDCCH that schedules a PDSCH in the same slot cannot indicate a new transmit beam for that PDSCH if it requires the UE to switch to a new UE receive beam. The reason is, due to decoding time, when the UE has finalized the decoding of the PDCCH that indicates a "new" beam, the PDSCH has already begun to be transmitted, and consequently, the UE has used the "old" receive beam for that PDSCH and its decoding may fail.

To resolve this, the DL beam indication in PDCCH in a slot [i.e., the transmission configuration indicator (TCI) state, see Section 9.3.3.1], refers to a future PDSCH transmission, for example, the next slot. This latency is a consequence of the use of analog beamforming in the UE that typically restricts the UE from receiving from two directions simultaneously. This kind of consideration is new to NR and was not needed in LTE due to the underlying assumption of all digital receivers.

The issues and difficulties to evolve LTE that arose from presence of CRS implied an aim of lean design for NR, where no other always-on signals are present except for the synchronization block (SSB) and the tracking reference signal (TRS) (see Section 9.3.1.1.1). The periodicity of "mandatory" transmitted signals was significantly increased from every subframe in LTE

(<0.5 ms) to 20 ms or longer in NR. In addition, NR supports configuring carriers without an SSB, where only TRS is present. Since TRS periodicity can be configured to 80 ms, this allows for an even "leaner" carrier. However, the lack of SSB on this type of carrier implies that it cannot be used for initial access, only as a part of carrier aggregation (CA) together with a carrier with SSB.

Also, the principle of "self-contained" transmissions was adopted both for PDCCH, physical uplink control channel (PUCCH), and PDSCH/PUSCH, which implies that demodulation reference signals are only present in the used RBs and used symbols for these channels, not outside. This allows forward compatibility as future new channels can be placed adjacent to release 15 PDSCH for example. In addition, it allows for beamforming any channel, even PDCCH and PUCCH, since the channel comes with "its own" reference signal, beamformed in the same way, to be used for demodulating that channel. Hence, transmission is agnostic to the number of used antennas in gNB or UE, respectively. However, complete self contained operation is not acheived in NR, since the specification requires a tracking reference signal (TRS) to be configured which is not transmitted together with the PDSCH.

CSI reporting and PDSCH transmission are decoupled in NR and it is up to the network to decide how to configure and utilize the CSI report for link adaptation of the DL transmission. This allows for use of either reciprocity-based operation or codebook-based feedback operation, or both, depending on the network or site deployment choice, without changing the PDSCH transmission scheme. Due to the use of UE–specific DM-RS, for example, LTE TM 10 like operation, there is only a single transmission mode in NR release 15 as this mode allows for arbitrary gNB precoding and thus covers a wide range of use cases (e.g., noncodebook-based transmissions as well as specification transparent antenna diversity transmissions). Hence the use of DM-RS allows the decoupling of CSI reporting from actual transmission since the network can freely use proprietary algorithms to determine any desired precoder.

To support a large number of coscheduled UEs by MU-MIMO scheduling, 12 orthogonal DM-RS ports are specified for both DL and UL and hence the signaling supports scheduling of up to 12 UEs simultaneously in MU-MIMO with orthogonal DM-RS. (Although even more can be scheduled with nonorthogonal DM-RS) This is a large improvement compared to LTE.

9.3 NR PHYSICAL LAYER SPECIFICATIONS FOR AAS

This section describes in more detail the specified functionality for the NR physical layer where focus has been given to parts that relate to AAS.

9.3.1 ANTENNA PORTS, REFERENCE SIGNALS, AND QUASICOLOCATION

The definitions of antenna ports, reference signals, and QCL are inherited from LTE; hence refer to Section 8.3.1 for an introduction.

Each reference signal in NR has primarily a single purpose and is thus optimized in, for example, density for that purpose. The RS density in the time–frequency grid is configurable. In addition, all RSs are UE-specifically configured although there are also default configurations of DM-RS. A default is needed to be able to communicate with the UE prior to radio resource control

(RRC) configuration, to receive, for example, system information in PDSCH and for the random access procedure using PUSCH.

Due to the absence of CRS in NR, a wide bandwidth *tracking reference signal* (TRS) was specified to assist the UE in performing fine time–frequency synchronization as the detected SSB cannot ensure good enough performance when MIMO and higher-order modulation is used. The synchronization by using SSB is only accurate enough to receive PDSCH and PDCCH with low-order [e.g., quadrature phase shift keying (QPSK)] modulation since the SSB is relatively narrow in bandwidth and only spans four OFDM symbols. In initial access, the UE must, however, rely on SSB-based synchronization as it occurs before the TRS has been configured.

Moreover, when NR is used in FR2, a *phase tracking RS* (PT-RS) can be configured in both UL and DL, which assists the receiver in performing demodulation in the presence of phase noise the effects of which impact the performance at high-bands. Hence, support of PT-RS is mandatory for a UE supporting FR2 operation.

In Table 9.2, the antenna ports for different reference signals are given, including the validity regions and the port numbering. Note that if the same port is used in two different regions, then the two estimated channels may be completely different, and the UE shall not interpolate or average these estimates across different regions. See the discussion in Section 8.3.1.1 for a more thorough description of "validity region" of an antenna port.

Note that a PT-RS port is oddly enough not specified to have a unique port number but is instead said to be *associated with* one of the used DM-RS ports for the PDSCH transmission.

For all channels, in particular PDCCH DM-RS, the validity region implies a self-contained transmission; hence the DM-RS used to demodulate a certain PDCCH is embedded among the resource

Table 9.2 Channels, Antenna Ports, and Region of Validity for NR Downlink and Uplink.

Channel	Port	Validity Region
PDCCH DM-RS	2000	Within a specified subset of RE used by the associated PDCCH
PDSCH DM-RS	1000–1011	One PRG (2 PRB, 4 PRB, or all scheduled RBs) and the PDSCH duration, that is, at most one slot
CSI-RS	3000–3031	• Configured bandwidth in one slot with measurement restriction enabled or for aperiodic CSI-RS • Configured bandwidth across all slots with measurement restriction disabled
PBCH DM-RS	4000	Within the RE used for the SSB with a certain SSB index
PT-RS	Associated to one of 1000–1005 (for DL)	Same as the associated DM-RS port for PDSCH or PUSCH respectively
PUSCH DM-RS	0–11 or 1000–1003	Whole scheduled PUSCH bandwidth and PUSCH duration, that is, at most one slot
SRS	1000–1003	Whole scheduled SRS bandwidth and SRS resource duration, that is, at most four OFDM symbols
PUCCH DM-RS	2000	Within a specified subset of RE used by the associated PUCCH
PRACH	4000	Not applicable (physical channel)

elements used for that PDCCH. Hence, the validity region is as compact it possibly can be. This should be compared to LTE where CRS is used for PDCCH demodulation and where CRS extends throughout the whole subframe and outside the PDCCH, that is, not a self-contained transmission. The self-contained design in NR allows for beamforming of the PDCCH/PDSCH/physical broadcast channel (PBCH) and therefore better utilization of AAS for transmissions compared to LTE.

For UL, NR uses DM-RS and PT-RS for demodulation in the same way as for DL, and sounding reference signal (SRS) is used for UL channel measurements.

9.3.1.1 Channel State Information-Reference Signal (CSI-RS)

The purpose of CSI-RS is to provide a reference signal for DL measurements. As in LTE, each CSI-RS is defined as a CSI-RS resource, where a resource can have one to 32 CSI-RS antenna ports.

In NR there is also the notion of *CSI-RS resource sets*, as a collection of CSI-RS resources. There are several occasions when it is useful to refer to a set of CSI-RS resources as will be exemplified below, but to take one example: in the case each CSI-RS resource is transmitted in a different beam, then the set represents a set of beams, and it is possible to configure the UE to report the best beam from this set. This then mimics the Class B CSI feedback in LTE (see Section 8.3.2.3).

The CSI-RS resource in NR is very flexible in its possible configuration and is UE-specifically configured by dedicated signaling. This means that only connected UEs can be aware of a CSI-RS transmission. It is a measurement signal that can be configured to be used for

- CSI measurements of channel and/or interference
- tracking of channel (configured as a TRS)
- beam management measurements
- beam failure detection
- mobility measurements
- radio link monitoring
- reservation of resources elements from PDSCH mapping

Evidently, CSI-RS resources are used for many purposes and each configured CSI-RS resource is tailored (optimized) for a single purpose, that is, a CSI-RS resource cannot in general be used for multiple purposes simultaneously. Therefore a UE will typically be configured with many CSI-RS resources simultaneously and as many as 192 CSI-RS different resources can be simultanously configured to a UE.

As mentioned earlier, a CSI-RS resource can have from one up to 32 antenna ports and is designed by a modular approach using aggregation of groups of resource elements [known as *code division multiplexing* (CDM) groups]. A CDM group is a group of $n \times m$ adjacent resource elements per PRB used for the CSI-RS resource, that is, using n adjacent subcarriers and m adjacent OFDM symbols. For CSI-RS, CDM groups of sizes 1×1, 2×1, 2×2, and 2×4 are used to aggregate CSI-RS resources from 1 to 32 ports. The possible sizes and locations of the CDM groups that constitute a CSI-RS resource with a certain number of ports are found in Ref. [2]. Note that the channel should ideally be flat within the RE of a CDM group to maintain orthogonality between the ports. Hence, larger CDM groups than 2×4, that is, eight REs, are not used. See some examples of CSI-RS resources are shown in Fig. 9.2.

A CSI-RS resource can be configured to start in any of the 14 symbols in the slot and the resource is mapped out across one, two or four OFDM symbols. The number of used symbols

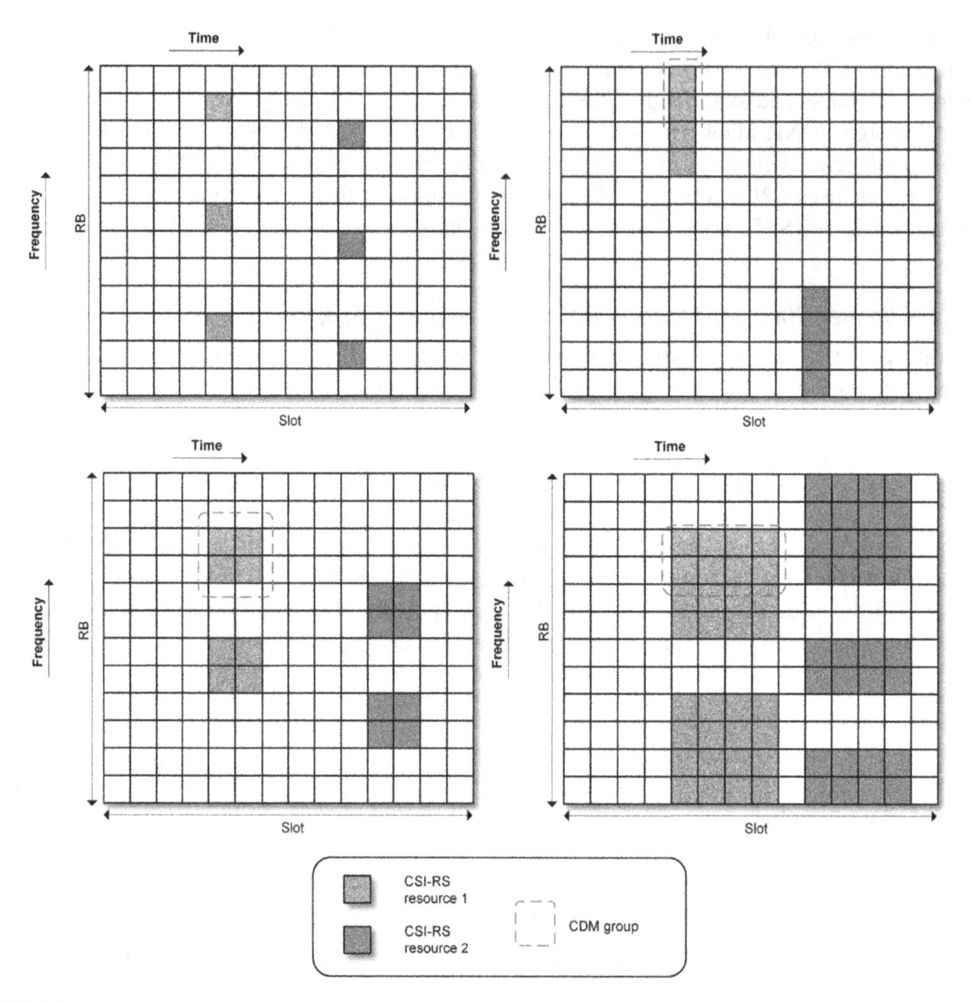

FIGURE 9.2

Four examples of CSI-RS resource configurations where upper left, upper right, lower left, and lower right figures correspond to 1, 4, 8, and 32 CSI-RS ports, respectively. For each example two (out of many) different CSI-RS resource configurations are illustrated, referred to as CSI-RS resource 1 and 2, respectively. In the 4, 8, and 32 port examples, the CDM group size is 2×1, 2×2, and 2×4 REs, respectively. Note that the eight-port CSI-RS resource is in this example mapped to two "islands" or REs and the 32-port CSI-RS resources are mapped to two and three "islands," respectively. This flexibility provides some flexibility to configure a resource, so it does not collide with other channel reference signals (in current or in future releases of NR).

depends on the number of ports the resource has, that is, one, two, or four symbols for 1–12, 4–16, or 24–32 ports, respectively.

Nominally, the CSI-RS density is 1 RE/port/PRB but it is also possible to configure with a lower density of 1/2, in case of a resource with a large number of ports in order to reduce overhead.

However, a lower density may reduce channel estimation performance. For the CSI-RS resource with a single port, there exists a special configuration that has a high-density 3 RE/port/PRB using a frequency comb structure where every fourth subcarrier carries the CSI-RS. This configuration is used for the TRS (see Section 9.3.1.1.1).

In Fig. 9.2, some examples of CSI-RS resource mapping to resource elements (known as *patterns*) are shown where each of the four subplots shows two possible configurations of a CSI-RS resource for the same number of ports.

The top left figure shows two instances of the special single-port CSI-RS resource that has density 3 RE/PRB. The CSI-RS resources in one and the same slot can be flexibly configured to be mapped to different subcarriers as the examples seen in Fig. 9.2 also show. The mapping pattern flexibility is thus very large to allow for forward compatibility as one may need to configure the CSI-RS resource to avoid collisions with reference signals or channels introduced in future releases.

The CSI-RS transmission bandwidth is flexibly configured from 24 PRB (or the used BWP if it is smaller than 24 PRB) up to the full BWP bandwidth in steps of four PRB. The CSI-RS periodicity can be configured from 5 slots up to 640 slots. This large range is used to support different subcarrier spacings since for high subcarrier spacings a large number of slots is needed to obtain a long periodicity, if desired. The maximal 640 slots corresponds to 80 ms CSI-RS periodicity at 120 kHz SCS . A CSI-RS resource can also be configured without periodicity when used as an aperiodic CSI-RS.

As mentioned in the introduction to this section, the CSI-RS resources are organized into *CSI-RS resource sets*, where a set can contain up to 64 CSI-RS resources. A certain resource may belong to multiple sets configured to the same UE. The use of CSI-RS resource sets will be explained in the following subsections where each CSI-RS use case is discussed separately.

Since a certain CSI-RS resource set can be configured for one out of many different purposes, this is indicated in the set configuration message. Table 9.3 shows the possible configuration parameter of a CSI-RS resource set and may seem cryptic to the reader. For example, the use of a set for beam management is indicated by the presence of a parameter "repetition." However, the rationale for this naming will be clarified in Section 9.3.1.1.2. In addition, the CSI-RS used for

Table 9.3 The Configuration of a CSI-RS Resource Set Determines the Use of the Set and Is Controlled by Whether the Parameter Is Present or Not in the RRC Configuration of the Resource Set.

CSI-RS Resource Set Configuration Parameter	Use Case
"trs-info"	If parameter is present, the CSI-RS resources in the set can be used for channel tracking and fine synchronization (i.e., a TRS)
"repetition"	If parameter is present, the CSI-RS resources in the set are used for beam management–related measurements and associated beam CSI reporting. The parameter value can be "on" or "off"
Neither of "trs-info" or "repetition" is present	If both parameters are absent, then the CSI-RS resources in the set are used for CSI acquisition

mobility measurements are configured separately and thus have none of the parameters as in Table 9.3.

Finally, as in LTE, a CSI-RS resource can be configured for channel measurements, that is, it has a *non-zero transmit power*, known as NZP CSI-RS. Alternatively, it can be configured as a *zero-power* CSI-RS (ZP CSI-RS), in which case the resource elements used by the resource are empty. The use cases for a ZP CSI-RS are elaborated in the subsections below.

9.3.1.1.1 CSI-RS configured as TRS

The aim of the TRS is to provide a periodic reference signal used for fine synchronization and channel analysis for the UE. This is necessary for both DL and UL reception since it allows the UE to synchronize the local oscillators in the gNB and UE. When a UE is connected, the 3GPP specification states that *it expects* to be configured with a TRS transmission from the network, to be able to perform this synchronization. Hence the TRS is an "always-on" transmitted signal although it can be disabled if there are no UEs connected to the cell, as only connected UEs are aware of TRS. Idle UEs perform synchronization solely on SSB when accessing the cell.

The TRS is configured as a CSI-RS resource set of two or four single-port CSI-RS resources across one or two adjacent slots (known as a TRS burst), respectively, and where "trs-Info" is set to "true" (see Fig. 9.3).

The special single-port CSI-RS resource configuration with density 3 RE/PRB (every fourth subcarrier) is used for the TRS. This is a comb structure in frequency direction and the distance between the two CSI-RS resources in each slot is fixed to four OFDM symbols (see Fig. 9.3). The absolute symbol position of the TRS in the slot can be configured in order to avoid collisions with other configured reference signals, for example, the PDCCH or SSB.

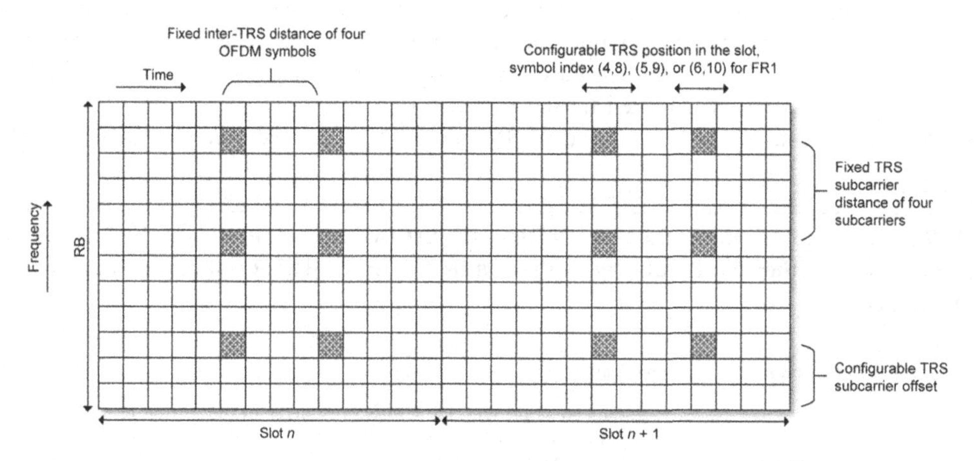

FIGURE 9.3

One possible configuration of TRS with a two-slot burst where TRS is mapped to symbol indices (5,9) in each of two adjacent slots (where first symbol in the slot has index 0). There are four possible subcarrier offsets and three possible symbol positions (in FR1) allowing for some configuration flexibility to avoid collisions with other configured channels and reference signals.

The TRS has a regular frequency pattern (i.e., comb) that is suitable for time offset estimation and a time extent over two slots that can be used for Doppler estimation. The choice of four symbol separation was selected as a good trade-off between accuracy and coping with large Doppler offsets and frequency errors. Moreover, although four single-port CSI-RS resources are used in the set that defines the TRS, these are in this special case defined and unified as a single port, that is, the UE can safely assume that they are coherently transmitted from the gNB so that the UE can jointly use all four CSI-RS resources of a TRS when estimating large-scale channel parameters.

For FR2, there is also an additional possibility to configure a *single-slot* TRS burst with only two OFDM symbols carrying the CSI-RS. The motivation was that for FR2 with analog beamforming where the gNB can only transmit in one direction at a time, a two-slot TRS would occupy too much resources just for the sake of TRS transmissions. Hence, a single-slot TRS can be used in FR2 at the risk of slightly worse channel tracking performance.

The TRS periodicity can be configured to 10, 20, 40, or 80 ms where the motivation for the very short 10 ms configuration was to support the case of high-speed train scenario. The UE can utilize and combine measurements from multiple TRS bursts if desired. Note that the SSB has periodicity of 20 ms and together with TRS, which is configured when the gNB is serving a UE, these are the mandatory always-on transmissions from the gNB. This enables a significantly lower overhead of "always-on signals" in NR compared to LTE.

In addition to the periodic TRS, which the UE expects the network to configure, there is also optional UE support for configuring an aperiodic TRS. This can be triggered by the gNB to assist the UE with a TRS measurement directly after discontinuous reception wake up or a handover, so the UE can quickly fine synchronize its receiver. This would improve subsequent reception and transmissions without needing to wait for the periodic TRS transmission, which may be rather long, up to 80 ms.

The TRS bandwidth can be configured as either 52 PRB (if the BWP bandwidth is large enough) or the full BWP bandwidth as a gNB choice.

Note that the CSI-RS resources configured as TRS cannot simultaneously be used for CSI feedback-based measurements, as the specification forbids such double functionality configuration, that is, a CSI report based on TRS cannot be configured. This restriction was introduced to keep UE complexity down. Hence, a TRS cannot be used for CSI feedback but it can, however, be used for RLM.

9.3.1.1.2 CSI-RS configured for beam management

The purpose of CSI-RS for beam management is aimed at providing a low overhead per resource and many resources as the number of beams can be large. Hence, for beam management–related measurements (discussed in Section 9.3.6.2), a CSI-RS resource set can be configured with multiple CSI-RS resources of one or two ports each, for received power (reference signal received power, RSRP) measurements. The intention of using a set of resources is that each CSI-RS resource can be transmitted in a different beam, that is, a beam sweep.

The single-port configuration implies a comb-based CSI-RS pattern with density every 12th or every 4th subcarrier. When the repetition parameter in the CSI-RS configuration has the value set as "on," then the UE can assume that the same spatial transmission filter (e.g., beam) is used by the gNB for all transmitted CSI-RS resources in the set, commonly referred to as Procedure 3 (P3) (see Section 9.3.6). The P3 operation allows the UE to use these repeated transmissions to evaluate different receive beams. If the repetition is "off," then the UE cannot make such assumption and thus the gNB may perform a beam sweep using the multiple CSI-RS resources in the set and the UE is thus measuring the received power for multiple gNB transmit beams. This is known as P2 during standardization. For further discussions of P1, P2, and P3, see the beam-based operations in Section 9.3.6.

Measurement on CSI-RS configured for beam management involves only received power (RSRP) measurements but contrary to RSRP measurements for mobility, the related measurement report is feedback using layer 1 signaling (i.e., carried by the CSI feedback), hence the notion of L1-RSRP. Reporting of L1-RSRP can be configured for simultaneous reporting of up to the four best CSI-RS resources (i.e., four beams). Note that interference is not considered in these beam management measurements but will be included in release 16 NR enhancements of beam management.

If a CSI-RS resource set is configured for beam management (with "repetition" parameter present), then the specification prevents that this set is used for the "normal" CSI acquisition, as described in the next section. Hence, the network needs to choose one usage only for each configured CSI-RS resource set.

9.3.1.1.3 CSI-RS configured for CSI acquisition

The use of CSI-RS for CSI acquisition is the "normal" use of the CSI-RS since its first introduction in LTE, aimed to provide a measurement signal for CSI reporting.

The CSI-RS resource set is used for this "normal" CSI measurements when neither *"repetition"* or *"trs-Info"* parameters are present in the CSI-RS configuration. A configuration of a resource set with multiple NZP CSI-RS resources can mimic the Class B operation in LTE (see Section 8.3.2.3), where the gNB can transmit the different CSI-RS resources using different precoding (e.g., vertical beams) and let the UE select one of them for computing the full CSI feedback [i.e., transmit precoding matrix (PMI), rank indicator (RI), channel quality indicator (CQI)].

The difference of this configuration compared to a set of CSI-RS resource configured for beam management is that a full CSI report is computed while if the set is configured for beam management, only L1-RSRP can be feedback. As the computation is more demanding for a full CSI report and since a CSI-RS in this case can have up to 32 ports, only a single CSI-RS resource from the set is selected for CSI feedback.

If a single CSI-RS resource is configured in a set, then this mimics Class A operation from LTE, where nonprecoded CSI-RS is used. Refer to the discussion in Section 8.3.2.3. Note that the NR CSI framework flexibly supports these (e.g., Class A and B as in LTE) and other different methods, without explicitly specifying these as independent "classes." This was one of the intentions with the CSI framework design in NR, that is, to avoid specifying a different CSI feedback class for each use case.

For channel measurements, the NZP CSI-RS is used and for interference measurements, the interference measurement resource (CSI—interference measurement, CSI-IM) or the NZP CSI-RS can be configured. The CSI-IM is a ZP CSI-RS with a size of 2×2 or 4×1 resource elements in adjacent subcarriers times OFDM symbols (per PRB). This is further discussed in the CSI configuration (Section 9.3.5.2).

9.3.1.1.4 CSI-RS configured as reserved resources

A ZP CSI-RS resource can be configured as a reserved resource, in case some resource elements need to be protected from transmissions, that is, these are not used for PDSCH transmission from the gNB and the UE can assume that the PDSCH to RE mapping skips these REs. The use cases for such configuration is the same as in LTE (refer to Section 8.3.1.4 for further elaboration).

9.3.1.1.5 CSI-RS configured for mobility

CSI-RS can also be configured for mobility measurements. Each of these NZP CSI-RS resources is UE-specifically configured single-port resources and hence not tied to the cell ID. This allows a large flexibility in the use of mobility measurements compared to LTE where signals used for L3 mobility are associated with a cell ID. However, the UE may still need an SSB to perform coarse synchronization before being able to receive the CSI-RS, so some dependence to a cell still exists even if CSI-RS is configured for L3 mobility. The reason is that SSB [e.g., primary synchronization signal (PSS) and secondary SS (SSS)] sequences depend on the cell ID. A CSI-RS resource is thus always associated with either the SSB of the serving cell or another configured SSB.

A distinguishing fact of a configured CSI-RS for mobility is that even if a UE is configured with such, it should not assume that its simultaneously received PDSCH is mapped around the REs occupied by any CSI-RS configured for mobility. This is different from all other use cases of CSI-RS in NR where PDSCH is mapped around a configured CSI-RS resource. The reason is that a UE can be configured with many such mobility resources (e.g., associated to neighbor cells and/or transmission points) and the overhead impact to PDSCH would be too large. Also, even if a UE is configured with many resources, it can only detect and receive a few of them as the path loss to most of them is too large for a given geographical UE position in the network.

9.3.1.2 PDSCH and PUSCH DM-RS

For demodulation of data channels, UE-specifically configured reference signals (DM-RS) are used. This is in line with the self-contained principle of NR, meaning that irrespectively of the number of scheduled OFDM symbols for data transmission, the DM-RS associated with the scheduled PDSCH are only transmitted within the scheduled resources (where a *scheduled resource* is the scheduled RBs and the scheduled OFDM symbols for the PDSCH or the PUSCH).

The DM-RS density is configurable in time and frequency and the number of OFDM symbols used for DM-RS in the scheduled resource can be 1, 2, 3, or 4. The DM-RS is designed by targeting an equal-distance symbol spacing between each DM-RS symbol, whenever possible.

In the following, the DM-RS symbol positions in a slot will be discussed first, followed by a description of the DM-RS structure within a DM-RS symbol.

9.3.1.2.1 DM-RS positions in the slot

A need to support flexible scheduling with shorter transmissions than a slot was identified for NR in order to have very low latency. Hence, two scheduling types, A and B, are defined where Type A is known as slot-based scheduling and Type B is known as nonslot-based scheduling (also referred to as mini-slots).

In Type A scheduling, the PDSCH is scheduled to be mapped to OFDM symbols from the beginning of the slot, where the start is flexible, in any of the four first symbols. A delayed start in the slot is used when PDCCH occupies the first symbols of the slot. Nevertheless, the DM-RS positions are the same irrespectively of the used start position. Scheduling of Type A can also end early in the slot, which is used in the cases the end of the slot contains other transmissions such as CSI-RS or if a switch to a UL is scheduled or configured.

Type B scheduling was introduced to support, for example, small packets with low-latency requirements and operation in unlicensed bands. It can also allow for a more efficient use of analog beamforming for small packets since in a given OFDM symbol, only one beam can be transmitted. Hence, using Type B scheduling, two different beams can be transmitted to two different UEs in the same slot (time multiplexed in different OFDM symbols), if the packets are small enough to fit into a few OFDM symbols. For Type B scheduling the PDSCH can be placed anywhere in the slot (hence the name nonslot-based scheduling) as long as the transmission end symbol is within the same slot as the start symbol.

The possible lengths of Scheduling Type A are shown in Table 9.4 where in the figure it is assumed that the data always start at the first symbol of the slot. The "duration" is for Type A always measured from the first symbol to the last symbol containing data in the slot. Note that even if for the PDSCH case, the actual start symbol may be any of 1–4 (see Table 9.4), the duration is measured from symbol 1.

The DM-RS symbol positions depend on the duration and the higher-layer configuration of the number of DM-RS symbols (see Fig. 9.4). The position of the DM-RS is the same for PUSCH and PDSCH, assuming duration and configuration are the same.

Note that when the duration is shorter and shorter, the actual number of DM-RS for the PDSCH is less than the configured number since there is simply no room to transmit the configured number. For example, with two additional DM-RS, there are three DM-RS symbols for durations 10–14 but only two for duration 8 and 9, and one for duration 4–7 (see Fig. 9.4). Hence, the actual number changes dynamically from slot to slot, depending on the scheduled duration.

For Type A scheduling, the location of the first DM-RS in the slot (known in 3GPP specifications as the *front-loaded DM-RS*) is either the *third or fourth* symbol and this position is configured by the network. The same symbol (third or fourth) is used for PDSCH and PUSCH Type A scheduling. Whether to use the third or fourth depends on the number of PDCCH symbols that is expected to be needed as PDCCH is transmitted prior to the first PDSCH DM-RS in the slot.

The first DM-RS in the slot is always present when PDSCH is scheduled, and an additional one, two, or three DM-RS symbols in the slot can be higher-layer configured using the parameter *DM-RS-add-pos*. Whether to configure these additional DM-RS depends on the UEs mobility, in order to cope with a time-varying channel within the slot.

Table 9.4 Possible Scheduling Types in NR for PDSCH and PUSCH and Their Positions in the Slot Where it Has Been Assumed That Symbol 1 Is the First Symbol in the Slot.

Scheduling Type	Start Symbol in Slot	Length in Symbols
PDSCH Type A	1, 2, 3, or 4	3–14
PDSCH Type B	1–13	2, 4, 7
PUSCH Type A	1	4–14
PUSCH Type B	1–14	1–14

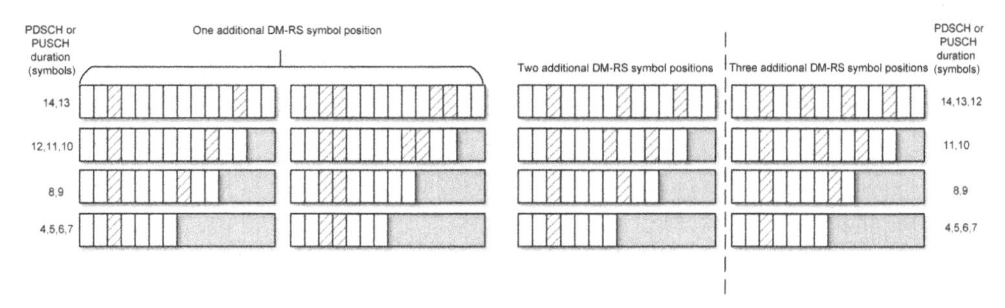

FIGURE 9.4

DM-RS positions for Type A PDSCH and PUSCH. It is in this figure assumed that the first DM-RS is in the third symbol (an alternative configured position is in the fourth symbol). The configurations with one, two, or three additional DM-RS positions are illustrated here, while the single DM-RS symbol configuration is not shown. The gray parts are OFDM symbols that are outside the scheduled PDSCH or PUSCH transmission duration and can be used for other downlink transmissions such as CSI-RS or can be used for an uplink transmission.

Using only one single DM-RS symbol (corresponding to *DM-RS-add-pos* = *0*) in the slot can be used in very static channel environments such as indoor office, while the other extreme of three additional DM-RS is beneficial only for very high-speed UEs exceeding 300 km/h. The most typical DM-RS configuration for most cases is using one additional DM-RS symbol (*DM-RS-add-pos* = *1*) in the scheduled resource.

In the case of zero or one additional DM-RS symbol position (*DM-RS-add-pos* = *0 or 1*), it is possible to further configure the use of a "double-symbol" where each of the two symbol positions uses two adjacent DM-RS symbols, in order to support up to 12 orthogonal DM-RS ports. See the second figure from left in Fig. 9.4.

For Type B scheduling, the first DM-RS is positioned in the first OFDM symbol of the scheduled resource and Type B can as in Type A be configured with one, two, or three additional DM-RS symbols. See corresponding Fig. 9.5 for DM-RS positions for Type B PUSCH scheduling and Fig. 9.6 for Type B PDSCH scheduling. The DM-RS positions for PUSCH depend on the higher-layer configuration of the parameter (*DM-RS-add-pos*) and the scheduling duration. Note that *DM-RS-add-pos* = *0* (only DM-RS in the first OFDM symbol) cases are not shown in these figures.

For Type B scheduling, the "duration" is defined as measured from the first to the last symbol of the actual PDSCH or PUSCH transmission.

As seen in Fig. 9.5, when the DM-RS has zero or one additional DM-RS symbols in the slot (*DM-RS-add-pos* = *0 or 1*), it is possible also in Type B as for Type A, to double the number of DM-RS symbols by enabling a configuration of two adjacent symbols at the position of each of these one or two DM-RS.

The use of either single or double adjacent DM-RS symbols in a scheduled PDSCH or PUSCH for both Type A and B is dynamic and indicated by the scheduling downlink control information (DCI). The double configuration increases received DM-RS energy and thus improve the robustness of the channel estimation, at the expense of increased overhead. Moreover, another use case for double adjacent symbol configuration is the doubling of the maximum number of orthogonal

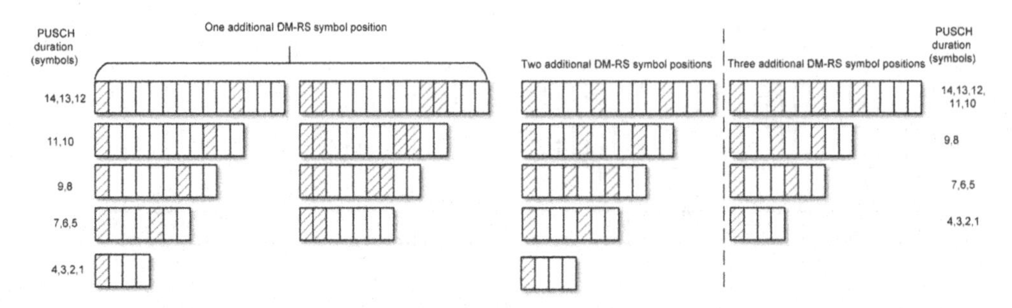

FIGURE 9.5

DM-RS configurations (i.e., positions in the scheduled resource) for Type B PUSCH scheduling (the single OFDM symbol DM-RS configurations are not shown) where for Type B, the PUSCH can be scheduled to be placed anywhere in the slot.

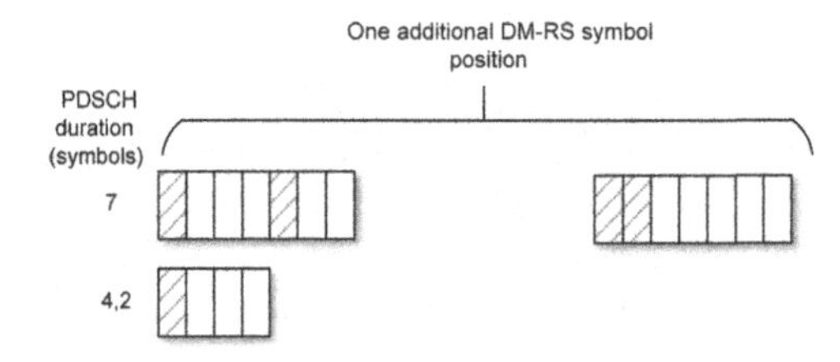

FIGURE 9.6

DM-RS positions for PDSCH Type B scheduling. Note that if the DM-RS in the first OFDM symbol collides with PDCCH, then the DM-RS is moved to the next available symbol. For PDSCH, only duration 2, 4, and 7 OFDM symbols are supported in NR release 15 and maximally one additional DM-RS symbol can be configured. The Type B PDSCH can be scheduled to be placed anywhere in the slot.

DM-RS antenna ports. This can be used in case of scheduling a high number of layers (larger than four) to the same UE in single-unit MIMO (SU-MIMO) or from/to different UEs in MU-MIMO scheduling.

For Type A scheduling, as mentioned above, the position of the first DM-RS symbol of the slot is either in the third or fourth symbol and this is indicated in the PBCH. This broadcast of a configuration is necessary to be able to receive PDSCH prior to the UE–specific configuration of the DM-RS. Also, two additional DM-RS symbols (*DM-RS-add-pos* = 2) are the default configuration unless configured otherwise by dedicated control signaling to the UE and this DM-RS pattern is therefore used to receive and transmit PDSCH and PUSCH during initial access.

9.3.1.2.2 DM-RS structure within a symbol

To allow for a DM-RS overhead that scales with the number of transmitted layers, the frequency density can also be configured. This facilitates support for a single layer up to 12-layer transmission with orthogonal DM-RS ports and with one DM-RS port per layer. Hence, two DM-RS types have been introduced and the type is higher-layer configured to the UE for PUSCH and PDSCH independently:

- DM-RS Type 1
  - supports up to four ports with a single DM-RS symbol; and
  - supports up to eight ports with two adjacent DM-RS symbols.
- DM-RS Type 2
  - supports up to six ports with a single DM-RS symbol; and
  - supports up to 12 ports with two adjacent DM-RS symbols.

For example, DM-RS Type 2 could be configured if MU-MIMO with many layers is foreseen, while DM-RS Type 1 could be configured for SU-MIMO and occasional MU-MIMO with up to eight layers. As will be discussed below, the DM-RS density is higher with DM-RS Type 1 compared to Type 2, and it is comb-based, which means it has potential for better channel estimation performance.

Within a symbol containing DM-RS, Fig. 9.7 shows how antenna ports are mapped, where the default (Type 1) use a "comb" of reference signals, that is, a DM-RS port is mapped to every other subcarrier. If resource elements in a DM-RS symbol are not used for DM-RS (e.g., a single comb is used), then it is possible to map PDSCH or PUSCH to these REs. An exception is the case when DFT-precoded OFDM is configured for a PUSCH transmission, then REs not used for DM-RS are always empty. Otherwise the low peak-to-average power ratio (PAPR) property of the DM-RS would be destroyed if frequency multiplexed with PUSCH data.

For both Type 1 and 2, an orthogonal cover code (OCC) of length 2 is applied in frequency (subcarrier) directions on top of these resources, which then enables two or three DM-RS antenna ports per OFDM symbol for Type 1 and 2, respectively, and thus in total four or six DM-RS ports.

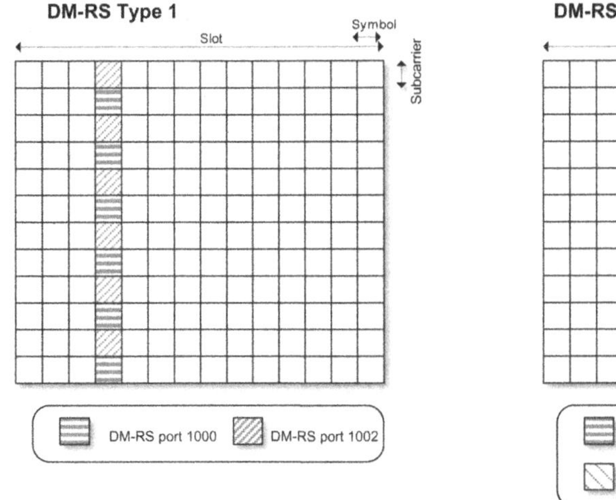

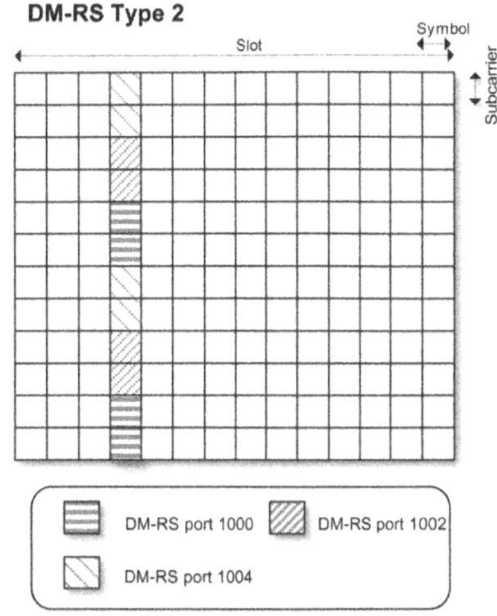

FIGURE 9.7

DM-RS Type 1 and Type 2 in case of a single OFDM symbol DM-RS configuration in a slot. Each marker represents a CDM group. Type 1 has two CDM groups (i.e. two combs), while Type 2 has three CDM groups, allowing for a higher number of orthogonal DM-RS ports at the expense of lower DM-RS density.

If double adjacent symbol DM-RS is configured, then the number of ports is doubled since a time-domain OCC is additionally used across the two adjacent OFDM symbols. The group of two or four REs for which time and/or frequency direction OCC is applied is called a CDM group and for DM-RS Type 1 and 2 there is thus two and three CDM groups, respectively.

However, in FR2, where phase noise is an issue, OCC to separate ports across time is not utilized, and the UE is thus not expected to receive antenna ports that are separated in time using OCC. This implies that at most six orthogonal DM-RS antenna ports can be used for FR2 and the doubling of the number of DM-RS ports by double adjacent symbols is only achieved in FR1. However, the double-symbol configuration still improves channel estimation.

The reason for this restriction not to use OCC is that the phase noise makes it problematic to de-spread the OCC by the receiver; hence only one of the two codes of the length 2 OCC can be used. This limits the number of orthogonal ports for MU-MIMO scheduling in FR2. On the other hand, in FR2, beamforming is likely used, and nonorthogonal ports may be possible to use successfully since the spatial separation of users by this beamforming isolation is high.

Type 2 DM-RS use two adjacent REs in the frequency direction instead of the comb structure used in Type 1, but with the same use of OCC structures in frequency of length 2 and in time of length 2 across adjacent OFDM symbols. For DM-RS Type 2, there are three CDM groups instead of two, thereby lower DM-RS density than Type 1 (but higher port capacity).

Although the DM-RS has a great configuration flexibility, it is worth to note which configurations are mandatory for a release 15 UE to support and which are optional. In Table 9.5, an overview of the UE capabilities is given where the different capability signaling classes are explained

Table 9.5 NR Release 15 UE Capabilities Related to DM-RS Configuration.

Scheduling	Mandatory UE Support	Mandatory UE Support with Capability Signaling	Optional UE Support with Capability Signaling
PDSCH Type A	• 1 DM-RS symbol • 2 DM-RS symbols • 3 DM-RS symbols for a single DM-RS port only • Type 1 DM-RS	• 3 DM-RS symbols for multiple DM-RS ports	• Two adjacent DM-RS symbols • 4 DM-RS symbols • Type 2 DM-RS
PDSCH Type B	• 1 DM-RS symbol • 2 DM-RS symbols		• Two adjacent DM-RS symbols • 4 DM-RS symbols
PUSCH Type A	• 1 DM-RS symbol • 2 DM-RS symbols • 3 DM-RS symbols for a single DM-RS port only • Type 1 DM-RS	• 3 DM-RS symbols for multiple DM-RS ports • Type 2 DM-RS • Two adjacent DM-RS symbols	4 DM-RS symbols
PUSCH Type B	• 1 DM-RS symbol • 2 DM-RS symbols • Type 1 DM-RS	• Type 2 DM-RS • Two adjacent DM-RS symbols	4 DM-RS symbols

Notes: *One DM-RS symbol implies the higher-layer parameter DM-RS-add-pos = 0, two DM-RS symbols imply DM-RS-add-pos = 1, and so on.*

in Section 10.1.3. The column with mandatory support in Table 9.5 implies that the UE will support this configuration without such capability signaling, which is the case for three DM-RS symbol configuration (*DM-RS-add-pos* = 2). This is because it is used for initial access, prior to dedicated DM-RS configuration. Hence, all NR UEs must support this configuration.

It is worth noting that two adjacent DM-RS symbols and Type 2 DM-RS are optional for UE to implement for PDSCH, which means that if MU-MIMO operation is desired, only the configuration with a single DM-RS symbol and Type 1 DM-RS is mandatory for UE to support. This then means that at most four orthogonal PDSCH DM-RS ports can be used, limiting MU-MIMO scheduling for PDSCH to at most four (single layer) UEs simultaneously, unless nonorthogonal DM-RS is used by means of different DM-RS sequences (as in LTE, see Section 8.3.1.5 for which case there is basically no theoretical limitation on the number of coscheduled DM-RS ports).

For PUSCH on the other hand, the Type 2 DM-RS and two adjacent DM-RS symbols are mandatory (although with capability signaling), which means up to 12 UEs can be coscheduled with MU-MIMO with orthogonal DM-RS ports.

This difference in UE capabilities for PDSCH and PUSCH DM-RS reflects the fact that MU-MIMO for UL is likely to be more beneficial than for DL, since for PUSCH, the receiver algorithm in the gNB can separate all the coscheduled users as it has full knowledge of the used DM-RS.

As in LTE PDSCH, two different DM-RS pseudo-random sequences can be configured for the UE and when scheduling, the DCI contains information on which of the two sequences to use. This allows for scheduling of nonorthogonal DM-RS (pseudo-orthogonal) by assigning the same DM-RS port to two different users but with different sequences. This can be used to increase the number of coscheduled UEs as a given DM-RS port can be reused by more than one UE provided that the spatial transmit filter (e.g., beamforming) suppresses the transmitted power in the unwanted (interfering) directions.

For DFT-precoded OFDM-based PUSCH, only DM-RS Type 1 is supported (comb-based mapping) and so-called Zadoff–Chu sequences are used for longer sequence lengths while computer-generated sequences are used for shorter sequence lengths (see Ref. [1] for more details of these sequences). In the case of DFT-precoded OFDM, only a single-layer PUSCH DM-RS is supported.

9.3.1.3 PDSCH PT-RS

Phase noise becomes an issue when operating NR at millimeter-wave bands (see Section 10.6.2.2). The consequence of phase noise is degraded link performance when using higher-order modulation for PDSCH such as 16-QAM, 64-QAM, and above. To mitigate phase noise and to reach high throughput also in FR2, a supporting reference signal for PDSCH demodulation can be configured, the *phase tracking RS* (PT-RS). Enabling PT-RS implies that a reference signal can be configured to be present in every OFDM symbol, and thus allowing for tracking the impact from the phase noise in the receiver, symbol by symbol.

The PT-RS and DM-RS can together be used by the receiver for phase noise tracking but also for other purposes such as frequency offset estimation. The PT-RS occupies at most a single subcarrier per PRB and is only present in the scheduled RBs, just as the DM-RS. As mentioned above, it is possible to configure the PT-RS to be present in every OFDM symbol, except in the OFDM symbols already occupied by DM-RS where PT-RS is unnecessary as DM-RS can also be used for the phase noise tracking. For PDCCH, the PDCCH DM-RS is present in every PDCCH symbol and there is no need for an associated PT-RS to receive PDCCH.

There is a trade-off between the PT-RS overhead and performance benefit, as enabling PT-RS removes available resource elements from the PDSCH. Evaluations showed that for QPSK modulation, PT-RS improves phase noise tracking but does not provide a net throughput gain when taking the PT-RS overhead into account, due to the inherent robustness of QPSK modulation. Hence, it is also possible to disable PT-RS for QPSK, and to configure a lower PT-RS time density for higher-order modulation, where every other or every fourth symbol contains PT-RS. There is a sweet spot in the PT-RS density that maximizes the throughput for a given modulation and coding scheme (MCS) and for a given scheduling bandwidth. This sweet spot also depends on the phase noise characteristics of the transmitter and receiver. If PT-RS density is lowered, the net performance may still be increased, especially in case the phase noise is correlated across multiple symbols as then there is no need for the highest time density.

See Fig. 9.8 for an illustration of a PT-RS which is mapped to every other OFDM symbol, except for the two OFDM symbols carrying the DM-RS. In these OFDM symbols, the DM-RS can be used for phase tracking instead of the PT-RS since the same antenna port is used.

Moreover, in the frequency domain, the PT-RS density can also be lowered and configured to be present in every second or every fourth scheduled PRB with a configurable PRB offset. Note

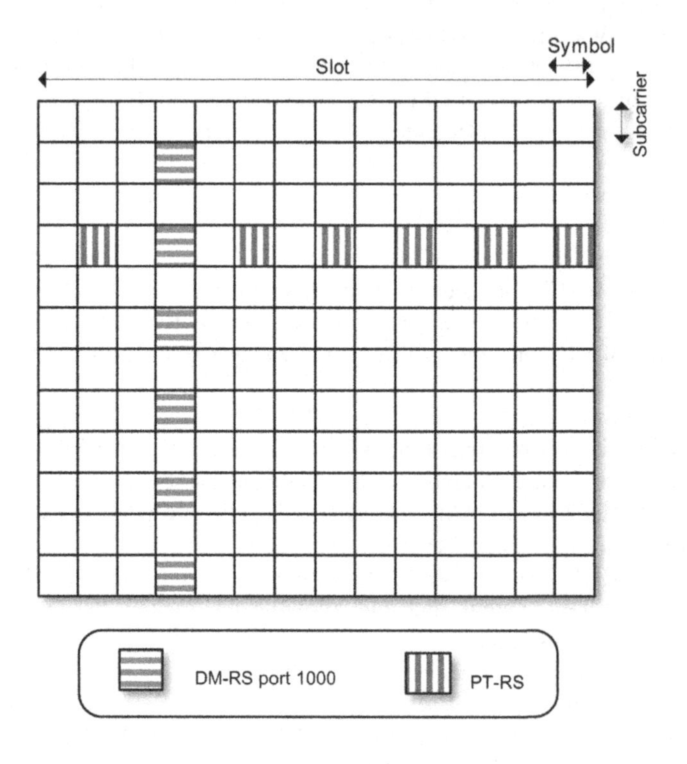

FIGURE 9.8

A resource block with DM-RS Type 1 and PT-RS with PT-RS density of every other OFDM symbol. Note that not every PRB scheduled for data transmission will have PT-RS, the frequency-domain density can be every other or every fourth PRB.

that the whole OFDM symbol is subject to the same common phase error due to phase noise; hence the noise is wideband in its characteristics and a higher PT-RS density in frequency thus only provides a better signal-to-noise ratio (SNR) for phase noise estimation due to more samples.

As mentioned, the used PT-RS density in time is designed to depend on the selected MCS of the scheduled PDSCH, and for highest MCS (e.g., 256-QAM), the PT-RS density is also highest. For the lowest MCS, that is, QPSK, the PT-RS is by default not present. In addition, the frequency density depends on the scheduled bandwidth of the PDSCH as the number of available PT-RS samples in an OFDM symbol scales with the scheduled bandwidth, where for the largest density it is enough to have PT-RS in every fourth PRB.

The subcarrier position of the PT-RS in a scheduled PRB is the same as one of the subcarriers used for the DM-RS of the lowest port number of the DM-RS ports used by the scheduled PDSCH. Moreover, which subcarrier among the multiple subcarriers used by that DM-RS in the PRB can be configured by RRC.

The PT-RS port is associated with a DM-RS port. The reason for association (instead of using precisely the same port) is that even though DM-RS and PT-RS are transmitted with the same precoder; the PT-RS power may be different than the DM-RS power, hence it is not exactly the same antenna port, according to the definition of an antenna port.

The MCS and scheduled bandwidth thresholds that are used to determine the switch between frequency and time densities for PT-RS have default values, but the UE can be configured with arbitrary thresholds if the gNB desires. The UE can also report recommended thresholds for both DL and UL by RRC signaling. This can be used to tailor the switching thresholds to the phase noise performance of the actually used local oscillators in the UE device (instead of the default thresholds). The gNB decides whether to adopt this threshold recommendation from the UE.

9.3.1.4 PUSCH PT-RS

The UL PT-RS is used to compensate for phase noise when demodulating PUSCH. The PT-RS design and procedures for PUSCH when CP-OFDM is used are the same as for PDSCH PT-RS as described in the previous section, but with the exception that PT-RS is present also for QPSK modulation in the default configuration. The reason for the difference with the DL is that gNB may need to use the PT-RS for frequency offset estimation, especially when only one DM-RS symbol is configured for PDSCH reception. The use of PT-RS for this purpose is not needed for the DL since TRS is present and can be used for such finer synchronization.

For PUCCH, the modulation is $\pi/2$-BPSK or QPSK, hence low spectral efficiency and robust against phase noise, and there is no need for a PT-RS.

For PUSCH configured with the DFT-precoded OFDM waveform, the PT-RS design is different. In this case, groups of PT-RS samples of $\pi/2$-BSPK modulated symbols are inserted prior to the DFT, that is, among the time-domain PUSCH data samples. The number of such PT-RS sample groups and the number of PT-RS samples within each group depend on the scheduled PUSCH bandwidth, to have sufficiently many groups and samples for large resource allocations and to avoid too large overhead for small PUSCH resource allocations. The thresholds for this bandwidth dependence, as well as time density (as which OFDM symbols PT-RS is mapped), are configured by higher-layer signaling to the UE.

For the UL, it is possible to configure two PT-RS, simultaneously transmitted from the UE. Note that for the DL, only a single PT-RS is available as the scope of release 15 was toward the end of standardization reduced by excluding multi-transmission point (multi-TRP) and multipanel operation. This was done in order to finish NR standardization on time. Two PT-RS ports for UL can, for example, be used when the UE has two antenna panels (e.g., two planar antenna arrays mounted on each side of a UE, facing in opposite directions) and each panel has its own local oscillator. In this case there may be two independent sources of phase noise, and thus different PUSCH MIMO layers may need to be compensated differently to track the phase noise or frequency offset.

The number of PT-RS ports transmitted can be zero, one or two for a PUSCH transmission using CP-OFDM waveform and with noncodebook-based PUSCH. To indicate the number of ports, an SRS resource is configured with an associated number of PT-RS ports, so when that SRS resource is selected and indicated by the UL scheduling grant, then the actual number of PT-RS ports the UE shall transmit together with the PUSCH is obtained from this grant.

Similarly, for codebook-based PUSCH, the actual number of PT-RS ports is indirectly indicated by the UL scheduling grant for PUSCH unless coherent codebook is used (see Section 9.3.4), in which case a single PT-RS port is always used whenever PT-RS is enabled.

9.3.1.5 Sounding reference signal

To perform UL channel measurements, SRSs are defined. It was observed during standardization that more freedom should be given to the UE to self-determine how to transmit an SRS, for example, transmitting SRS in a certain beam direction. Also, noncodebook-based precoding was introduced in NR, which is based on a similar principle where UE determines the precoder on its own.

To support these cases, it is useful for the gNB to maintain some control on how UL is transmitted, as the gNB needs to handle the interference on the UL. Therefore it is specified that the UE can transmit multiple different SRS resources (e.g., in different beams) and let the gNB select one SRS resource among these resources to be used for the subsequent PUSCH and/or PUCCH transmission, that is, in UL beamforming. How the UE performs the actual beamforming remains a UE implementation issue, that is, UL beams are not specified in standard.

To support such gNB-based SRS resource selection and indication, *SRS resource sets* were introduced as a collection of *SRS resources*. To summarize:

- *An SRS resource*
  - can be configured to 1, 2, or 4 SRS ports;
  - can span 1, 2, or 4 adjacent OFDM symbols in one slot;
  - all SRS ports are present in each OFDM symbol of the resource;
  - can have a repetition factor or 1, 2, or 4 in the resource, in which case the same SRS subcarriers are used in each repetition;
  - can be transmitted anywhere in the last 6 symbols of a slot.
- *An SRS resource set*
  - can have up to 64 SRS resources;
  - can have multiple SRS resources in the same slot where resources are adjacent or nonadjacent in time;
  - can have SRS resources in different slots;
  - is configured with a single "usage" (see Table 9.6).

The motivation is analogous to CSI-RS resource sets as introduced in Section 9.3.1.1, for example, used for the receiver to select a preferred beam. A set is restricted to be configured for only

Table 9.6 The Different Uses That Can Be Configured for an SRS Resource Set.

SRS Resource Set Configuration Parameter "Usage"	Use
"Antenna switching"	Reciprocity-based measurements to acquire downlink CSI
"Beam management"	To assist in finding a suitable uplink transmit beam for UE or a suitable gNB receive beam
"Codebook"	For uplink link adaptation for PUSCH, for example, for gNB to determine PUSCH precoder (and PUCCH)
"Noncodebook"	For uplink link adaptation for PUSCH for noncodebook-based uplink transmissions

one purpose at a time, and since SRS is used for multiple functionalities (see below), the UE can be configured with multiple SRS resource sets simultaneously. For example, one set is used to assist gNB in selecting a precoder for PUSCH, while another set is used for reciprocity-based operation to assist PDSCH precoding.

An SRS resource can in release 15 be transmitted anywhere in the six last symbols of a slot (see Fig. 9.9). This allows for a DL transmitted PDCCH in the beginning of a slot, a guard period (GP) for the downlink to UL switch (in case of TDD) and the SRS transmission in the end of the slot. An SRS resource spans either one, two, or four adjacent OFDM symbols; hence the last six symbols of a slot can be used for transmitting more than one SRS resource, for example, the multiple SRS resources of an SRS resource set.

When an SRS resource is mapped to more than one OFDM symbol, then each SRS port of the SRS resource is present in every symbol and across the whole configured SRS bandwidth of the resource. An SRS antenna port can thus be repeatedly transmitted by the UE in two or four symbols in a slot, which can be used to extend SRS coverage.

An SRS port transmission is mapped to every second or fourth subcarrier in the OFDM symbol (i.e., a comb structure is used). SRS uses the same low PAPR sequence as in LTE, a Zadoff–Chu sequence, where up to 12 different cyclic shifts (CS) of the sequence are used to define different SRS antenna ports.

The different configuration alternatives of mapping of SRS ports of an SRS resource to subcarriers in an OFDM symbol can be seen in Fig. 9.9 using either a comb-4 or a comb-2 structure. To summarize:

- For a single SRS port resource, the port can be mapped to any of the combs and a CS can be applied (to separate the SRS port from another UEs transmission by using different CS and/or different comb);
- For a two SRS port resource, both ports are mapped to the same comb and separated by CS. Any of the combs can be configured for this SRS resource;
- For a four SRS port resource, either all four ports are mapped to the same comb and separated by CS, or, groups of two ports are mapped to either of two configured combs and separated by CS within the group.

Note that it is not possible to map a four-port SRS resource to four different combs, CS must be used to separate at least two ports. In addition, for the four-port and four-comb case when two

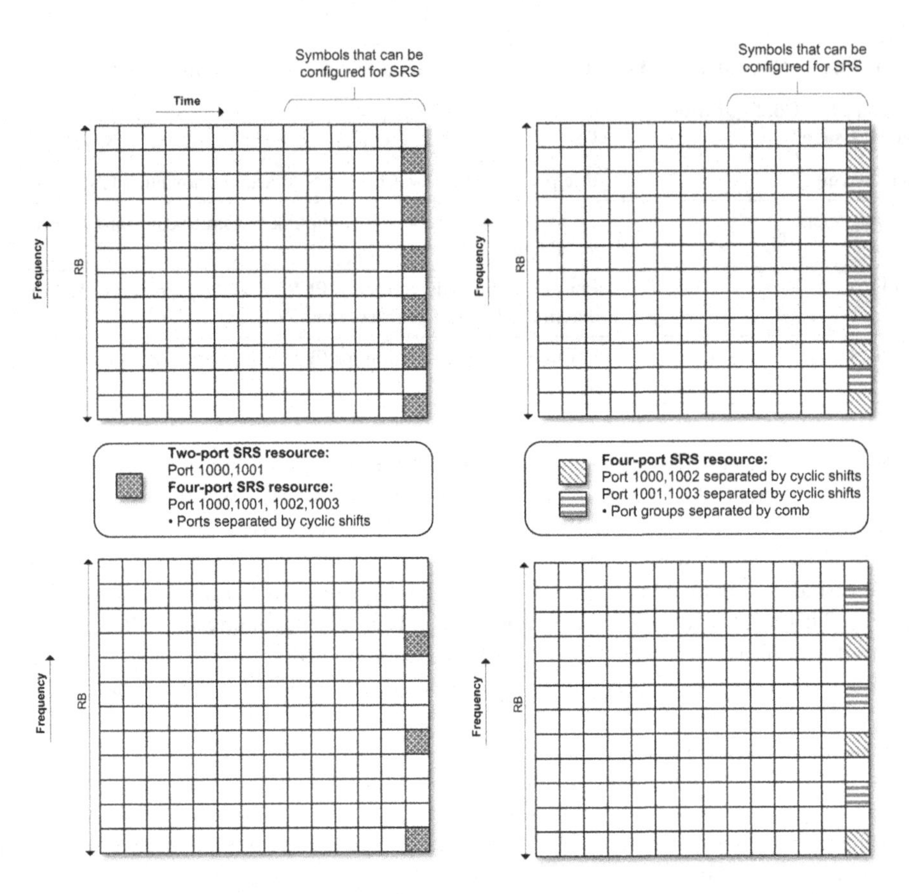

FIGURE 9.9

An SRS resource can be configured within the last six OFDM symbols of a slot and use a comb-2 structure (upper plots) or a comb-4 structure (lower plots). The SRS resource has one, two, or four SRS ports that are present in all OFDM symbols of the SRS resource (the figure shows only the single OFDM symbol case).

combs are configured, the two combs cannot be adjacent since that would prevent multiplexing another SRS resource in the same OFDM symbol.

To randomize SRS interference between users transmitting SRS within the same bandwidth, in the same cell and in different cells, a time-dependent sequence randomization (known as sequence hopping) can be configured for the SRS sequence. The sequence used for SRS depends pseudo-randomly on both slot index and symbol index within a slot. In addition, the used SRS sequence initialization is UE-specifically configured by RRC.

An SRS resource can be periodic (in which case it is only configured by RRC), aperiodic (configured by RRC but an actual transmission is triggered by DCI), or semipersistent (SP) [configured by RRC and activated/deactivated by medium access control-control element (MAC CE)]. The use of SP SRS allows to start and stop the periodic SRS transmission using MAC signaling from the gNB to the

UE, which is faster than the RRC-based control. It thus provides some means to trigger periodic SRS transmissions when needed only, to avoid interference and unnecessary transmissions from the UE.

The periodic SRSs can have a periodicity ranging from 1 slot to 2560 slots in 17 different periodicity steps (i.e., from 125 μs for 120 kHz SCS to 2.56 seconds for 15 kHz SCS). This large configurability range aims to allow for the same maximum periodicity, when expressed in milliseconds, irrespectively of the used subcarrier spacing (as the number of slots per second increases with the subcarrier spacing).

As discussed for LTE in Section 8.3.1.7, there is a trade-off between coverage and SRS transmission bandwidth. Since the total transmit power of a UE is limited, one way to improve SRS coverage is to configure the UE to transmit the SRS over a small bandwidth. Given a constant transmit power, the *power density* of the SRS transmission is then increased and thus SRS coverage improves. The drawback is that only a part of the bandwidth is sounded.

SRS frequency hopping is therefore specified, where SRS transmissions at different times occupy different parts of the band. The smallest SRS transmission bandwidth in NR is four RBs and the maximum is 272 RBs, and intermediate bandwidths can be configured in 64 possible bandwidth steps. The part of the band that is used for SRS transmission in a given OFDM symbol is pseudo-randomly determined by a formula in specifications, thereby the common term SRS frequency hopping. All parts of the total configured SRS bandwidth are sounded before the randomizations restart with the first part again.

It is important for reciprocity-based operation to measure as much of the UL channel bandwidth as possible in as short time as possible, to get the instantaneous snapshot of the channel. Ideally, the whole system bandwidth should be measured in a single OFDM symbol, for all SRS ports. However, this is only possible if the UE is close to the receiving base station, as the power spectral density is low when the UE power is used in a full bandwidth transmission. Hence, to achieve full bandwidth sounding in a short time span, NR introduces intra-slot hopping, where in the same slot but different OFDM symbols, different parts of the bandwidth are sounded in the UL. Interslot hopping, as used in LTE, is also possible. In NR, a combination of inter- and intra-slot hopping is also possible. See some examples in the following paragraphs.

With hopping enabled and without repetition enabled, the SRS is transmitted in different parts of the bandwidth in the different OFDM symbols of the SRS resource. This is used to focus the available transmit power to a smaller bandwidth (higher power spectral density) and obtain measurements for a large part of the bandwidth. With repetition enabled, the SRS is transmitted in the same part of the band for two or four OFDM symbols in the SRS resource.

Figs. 9.10–9.12 illustrate some examples of configurations of SRS hopping and repetition for the case of a periodic SRS resource.

An alternative use of SRS repetition is when the SRS is beamformed, as in FR2, to allow the gNB to perform a gNB receive beam tuning. Since the UE repeats the SRS transmission using the same transmit beam multiple times, the gNB can evaluate the performance of several gNB receive beam candidates. The performance of these different gNB receive beams can be directly compared since it is known that the UE keeps its transmit beam constant for each transmission. This is like the P3 procedure for CSI-RS repetition as discussed in Section 9.3.1.1.2. This is also used for the UL beam management procedures (see Section 9.3.6.4).

The SRS is configured to the UE by one or more SRS resource sets that each contains one or more SRS resources. Each set is designated for a certain "usage" (see Table 9.6). These different

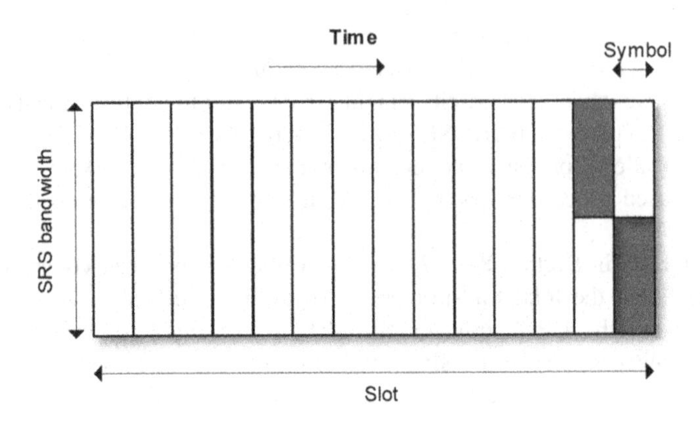

FIGURE 9.10

Intra-slot SRS hopping for a two OFDM symbol SRS resource, where in each transmission half of the configured SRS bandwidth is sounded so that the full SRS bandwidth is sounded in each slot. The power density of the transmission can be increased by 3 dB compared to sounding the full SRS bandwidth in a single slot, Note that the SRS bandwidth (which is multiple PRBs) can be configured smaller than the total bandwidth part bandwidth.

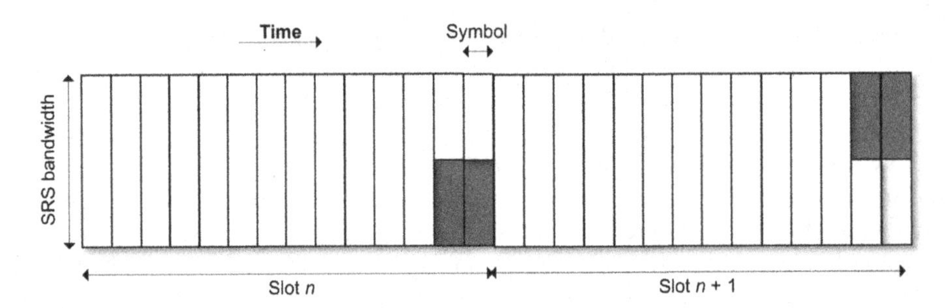

FIGURE 9.11

Interslot SRS hopping with two times repetition for a two OFDM symbol SRS resource. In this case, only half of the SRS bandwidth is sounded in each slot, so that power density is increased by 3 dB compared to a full bandwidth sounding, plus the use of repetition, which increases the coverage by an additional 3 dB. Adjacent slots are shown in the figure, which is one configuration possibility.

uses will be explored in the following subsections. Note that an SRS resource transmitted in a given set cannot be used for other use than it has been configured for, for example, an SRS transmission for "antenna switching" cannot simultaneously be used for "codebook"-based usage and vice versa. This is a weakness in release 15 NR as it leads to unnecessary configuration of multiple SRS resource sets and increased SRS overhead.

Since each configured SRS resource set can only have one "usage" there is a need for configuring several resource sets simultaneously to the UE. A UE can therefore be configured with up to 16 SRS resource sets per BWP (although limited to four sets in FR1) and each set can contain maximally 64 SRS resources. The SRS resources belonging to a set can be in the same slot or can be

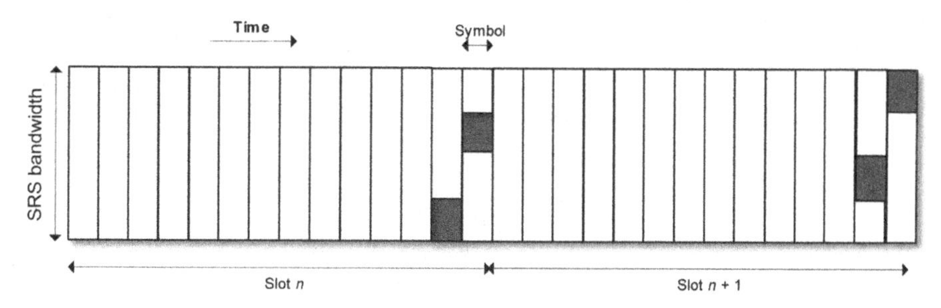

FIGURE 9.12

Intra- and interslot hopping for a two OFDM symbol SRS resource. In each symbol, only a quarter of the SRS bandwidth is sounded, and it takes two slots to sound the whole SRS bandwidth (which still may be configured as smaller than the bandwidth part bandwidth). The power density per SRS transmission is increased by 6 dB compared to full bandwidth sounding.

distributed across different slots. In FR1, the total number of SRS resources is limited to 10 as the larger number of resources is needed only when SRS resources are transmitted in different UL beams.

Aperiodic SRS can be triggered by either the UL DCI (that schedules PUSCH) or the DL DCI (that schedules PDSCH). Using the DL DCI to trigger SRS is useful for reciprocity-based operation as the SRS in this case is related to the DL performance and channel acquisition intended for the DL.

9.3.1.5.1 SRS for reciprocity-based operation

The purpose of using SRS for reciprocity is to assist the DL PDSCH transmission by allowing the gNB to measure the UL MIMO channel (see Section 6.4.2 about reciprocity-based CSI). When configuring an SRS for such UL channel sounding purpose, then the "usage" parameter in the SRS resource set configuration is set to "AntennaSwitching." This name may sound confusing if full channel sounding is used (and hence no antenna switching is performed) because even in this case the usage parameter is set for "AntennaSwitching."

In Section 6.7.1.2, the partial channel problem is introduced and to remedy this, the use of antenna switching can be configured to enable reciprocity operation. If the UE has the same number of TX as RX, then it is capable of providing "full" channel sounding without actual switching of antennas, which provides the ideal case for reciprocity-based operation. See also the same discussion for LTE UE in Section 8.3.1.7. The UE reports its capability of supporting antenna switching to the network (or rather it is an incapability of supporting full channel sounding), and this capability reporting includes whether the UE supports full channel sounding. In the latter case, actual antenna switching is not needed for the UE but as mentioned above, still the "usage" parameter of the SRS resource set configured to perform the full channel sounding will be "AntennaSwitching," which is a bit misleading.

If the UE has reported capability for full sounding operation, it means it can for each receive antenna also perform a simultaneous transmission on all these antennas, without the need for antenna switching. Hence, it can transmit an SRS from all ports in one and the same OFDM

Table 9.7 Possible SRS Configurations for Antenna Switching Used for DL CSI Acquisition.

Use case	SRS Resource Sets	SRS Resources per Resource Set	SRS Ports per SRS Resource
1T2R	1 or 2	{2, 2}	1
1T4R	1	4 (periodic or semipersistent SRS)	1
1T4R	1 or 2	{2, 2} or {1, 3} (for aperiodic SRS, ports transmitted in two slots)	1
2T4R	1 or 2	{2, 2}	2
1T = 1R	1 or 2	1	1
2T = 2R	1 or 2	1	2
4T = 4R	1 or 2	1	4

symbol. In this case, the UE can be configured with an SRS resource set with a single SRS resource having 1, 2, or 4 ports. The 3GPP terminology for full sounding is "1T = 1R," "2T = 2R," and "4T = 4R" for the case of one, two, and four transmit/receive antennas, respectively.

Unfortunately, UEs may not have the same number of transmit branches as receive branches and then the UE needs to indicate partial sounding support and antenna switching is needed. The different flavors of antenna switching are indicated from UE to the network as "1T2R," "1T4R," and "2T4R," respectively (see also Section 8.3.1.7 for the similar antenna switching functionality in LTE).

An overview of these different configurations for SRS antenna switching is given in Table 9.7. A UE that is not capable of antenna switching although it has >1 RX antenna, needs to indicate 1T = 1R, then the gNB would configure an SRS resource set with one single-port SRS resource. Note that the different SRS resources in a set configured for "AntennaSwitching" must be separated with at least one or two OFDM symbols for subcarrier spacings less than 120 kHz and 120 kHz respectively, to give enough time for the UE to perform the actual antenna switching. This is known as the GP, and during the GP, the UE is not transmitting.

In the cases of 1T2R, 1T4R, and 2T4R, if two SRS resource sets are configured, then the two sets must have different time-domain behavior, for example, if one is configured as periodic, then the other can be aperiodically triggered. The rationale is that a periodic antenna switching is configured using the first set and with a long periodicity and an aperiodic set for antenna switching can be triggered on demand to get a "fresh" channel sounding estimate. Note that this restriction of different time behavior of the two sets does not apply for the full sounding cases (1T = 1R, 2T = 2R, and 4T = 4R).

For FR1, the SCS is always less than 120 kHz and assuming an example of 2T4R, the intra-slot antenna switching can be configured as in Fig. 9.13, where an SRS resource set with two SRS resources is used with a guard symbol in between. There may also be a guard symbol before the first SRS in this example as there may be a DL to UL switch if the first part of the slot was used for DL reception by the UE.

If even larger coverage for the SRS is needed with antenna switching, then an SRS resource of four symbols can be used, although then there is no room for two SRS resources within a single slot as it can maximally be six; hence the antenna switching must be configured across two slots.

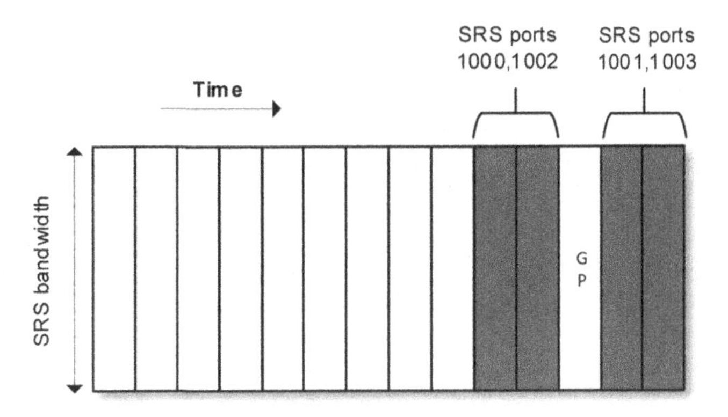

FIGURE 9.13

Intra-slot SRS antenna switching for 2T4R where a set of two SRS resources are used with two symbols each (repetition of factor two). In between the group of two symbols, a guard period (GP) is introduced.

See Fig. 9.14 for an example of this configuration using 2T4R. Here, the SRS resource set of two SRS resources is split across two slots, which is possible for periodic and SP SRS configuration since in this case the SRS slot offset is configured per SRS resource (and thus allows mapping of resources in the same set to different slots).

For an aperiodic triggered SRS resource set, on the other hand, the slot offset is configured per resource set; hence the split of the two resources in the same set across different slots as for periodic or SP SRS is not possible. For the special case of 1T4R antenna switching with aperiodic SRS requiring two slots to complete, an exception is made such that two aperiodic sets may be configured, thus allowing a split of the four resources across two different slots (either as a $2+2$ split or a $1+3$ split).

For the use of reciprocity, it is also important to consider CA use case. Typically, in CA, the number of DL carriers is larger than the number of UL carriers and fast carrier switching is therefore supported in LTE and NR in order to transmit SRS on any UL carrier (even though such a carrier does not have possibility for PUSCH/PUCCH transmission).

The principle is that the UE temporarily turns off the UL transmission on a carrier, performs RF retuning and transmits SRS on another PUSCH-less carrier, and then again performs RF retuning to return to the "normal" PUSCH carrier. However, when performing this carrier switching, resources for PUSCH transmission are lost, which degrades the UL performance in order to benefit the DL, PDSCH, performance. See also Section 6.7.7 for further discussion on fast SRS carrier switching

It should be kept in mind that the SRS carrier switching feature is an optional UE feature as it complicates the UE design. There is an associated UE capability that indicates the needed time gap to perform the switching. The switch refers to the RF retuning between one band pair, that is, between one band and another band that does not have PUSCH. The UE can indicate a gap in several steps between 0 and 900 μs, depending on its implementation. The network can request this capability from the UE. The SRS carrier switching can be configured to be periodic or triggered aperiodically. Also, carrier switching can be combined with the antenna switching feature if the UE also supports that.

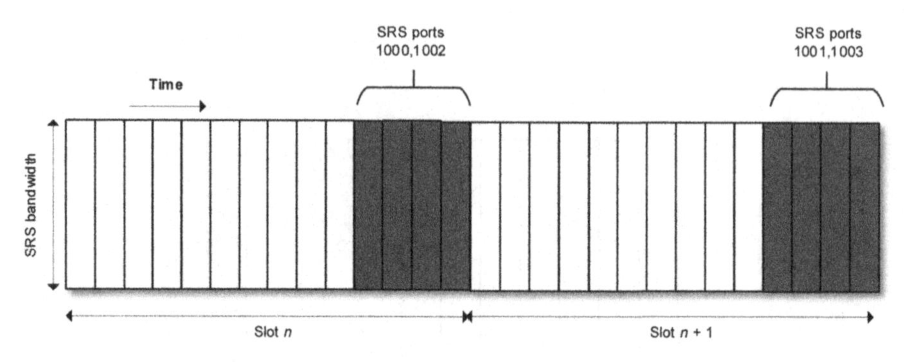

FIGURE 9.14

Example of interslot SRS antenna switching for 2T4R where a set of two SRS resources are used with four symbols per SRS resource that achieves a four-time repetition of each SRS port.

9.3.1.5.2 SRS for beam management

When the "usage" parameter in the SRS resource set configuration is set to "BeamManagement," the purpose is to assist finding beam directions, both for UL transmission at UE and UL reception at gNB. In this case, an SRS resource set with different SRS resources in different OFDM symbols may correspond to different (or same) transmit beams at the UE. More than one set can be simultaneously configured, to support the case a UE has more than one antenna panel to transmit a UL beam. An antenna panel can be seen as a planar array on the UE, and multiple such panels can be used to achieve omnidirectional coverage at the UE. In 3GPP, a model for such UEs was established (see the technical report describing the 5G channel model [9]). A UE may then choose to transmit the SRS resources belonging to a set from a given antenna panel only, although the mapping of a resource set to a panel is up to UE implementation

Hence, the gNB can use this to evaluate different transmit beams for the UL and select which of the transmitted SRS resources (i.e., beams) is preferred and should be used for the scheduled PUSCH transmission. To use the beam for PUSCH, the selected SRS resource is then configured as the *spatial relation* for the SRS used for codebook or noncodebook-based PUSCH. See Section 9.3.4.1 for the definition of spatial relations introduced to enable that a beam found from beam management can subsequently be used for PUSCH transmissions.

Note however that the NR specifications are a bit vague on the point of the intended behavior that the UE selects *different transmit beams* for each SRS resource. This behavior is not clear in specifications, although it was the intention when specifying this for beam management. There was a difficulty in standardization on how to specify that a UE shall change the transmit beam since "UE beams" are not defined in the physical layer specifications. However, it is possible to configure a spatial relation (s) (Section 9.3.4.1) to control the direction where the UE transmits an SRS, while if such a spatial relation(s) is not configured, then the UE behavior is not clearly defined.

9.3.1.5.3 SRS for codebook-based PUSCH

When the "usage" parameter in the SRS resource set configuration is set to "Codebook," the SRS resource is used to measure the UL MIMO channel to determine the UL precoder from the codebook, as well as rank, and other transmission parameters used in the UL link adaptation. See Section 9.3.4.2 for codebook PUSCH transmission procedures.

An SRS resource set configured for this usage can have one or two SRS resources. In case of two resources in the set, they have the same time configuration, for example, both are configured as "periodic" or "aperiodic" and both have the same number of SRS ports.

The use of two SRS resources can be utilized in case the UE has an implementation of two antenna panels (e.g., two planar antenna arrays on opposite side of the UE). The UE may be implemented to transmit one SRS resource per panel; then the gNB can measure the SRS to find the best panel. The gNB can then indicate a UE panel (although only in principle, since use of panels is a UE implementation issue) in the PUSCH scheduling grant using the one-bit *SRS resource indication* (SRI). The UE shall then transmit the PUSCH using the same panel as it previously used to transmit the SRS resource that was indicated by SRI.

Supporting two SRS resources for the SRS resource set when in codebook-based operation is a UE capability. Moreover, the UE can only be configured with one SRS resource set with "usage" parameter "Codebook" (although it can be simultaneously configured with sets with other usage parameter values).

9.3.1.5.4 SRS for noncodebook-based PUSCH

In NR, noncodebook PUSCH transmission has been introduced, which does not exist in LTE. The intention is that the UE autonomously decides the precoding to use for each SRS port. The UE may therefore utilize DL-UL reciprocity and then precode an SRS transmission in a preferred direction. Furthermore, the UE guarantees that the SRS port and subsequently scheduled PUSCH DM-RS port are precoded in the same way. See Section 9.3.4.3 for noncodebook PUSCH transmission procedures.

When the "usage" parameter in the SRS resource set configuration is set to "nonCodebook," this set can have one, two, or four SRS resources depending on UE capability and each resource has only a single SRS port. Each SRS resource then represents a PUSCH layer; hence if the UE supports only two-layer PUSCH transmission, then an SRS resource set with two SRS resources is configured.

The SRI in the PUSCH scheduling grant selects a subset of, or all, these SRS resources for transmission of PUSCH and the UE then transmits one PUSCH layer for each indicated SRS resource. If a rank 1 PUSCH transmission is scheduled, then the SRI selects a single SRS resource out of the up to four SRS resources, and so on. The UE should then transmit the PUSCH layer in the same way as it transmitted the indicated SRS resource in the most recent SRS resource transmission.

9.3.1.6 The QCL framework in NR

The aim of introducing the *quasi-co-location* (QCL) framework is to improve the UE reception performance by indicating to the UE which sets of DL antenna ports can be expected to have the same large-scale channel properties (e.g., channel delay spread). See Section 3.5 on large-scale channel parameters. Thereby, the UE can use one port for synchronization and measurements to receive a transmission that uses another antenna port if they are QCL. If QCL is not indicated between two ports, then the UE cannot make such an assumption. This framework allows the use of more advanced network deployments where ports are transmitted to the same UE from physically noncolocated transmission points. It can also relax network implementation in case different transmission points are not perfectly synchronized in which case these ports are not indicated as QCL to the UE.

It is recommended to be familiar with the concepts outlined in Section 8.3.1.2, the introduction of QCL in LTE release 11, which content is also highly relevant for NR. This is because the QCL framework reuses the same definition of time and frequency large-scale parameters as in LTE, although there is an NR-specific extension with a new, spatial, large-scale parameter.

The spatial parameter was added in NR to assist in certain UE and/or gNB implementations where precoding or beamforming is performed in time domain, often implemented by analog beamforming (see Chapter 7). Time-domain beamforming means that the beamforming is wideband, and it must be adjusted to a certain radiation pattern or reception pattern *prior to* receiving the signals. In other words, the transmitter and/or receiver cannot transmit or receive to/from all directions simultaneously, an implementation model that was not assumed in the standardization of LTE.

Typically, time-domain beamforming is used in FR2, at millimeter-wave frequencies. In 3GPP it was then decided that such new functionality is handled under the QCL framework by the introduction of a new large-scale parameter referred to as "*spatial RX*." Hence, a QCL relation with respect to "spatial RX" can in NR be defined between two antenna ports.

The QCL parameter of "spatial RX" is used to configure or indicate the network's intent to the UE of a transmit beam that will be used for PDSCH (or PDCCH) transmissions. This is further elaborated in detail in Section 9.3.6.2 on beam management procedures but here follows a brief example of how this can be used:

Assume that a UE is configured to receive a CSI-RS resource transmitted periodically by the gNB and where the gNB uses the same beam every time it transmits the CSI-RS resource. The UE has over time tuned the receive beam to receive this periodic RS, possibly with some receive beam refinement every time the periodic CSI-RS is transmitted. The gNB then occasionally schedules the UE to receive a PDSCH, and the gNB intends to use the same transmit beam for that PDSCH as it uses when it transmits the periodic CSI-RS. The gNB then indicates in the PDCCH that the CSI-RS port is QCL with the PDSCH DM-RS port with respect to "spatial RX." This informs the UE that the same RX beam as used to receive the CSI-RS can be used also to receive the PDSCH.

In the NR QCL framework, a QCL relation is defined between a source antenna port (e.g., a CSI-RS) and a target antenna port (e.g., a PDCCH DM-RS), both transmitted in the DL. The available large-scale parameters are grouped into four QCL types:

- QCL Type A: {Doppler shift, Doppler spread, average delay, delay spread}
- QCL Type B: {Doppler shift, Doppler spread}
- QCL Type C: {Doppler shift, average delay}
- QCL Type D: {Spatial RX parameter}

Note that QCL Type D is optional to support by the UE, but UEs operating in FR2 will typically indicate support for Type D. QCL Type A contains all time/frequency large-scale channel parameters, which basically means the two RS are transmitted from the same transmission point and similar coverage (i.e., beamwidth of the transmission). QCL Type C can be used in case the beamwidth of the transmission of the two RSs are different in which case the main path is the same (same average delay) but the delay spread not necessarily is the same. QCL Type B can be used when the transmission point is different for the two RSs; since then the average delay is not the same.

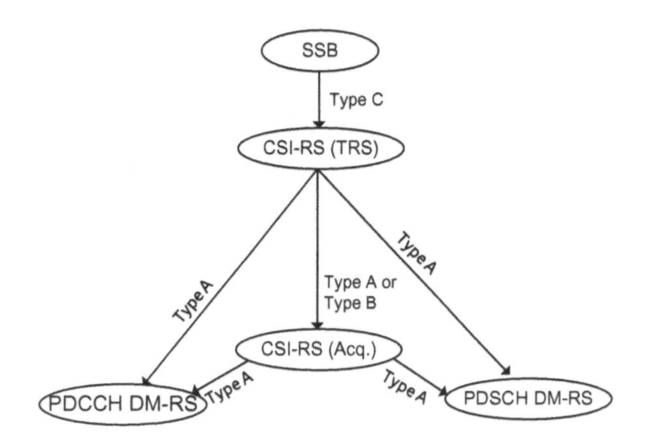

FIGURE 9.15

Possible QCL relations when QCL Type D is not applicable (e.g., FR1).

One can view the QCL framework configured to a certain UE as a tree structure with the links between the source and target reference signals. This is illustrated in Fig. 9.15 where SSB is the source for a target TRS, while the TRS in turn is the source for a target DM-RS. Each link has a QCL type and for some links multiple types are possible and the network can configure which type applies, depending on how the network is deployed.

In Fig. 9.15, all the possible configurations of QCL relations are illustrated when QCL Type D is not applicable, for example, for typical FR1 operation. Note that all links may not be configured simultaneously as a target antenna port can only have one source antenna port configured at a given time, for a given QCL Type. The following discussions and Fig. 9.15 assume the UE is in connected state and has been RRC configured with these QCL relations, while the default QCL assumptions used during, for example, initial access, are discussed in Section 9.3.6.1.

For example, in Fig. 9.15 a PDCCH DM-RS or a PDSCH DM-RS can be configured to either be QCL with TRS as source, or with a CSI-RS for CSI acquisition as source, in any case using QCL Type A. Hence to receive the control and/or the data channels, the UE can use this source for all the time and frequency synchronization parameters.

In the most basic deployment, assuming a cell with an SSB and a single TRS, the TRS is used to perform fine time/frequency synchronization as the basis for the PDCCH and PDSCH reception. A CSI-RS for CSI acquisition can also be configured using the TRS as the source by QCL Type A. Using the CSI-RS for CSI acquisition as a QCL source for DM-RS is typically not used in a basic deployment case (the link will not be configured), since the TRS provides a more reliable source for synchronization due to its time and frequency structure (see Section 9.3.1.1.1).

In a deployment where *dynamic point selection* (DPS) is used (see also Section 8.3.4.3), the UE can be configured with multiple TRSs. The QCL source for the PDSCH DM-RS reception can then switch dynamically between different TRSs. For each such additional transmission point, a CSI-RS for CQI acquisition is also configured. Note that NR release 15 does not support a configuration where these TRSs have an SSB of another cell as a QCL source. Hence, DPS between transmission

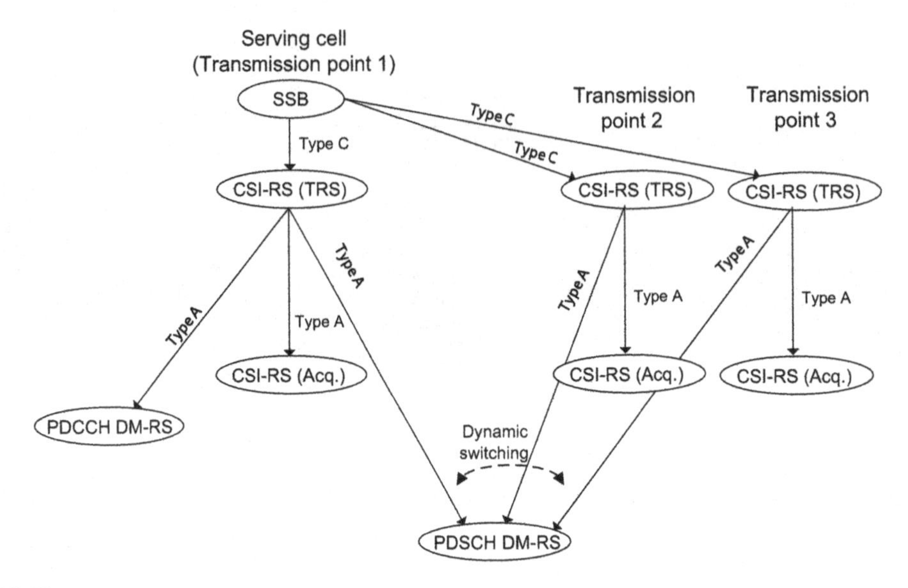

FIGURE 9.16

Example of configured QCL relations for dynamic switching of the transmission point used for PDSCH transmission to the UE (in the case of no QCL Type D, e.g., FR1 operation) where the three transmission points belong to the same cell. The PDCCH is always QCL with the serving cell and transmission point 1 but the PDSCH can be QCL with either of the two additional transmission points and this is indicated by the PDCCH.

points belonging to different cells (i.e., different physical cell ID) is not possible in release 15 but is introduced as a release 16 enhancement.

Fig. 9.16 illustrates an example of a cell with three transmission points where DPS between these is used. Hence, some degree of "local mobility" is achieved, since the UE can roam around within the cell and receive PDSCH from the currently "best" transmission point without involving higher-layer mobility and handover procedures. In this example, the PDCCH is transmitted from one point only, but it is possible to move the PDCCH to another transmission point, although it is a somewhat slower process than the dynamic switching, as it involves MAC protocol signaling.

For FR2 and when QCL Type D is applicable, things get more intricate when it comes to the number of QCL configuration possibilities. The NR specifications support several different options to deploy a beam-based system as multiple SSBs and multiple TRSs can be configured and individually beamformed to enhance the coverage.

Moreover, CSI-RS for beam management can be configured in addition to these beamformed SSBs, typically with more narrow beamwidths and thus larger gains compared to the SSB or TRS beams. See further Section 9.3.6 on beam-based operation.

Fig. 9.17 illustrates all the *possible* QCL relations in connected mode, when QCL type D is applicable and when CSI-RS for beam management is included. Note again that for a given QCL Type, a target RS can only have a single source RS configured at a given point in time.

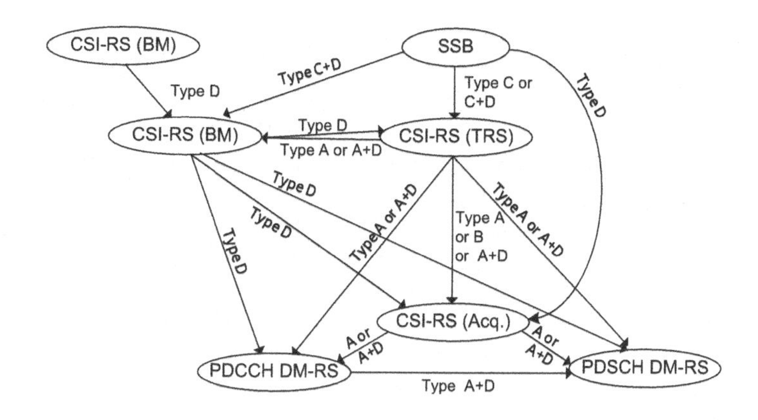

FIGURE 9.17

Possible QCL relations when QCL Type D is applicable (e.g., FR2). Note that a CSI-RS for beam management can also have another CSI-RS for beam management as QCL Type D source. At a given point in time, each target RS can only have one source RS for a given QCL Type and this is controlled by indicating/configuring TCI states for the target RS.

As an example of the configuration flexibility, a PDSCH DM-RS can be QCL Type D with either a beam management CSI-RS, a CSI-RS for CSI acquisition or a TRS, depending on the desired beam management implementation. In Sections 10.5 and 10.6, two use cases of QCL configuration for beam management is exemplified.

To perform the book-keeping of the many configurations of QCL relations between RSs, a list of *transmission indicator states* (TCI) is introduced in NR, where each TCI state contains a source RS that can be used for indicating a QCL relation to a target RS. Hence, when configuring a target RS, a TCI state can be indicated in the configuration and then the QCL relation to the source RS (and QCL type) is thereby established. See further in Section 9.3.3.1.

There are three different configuration mechanisms used to associate a TCI state to a target RS:

- RRC (e.g., for periodic CSI-RS)
- MAC CE (e.g., for PDCCH DM-RS or SP CSI-RS)
- DCI (e.g., for PDSCH DM-RS and for aperiodic CSI-RS)

In addition, before RRC configuration of a list of TCI states, there exists a default QCL relation between PDCCH DM-RS, PDSCH DM-RS, and an SSB. These default QCL relations are used in the initial access procedure (see further Section 9.3.6.1).

9.3.2 INITIAL ACCESS

Initial access is the procedure for a UE to find a cell, acquire system information, and perform the random access procedure. In this section, initial access is briefly described under the assumption of a nonbeam-based system (i.e., a single SS/PBCH). See Section 6.2.1 on UE to cell association background concept. In Section 9.3.6.1, the initial access for a beam-based system, with multiple SS/PBCHs, is described as an extension to the single SS/PBCH description here.

When a UE attempts to access an NR cell, it searches for the PSS. When the PSS is detected, it can also detect the SSS and read the basic system information carried by the PBCH.

The group of synchronization signals (SS) and PBCH is specified as an SSB, typically referred to as an SS block (SSB) in NR. When detecting SS, the UE obtains the physical cell identity, achieve DL synchronization in both time and frequency, and acquires the subframe number and the frame, slot, and symbol synchronization by decoding the PBCH. Hence, PBCH carries a small payload containing only the basic system information. The PBCH also contains information on how to find the *system information block 1* (SIB1) that contains the remaining system information needed to access the cell.

The SSB is mapped to four OFDM symbols in the time domain and 240 contiguous subcarriers (20 RBs) in the frequency domain, as shown in Fig. 9.18. The SSB has a default periodicity of 20 ms (as assumed by the UE in initial access) and the UE can average measurements across each transmission of the same SSB (20 ms separation).

The UE then finds information in SIB1 on how and when to transmit the random access channel (physical random access channel, PRACH) that initiates the access to the cell and the transfer from idle mode to connected mode for a UE.

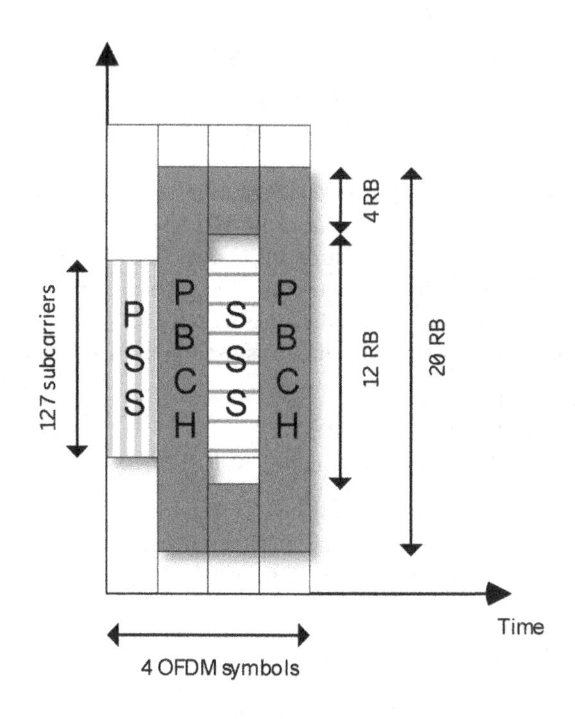

FIGURE 9.18

Illustration of 5G NR SS/PBCH (i.e., an "SSB") where PSS and SSS are the primary and secondary synchronization channels, respectively, and PBCH is the physical broadcast channel containing the master information block (MIB).

9.3.3 PDSCH AND PDCCH TRANSMISSION PROCEDURES

To transmit control and data to the UE, procedures and functionality related to PDCCH and PDSCH transmissions are defined which will be discussed briefly here.

In NR there is only a single transmission scheme for PDSCH. It resembles the transmission scheme of LTE TM 10 and supports up to eight-layer MIMO reception by the UE. In LTE, a transmit diversity scheme using the Alamouti encoding was specified; however, in NR standardization it was decided to not specify a transmit diversity scheme as this can be achieved by implementation-based spatial diversity methods, transparent to the UE. Another reason is that an Alamouti encoded signal is behaving as a rank 2 interferer and since the receiver is not aware of that the interfering signal is Alamouti encoded, an advanced receiver cannot take that into account. Also, a spatial diversity scheme such as Alamouti (based on orthogonal designs) does not scale well with increasing number of antennas and implementation-based methods are anyway needed for AAS. Therefore there is only a single transmission scheme for PDSCH and PDCCH and the PBCH has only a single antenna port; hence diversity is specification transparent.

NR uses *low-density parity-check codes* as the channel code for data. A set of encoded bits is in the physical layer denoted a *code word* (CW) and a single CW is used for one- to four-layer PDSCH transmission to a UE. If the number of transmitted layers is between five and eight, then two CWs are encoded and transmitted by the PDSCH. This is different compared to LTE where two CWs are used already when PDSCH contains two layers or more. See Fig. 9.19 for an illustration comparing how LTE and NR differ in the CW to layer mapping.

So, why is the mapping different in LTE and NR? In LTE, the motivation for two CWs already for two-layer transmission was the possibility to perform per layer link adaptation of MCS plus the possibility to use an advanced receiver which in principle detects one CW first, subtracts the detected CW from the incoming data stream to enhance the detection performance of the second CW. The drawback of two CW is additional CSI and hybrid automatic repeat request-acknowledgment (HARQ-ACK) feedback overhead since these are defined per CW.

When NR specifications were developed, it was observed that even if the LTE specification had allowed implementation of such iterative receivers for more than a decade, they were not used in practice. Furthermore, more detailed evaluations show that the performance benefits of having two CWs compared to a single CW for, for example, two-layer transmissions are negligible when considering impairments due to implementation, quantization and feedback delays. Hence, NR adopted a scheme with a single CW for one- to four-layer transmission and two CWs for five- to eight-layer transmission.

The modulated symbols of a CW are mapped across layers first, then across the subcarriers of the scheduled resource within an OFDM symbol and lastly across OFDM symbols. This mapping across time dimension last means that the UE may start demodulating the first part of the CW already after receiving the first few OFDM symbols of data. This enables a system with low latency as the ACK/negative ACK can be made available earlier compared to if time first mapping would have been supported since in this case the whole slot needs to be received before decoding can start.

Another introduced feature to support a physical layer with very low latency is the possibility to schedule a very short PDSCH or PUSCH transmission, much shorter than the slot duration. Hence, both slot-based and nonslot-based scheduling are specified in NR, where slot-based (known as

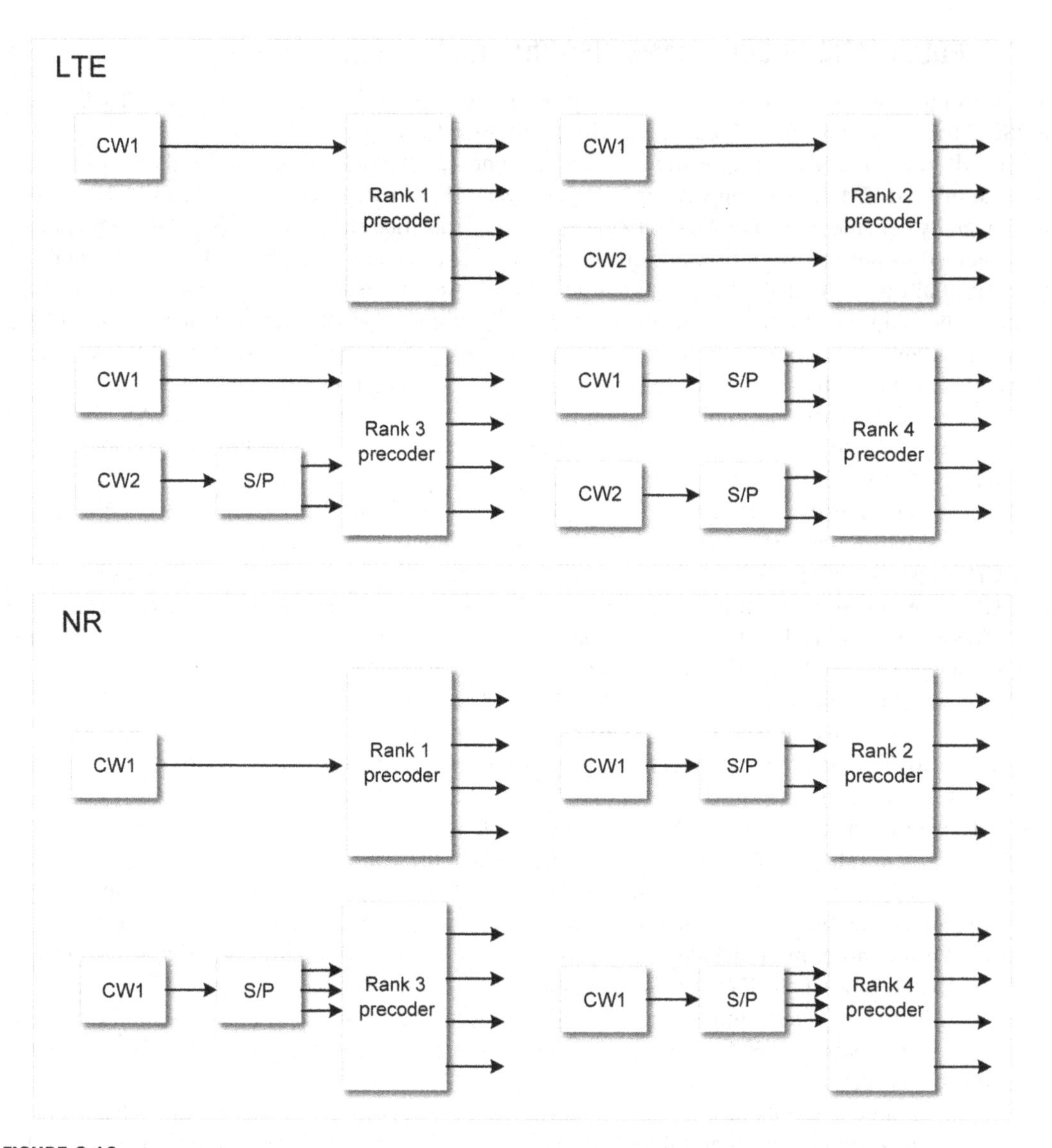

FIGURE 9.19

Codeword to layer mapping for LTE and NR for one- to four-layer PDSCH or PUSCH transmission. For more than four-layer PDSCH, the same codeword to layer mapping is used in LTE and NR (not shown in this figure). The serial-to-parallel (S/P) maps a stream of encoded and modulated symbols on to multiple parallel streams, knows as the layers, thereby implementing the multilayer transmission.

PDSCH mapping Type A) resembles LTE, where PDSCH starts in the beginning of the slot and can end at the end of the slot, or earlier. Nonslot-based scheduling (Type B) can span two, four, or seven OFDM symbols and can basically start and end anywhere in a slot (as long as a transmission does not cross the slot boundary). See further the discussion about DM-RS in Section 9.3.1.2.1 and Table 9.4 about scheduling Type A and B and the supported length in number of symbols, of these scheduling types.

A PDSCH can be scheduled by a PDCCH that carries DCI using a small payload PDCCH (DCI format 1_0) or a larger payload PDCCH (DCI format 1_1) where the smaller DCI format provides limited functionality, for example, it only supports scheduling of a PDSCH transmission using a single DM-RS port (single layer). The smaller DCI format is a default DCI format and thus schedules a default PDSCH configuration. The default configuration is used during initial access communication and can also be used as a fallback to a known transmission configuration, if additional transmission modes or schemes are introduced in NR in the future. Such fallback mechanism is used in LTE, see the LTE PDCCH transmission procedures in Section 8.3.3.

Increasing the bandwidth that the UE can utilize when measuring and computing channel estimates generally leads to better channel estimation performance. However, allowing the UE to use the full bandwidth measurement would prevent the network to use frequency selective precoding. The reason is that changing the precoding matrix from one PRB to another can abruptly change the effective channel making interpolation difficult. To balance this trade-off between UE channel estimation performance and network-based frequency selective precoding, PRB bundling is introduced. A bundle of RBs is defined, over which the UE can safely filter or average the DM-RS-based channel estimates. The gNB in turn keeps the MIMO precoder fixed across the subcarriers within each PRB bundle in order not to cause any abrupt change in the effective channel.

In NR, PRB bundles of two and four RBs are supported as well as a configuration called "wideband," which means that the UE can use the DM-RS of the full scheduled bandwidth to estimate the channel. Wideband is only applicable if the scheduled bandwidth is more than half of the BWP. A use case for "wideband" is to allow for using time-domain channel estimation algorithms in the UE that can provide improved channel estimation performance compared to frequency-domain channel estimation. A wideband PRB bundle is also useful for reciprocity-based operation, as quantization of the CSI feedback into sub-bands is not applicable.

In addition, the PRB bundling can be dynamically indicated, where the scheduling PDCCH contains an indicator of either two PRB or four PRB (depending on higher-layer configuration) bundling and the "wideband" option. The motivation for supporting dynamic PRB bundle size indication is that the gNB may utilize both SRS-based and codebook-based CSI acquisition and may need to dynamically switch between them in the precoding (where in case of codebook-based CSI, typically PRB bundles of two or four RBs is used for PDSCH precoding while for reciprocity-based CSI the PDSCH precoding may use "wideband").

9.3.3.1 TCI states

As described in Section 9.3.1.6, the QCL framework is central in NR. Between two configured reference signals there can be a QCL relation (see one example of such relations in Fig. 9.15). In the case of beam management in particular, when the UE moves in the cell, there is a need to change the configured relations, for example, to reassociate the relation between PDCCH DM-RS and the SSB.

In 3GPP, it was seen beneficial to be able to preconfigure many potentially useful QCL relations using RRC signaling, and then activate/deactivate a given relation as needed. This preconfiguration is useful because RRC-based signaling is a rather slow and there is a need to be able to quickly switch QCL relations. Moreover, there was a need to keep track of all such relations between configured reference signals, and for this reason, the *transmission configuration indicator* (TCI) states were introduced. These states provide a book-keeping framework for the multiple and simultaneously configured QCL relations between RS that the UE can or need to use for measurements and demodulation.

Hence, TCI states are configured first by RRC, and when subsequently configuring a DL RS such as a DM-RS or a CSI-RS, an association with a TCI state is indicated in the DL RS configuration. There are also default QCL relations used in initial access which do not use TCI states. Details of these mechanisms will be described in this section.

To recap the QCL framework for NR as introduced in Section 9.3.1.6: The framework is needed as the UE needs to know which DL signal to perform synchronization toward, before receiving any DL signals such as a measurement reference signal, a control channel and/or a data channel.

If there is a single transmission point that serves the UE, there would be less need for a QCL framework but as deployments get more advanced, and if beam-based operation is used (see Chapter 7), then a QCL framework is necessary.

To give an example on how TCI states are used, each beam transmitted from a TRP (or that potentially can be transmitted in some point in time) is defined by the beamformed transmission of a configured unique reference signal, for example, a periodic CSI-RS (e.g., TRS). Each configured TCI state thus contains a periodic CSI-RS which then represents a certain beam.

One may therefore in this context interpret each TCI state interchangeably as either a "beam" or a proxy for a certain configured reference signal.

A periodic CSI-RS is then a source RS for a QCL relation to receive PDCCH and PDSCH in the same beam as the periodic source RS. The UE can use this periodic CSI-RS to adjust its RX beam and to perform synchronization and channel analysis.

When PDCCH and/or PDSCH transmissions should be switched to a new beam, then the UE must first be informed that a new source periodic CSI-RS should be used for the QCL assumption when receiving PDCCH and/or PDSCH. A new TCI state for the new beam is therefore indicated to the UE, to be used for PDCCH and PDSCH reception in the new beam. How to make this indication is discussed in further detail below.

In the RRC configuration of PDSCH related parameters, the list of TCI states is included. Each TCI state in this configuration contains a pointer to a source RS which can be either a CSI-RS resource, a TRS, or an SSB. The TCI state configuration also contains the associated QCL type (A, B, or C). See the left part of Fig. 9.20 where RSs are configured to TCI states.

For FR2, a TCI state can optionally be extended to contain a second pointer to one additional source RS, where CSI-RS, TRS, or SSB are valid as the source RS for QCL Type D in that TCI state. The TCI state definition also contains a pointer to a cell ID and a BWP ID. Hence, it is possible to configure the UE to derive its large-scale parameter (or controlling its RX beam in FR2) based on measurements on RSs in another cell.

A UE is preconfigured with a list of up to 128 TCI states by RRC signaling and these states then act as labels for different source RSs (or equivalently as labels for different beams). Now, when scheduling data or control, or when triggering a transmission of a CSI-RS, then a TCI state is

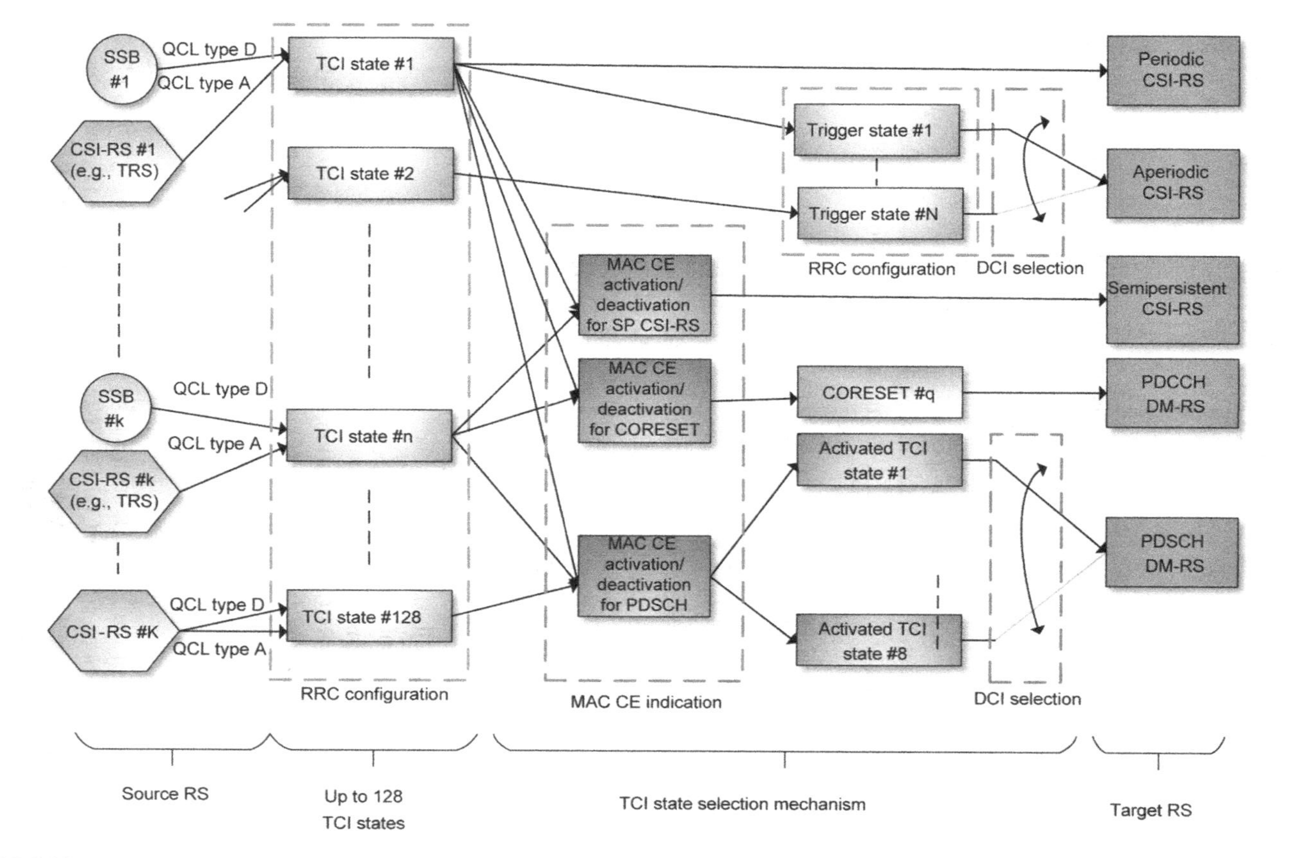

FIGURE 9.20

Framework for TCI states used to assign QCL relations between CSI-RS (e.g., TRS or CSI-RS for beam management or CSI for CSI acquisition) or SS/PBCH and PDCCH DM-RS or PDSCH DM-RS. The figure shows FR2 where each TCI state has one RS for QCL Type D and one more RS for QCL Type A, B, or C. (These two RS may also be the same RS as for TCI state #128 in this example.) Here, the maximal 128 TCI states have been configured to the UE, by RRC. A periodic CSI-RS, for example, used for CSI acquisition is directly associated with a TCI state (i.e., configured by RRC). For PDCCH DM-RS, MAC CE is used to select and activate one of the RRC configured TCI states for the CORESET used by the PDCCH. For PDSCH DM-RS, MAC CE is used to select and activate TCI states for PDSCH DM-RS, either a single TCI state or up to eight TCI states, from the list of RRC configured TCI states. In the latter case, the scheduling DCI selects one of the eight activated TCI states for PDSCH DM-RS. It is also possible to configure PDSCH DM-RS using the same TCI state as the PDCCH that scheduled the PDSCH (i.e., same beam is used for PDCCH and PDSCH).

tied to the target RS of PDSCH DM-RS, PDCCH DM-RS, or a CSI-RS, respectively, thereby completing the link between source RS and target RS.

Depending on what the target RS is used for, there are different mechanisms used to assign and make the connection between the source RS to the target RS (RRC, MAC CE, or DCI) (see Fig. 9.20).

9.3.3.1.1 TCI states for CSI-RS

For periodic CSI-RS, for example, TRS, the TCI state is RRC configured (see Fig. 9.20). If more flexibility is desirable in changing TCI states for CSI-RS, then SP CSI-RS can be used, which is activated/deactivated by MAC CE. This MAC CE message also contains selection of a TCI state to be used, so it is possible to change TCI state for an SP signal without the need for using RRC. For aperiodic CSI-RS, up to 63 trigger states can be configured, each of them with a different RRC configured TCI state, if desired. The DCI that triggers an aperiodic CSI-RS thus selects a trigger state and hence also an associated TCI state. So, for aperiodic CSI-RS there is a large flexibility available in selecting TCI state dynamically for a given CSI-RS resource by configuring the same CSI-RS resource in many trigger states, but with different TCI states.

9.3.3.1.2 TCI states for PDSCH DM-RS

For PDSCH DM-RS the TCI state (beam) switching can be done from one scheduled PDSCH to another (see lower part of Fig. 9.20). This is obtained by configuring the presence of a 3-bit TCI state indicator filed in the scheduling DCI. Hence, for each scheduled PDSCH, a TCI state is indicated dynamically by DCI. However, not any (up to 128) RRC configured state can be selected dynamically by DCI as that would imply a too large indicator field in the DCI. It would also imply a huge UE complexity as it must be prepared to receive (i.e., to be synchronized to) all these TCI states (source RSs) simultaneously. Instead MAC CE is used to downselect and reconfigure to a subset of at most eight TCI states from the possibly larger list of up to 128 TCI states. The DCI can then dynamically select from this subset of eight only when scheduling PDSCH. Hence, dynamic switching between at most eight beams and/or transmission points is supported (also known as dynamic point or dynamic beam selection).

Each such TCI state thus defines different source RS for QCL purpose and NR can in principle support transmission from up to 128 different transmission points, antenna panels, or 128 different transmit beam directions without a need for RRC reconfiguration of the DL parameters. Only MAC CE signaling is needed, which is significantly faster than RRC and avoids involving layers above MAC layer.

9.3.3.1.3 TCI states for PDCCH DM-RS

For PDCCH reception, control resource sets (CORESETs) are defined, which collects physical layer parameters related to the detection of the PDCCH, such as the number of OFDM symbols (1, 2, or 3) and the configured frequency resources. Each CORESET thus contains PDCCH candidates and is configured with a TCI state, so if the UE receives a PDCCH, the PDCCH DM-RS is QCL with the source RS as indicated by the TCI state of the CORESET for which the PDCCH belongs.

For a CORESET, the used TCI state can either be configured by RRC only (if the TCI state list is only of length 1) or by RRC plus MAC CE selection of a single TCI state from the list. Hence, since DCI is not used to select the TCI state, the beam for PDCCH cannot be changed as fast as the PDSCH beam since switching is done by MAC CE only.

A UE can be configured with up to three CORESETs and it is thereby possible to configure one CORESET per transmission point or beam, each with a unique TCI state. After the list of TCI states has been configured by RRC, one TCI state is activated per CORESET by MAC CE signaling. The new active TCI state for the CORESET should be valid no later than 3 ms after the HARQ-ACK for the PDSCH that carried the MAC CE. Note, however, that the "old" TCI state is still valid until the switching point 3 ms later, so communication is not interrupted during the time of new TCI state indication and the switching point.

A possible source RS for a TCI state that is used for PDCCH is an SSB; hence each SSB used in the cell is configured with a unique TCI state and when the UE moves around in the cell, MAC CE signaling is used to update the TCI state for the configured CORESET. Since a cell in NR can have multiple SSBs (as opposed to the single PSS/SSS/PBCH per cell in LTE), the MAC CE reconfiguration of the active TCI state for a CORESET can be viewed as an intra-cell handover command, except without higher-layer (Layer 3) involvement.

9.3.3.1.4 A baseline use of TCI states for PDCCH and PDSCH

In the baseline configuration, the TCI state used for CORESET/PDCCH is also inherited for PDSCH; hence the same transmission point or the same beam is used to transmit both PDCCH and PDSCH to the same UE. The CORESET is then configured without the RRC parameter *tciPresentinDCI* and the DCI does not have the 3-bit field to select TCI state dynamically. This mode of operation is reasonable since if, for example, a beam is found that maximizes the SINR at the receiver, that beam should be used for transmission of *both* control and data to a UE. Hence, the UE uses the same source RS for time/frequency synchronization for PDCCH and PDSCH reception in this case.

In FR2 and if QCL Type D is enabled, the UE additionally assumes that it can use the same RX beam to receive PDCCH and PDSCH. Note that PDSCH beam switching is in this case as slow as for PDCCH, since the dynamics are dictated by the PDCCH beam selection behavior that is reconfigured using MAC CE as discussed above.

9.3.3.1.5 More advanced use of TCI states for PDCCH and PDSCH

In a more advanced configuration, the active TCI states for PDCCH (i.e., the CORESET) and PDSCH can be configured to be different. The CORESET configuration then enables the RRC parameter *tciPresentinDCI*, which introduces an information element to the DCI of at most 3 bits, used to select the TCI state for the scheduled PDSCH. This enables transmitting the PDCCH and PDSCH from different transmission points. As mentioned earlier, MAC CE signaling is used to activate the up to eight TCI states from the list, of which one is then further dynamically selected by DCI for receiving the PDSCH (see Fig. 9.20).

Active tracking of the channels by measuring RSs (e.g., TRS) in multiple TCI states increases UE complexity and it is thus a UE capability on how many active TCI states the UE can track simultaneously. In addition, even if the UE is actively tracking up to eight TCI states, there is a delay (when in FR2) from the indication by DCI to when the UE has applied the indicated TCI state, since the UE may have to adjust the RX beam (if it uses RX beams). The configuration delay from DCI indication to application of the DCI indicated TCI state is a UE capability (1/2, 1, or 2 slots). During this time the "old" TCI state is assumed.

To exemplify one use of the more advanced configuration in FR1: control messages can then be transmitted from a main gNB transmission point, while the data transmission is dynamically selected between two or more other transmission points. This is known as DPS and is also supported in LTE by the use of TM 10.

Another example of an advanced configuration is in FR2, where multiple fixed beams are used and each beam is associated with a TCI state. The beam used for PDCCH may be wider, with lower beamforming gain (e.g., same as the beam used for a SS/PBCH), while the beam used for PDSCH transmission may be a high-gain narrow beam (e.g., a beam found by the network using a recent CSI-RS-based beam sweep). The UE could thus benefit from using the SS/PBCH as source RS for PDCCH and the CSI-RS as source RS for the PDSCH in this case, even if the wide beam is spatially overlapping with the narrow beam. However, if the UE RX beamwidths are very wide, the benefit of this advanced configuration compared to the baseline is likely marginal since the same RX beam will be used for PDCCH and PDSCH, that is, irrespectively of the beamwidth of the gNB TX beam (assuming the wide beam is overlapping the narrow beam).

Finally, note that the maximum number of TCI states a UE can be configured with (by RRC) is a UE capability, where the smallest number a UE is mandated to support is 4 in FR1 and 64 in FR2.

9.3.4 PUSCH TRANSMISSION PROCEDURES

The NR UL supports up to four-layer MIMO codebook-based transmission for PUSCH as in LTE. In addition, both noncodebook-based operation and beamformed UE transmissions are supported. The NR UL supports CP-based OFDM as a baseline but a DFT-precoded OFDM can be configured to reduce the PAPR and increase the coverage of the transmission (see Section 5.1.5). However, when DFT precoding is enabled, only a single-layer PUSCH transmission is supported.

UL MIMO (multilayer transmission) transmission is supported by indicating to the UE an UL transmission grant in DCI format 0_1. The DCI contains an SRI and "Precoding information and number of layers" bit fields. Note that these fields are not present in the fallback DCI format 0_0 which means only single-port PUSCH transmission is supported when the UE is scheduled with the DCI format 0_0.

9.3.4.1 Spatial relation

Introducing noncodebook-based transmission and UL beamforming implies that the direction of the transmitted energy is no longer completely controlled by the gNB, as in the codebook-based PUSCH. Hence, for the network to maintain some control over how the UE directs the transmitted energy, for example, for interference management purpose and for FR2 to allow the gNB to prepare a good RX beam that corresponds to the UE transmission, there was a need to introduce a new functionality referred to as *spatial relation*.

A spatial relation is different from a QCL relation in that it is defined on the UE side while QCL is defined on the gNB side (as observed at the UE). A spatial relation can be configured to hold between a received DL reference signal and a transmitted UL signal, whereas a QCL relation is always between two transmitted DL signals (observed at a receiver). Hence, the spatial relation framework provides a method for the network to steer the UL transmission so that it improves the reception quality at a certain receiving network node, that is, gNB.

The use of the spatial relation features relies on UE support for beam correspondence, where the receive and transmit chains at the UE are calibrated so that the UE can transmit back in the same direction as it has received a DL signal.

A spatial relation can alternatively by configured between two UL transmissions at the UE, such as an SRS and a PUCCH. If a spatial relation is configured between these, it means the UE shall transmit the PUCCH with the same antenna pattern, for example, beam, that it previously transmitted the associated SRS. Thereby, the network receiver can infer the quality it would receive PUCCH with by measuring the SRS. In addition, the gNB can in this case use the same receive spatial filter/beam to receive PUCCH and SRS.

Hence, like configuring QCL relations by TCI states for the DL transmissions, a list of spatial relations can be configured for the UL transmissions.

The spatial relation for SRS is configured per SRS resource; hence a target SRS resource may have a spatial relation to a source CSI-RS, an SS/BPCH (on same or different cell) or another SRS resource (on same or different UL BWP).

For periodic or aperiodic SRS resources, the spatial relation is configured by RRC signaling but for SP SRS transmissions (periodic transmissions activated and deactivated by MAC CE) the MAC CE is used to activate one spatial relation from a list of RRC configured spatial relations, providing somewhat better flexibility.

Note that configuration of a spatial relation to an SRS resource is optional for gNB. If not configured, the UE can select the "direction" of the transmission, that is, the transmit precoder, freely. One use case for this is a UL SRS beam sweep, where an SRS resource set is triggered without spatial relations assigned and the UE can then transmit multiple SRS resources in different directions to "probe" possible new UL directions for data transmissions. It is also possible to have a gNB-assisted SRS beam sweep where the SRS resources have a spatial relation to, for example, different CSI-RS transmitted from different TRPs. Then the network can use these multiple SRS transmissions to evaluate the reception quality of the UL at each of multiple TRPs.

When a good UL direction is found by an SRS resource transmission, the gNB can configure that SRS resource as the reference RS (using RRC/MAC CE) for the spatial relation to the SRS resource used for, for example, codebook-based PUSCH transmissions. Hence, the SRS used for probing the UL is reused as the reference for SRS-based link adaptation measurements intended for PUSCH transmission and this is made possible by using a spatial relation configuration between two SRS resources. In this case, the source SRS resource may belong to an SRS resource set used for beam management (UL beam sweeps) and the other, target, is an SRS resource used for codebook-based PUSCH.

Note that using the spatial relation between SRSs observed from an SRS beam sweep is also useful if the UE does not support beam correspondence, in which the gNB can select a preferred UL SRS transmission direction from such "blind SRS beam sweep" and then apply the preferred direction for the data and control transmission.

For PUCCH, a list of up to eight spatial relations can be configured by RRC where the reference RS is a CSI-RS or a SS/PBCH or another SRS. From these eight spatial relations, MAC CE is used to activate one of them for each PUCCH resource. Note that if an SS/PBCH is used as the reference RS, and the serving node is using more than eight SS/PBCHs (e.g., 64), then RRC reconfiguration of PUCCH spatial relations is needed whenever the UE enters the coverage area of a new SS/PBCH in the cell, which is a drawback. This issue is addressed in NR release 16 where up to 64 spatial relations can be configured to the UE for PUCCH.

Alternatively, a CSI-RS can be configured that "follows" the UE movements, that is, the precoding of the transmission of this CSI-RS which is used for PUCCH spatial relation is adapted dynamically if side information is available. Side information in this case could be obtained from UL measurements. In this case there is no need to reconfigure the spatial relation reference RS, but it comes at the cost of increased gNB complexity.

9.3.4.2 Codebook-based PUSCH

For PUSCH transmissions, the UE is configured for either "Codebook" or "Noncodebook"-based transmissions.

The codebook-based transmission is like LTE PUSCH UL MIMO introduced in release 10, with the extension that one of two configured SRS resources in an SRS resource set, configured to *usage = "codebook,"* can be indicated as the reference for the PUSCH.

The DCI contains an SRI bit that identifies one of two SRS resources, a transmit precoding matrix indicator (TPMI), and a rank and MCS for the PUSCH transmission. The UE shall transmit the PUSCH using the precoder indicated by TPMI and using the same antenna virtualization as it used of the SRS ports for the latest transmission of the SRS resource indicated by the SRI.

Another difference of NR codebook-based UL from LTE operation is the introduction of different UE coherence capabilities defined between different SRS antenna ports. In LTE, it is assumed that all SRS ports are coherent, which allows for up to 6 dB coverage gain (coherent addition of four antennas) but has large impact on UE implementation complexity. Coherent means that the phase relation between SRS antenna ports is maintained over time, so the gNB can measure SRS in one slot and indicate a precoder to use for PUSCH transmission in another, later slot, where the precoder contains cophasing of SRS antenna ports.

Supporting this coherency may be complex to implement for UE vendors, and this was addressed explicitly in NR by introducing support also for less complex UE implementations together with an associated capability signaling. Precoding matrices that do not use cophasing of antenna ports are thus specified in NR, which do not require coherence.

Hence, the UL precoding codebook for a given rank is divided into three overlapping subsets. If the UE indicates full coherence capability, then all SRS ports can be transmitted coherently, and it is thus possible to cophase these multiple ports to achieve spatial precoding gains. The full codebook of all available precoding matrices can be used for such a UE.

If the UE declared partial coherence capability, applicable for the case of four SRS ports, only pairs of SRS ports (port {0,2} and port {1,3}, respectively) are coherently transmitted, and thus precoding matrices that cophase across these pairs (e.g., across port 0 and 1) are excluded from this UE's UL codebook. A typical implementation could be that the UE uses dual-polarized antennas and the two ports corresponding to different polarizations of the same antenna element can be assumed to be coherent while ports on different, spatially separated antennas on the device are not coherent.

Lastly, a UE can also declare noncoherence, which means that no SRS ports are implemented as coherent and only antenna selection is then used, per transmitted PUSCH layer. In other words,

TPMI	Type	Precoding matrix W							
0–5	Non coherent	$\frac{1}{2}\begin{bmatrix}1 & 0\\0 & 1\\0 & 0\\0 & 0\end{bmatrix}$	$\frac{1}{2}\begin{bmatrix}1 & 0\\0 & 0\\0 & 1\\0 & 0\end{bmatrix}$	$\frac{1}{2}\begin{bmatrix}1 & 0\\0 & 0\\0 & 0\\0 & 1\end{bmatrix}$	$\frac{1}{2}\begin{bmatrix}0 & 0\\1 & 0\\0 & 1\\0 & 0\end{bmatrix}$	$\frac{1}{2}\begin{bmatrix}0 & 0\\1 & 0\\0 & 0\\0 & 1\end{bmatrix}$			
6–13	Partial coherent	$\frac{1}{2}\begin{bmatrix}1 & 0\\0 & 1\\1 & 0\\0 & -j\end{bmatrix}$	$\frac{1}{2}\begin{bmatrix}1 & 0\\0 & 1\\1 & 0\\0 & j\end{bmatrix}$	$\frac{1}{2}\begin{bmatrix}1 & 0\\0 & 1\\-j & 0\\0 & 1\end{bmatrix}$	$\frac{1}{2}\begin{bmatrix}1 & 0\\0 & 1\\-j & 0\\0 & -1\end{bmatrix}$	$\frac{1}{2}\begin{bmatrix}1 & 0\\0 & 1\\-1 & 0\\0 & -j\end{bmatrix}$	$\frac{1}{2}\begin{bmatrix}1 & 0\\0 & 1\\-1 & 0\\0 & j\end{bmatrix}$	$\frac{1}{2}\begin{bmatrix}1 & 0\\0 & 1\\j & 0\\0 & 1\end{bmatrix}$	$\frac{1}{2}\begin{bmatrix}1 & 0\\0 & 1\\j & 0\\0 & -1\end{bmatrix}$
14–21	Fully coherent	$\frac{1}{2\sqrt{2}}\begin{bmatrix}1 & 1\\1 & 1\\1 & -1\\1 & -1\end{bmatrix}$	$\frac{1}{2\sqrt{2}}\begin{bmatrix}1 & 1\\1 & 1\\j & -j\\j & -j\end{bmatrix}$	$\frac{1}{2\sqrt{2}}\begin{bmatrix}1 & 1\\j & j\\1 & -1\\j & -j\end{bmatrix}$	$\frac{1}{2\sqrt{2}}\begin{bmatrix}1 & 1\\j & j\\j & -j\\-1 & 1\end{bmatrix}$	$\frac{1}{2\sqrt{2}}\begin{bmatrix}1 & 1\\-1 & -1\\1 & -1\\-1 & 1\end{bmatrix}$	$\frac{1}{2\sqrt{2}}\begin{bmatrix}1 & 1\\-1 & -1\\j & -j\\-j & j\end{bmatrix}$	$\frac{1}{2\sqrt{2}}\begin{bmatrix}1 & 1\\-j & -j\\1 & -1\\-j & j\end{bmatrix}$	$\frac{1}{2\sqrt{2}}\begin{bmatrix}1 & 1\\-j & -j\\j & -j\\1 & -1\end{bmatrix}$

FIGURE 9.21

An example of the codebook for uplink codebook–based transmission for PUSCH; the table shows the rank 2 codebook for four SRS antenna ports and the grouping by table rows into precoding matrices for noncoherent, partial coherent, and full coherent transmission, respectively.

no cophasing at all among the UE's TX chains can be achieved and hence no coherent precoding gains can be achieved.

An example of the UL codebook for four SRS ports and two-layer transmission can be seen in Fig. 9.14. Here, the precoder that a UE shall use for PUSCH is indicated in the DCI with a transmit precoding matrix indicator (TPMI).

A UE that declared fully coherent may be scheduled with TPMI = {0,. . .,21} while if partial coherence has been declared, the UE may only be scheduled with TPMI = {0,. . .,13} and for a noncoherent UE only TPMI = {0,. . .,5} can be used (Fig. 9.21).

9.3.4.3 Noncodebook-based PUSCH

The noncodebook-based transmission is introduced to allow reciprocity-based PUSCH transmissions. This is a new functionality compared to LTE and is an optional feature for the UE to support. In this case, the UE transmits up to four SRS resources of an SRS resource set where each such SRS resource is restricted to be configured to have a single SRS port only. The UE transmits the SRS resources using transmit precoding (or beamforming) that is not known to the gNB. The gNB then measures the channel for each SRS resource and then selects which to use for the PUSCH transmission. The number of resources the gNB selects equals the rank of the PUSCH transmission since each resource has a single port only.

The gNB then indicates in the scheduling DCI that which set of SRS resources to use for the PUSCH transmission. This is done using the SRI field in the DCI, which in this noncodebook-based case can indicate multiple SRS resources.

The UE transmits one PUSCH layer per indicated SRS resource using the same precoder/beam as it used in the most recent transmission of that SRS resource. Hence, for noncodebook-based PUSCH, the number of SRS resources selected by SRI equals the scheduled rank for PUSCH.

A straightforward implementation of noncodebook-based operation is based on antenna selection. SRI can identify a subset of up to four antennas for the UE to transmit on. This can provide diversity and/or select the antenna(s) pointing most directly at the gNB. When multiple layers are transmitted, the UE is not required to maintain coherent transmission among the transmitting antennas, which enables a low-complexity UL MIMO design like noncoherent codebook-based operation.

It is also possible in the noncodebook-based transmission scheme to configure an associated NZP CSI-RS for an SRS resource set, which means the UE can calculate the precoder used for the transmission of SRS based on measurement of this NZP CSI-RS resource. Hence, this is reciprocity-based PUSCH transmission, and it is also applicable for FR1. The use of reciprocity-based precoding in the UE requires not only phase coherence between the UE's TX chains but also calibration between its TX and RX chains. Consequently, this is a more advanced UE implementation, and the use of an associated NZP CSI-RS for noncodebook-based operation is also an optional UE capability. The similarities with spatial relation (Section 9.3.4.1) are large (using a DL RS to guide the UE in performing a directive UL transmission), although the associated DL RS is in this case configured on an SRS resource set level instead of the SRS resource level as it is for the spatial relation.

9.3.5 CSI FRAMEWORK

The purpose of the CSI framework is to provide the network with DL measurements; hence it includes configuration of reference signals for these measurements and the associated reporting. The NR CSI framework is significantly more flexible than the corresponding LTE framework. The guiding principle during standardization has been to keep a modular approach where CSI measurement reference signals and CSI reporting are configured independently and only tied together based on the desired CSI reporting functionality. In addition, the principles of implicit CSI reporting have been used (see Section 6.4.1.2).

Hence, in NR, the CSI measurements, the CSI reporting, and the CSI-RS transmission are independent configurations. This goal was achieved by support for configuration of one or more CSI report settings used to define the content of a CSI report and one or more CSI resource settings, used to define the RS(s) used to perform CSI measurements.

Hence, these two settings and their combinations provide a huge flexibility in the configuration, and it should be noted again that this reporting is not connected to how the actual transmission is performed; it is only a measurement that the UE reports to the gNB (cf. LTE where CSI reporting is closely tied to the transmission mode).

In Fig. 9.22, it is illustrated how a CSI report configuration can be flexibly tied to two configured CSI resources used for measurements of the channel (desired signal) and the interference, respectively. Note that it is also possible to have a single channel measurement resource or even two interference measurement resources in the CSI report configuration.

In addition, the UE can be configured with a list of *aperiodic trigger states*. Each state points to one or more CSI report configurations and a state is triggered dynamically by PDCCH to obtain an aperiodic CSI report from the UE.

In the following the details of the CSI report setting and the CSI measurement setting will be further elaborated.

9.3.5.1 CSI report settings

The report setting configures how the UE shall generate a certain CSI report. It is linked to one or more CSI resource configurations which configure how the UE should make measurements for the report.

The gNB can configure several different types of reports simultaneously. Hence, the UE can be configured with one or more CSI report settings, by the higher-layer parameter *CSI-ReportConfig* carried by dedicated RRC signaling. Each such setting configures either a periodic CSI report using

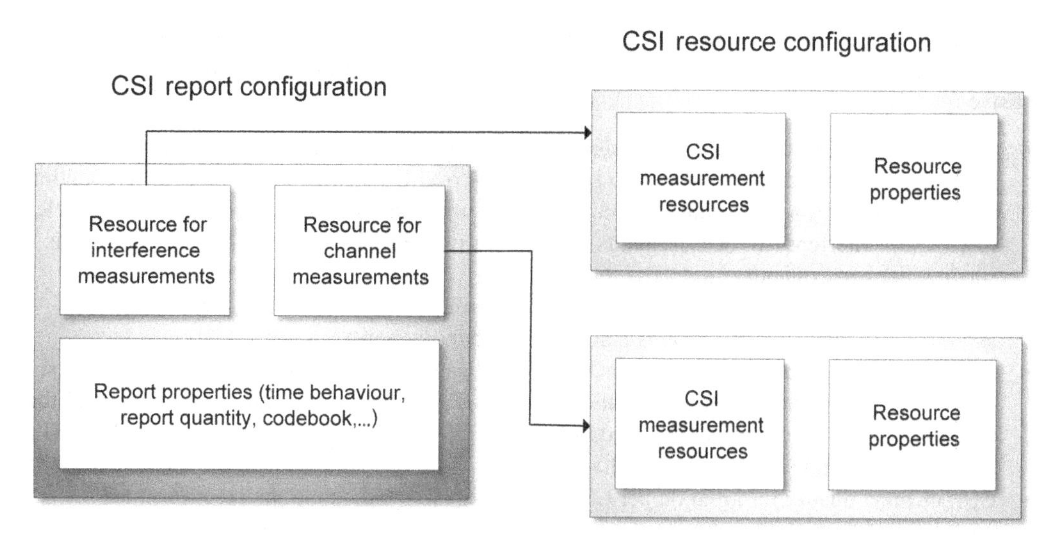

FIGURE 9.22

Illustration of the flexibility in the NR CSI feedback framework where configuration of reports and measurements are separate but can be linked together depending on the desired CSI reporting.

PUCCH or an aperiodic CSI report using PUSCH. There is also a possibility to configure an SP CSI report using either PUCCH or PUSCH.

The PUCCH-based SP CSI report is similar to a periodic CSI report but can be enabled or disabled via MAC CE signaling, which is faster than the RRC signaling–based periodic report configuration.

The CSI report setting contains many parameters of configuration, for example:

- An identifier of resources for channel measurements;
- An identifier of resources for interference measurements;
- Whether reporting is periodic, aperiodic, or semipersistent;
- The reporting quantity (see below);
- The bandwidth for which the CSI report should be estimated;
- Whether CQI is reported as a single wideband value or per sub-band;
- Whether PMI is reported as a single wideband value or per sub-band;
- The sub-band bandwidth (4, 8, 16, or 32 RBs are possible, available values depend on the bandwidth of the BWP);
- Whether measurement restriction applies (i.e., a single shot measurement or time-averaged measurements) for channel and interference, respectively;
- Whether codebook Type I or Type II is used and related codebook subset restriction;
- Which CQI table and thus block error rate (BLER) target to use;
- Number of measured RS to be reported (1, 2, or 4) for beam reporting;
- The PUCCH resource, when applicable;
- Power control for the PUSCH transmission in case of SP-CSI reporting.

The different CSI report quantities from the UE to the network are the following, where CQI, RI, PMI, and CSI-RS resource indicator (CRI) are also used in LTE, while the remaining ones are exclusive for NR:

- *CRI*, CSI-RS resource indicator. In case the configured measurement contains a set of CSI-RS resources, then the CRI is used to select a preferred CSI-RS resource (e.g., corresponding to a beam out of a set of beams) for channel measurement and CRI can also be used to select CSI-IM for interference measurements. CRI is not used for Type II codebooks. The CRI is a single, wideband quantity;
- *RI*, rank indicator, represents the recommended number of PDSCH layers, calculated assuming the selected CRI. This is a single, wideband quantity;
- *PMI*, precoding matrix information, is used to indicate the recommended precoding matrix, consisting of an index *i1* and one or more indices *i2* (i.e., per sub-band) to select each of the two matrices respectively in the precoding matrix $W = W^{(1)}W^{(2)}$ (see Section 9.3.5.3). PMI is calculated assuming the selected CRI and RI;
- *CQI*, channel quality information per CW, is a 4-bit value indicating the recommended modulation scheme and code rate, assuming the selected PMI, RI, and CRI. The CQI can be either wideband or per sub-band. The CQI shall be reported for a configurable BLER target of either 10% for the mobile broadband use case or 0.001% for ultra-reliable transmission use case;
- *SSBRI*, SSB indicator, is like CRI but indicates instead which SSB the accompanying CSI report is valid for. This is a wideband quantity and a report for beam management can contain up to four SSBRIs;
- *LI*, layer indicator, in case of $RI > 1$, indicates the strongest layer, that is, column, of the selected precoding matrix, assuming the selected CQI, PMI, RI, and CRI. LI can be used by gNB to ensure that PT-RS is transmitted on the strongest layer, and is a single, wideband quantity;
- *L1-RSRP*, carries a single or multiple RSRP measurements. For the case of a single L1-RSRP, a 7-bit value is used using the range $[-140, -44]$ dBm with 1 dB step size. The measurement RS could be CSI-RS, SS/PBCH, or both. L1-RSRP is a wideband quantity and such reporting can be configured in both FR1 and FR2, although it is only mandatory for a UE to support L1-RSRP in FR2.

Given these measurement quantities, a CSI report configuration can be configured to feedback one of the possible combinations shown in Table 9.8. Note that if only a single RS is configured for measurement, then the CRI field has zero bits even if the quantity name includes the term "cri."

9.3.5.2 CSI resource configuration

The UE can be configured with one or more CSI resource settings, by the higher-layer parameter *CSI-ResourceConfig* carried by dedicated RRC signaling. If multiple CSI-RS resources have been configured, the CSI measurement indicates which of these RSs to be used in a particular measurement.

Hence, each such resource setting contains a list of CSI resource sets that may include NZP CSI-RS resource sets, SSB sets or resources for interference measurements, and the CSI-IM resource sets.

Table 9.8 CQI Report Quantities in NR, used to configure the content of a CSI report.

Quantity	Description	Typical Application
None	No CSI report	Beam management with "repetition" on (see Section 9.3.6.2), where a report is not needed.
cri-RI-PMI-CQI	Full CSI report, either wideband or per sub-band	Basic CSI reporting for CSI acquisition
cri-RI-LI-PMI-CQI	Full CSI report, with LI, either wideband or per sub-band	Basic CSI reporting for CSI acquisition with LI, typically for FR2 to determine the best DM-RS layer for PT-RS mapping
cri-RI-i1	A wideband partial PMI, for $\boldsymbol{W}^{(1)}$ only, is reported together with rank and CRI	Used to obtain information about a preferred long-term and wideband channel property. Needs to be combined with another report quantity to obtain CQI
cri-RI-i1-CQI	A wideband partial PMI, for $\boldsymbol{W}^{(1)}$ only, is reported together with rank, CRI, and a wideband or per sub-band CQI. For CQI calculation, the UE assumes a randomly selected $\boldsymbol{W}^{(2)}$ from the codebook, per group of RBs	Type I reporting and single panel only. Can be used for higher speed UE where the optimal $\boldsymbol{W}^{(2)}$ anyway will be outdated if feed back to the gNB, due to feedback delays
cri-RI-CQI	CSI report without PMI, assuming a scaled identity matrix as a precoding matrix. Wideband or sub-band CQI	CQI report for reciprocity-based operation
cri-RSRP	CSI-RS resource selection and RSRP	Beam selection and RSRP report per beam, CSI-RS-based beams
ssb-Index-RSRP	SSBRI and RSRP	Beam selection and RSRP report per beam, SS/PBCH-based beams

To measure interference, either NZP CSI-RS (i.e., actual channel estimation of interferer) or CSI-IM (power measurement in "empty" resource elements) can be configured. Using NZP CSI-RS for interference measurement can be used for CSI feedback in MU-MIMO scheduling, where two NZP CSI-RS are configured, and UE A uses the first for channel and the second for interference measurements and UE B uses the second for channel and first as interference measurements. In this way, the mutual interference due to MU-MIMO is captured in the CSI report. The use of NZP CSI-RS measurements for interference is available only for the aperiodic CSI report, triggered by PDCCH.

The number of resource settings (i.e., measurement RSs) linked to a report setting (i.e., a certain CSI report) can be configured according to the alternatives in Table 9.9.

9.3.5.3 Overview of Type I codebooks for single and multipanel

As in LTE, implicit CSI feedback is used (see Section 6.4.1.2) and a codebook of precoding matrices has been defined and the design approach is similar to the LTE release 13 codebook for

Table 9.9 Four Different Configurations for a CSI Report Are Available, Using One, Two, or Three Different Measurement RSs (By Linking to One, Two or Three Resource Settings).

Number of Resource Settings Linked to a Report Setting	These Measurements Are Used for	Applicable Time Behavior of the CSI Report
One	• Channel for L1-RSRP	Aperiodic, semipersistent, or periodic CSI reporting
Two	• Channel (NZP CSI-RS) • Interference (CSI-IM)	Aperiodic, periodic, or semipersistent
Two	• Channel (NZP CSI-RS) • Interference (NZP CSI-RS)	Aperiodic only
Three	• Channel (NZP CSI-RS) • Interference (NZP CSI-RS) • Residual interference (CSI-IM)	Aperiodic only

one-dimensional (1D) and two-dimensional (2D) antenna port layouts (see Section 8.3.2.2 for the LTE codebook discussion). See also Chapter 6, in general for an introduction to MIMO precoding matrices and codebook design fundamentals.

In LTE, an advanced codebook was defined in release 14 to improve the MU-MIMO performance (see Section 8.3.2.4). In NR, these two categories of codebooks also exist and are denoted Type I and Type II, respectively. In addition, for Type I codebooks, NR also supports CSI feedback for multipanel antenna arrays.

The Type I codebook in NR is designed for SU- or MU-MIMO operation having moderate CSI feedback overhead and relatively low UE complexity to decide the preferred precoder. Both 1D and 2D antenna port layouts are supported with 2, 4, 8, 12, 16, 24, and 32 CSI-RS ports. As mentioned, there are in NR both single and multipanel codebooks, where multipanel implies the use of more than one antenna array panel where the panels are physically separated from each other and where both are simultaneously used for the same PDSCH transmission.

The multipanel codebook was introduced primarily for FR2 band operation since it was argued from an implementation perspective that such multipanel gNBs would be of interest to vendors using AAS panels placed next to each other, that is, cosited. When a PDSCH transmission to a UE uses more than one AAS panel there is a need to cophase the transmission of a PDSCH layer across these two or more panels, to achieve spatial combining gains.

Hence the CSI feedback framework for the multipanel case is an extension of the single-panel codebook where an unknown distance between any two such panels is assumed. This arbitrary distance implies an arbitrary phase shift, so the uniform linear array model that has been used for the codebook within a single panel does not hold across different panels.

Such arbitrary phase jump is assumed in the Type I multipanel codebook design which supports a configuration of two or four panels and where the cophasing can be configured to be wideband or per sub-band.

When multipanel CSI feedback is used, the reported rank is restricted to a maximum of 4 and the number of CSI-RS ports per panel is at most eight.

Regarding the codebook within a panel (or for the single-panel configuration), the codebook for two ports is the same as in LTE and for four ports or more it has the same matrix product structure as the LTE release 10 codebook (see Section 8.3.2.2).

$$\boldsymbol{W} = \boldsymbol{W}_1 \boldsymbol{W}_2,$$

where the first, inner, $N_T \times \tilde{N}_T$ precoder matrix $\boldsymbol{W}^{(1)}$ is block diagonal with a matrix $\boldsymbol{B}$ repeated twice in the matrix diagonal; hence

$$\boldsymbol{W}_1 = \begin{bmatrix} \boldsymbol{B} & \boldsymbol{0} \\ \boldsymbol{0} & \boldsymbol{B} \end{bmatrix}$$

and where $\tilde{N}_T$ is a design parameter, which is related to the number of remaining beams after the channel dimension reduction by $\boldsymbol{W}_1$. Here, $\boldsymbol{W}_1$ is reported wideband, while $\boldsymbol{W}^{(2)}$ can be reported per sub-band or also wideband.

The dual-polarized uniform planar array (see Section 4.5.2.1) was assumed in this codebook design, as in LTE release 10. Compared to LTE, the NR codebook configurability is simplified, and $\boldsymbol{B}$ contains either $S = 1$ or $S = 4$ DFT vectors where S is a configurable parameter.

For rank 3 and above, the codebook is only defined for $S = 1$; hence $S = 4$ and sub-band beam selection are only an option for rank 1 and 2. See Chapter 6, on the definition of DFT vectors and the relation to a "beam." Note that in NR specifications, $S = 1$ and $S = 4$ are termed Codebook Mode 1 and 2, respectively.

The matrix $\boldsymbol{W}_2$ then performs polarization cophasing and if $S = 4$, then $\boldsymbol{W}_2$ also performs beam selection, for example, in the rank 1 case, one out of four columns of $\boldsymbol{B}$ is selected, per sub-band. This increases the feedback overhead significantly since two additional bits are needed per sub-band compared to S = 1, used to select a beam from the four possible beams.

The spatial oversampling factor in NR is fixed to four and thus not configurable as in LTE. The oversampling reduces the straddling loss, that is, the beamforming gain loss, that occurs when the UE is in between two beam peaks.

9.3.5.4 Details of Type I single-panel codebook

For the interested reader, this section contains more details of the NR single-panel Type I codebook design and resulting beam structure. The codebook will be described for the $S = 1$ configuration since it has the lowest PMI overhead and basically the same performance as the $S = 4$ configuration.

In Table 9.10, the supported 1D and 2D antenna port layouts are shown. For example, a (4,2) codebook can be configured for an antenna array having four ports in a first spatial dimension (e.g., columns in a horizontal array) and two ports in the second spatial dimension (e.g., vertical, two rows per column) plus the two ports in the polarization dimension. Hence, a CSI-RS resource with $2N_1N_2$ antenna ports needs to be configured for the measurements, in this example a 16-port resource.

The two-port codebook (1,1) for rank 1 and 2 is the same as in LTE, except that for rank 2, the identity matrix is not included in NR. See the description of the LTE 2 port codebook in Section 8.3.2.2.

Table 9.10 Supported (Indicated by X) NR Release 15 Codebooks for Combinations of (N_1, N_2), the Number of Antenna Ports for Each Dimension in the Two-Dimensional Port Layout.

		N_2 Ports for Dimension Two			
CSI-RS Ports		**1**	**2**	**3**	**4**
N_1 ports for dimension one	1	X	–	–	–
	2	X	X	–	–
	3	–	X	–	–
	4	X	X	X	X
	6	X	X	–	–
	8	X	X	–	–
	12	X	–	–	–
	16	X	–	–	–

For four ports or above, the $\mathbf{W} = \mathbf{W}_1\mathbf{W}_2$ structure is used with

$$\mathbf{W}_1 = \begin{bmatrix} \mathbf{B} & \mathbf{0} \\ \mathbf{0} & \mathbf{B} \end{bmatrix},$$

and where $\mathbf{B}$ is composed of S oversampled 2D DFT beams with an oversampling factor of 4. Hence in the $S=1$ case, the UE selects one beam from $4N_1$ and $4N_2$ beams per dimension, respectively.

In Fig. 9.23, 128 beam directions are shown, as each circle represents a DFT beam main direction and the example given is an antenna port layout of (2,4), that is, the (4,2) port layout rotated 90 degrees. There are N_1 and N_2 orthogonal beams per dimension and hence, the striped circles in the figure represent one set of orthogonal beam directions. Note that in this illustration, the circles represent the "tip" of each beam, the main beam direction, to illustrate the concept, but certainly there are sidelobes as well.

For rank 1, the UE selects $\mathbf{W}_1$ by selecting one out of these *$16N_1N_2$* different beam directions, valid for the whole bandwidth of the configured CSI-RS resource. Hence, $\mathbf{W}_1$ is determined by two indices, one per dimension, respectively. The cophasing of the two polarizations is then determined by selecting $\mathbf{W}_2$. Hence this is the same as the LTE two-port codebook for rank 1, and it is illustrative how $\mathbf{W}_1$ reduced the dimensions to two. The feedback of $\mathbf{W}_2$ can be configured to be reported as a single wideband value or per sub-band.

$$\mathbf{W}_2 = \frac{1}{\sqrt{2}}\begin{bmatrix} 1 \\ 1 \end{bmatrix}, \frac{1}{\sqrt{2}}\begin{bmatrix} 1 \\ -1 \end{bmatrix}, \frac{1}{\sqrt{2}}\begin{bmatrix} 1 \\ j \end{bmatrix}, \frac{1}{\sqrt{2}}\begin{bmatrix} 1 \\ -j \end{bmatrix}.$$

For rank 2, the UE selects $\mathbf{W}_1$ by selecting a beam for the first layer from the *$16N_1N_2$* different beam directions as in the rank 1 case, and for the second layer it can choose to select the same beam as for the first layer, or it can select an orthogonal beam to the beam of the first layer, in either of the two dimensions. (For rectangular port layouts, $N_1 = N_2$, it is also possible to select a beam that is orthogonal to the first layer beam in both dimensions simultaneously) Hence, two

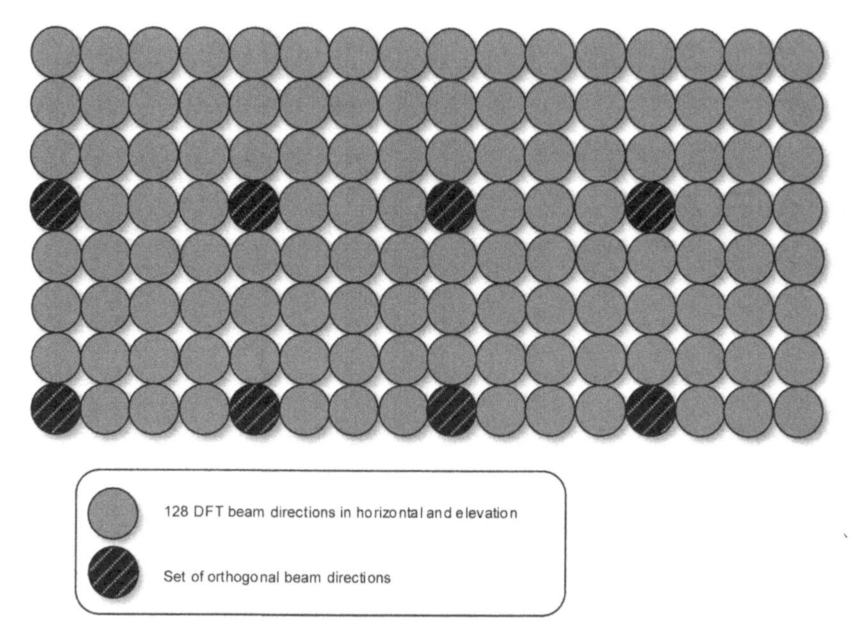

FIGURE 9.23

The 128 mean beam directions resulting from a two-dimensional port layout (2,4) where due to spatial oversampling of four, there are 8 vertical beams and 16 horizontal beam directions. The marked beams form a set of orthogonal DFT beams; hence every fourth beam in the horizontal and vertical direction forms a set of orthogonal beams.

additional bits are added to the determination of $\boldsymbol{W}_1$ compared to the rank 1 case. For rank 2, the specification uses three indices to select $\boldsymbol{W}_1$, the beam of the first and second dimensions, respectively, plus an index to select the beam of the second layer relative to the first layer (the two additional bits), see (Fig. 9.24).

Note that in the 1D port layout cases, the three closest orthogonal beams in one dimension are the alternative beams for the second layer as there is only one dimension. In addition, in the four-port codebook case, only the closest orthogonal beam can be selected as there are only eight beams to choose from by $\boldsymbol{W}_1$.

The cophasing is done per sub-band or wideband (configurable) using 1 bit by

$$\boldsymbol{W}_2 = \begin{bmatrix} 1 & 1 \\ 1 & -1 \end{bmatrix}, \begin{bmatrix} 1 & 1 \\ j & -j \end{bmatrix}$$

For *rank 3 and 4 with less than 16 ports*, the inner precoder $\boldsymbol{W}_1$ is selected as in the rank 1 and 2 cases, but where the first and second layer share a (main) beam. The third and possibly the fourth layer share a second beam where this second beam is orthogonal in either or both vertical and horizontal dimensions relative to the main beam used for the first and second layer. Hence, the matrix $\boldsymbol{B}$ has two columns in this case.

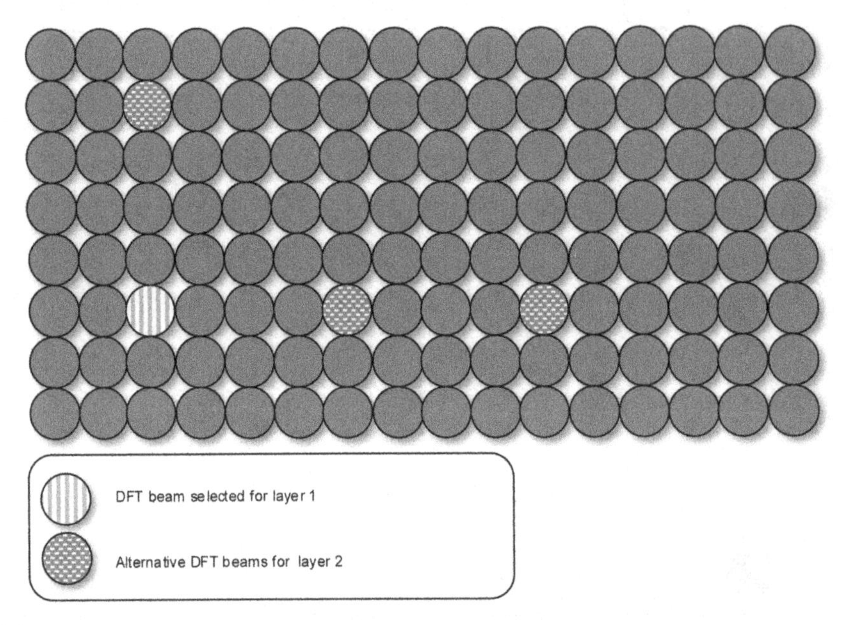

FIGURE 9.24

Example for rank 2 beam selection. For the second layer, the UE can select the same beam as for the first layer, or an orthogonal beam, from the set of beams shown in this example for the two-dimensional port layout (2,4). Note that some of the beams orthogonal to the layer 1 beam (cf. Fig. 9.23) cannot be selected by the layer 2 beam which is a restriction in order to use only 2 bits for the beam selection.

The cophasing is done per sub-band or wideband (configurable) using 1-bit indication of one of the following precoders.

$$\mathbf{W}_2 = \begin{bmatrix} 1 & 1 & 0 \\ 0 & 0 & 1 \\ 1 & -1 & 0 \\ 0 & 0 & 1 \end{bmatrix}, \begin{bmatrix} 1 & 1 & 0 \\ 0 & 0 & 1 \\ j & -j & 0 \\ 0 & 0 & j \end{bmatrix}$$

for rank 3 or

$$\mathbf{W}_2 = \begin{bmatrix} 1 & 1 & 0 & 0 \\ 0 & 0 & 1 & 1 \\ 1 & -1 & 0 & 0 \\ 0 & 0 & 1 & -1 \end{bmatrix}, \begin{bmatrix} 1 & 1 & 0 & 0 \\ 0 & 0 & 1 & 1 \\ j & -j & 0 & 0 \\ 0 & 0 & j & -j \end{bmatrix}$$

for rank 4.

For *rank 3 and 4 with 16 ports or more*, the antenna ports of the first dimension are split into two equal parts, hence having $N_1/2$ ports. The reason is that the number of ports of this dimension is getting large and the beamwidth in this dimension may be too narrow for some propagation channels. Therefore a mechanism to widen the beam is introduced as follows: A DFT beam is selected as to be transmitted from only half ($N_1/2$) of the antenna ports. The same selected DFT beam is used for both parts but the beam direction of the second part can be shifted relative to the beam direction of the first part. This phase shift is selected as a wideband parameter using 2 bits. Note that the indicator for selection of a beam for the first dimension has been reduced by 1 bit compared to if N_1 ports would have been used, since the DFT beam of the first dimension is mapped to half of the number of ports ($N_1/2$) only (and hence the number of beams is reduced from $4N_1$ to $2N_1$).

For the third and fourth layer, a beam orthogonal to the beam of the first and second layer is used, as for the codebook of less than 16 ports. The cophasing of layer within the same beam also follows the same principle, using 1 bit for wideband or per sub-band, depending on configuration of the CSI report.

For *rank 5 and 6*, the first two layers share the same main beam, and the third and fourth layer share the same beam direction which is orthogonal to the main beam direction where orthogonality is achieved by selecting an orthogonal beam to the first layer beam, in the first dimension (Fig. 9.25). The fifth and sixth layer share the same beam that is yet another orthogonal beam to the main beam by selecting an orthogonal beam in both first and second dimensions. Cophasing of layers within the same beam is done as for previous ranks by 1 bit per sub-band.

For *rank 7 and 8*, the same principle is further extended, by selecting a main beam direction for the first and second layer, and then orthogonal beam directions are selected in one or both dimensions, relative to the main beam, in order to get four orthogonal beams. Cophasing of layers within the same beam is done as for previous ranks by 1 bit per sub-band.

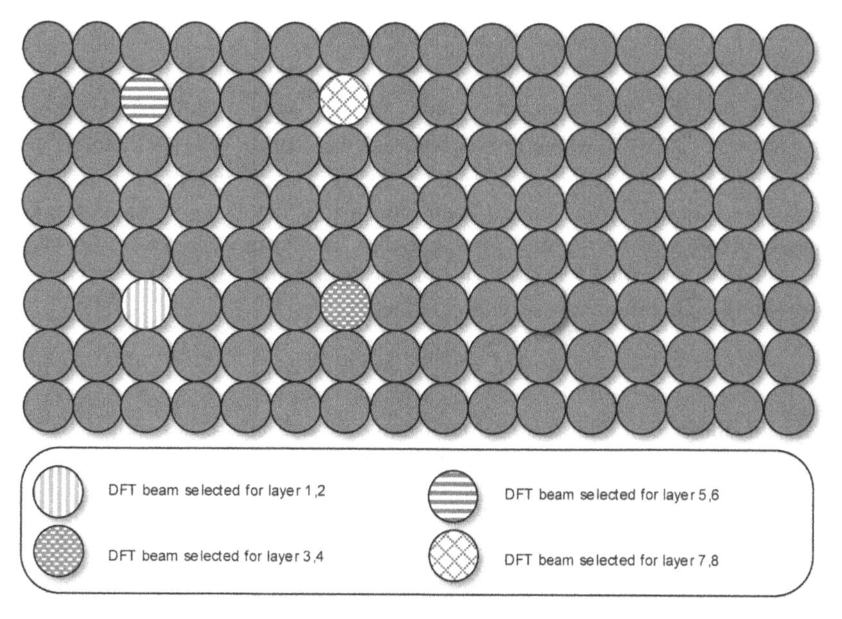

FIGURE 9.25

Example for higher rank beam selection of 5–8-layer transmission where orthogonal DFT beams are utilized. Every pair of two layers uses the same beam. The set of beams shown in this example is for the two-dimensional port layout (2,4).

9.3.5.5 Type II codebooks

In LTE, the advanced CSI feedback was introduced in release 14, mainly in order to improve performance for MU-MIMO. NR has a similar, but even higher-resolution CSI feedback mode, the "Type II" codebook. It is specified for single-panel CSI feedback only and for rank 1 and 2 feedback. From 4 to 32 CSI-RS ports are supported and there exist two Type II codebooks, the standard Type II and the "port selection" Type II codebook. In the following, the standard Type II codebook is described.

A difference with LTE advanced CSI feedback is that up to four beams can be configured for the advanced feedback (in cases with more than four CSI-RS ports), where more beams are useful in rich scattering channels. A multi-antenna channel that has several nonnegligible eigenvalues requires also more eigenvectors (span a higher dimensional subspace) to minimize the error in representing a MIMO channel with as few bits as possible without losing the essential channel information (see Section 6.4.1.2).

The codebook for rank 1 feedback can be expressed as

$$\boldsymbol{W} = \begin{bmatrix} \tilde{\boldsymbol{w}}_{0,0} \\ \tilde{\boldsymbol{w}}_{1,0} \end{bmatrix} = \boldsymbol{W}_1 \boldsymbol{W}_2,$$

where the precoding vector now is a weighted and cophased sum over $L = 2$, 3, or 4 beam directions instead of $L = 1$ in the Type I codebook. The value L is configurable and represents a trade-off between overhead and CSI accuracy. The expression for the weight vector is given by

$$\tilde{\boldsymbol{w}}_{r,l} = \sum_{i=0}^{L-1} \boldsymbol{b}_{k_1^{(i)} k_2^{(i)}} \cdot p_{r,l,i}^{(\mathrm{WB})} \cdot p_{r,l,i}^{(\mathrm{SB})} \cdot c_{r,l,i}$$

Here, $\boldsymbol{b}_{k_1^{(i)} k_2^{(i)}}$ is a 2D DFT beam selected by index pair $k_1^{(i)}$ and $k_2^{(i)}$ and there are two real-valued power scaling factors, $p_{r,l,i}^{(WB)}, p_{r,l,i}^{(SB)}$ where one is reported wideband (WB) and another per sub-band (SB). Whether to use a per sub-band power scaling factor is configurable as it increases feedback overhead. Lastly, there is a complex-valued beam cophasing factor $c_{r,l,i}$ that can be configured to belong to either a QPSK alphabet or an 8-PSK alphabet. The beams are selected from an orthogonal basis in two dimensions with a spatial oversampling factor of 4.

The corresponding rank 2 codebook is expressed as

$$\boldsymbol{W} = \begin{bmatrix} \tilde{\boldsymbol{w}}_{0,0} & \tilde{\boldsymbol{w}}_{0,1} \\ \tilde{\boldsymbol{w}}_{1,0} & \tilde{\boldsymbol{w}}_{1,1} \end{bmatrix} = \boldsymbol{W}_1 \boldsymbol{W}_2$$

In Fig. 9.26, the illustration shows how four beams are combined and weighted. Note that the figure shows one polarization only for simplicity of illustration. The wideband power scaling is one of

$$p_{r,l,i}^{(\mathrm{WB})} \in \{1, \sqrt{0.5}, \sqrt{0.25}, \sqrt{0.125}, \sqrt{0.0625}, \sqrt{0.0313}, \sqrt{0.0156}, 0\}$$

and the sub-band power scaling is $p_{r,l,i}^{(\mathrm{SB})} \in \{1, \sqrt{0.5}\}$.

In Section 13.6.2, a comparison of the performance of Type I and Type II feedback is given.

9.3.5.6 CSI computation limits

The large flexibility of the CSI framework in NR puts large demands on UE implementation. To be able to cope with the CSI computation complexity and to handle aperiodic CSI requests on top of

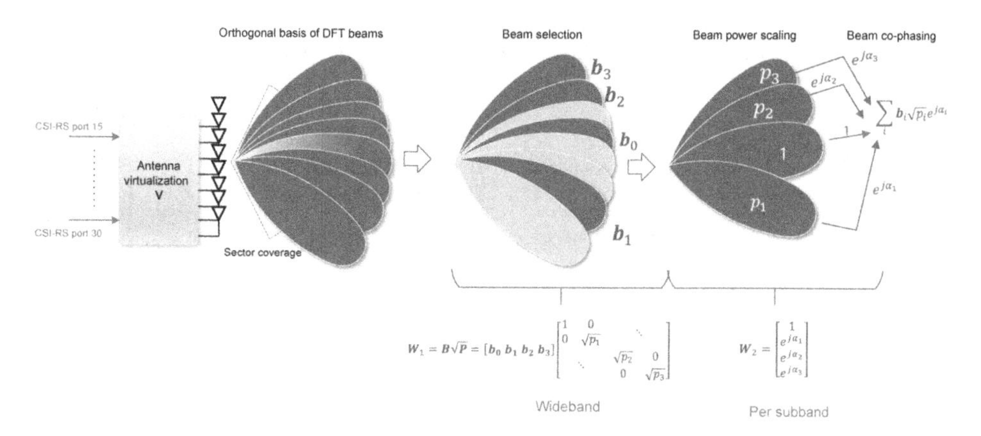

FIGURE 9.26

CSI feedback type II in NR, where the UE can be configured to select up to four beams to represent the MIMO channel. The illustration only shows the cophasing of beams for one of the two polarizations (for simplicity). Also, wideband-only power scaling is assumed in this figure.

periodic CSI reporting, the UE reports the number of simultaneous ongoing CSI calculations, known as CSI processing units (CPUs), as a UE capability. The capability values that can be reported is between 5 and 32 CPU:s (where the indicated CPU value should be across all carriers in case of CA) and a CPU value between 1 and 8 when counted per carrier.

The CPUs are thus a pool of generic computational resources for CSI. As an example, the UE can indicate support for two configured CSI report settings but only a single CPU. This means that the gNB can trigger any of the two configured CSI report configurations, but gNB must time multiplex these different CSI reports since this UE has only support for a single CPU.

If there are not enough CPUs available due to that the UE is already processing other CSI reports, the allocated CSI reporting does not have to be calculated by the UE and it can instead report stale CSI, such as a previously calculated CSI report stored in memory or simply padding the CSI report with dummy bits.

Each CSI report occupies a number of O_{CPU} CPUs from a starting allocation time until the last symbol of the physical channel (i.e., PUCCH or PUSCH) carrying the CSI report has finished transmitting, whereby the O_{CPU} CPUs are released.

The number of O_{CPU} CPUs occupied by a certain CSI report calculation depends on the content of the report. For nonbeam management reports, the number of occupied CPU is the number of CSI-RS resources in the CSI-RS resource set for channel measurement. Hence, if CRI selects among eight CSI-RS resources, this occupies $O_{CPU} = 8$. For beam management (e.g., L1-RSRP) reports, only a single CPU ($O_{CPU} = 1$) is occupied, even if multiple CSI-RS resources are included in the CSI-RS resource set for channel measurement.

For aperiodic reporting, the CSI reports are classified into different latency classes, each with different timing requirements (see Table 9.11). The minimum number of OFDM symbols between the last symbol of the PDCCH triggering the aperiodic CSI report and the first symbol of the triggered PUSCH that carries the CSI report is denoted as the parameter Z.

Table 9.11 CSI Reporting Latency Classes for Aperiodic Reporting Depending on Report Content.

Ultralow-Latency CSI (30 kHz SCS Assumed for Z, Z′)	Low-Latency CSI (30 kHz SCS Assumed for Z, Z′)	Beam Reporting Latency (120 kHz SCS Assumed for Z, Z′)	High-Latency CSI (30 kHz SCS Assumed for Z, Z′)
• Wideband frequency granularity • A single CSI (i.e., no CRI reporting) • At most four CSI-RS ports measured • PMI reporting with Type I Single-Panel codebook or non-PMI reporting • Zero CPU currently occupied in UE • No PUSCH transmission containing data or HARQ-ACK scheduled together with CSI	• Wideband frequency granularity • A single CSI-RS resource (i.e., no CRI reporting) with at most 4 CSI-RS ports • PMI reporting with Type I Single-Panel codebook or non-PMI reporting	• L1-RSRP reporting (i.e., beam management) • *Note 1: X_4 is a UE capability on beam reporting timing and can take values {14,28,56} OFDM symbols* • *Note 2: KB_2 is a UE capability on beam switching timing (when PDCCH and CSI-RS have different beams/QCL Type D) and can take values {14,28,48,224,336}. OFDM symbols where the latter includes antenna panel activation delay.*	All other cases
$Z = 13,\ Z' = 11$	$Z = 33,\ Z' = 30$	$Z = \min(97, X_4 + KB_2),\ Z' = X_4$	$Z = 72,\ Z' = 69$

Notes: *The values of Z and Z′ are measured in number of OFDM symbols and the table shows values 30 kHz subcarrier spacing (SCS) except for beam reporting where 120 kHz SCS values are used. See Ref. [5] for Z, Z′ values for other SCS.*

If aperiodic CSI-RS or CSI-IM is used with the aperiodic CSI report, the aperiodic CSI-RS could potentially be triggered close to the triggered PUSCH transmission. Therefore the minimum number of OFDM symbols Z' between the last symbol of the aperiodic CSI-RS/IM used to calculate the report and the first symbol of the PUSCH that carries the CSI report is also specified.

If the Z-criterion and Z'-criterion are not both fulfilled, that is, the gNB triggers the PUSCH too close to the PDCCH (or the aperiodic CSI-RS/CSI-IM too close to the PUSCH), the UE can simply ignore the scheduling DCI for the aperiodic CSI report.

9.3.6 BEAM-BASED OPERATION

NR is inherently designed to support beam-based operation, which implies that the network and UE may choose to use transmit and receive spatial filtering (i.e., beamforming) for all signals and channels. The use of a beam from gNB transmission means that a transmission is not covering the whole served cell from a horizontal and/or vertical angle perspective.

To recap Chapter 7, when a beam is used for reception, the whole channel cannot be observed at one measurement, since the beam implies that some directions are spatially filtered out. This leads to an observability problem that needs special techniques and functionality to manage.

It is important to stress that all these introduced beam management features are mainly targeting FR2 operation. Some of these functionalities may not be needed in FR2, depending on the actual used implementation in gNB and/or UE. Additionally, some of these beam-based features can also be useful in FR1 operation although the support of these features is typically optional for the UE in FR1.

The key components of beam-based functionality are:

- *Initial access*—How to design network access procedures if the access signals such as synchronization and system information are beamformed, that is, do not cover the whole cell? Also, how do the gNB and UE transmit and receive beams align at initial access without using dedicated beam finding procedures?
- *Observability*—How to find good transmit and receive beams for both gNB and UE, that are aligned, that is, the receive beam depends both on the propagation channel and how the corresponding transmit beam is directed;
- *Indication of intent*—As processing of received control and data is not instantaneous, but extends over several symbols, the activation of a new beam cannot be instant. Hence, the beam manager's intent to switch to a new beam must be signaled in advance to the UE. This introduces some delays in the beam management;
- *Beam recovery*—It happens that the transmit-receive beam alignment fails. Hence, procedures must be specified for how the UE and gNB behave when this happens, without the need for the full higher-layer procedures of declaring radio link failure (RLF).

To solve these tasks, the beam management procedures are specified, comprising the following components that have been introduced earlier in the chapter:

- QCL relation using Spatial RX (Section 9.3.1.6)
- TCI states (Section 9.3.3.1)
- Spatial relations (Section 9.3.4.1)
- P1, P2, and P3 procedures (Section 7.4.2)
- L1-RSRP measurements (Section 9.3.5.1)

The beam management toolbox is quite extensive, and not all components need to be used. In fact, it is possible to use a very basic beam management framework with only SSB beams, as these beams define the coverage of the cell. Then, to optimize the performance, CSI-RS for beam management can additionally be configured, allowing for narrower and thus higher-gain beams. See Section 10.6 for some examples of this.

In this section, it will be described how these different components are used to perform the beam management and solve the four tasks outlined above.

Beams can in NR be used for both broadcast signals and for dedicated data channels; hence NR can support a fully beam-based network operation in contrast to LTE where broadcast signals are covering/defining the cell (i.e., nonbeam-based broadcasting). The UE may also use beamforming for transmission and reception.

A central concept is *beam correspondence*, meaning that transmission and reception beams are aligned (within some margins of error). If a UE and gNB support beam correspondence, procedures are simplified and access to network is faster. For example, the UE can transmit the random access channel in the same beam as it previously has detected an SSB.

Another central concept in beam-based operation is the "QCL Type D," which indicates "Spatial RX" relation between two reference signals transmitted from the gNB (see Section 9.3.1.6). Signaling or configuring the relation between two RSs of QCL Type D ensures that the UE can use the same receive beam to receive the RSs even if they are transmitted in two different time instances. Hence, a "beam memory" is introduced by the specified beam-based procedures.

The use of beam indication in the scheduling DCI is helpful in UE implementations where the UE cannot receive signals omnidirectionally and thus a beam must first be formed in time domain, that is, in advance, before starting to receive the actual signal. Therefore there is a need to label different gNB beams since indication of an intent to switch to a new gNB beam requires some method for such labeling. However, gNB beams are not explicitly numbered in NR; instead a reference is made to the index of the RS that is transmitted with a certain beam, and the TCI states thus serve the purpose of labeling beams.

As discussed previously in Section 9.3.2, the TCI states can be used to indicate QCL Type D, and one can thus interpret different TCI states as equivalent to different transmit beams and/or transmission points of the gNB. Hence, for each active TCI state, the UE should maintain a suitable receiver configuration, such as a preferred receive beam (if the UE uses time-domain beamforming) and channel estimation parameters based on estimated delay spread, Doppler shift, etc.

The observability in beam management revolves around the problem of finding one or more suitable gNB transmit and UE receive beams for the gNB to UE link and to indicate to the UE the intention of the gNB transmitter to use a certain beam. Sometimes terms as *beam alignment* or *beam pairs* are used where the transmit and receive beams form the pair.

In designing the beam shapes, assuming classical beamforming is used (see Chapter 6), the gNB must choose the beamwidth. The choice to make a beamwidth wider to cover a larger sector angle in horizontal and/or vertical (i.e., a larger *solid angle*) will sacrifice the coverage.

For some channels a low spectral efficiency [a low modulation order (QPSK) and a low code rate] is commonly used, for example, those used to broadcast system information or random access responses. These are intended for multiple UEs simultaneously and to UEs in unknown directions and for these, lower directivity still gives acceptable coverage. As the UE position is unknown, a beam sweep is needed to cover the whole served cell volume. Hence, the network can for these types of broadcast channels trade directivity for a larger solid angle per transmission (e.g., a sector of the cell) and thus fewer such repeated beam sweep transmissions in different directions are needed to cover the whole served volume. Using fewer such beams will reduce the overhead.

The data transmission is intended for a single UE and as high spectral efficiency as possible is targeted. For this transmission it is therefore beneficial to maximize the receive SNR at the UE while at the same time avoid transmitting in undesired directions to minimize interference towards other UEs.

Hence, there is a need to utilize the potential of the full antenna array coherently for these dedicated transmissions to the UE, and this requires more elaborate fine-tuning mechanisms in the beam management. With narrower beams, the beam direction is more sensitive to beam pointing errors.

For these reasons, the NR specification supports two types of reference signals for beam management measurements and either of them can be used when a reference to a beam is needed:

- The SSB is a periodic signal transmitted in a wider beamwidth with medium coverage and thus medium directivity, shared by all users in the cell. Up to 64 different SS/PBCH beams can be transmitted in a 20 ms period, and these are static beam directions;

- The CSI-RS for beam management, typically used as an aperiodic reference signal using dedicated beamforming to one UE. It is typically using a narrow beamwidth with high directivity.

In the following section, the use of these two different signals for beam management is discussed, where SSB beams are typically coarse beams, suitable for common control signaling, beam recovery, and for initial access, and CSI-RS for beam management can additionally be configured to fine-tune high directivity beams used for the high-performance data transmission.

When a suitable beam has been found, then CSI reporting for link adaptation purpose can also be triggered (to feed back CQI, PMI, and RI). Hence in addition to these beam management reference signals, a CSI-RS for CSI acquisition (Section 9.3.1.1.3) can be configured. See how this is used in the use case examples 4 and 5 in Chapter 10.

9.3.6.1 Initial access and beam-based operation prior to connected state

To support beam-based cell search operation (see Sections 6.2.1 and 9.3.2), the network transmits multiple SSBs that are beamformed in different directions (Fig. 9.27). Up to a maximum of $L_{\max} = 4, 8,$ or 64 such SSBs can be transmitted in a 5 ms burst within the 20 ms period, hence supporting an SSB beam sweep in up to $L_{\max}$ different beam directions. The localization into 5 ms out of 20 ms allows minimization of transmission of these SSBs that are "always-on" signals, since if there is no traffic at all in the cell, at least 75% of all slots will be completely empty as opposed to LTE where transmission of CRS is mandatory in all subframes.

The used value $L_{\max}$ depends on the frequency band (see Table 9.12).

A UE performing cell search and initial access to a network can thus utilize the fact that an SSB is transmitted with a 20 ms periodicity. If the UE is using receive beamforming using an analog implementation, then the UE may change its received beam every 20 ms to probe different receive beam directions in order to improve the link quality.

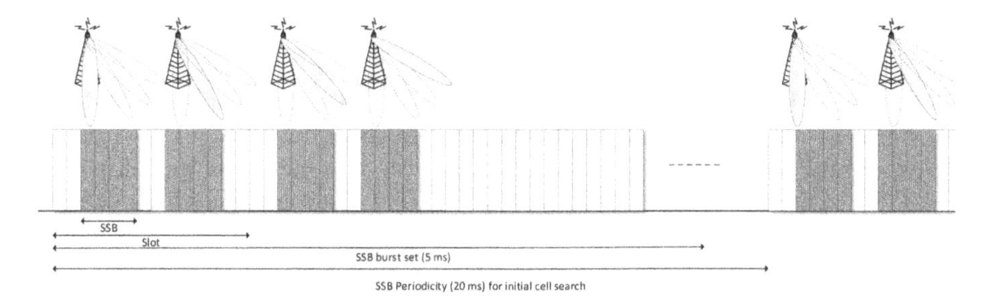

FIGURE 9.27

The SS/PBCH (SSB) period is 20 ms and is in a burst set of 5 ms in the beginning of the period. Each SSB within the period is transmitted in a different beam direction to facilitate a beam sweep that covers the whole cell. The figure shows an example of four SSBs where the illustrated pattern is configurable for 30 kHz subcarrier spacing case. Note that the burst set length and periodicity (5 and 20 ms) are independent of the subcarrier spacing. With increased subcarrier spacing, there are more slots within the burst set and thus room for more SSBs with a maximum of 64 SSBs for 120 kHz subcarrier spacing.

Table 9.12 The Dependence of the Value L_{max} (Maximum Number of SSBs and Thus Different SSB Beams) on Different Subcarrier Spacing (SCS) and Frequency Ranges.

SCS (kHz)	Frequency Range	L_{max}
15	≤ 3 GHz	4
15	3–7.125 GHz	8
30	≤ 3 GHz (paired spectrum)	8
30	≤ 2.4 GHz (unpaired spectrum)	4
30	2.4–7.125 GHz (unpaired spectrum)	8
120 or 240	24.25–52.6 GHz	64

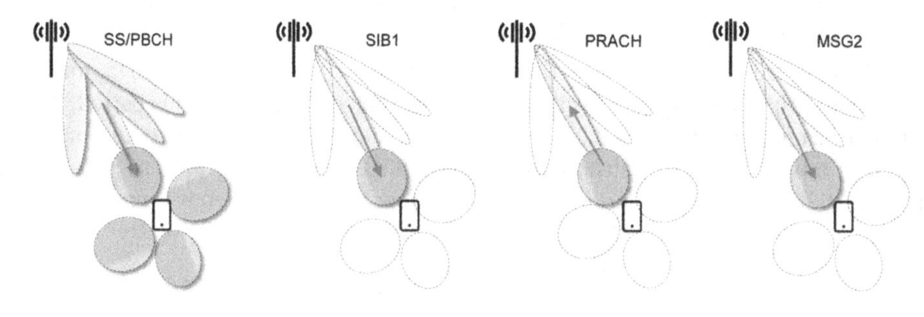

FIGURE 9.28

Initial access in a beam-based system. The UE detects a preferred SSB using a suitable RX beam. It receives further system information by SIB1 and then transmits PRACH in the same beam as the receive beam. The network can then use the same beam to receive PRACH as it used to transmit the SSB.

The UE then detects an SSB and stores the information related to the RX beam used to receive the block. As the periodicity is known as 20 ms, the UE may also periodically refine the RX beam for a detected SSB or alternatively improve the reception quality of the PBCH by averaging.

After detecting the SS/PBCH and decoding the master information block (MIB) carried by PBCH, the UE has acquired the frame, slot, and symbol timing information of the cell and knows which SS/PBCH out of the up to $L_{max} = 64$ possible SS/PBCHs it has detected. The MIB also contains information on how to find system information block 1 (SIB1).

A PDCCH scheduling a PDSCH carrying SIB1 is periodically transmitted and contains the necessary parameters for configuring the random access procedure. These PDCCH and PDSCH DM-RS are QCL with the detected SSB/PBCH block with respect to Type A and D. This means that the UE may use the same RX beam to receive SIB1 as it used to receive the detected SSB (see Fig. 9.28). It also means that for each beam used for a SS/PBCH, the network also needs to transmit an SIB1. Furthermore, in FR2 it is possible to configure the SIB1 in the same four OFDM symbols as the associated SS/PBCH (i.e., frequency multiplexed) to save overhead in transmission of these periodically transmitted signals and channels.

Based on the index of the detected SS/PBCH (out of the up to 64 possible SS/PBCHs), there is an associated set of PRACH preambles and PRACH time/frequency location and this association is configured by SIB1. Due to this association between a detected DL beam used for SS/PBCH and a time and preamble for the UE to perform PRACH transmission, the gNB can, if it uses time-domain/analog beamforming, be prepared to receive in the same beam direction to detect the PRACH as the beam direction it used when transmitting the associated SS/PBCH.

If the UE supports beam correspondence, it can similarly as the gNB use the same beam to transmit PRACH as the associated receive beam it used to receive the detected SS/PBCH.

When the gNB detects a PRACH from a UE, it simultaneously knows the SS/PBCH index the UE detected and thus also the associated gNB TX beam. Hence, the transmit and receive beams for both the gNB and the UE have been established based on this initial access procedure.

The TX and RX beams established in the initial access procedure are then used by the gNB and the UE in the following PDSCH and PUSCH transmissions required to complete the random access procedure, to configure security parameters, to convey UE capabilities, and to convey the dedicated RRC configuration. The UE can thus assume that subsequent PDSCH transmissions from the gNB are QCL with the SS/PBCH found during this initial access unless configured otherwise.

In case the UE does not support beam correspondence, the UE must repeat the PRACH transmissions in different directions and listen for the network response using the same beam used to detect the SS/PBCH. After successful connection establishment, it will have a transmit beam direction and a receive beam direction. Access without beam correspondence leads thus to an increased latency in the initial access procedure and increased interference due to excessive PRACH transmission attempts, compared to a UE supporting beam correspondence.

9.3.6.2 Beam management procedures and measurements

After initial access, dedicated RRC signaling is used to configure the UE with beam management–related features such as TCI states, CORESET for beam failure recovery (BFL) monitoring and if desired, CSI-RS for beam management. The beam management procedures are primarily designed for FR2 although some functionality can also be used in FR1 bands, if the UE supports the required features.

Beam management measurements and beam finding can be categorized into three procedures for the DL, procedure 1 (P1), P2, and P3. See a more elaborate discussion in Chapter 7.

- *P1—TX and RX beam finding*: This procedure is used in initial access, where the UE finds a preferred network TX beam and a corresponding UE RX beam. It can also be used by connected UEs on, for example, nonstandalone NR carriers, where the UE measures on SSBs (or configured periodic CSI-RS for beam management) and identifies a set of preferred SSB beams and reports L1-RSRP and the SSBRI to indicate this to the network. In this case the UE also finds and stores in memory the preferred RX beam for the single or set of preferred gNB TX beams;
- *P2—TX beam refinement*: gNB triggers a measurement of a gNB TX beam sweep of multiple beams using CSI-RS for beam management. Typically, aperiodic CSI-RS for beam management is used. The UE reports its preferred beam and L1-RSRP. The UE may also store the RX beam it used for these beams;

- *P3—RX beam refinement*: gNB triggers a measurement of a single gNB TX beam, repeatedly transmitted, so that the UE can have multiple opportunities to refine its RX beam for the gNB TX beam.

As will be discussed below, a system can work with only P1 procedures implemented, but better performance can be achieved if P2 is added and potentially also P3. Whether P3 is beneficial is unclear as it depends on how the UE has implemented its RX beam selection procedures, for example, how well can the UE select RX beam without P3-based network assistance.

One typical way of operating beam management is a two-step approach where the first P1 is used to find a coarse beam, that is, an SSB beam, and then gNB triggers a P2 beam sweep with multiple narrower beams within the beam coverage of the SSB beam found in P1. The purpose of P2 is to find a suitable high-gain beam to be used for PDSCH transmission. The PDCCH may still be transmitted using the wider beam found during the P1 procedure as antenna gain is not as important, since PDCCH coverage may be enough anyway.

In RRC connected state, the *active* TCI states are central in the beam management since these are potential candidates for being used for a scheduled PDSCH, a PDCCH, and/or CSI-RS. Also, dynamic switching between these beams corresponding to active TCI states can be performed. Switching to a beam related to an inactive TCI state requires the first MAC CE activation of that TCI state.

How to select which TCI states to enabled as active for a given UE is equivalent to selecting the set of beams a UE may be scheduled on, particularly if static beams are used (static beams means that a given RS resource is periodically transmitted in the same beam so there is a one-to-one mapping between beam and resource identifier). This is achieved with the P1 and P2 procedures.

Each configured TCI state contains a pointer to a configured source reference signal with respect to QCL Type D, which has previously been transmitted by the network. When referring to that TCI state, if it is *active*, the UE knows which receiver configuration it should apply (e.g., RX beam). If this RS is a periodic RS, such as the SS/PBCH, a TRS, or a periodic CSI-RS, then the UE can update and refine its receiver configuration/RX beam periodically.

In Fig. 9.29, beam management procedures are illustrated by five steps, where it has been assumed four SSB beams are transmitted by the network periodically. It has also been assumed that dynamic TCI state indication is enabled, in order to illustrate the switching of TCI states. As discussed in Section 9.3.3.1, it is also possible to configure the TCI state for PDSCH to follow the TCI state configured by MAC CE for the PDCCH. In this case, MAC CE (carried by a PDSCH) must instead be used to perform the indication and switching of TCI states in Fig. 9.29 instead of PDCCH-based switching.

Hence, refer to Fig. 9.29 in the following steps:

1. During initial access, the UE detects a preferred SSB beam and a related preferred RX beam. From the initial access procedure, a default beam pair is established.
2. The default beam pair is used for further communication, for example, to convey UE capabilities to the network and for establishing the RRC connection. This is used to configure TRS for better synchronization, TCI states, and to enable measurements and beam reporting (e.g., periodic L1-RSRP reporting of the SSB beam measurements).
   a. The UE is configured with a list of four TCI states. For FR2 operation, each TCI state contains two source RSs for QCL Type A/C and Type D, respectively.

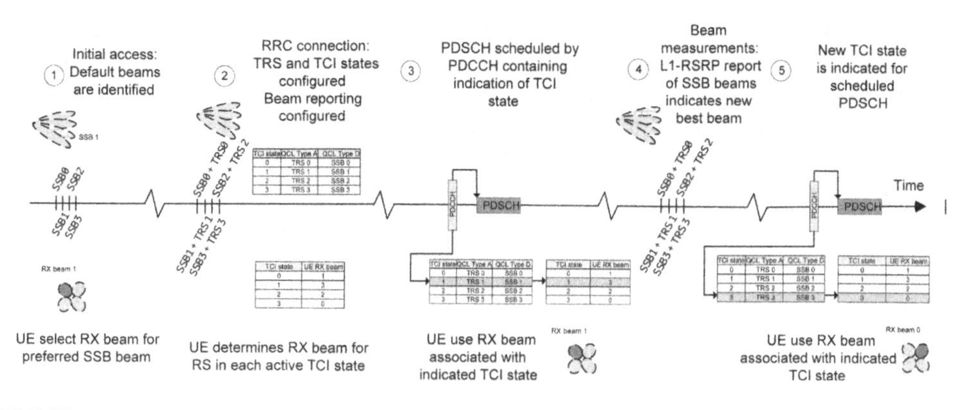

FIGURE 9.29

Illustration of dynamics in a beam management procedure where SSB beams are used and where four active TCI states are configured for PDSCH and where the PDCCH is used to select the TCI state used for PDSCH.

  b. The UE determines the preferred RX beam for each active TCI state and there can be 1, 2, 4, or 8 active TCI states. Since the SSBs are periodic, the UE can successfully refine the RX beam for each active TCI state, over time. This is particularly important when UE rotates as the RX beam may change for a given SSB beam in that case.

3. UE is scheduled a PDSCH by the PDCCH and the PDCCH indicates a TCI state to use for PDSCH reception. The UE uses the RX beam previously determined for that TCI state to receive the PDSCH.
4. UE periodically feeds back L1-RSRP reports per SSB beam and when UE moves in the cell, a better SSB beam is found and reported.
5. When UE is scheduled again, a different TCI state is selected which now corresponds to the new found SSB beam and hence PDSCH is transmitted in the new beam and the UE uses the RX beam associated with that new TCI state.
   a. Note that in FR2 there is a delay until the newly indicated TCI state is applied as discussed in Section 9.3.3.1. Hence, the PDSCH received in the same slot as the PDDCH carrying the new TCI state may be received using the "old" TCI state and the new TCI state is valid for PDSCH received in the next slot.

The above measurements and reporting procedures are used to maintain a set of "good" beams for transmission to (and reception from) each UE.

Note that if CSI-RS for beam management is used, for example, as in the P2 procedure, then there is an associated TCI state configured for each such CSI-RS resource, and this TCI state then typically contains an SS/PBCH as the source for QCL Type D. Hence, when performing measurements on CSI-RS for beam management, the UE will use the same RX beam as it used to receive the SS/PBCH of the TCI state. The SS/PBCH, on the other hand, does not have any TCI state associated with it. It is on the top level of the "QCL hierarchy" and the UE must thus autonomously train a suitable RX beam for each SS/PBCH, that is, the P1 procedure. See also the examples in Chapter 10.

When the network has found a new beam to use for the UE based on the RSRP reporting from the UE (or alternatively, based on measurements on UL signals transmitted from the UE such as SRS), it must update the active TCI state(s) for the UE, so that the active TCI state is a state that contains the RS that corresponded to the best beam report. When this has been accomplished, the newly found beam can be used for PDCCH and/or PDSCH transmission. The MAC CE signaling is used to reconfigure active TCI states for the PDCCH and/or PDSCH out of the up to 128 configured TCI states.

In addition to the above mechanism to find and configure a transmit beam for the network, the P3 procedure can be added. In this case, the network triggers a CSI measurement for beam management (i.e., RSRP based) using a CSI-RS resource set where the "repetition" parameter is set to "on." This means that gNB will transmit all the CSI-RS resources in this set with the same spatial filter, that is, using the same transmit beam. This repetition comes with an overhead cost but allows the UE to probe different RX beams for the fixed transmit beam and the network can configure the UE to report an RSRP measurement assuming the best-found RX beam.

The use of *aperiodic CSI-RS* transmissions is enabled by configuration of a list of aperiodic *trigger states* and an associated indicator in DCI to select a trigger state. Moreover, a list of TCI states is configured to be associated with each trigger state, where the RSs of the TCI states in the TCI state list are the CSI-RS resources of an aperiodic CSI-RS resource set. In this way, the source QCL for an aperiodically triggered CSI-RS resource can be indicated dynamically by DCI.

For each set of aperiodic CSI-RS resources, repetition can be set as either "on" or "off" where "off" indicates that the network may perform a beam sweep using the different CSI-RS resources of the set.

9.3.6.3 Beam failure recovery

If the UE loses the connection to the serving node, it can begin an RLF procedure and a new connection can be established to a new cell or carrier. When using a beam-based network, a need for a lower layer failure handling was seen, since there may be many beams in one and the same cell and it would cause too much impact on higher layers if a UE begins RLF procedure whenever the connection is lost due to, for example, blocking. Such "beam failures" may be frequent if narrow beams are used, and this could be handled by lower layers without triggering an RLF. For this reason, BFR procedures were introduced in NR.

A UE uses the CSI-RS or the SSB of the TCI state for a configured BFR-specific CORESET as the reference signal for beam failure detection. It is also possible to configure an alternative periodic CSI-RS for this purpose. If the BLER of a hypothetical PDCCH received using the configured RS is above a threshold, then a beam failure instance is counted and if the number of such instances exceeds a configured threshold, then "beam failure" is declared.

In the event of a beam failure, the UE begins to measure on beam identification RSs, to find a new network transmit beam and a suitable UE receive beam (i.e., a TX-RX beam pair). These RSs may be SSBs or configured periodic CSI-RS for beam management. If the L1-RSRP of a new beam pair is above a threshold, then a new beam is identified as detected, and the UE sends a *beam failure recovery request* (BFRQ) to notify the network.

The BFRQ can be carried by a configured PRACH resource that can be associated with the beam identification RS; hence the network knows in this case for each PRACH resource the suitable RX beam. When the network detects the BFRQ, it can respond in DL by a PDCCH from the BFR-specific CORESET, with a PDSCH that uses the same beam as the beam identification RS.

If the UE does not get a response from the network within a configurable time window, it will increase the TX power and perform a retransmission of the BFRQ.

As an alternative mechanism to protect against beam failures, the UE can be configured with more than one CORESET, where each CORESET is associated with its own TCI state. Hence, different PDCCH candidates can be transmitted from different beams and/or different TRPs. This will increase the probability that at least one link has sufficient coverage for communication.

9.3.6.4 Beam-based uplink transmissions

For the UL, UL beam sweeps of SRS transmission can be configured to find a suitable UL transmit beam. It is also possible to configure the UE to transmit SRS in the same beam repeatedly, for the gNB to tune its RX beam. This was discussed in Section 9.3.1.5.2 on SRS.

These two procedures are similar to the P2 and P3 procedures for the DL and were also referred to as U2 and U3, respectively, during the standardization work in 3GPP.

When a suitable beam has been found, then the SRS used to transmit that beam can be configured as the spatial relation for PUSCH and PUCCH, so these channels will use the same beam (see Section 9.3.4).

9.4 NR SUMMARY AND EVOLUTION

The first release of NR specifications is now available and provides an enormous and flexible toolbox of features and configurations. The number of RRC parameters used to configure an NR UE is significantly increased compared to LTE. It has been mentioned that NR has at least six times more rows in the RRC parameter list compared to LTE RRC, and then it should be observed that a lot of these NR parameters are lists of further parameters. The number of combinations and hence alternatives to configure an NR network and UE is a monstrous number.

With all this configurability, there is a large room for network implementation creativity to enhance network performance in the coming years. Nevertheless, some typical NR configurations based on the toolbox are discussed in Chapter 10, to aid the reader in understanding on a higher level how to configure NR for some basic scenarios and deployments.

NR specifications are a great success and many shortcomings observed in LTE have been removed. There is no need to frequently transmit signals when there is no traffic (i.e., CRS) and hence the carrier is lean and mandatory transmissions from the gNB have been minimized. Overhead is configurable depending on the scenario. Advanced antenna systems are well supported and it has been ensured that tools are specified that support a huge range of frequency bands from below 1 GHz to above 50 GHz.

The scope of the multi-antenna evolution for NR standardization is first to correct and improve obvious shortcomings of this first release. One issue is the PAPR for DM-RS, which due to a mistake in the DM-RS agreements is designed to be higher than the PAPR for PDSCH and PUSCH, respectively, for rank 3 and above. This is very undesirable but has already been corrected in release 16. Also, beam management suffers from some inefficiencies, especially the configuration of spatial relations for PUCCH, and this is also in the scope of release 16 enhancements. Another area is improvements in UL coverage, for the UE to be able to utilize all power in UL transmissions. Also, DPS between TRPs from different cells is a topic for release 16 enhancements.

For release.17, more forward-looking enhancements by increasing efficiency and effectiveness are considered and specified if they are commercially justified. Creating a gap between standardized features and used features in real products should be avoided (as happened in the later releases of LTE). Enhancements are to be considered in the areas of reciprocity-based operation, for example, the partial reciprocity problem, to enhance UL coverage and reduce overhead. In addition, whether MIMO can provide further latency reduction for low-latency services is another interesting evolution track. Furthermore, an interesting area for the future is true self-contained demodulation, where the UE use the DM-RS to perform fine synchronization, and the need for TRS to assist demodulation is then removed.

Machine learning and artificial intelligence (AI) is gaining a lot of interest to enhance network performance and some components to better support this functionality may require standardization. For example, the CSI feedback is an interesting area where AI solutions may provide benefits.

Also, there are discussions to extend the FR to in between FR1 and FR2 and above FR2. This may need further multi-antenna adaptations.

REFERENCES

[1] E. Dahlman, S. Parkvall, J. Sköld, 5G NR: The Next Generation Wireless Technology, Academic Press, 2018.
[2] 3GPP TR 38.211, V15.7.0 (2019-09) NR, Physical Channels and Modulation.
[3] 3GPP TR 38.212, V15.7.0 (2019-09) NR, Multiplexing and Channel Coding.
[4] 3GPP TR 38.213, V15.7.0 (2019-09) NR, Physical Layer Procedures for Control.
[5] 3GPP TR 38.214, V15.7.0 (2019-09) NR, Physical Layer Procedures for Data.
[6] 3GPP TR 38.216, V15.5.0 (2019-69) NR, Physical Layer Measurements.
[7] 3GPP Tdoc, R1−163215, Overview of NR, Ericsson.
[8] O. Liberg, et al., Cellular Internet of Things: Technologies, Standards, and Performance, Academic Press, 2019.
[9] 3GPP TR 36.901, V14.3.0 (2017-12), Study on Channel Model for Frequencies from 0.5 to 100 GHz.

CHAPTER

END-TO-END FEATURES

10

10.1 INTRODUCTION

In Chapters 6 and 7, the fundamentals of multi-antenna functionality have been outlined, and in Chapter 9, it is shown how AAS support has developed in the 3GPP standardization of NR. The 3GPP toolbox of specified features gives a very large optimization problem on how to select features and configurations to achieve the best possible AAS performance for a given deployment. The physical layer specifications mainly describe the expected UE behavior and not so much about the corresponding network feature design. Hence, a lot of implementation freedom for the AAS base station is left for further optimization and tuning. Also for the UE there is significant room for implementation specific solutions and algorithms. To sustain a well-functioning and high-performing network, operators must continuously make decisions regarding what 3GPP features to enable in their networks.

The purpose of this chapter is to give some examples of the process and thinking when making such selections, and to illustrate end-to-end features using AAS. It will be discussed what is required by the network side of RAN (gNB) and in the UE, respectively, to make an overall solution that can be applied in a real live network. The examples illustrate how some of the AAS technologies outlined in Chapters 6 and 7, can be implemented using the 3GPP standards for NR. Examples will be provided for FDD and TDD using AAS, and the focus here is the downlink.

Note that the intention is not to provide an exhaustive description or a final recipe for how to configure a network, but to give the reader a sense of the challenges, issues, and possible solutions in this process. Real implementations in commercial systems are considerably more complex. Comments on the resulting network performance and the effects of making different choices with respect to both 3GPP and design will also be made. These comments should be seen only as illustrations, as a full discussion taking in all degrees of freedom, based on both the 3GPP toolbox and the design choices, is beyond the scope of this book.

10.1.1 GENERAL CONSIDERATIONS

When designing AAS features or, for that sake, any system features, some high-level system design considerations and aspects need to be made that may influence the feature choices and the design of the features, respectively. Some examples are listed below:

- what technology to be used, FDD or TDD, and whether to use LTE or NR;
- what spectrum to use and aspects related to joint operation between different frequency bands;
- what AAS configuration to use (e.g., number of ports, 1D, or 2D port layouts);
- deployment scenario, for example, dense urban with high-rise buildings, urban with mostly low-rise buildings, suburban, rural, or similar;

Advanced Antenna Systems for 5G Network Deployments. DOI: https://doi.org/10.1016/B978-0-12-820046-9.00010-1

- the expected traffic density in the intended AAS coverage areas;
- performance requirements
  - which multi-antenna technology features to use (beamforming-based, SU-MIMO or MU-MIMO);
  - which uplink and downlink features to use;
- UE capabilities of UEs already available in field and UE suppliers plans for enabling new AAS feature introduction;
- interaction with already existing system features;
- cost versus performance tradeoffs;
- site constraints, such as installation height, site permits, etc.

The list can be made much longer and, in this chapter, these choices will be further discussed using case examples. In addition, Chapters 13 and 14 will also discuss related aspects that operators, network vendors and UE vendors must take into account.

Some features are mandatory to be implemented in a UE while whether a network vendor choose to implement the feature is optional unless the feature is critical for network performance. For optional features, negotiations between network and UE vendors is needed, as there is no point in implementing these on one side of the link if the other side does not commit to make the associated implementation. To verify implementations, there is *interoperability and device testing* (IODT) for features and the UE may signal to the network that it supports a certain feature only if it has passed the related IODT with network vendors.

10.1.2 NETWORK CONFIGURATIONS BASED ON 3GPP SPECIFICATIONS TO SUPPORT AAS

For a network deployment to support AAS, several choices must be made on both network (gNB) and UE sides, respectively, based on the 3GPP specifications. These choices can be classified into some fundamental areas, as described next. Depending on the use case, additional selections might be required as well to support more advanced features. These will be discussed in relation to each specific use case in upcoming sections.

For NR, the AAS configuration needs to decide on parameters in the following fundamental areas (see Section 8.7 for a description of them):

- the SS/PBCH configuration
- the TRS and TCI configuration
- the DM-RS configuration
- codebook configuration
- CSI reporting configuration
- CSI measurement configuration.

Examples of how these configurations can be used will be illustrated in the use cases outlined in Sections 9.2–9.5.

10.1.3 USER EQUIPMENT CAPABILITIES RELATED TO AAS

As it is way too complex to implement all functionalities as specified by 3GPP in a UE, UE capability signaling has been introduced. The general principle is that the UE signals its capabilities to the network, the network chooses which of the related features to use and then configures the UE with those features accordingly. Hence, only features and parameters that the UE supports can be configured. Otherwise the UE will reject the entire configuration message and perform an RRC reestablishment. Capabilities are stored in the core network when the UE is in the IDLE mode to avoid frequent signaling of capabilities if a UE switches between IDLE and CONNECTED states.

The 3GPP capability signaling from the UE to the network is rather complicated as capabilities can be per frequency band and per combination of bands (e.g., carrier aggregation case). Also, capabilities can depend on the FDD or TDD duplex mode. In addition, LTE and NR dual connectivity requires a joint evaluation of LTE and NR capabilities.

The most straightforward capability for a feature is if the capability is indicated "per-UE", that is, if indicated, then the UE supports the feature irrespectively of the frequency band, duplex mode or number of carriers in carrier aggregation. But there are also capabilities that are specific and separate for e.g. TDD or FDD, for certain frequency bands, etc.

The intention of this chapter is not to describe all details but to give a high-level understanding of different types of capability signaling, particularly for AAS-related features.

In NR, features subject to a capability are classified as follows:

- **(M) Mandatory without capability signaling;** An NR UE must be implemented with this feature to be compliant with NR specifications. All networks may assume that all UEs support it and have tested it successfully;
- **(MC) Mandatory with capability signaling;** An NR UE must be implemented with this feature; however, to avoid the lack of enough opportunities for interoperability device testing (IODT) that prevents UE vendors from releasing their devices, the capability signaling is used to indicate whether the UE has performed a successful IODT of the feature. Hence, as soon as the network vendors offer testing opportunities, such features are expected to be widely supported by UEs. Nevertheless, a network implementation should verify, based on the capability signaling, whether it is possible to go ahead and configure the feature for a given UE;
- **(OC) Optional with capability signaling;** An NR UE may be implemented with the feature, and if this is done and a successful IODT has been performed, it will signal this to the network. Networks may configure the feature only if the UE supports it according to its UE capabilities;
- **(O) Optional without capability signaling;** An NR UE may be implemented with the feature, and whether this is done or not will not be indicated to the network as the network does not need to know. This is typically the case of a functionality applicable to the IDLE mode. However, this type is not used for AAS-related features;
- **(C) Conditional capability;** Used for example if a UE supporting features A and B must automatically support feature C.

The basic UE capabilities related to MIMO and AAS in NR that are mandatory without capability signaling (M) are shown in Table 10.1, where the reader is expected to be familiar with the concepts introduced in Chapter 9.

Furthermore, the basic AAS-related capabilities in Table 10.2 are those that are mandatory with capability signaling (MC) or optional with capability signaling (OC). Hence, a successful IODT must have been performed and the UE must indicate to the network if this has been performed

Table 10.1 Mandatory UE Capabilities (without Capability Signaling) (M) Related to AAS for an NR Release15 UE.

UE Capability	Content	Type
Waveform	CP-OFDM for DL and UL plus DFT precoded OFDM for UL	M
PDSCH reception	Single layer	M
PUSCH transmission	Single layer (single Tx) and single SRS resource with one port configured for codebook-based PUSCH. Scheduling Type A and Type B.	M
UCI feedback	Periodic CSI on PUCCH and aperiodic CSI on PUSCH	M
Codebook	Type I single panel codebook in FR1 for two, four, and eight CSI-RS ports (in the eight-port case, only wideband CSI report)	M
TRS	All periodicities and both TRS bandwidth configurations	M
SRS	One-port SRS, at least one periodic and one aperiodic SRS resource per component carrier	M
TCI states	Single active TCI state for PDSCH reception	M
PDSCH DM-RS	Type A scheduling: one, two, or three DM-RS symbols (for single DM-RS port) Type B scheduling: one or two DM-RS symbols (conditional on that Type B is supported)	M
PUSCH DM-RS	Type A scheduling: one, two, or three DM-RS symbols (for single DM-RS port) Type B scheduling: one or two DM-RS symbols	M

before the network can enable the feature. In addition, for some of these features, the UE reports an associated parameter value when indicating the capability.

When comparing the UE capability signaling to what the network has implemented, a mismatch may occur which needs to be further handled by the network. For example, in case where not all UEs indicate support for the 32-port codebook and gNB has an AAS with 32 antennas, the NW must perform antenna virtualization (see Section 8.1.1, where virtualization is implemented either per UE, or per cell).

Assume, for example, that the first group of served UEs has indicated support for at most the 16-port codebook, while the second group of served UEs supports up to 32 ports. Then antenna virtualization must be used for the first group to map the 16 CSI-RS ports to the 32 antennas of the AAS. This group will thus configure a CSI-RS resource of 16 ports. For the second group, no antenna virtualization is needed, and they are thus configured a 32-port CSI-RS resource. Hence, from a network perspective, two separate CSI-RS resources must be configured which increase overhead.

Alternatively, the network uses antenna virtualization also for the second group of UEs, so all UEs in the cell measures CSI-RS using a single 16-port CSI-RS resource. Then the performance may not be as good as what could have been achieved for those UEs that support 32 ports, but, on the other hand, only a single CSI-RS resource is needed which is beneficial from an overhead point of view. It may also simplify the gNB processing since all UEs are using the same codebook.

This is one example of a tradeoff and alignment that the network must do after receiving UE capabilities. It may get even more complicated and differentiated when the number of possible values in a UE capability signaling is larger.

Table 10.2 Mandatory UE Capabilities (with Capability Signaling) (MC) and Optional with Capability Signaling (OC) Related to AAS for an NR Release 15 UE.

UE Capability	Content	Type
PDSCH reception	The maximum number of received layers the UE can handle, either one, two, four, or eight layers. At least four MIMO layers are MC in the bands where 4Rx is specified as mandatory for the given UE.	MC
PUSCH transmission	Codebook-based PUSCH with the indication of a maximum of one, two, or four layers	OC
L1-RSRP	The maximum number of L1-RSRP values per CSI report for both FR1 and FR2, either one, two, or four	MC
Type I single panel codebook	The maximum number of CSI-RS ports for the codebook, either 8, 12, 16, 24, or 32 ports	MC
Type I multipanel codebook	Support for maximum 8-, 16-, or 32-port multipanel codebook Support for either up to two or four antenna panels	OC
Type II codebook	Support for maximum 4-, 8-, 12-, 16-, 24-, or 32-port codebook for Type II Support of amplitude scaling, either wideband or wideband/sub-band	OC
PDSCH PT-RS	Support for one PT-RS port for PDSCH reception	MC (FR2) OC (FR1)
PUSCH PT-RS	Support for one PT-RS port for PUSCH transmission	MC (FR2) OC (FR1)
TRS	TRS burst length, either one slot or both one and two slots The maximum number of simultaneously tracked TRS per CC, a value from 1 to 8 The maximum number of simultaneously configured TRS per CC, from 1 to 64	MC
SRS	The maximum number of aperiodic SRS resources per BPW, either 1, 2, 4, 8, or 16 The maximum number of aperiodic SRS resources per BPW per slot, either 1, 2, 3, 4, 5, or 6 The maximum number of periodic SRS resources per BPW, either 1, 2, 4, 8, or 16 The maximum number of SRS ports per resource, either 1, 2, or 4 Support for semipersistent SRS	MC
TCI states	Support for maximum 1, 2, 4, or 8 active TCI states and either 4, 8, 16, 32, 64, or 128 configured TCI states per BWP per CC. In FR2, UE is mandated to signal at least 64 configured states. In FR1, UE is mandated to report at least the same as the number of allowed SSB in that band (i.e., at least four).	MC
PDSCH DM-RS type	DM-RS Type 1 or Type 1 + Type 2	MC (Type 1) OC (Type 2)
PDSCH DM-RS	Two adjacent DM-RS symbols (to support up to 12 ports)	OC
PUSCH DM-RS type	DM-RS Type 1 + Type 2	MC
PUSCH DM-RS	Two adjacent DM-RS symbols (to support up to 12 ports)	MC

10.1.4 EXAMPLES OF AAS END-TO-END USE CASES

Below is the selection of AAS use cases for downlink transmissions that will be presented in the following sections. The rationale for this selection is to cover use cases that are commonly deployed in practice and to also include AAS on a high-band in two of the cases.

The use cases that will be covered in this chapter are:

- **Use case 1**: SU-MIMO with normal CSI feedback
- **Use case 2:** MU-MIMO with advanced CSI feedback
- **Use case 3:** MU-MIMO with reciprocity-based precoding
- **Use case 4:** Beam management (BM) based on SSB
- **Use case 5**: BM based on SSB and CSI-RS.

10.2 USE CASE 1: SU-MIMO WITH NORMAL CSI FEEDBACK

10.2.1 INTRODUCTION

The first use case is a baseline SU-MIMO configuration in FDD or TDD. Deployments target lower frequency bands, hence FR1, and in this example use case, we assume FDD for simplicity. Both codebook-based CSI feedback and PDSCH transmissions are used. Hence, "beams" are predefined in the standardized codebook. The complexity at the network side to determine a precoder for SU-MIMO transmission is negligible, since the base station follows the recommendation of a precoder from the UE.[1] This scheme is robust and can be applied in TDD as well as in FDD. There is no reliance on reciprocity and the target is SU-MIMO scheduling only.

The codebook can be interpreted as it defines multiple possible beams (see Chapter 8). Spatial oversampling is used for these beams, where additional beams with beam directions in between two orthogonal beam directions are defined and the spatial oversampling factor is fixed at 4 in NR. Hence, the n:th and (n + 4):th beam in one dimension (e.g. in horisontal plane) are orthogonal in the NR codebook.

Using MIMO precoding (or, equivalently, beamforming) for SU-MIMO will give performance benefits in most scenarios. Different scenarios (including spatial traffic distribution) call for different antenna configurations, and this has implications on what codebook and antenna port layout to choose. One example is whether to use one-dimensional (e.g., horizontal antenna port layout only) or two-dimensional beamforming. Different scenarios impact also how to implement antenna virtualization in the AAS and how to select the cell shape, that is, the broadcast of cell-defining signals.

In low-rise urban and suburban environments, the UEs are distributed over the whole horizontal area, but only to a limited extent in the vertical domain. In this case, horizontal beamforming will give significant benefits, whereas adaptivity in using vertical beamforming will provide very limited additional benefits. How the performance of different multi-antenna schemes is affected by the

[1]In NR, the system is free to use any precoder it prefers for transmission of data, since the precoder for data transmission is transparent to the UE because DM-RS is precoded in the same way as the PDSCH. However, it is, in practice, difficult in SU-MIMO to motivate the use of any other precoder than the one from the codebook, as recommended by the UE.

deployment scenario and the choice of antenna configuration is thoroughly discussed in Chapter 13.

In dense urban high-rise scenarios, however, the UEs will be distributed in both horizontal and vertical dimensions and hence there are benefits in this case of two-dimensional beamforming.

10.2.2 SELECTING FEATURES FROM THE 3GPP TOOLBOX

For a network to support codebook-based precoding, several decisions must be made on both network (gNB) and UE sides. These choices are not only based on the available 3GPP toolbox but also implementation decisions. This section contains the selections from the 3GPP toolbox and Section 9.2.3 will discuss the associated implementation choices.

For an NR UE, the mandatory functionality (without capability signaling) is surprisingly not enough to support multilayer MIMO (see Section 9.1.2). To recap, the baseline that any NR UE must support is limited to:

- single-layer PDSCH reception
- measurements of a CSI-RS with two, four, and eight ports
- type I single-panel codebook.

A UE vendor that intends to provide a UE design that supports MIMO reception must therefore implement and perform IODT for at least the first, but preferably both, of the following features:

- More than single-layer PDSCH reception (two, four, or eight layers);
- Type I codebook based with up to 32 ports (12, 16, 14, or 32 ports).

These features are mandatory with capability signaling, and both are necessary to support codebook-based AAS in FR1.

10.2.2.1 Example of choices made for suburban or rural deployments with low to medium traffic load

The suburban and rural deployment are typical environments where channel reciprocity may not be used, since the probability of MU-MIMO scheduling is low due to low traffic load and a typical low signal-to-noise ratio (see Chapter 13). Keep in mind that a reciprocity-based operation (Section 6.4.2), when it actually can be used, is mainly useful for MU-MIMO.

An exception is fixed wireless access, which would benefit from reciprocity and MU-MIMO even in a suburban environment due to the high traffic load. The example provided here, nevertheless, assumes that reciprocity is not used and that the traffic load of mobile users is medium to low. Hence the operator does not see a strong need for MU-MIMO scheduling. Excluding the reciprocity-based operation means that the gNB complexity in terms of uplink sounding and calculation of a precoder can be kept relatively low.

Also, a suburban or rural deployment implies that there is less need for elevation beamforming unless the terrain is hilly. Hence all adaptable, or nonadaptable antenna ports are placed in the horizontal dimension in this example, that is, a one-dimensional horizontal port layout can be used.

As the target in this use case is SU-MIMO only, there is no benefit of a very detailed high-resolution (HR) CSI feedback. Hence, the CSI feedback overhead can be kept moderate by selecting the appropriate feedback mode. A wideband CSI report is, in many cases, enough but can be complemented with an aperiodically triggered CSI report with per sub-band CSI. The latter would be used in case a large downlink data packet taking several subframes to transfer is expected, since triggering and feedback take several slots. If the packet is small, then there is no incentive to trigger the aperiodic detailed CSI report as the packet would have been delivered before the detailed CSI report has been received.

When using SU-MIMO and with the codebook-based operation, the gNB would simply follow the rank and PMI recommendation from the UE by the periodic CSI report, which can be a wideband report. A periodicity of 40 ms is typically used for CSI reporting in this case but can be even longer if the propagation channel and interference variations are expected to be constant over a longer period (i.e., depending on the scenario).

Table 10.3 shows an overview of suitable configurations. The periodic CSI-RS configured for CSI acquisition (see Section 9.3.1.1.3) can preferably be configured to be transmitted in the same slot as the SSB, which means that remaining slots are completely empty if there is no traffic in the cell. This allows for power amplifier power-down for a few milliseconds which enables an energy-efficient operation at low traffic hours, for example, during night time.

A CSI reporting configuration (see Section 9.3.5) uses a wideband periodic report complemented by sub-band PMI reporting for the aperiodic reporting based on aperiodic NZP CSI-RS measurements. An interference measurement resource (IMR) is configured as a ZP CSI-RS to capture the inter-cell interference in the CSI report. The Type I codebook for 1D port layout is suitable. The higher resolution Type II codebook does not provide enough benefits in SU-MIMO scheduling to motivate its larger CSI reporting overhead.

The DM-RS Type 1 can be used, and it is assumed here that a UE can receive at most four PDSCH layers. If a served UE supports up to eight-layer PDSCH, then it will indicate that in capability signaling and also indicate support for eight DM-RS ports.

Table 10.3 Configuration of 3GPP Features for a Typical Codebook-Based CSI Feedback System such as FDD or TDD Without Using Any Channel Reciprocity Information (See Section 9.3.5 for Details of CSI Framework).

Feature	Configuration
CSI measurement	• CSI-RS configured for CSI acquisition (Section 9.3.1.1.13) • Periodic CSI-RS with 40 ms periodicity • Aperiodic CSI-RS • No measurement restriction configured
CSI reporting	• Wideband report for periodic report • Wideband CQI, sub-band PMI for aperiodic report • Interference measured based on ZP CSI-RS
Codebook	Type I codebook for 1D antenna port layout (see Section 9.3.5.3)
DM-RS	Type 1, single symbol, and with one additional DM-RS in the slot (see Section 9.3.1.2)
SS/PBCH	A single SS/PBCH is configured
TRS and TCI	• A single TRS is configured. • A single TCI state used for CSI-RS, PDCCH, and PDSCH and one TCI state for SSB. In total, two TCI states configured.

Since this is assumed to be an FR1 operation, it is, in principle, enough with a single TRS and a single TCI state to indicate the TRS as the QCL source for the PDCCH, PDSCH, and the CSI-RS for CSI acquisition.

Note, however, that at least two TCI states must always be configured since the single TRS needs an SSB as source for its QCL. As QCL relations are configured via TCI states, the SSB consumes one additional TCI state (see Section 9.3.1.6 for a further description of the QCL framework in NR).

10.2.3 IMPLEMENTATION ASPECTS

10.2.3.1 Network side

For the implementation of an AAS, it is recognized that the number of antenna ports is limited to one for some channels while the number of adaptable antennas in the AAS can be significantly larger. In this case, the implementation of an antenna virtualization is needed (see Figs. 8.6 and 8.7 in Section 8.1.1). The purpose of antenna virtualization in this case is not to create a narrow horizontal beam for increasing the coverage in some direction, but rather to ensure that all ports have full horizontal cell coverage. Hence, the antenna virtualization needs to be designed to provide a broad beam horizontally (while the beam width in the elevation angle is rather narrow to provide a high AAS antenna gain).

For NR, the SSB, TRS, and some system information need to have full cell coverage, so these channels need to be transmitted using an antenna virtualization like the LTE case, assuming here that a single SSB and TRS is configured.

For any served UE, a CSI-RS resource with the same number of ports as the number of adaptable antennas (i.e., the 1D horizontal port layout) is configured unless the number of AAS radio branches is so large that virtualization is needed (taking into account that NR supports at most 32 ports in a CSI-RS resource). All UEs can be thus configured to share the same CSI-RS resource, as each CSI-RS port covers the entire cell (i.e., using nonprecoded CSI-RS, also known as the Class A operation in LTE).

After the configuration, the operations in the gNB are rather straightforward. The UE selects the preferred precoder and the gNB follows the UE recommendation as there is no benefit for the network to re-compute the precoder, since only SU-MIMO scheduling is used.

10.2.3.2 User equipment side

There are a few UE implementation aspects worth to mention for this use case. The UE must measure the downlink channel for CSI feedback on all CSI-RS ports. Since no measurement restriction is configured, the UE can utilize (e.g., average) the channel estimates from multiple CSI-RS transmissions in time. The interference estimates can also be averaged over time. Alternatively, the UE estimates an intermediate parameter, such as the SINR, and averages this parameter across time and computes the CSI report based on such intermediate parameter.

For CSI reporting, the UE then decides on the preferred rank (RI), a wideband precoder from the codebook (PMI) and a wideband CQI. If the UE is triggered to feedback an aperiodic CSI report, the UE needs to estimate one precoder per sub-band as well (sub-band PMI).

There are many ways to search for the preferred RI, PMI, and CQI, and it is up to the UE's implementation choice how to perform these operations as long as the tests defined by RAN4 are passed.

10.2.4 END-TO-END FEATURE DESCRIPTION

The codebook-based SU-MIMO feature in FR1 can be summarized in the following way (see also Fig. 10.1):

- During access to the cell, the UE indicates its capabilities to the network;
- The gNB decides on all necessary higher-layer configurations and configures the UE. The CSI-RS is configured for CSI acquisition;
- The UE estimates CSI using CSI-RS measurements;
- The UE periodically estimates a coarse CSI, using wideband (PMI, CQI, RI), and reports it to the system periodically using PUCCH;
- If a small data packet arrives from the network to the UE, the coarse CSI is used for link adaptation and the PDSCH can be scheduled directly. Hence, the PDCCH schedules a PDSCH;
- If, however, a large data packet arrives, for which the transmission will take many subframes/ slots, it is worth to trigger an aperiodic and more detailed CSI report; hence, an aperiodic CSI report is requested from the UE;

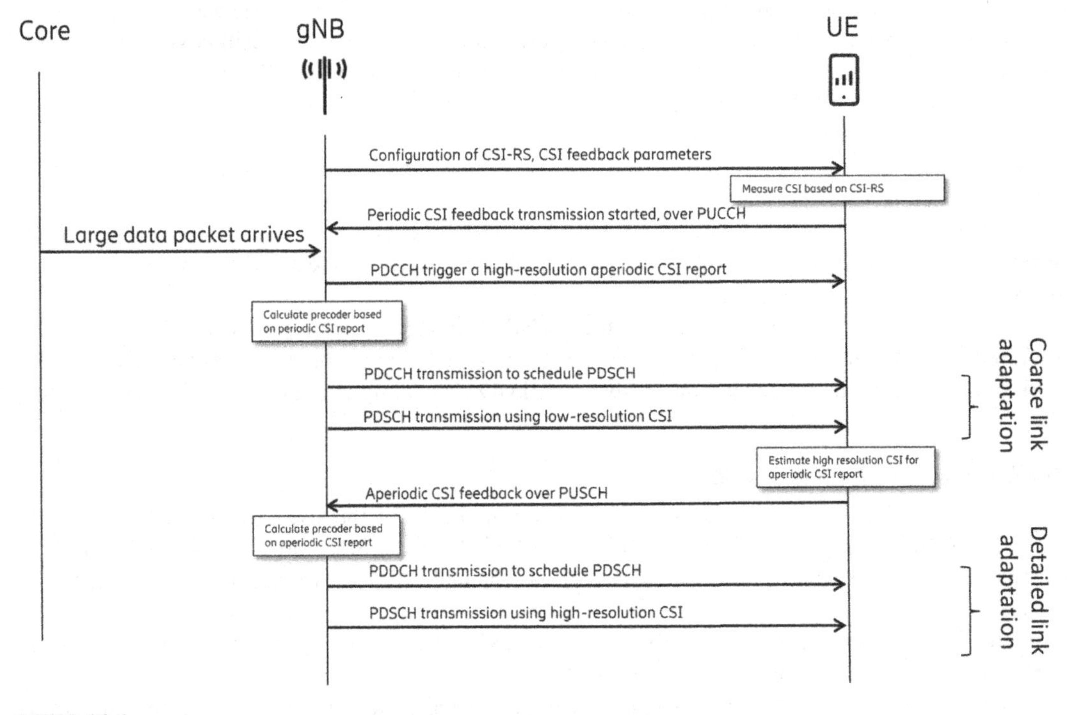

FIGURE 10.1

Sequence diagram for use case 1: Codebook-based precoding for SU-MIMO in FDD or TDD in NR FR1. The functionality in the boxes on gNB and UE sides are subject to vendor-specific implementation.

- The UE estimates wideband (CQI, RI) and per sub-band PMI and reports aperiodic CSI to the network using PUSCH;
- After receiving the detailed CSI report, the large data packet is scheduled by the PDCCH and is transmitted, which schedules a PDSCH. The link adaptation utilizes the aperiodic CSI report, that is, a more detailed link adaptation compared to the periodic CSI report;
- Note that the network can begin transmitting the large packet directly based on the coarse CSI report while waiting for the detailed CSI report to be used to enhance the throughput in the remaining subframes/slots.

10.2.5 PERFORMANCE

10.2.5.1 Codebook impacts

Codebook Type I (see Section 9.3.5.3) is used when only the SU-MIMO operation is targeted, and the spatial oversampling factor is in NR fixed to 4. Hence there are $4N_1$ horizontal beam directions, where $2N_1$ is the total number of antenna ports in the horizontal port layout (counting both polarizations). This results in 16 DFT-based physical beam directions in the horizontal plane in case an eight-port CSI-RS resource is configured.

10.2.5.2 Overhead

For NR, the overhead versus performance tradeoff has the possibility to be adjusted more accurately compared to LTE due to the large configuration flexibility in NR. As there is no CRS in NR, the overhead can be made very small when using only a periodic CSI-RS transmission plus the SSB and TRS overheads.

10.3 USE CASE 2: MU-MIMO WITH CSI FEEDBACK TYPE II

10.3.1 INTRODUCTION

The underlying assumption of this use case example is an FDD or TDD deployment in NR where MU-MIMO scheduling is expected to be common and Type II CSI feedback is used. A lower band deployment is assumed, hence an FR1 band.

Implementing support for MU-MIMO is motivated in deployments with high load and good enough SINR plus a high probability to have data available to several users simultaneously. For MU-MIMO transmission, it is important to minimize the interference from the base station to each co-scheduled user, that is, the cross-layer interference. This is achieved by designing the MIMO precoder to each scheduled UE based on recent information of the channel to all co-scheduled UEs. Such detailed multiuser channel information can be obtained by using advanced CSI feedback, which is the assumption in this section. It can alternatively be obtained by reciprocity measurements which is the assumption in Section 10.4 (see also the discussion in Chapter 6).

Support for Type II CSI feedback is optional for an NR UE, as is discussed more below.

10.3.2 SELECTING FEATURES FROM THE 3GPP TOOLBOX

The network needs to determine the same configuration parameters as in the SU-MIMO case. However, as will be discussed below, the values of the parameters will not be the same in this MU-MIMO case as in the SU-MIMO case.

For an NR UE, only single-layer PDSCH is mandatory (see Table 10.1); hence, it is beneficial if the NR UE supports multilayer PDSCH reception as a rank 2 transmission per UE is common also for MU-MIMO. This feature is mandatory with capability signaling:

- More than single-layer PDSCH reception (two, four, or eight layers).

  In addition, these following UE capabilities would enhance the performance of MU-MIMO with HR CSI feedback. These are all optional for a UE to support:
- Type II codebook based with maximum (4, 8, 12, 16, 24, and 32) ports;
- DM-RS Type 1 with two adjacent DM-RS symbols (support up to eight co-scheduled UEs/layers);
- DM-RS Type 2 with a single DM-RS symbol (support up to six co-scheduled UEs/layers);
- DM-RS Type 2 with two adjacent DM-RS symbols (support up to 12 co-scheduled UEs/layers).

The UE needs to implement at least the first of these four optional features to enable the Type II feedback described in Section 9.3.5.5. If MU-MIMO scheduling of more than four co-scheduled UEs (or more than four layers transmitted by gNB), with orthogonal DM-RS, then either DM-RS Type 2 or DM-RS Type 1 with two adjacent DM-RS symbols needs to be implemented by the UE (see Section 9.3.1.2 on DM-RS configurations). However, gNB can always use non-orthogonal DM-RS to cos-chedule UEs, as mentioned also in the previous use case.

Note also that a UE can expect that a co-scheduled UE uses the same DM-RS configuration (to allow for the suppression of interference from co-scheduled users). Hence if the network chooses to use DM-RS Type 2 for the UEs, then a significant number of served UEs in the cell need to support the optional DM-RS Type 2, to be able to perform MU-MIMO scheduling using Type 2. Otherwise the network configures the use of the mandatory DM-RS Type 1 for all UEs served by the cell.

10.3.2.1 Example of choices made for a dense urban deployment with high traffic load

The related configuration and parameter choices are provided for this use case of a dense urban environment and with high traffic load. A dense urban deployment may imply a large benefit in exploiting the elevation domain as well as azimuth domain to separate users for MU-MIMO (see Chapter 13). This is due to the three-dimensional distribution of users. Hence, an AAS with a 2D port layout is the preferred choice, where antenna ports are placed in the horizontal antenna dimension as well as the vertical dimension.

The advanced CSI feedback is useful to support MU-MIMO operation in both FDD and TDD as there are limits on the applicability of reciprocity-based operation, as discussed in Chapter 6. With CSI Type II feedback, the partial reciprocity problem is avoided and the need for sounding reference signal (SRS) antenna switching and SRS carrier switching is also avoided.

Even though all UEs support SRS transmission, antenna switching (see Section 9.3.1.5.1) is a mandatory feature but with capability signaling. Moreover, CSI feedback Type II is optional for UE to support. Hence support for MU-MIMO using either reciprocity or CSI Type II depends highly on the UE implementation, and to make MU-MIMO possible, a large fraction of the UEs need to support at least one of antenna switching or Type II CSI feedback.

Table 10.4 Selection of 3GPP Features for a Type II Codebook-Based CSI Feedback System Such as FDD or TDD without Reciprocity-Based Operation in a Dense Urban Scenario.

Feature	Choice	Comment
CSI measurement	• CSI-RS configured for CSI acquisition (see Section 9.3.1.1.3) • Periodic CSI-RS with 40 ms periodicity • Aperiodic CSI-RS	Rather long periodicity to keep overhead low and only wideband PMI and CQI. Periodic CSI-RS can be in same slot as SS/PBCH for energy efficiency.
CSI reporting	• Wideband report for periodic report, • Sub-band CQI, sub-band PMI for aperiodic report • Interference measured based on ZP CSI-RS	Sub-band CQI and sub-band PMI beneficial for MU-MIMO
Codebook	Type II codebook for 2D antenna port layout	High-resolution codebook beneficial for MU-MIMO. 2D codebook needed since elevation domain is exploited in dense urban.
DM-RS	• Type 1 with one additional DM-RS symbol and two adjacent DM-RS symbols • If traffic load is very high, Type 2 is used	• Type 1 with two adjacent DM-RS symbols supports up to eight-layer MU-MIMO • If the load is even higher, Type 2 can be used to support 12 layers (if majority of served UEs support Type 2)
SS/PBCH	A single SS/PBCH is configured	The only SS/PBCH is cell wide and hence defines the coverage of the cell
TRS and TCI	A single TRS is configured. A single TCI state used for CSI-RS, PDCCH, and PDSCH	The TRS is cell covering and serves as the source for fine synchronization to receive all channels and reference signals

Table 10.4 shows an overview of the choices in the 3GPP toolbox to support MU-MIMO for NR with Type II codebook.

10.3.3 IMPLEMENTATION ASPECTS

10.3.3.1 Network side

To operate using Type II CSI feedback and MU-MIMO scheduling, the gNB needs to decide on a scheduler strategy with associated precoding algorithm that uses the HR CSI feedback and computes a suitable precoder for each co-scheduled UE.

This is a non-trivial task with multiple dependent factors and many degrees of freedom, for example,

- the criteria used to determine on a SU-MIMO or an MU-MIMO scheduling;
- the different users reported capabilities;
- the maximum number of UEs to schedule simultaneously;
- the maximum total number of layers to be transmitted from gNB or eNB;
- the maximum number of layers to be scheduled per UE;

- whether to use orthogonal DM-RS only or to use a combination of orthogonal and nonorthogonal DM-RS;
- the algorithm to compute the precoder for MU-MIMO scheduling to reduce the cross-interference between the co-scheduled UEs;
- the method to provide accurate link adaptation per co-scheduled UE in MU-MIMO, considering the re-computed precoder.

For broadcast signals, synchronization and common channels, control channels, etc., the same issue as the SU-MIMO case of Section 10.2 needs to be solved, that is, how to produce cell covering transmissions using a MIMO antenna array. A difference in implementation is that, in this case, a 2D antenna array of adaptable antennas (2D port layout) is used.

A CSI-RS resource configured for CSI acquisition with the same number of ports as the number of adaptable antennas is configured, to be used for the advanced CSI feedback, and, alternatively, antenna virtualization is used (see Section 8.1.1) if the number of ports is fewer than the number of adaptable antennas. All UEs can be configured with the same CSI-RS resource, as each CSI-RS port covers the entire cell. However, as in the SU-MIMO case, different UEs may support different maximum number of CSI-RS ports which may be smaller than the number of adaptable antennas at the gNB. The gNB then needs to use antenna virtualization to map the smaller number of CSI-RS ports to a larger number of antennas (see Section 8.1.1).

The gNB needs to have complex algorithms implemented that compute the MIMO precoder for each UE based on the joint CSI feedback from all co-scheduled UEs. This is clearly more demanding compared to the SU-MIMO-only case of Section 10.2, where gNB can directly use the CSI feedback from the UE. The cross-user MU-MIMO interference must be considered by the algorithm (see Section 6.3.3 for more discussion on precoder algorithms).

10.3.3.2 User equipment

The UE must measure the radio channel for Type II CSI feedback on all CSI-RS ports of the CSI-RS resource to compute the CSI feedback. This may be more demanding for the UE compared to normal/Type I codebook feedback.

10.3.4 END-TO-END FEATURE DESCRIPTION

The feature can thus be summarized in the following way, assuming two co-scheduled UEs in this example (see Fig. 10.2):

- During access to the cell, the UE indicates its capabilities to the network;
- The gNB decides on all necessary higher-layer configurations and configures the UE with Type II codebook (Section 9.3.5.5), and a CSI-RS for CSI acquisition (Section 9.3.1.1.3);
- When a data packet arrives, an aperiodic CSI report is triggered;
- The UE measures CSI on CSI-RS Type II codebook;
- The UE reports CSI to the system;
- The gNB decides which UEs to be co-scheduled and computes the precoders $\boldsymbol{W}^{(1)}$ and $\boldsymbol{W}^{(2)}$ for each of the two co-scheduled UEs, respectively. This includes estimating the cross-interference and thus estimating the effective CQI, essentially performing the link adaptation;
- The gNB transmit the two PDSCH using precoders $\boldsymbol{W}^{(1)}$ and $\boldsymbol{W}^{(2)}$ for the two UEs, respectively.

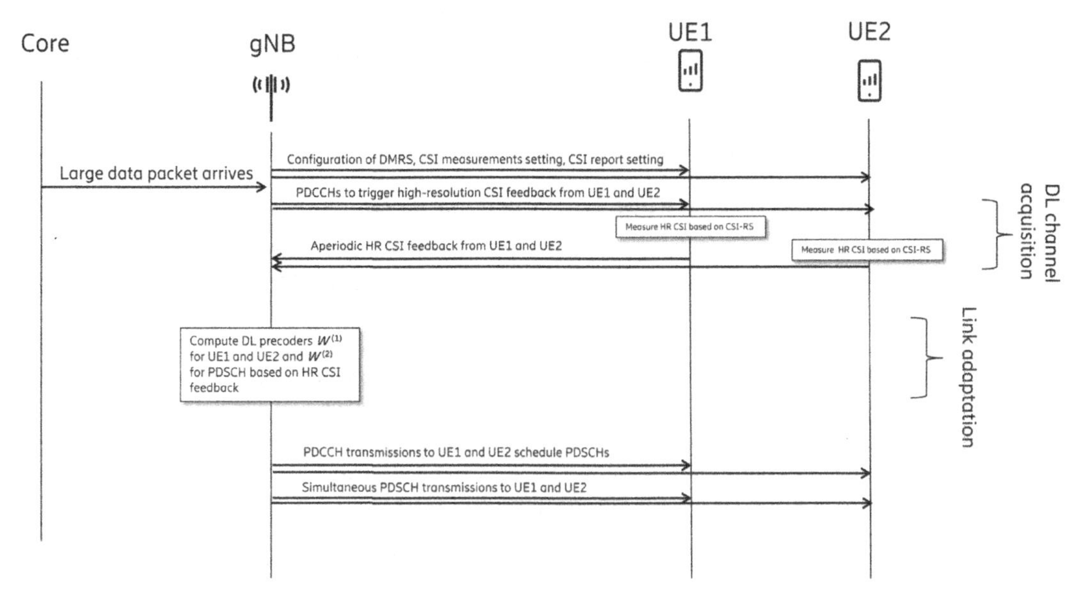

FIGURE 10.2

Sequence diagram for use case 2: The steps of MU-MIMO scheduling using CSI feedback as the Type II codebook-based feedback in NR which has high resolution (HR) in the spatial domain.

10.3.5 PERFORMANCE

Type II feedback implies an increase in the feedback signaling overhead. Therefore the measurement reports can be quite large and take substantial amounts of RBs from uplink data transmission. The gains from MU-MIMO in terms of throughput and/or latency need to be substantial in order to motivate the increased uplink overhead from these CSI measurement reports. In Chapter 13 the performance discussion will be elaborated in more detail.

10.4 USE CASE 3: MU-MIMO WITH RECIPROCITY-BASED PRECODING

10.4.1 INTRODUCTION

This use case assumes a TDD deployment in NR where channel reciprocity is utilized and MU-MIMO scheduling is expected to be common. The lower band is assumed for NR, hence FR1 operation.

In TDD, the same frequency band and thus the same propagation channel is used for both UL and DL, as explained in Chapter 6. This enables the system to make the DL channel estimate based on transmissions in the UL. The eNB/gNB uses a reference signal in the UL for channel estimation and the SRS can be used for this purpose. The UE can either transmit one SRS port per receive antenna, or use antenna switching in case the number of receive antennas is larger than the number of transmit antennas. The 3GPP specifications have support for antenna switching (see Section 9.3.1.5).

For reciprocity-based beamforming, the gNB can make the channel estimates on the UL and select the precoder for the DL without restrictions to a codebook, that is, generalized beamforming (see Chapter 6).

10.4.2 SELECTING FEATURES FROM THE 3GPP TOOLBOX

The same configuration parameters need to be determined as in the SU-MIMO case and outlined in Section 10.2, although a codebook does not need to be configured. The SRS is configured for the reciprocity-based operation (known as "antenna switching" in NR specifications of SRS configuration).

The mandatory functionality (without capability signaling) is not enough to support multilayer MIMO (see Table 10.1), since only single-layer PDSCH and a single SRS port are included in the baseline. In most frequency bands in FR1, it is mandatory for a UE to support four receive antennas, but a UE with four transmit branches is likely not supported for these UEs, at least not initially. Hence SRS resources with more than one SRS port and SRS antenna switching are likely to be implemented by the UE for the reciprocity-based operation. Otherwise, the network vendor will face the partial channel problem (see Chapter 6) and performance is degraded.

The NR UE must therefore be implemented with and have performed IODT of the following features that are mandatory with capability signaling:

- more than single-layer PDSCH reception (two, four, or eight layers);
- an aperiodic SRS resource with more than one antenna port (two or four);
- DM-RS Type 1 with two adjacent DM-RS symbols (support up to eight co-scheduled UEs/layers with orthogonal DM-RS);
- DM-RS Type 2 with a single DM-RS symbol (support up to six co-scheduled UEs/layers with orthogonal DM-RS);
- DM-RS Type 2 with two adjacent DM-RS symbols (support up to 12 co-scheduled UEs/layers with orthogonal DM-RS).

In addition, the additional capabilities in Table 10.5 are beneficial to efficiently support the reciprocity-based operation in NR.

Table 10.5 Additional Capabilities Relevant for Scheduling Based on Channel Reciprocity. Here OC Is Optional with Capability Signaling and MC Is Mandatory with Capability Signaling.

UE Capability	Content	Type
CSI report without PMI	If supported, a codebook does not need to be configured to the UE as there is no PMI feedback	OC
SRS TX switch	The UE reports which of these antenna switching cases it supports ("1T2R," "1T4R," "2T4R," "1T4R/2T4R," "1 T = 1 R," "2 T = 2 R," "4 T = 4 R") (see Section 9.3.1.5.1).	MC
SRS carrier switching	Whether the UE supports carrier switching for SRS transmission on an uplink carrier that does not have PUSCH/PUCCH transmission, in order to enable a reciprocity-based PDSCH transmission in downlink on that carrier	OC

10.4.2.1 Examples of choices made for reciprocity-based scheduling

In a TDD system where the traffic load varies, the likelihood of MU-MIMO scheduling varies and at low traffic hours, SU-MIMO scheduling can be used.

Whether elevation beamforming is useful depends on the scenario. If there are UEs distributed in both vertical and horizontal dimensions, then a massive MIMO array with adaptable antennas in two dimensions is beneficial.

However, as no codebook is configured, there is no difference for the UE whether the massive MIMO antenna array use a one-dimensional or two-dimensional port layout. The UE simply transmits SRS, and it is up to the eNB or gNB to measure the uplink channel using all receive antennas.

For the SU-MIMO and reciprocity-based operation, the gNB would simply perform a relatively low complexity operation on the received uplink channel to determine the precoder for PDSCH transmissions.

For MU-MIMO scheduling, on the other hand, the computations in gNB can be quite elaborated as the cross-user interference needs to be handled when computing the precoders. This is similar to the CSI feedback Type II case of the previous section.

Once a co-scheduling precoder has been determined per UE, the network needs to determine the CQI and the rank per UE. Hence, an aperiodic CSI report may be triggered to be reported from each of the co-scheduled UEs, where the aperiodic CSI-RS for each UE is precoded according to the derived co-scheduling precoder. In this case, PMI reporting has been disabled. The report thus contains the CQI and RI, taking into account the MU-MIMO scheduling and the transmit precoding.

Hence a two-step approach is needed when reciprocity is utilized, where, in the first step, the precoder is determined and, in the second step, the CQI and RI are determined. This may be a drawback with the SRS based reciprocity operation compared to the Type II codebook-based operation in the previous section where the CSI report contained a CQI. However, in the codebook-based case, this CQI needs to be re-computed by the base station as it does not contain the cross-user interference.

Table 10.6 shows the overview of the corresponding choices in the 3GPP toolbox for NR. The periodic CSI feedback can be used as a fallback, and to be able to perform link adaptation to schedule short packets for which it is not worth, the overhead and latency are induced by triggering an aperiodic CSI.

10.4.3 IMPLEMENTATION ASPECTS

10.4.3.1 Network side

The network side implementation needs to decide on the same issues as in MU-MIMO using advanced CSI feedback (see Section 10.3.3.1). The algorithm used to compute precoders may be different compared to the advanced CSI feedback case. It is also possible that there is a mix of UEs served by the network where some use advanced CSI feedback Type II and others use SRS.

10.4.3.2 User equipment

A UE must implement antenna switching to fully benefit from reciprocity-based operation. If carrier aggregation is used, and the UE supports fewer uplink carriers than downlink carriers, then SRS carrier switching also needs to be implemented.

Table 10.6 For the Case of NR, the Selection of 3GPP Features for a Typical Reciprocity-Based Operation in TDD.

Feature	Choice	Comment
Periodic CSI	• CSI-RS configured for CSI acquisition (Section 9.3.1.1.3) • Periodic CSI-RS with 40 ms periodicity and periodic reporting	Rather long periodicity to keep overhead low and only wideband PMI and CQI. Periodic CSI-RS can be in same slot as SS/PBCH for energy efficiency.
Aperiodic CSI	• CSI-RS configured for CSI acquisition • Aperiodic CSI measurements and aperiodic CSI reporting with PMI disabled	To obtain CQI and RI after precoder has been determined by SRS measurements
CSI report	• Wideband report for periodic report, sub-band reporting for aperiodic reporting • Interference measured using ZP CSI-RS	Wideband enough for periodic reporting and for SU-MIMO scheduling Sub-band reporting for MU-MIMO scheduling for better accuracy
SRS	Aperiodic SRS configured for the reciprocity-based operation, for example, with antenna switching (Section 9.3.1.5.1)	Note that additional SRS may be configured for PUSCH transmissions (Section 9.3.1.5.3)
DM-RS	Type 1 with one additional DM-RS and two adjacent DM-RS symbols If traffic load is very high, Type 2 is used	Type 1 with two adjacent DM-RS symbols supports up to eight-layer MU-MIMO. If the load is even higher, Type 2 can be used to support 12 layers (if UE supports Type 2). Alternatively, a nonorthogonal DM-RS is used
SS/PBCH	A single SS/PBCH is configured	The only SS/PBCH is cell wide and hence defines the overage of the cell
TRS and TCI	A single TRS is configured. A single TCI state used for CSI-RS, PDCCH, and PDSCH and one TCI state for SSB.	The TRS is cell covering and serves as the source for fine synchronization to receive all channels and reference signals

Compared to the advanced CSI feedback case, the computations in the UE is expected to be less demanding as the task of the UE is to transmit SRS. Hence the complexity to support MU-MIMO is mainly transferred to the network side in this use case.

10.4.4 END-TO-END FEATURE DESCRIPTION

The feature can thus be summarized in the following way, assuming two UEs to be co-scheduled in MU-MIMO (see also Fig. 10.3):

- The UE indicates its capabilities to the network;
- The eNB or gNB decides on all necessary higher-layer configurations and configures the UE with an SRS for the reciprocity-based operation (Section 9.3.1.5.1) and CSI feedback using CSI-RS for CSI acquisition (Section 9.3.1.1.3);
- When a data packet arrives, an aperiodic SRS is triggered for both UEs;

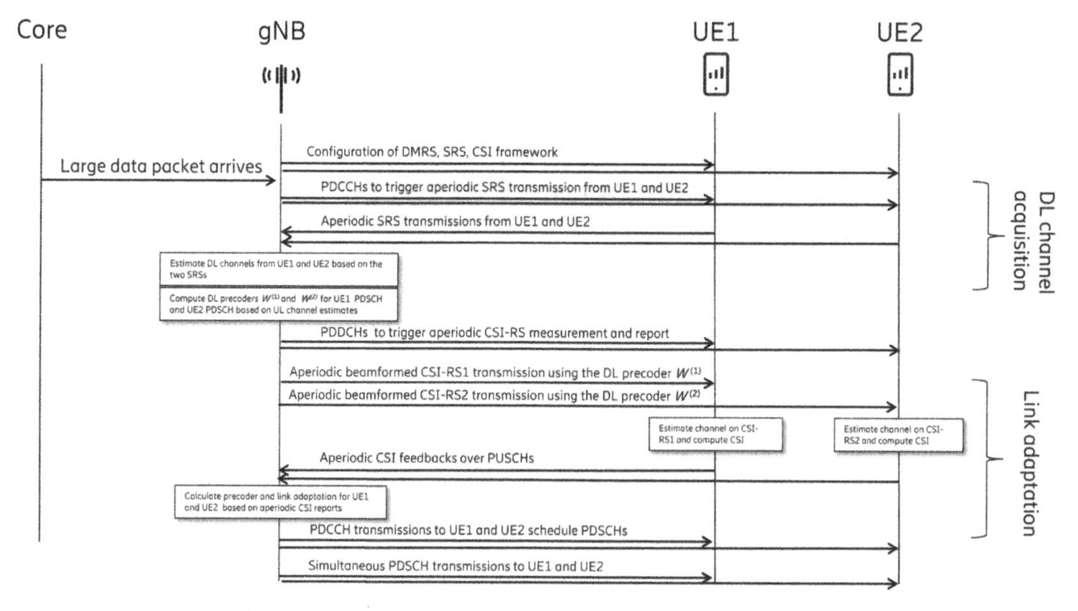

FIGURE 10.3

Sequence diagram for Use case 3: Reciprocity-based precoding for MU-MIMO in TDD; NR. The SRS shown here is intended for reciprocity-based operation. Note that the UE is likely also configured with SRS for uplink transmissions, for example, codebook-based PUSCH (see Section 9.3.1.5.3).

- The UEs transmit SRS for a reciprocity-based operation;
- The network estimates the UL channels and computes the precoders $\boldsymbol{W}^{(1)}$ and $\boldsymbol{W}^{(2)}$ for the two UEs, respectively;
- An aperiodic CSI report is triggered where the transmitted CSI-RS to the two UES are precoded with $\boldsymbol{W}^{(1)}$ and $\boldsymbol{W}^{(2)}$, respectively;
- The UEs calculate the CQI and RI, including the interference measurements using IMR (to capture cross-interference);
- The UEs feedback CSI reports to the network;
- The network recalculates the CQI and adjusts the precoders for the reported ranks;
- The gNB or eNB transmits the two PDSCH using precoders $\boldsymbol{W}^{(1)}$ and $\boldsymbol{W}^{(2)}$ for the two UEs, respectively.

10.4.5 PERFORMANCE

The performance of the reciprocity-based beamforming relies on the quality of the channel estimates on the received SRS. Therefore, the coverage of the SRSs largely determines the performance and the applicability of reciprocity-based beamforming. When the UL coverage is good, reciprocity-based beamforming can provide better performance than codebook-based beamforming, since the actual channel estimate can be used rather than a precoder in a codebook, which will only approximate the true channel.

When the coverage is poor, the channel estimate deteriorates, and codebook-based beamforming/SU-MIMO tends to perform better (for further details, see Chapter 13).

10.5 USE CASE 4: BEAM MANAGEMENT BASED ON SSB

10.5.1 INTRODUCTION

In this section, NR is assumed, operating in FR2 with a beam based approach for all transmissions from the gNB. In this example, the simplest approach of BM is used, which relies entirely on measurements of SSB beams. Hence after a suitable SSB beam has been found, the same beam is then used for PDCCH and PDSCH transmissions to a UE and that the beam direction may then also be used to receive PUSCH and PUCCH from the same UE.

Also, when a "good" beam has been found, the gNB may trigger a CSI-RS configured for CSI acquisition, where this CSI-RS is using the beam found in the BM procedure. This is done in order to obtain the CQI, PMI, and rank for PDSCH link adaptation when using this beam.

To ensure good coverage for data transmissions, high beam directivity is needed and the SSB beam directivity must thus also be high as it is assumed to be the same beam. Therefore many SSB beams are needed to cover the entire cell coverage area. One typical value for such operation can be 48 SSB beams, which is assumed in the following. A major drawback with this approach is the SSB overhead, as each of these 48 SSB is transmitted with a 20 ms periodicity. These SSB transmissions consume resources that otherwise would be available for other transmissions like PDSCH. In addition, there will be a TRS overhead, as one TRS per beam is configured (see below). The benefit, on the other hand, may be simplicity, as a beam useful for PDSCH transmissions is directly found from the SSB measurement procedure.

In Section 10.6, an alternative approach using a hierarchical method for BM is exemplified, where the SSB beams + CSI-RS beams are used (i.e., CSI-RS configured for BM is additionally used in Section 10.6).

10.5.2 SELECTING FEATURES FROM THE 3GPP TOOLBOX

The reader is assumed to be familiar with relevant concepts in Section 9.3, especially terms for BM such as QCL, TCI states, and SSB.

10.5.2.1 Network side

For BM, in this example, the network will configure 48 SSB beams where each of these beams will contain a TRS as well, hence, in total, 48 TRS configurations. The QCL relations are thus configured to the UE by configuring a list of 96 TCI states as follows:

- 48 TCI states numbered 0–47, each containing a unique SSB which is the source QCL RS for Type C and Type D;
- 48 TCI states numbered 48–95, each containing a unique TRS which is the source QCL RS for Types A and Type D.

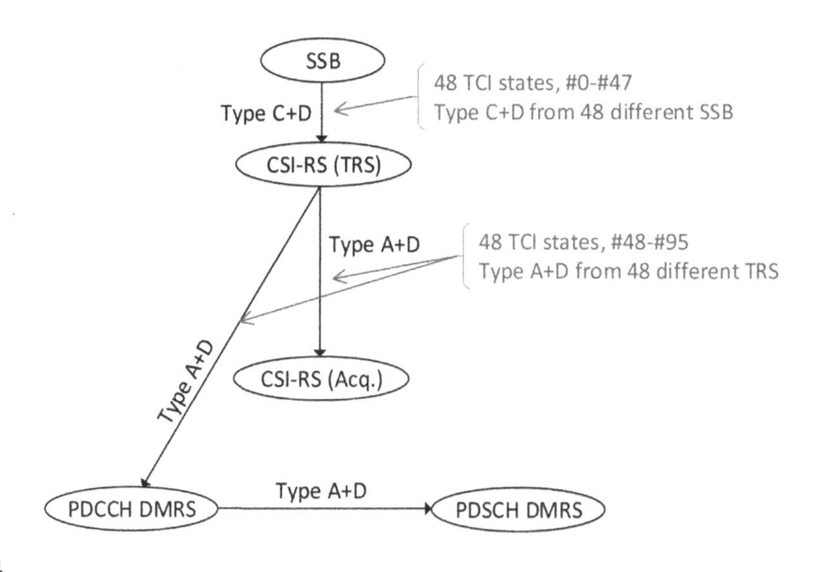

FIGURE 10.4

Configured QCL relations and TCI states for SSB-based beam management.

When configuring a TRS, a TCI state from the range 0 to 47 is indicated for each TRS, which contains the SSB transmitted by the same transmit beam as the TRS. Hence the TCI states 0–47 containing SSB are only used to aid the reception of the TRS.

When configuring the PDCCH DM-RS, one of the TCI states 48–95 is used, and hence the UE uses that associated TRS to receive PDCCH DM-RS (see Fig. 10.4). Moreover, the network will most likely choose to transmit PDCCH and PDSCH in the same beam. Hence, the PDCCH DM-RS and the PDSCH DM-RS are associated with the same TCI state, that is, the same TRS is used as source for QCL.

In addition, a CSI-RS for CSI acquisition (Section 9.3.1.1.3) is also configured with one state from the same set of TCI states 48–95. The gNB can then use the same beam to transmit the triplet SSB, TRS, and CSI-RS for CSI acquisition which the UE uses for synchronization and CSI measurements. Note that the CSI-RS for CSI acquisition can be transmitted on demand only, to save overhead. Also, the number of CSI-RS ports of this CSI resource is typically small, for example, two ports only, since it reflects the maximum number of PDSCH layers the UE can receive (and gNB can transmit) per beam. The PMI feedback in the CSI report is in this case using the two-port codebook and rank indication is either indicating rank 1 or 2.

The CSI-RS is typically aperiodic, and only transmitted in a beam when the UE is triggered to perform a measurement. The reporting is also aperiodically triggered and used when a report is needed to adjust the link adaptation for PDCCH or PDSCH.

Additionally, for the BM reporting, the network will configure a periodic L1-RSRP reporting of the best (1, 2, or 4) SSB beams that the UE finds, to continuously have an update of the preferred transmit beam for the UE. Hence, this is the P1 procedure (see Chapter 7), and the UE is expected to autonomously find a suitable RX beam for each SSB beam (see Fig. 10.5 for an illustration of

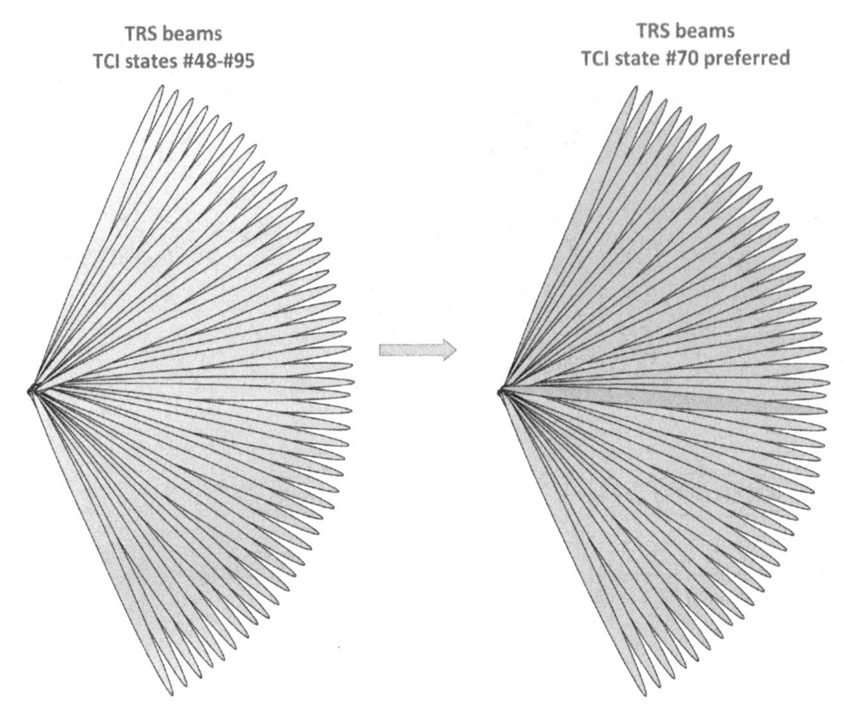

FIGURE 10.5

An example where a larger number of high-gain TRS beams (or equivalently SSB beams) is used (P1) and where one TRS beam has been found. This illustration assumes all beams are in the horizontal plane only, which typically is not the case.

the many SSB beams used in this approach but note that the beams are usually not all in the same plane as the figure illustrates but pointing in both a horizontal and azimuthal angle).

Beam switching (i.e., changing the active TCI state) for PDSCH and PDCCH is performed by MAC CE signaling, where one of the TCI states defined for PDCCH DM-RS would be activated using MAC CE, and as the UE moves, the active TCI state would be updated using MAC CE. The delay is 3 ms for this update, from the time when the ACK of the MAC CE message is transmitted from the UE.

Note that only a single active TCI state is used and selected by MAC CE in this simplest configuration, although 96 TCI states are configured by RRC. Hence, a single SSB beam is used at a time, for both data and control communication.

It is possible to activate more than one TCI state by MAC CE and dynamically switch the SSB beam used for PDSCH by using DCI. This increases, however, the complexity and is also dependent on the supported UE capability to have more than a single TCI state active simultaneously. Even in this case of dynamic PDSCH beam switching, the switching of beam for PDCCH must use MAC CE, and hence the delay is at least 3 ms.

Table 10.7 Additional UE Capabilities Related to SSB-Based Beam Management Operations.

UE Capability	Content	Type
TCI states	UE indicates maximum one, two, four, or eight active TCI states per carrier UE indicates maximum 64 or 128 configured TCI states per carrier	MC
PDSCH beam switching	Time duration between PDCCH indicating a beam switch and corresponding PDSCH. The UE indicates one value from (14, 28) symbols for SCS = 120 kHz	MC
Beam correspondence	See Section 7.4.1.1 on the definition of beam correspondence	MC
Periodic beam report	Support beam reporting on PUCCH	MC
L1-RSRP report	The maximum number of L1-RSRP the UE can include in one report from 1, 2, or 4	MC
PT-RS	Supported, with a single PT-RS port	MC
PDSCH reception	Reception of more than one single-layer PDSCH, an indication of at most two, four, or eight layers	MC

10.5.2.2 User equipment side

In the SSB-only-based BM case, the features related to the UE capabilities in Table 10.7 are of interest to be implemented in the UE. For a simple operation, a single active TCI state may be enough, but the UE may face problems if the number of SSB beams is very large and hence narrow in beam width. A problem is that it may, in this case, require more frequent MAC CE reconfiguration of the TCI state and thus transmit the beam when the UE moves in the cell.

UE support for beam correspondence is also crucial to allow, for example, fast UE access to the network. If not supported by the UE, then the UE cannot rely on downlink measurements to determine the beam to use for uplink transmissions. It means the access procedure is "blind" and the UE needs to probe by transmitting PRACH in multiple directions until it gets a positive response from the gNB. This incurs a delay in cell access. In addition, the gNB must configure dedicated procedures for uplink BM in order to find a good beam for PUSCH and PUCCH transmissions. If beam correspondence is supported, the UE can simply follow the same beam direction to transmit as it used to receive the SSB beam.

Beam reporting on PUCCH needs to be implemented to be able to perform BM, or, alternatively, beam reporting is only performed when triggered, using PUSCH. In addition, simultaneous L1-RSRP reporting using multiple SSB beams can be useful to improve the network performance, especially if two or more beams are close to each other in received RSRP, since it gives the network a possibility to choose from a set of beams when transmitting to the UE.

Note that if two UEs share the same beam (same TCI state), and the gNB uses time domain beamforming (e.g., analog beamforming implementation), then PDSCH transmission to the two UEs can be simultaneous, for example, frequency multiplexed in the same slot and beam.

In addition, the UE must support PT-RS reception and transmission; otherwise, the performance at high spectral efficiencies (16QAM and above) will be severely degraded.

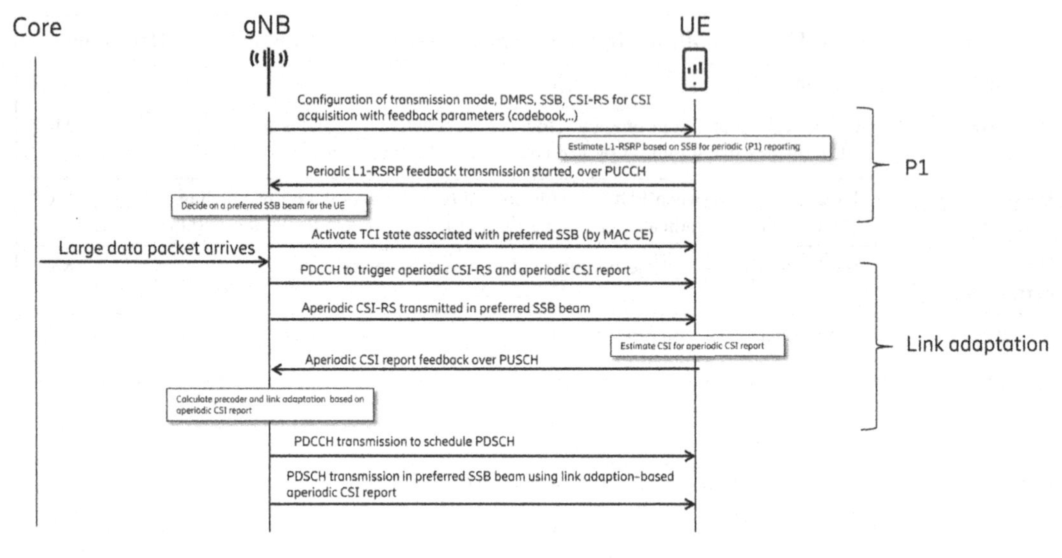

FIGURE 10.6

Beam management based on SSB in NR FR2 (based on P1 only) and using CSI-RS for CQI acquisition to provide accurate link adaptation in the found beam.

10.5.3 IMPLEMENTATION ASPECT IN THE USER EQUIPMENT

As there is no P3 procedure configured by the network in this example (i.e., to assist UE to find RX beam), the UE must continuously find a suitable RX beam by measuring on the SSB and/or the TRS that is QCL with the SSB.

10.5.4 END-TO-END FEATURE DESCRIPTION

The feature can thus be summarized in the following way (see also Fig. 10.6):

- During access to the cell, the UE indicates its capabilities to the network;
- During access procedures, an SSB beam is found and used for further communication to the UE until the BM framework and SSB beam reporting has begun;
- The gNB decides on all necessary higher layer configurations and configures the UE with TRS, TCI states, DM-RS, and CSI framework, where the CSI framework includes configuring both BM reporting (L1-RSRP) and CSI-RS for CSI acquisition and associated CSI reporting;
- The UE measures L1-RSRP on SSBs, selects a preferred SSB, and begins periodic L1-RSRP reporting to the network using PUCCH;
- The UE is configured an active TCI state for PDCCH DM-RS, using MAC CE;
- If small data packet arrives from network, to the UE, the PDCCH is transmitted which schedules a PDSCH directly using coarse link adaptation based on L1-RSRP only. The TRS beam indicated by the active TCI state is used for transmitting PDCCH and PDSCH;

- If a large data packet arrives, for which transmission will take many subframes, it is worth to trigger an aperiodic and more detailed CSI report based on CSI-RS configured for CSI acquisition. In this case, an aperiodic CSI report is requested from the UE. The same active TCI state is used, and hence the same TRS beam is used;
- The UE estimates RI, wideband PMI, and per sub-band CQI, and reports to the system using PUSCH;
- After receiving the detailed CSI report (including PMI, CQI, RI), the large data packet is scheduled by the PDCCH followed by the PDSCH where link adaptation is done using the aperiodic CSI reporting, that is, using detailed link adaptation.

10.6 USE CASE 5: BEAM MANAGEMENT BASED ON SSB AND CSI-RS

10.6.1 INTRODUCTION

This section also relates to NR FR2, where beams are used for all transmissions from the gNB in order to extend coverage. Here we assume a more advanced BM approach with less overhead, relying on BM using both SSB and CSI-RS configured for BM. The SSB beams are used to find suitable beam directions which can be used for PDCCH transmission. After an SSB beam has been found, a second step using a beam sweep of narrower, higher gain beams is performed, by using CSI-RS for BM. These high-gain beams can then be used for PDSCH transmission to maximize the spectral efficiency.

After a "good" high-gain beam has been found for a UE, a CSI measurement for CSI acquisition (Section 9.3.1.1.3) can be triggered where the CSI-RS is using the beam found in the BM procedure. This is done in order to obtain the CQI, PMI, and rank for PDSCH link adaptation when using this beam.

Hence, this is a hierarchical method for BM, where wider SSB beams together with narrower CSI-RS beams are used. These are the P1 and P2 procedures, as defined in Chapter 7. In addition, a P3 procedure can optionally be configured to further improve the UE in tuning a suitable UE RX beam for the active network Tx beam.

In this case, a number of SSB beams are much fewer, have wider beam widths, but also less gain compared to the example in the previous section. The high-gain CSI-RS reference signals for BM and reporting are triggered on demand, that is, aperiodically. Hence, less SSB overhead compared to the SSB only BM approach of the previous section, but it requires the two-step, hierarchical BM procedure. The coverage of each SSB beam is lower in this case; however, the uplink is typically the bottleneck for coverage so the impact of fewer SSB beams and less downlink coverage may not be significant.

10.6.2 SELECTING FEATURES FROM THE 3GPP TOOLBOX

10.6.2.1 Network side

For BM, the network will in this example configure eight SSB beams, and in each of these beams a TRS, and hence, in total, eight TRS configurations are needed to be configured to the UE. The same beam is used for each pair of SSB and TRS transmission.

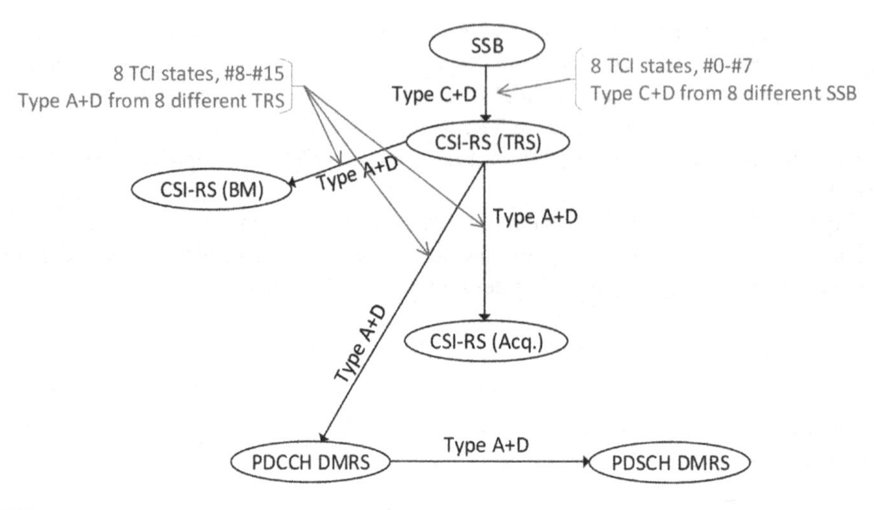

FIGURE 10.7

Configured QCL relation and TCI states for SSB plus CSI-RS for beam management (BM) based beam management.

As the SSB beams can be wider, due to which CSI-RS for BM is used for the high-gain BM, eight SSB beams are enough to keep the overhead of periodic signals at a minimum.

The needed QCL relations are thus configured to the UE by using a list of 16 TCI states:

- 8 TCI states numbered 0–7, each containing a unique SSB which is the source QCL RS for both Type C and Type D;
- 8 TCI states numbered 8–15, each containing a unique TRS which is the source QCL RS for both Type A and Type D.

When configuring a TRS, one of the eight possible TCI states (0–7) is indicated, and this state contains the corresponding SSB transmitted by the same transmit beam. The QCL relations needed for this configuration is illustrated in (Fig. 10.7).

When configuring PDCCH DM-RS, one of the TCI states (8–15) is activated. Hence, the UE uses the TRS beam of the activated TCI state as a reference to receive PDCCH DM-RS and thus uses the same beam to receive PDCCH as it is used to receive the TRS. Moreover, the network can choose to transmit PDCCH and PDSCH in the same beam. Hence, the PDCCH DM-RS and the PDSCH DM-RS are associated with the same active TCI state, that is, the same TRS. To switch the active TCI state, that is, the beam, when the UE moves around in the cell, MAC CE signaling is used to select another of the eight configured TCI states to be the active TCI state.

To further enhance the PDSCH performance, measurements to identify high-gain beams can be triggered, by using CSI-RS for beam management. Hence, within the wider beam that has been found to be best for the UE, a beam sweep of, for example, four different high-gain beam directions can be triggered for measurements and reported aperiodically. Aperiodic triggering is important here to not introduce too much CSI-RS overhead; hence, this beam sweep is only triggered when needed. This is also known as the "P2" procedure (see Chapter 7).

The gNB then finds a good high-gain beam for PDSCH transmission. The network can also choose to use this high-gain beam for the PDCCH transmission to the same UE. Alternatively, the gNB can continue to use the wider TRS beam for PDCCH transmissions to maintain somewhat better robustness against UE mobility at the expense of a lower beamforming gain.

Note that in this example, the UE RX beam is tuned based on the wider TRS beam of the activated TCI state. So even if a high-gain PDSCH beam is found by the beam sweep, the reference for RX beam adjustment is still the wide beam. This may be suboptimal, since the UE could optimally tune its RX beam towards the found high-gain beam instead to improve link performance further.

This further refinement can be accomplished in two ways; either the UE can use a smart implementation that recognizes that a new better RX beam can be used after the P2 beam sweep, even though the wide TRS beam is used as the reference in the TCI state signaling from the network. The network can also take control and trigger a P3 procedure to tune the RX beam for the found high-gain beam. However, whether this would improve performance is uncertain as it depends on UE implementation of RX beam finding algorithms.

In addition, a CSI-RS for CSI acquisition is also configured with the same TCI states as the TRS, but the gNB will use the high-gain beam to transmit the CSI-RS for CSI acquisition to get correct measurement for the CSI report and thus the link adaptation. Since the high-gain beam is within the beam width of the TRS beam, the UE will still be able to receive the CSI-RS for CSI acquisition (and the PDSCH), although, as mentioned above, the RX beam may in this case not be optimized for the high-gain beam. Note that the CSI acquisition can be triggered to save overhead. Hence, the CSI-RS is aperiodic, and only transmitted in a beam when the UE is triggered to perform a measurement and when a report to perform link adaptation for PDCCH or PDSCH is needed (Fig. 10.8).

An alternative approach is also possible, in which the UE behavior is more predictable and also avoids uncertainties in UE implementation of searching RX beam towards high-gain beams. This alternative further utilizes the available tools in the specification and is enabled by configuring

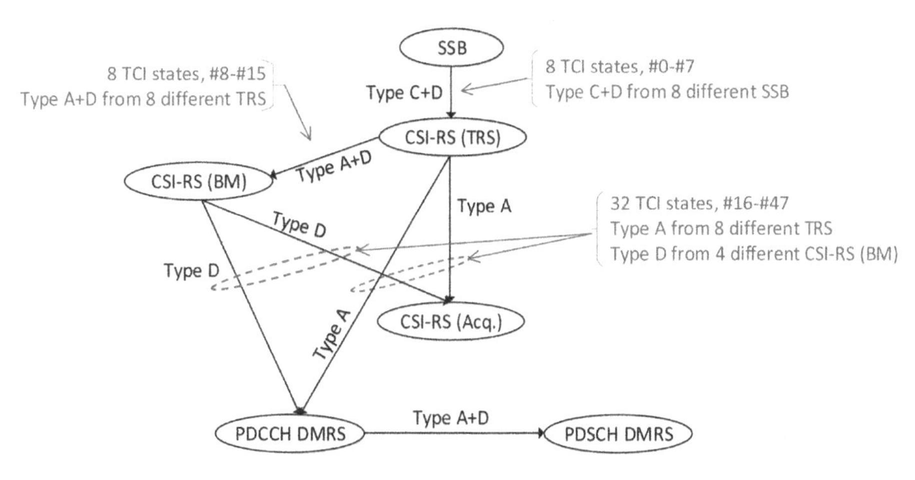

FIGURE 10.8

Configured QCL relation and TCI states for SSB plus CSI-RS for beam management (BM) based beam management where TCI states are also included for the high-gain beams found in the CSI-RS (BM) beam sweep.

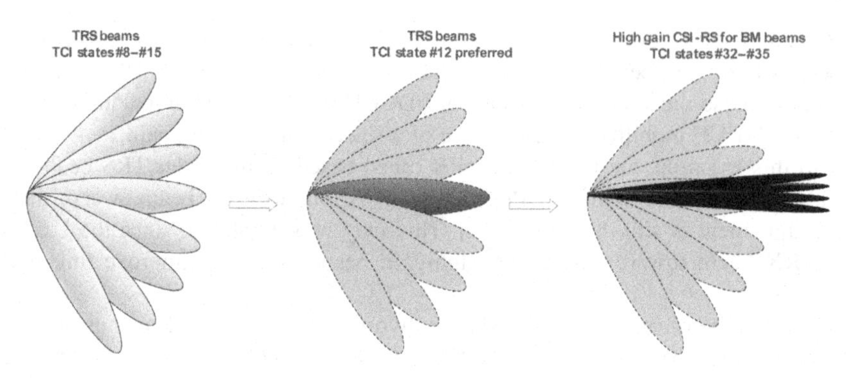

FIGURE 10.9

An example where wider TRS beams (or equivalently SSB beams) are used in the first step (P1). From this step, one TRS beam is found. In the second step, an aperiodic beam sweep is used to further identify a high-gain beam within the beam width of the wider TRS beam.

Table 10.8 Additional UE Capabilities Related to SSB + CSI-RS-Based Beam Management Operations.

UE Capability	Content	Type
PDSCH beam switching	Time duration between PDCCH and PDSCH indicating a beam switch. The UE indicates one value from (14, 28) symbols for SCS = 120 kHz	MC
TCI states	UE indicates at most one, two, four, or eight active TCI states per CC UE indicates at most 64 or 128 configured TCI states per CC	MC
Beam correspondence	See Section 7.4.1.1 on the definition of beam correspondence	MC
Periodic beam report	Support beam reporting on PUCCH	MC
Aperiodic beam report	Support triggered beam reporting on PUSCH	MC
SSB/CSI-RS for beam measurement	The UE reports the maximum number of SSB/CSI-RS resources across all CCs for L1-RSRP measurements as one value from (8, 16, 32, 64)	MC
P3 procedure	Support Rx beam switching procedure using CSI-RS resource repetition "ON." The UE reports a recommended number of repetitions from 2, 3, 4, 5, 6, 7, or 8.	MC
L1-RSRP report	The maximum number of L1-RSRP the UE can report in one report from 1, 2, or 4	MC

additional TCI states also for the high-gain beams (see Fig. 10.9). This allows the network to indicate to the UE that the RX beam for a certain high-gain beam should be used to receive PDCCH and PDSCH and further to measure CSI-RS for CSI acquisition (see Fig. 10.8 for such configuration by the network).

In this example, for each TRS beam, there are four underlapping CSI-RS beams with higher a beamforming gain (see Fig. 10.9). These can be triggered aperiodically, to save the overhead.

Hence, TCI states 16–47 represent the high-gain beams and can be used as an active TCI state for PDCCH DM-RS and for CSI-RS, for CSI acquisition.

A P3 procedure can additionally be triggered for the found high-gain beam to further tune the RX beam of the UE. With this approach, the TX and RX beam pairs are very well tuned and the CSI for link adaptation is also defined accurately. A drawback is larger sensitivity to UE mobility, since if the UE moves out of a high-gain beam, then a new TCI state needs to be activated by the MAC CE signaling. If UE mobility is high, then it may be better to use the robust configuration of Fig. 10.7.

10.6.2.2 User equipment

In this mode of operation of the network, the UE capabilities in Table 10.8 are of interest to implement in the UE. In addition to the capabilities already discussed in Section 10.1.3, there is also a need to support the aperiodic beam reporting since it is needed for the CSI-RS based beam sweep. This mode of operation requires also a larger number of configured TCI states, and hence there may be a need for more than 64 TCI states to have full flexibility from the network side, especially if the number of SSB beams is moderately high, as they consume two times the number of SSB beams (for each SSB TCI state, one TCI state is also needed for each TRS).

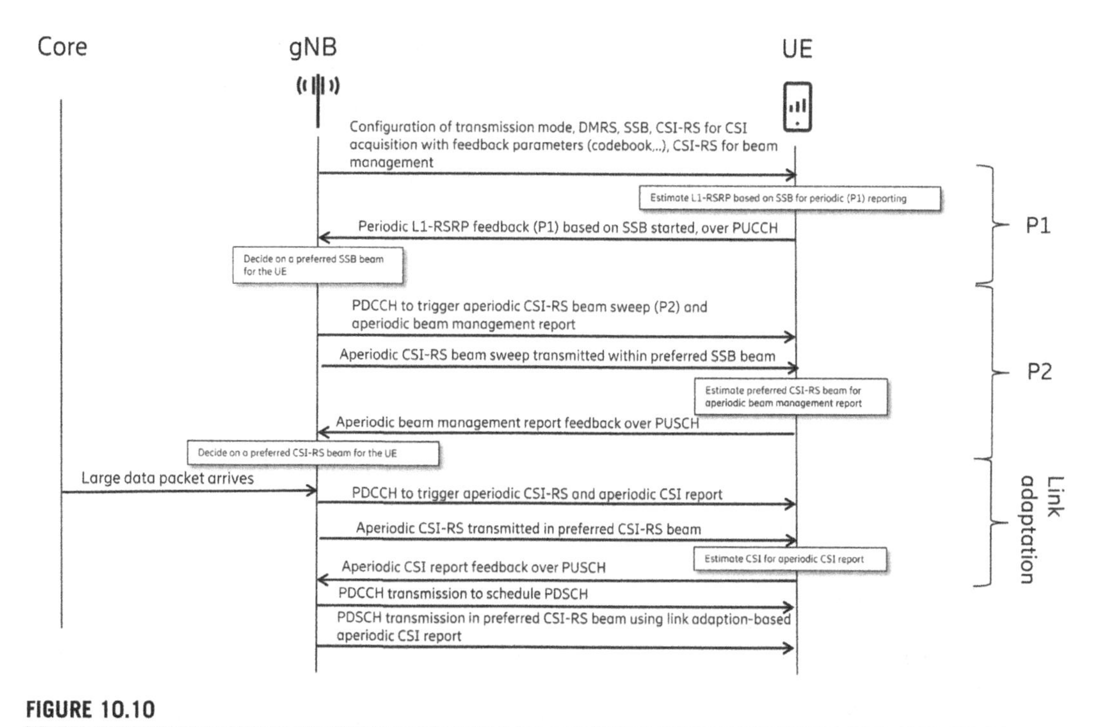

FIGURE 10.10

Case 5: Beam management in NR FR2 based on SSB and aperiodic CSI-RS for beam management (P1 + P2) as well as CSI-RS for CSI acquisition.

10.6.3 END-TO-END FEATURE DESCRIPTION

The feature can thus be summarized in the following way (see also Fig. 10.10):

- During access to the cell, the UE indicates its capabilities to the network;
- During access procedures, an SSB beam is found and used for further communication to the UE until the BM framework and SSB beam reporting has begun;
- The gNB decides on all necessary higher layer configurations and configures the UE with TRS, TCI states, DM-RS, and CSI framework, where the CSI framework includes configuring both BM reporting (L1-RSRP), CSI-RS, and CSI reporting for BM, and CSI-RS for CSI acquisition and the associated CSI reporting;
- The UE measures L1-RSRP on SSBs, selects a preferred SSB, and begins periodic L1-RSRP reporting to the network using PUCCH;
- The network triggers an aperiodic beam sweep within the beam width of the preferred SSB beam, using high-gain beams;
- The UE reports CSI report from the beam sweep;
- Based on the P1 and P2 procedures, the UE is configured with an active TCI state for PDCCH DM-RS, using MAC CE;
- If a small data packet arrives from the network to the UE, the PDCCH is transmitted which schedules a PDSCH directly using coarse link adaptation based on L1-RSRP only;
- If a large data packet arrives, for which transmission will take many subframes, it is worth to trigger an aperiodic and more detailed CSI report based on CSI for CSI acquisition. In this case, an aperiodic CSI report is requested from the UE. The same active TCI state, as found in the P1 and P2 procedures, is used, and hence the same beam is used for CSI acquisition;
- The UE estimates wideband (CQI, RI) and per sub-band PMI and reports to the system using PUSCH;
- After receiving the detailed CSI report, the large data packet is scheduled by the PDCCH and is transmitted by the associated PDSCH directly where link adaptation is done using the aperiodic CSI reporting, that is, using detailed link adaptation.

10.7 SUMMARY

In this chapter, it has been shown how a selection of concepts described in Chapters 6, and 7, can be implemented using the 3GPP NR standard described in Chapter 9. To make a feature work end to end, certain choices from the 3GPP standard must be made both on the system side and on the UE side. In addition to this, complementary proprietary implementations in the system and the UE are required to make a complete end-to-end feature. Here, an alignment in the industry is necessary since many features are optional or mandatory with capability signaling.

The choices to be made depend on a number of factors that influence the requirements and the constraints of the feature and its performance in a field environment. Examples include what AAS structure to use and the corresponding subarray sizes, the deployment environment, performance requirements, UEs available in field, multiband requirements, etc. These factors influence, for example, the appropriate parameter choices from the 3GPP standard.

To sustain a well-functioning and high-performing network, operators must continuously make decisions regarding what 3GPP features and vendor-specific complementary algorithms to enable in their networks.

CHAPTER

RADIO PERFORMANCE REQUIREMENTS AND REGULATION

11

11.1 INTRODUCTION

Deployment of radio network equipment and in particular advanced antenna system (AAS) base stations requires that a number of 3GPP and regulatory requirements are fulfilled. The introduction of AAS architecture and beamforming capability over the past few years has necessitated a new paradigm for specifying and assessing such requirements, based on *over-the-air* (OTA) requirement definition and testing. The radio requirements, which typically arise from the need to guarantee performance and co-existence in the radio spectrum with other systems, impose technology opportunities, benefits, and constraints for AAS systems.

While 3GPP physical layer AAS aspects have been discussed in Chapter 8, for LTE and the evolution towards NR, and Chapter 9, for NR, this chapter will provide an overview of 3GPP radio and some regulatory requirements and their impact on AAS implementations. Section 11.2 elaborates on the purpose of such requirements and describes briefly general categories of requirement, applicable to both non-AAS and AAS. Section 11.3 provides a description of the radio requirements. Section 11.4 then outlines issues that arise when applying the requirements to AAS technologies and serves as a background for Section 11.5, which describes the conceptual approaches developed to solve these issues. Apart from the conceptual approach, there is also a need to set thresholds and limits for OTA requirements, and the background and technology constraints driving the agreed limits are described in Section 11.6.

Having elaborated on the requirements themselves, Section 11.7 discusses the impact of the need to meet 3GPP and regulatory requirements on potential hardware implementation gains from massive multiple-input multiple-output (MIMO) (massive MIMO is defined in Chapter 1).

Finally, Section 11.9 outlines some practical OTA testing methods for assessing requirement compliance.

11.2 PURPOSE OF RADIO REQUIREMENTS

Spectrum is a shared resource that is scarce and expensive. Only a small fraction of the spectrum is available to cellular technologies, necessitating extensive investments for spectrum acquisition, and thus efficient spectrum usage is essential in order to provide a sufficient return. Apart from efficiency, investments in spectrum require confidence that the spectrum can be used for providing a wide variety of services with good quality whilst sharing the spectrum resource with other users. Interference and compatibility between incumbents (cellular operators or other

Advanced Antenna Systems for 5G Network Deployments. DOI: https://doi.org/10.1016/B978-0-12-820046-9.00011-3

services) must be carefully managed considering impacts to all parties, fairness, and technology limitations. Standardized radio requirements are efficient means to satisfy the need for clean and useful usage of the spectrum by prescribing expectations for interference management.

Radio requirements are diverse and are specified in several standardization communities (the main one in the case of the cellular industry being 3GPP) and regulatory bodies. 3GPP requirements represent an industry consensus on parameters needed to ensure efficient usage of the spectrum and good co-existence properties between different spectrum users. Regulatory requirements are legal constraints that must be met in order to lawfully deploy cellular systems. Generally, commercially deployed base stations have to be designed to meet both 3GPP defined and regulatory requirements.

3GPP requirements cover transmitter unwanted emissions that limit the harmful interference, receiver requirements limiting susceptibility to interference from other systems, requirements that relate to efficient spectrum usage, and requirements that enable predictable user equipment (UE) behavior. The requirements are standardized in 3GPP Working Group 4 (WG4) and cover base station radio frequency (RF) (core and conformance testing) [1,2], UE RF [3], electromagnetic compatibility (EMC) [4], and radio resource management (RRM) [5]. The radio requirements relate to key radio parameters, whilst the intention with RRM requirements is to create predictability and performance boundaries when certain UE procedures or functions are executed, for example, mobility.

In addition, WG4 has also developed base station radio requirements that handle multistandard operation [6,7], in which different access technologies operate simultaneously through a common radio, for example, GSM + LTE or NR + LTE + NB-IoT. The capability to operate multiple standards (whilst meeting all relevant requirements) enables smooth migration between different generations because over time spectrum can be re-farmed gradually from one radio access technology to the next using the same deployed radio equipment.

Regulatory requirements in all regions cover at least transmitter unwanted emissions, whilst in some regions, for example, Europe, some key receiver requirements are also regulated.

The regulatory requirements are often based on extensive co-existence studies between different systems performed either in 3GPP or regulatory bodies to ensure operational compatibility between coexisting cellular and other systems including systems deployed in adjacent or close frequency bands.

In addition, there are requirements that capture EMC and electromagnetic field (EMF) aspects. The EMC requirements that are related to cellular radio systems encompass radiated emission (apart from the intended emissions) and immunity. Radiated emission for a non-AAS type of base station comprises the radiated emissions from the base station enclosure that do not arise in the radio transmitter itself, but rather other electronic components within the base station. EMC emissions are tested by means of measuring the emissions of the BS enclosure with the antenna ports of the base station terminated.

For AAS base stations there is no possibility to distinguish the EMC radiated emission (i.e., emissions from nonradio transmitter related electronics) and unwanted emissions caused by radio transmitter nonlinearities (since both are unstructured noise power) and thus a combined requirement is applied for both. EMC immunity captures the performance of equipment when it is exposed to high level of radiated interferer and for a large frequency range and is more stringent than the BS receiver radio requirements.

Although EMF is not covered in 3GPP standards, all base stations including AAS base stations need to comply with applicable standards and regulations related to human exposure to RF EMF. EMF aspects are further elaborated in Section 11.8.

In addition to providing a framework for managing co-existence, the standardized radio requirements establish a common ground for comparing different equipment from different vendors in similar or the same circumstances, enabling validation of spectrum compatibility between equipment and simple performance benchmarking. This is possible since conformance specifications contain well-defined test models and test configurations that are based on equipment declarations.

Section 11.3 provides a description of base station radio requirements.

11.3 RADIO REQUIREMENT DESCRIPTION

11.3.1 GENERAL OVERVIEW

Each type of radio requirement (as described in Section 11.2) serves a specific individual purpose, but the impacts of the requirements on design and operation are interrelated. In this section, the different base station requirements are elaborated in terms of purpose and interrelation.

It should be noted that the 3GPP specifications cover a large number of frequency bands, including both paired and unpaired spectrum. Many requirements are common for paired and unpaired bands, but there are some specific requirements for unpaired bands, which are described in this section.

For non-AAS base station architectures, the reference point for radio requirements was called the antenna reference point (ARP) or antenna connector, at which a physical test point was assumed to be available. For AAS base stations the reference point for radio requirement may be over the air (OTA) as in many cases no physical connectors are available or feasible to implement.

Four different classes of base station are specified in 3GPP covering deployments above the roof-top (wide-area BS), below the roof-top (medium range), hotspot/pico (local area), and in homes. The radio requirements differ between different classes and relate to the intended scenario.

The 3GPP base station radio requirements can be grouped as follows:

- in-band requirements (both transmitter and receiver);
- out-of-band requirements (both transmitter and receiver);
- co-location requirements (both transmitter and receiver);
- receiver demodulation performance requirements.

The in-band requirements manage co-existence between different cellular network operators occupying spectrum within the same frequency band and also regulate some basic network performance aspects such as coverage, available power and accuracy, and signal quality. The in-band requirements are applicable over a frequency range that is the operating band extended by 10 MHz on each side for transmitter requirements and 20 MHz for receiver requirements for a non-AAS base station. The extension of operating band over which the in-band requirements are applicable depends on the size of the band and the type of base station, non-AAS or AAS. Since the in-band requirements cannot be met by means of analog RF filters (unlike the out-of-band requirements), they are generally less stringent than the out-of-band requirements.

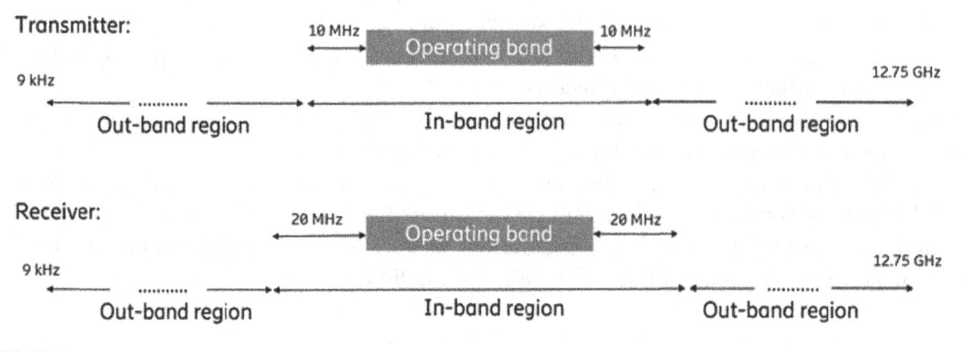

FIGURE 11.1

Example of relation between in-band and out-band requirements.

The out-of-band requirements generally regulate co-existence to other systems by means of general spurious emission or receiver out-of-band blocking and also include some specific interference and blocking requirements for specific systems. Other systems in this context mean non-cellular systems and also cellular systems operating on other frequency bands. The transmitter spurious emission is the unwanted emissions due to, for example, transmitter nonlinearity, harmonics, or parasitic emissions, while out-of-band blocking is the receiver's ability to receive a wanted signal in the presence of an interferer outside the operating band as shown in Fig. 11.1.

Co-location requirements, which in practice are a subset of out-of-band requirements, address deployments in which a site is reused by means of colocating base stations operating in different bands. The requirements specify maximum interference between bands on the same site (regardless of whether the base stations associated with each of the bands belong to same or different operators) and are generally much stricter than the general out-of-band requirements due to the assumed close proximity of the co-located base stations.

The base station receiver demodulation performance requirements specify in general the minimum required user throughput (or in some cases, other criteria) for specifically defined uplink (UL) physical channel settings given certain signal-to-noise ratio (SNR) conditions in different fading conditions.

11.3.1.1 Simplified description of base station transmitter and receiver

A very simplified schematic of the BS transmitter and receiver is provided in this section to facilitate the descriptions in the following sections in relation to requirements and the implementation aspects that requirements are intended to regulate. The schematics are of course by no means intended to completely describe the highly complex implementation of a base station transmitter and receiver rather should be seen as simplified examples.

For non-AAS base station a simplified schematic over the radio transmitter chain is presented in Fig. 11.2.

The blocks before the D/A converter are located in the digital part of radio and fulfill the following purposes:

Digital channel filtering is applied to individual carriers within the band and ensures that the parts of the modulation spectrum that overlap other carriers are sufficiently suppressed before the

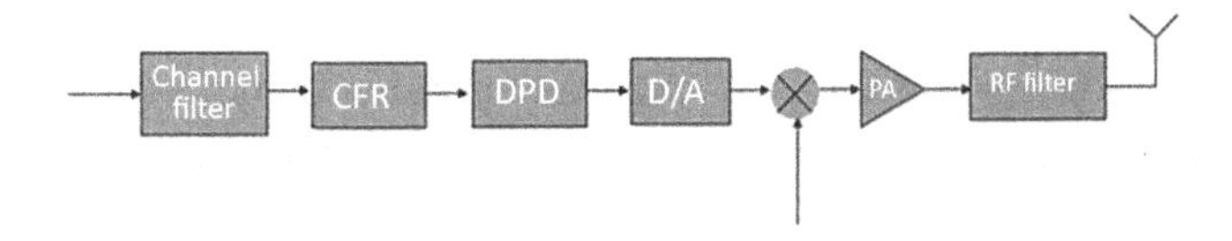

FIGURE 11.2

Simplified radio transmitter chain for non-AAS.

signal enters the analog domain (where additional impairments occur and are also handled). The filtering ensures that in-band requirements between different carriers (such as operating band unwanted emissions (OBUE) masks and adjacent channel leakage ratio (ACLR)) are met.

The modulation spectrum for orthogonal frequency division multiplexing (OFDM)−based 3GPP technologies is worse than the base station transmitter emission requirements and thus a digital channel filter is needed. Digital filtering alone is not sufficient for meeting all of the requirements due to the impact of the analog processing.

When transmitting an analog signal, the linearity of the amplifier system and peak-to-average power ratio (PAPR) of the transmitted signal are of critical importance. To remain linear, the amplifier system must always be kept within certain output power boundaries. The efficiency with which this can be achieved depends on the PAPR of the signal, which is an intrinsic quality of the waveform. A large PAPR implies that a large power headroom above the average signal level needs to be reserved in the power amplifier (PA) to avoid distortion of the peaks. Operating with a large power headroom above the average power generally implies poor amplifier efficiency. Unfortunately, the PAPR of OFDM-based waveforms is large. Crest factor reduction (CFR) is a time-domain nonlinear algorithm that reduces the PAPR of the wanted signal to a more reasonable level, enabling design of more efficient PAs.

CFR introduces distortion to the signal, degrading the transmitter signal quality. The additional distortion introduced by the CFR must therefore be carefully traded against the benefit of improved PA efficiency due to reduced PAPR.

Even after processing using algorithms such as CFR that reduce the PAPR, PAs are still generally not linear and introduce additional distortion. Distortion in the PA has two effects; firstly, the wanted signal itself is distorted, contributing to the transmitter error vector magnitude (EVM). Secondly, PA distortion creates unwanted emissions at frequencies outside of the wanted signal, which must not exceed the allowed unwanted emissions. Digital predistortion (DPD) is a method to improve the linearity of the PA and ensure that the unwanted emission requirements are fulfilled, in particular in-band emission requirements, which are quite stringent. DPD operates by means of measuring the distortion caused by the amplifier, and predistorting the input signal with an inverse of the distortion, such that the output signal appears to be less distorted. The DPD often contains a receiver at the PA output to measure the actual transmitted signal, which is compared with the input to compute the PA distortion characteristics. Knowledge of the PA distortion characteristics is used to provide adaptation parameters to predistort the input signal. The cost of DPD is spending additional power on the nonlinear signal processing algorithms that calculate the distortion. The alternative to predistortion is operating the PA with low power, that is, large power back-off and low efficiency. In designing the transmitter, the power penalty of the predistortion must be outweighed by the benefit of increasing PA power efficiency.

The RF filters are analog filters that have the purpose of reducing the out-of-band unwanted emissions either toward the own receiver (if frequency division duplex (FDD)) or other bands to facilitate co-existence and co-location. The analog filter passband covers a large part of, or the whole, operating band, including all carriers. Thus the analog filters do not contribute to meeting in-band requirements. It is generally not possible to provide analog filters for individual configurations of carriers within the operating band, because an individual filter would be needed for each possible configuration of carriers, and many possible carrier configurations exist. This architectural constraint is a strong motivating factor for defining in-band requirements within a band and out-of-band requirements outside the band; out-of-band requirements can be dimensioned considering analog filtering possibilities but in-band requirements cannot.

Many more stages of amplification, filtering, and possibly upconversion with corresponding filtering can be included. In addition, direct upconversion to RF can be done in some possible transmitter implementations.

For AAS base stations with beamforming and multi-antenna transmission capability, Fig. 11.3 depicts a simple example of a transmitter schematic, which includes several RF chains as well as a beamformer, inverse fast Fourier transform, and cyclic prefix addition–related functionality.

A very simplified example schematic of a base station receiver is presented in Fig. 11.4.

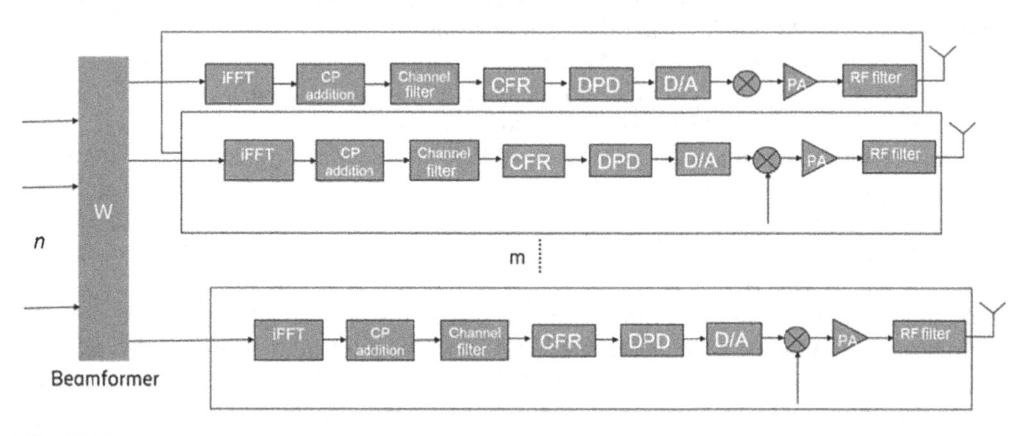

FIGURE 11.3

Simplified schematic of an example AAS transmitter.

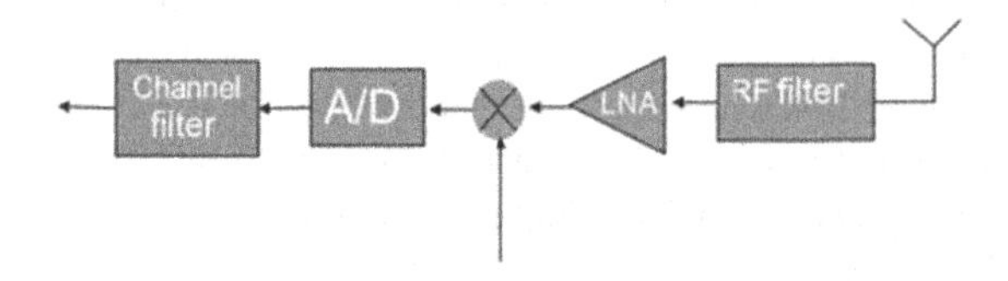

FIGURE 11.4

Simplified example schematic of a base station receiver.

For the receiver, the analog RF filter has the purpose of suppressing strong interferers, either from the base station's own transmitter (if FDD) or from powerful transmitter signals from other systems in other bands. The digital channel filter provides enough attenuation to mitigate the in-band interference often caused by transmissions on other carriers from uncoordinated terminals of other operators. Similarly to the transmitter, the analog filter passband covers the operating band. Thus the base station is able to filter away strong out-of-band interferers, whilst in-band interferers must be handled by means of providing sufficient dynamic range in the analog system so that the interferers do not overload or saturate the low-noise amplifier (LNA) or A/D converters (ADC). Digital channel filtering after analog-to-digital conversion is also required to achieve sufficient selectivity for the wanted signal over the adjacent interfering signals.

The LNA close to the RF filter is designed to have a very low noise figure (NF) (noise figure is defined as the ratio of output SNR to input SNR). Additional amplification, downconversion steps, and filtering can also be embedded in the receiver chain. In addition, the receiver chain usually contains an automatic gain control, which adapts the input signal level to the ADCs to ensure that both low and high wanted signal levels can be received. The ADCs must be dimensioned such that low-level wanted signals can be received in presence of strong interferers, in particular for in-band interferers that are not subject to analog filtering. Direct downconversion is also a possible approach for some receiver implementations.

For non-AAS base stations with few transceivers and high power from each transceiver, RF filters with low passband losses (i.e., with high so-called Q-values that implies larger size) and many poles and zeroes are feasible. Q-value or quality factor is measure for ratio of the peak energy stored in the resonator in a cycle of oscillation to the energy lost per radian of the cycle. Higher Q-value indicates a lower rate of energy loss.

Even though such filters are large and bulky, they can be fitted into base station mechanical structures. For AAS base stations with many subarrays and a transceiver per subarray, individual filters are needed per transceiver. Fitting many and bulky filters into the limited size and volume is not feasible and thus the AAS RF filters need to have compact size and be miniaturized. This means that the achievable filter attenuation for AAS base stations is more limited, and the insertion losses in the passband higher than for non-AAS when applying the same attenuation requirement. Consequently, whereas for non-AAS BS the in-band requirements are applicable over a frequency range that is the operating band extended by 10 MHz on each side for transmitter requirements and 20 MHz for receiver requirements, for AAS base stations, the range of frequencies considered for in-band requirements is extended depending on the type of AAS base station and the operating band limitations.

New radio (NR) is more versatile and flexible than long-term evolution (LTE) as multiple numerologies are supported, and it is a deployment choice which numerology to choose. For sub-6 GHz (frequency range 1, FR1), NR supports numerologies based on 15, 30, and 60 kHz subcarrier spacing (SCS) both for physical downlink shared channel (PDSCH) and SS/PBCH block (SSB), while LTE only supports 15 kHz SCS. In addition, NR supports large carrier bandwidths up to 100 MHz, while LTE is limited to a maximum of 20 MHz. The number of combinations of supported carrier bandwidths and SCS is significantly larger for NR compared to LTE. For millimeter-wave (mm-wave) frequency bands, carrier bandwidths of up to 400 MHz and SCS of 60 and 120 kHz are supported for the PDSCH, while SSB can support 120 and 240 kHz SCS.

The flexibility leads to a larger number of permutations of requirements and therefore specifications for NR base stations and NR UE are more complex compared to previous generations.

NR also supports higher spectrum utilization for many bandwidth and SCS combinations, which makes the digital channel filtering more challenging than for LTE. A digital channel filter needs to be implemented for every supported carrier. For non-AAS base stations, which have a small number of transceivers, the total channel filter processing resources are significantly less than for AAS base stations, which have many subarrays/transceivers, each needing channel filtering resources.

11.3.2 IN-BAND TRANSMITTER REQUIREMENTS

The in-band transmitter requirements contain a number of requirements relating to output power.

The main output power requirement is on the output power accuracy. The output power of the base station is declared by the vendor. 3GPP conformance requirements state that the base station must be capable of achieving the declared power to within an allowed accuracy.

There is no output power limit specified for the wide-area base station class, but for other base station classes output power restrictions are specified due to the need for ensuring co-existence between different operators and/or heterogeneous deployment layouts operating in the same geographical area.

Other output power requirements relate to so-called transmitter dynamics (i.e., handling of variations of transmit power in time or frequency). These comprise two requirements. The first is maximum expected power variation between different resource elements in the frequency domain, which is both physical channel and modulation scheme–dependent (resource element power may be boosted and de-boosted, depending on the channel and modulation scheme). The second requirement is the transmitter capability to vary the total transmit power from the power needed for single physical resource block (PRB) allocation to the power needed for full PRB allocation. As an example, for a 20 MHz LTE carrier supporting up to 100 PRBs, the total power dynamic range is 20 dB (i.e., a factor 100 between single PRB and 100 PRB) and the required supportable power difference between resource elements is 6 dB (which is considered sufficient to accommodate variations between physical channels). The transmitter implementation, for example, the digital-to-analog converter effective number of bits should be designed to handle all of the power dynamics requirements.

The signal quality requirements capture several aspects of transmitter performance. Modulation quality refers to the accuracy with which signal constellations are represented in the transmitted signal. The so-called EVM is used as metric. The EVM is a relative requirement for the transmitter and depends on the maximum supported modulation order. The higher the modulation order, the stricter the EVM requirements. EVM is expressed in the 3GPP specifications as a maximum allowed percentage distortion in the transmitted waveform to the ideal waveform constellations (see Fig. 11.5) and is calculated as follows:

$$\mathrm{EVM}_{(\%)} = \sqrt{\frac{\text{Error vector}}{\text{Reference signal}}} \times 100\% \tag{11.1}$$

The EVM expressed in percentage can be translated to a transmitter signal-to-interference-and-noise ratio (SINR) ceiling simply by $\mathrm{SINR}_{(\mathrm{dB})} = -20\log_{10}(\mathrm{EVM}_{(\%)})$ where EVM is the ratio

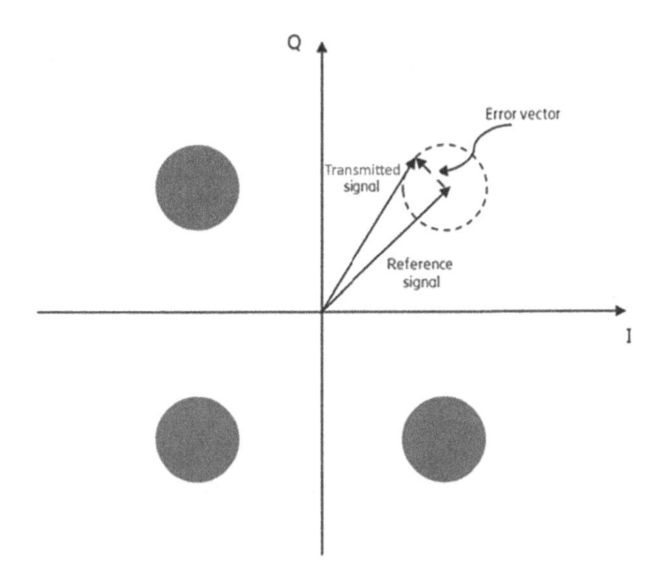

FIGURE 11.5

Error vector magnitude description for QPSK constellation.

between error vector to reference signal as in Fig. 11.5. As an example, a BS EVM of 10% corresponds to SINR of 20 dB.

The transmitter SINR sets a ceiling on the maximum achievable throughput, regardless of the radio conditions. The lower the EVM percentage, the higher the achievable SINR at the transmitter and corresponding maximum achievable user throughput.

It is noteworthy that EVM contains many different transmitter impairments such as distortion induced by peak reduction algorithms, phase noise (PN), frequency error, etc.

In addition to EVM, frequency error (which is also a regulatory requirement in some regions) is a part of the signal quality requirement set. The frequency error regulates the base station ability to transmit accurately on the assigned carrier frequency. Apart from correct use of spectrum, the frequency error also has impact on demodulation performance at the UE as the receiver must regenerate the carrier frequency from the received signal. Error arising from BS frequency variation will combine with other effects caused by the channel due to Doppler shift. A third type of signal quality requirement is the so-called time alignment error, which sets a maximum time variation between different transmission chains in a MIMO capable BS. The time alignment error requirement avoids UE demodulation degradation due to excessive time offsets between different reference signals.

The in-band unwanted emission requirements regulate the acceptable level of emissions arising from factors such as transmitter nonlinearities. They are specified based on extensive co-existence studies and consideration of regulation in different regions, in order to allow for co-existence between cellular systems in the same operating band in a multi-operator and multi-radio access environment.

A relative requirement called adjacent channel leakage ratio (ACLR) specifies a limit on the ratio between power levels of the wanted signal and emissions in the adjacent channel (both wanted and unwanted emissions are measured with same filter bandwidth) of 45 dB for below 6 GHz or 28 dB for above 24 GHz; that is, the emission in adjacent channels should be suppressed by at least 45 (or 28) dB compared to the wanted signal power. The same ACLR requirement is applicable for all base station classes as shown in Fig. 11.6.

Degradation to other systems is caused by two effects: leakage into their receiver band, which is regulated by ACLR, and their own ability to suppress interference, regulated by receiver adjacent channel selectivity (ACS) and described in Section 11.3.3. Co-existence and compatibility studies jointly assess the impact of ACLR in conjunction with receiver selectivity in the adjacent channel to provide a measure of joint interference level, in which both transmitter and receiver imperfections are taken into account. ACLR impacts the base station implementation and ACS the UE impementation. The BS ACLR and UE ACS requirements must ensure that the composite effect is within an acceptable limit and that implementation complexity is reasonable.

In addition to ACLR, absolute unwanted emission masks are specified as so-called OBUE. Whilst ACLR is defined in terms of total power over the adjacent channel, OBUE has a fine frequency resolution. The ACLR and OBUE are complementary in that ACLR regulates total emissions, whilst OBUE ensures that the distribution of emissions over frequency is acceptable. The OBUE is applicable across the whole transmit operating band extended with 20 MHz on each edge for frequency bands smaller than 100 and 60 MHz for operating bands larger than 100 MHz respectively whereas ACLR covers only the adjacent channels. The OBUE mask differs between base station classes and between regions. LTE (for carrier bandwidths equal to or larger than 5 MHz) and NR share common requirements on OBUE. There are also regulatory in-band emissions requirements in some regions (Fig. 11.7).

The transmitter intermodulation requirement is yet another emission requirement related to co-located base station scenarios and is described further in Section 11.3.5.

Some important transmitter in-band requirements are summarized in Table 11.1.

11.3.3 IN-BAND RECEIVER REQUIREMENTS

The in-band receiver requirements relate to:

- in-channel properties such as receiver sensitivity, dynamic range, or in-channel selectivity (ICS);
- the selectivity of the receiver such as ACS, in-band, and narrowband blocking;
- the receiver linearity such as receiver intermodulation.

Receiver sensitivity is a measure of the receiver's capability to receive a low mean power at the ARP whilst maintaining a throughput requirement for certain defined conditions. The conditions are specified as a so-called "reference measurement channel" in a non-fading environment. A "reference measurement channel" refers to a physical channel and modulation configuration used for sending random data, based on which the achieved throughput of the receiver when subject to a test signal at the sensitivity level is assessed.

The reference sensitivity measured in dBm depends on the base station NF, bandwidth and SNR needed to achieve the target throughput with the reference measurement channel and is calculated as:

$$\text{Receiver sensitivity}_{(\text{dBm})} = -174 + \text{noise figure} + 10\log_{10}\left(\text{bandwidth}_{(\text{Hz})}\right) + \text{SNR}_{(\text{dB})} \qquad (11.2)$$

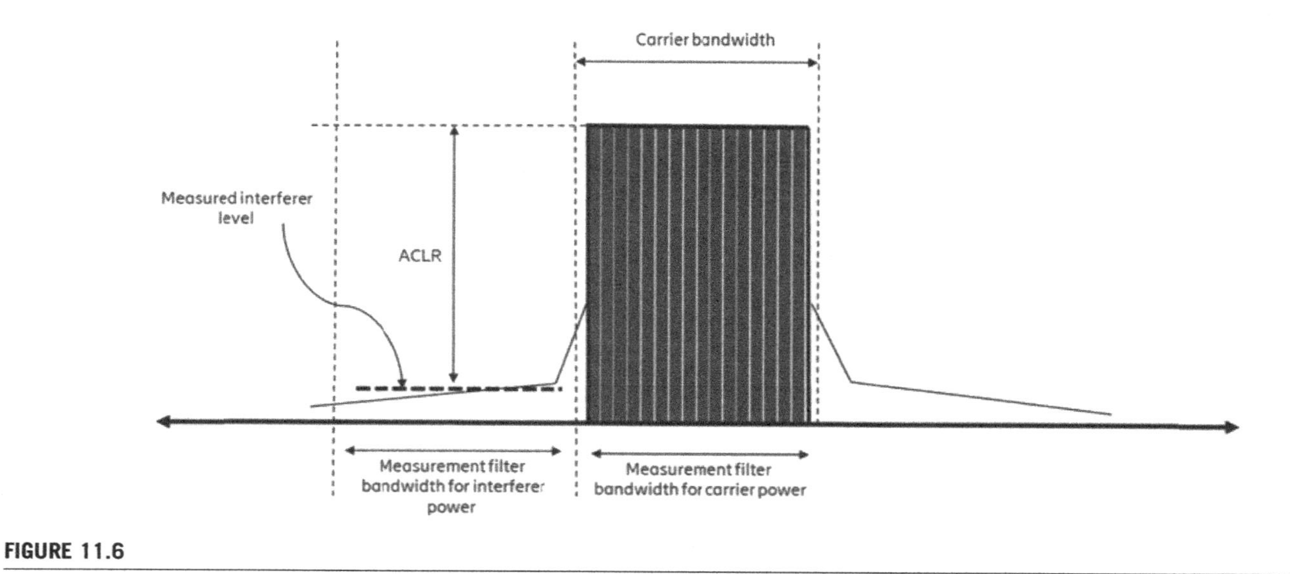

FIGURE 11.6

ACLR illustration.

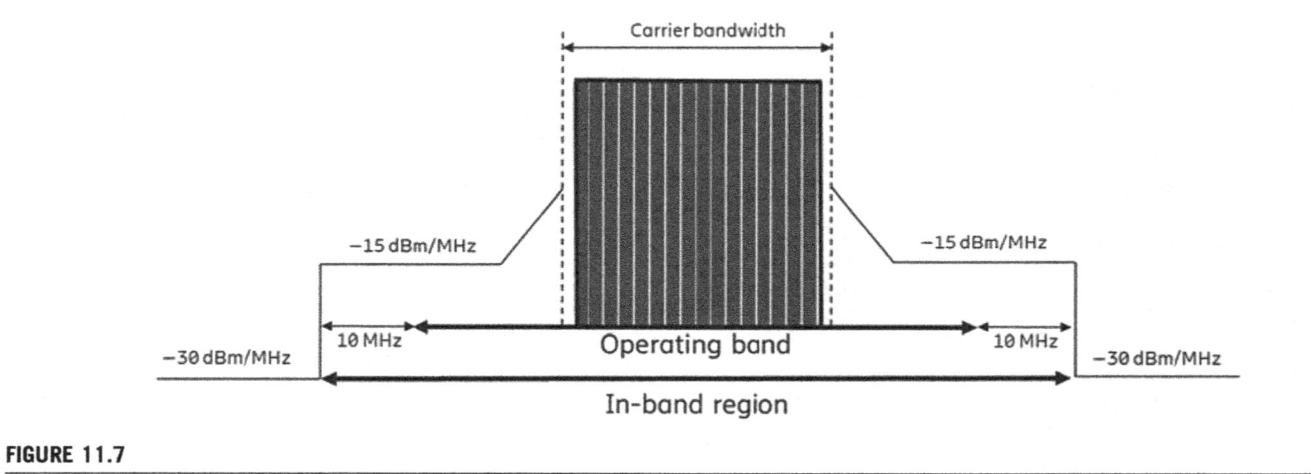

FIGURE 11.7

Operating band unwanted emission illustration.

Table 11.1 Summary of Transmitter In-Band Requirement.

Requirement	Description	Purpose
Base station output power	Measure of accuracy of the base station output power	Accurate planning of network; the accuracy of output power defines the accuracy of the cells and handover borders toward other cells
Output power dynamics	Measure of minimum supported power difference between the resource element (RE) power and average RE power for various physical channel and modulation formats Also, the ability of the transmitter to handle the power difference between full physical resource block (PRB) allocation at maximum power and single PRB allocation with the same power spectral density (PSD) as the full power allocation	This requirement in conjunction with in-band emission–related requirement captures a minimum requirement transmitter dynamic range including sufficient implementation of D/A converters
Transmitter ON/OFF power and transient time	An unpaired spectrum (time division duplex, TDD)-specific requirement for setting a boundary on the transient time between the PA being on for transmission and powered down for reception. Also, the maximum allowed output power from the transmitter when temporarily powered down (during reception)	It is essential to ensure low enough guard time between transmission and reception so that overhead is not sacrificed unnecessary (as guard symbols), whilst minimizing the impact of transmitter power interfering with receiver time slots, as well as handling interference due to transmitter power that propagates to other BS receivers. The power level and transition time are related to the maximum supportable cell size
Signal quality	Captures the modulation quality, frequency error, and time alignment error for any multi-antenna implementation such as TX diversity or MIMO	Transmitter signal quality limits the achievable SINR and thus the user throughput while frequency error also indirectly implies the need for proper frequency shift compensation in the receiver when the signal is demodulated
Adjacent channel leakage power ratio (ACLR)	The ratio between output power of the BS to the power of the unwanted emissions in the adjacent channel	ACLR is a relative emission requirement that limits the unwanted emission impact to an adjacent operator, and in conjunction with UE receiver selectivity relates to the allowable degradation
Operating band unwanted emission (OBUE)	Limits on in-band absolute unwanted emission mask with higher resolution than ACLR	Limits the distribution of unwanted emissions over frequency. OBUE is regulatory in many regions
Transmitter intermodulation	Ensures the fulfillment of unwanted emission requirements in the presence of an interferer with 30 dB lower level compared to wanted signal. The transmitter intermodulation requirement may also be considered as belonging to co-location type of requirements	Ensures that the BS can tolerate the reverse interference coming from co-located systems on the same site assuming antenna isolation of 30 dB (which is a common scenario in reality) whilst not generating additional interference due to intermodulations

The NF of the BS receiver is defined as a ratio of output SNR to input SNR of the receiver capturing the active components as well as losses in the passive components.

The reference sensitivity relates to the noise floor in the receiver, considering the analog components. It is proportional to the UL link budget and hence the coverage that can be supported by the BS.

Apart from coverage, the receiver sensitivity affects the amount of transmit power needed for UEs to reach an SINR target at the base station receiver, which in turn impacts neighbor operators since increased transmit power will cause increased unwanted emissions from UEs. Thus sensitivity is also indirectly related to co-existence between operators.

The sensitivity requirement has different values depending on the base station class. Wide-area BS have the most stringent sensitivity requirements, because the UL coverage needs to be large. The sensitivity requirement for a medium-range base station is less stringent (i.e., a higher BS NF is allowed) and local area BS sensitivity requirements are less stringent still. The difference in sensitivity is related to the fact that the micro and pico scenarios do not require large cell radii, and that the cost for smaller BS sizes needs to be proportionally lower than for a large macro site. The minimum sensitivity requirement for these classes is bounded by the need to avoid small cells increasing UE transmit power (due to poor sensitivity) to the extent that surrounding macro cells would suffer interference, and the sensitivity bounds for micro and pico scenarios have been derived by means of co-existence scenario analysis.

Receiver sensitivity and other receiver requirements are defined as a regulatory requirement in some regions, for example, Europe.

The so-called receiver dynamic range requirements capture the receiver's ability to receive a signal either at high SINR level or in the presence of strong co-channel interference. The combination of both strong interference and high SINR implies a large input power level to the BS receiver, and thus the requirement sets an expectation on the largest input power level for a wanted signal. Together with the sensitivity requirement (which corresponds to the lowest expected input signal level), this requirement partly affects the design of ADCs in terms of the needed ENOB.

The ICS requirement is based on having a wanted signal that is allocated within a subportion of the carrier on one side of the direct current (DC) carrier and an "interfering signal" with a significantly higher level, which is allocated within the other subportion of the carrier at the other side of DC and symmetrically opposite to wanted signal. This requirement captures the ability of the receiver to handle signals from different UEs arriving at the BS with different power levels depending on scheduled data rate and associated SINR. Since the signals are within the same channel, filtering will not suppress either signal. In principle, the signals are orthogonal but in practice effects such as PN jitter will cause interference from the strong signal to the weaker one.

The ACS requirement is a measure of the receiver's ability to receive a wanted signal in the presence of an interferer in the adjacent channel. The requirement is related to the ability of the BS receiver to suppress power from transmissions of an adjacent operator's UEs. ACS sets the requirement on the ability of the receiver to suppress interference in an adjacent channel that is significant, but not strong enough to cause the receiver to saturate. The ACS is specified as an absolute interferer together with a wanted signal whose power is increased by an allowed degradation (generally 6 dB in 3GPP) compared to reference sensitivity. The interferer is modulated so that it represents the actual power spectral density (PSD), bandwidth, and peak-to-average of an interferer of an adjacent operator UE using the same bandwidth as wanted signal. The interferer level for the wide-area BS class is −52 dBm, while for other BS classes, the interferer level is adjusted in relation to corresponding BS class-specific sensitivity level/noise floor requirement. The selectivity requirements affect amongst other things the digital filtering, linearity, and the design of ADCs in the receiver chain.

In addition to ACS, another requirement is defined on receiver in-band blocking. The in-band blocking requirement also defines an interferer that is modulated and has bandwidth of 5 MHz (e.g., for LTE). However, the interferer is not allocated at the adjacent channel but instead at larger frequency offset and has a much greater power level than the ACS interferer. Similar to ACS, the allowed degradation in receiver performance compared to reference sensitivity is generally 6 dB. The interferer level for the wide-area BS class is −43 dBm, while for other BS classes, the interferer level is adjusted in relation to the corresponding BS class-specific sensitivity/noise floor requirement. For some frequency bands, a further requirement based on a narrowband interferer close to the carrier is specified, for which the interferer is modulated and has a bandwidth corresponding to 1 PRB. The in-band blocking requirements set a minimum expectation on the ability of the BS to receive a strong signal from an adjacent operator's UE without saturating the receiver electronics. Whilst the ACS tests the ability of the BS receiver to suppress moderate power interference that is close enough to cause interference but not strong enough to cause the receiver operation to saturate, the blocking is related to avoiding saturation due to occasional strong interference. The intention of the narrowband requirement is to mimic high PSD, narrow bandwidth interferers such as GSM UEs.

The receiver intermodulation requirement sets an expectation on the receiver linearity by means of subjecting the receiver to a wanted signal in the presence of two interferers at different frequency offsets (one continuous wave (CW) and the other modulated). The interferer frequency offsets are chosen to ensure that unwanted intermodulation products due to receiver non-linearities fall within the frequency range of the wanted signal. In addition to linearity, the selectivity of the receiver is also stressed. Similar to other receiver requirements, the allowed degradation of the wanted signal compared to the reference sensitivity level is generally 6 dB. The interferer level for receiver intermodulation is BS class-dependent, similarly to ACS and blocking.

Some important receiver requirements are summarized in Table 11.2.

11.3.4 OUT-OF-BAND REQUIREMENTS

In addition to in-band requirements, several out-of-band requirements are specified both in 3GPP and in regulation. Some out-of-band requirements are generic and relate to regulation. Examples of general out-of-band requirements include transmitter spurious emissions or receiver spurious emissions, which specify the maximum unwanted emission level across a very large frequency range. In 3GPP the general transmitter spurious emission requirement is divided into two categories. The first, category A requirements have a level of −13 dBm/MHz and are adapted to regulatory requirements in some regions such as United States or Japan. The second, category B spurious emission requirements are applicable in Europe and specify a −30 dBm/MHz level. For either category, the general spurious emission level applies for the range of 30 MHz to either 12.75 GHz or the 5th harmonic (occurring at frequencies five times the operating band) of the upper frequency

Table 11.2 Summary of In-Band Receiver Requirements.

Requirement	Description	Purpose
Receiver sensitivity	A measure of the BS ability to receive and properly demodulate a signal at a specified minimum power level to achieve a specified throughput	The receiver sensitivity sets minimum expectations for both the noise figure of the receiver and demodulation performance of the baseband for static channel conditions and provides an expectation for achievable UL coverage
Dynamic range	A measure of the BS ability to receive a signal at a high SINR level in the presence of strong interference	Ensuring proper implementation of the receiver to secure the performance when the receiver level is high due to high SNR and/or strong interference
Adjacent channel selectivity (ACS)	A measure of the BS ability to receive a wanted signal in the presence of a modulated (CDMA or OFDM) interferer in the adjacent channel, where the interferer has same channel bandwidth as wanted signal	Sets an expectation on the receiver selectivity toward the adjacent channel, that is, to what extent the interferers at the adjacent channel frequency (which mostly are generated from uncoordinated UEs belonging to another operator) can be suppressed to ensure proper receiver performance in a multi-operator environment
In-band blocking	A measure of BS ability to receive a wanted signal in the presence of a modulated (CDMA or OFDM) interferer allocated at a larger frequency separation than is the case for the ACS requirement	Sets an expectation on the receiver dynamic range and selectivity at larger frequency offsets compared to ACS and to what extent the interferer from, for example, uncoordinated UEs can be suppressed to ensure proper receiver performance in a multi-operator environment
Narrowband blocking	A measure of the BS ability to receive a wanted signal in the presence of a narrowband modulated (GMSK or OFDM) interferer allocated at small frequency separation	Sets a minimum expectation on the receiver selectivity toward narrowband high PSD interferers at small frequency offsets with high resolution compared to ACS and to what extent interference from, for example, uncoordinated UEs either a GSM system or LTE UEs transmitting with a low number of PRBs and high PSD can be suppressed
Receiver intermodulation	A measure of the BS ability to receive a wanted signal in the presence of two interferer signals, one CW, and the other modulated (considering both wide or narrow bandwidths). The interferer frequency separations are allocated such that the intermodulation products generated due to receiver nonlinearities fall into the frequency range of the wanted signal	The receiver intermodulation, apart from also regulating the minimum selectivity of the receiver sets an expectation on receiver linearity. The intention of the requirement is to ensure that the receiver linearity is sufficient to suppress the intermodulation products in the receiver level for each of the various interferer types
In-channel selectivity (ICS)	A measure of receiver's ability to receive a wanted signal with a very low power occupying a limited amount of PRBs on one side of DC subcarrier in the presence of a strong signal on the other side of DC subcarrier	Sets an expectation on the receive ability to suppress the DC leakage scenarios in which the received power from different UEs served by same carrier is high

edge of the transmit part of the operating band but excluding the frequency range covered by the in-band requirements.

In some regions, there exist specific spurious emissions requirements related to so-called co-existence in the same geographical area. These requirements are designed for specific cases in which different operators and bands do not share the same site and thus are not co-located. The requirements apply within specified frequency ranges, which correspond to other operating bands. These types of requirements cover the UL and downlink (DL) part of the bands, implying that the BS (UL) and UE (DL) receive bands are protected. In addition, there are regional requirements that are similar to the co-existence requirements protecting other systems (non-3GPP) such as personal handy-phone system in Japan or digital terrestrial television in Europe. The co-existence requirements are more stringent than the general spurious emissions requirements because the systems operating within their frequency ranges require increased protection.

The receiver spurious emission requirements cover the same frequency range as the transmitter spurious emission requirements and are intended for nonduplexed antenna ports or for unpaired bands for which UL is switched with DL.

Another out-of-band requirement that is general is the out-of-band blocking requirement. The out-of-band blocking requirement is a receiver requirement and is similar to the in-band blocking requirement in that the BS is subject to a wanted signal, with an allowed additional power margin to account for degradation, and an interferer. In this case, the interferer is a CW at a level of −15 dBm regardless of BS class. The interferer is swept with 1 MHz step within a range from 9 kHz for conducted and 30 MHz for OTA up to 12.75 GHz, but excluding the range covered by in-band requirements.

There are also specific requirements originating from co-location scenarios. For such scenarios, it is assumed that the isolation between the co-located antennas of different base stations (at the antenna connectors) is 30 dB, which is seen as a reasonable worst-case estimate. This is further discussed in the next section.

11.3.5 CO-LOCATION

Due to the cost and complexity of acquiring sites, it is very desirable to be able to install multiple base stations covering different bands at the same site. The base stations may belong to the same operator, or it may be that the site is shared, and the base stations belong to different operators. It is also possible that different operators may wish to locate base stations operating in the same band (but on different carriers within the band) at the same site. For these cases, it is important to ensure that unwanted emissions from the transmitter of one base station do not degrade the receiver of another base station, and that the receiver of one base station is not degraded by the wanted signal transmission of another base station.

To provide for the possibility to co-locate, 3GPP has created a number of transmitter and receiver requirements specific to co-location. These requirements are specific unwanted emissions and receiver blocking requirements. Since the base stations are intended to be in close proximity, the co-location requirements are stricter than general unwanted emissions/blocking requirements.

Co-location requirements are band-specific and conformance to co-location requirements is not mandatory. The base station vendor declares which co-location requirements the BS meets, which indicates the bands with which the BS can be co-located.

To ensure that co-located base stations operating at different bands do not create unwanted interference that is greater than an acceptable level, a set of band-specific co-location spurious emission requirements are defined in 3GPP; they apply to co-located bands and are either −98 dBm/100 kHz or −96 dBm/100 kHz for a macro base station (or wide-area BS). The level is different depending on the BS class that relates to the concerned BS class noise floor.

The level of the unwanted emissions requirement (as well as other co-location requirements) has been derived based on an assumption of a 30 dB isolation between the antenna connectors of each base station. The origin of the 30 dB assumption is not well documented but is related to the lowest expected isolation for co-located BS antennas that are expected to be encountered in practice.

The receiver co-location–related requirements are a set of band-specific co-location blocking requirements that state that the receiver should be able to cope with high power transmitted in the DL of the other band. (For FDD operation, there is also a requirement on the required blocking rejection of the receiver from DL transmissions from the own transmitter.) For the wide-area BS class, the requirement is set to +16 dBm. The co-location blocking level is also BS class-specific so other BS classes have different requirements. As for the transmit case, the receiver requirement assumes 30 dB antenna isolation, which means that a carrier transmitting at full power of up to 46 dBm can be handled if the requirement is met. Meeting the co-location requirements in 3GPP is achievable by using sophisticated RF filters. The requirements provide a powerful toolbox to allow for adding new bands and new equipment to a site without posing any degradation on existing systems and bands (or the newly deployed band).

Transmitter intermodulation is also another co-location-related requirement. All of the unwanted emission requirements are applied, whilst an interferer with level of 30 dB lower than the total power of wanted carriers is applied in reverse direction. The interferer is swept in frequency. This requirement emulates the impact of strong power from another BS antenna assuming 30 dB of co-location isolation leaking into the transmitter. Such reverse leakage should be handled in a manner that does not prevent all existing unwanted emission requirements from being fulfilled.

For AAS base stations, due to limitations of OTA chambers in terms of dynamic range and minimum detectable level, the OTA co-location requirements are tested based on a proximity principle. The assumed 30 dB of isolation for non-AAS products is translated to placement of a reference antenna at 10 cm distance to base station under test. In addition, the requirements levels are adapted and scaled to map the conducted co-location levels to OTA ones. Fig. 11.8 visualizes co-location OTA testing for AAS base station.

It should be noted that management of passive intermodulation (PIM) has been a challenge since the beginning of cellular systems and has become more accentuated due to support for wider bandwidths, noncontiguous and multi-band operations. PIM arises due to nonlinearities in passive components and with wideband modulated operation resulting in wideband interference. If PIM products hit a receiver band, the induced interference may degrade the sensitivity of the receiver. PIM is not only generated in the base station itself but also in the feeder cable, jumpers, external filters, site equipment, and in some cases, the installations surrounding sites. Limitations on PIM are not directly specified in 3GPP base station specifications as many contributions to PIM come from external sources. Since PIM affects the receiver sensitivity and coverage in UL, significant effort is made to keep the base station PIM to an acceptable low level. Apart from low PIM BS design and manufacturing, careful site planning and high-quality site equipment and installation are

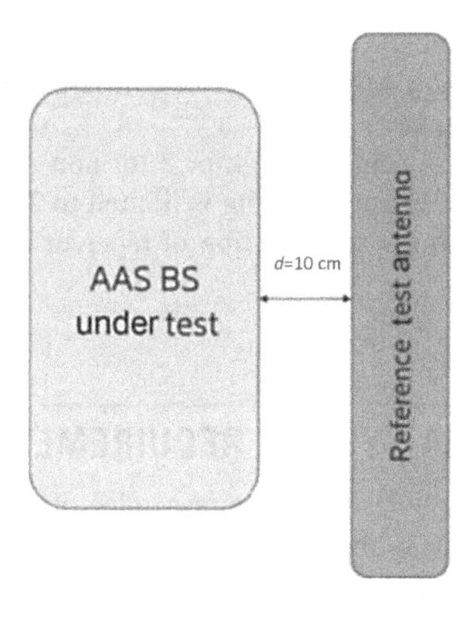

FIGURE 11.8

OTA co-location testing based on the proximity principle.

also important to mitigate PIM. An additional advanced approach is to deploy PIM cancellation schemes and equipment.

11.3.6 RECEIVER DEMODULATION PERFORMANCE

In addition to RF requirements, 3GPP BS specifications also incorporate a set of so-called receiver demodulation performance requirements (sometimes called receiver baseband performance requirements). The intention with receiver demodulation performance requirements is to capture the robustness of the receiver signal processing algorithms in the base station. Properties such as channel estimation for different fading environments, frequency error, and Doppler shift handling are subject to minimum performance requirements. The receiver demodulation requirements are expressed as either probabilities of expected outcome on signaling channels or as throughput. They are designed such that the testing exercises the baseband algorithms, but the outcome is independent of the analog radio implementation. For testing, a wanted signal that is expected to be demodulated is passed into the antenna connectors. The wanted signal is in many cases subjected to emulated channel fading. The signal will pass through the RF components of the receiver and in order to avoid that the outcomes are influenced by the radio performance, *additive white Gaussian noise* (AWGN) at a level of 20 dB higher than the base station noise floor is added to the wanted signal that is input to the antenna connectors. The addition of the artificial noise ensures that the base station noise floor becomes negligible as the linear sum of BS noise floor and AWGN will be more or less equal to AWGN level.

For NR/5G, the receiver demodulation performance requirements cover the NR UL physical channels of physical uplink shared channel, physical uplink control channel, and physical random access channel. UL physical channels are elaborated in Section 9.2. Each of these channels carries different user or control information to enable initial access to the network, including necessary information for channel quality, hybrid automatic repeat request processing, and user data. The requirements cover different use cases by defining multiple channel formats with different densities of relevant reference signals such as DM-RS and PT-RS, different modulation and coding schemes, as well as various fading environments.

The number of receiver antennas could be 2, 4, or 8 for non-AAS type of BS but for AAS and OTA, the number of RX branches subject to testing is limited to 2 since the OTA test chambers are limited to two orthogonal polarizations. OTA testing of receiver demodulation performance is further discussed in Section 11.5.3.

11.4 ADVANCED ANTENNA SYSTEM REQUIREMENT APPROACH

The previous Sections 11.2 and 11.3 described the principles of 3GPP requirements that are applicable to cellular base stations. Prior to the development of AAS specifications, radio requirements have always been defined and applied at antenna connectors. This section describes the rationale behind applying requirements at antenna connectors for non-AAS. For AAS BS, a number of conceptual challenges arise with respect to how to define requirements. These challenges are outlined in this section, while the following Section 11.5 describes the OTA-based solutions developed for the 3GPP specifications and regulation.

11.4.1 REQUIREMENT APPROACHES BEFORE ADVANCED ANTENNA SYSTEM

BS architectures from 2G until the development of AAS base stations within the context of 4G consisted of subsystems for handling higher layer protocols, baseband, radio, and antenna. The antenna was a physically separate component from the unit containing the active radio circuits. In many installations, the radio unit would be separated from the antenna (located at the top of the tower) by a long cable. The base station and antenna units were commonly procured separately, possibly from different companies. BS architectures based on separate radio unit and antenna will continue to form an important part of 5G infrastructure, in particular for lower bands. For wide-area/macro scenarios, the antenna part of such a non-AAS system is generally a fixed beam, passive antenna that covers a conventionalsector.

Base station architectures evolved to the concept of remote radio unit, in which at least a portion of the active radio components is located physically closer to the antenna itself to reduce losses due to cabling. The evolution toward decreasing the distance from antenna to radio ultimately resulted in a class of BS that integrates one or more active radio and passive antennas into the same hardware. Integration of radio and passive antenna gives rise to the need to solve complex problems such as optimizing weight, form factor, and PIM but does not change the basic paradigm of one or a small number of active radio transmitters connecting to sector-wide passive antenna(s) with a fixed beam radiation pattern.

The impact of some of the key radio requirements, such as carrier power and unwanted emissions, differs across space because of the directional antenna pattern. For base station architectures involving a passive antenna, the variation in key antenna parameters (such as beamwidth and sidelobe suppression) between different antennas is sufficiently low that the spatial impacts of radiated radio power and imperfections can be modeled based on a generally assumed antenna pattern.

Modeling in simulations using a general antenna pattern assumption enables requirement levels to be derived and applied at the active radio output of a base station that is sufficient to specify and regulate co-existence aspects, when a compliant radio is deployed with a fixed beam antenna.

For relative requirements such as ACLR and EVM, since both the wanted signal and interference/distortion components are radiated through the same passive antenna, the ratio is the same at all points in space and so the ratio requirements can also be applied at the radio output.

The fact that separate radio and passive fixed beam antenna-based architectures enable conformance and regulatory requirements to be specified at the output of the active radio (referred as antenna port or ARP) is highly convenient, since this enables the base station and antenna components of the system to be separated and conformance testing to be performed on the base station in isolation. Antenna connectors are generally readily available, since base stations need to be connected to antennas via cabling.

Originally, conformance requirements were derived assuming a single radio transmitter. The introduction of two TX MIMO for 4G and later development of four TX and eight TX MIMO continued to be based on radio equipment where requirements are applied at the outputs of each individual radio transmitter. For some types of requirement, this implied an increase in the total output power from the BS, since two, four, or even eight transmitters meeting the same requirement on a power level (e.g., emissions) will output two to eight times as much power in total.

11.4.2 CHALLENGES INTRODUCED BY ADVANCED ANTENNA SYSTEM

In 3GPP RAN4 it is assumed that an AAS base station comprises a system in which active radio/transceivers and antenna components are integrated, which is capable of dynamically varying the radiation pattern in a controlled manner. The ability to dynamically adjust the radiation pattern contrasts with the systems described in Section 11.4.1, in which the radiation pattern is fixed. AAS base stations may be built around a number of different architecture concepts, ranging from fully digital, in which each individual antenna element is driven by an individual radio transmitter to hybrid, in which groups of elements (e.g., columns) are driven by a transmitter. Chapter 14, considers in more detail the application scenarios and benefits/drawbacks of each architecture option.

A first and very significant challenge for an AAS is that when there is a large number of radio transmitters and receivers, the feasibility of building antenna connectors for each individual radio transmitter together with a testing setup that enables testing of the combined response of individual radios is severely compromised. For large arrays and in particular for millimetric-wave systems, provision of test connectors becomes unfeasible. Even for systems for which connectors can be provided, since beamforming is a key component of the base station function it is desirable to set some requirements that encompass the antenna array as well as the radio/transceiver. Thus, for AAS base stations, it is necessary to specify requirements and corresponding relevant metrics and perform conformance testing (as well as other forms of testing such as research and development testing and manufacturing testing) over the air, that is, by means of analyzing signals radiated from the base station and/or exposing the base station to radiated signals in a controlled environment. OTA testing of base stations is a major paradigm shift for base station testing that has resulted in significant work in standardization and regulatory fora. However, OTA testing alone is not the only challenge to consider in setting regulatory requirements for AAS base stations.

The beamforming functionality of the base station is in general distributed between so-called analog beamforming, which takes place within the antenna element groups driven by each individual transmitter or receiver and digital beamforming, which takes place in the digital domain prior to conversion to analog transmission (as described in Section 7.3). The radiated beam pattern is the composite result of all digital and analog beam steering, antenna and subarray radiation patterns, and transceiver properties. For any conformance requirements that need to consider the radiation pattern of the base station, the requirements need to be placed on the AAS BS as a whole such that the impacts of all beamforming mechanisms in addition to transceiver performance are captured.

The radiation pattern of unwanted emissions from an AAS base station depends on several factors. Some unwanted emissions components are generated by random noise-like processes, which are uncorrelated between different transmitters. These components will not experience coherent beamforming and will be radiated according to the pattern of the element groups driven by each transmitter. Other components are related to distortions of the wanted signal, such as intermodulations. If the wanted signal contains a number of beams, and/or considering that the phase impact of intermodulations will vary in a frequency-specific manner, the pattern of emissions arising from such transmitter distortions is likely to be beamformed, but potentially with a different pattern to that of the wanted signal (See Fig. 11.9). A further category of unwanted emissions arises from specific components of the transmitter, for example, local oscillators (LOs) that give rise to narrowband interference. These components may also experience coherent beamforming gain through the array, and the beam pattern may or may not be aligned with the wanted signal depending on the array and transmitter architecture.

Fig. 11.10 depicts a simulation result illustrating the direction of unwanted emissions components in relation to two carriers. In the center, the spectra and angular distribution of the two carriers are observed. The carriers have different beam directions. To the left and right, the spectra

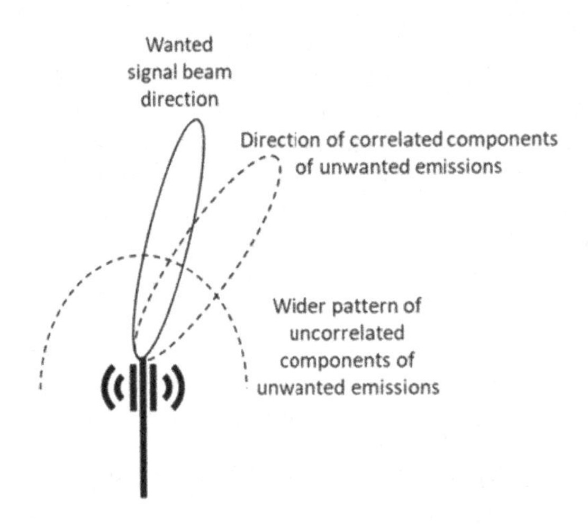

FIGURE 11.9

Example of correlated and uncorrelated unwanted emission components that are not aligned with the wanted signal direction.

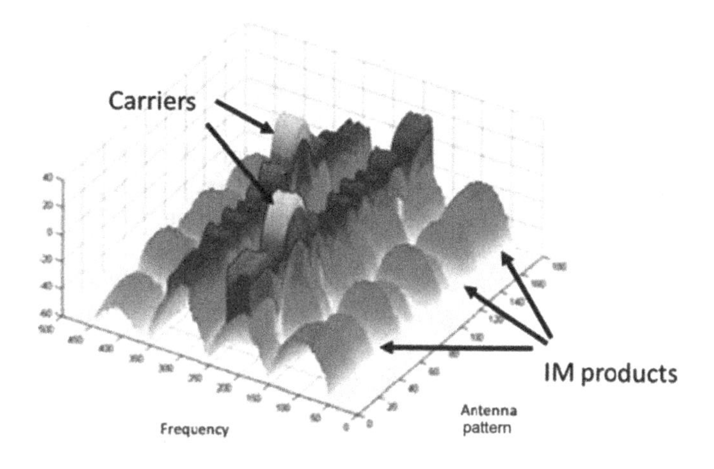

FIGURE 11.10

Simulation result of wanted and unwanted emission directions.

and angular distribution of the intermodulation products are visible. The IM has peaks of radiation, but they are not aligned with the peaks of either of the carriers.

The fact that both the wanted signal and the unwanted emissions are subject to time-varying beamforming implies that evaluation of interference and co-existence properties cannot assume a fixed antenna pattern as is the case for non-AAS systems. Consideration must be given to both the potentially wide variation in beam patterns of different implementations and to the implications of time-varying beam patterns on aspects regulated by conformance requirements such as co-existence with other systems. Section 11.5.2 outlines the approach and conclusions for setting emissions and power requirements for AAS base stations. Furthermore, the spatial pattern of the wanted signal and interference or distortion may differ for the reasons outlined above. For this reason, requirements based on quantities that are ratios of power levels, such as ACLR and EVM, will vary in space and will for many directions not be the same as the ratios achieved at the transmitting radios. Additional consideration is needed on the implications of the spatial variation of relative requirements.

In the receiver direction, a discrepancy between the portion of the antenna aperture related to each individual receiver and that of the whole array considering all types of beamforming necessitates careful evaluation. The sensitivity of the array as a whole depends on the combined beamforming within the array, the NF of the radio receivers, and the baseband implementation. Some of the receiver requirements are, however, designed to set limits on desensitization (degradation of sensitivity) caused by the introduction of interferers. For example, the receiver blocking requirement introduces a high-power interferer and regulates the power level at which receivers saturate and any other forms of desensitization that may arise. The blocking interferer causes distortions in individual radio receivers; in the worst-case overloading of the receiver ADCs that causes the receivers to cease to operate. The distortions may be nonlinear and uncorrelated between different receivers. A key problem when considering such effects and the associated radio requirements is that the impact of the interferer on individual receivers depends on the gain of subportions of the array aperture (i.e., subarray radiation patterns), whilst the sensitivity itself depends on the whole array architecture and radiation pattern. The implication of this discrepancy between which part of the aperture is relevant for which type of signal can differ depending on the receiver architecture, which is problematic when attempting to create a generic radio requirement that treats the base station as a black box (Fig. 11.11).

Section 11.4.1 mentioned that non-AAS radio requirements defined unwanted emissions on each individual radio. For an AAS base station, setting unwanted emissions requirements per individual radio becomes problematic for a number of reasons. Firstly, it is desirable to set requirements on the base station as a whole that do not depend on the individual AAS base station architecture, and it would be undesirable for different array architectures with different numbers of transmitters to present differing levels of interference to the outside world. Secondly, it would be unacceptable for an AAS BS with a large number of transmitters to present an unacceptably large level of total interference simply due to an emissions requirement being applied to each transmitter individually.

Apart from the radio requirements, the baseband demodulation requirements are potentially challenging for an AAS BS. The baseband demodulation requirements typically include a stimulus signal that is modulated based on a model of a multi-path fading environment. Without access to

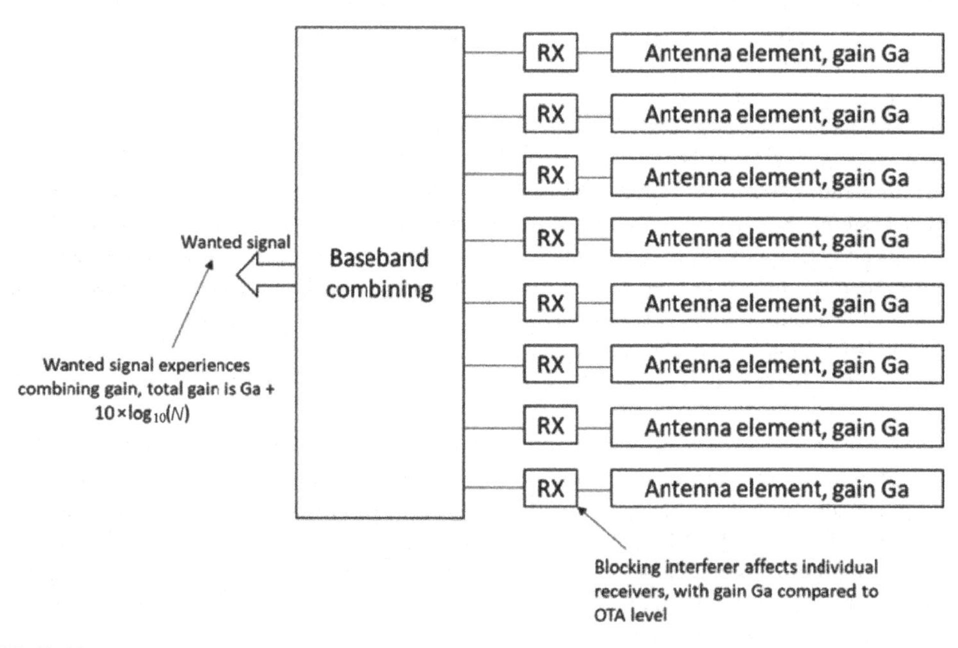

FIGURE 11.11

Blocking interference and wanted signal are relevant at different points in the architecture.

antenna connectors, it is highly challenging to create a controlled and reproducible multi-path fading environment in an OTA test.

In summary, the architecture and functionality of AAS base stations gave rise to several challenges for setting conformance and regulatory requirements. The need for OTA testing is a paradigm shift without precedent from previous base station technologies. The complexity in emulating real multi-path fading environments in an OTA test chamber is prohibitive and thus current state-of-the-art OTA testing without multi-path was considered. Apart from the new approach to testing, other aspects to consider included the impacts of beamforming, time-varying unwanted emissions, various types of architectures with different spatial behaviour and different responses to stimulus signals. Furthermore, the need to set requirements on the whole base station system (comprising antennas, radios, and digital processing) that do not depend on the architecture necessitated careful evaluation.

11.5 OVER-THE-AIR REQUIREMENT CONCEPT AND METRICS

The previous subsection described the challenges that arise when considering conformance and regulatory requirements for AAS base stations. This section describes the conceptual design of radio requirements that solve these issues, but not the actual requirement levels. Section 11.6 describes how requirement levels are set in 3GPP and Section 11.8 presents some practical approaches to OTA testing. Whilst much of the underlying work associated with this section has been performed in 3GPP RAN4, other regulatory bodies such as international telecomunication union, Federal Communications Commission (FCC), and ECC have also evaluated the issues and reached similar conclusions.

The 5G/NR base station RF specifications have been specified based on different so-called requirement sets.

The current 3GPP NR standard supports a large number of frequency bands from ~450 MHz to ~43 GHz. All frequency bands up to 7.125 GHz are classified as belonging to FR1, while frequency bands within ~24–~52 GHz are referred to as frequency range 2 (FR2).

For each frequency range, the concepts applied for defining the requirements are the same and the same set of requirements applies. However, between the frequency ranges there are some differences in requirement concept and applicability.

For FR1, three different requirement sets were created as follows:

- 1-C: All conducted
- 1-H: Hybrid requirements
- 1-O: All OTA.

A base station operating in FR1 is declared as conforming to just one of these three requirements sets; either 1-C, 1-H, or 1-O.

The requirement set 1-C describes a set of requirements for which requirement and conformance testing are specified at the antenna port of the BS and measured in a conducted fashion. This requirement set is intended for 5G base stations whose architecture is the same as non-AAS type base stations (i.e. without integrated antennas and with a low number of transceivers).

The requirement set 1-H is developed for AAS base stations but does not consist of OTA testing of all requirements. Conducted requirements are applicable for each transceiver in the AAS. The conducted requirements are not identical to 1-C, since limits to the total emissions are imposed. In addition, two OTA requirements, equivalent isotropic radiated power (EIRP) accuracy and OTA reference sensitivity, are applied. AAS base stations compliant to requirement 1-H require accessibility to connectors on each subarray. 1-H requirement set does not necessitate measurement of total radiated power (TRP) or any other OTA measurement at frequencies outside of the operating band, which in some circumstances simplify the testing (at the cost of needing to provide access to antenna connectors).

The requirement set 1-O is created for AAS base stations without antenna connectors, as the requirement definition and conformance testing are OTA for all requirements.

For FR2 covering the mm-wave frequency ranges, a single OTA requirement set 2-O is specified, because the AAS base station requires a high level of integration and it not feasible to implement connectors.

The FR2 AAS base station has slightly lower requirement coverage compared to FR1. Requirements such as transmitter intermodulation, co-location requirements, or receiver dynamic range are not specified for FR2 as analysis in 3GPP concluded that the requirements are not needed.

In coming standard releases when the regulatory situation around 7–24 and above 52.6 GHz in different regions is settled, new frequency ranges may be introduced in the specifications.

11.5.1 INITIAL OVER-THE-AIR REQUIREMENTS FOR ADVANCED ANTENNA SYSTEM

Since AAS base stations integrate antennas and introduce beamforming capabilities, it was decided early on in 3GPP that new OTA requirements should be introduced on beamformed radiated power in the transmit direction and beamformed sensitivity in the receive direction. It should be noted that from the perspective of 3GPP RAN4, radiated EIRP is not strictly necessary since it does not relate to co-existence, co-location, or any of the other usual reasons for introducing conformance requirements, and does not form the basis for declaring a rated power for the base station (TRP is used as rated power, which will be discussed in later sections). Furthermore, the transmit power requirement is not really a requirement at all, but rather a standardized means of verifying a declaration about basic beamforming. However, it is desirable to introduce some standardized tests relating to beamformed power in 3GPP, and furthermore maximum EIRP is a controlled parameter in some regulatory regimes, making a framework for declaring EIRP useful. OTA sensitivity, on the other hand, is a necessary part of other receiver requirements, as many receiver requirements are based on OTA sensitivity and so is a usual and necessary part of the conformance framework.

The 3GPP requirement on EIRP is based upon a declaration of a nominal range of beam directions for the base station and the EIRP achieved in each direction. The declaration of beam directions is very flexible such that all types of implementation can be accommodated. The specific details of how the declaration is made are discussed further in the annexes of the 3GPP technical report 37.842 [8]. There are no constraints on the beamforming and EIRP levels, and the "requirement" is simply that when tested, the declared EIRP is met in the declared directions to within a given accuracy.

It is very important to understand that the beamforming declarations made for the 3GPP EIRP requirement are nominal test patterns and may be quite different from the types of beamforming that may be applied by the base station during actual operation. Chapter 6, describes approaches to beamforming in more detail. In particular, for advanced forms of beamforming/multi-antenna transmission, there is no concept of a "beam" as such, but rather power is distributed in different directions according to the instantaneous state of the channel in a manner that can maximize received SINR at the receive antennas of the intended UE. The 3GPP-declared beamforming behavior does not test the intended beamforming performance of the base station in a deployment, but rather declares some test beam patterns that can be used to verify that the RF and antenna architecture of the base station is capable of coherently directing energy in an anechoic environment.

The OTA sensitivity requirement states that when the base station is exposed to a stimulus signal, a prescribed minimum throughput should be achieved. The conducted sensitivity requirement is an indirect requirement on the NF of the receiver. The OTA sensitivity requirement is an indirect requirement on jointly NF, antenna performance, and analog/digital combining. Two types of OTA sensitivity requirement are defined in the 3GPP specifications. One type of requirement is to meet a declared sensitivity level over a declared range of angles with respect to the base station receiver. The second type of OTA sensitivity requirement, referred to a reference OTA sensitivity, is to meet a minimum sensitivity over a declared range of angles. The ranges of angles over which sensitivity is achieved can differ for the two requirements. The first requirement is intended to serve as a standardized declaration and test of the best-case sensitivity achievable within a range of angles relating to a cell area. Unlike the EIRP requirement, the declaration does not assume beams or beamforming, and the receiver may use any type of combining to achieve the declared sensitivity. It is important to note though that the combining approach adopted in an anechoic test environment may differ significantly from a multi-path fading environment during real operation. The second type of OTA sensitivity requirement, OTA reference sensitivity relates the receiver performance to the existing conducted sensitivity requirement over a declared range of angles (i.e., ensures that the sensitivity performance is the same as for the conducted requirement within the range of angles). The declared range in this case may not relate to any cell coverage area, but rather is used as a parameter for mapping conducted to OTA requirements as described in Section 11.6. Both types of requirement are used for setting the wanted signal level for other OTA receiver requirements.

11.5.2 OTHER OVER-THE-AIR TRANSMITTER REQUIREMENTS

When developing requirements for unwanted emissions, as discussed in Section 11.5.1, it is important to consider that the unwanted emissions may be beamformed and that the beamforming pattern may be implementation-dependent and may vary with time. It is useful to differentiate between three types of unwanted emissions:

- in-band unwanted emissions, for which the victim of the emissions is another 3GPP network,
- out-of-band emissions (for which the victim of the emissions is likely to be other types of system), and
- emissions that are related to interference between co-located systems, generally between 3GPP base stations.

To understand the impact of in-band unwanted emissions on co-existence properties, Monte-Carlo statistical simulations on throughput loss in a victim network caused by an aggressor network causing emissions were performed with a large number of separate "drops" (i.e., instantaneous combinations) of UE positions. Throughput loss in a victim network depends on the instantaneous positions of UEs in the two networks. Monte-Carlo simulations provide a statistical distribution of throughput loss considering all of the modeled UE position combinations. The simulations modeled the aggressor network as using AAS base stations and beamforming. To emulate the effect of different spatial patterns, the unwanted emissions from the aggressor network were modeled as fully beamformed (i.e., the same time-varying pattern as the intended signal), non-beamformed (i.e., wide beam static patterns), and some variations in between. Simulations generally indicated that the relationship between the total power (i.e., sum of power in all directions) correlated, to within a few dBs to the statistical throughput impact of the interference in the victim network regardless of the spatial pattern assumption for the unwanted emissions. On the other hand, the relationship between EIRP of the unwanted emissions (i.e., maximum directional power) and the statistical throughput loss in the victim network depended on the assumption for the unwanted emissions spatial pattern. This implies that the TRP of the base station emissions is a good metric for setting an in-band unwanted emissions requirement that regulates interference toward other 3GPP networks on other frequencies. An explanation for this is that the instantaneous throughput impact depends on the relationship between the direction and intensity of the unwanted emissions and the position of the victim UE that is suffering the interference. Clearly if the victim UE is within a beam of unwanted emissions, then it will suffer throughput loss, whereas if it is outside of such a beam, throughput loss will be minimal and less than for non-AAS systems. When considering the average throughput loss in a victim network, all different relationships between unwanted emissions beam energy and victim UEs are considered and the average throughput impact is related to the "average" beam pattern, that is, the average radiated power.

This is illustrated in Fig. 11.12, which depicts two statistical simulations of throughput loss in a victim network compared to ACLR. ACLR is calculated based on conducted output emissions, TRP emissions, and EIRP emissions. Antenna patterns and beamforming are modeled in the simulation. In the top case, the array is assumed to have analog beamforming, whereas in the bottom, the array has digital beamforming and the emissions are assumed to be fully uncorrelated between transmitters. It can be observed that the TRP (and conducted) ACLR relate to throughput loss in the same manner for both implementations. However, the EIRP ACLR relation to throughput loss depends on array type. Hence, TRP ACLR (and more generically, TRP unwanted emissions) provides a more generic means to set emissions limits that lead to the same level of impact on neighbor systems for all array types.

As a result of these considerations, the in-band unwanted emissions requirements are expressed in terms of TRP. Also, the ACLR is expressed as a ratio of the TRP in-channel power to the TRP adjacent channel power. Since TRP corresponds very well to the power output of the radio transmitters, the unwanted emissions requirement levels are based on the conducted levels defined in previous specifications. However, as discussed in the previous section an AAS base station has a large number of transmitters, and the previous specifications set emissions requirements per transmitter. To set a consistent emissions requirement whilst taking into account the existence of multiple transmitters, the AAS in-band emissions requirements are set as TRP output from all transmitters simultaneously, but the levels are adjusted by 9 dB compared to the per transmitter

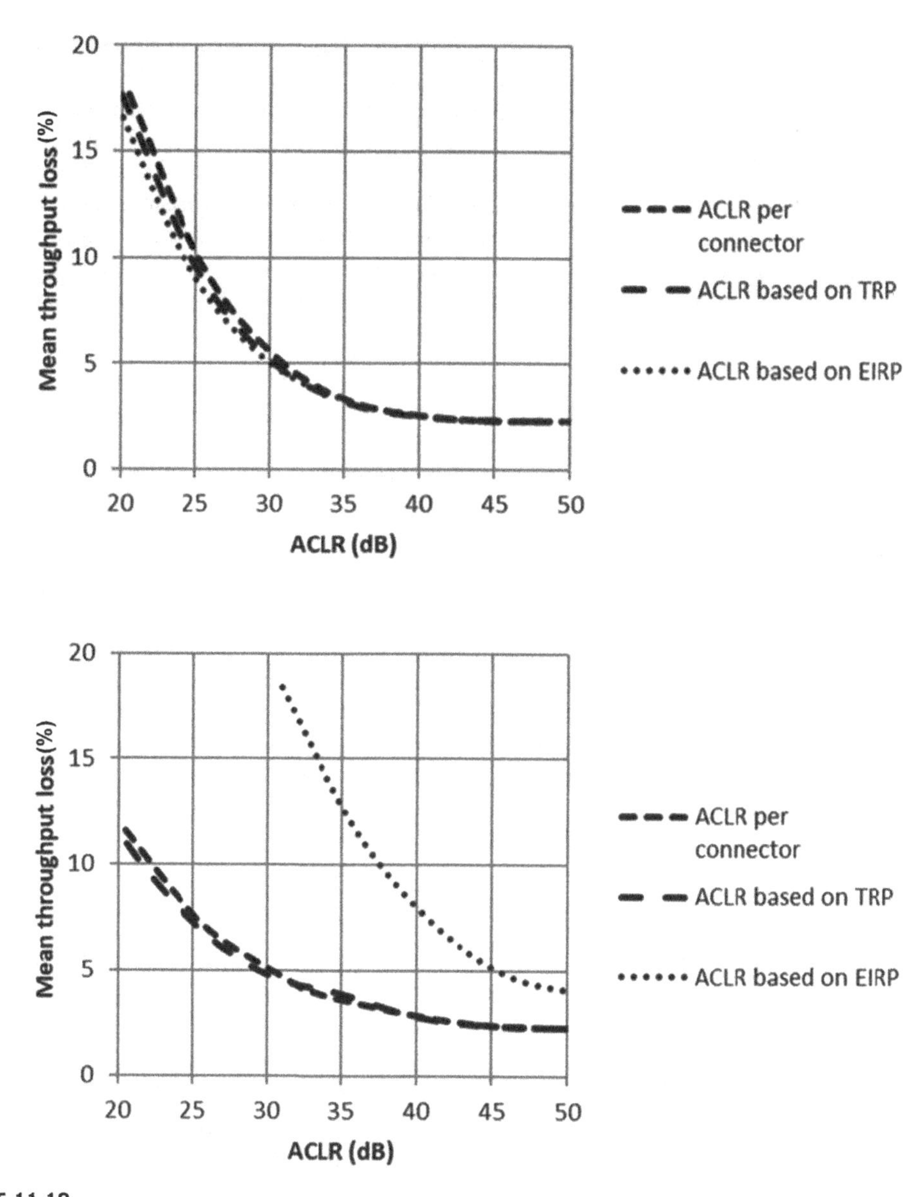

FIGURE 11.12

ACLR versus TRP/EIRP for two different types of arrays: analog beamforming (top) and digital beamforming (bottom).

requirements of the non-AAS specifications when AAS base station contains more than eight transmitters. The 9 dB adjustment was justified in 3GPP on the basis that functionality for implementing eight transmitter MIMO (with eight times, or 9 dB greater power) had already been included in the specifications prior to the development of AAS conformance requirements and so eight TX MIMO could potentially already have been implemented. Limiting the emissions adjustment to 9 dB imposes a ceiling on emissions that matches the release 13 functionality. The 9 dB adjustment generally applies to all unwanted emissions levels, except that in some regions it is not allowed for certain regulatory requirements (e.g. spurious emissions).

The out-of-band emissions requirements consist of a general spurious emissions limit and some specific additional requirements in other frequency bands protecting specific bands and services (as described in more detail in Section 11.3.4). Victim services in the out-of-band domain are diverse and the impact of AAS base stations may need a case by case examination. Nonetheless, some general principles can be considered:

- The group delay distortion of filters (a measure of how phase is changing over frequency) in the out-of-band domain may differ randomly between transmitters.
- The antenna spacing relative to wavelength changes.
- The antenna characteristics differ to in-band and the processes giving rise to spurious emissions may not be correlated.

It was observed that due to these factors, in general, out-of-band emissions tend not to be beamformed and are somewhat spatially white. This is especially true for full digital beamforming. For hybrid beamforming, the likelihood of de-correlation increases as the frequency relative to the carrier center frequency increases. Some factors such as LO leakage may be correlated between transmitters and less spatially white; however, such components are often narrowband in nature. Based on these observations and considering that previous spurious emissions requirements were defined as conducted, which is proportional to TRP, TRP was also adopted for spurious emissions as the most practical means of regulating emissions toward other systems. TRP is also convenient since for some frequency ranges, it is necessary to measure the unwanted emissions in the antenna near field, within which spatial patterns cannot be observed or measured.

Since TRP relates directly to output power from the radio transmitters, the conducted out-of-band emissions limits from the previous specifications were used for setting the OTA limits. In some cases, the 9 dB adjustment for OTA emissions limits compared to conducted was applied with the same reasoning as in-band emissions. However, in some regulatory domains (such as FCC and electronic communications committe (ECC)), the existing general spurious emissions limit is interpreted as applying to the whole base station and the adjustment is not allowed.

Although the concept of TRP-based unwanted emissions requirements can be applied for many of the in-band and out-of-band emissions requirements, there is a class of emissions requirements for which a TRP-based metric cannot be applied. These are the so-called "co-location" requirements, as described in Section 11.3.5. Due to the close proximity of the base stations in these scenarios, which implies near-field interaction (as described in Section 3.3) and relatively low isolation between the antennas, the allowed unwanted emissions levels are much lower than for other types of emissions requirements. In addition to the co-location emissions requirements, there are also transmitter intermodulation requirements, for which a high-power signal on another frequency is injected into the BS transmitter (modeling a co-located transmitter), in order to verify that any arising intermodulation products are sufficiently suppressed. Both the co-location emissions and the transmitter intermodulation requirements cannot easily be tested in the antenna far-field due to the very low emission levels in the one case and the need to deliver high power in the other respectively. To enable testing of the requirements, 3GPP developed the concept of co-location OTA testing, in which a test antenna is located in close proximity to the base station under test, physically mimicking co-location conditions. The size and location of the colocation test antenna are standardized in the 3GPP specifications.

Fig. 11.13 illustrates the setup for co-location measurements for transmitter intermodulation. For Co-location based emissions requirements, the roles of the spectrum analyzer and signal generator are reversed. For co-location blocking, the spectrum analyzer is replaced with a signal generator generating the wanted signal.

In addition to the EIRP and unwanted emissions requirements, a further declaration and requirement are placed upon the TRP of the base station. In the 3GPP specifications, an upper limit for the TRP of some BS classes is set. The reason that TRP is regulated rather than EIRP is that in the 3GPP specifications, the power limit is related to enabling co-existence between different frequency layers and operators. UEs on another frequency suffer receiver interference due to their ACS. The level of interference depends on the amount of radiated output power in the adjacent channel. Hence in order to prevent micro and pico base stations, which are deployed closer to UEs, from causing excessive degradation to UEs in an adjacent channel, the radiated power output from micro and pico BS is restricted. Similar to the ACLR and in-band unwanted emissions, the impact of in-channel power depends on the relative directions of the radiated energy and the victim UE. Thus again in a similar manner to unwanted emissions, over time and on average the impact of the set of different combinations of beam and victim UE direction relates better to the total power radiated in all directions than EIRP at any individual instant in time. For this reason, a requirement on TRP accuracy was created and the power limits were placed on TRP in 3GPP. It should be noted that

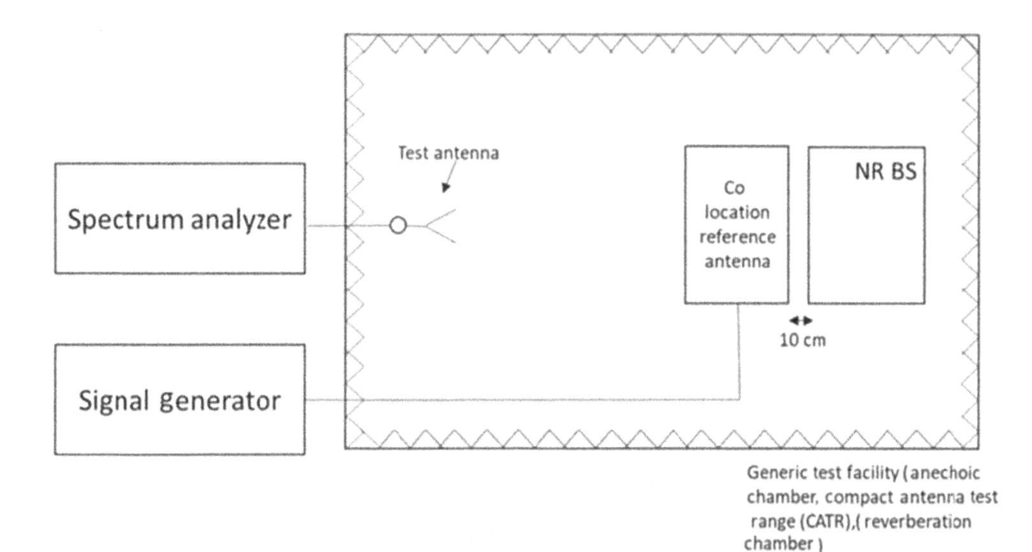

FIGURE 11.13

Illustration of co-location test setup for transmitter intermodulation.

some regulatory requirements (but not 3GPP) additionally place maximum EIRP limits to regulate EMF effects.

The remaining major transmitter conformance requirements in 3GPP relate to signal quality, including EVM, frequency error, and time alignment error. Since these requirements regulate properties relating to the structure of the transmitted signal rather than power levels, the requirement levels can be set to the same values as was the case for conducted conformance requirements. The signal quality requirements need only to be met at the receiving UE and not at other points in space. For this reason, the signal quality requirements are defined and tested at a number of directions in space at which it is expected that a UE could be scheduled. As for other requirements, it should be noted that the range of points in space at which the signal quality requirements are met are valid in an anechoic environment only, and in a real operating environment with complex radiation patterns the propagation of power to the UE does not have a one-to-one mapping with directional testing in an anechoic chamber. Nevertheless, the contrived anechoic environment and declared set of conformance directions are sufficient for an RF conformance test that verifies a minimum signal quality from the BS. It should also be noted that even in an anechoic environment, the requirements are not specified as applying for specific "beams," but rather as applying in specific directions. This distinction is important as it implies that also in the anechoic environment, the beamforming approach used to meet the signal quality requirements is implementation-dependent. For example, some implementations may meet an EVM requirement in a declared direction by pointing a beam in the test direction, whereas other implementations may for implementation-specific reasons meet the requirement away from the center of the main lobe or in a sidelobe of a beam pointing in a different direction (this may be the case, for example, for an implementation that does not include dynamic beamforming in one plane, for example, horizontal plane).

Four types of OTA requirement were thus created for transmitter conformance testing:

- EIRP and power requirements, based on nominal beam declarations and defined in the center of "beams,"
- directional signal quality requirements met in certain directions (with no related beam declarations),
- TRP requirements, including rated BS power and most unwanted emissions, and
- co-location requirements based on an assumed co-location antenna test setup.

Table 11.3 depicts how the 3GPP requirements map to different requirement types.

In summary, AAS transmitter requirements can be divided into several groups, for example, directional, nondirectional, and co-location based.

Directional requirements are specified and measured in specific directions or set of directions (in many cases declared) and often are related to network performance (i.e., key performance indicators). As examples, EIRP accuracy and signal quality (EVM) belong to directional requirements, for which each in one way or other impacts the network experience in terms of coverage or achievable bit rates. Some directional requirements allow for possibility of benefiting from spatial/beamforming gains. For example, with EVM any uncorrelated components of transmitter distortions will not be beamformed and hence will be spatially white compared to beamforming of wanted signal.

Table 11.3 Requirement Mapping to Different Requirement Types.

Requirement	Type of Requirement
EIRP accuracy	Directional beam related
Rated output power	TRP
TX power dynamic range	Directional beam related
EVM	Directional (non-beam related)
Frequency error	Directional (non-beam related)
Time alignment error	Directional (non-beam related)
TDD OFF power	Co-location (E-UTRA and NR FR1) TRP (NR FR2)
TDD OFF transient time	Co-location (E-UTRA and NR FR1) Directional (NR FR2)
Occupied bandwidth	Directional (non-beam related)
ACLR	TRP
Operating band unwanted emissions	TRP
General spurious emissions	TRP
Co-location-related spurious emissions	Co-location
co-existence with other systems spurious emissions	TRP
TX intermodulation	Co-location (and TRP for measuring emissions)

Unwanted emission requirements such as emission masks, ACLR and general spurious emissions, and also the TRP power accuracy requirement belong to a group of nondirectional requirements that are specified and measured as TRP.

Co-location-based requirements are related to a near-field type of impact considering a (measurement) reference antenna placed in proximity to the AAS BS. The distance between the AAS BS and the measurement reference antenna should emulate the worst-case expected antenna isolation ($d = 10$ cm). Co-location spurious emission requirement is one example requirement that illustrates the need for this kind of test setup. The conducted requirement level (−96 dBm/100 kHz) will lead to far-field power levels that are much lower than the minimum detectable level in any OTA measurement chamber measured either as EIRP or TRP and thus the need for the co-location measurement reference antenna setup.

Transmit intermodulation is also a co-location based requirement. An interferer is injected through a measurement reference antenna and then the emission requirements from base station under test are measured.

11.5.3 OVER-THE-AIR RECEIVER REQUIREMENTS

The OTA receiver sensitivity requirement was described in detail earlier in Section 11.5.1. The other receiver requirements consist of transmitting a wanted signal whose power has some relation to the sensitivity level and an interfering signal to the base station. The blocking requirements are described in more detail in Section 11.3.3. The interferer may model a high-power blocking signal (modeling a signal coming from a UE on another uncoordinated network), a lower power signal (modeling another in-channel UE) or multiple signals that interact.

In real operation, the wanted signal and blocking signals will come from different directions in space. Furthermore, as described in Section 11.4.1, the portion of the antenna aperture relevant to the wanted signal and the interferer may differ in an implementation architecture–dependent manner. (The wanted signal is combined across the entire antenna array, whereas the interferer impact is on the RF properties of individual receivers. The interferer level at each receiver depends on the subarray radiation patterns). This means that the worst-case combination of wanted signal and interferer may differ, depending on the AAS base station receiver architecture. These issues significantly complicate the design of OTA receiver requirements.

To simplify the receiver requirement definition and testing, 3GPP decided to specify requirements in which both the wanted signal and the additional interferer(s) come from the same direction in space. The practicalities of testing with several different source signal directions, and also the limitations of the number of combinations that can feasibly be tested implied that other approaches would be too complex. The level of the interfering signals was, however, investigated and simulated such that it would be statistically a reasonable representation of a blocking level to the receivers considering multioperator non-co-ordinated operation.

Blocking levels are defined in 3GPP based on statistical simulations of interference distribution. The 99.99:th percentile of the interference level observed by dropping interferer UEs at various positions is used to set the blocking level. The blocking level will relate to a worst case (possibly not the absolute worst case, but a case in which the UEs are infrequently located but cause severe interference).

For FR1 (below 7.125 GHz)operation, it was observed that for different architectures, whilst the direction of arrival of the blocking signal in the statistical worst case differs depending on the antenna pattern related to each individual subarray, the RF level experienced within the receiver is similar for all architectures, as illustrated in Fig. 11.14. For arrays with analog/passive combining, the 99.99:th percentile blocker tends to be in the main lobe of the beam or in a strong sidelobe. For a digital array, the element gain is lower, but the beamwidth is wider. Although the array gain differs between the two array types, the difference in array gain is generally compensated by a difference in pathloss to the 99.99:th percentile interferer positions.

Thus a single internal blocker level can be set as a minimum requirement, which is compensated by an assumed gain for the antenna pattern associated with the receiver in a test direction in order to obtain a test specific OTA level. The approach of mapping conducted to OTA receiver requirements is described in more detail in Section 11.6.1.

For mm-wave operation, statistical simulation revealed a divergence between the worst-case OTA interferer level experienced by hybrid or fully digital beamforming architectures and also dependent on the array size and corresponding beamforming gain. In order to set a single requirement, an OTA interferer level that is relative to the receiver sensitivity was set. Setting the interferer level in this manner has the effect that the impact of array size is partially compensated, and the level is somewhat lenient for a fully digital beamforming and somewhat strict for a fully hybrid beamforming. The design of the levels for mm-wave is intended to provide reasonable robustness whilst being implementation agnostic.

Receiver blocking requirements are separated into in-band and out-of-band. In-band requirements relate to co-existence with other 3GPP networks. Out-of-band regulates co-existence with other types of service. The existing conducted out-of-band blocking requirements of −15 dBm is

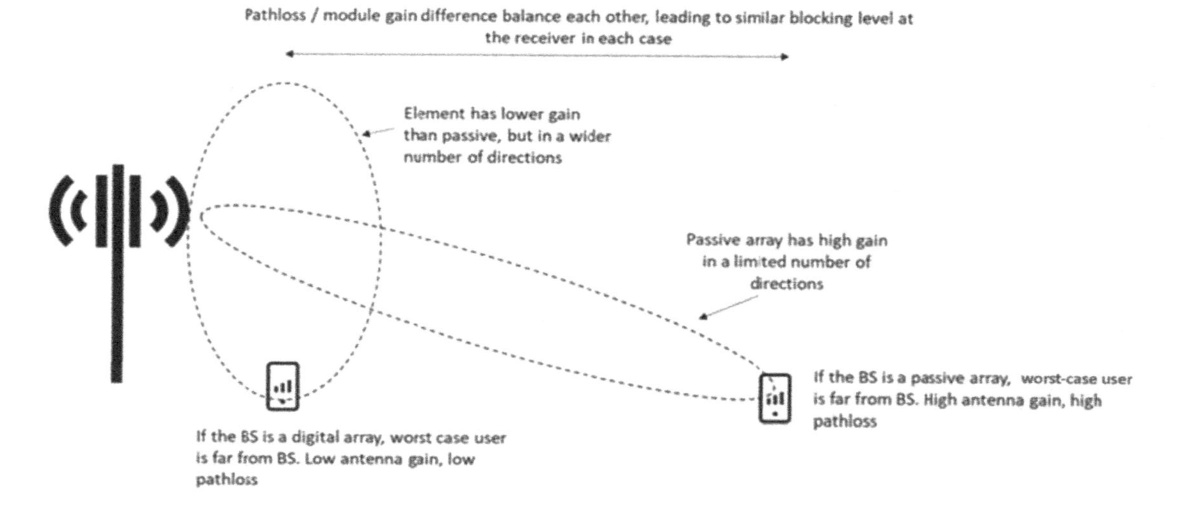

FIGURE 11.14

Illustration of how antenna gain and pathloss counteract one another for statistical worst-case blocking for different antenna arrays.

applicable over the whole spurious frequency range. When translated to OTA, the response of the base station antennas must be considered. With increasing frequency, the power absorbed by an antenna decreases and hence to maintain the same effect as the −15 dBm conducted requirement, in principle the power of an OTA blocking signal must increase with frequency. However, at higher frequencies, the required OTA signal power becomes too large to generate in a test chamber and does not correspond to any real-life blocking situation. In summary, the conducted −15 dBm requirement is overspecified. To resolve this inconsistency, out-of-band blocking is specified with constant field strength, corresponding to a varying OTA RX power level across all frequencies. This approach is practical, consistent with expected real interference situations and consistent with the approach for immunity testing as part of EMC.

Receiver requirements are thus mainly specified and are applicable in a declared set of directions, with both wanted signal and any interferers coming from the same direction.

For co-location blocking, the conducted interferer level is specified to +16 dBm for non-AAS. Due to link budget of typical chambers, the required far-field interferer power level that must be created to verify the requirement is not practical as an extremely high-power PA would be needed to emulate the conducted 16 dBm level at the base station under test. By applying a practical power level which is 16 dBm + 30 dB antenna isolation = 46 dBm, injected into measurement reference antenna at the specified physical separation (d = 10 cm) to the AAS base station with the set up described in Section 11.5.2, the co-location blocking of AAS can be feasibly tested.

A further class of requirements is the baseband demodulation requirements, as described in Section 11.3.6.

In Fig. 11.15, the terms ATT refers to attenuators that accurately adjust the power to the needed levels, while the term hybrid refers to analog networks that sum up the signals.

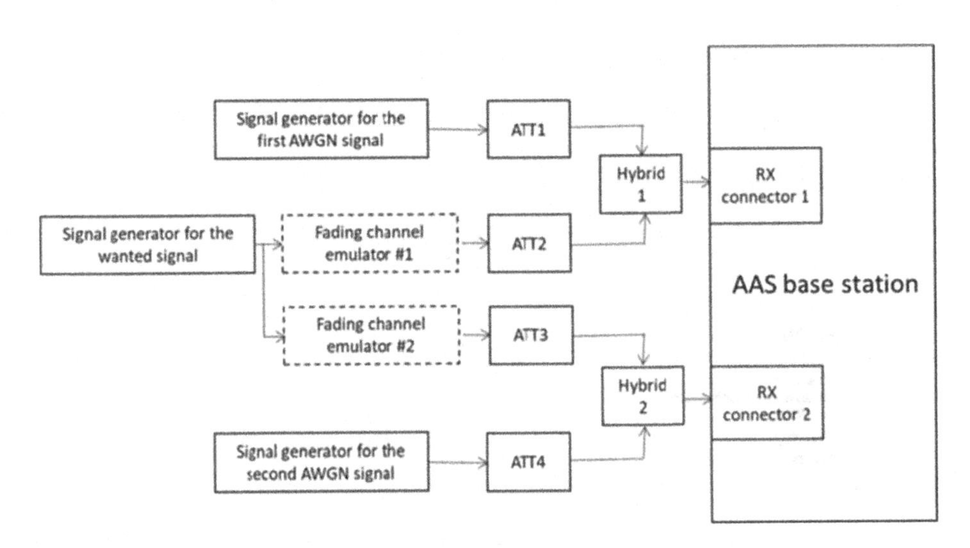

FIGURE 11.15

Example test setup for testing UL 2TX MIMO reception using conducted testing.

The demodulation requirements require modeling of an independent fading channel between every transmitter and every receiver. With current state-of-the-art testing technology, it is not possible to emulate fading channels within a test chamber in a controllable and reproducible manner.

OTA demodulation tests are defined in the current specifications in a simplistic manner. An anechoic OTA test chamber is assumed. The stimulus signal and AWGN are transmitted to the BS with independent fading channels for each polarization (the independent fading channels are emulated on the conducted signals by test equipment prior to transmission in the chamber). This approach relies on polarizations being orthogonal at the transmitter and receiver. It is not possible to create more than one independent fading channel per polarization because it is not possible to create more than one path to all co-polarized receivers within the chamber, and hence all receivers with the same polarization must receive the same signal with the same fading channel. Thus at most 2 RX diversity can be tested, even though the AAS base station may have many more receivers. Any number of UE transmitters can be modeled, since fading channels can be generated independently by the test equipment per transmitter and combined for each demodulation branch prior to transmission in the chamber. Thus for example, 2×2 MIMO can be tested as illustrated in Fig. 11.16.

11.5.4 BEAM QUALITY ASPECTS

As was discussed in previous sections, the 3GPP requirement on EIRP accuracy includes declaration of test beams that are transmitted and measured in an anechoic environment. The intention of the 3GPP requirement is to provide a basic test of the ability of the RF components of the base station to transmit coherently. The compliance testing does not include any standardized measurement or requirement relating to the correct operation of the beamforming, or "beam quality."

"Beam quality" is non-trivial to define and compare. A base station must transmit to at least two types of UE. Firstly, it must transmit to or receive from UEs whose position is not known. Such transmissions must cover the whole intended cell area, though not necessarily all at the same time. Examples of transmission to/from UEs with unknown position are SSB and RACH. Secondly, traffic is generally sent to/from a UE for which more information is known in terms of the channel state and direction. The propagation channel can vary with frequency and time and to maximize throughput; the base station must spread radiation optimally according to the instantaneous state of the channel. Figs. 11.17 and 11.18 depict an example of transmissions to an FDD UE using subband-specific precoding and a TDD UE using reciprocity beamforming, respectively. As can be seen in the figures, the radiation is non-trivial and will also vary rapidly in time. Even for base stations implementing a grid of beams type beamforming, the optimal beam direction(s) will vary in time and frequency.

It is necessary for the base station to both provide basic coverage for system information for UEs whose position is unknown and optimize the transmission of user data to known UEs. Definition and measurement of metrics suitable for assessing the effectiveness of beamforming is a complex task, and lies somewhat outside of the normal scope of conformance testing, since it does not relate to aspects such as co-existence and management of interference. For these reasons, to date the conformance framework does not include any requirements or metrics relating to beamforming performance.

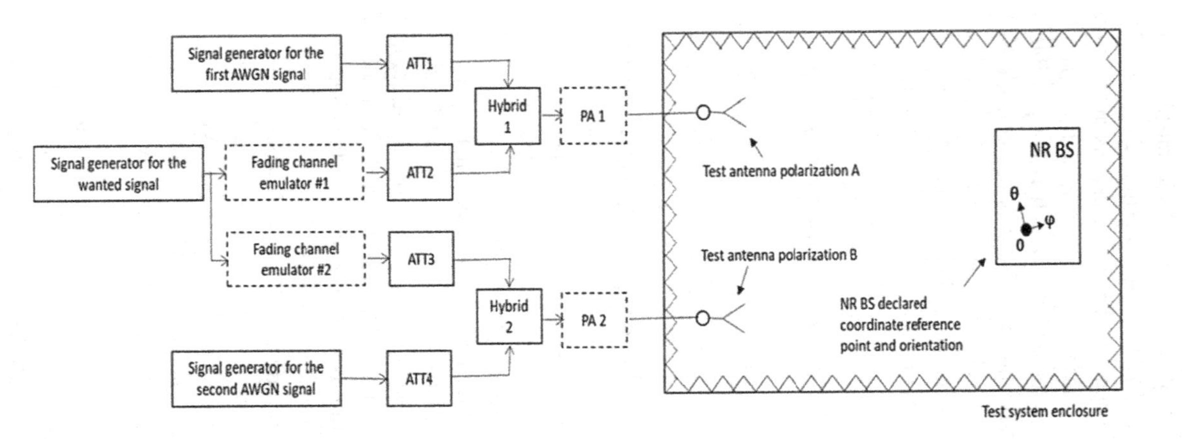

FIGURE 11.16

OTA test setup corresponding to the conducted setup of Fig. 11.15 assuming that the two BS receivers are on orthogonal polarizations.

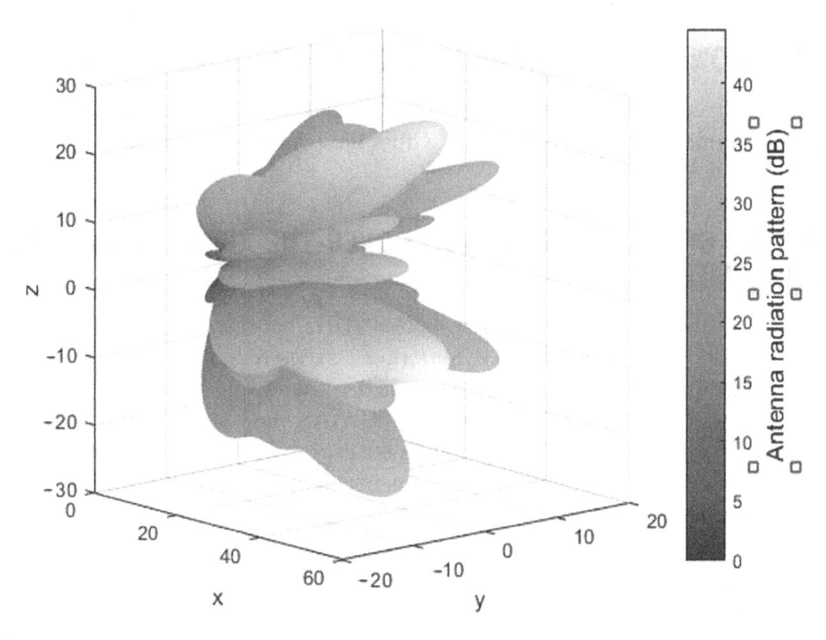

FIGURE 11.17

Example of instantaneous radiation pattern for reaching a single-user of an FDD sub-band precoding transmission.

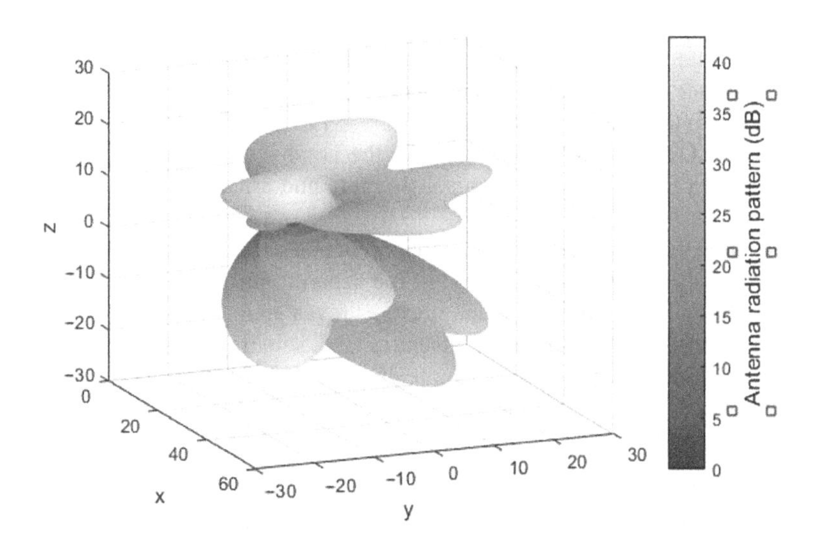

FIGURE 11.18

Example of instantaneous radiation pattern for reaching a single-user of a TDD transmission based on singular value decomposition and reciprocal beamforming.

11.6 DERIVATION OF REQUIREMENT THRESHOLDS AND LEVELS

The previous section described the concepts used for defining OTA requirements. This section describes how requirement levels have been set within the context of these conceptual approaches.

11.6.1 MAPPING OF FREQUENCY RANGE 1 OVER-THE-AIR TO CONDUCTED REQUIREMENTS

Systems such as 2G, 3G, and 4G operating at below 6 GHz have operated for many years and, as discussed in Section 11.3, a well-established conformance requirement framework exists. Deployed equipment is designed based on the assumed co-existence framework (in terms of emissions levels) behind the legacy non-AAS requirements. 4G and 5G AAS base stations will operate within the same bands and must co-exist with existing equipment and offer better network performance. It is necessary to take the existing conducted requirements as a baseline for deriving adapted AAS requirements in these bands.

The AAS requirements, as conceptually described in Section 11.5, are related to the existing requirements based on a number of general principles.

Transmitter unwanted emissions requirements that are expressed as TRP are in general related directly to the conducted transmitter unwanted emissions requirements. Thus the conducted requirement levels are used as a basis for the OTA levels. Since the unwanted emissions TRP requirements are set on the BS as a whole, for some of the emission limits an adjustment of 9 dB is applied to translate the conducted limit to an OTA limit, as described in Section 11.5.2.

The co-location unwanted emissions limits expressed at the output of the co-location test antenna are related to the conducted limits by assuming a 30 dB coupling loss between the BS output and the test antenna (i.e., the OTA limit is equal to the conducted limit reduced by 30 dB). Similarly, for transmitter intermodulation, the conducted intermodulation level applied at the input to the co-location test antenna is increased by 30 dB.

The signal quality OTA limits are the same as the conducted requirement limits, since these requirements are related to the transmitted signal structure and are not impacted by being measured over the air.

In the receiver direction, conducted receiver requirements are specified in terms of signal levels at the receiver antenna connector. Investigation of the subarray properties of different types of subarrays relating to different implementation architectures revealed that although the worst-case direction of arrival may differ for different architectures and subarray radiation patterns, the signal levels experienced at the radio receivers are quite similar. This implies that receiver requirements for FR1 can be verified by illuminating the base station in known directions and adjusting the strength of the OTA stimulus signal according to an assumed subarray gain.

The subarray gain assumption is derived based on declarations of the spatial behavior of the base station. A so-called "sensitivity range of angles of arrival (RoAoA)" is declared, which is a range of angles within which the sensitivity is no more than 3 dB different from the sensitivity in the boresight direction, assuming that all combining (both analog and digital) is active. The contour of the "sensitivity RoAoA" corresponds approximately to the envelope of the maximum receive power that can be achieved with combining, which is directly related to the element or subarray

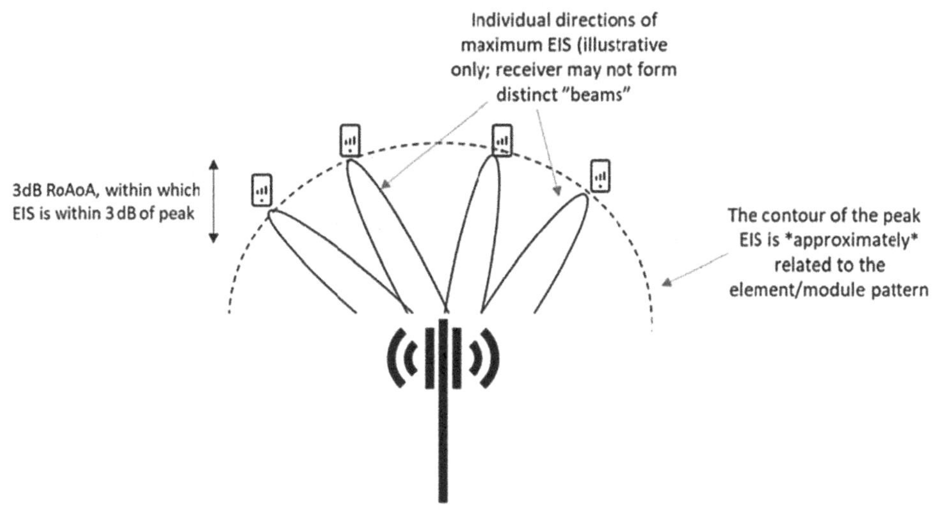

FIGURE 11.19

Illustration of the relation between sensitivity RoAoA and directions of maximum sensitivity.

pattern of each of the individual subarray receivers. From the declared envelope contour, an assumed subarray beamwidth is deduced. From the assumed subarray beamwidth, a nominal subarray gain is calculated, which relates beamwidth to gain.

Fig. 11.19 illustrates the relation between the directions at which maximum sensitivity is ≤ 3 dB lower than maximum sensitivity in boresight. ("beams" are included in this figure for illustration, but it should be noted that in practice, with minimum mean square error (MMSE) combining, UL radiation patterns can be more complex than simple beams). The contour of equivalent isotropic sensitivity (EIS) in different directions is approximately related to the element pattern. It was decided in 3GPP that this approximation is near enough to base a calculation of nominal element gain upon.

The calculated subarray gain is applied to the signal levels of the conducted requirement levels to derive OTA levels. In the 3GPP specifications, the nominal subarray gain is termed $\Delta_{OTAREFSENS}$ (for AAS and NR FR1). In this manner, based on declarations of the observable behavior of the base station, OTA receiver requirements have been defined (Figs. 11.20–11.22).

Using the above approach, OTA levels for both the wanted signal and the interfering signal can be derived. However, the assumed subarray gain for the wanted signal leads to a level much higher than the best achievable sensitivity, because the wanted signal will also experience gain from combining in the baseband. On the other hand, the impact of the interfering signal is on the individual radio receivers and the subarray gain is the correct transformation factor.

To provide greater requirement coverage, in parallel with the requirements based on reference sensitivity, an additional set of receiver requirements are set based on the declared best achievable, "minimum" sensitivity, minSENS, which includes baseband combining. For these parallel requirements, the interferer levels are also adjusted.

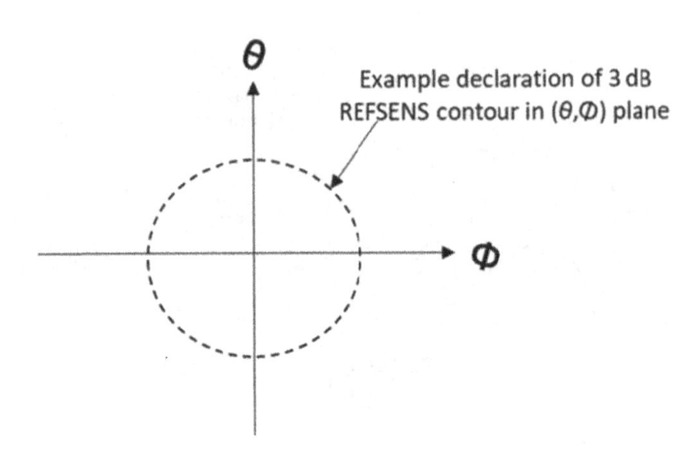

FIGURE 11.20

Example declaration of reference sensitivity (REFSENS) RoAoA in (θ,Φ) plane (i.e., zenith, azimuth plane).

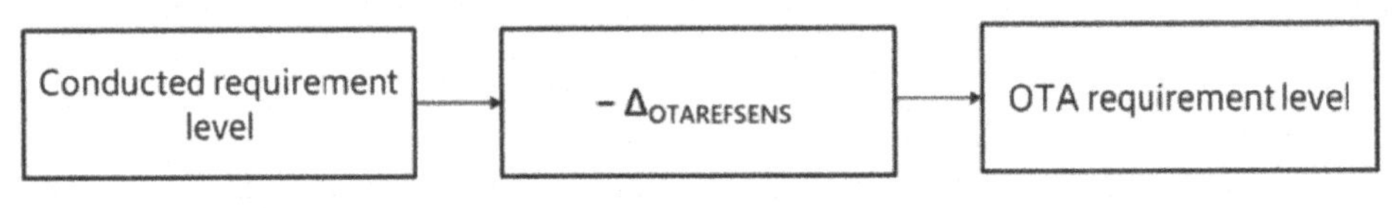

FIGURE 11.21

Conducted requirement levels are adjusted to OTA levels based on $\Delta_{\mathrm{OTAREFSENS}}$.

The first set of requirements based on reference sensitivity and the assumed subarray gain can be seen as testing the dynamic range of the receiver and its ability to handle high-power interferers. The second set of requirements based on minimum sensitivity tests the selectivity of the receiver, because they are applied with the lowest possible OTA wanted signal level and with an interferer whose power relative to the OTA level is the same as for the blocking requirement.

Out-of-band blocking is treated differently since, as described in Section 11.3.4, the interferer field strength is frequency-independent.

11.6.2 ADVANCED ANTENNA SYSTEM FOR MILLIMETER-WAVE FREQUENCIES

A new dimension for 5G is the inclusion of frequency bands allocated in mm-wave frequency ranges. Due to propagation conditions at higher frequencies high beamforming gain is a necessity to combat the propagation losses and thus an AAS type of base station is the only solution for these frequency ranges. The physical size of the antenna arrays in the mm-wave frequency range is relatively small due to the wavelength dependency of the radiating elements and the separation between the radiating elements being around half a wavelength (to avoid grating lobes). A high level of circuit integration and a compact building practice are needed to construct mm-wave

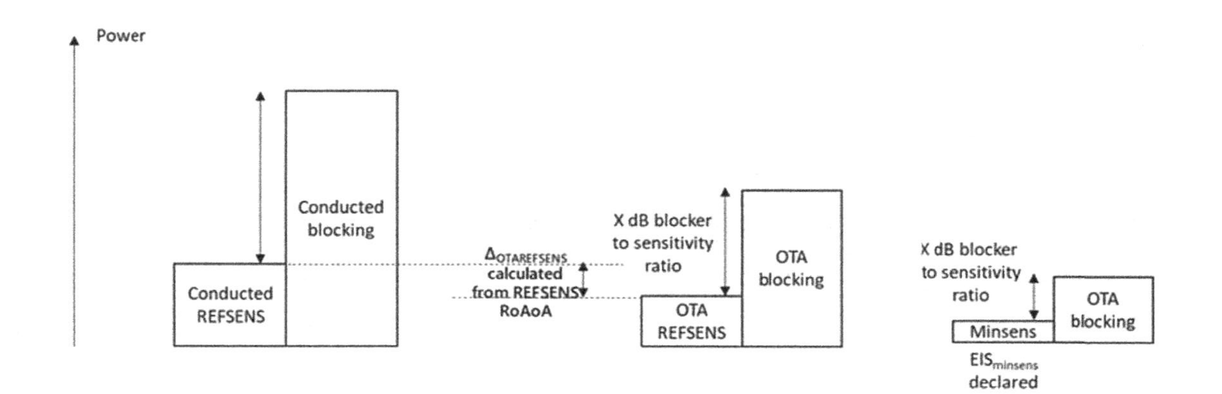

FIGURE 11.22

Illustration of the relation between the conducted RX requirement levels, OTA levels based on REFSENS, and OTA levels based on minSENS.

systems that incorporate a very large number of transceivers and radiating elements (antennas). In addition, the AAS base stations designed for mm-wave frequencies require careful and often complex considerations regarding power efficiency and heat dissipation within the small area/volume. These considerations are of increased importance in comparison with systems designed for lower frequencies such as FR1.

In general, RF components and subsystem performance degrade with increasing frequency. For example, PA output power and efficiency, LNA noise figure (NF) and linearity, phase noise (PN), etc. all degrade at higher carrier frequencies. In some cases, these are overcome by antenna array size, whereas in other cases, RF requirements have to be less stringent compared to those at lower frequencies [9,10].

The frequency bands specified for FR2 are quite wide and are in the range of few GHz. The minimum and maximum carrier bandwidth supported in 3GPP specifications are 50 and 400 MHz, respectively. Even larger carrier bandwidths may be added in future standards. In addition, carrier aggregation is also supported, implying the need to support larger RF bandwidth than maximum carrier bandwidth. The larger bandwidths available at mm-wave frequency bands in FR2 present challenges such as the performance of PAs and LNAs and also the scale of the data conversion interfaces between analog and digital domains in both the receiver and the transmitter chains.

Analog filtering for mm-wave is also a challenge considering the need for highly integrable filters in a highly size constrained design. The achievable attenuation and needed guard bands at the band edges are significantly higher compared to systems for lower frequency bands in FR1.

In addition, there are complex parameter interdependencies between the transmitter unwanted emissions and the achievable output power and efficiency. In the case of the receiver, the NF, linearity, dynamic range, and bandwidth are also interdependent. These interdependencies play an important role when specifying and designing a mm-wave frequency AAS system.

In the following subsections, some important performance parameters and their dependencies are further elaborated.

11.6.2.1 Power amplifier aspects

The output power capability of PAs for a given integrated circuit technology degrades by roughly 20 dB per decade, as shown in Fig. 11.23. The fundamental reason behind the degradation is the conflict between power capability and frequency, due to the so-called Johnson limit, which is technology-related. Higher operational frequencies require smaller transistor device geometries, which subsequently result in lower operational power capability in order to prevent dielectric breakdown from the increased electric field strengths.

mm-wave integrated circuits have traditionally been manufactured using so-called III–V materials, that is, a combination of elements from groups III and V of the periodic table, such as Gallium Arsenide (GaAs) and more recently Gallium Nitride (GaN). Integrated circuit technologies based on GaN are more expensive than conventional silicon-based technologies and offer limited integration options for more sophisticated circuits, for example, digital or mixed-signal circuits. However, GaN is now maturing rapidly and delivers power levels an order of magnitude higher than those achieved in mm-wave complementary metal oxide semiconductor (CMOS).

Fig. 11.24 shows the saturated power-added efficiency (PAE) as function of frequency for different technologies.

PAE is defined as $\text{PAE} = 100 \times \left([\text{P}_{\text{OUT}}]_{\text{RF}} - [\text{P}_{\text{IN}}]_{\text{RF}}\right)/[P_{\text{DC}}]_{\text{TOTAL}}$.

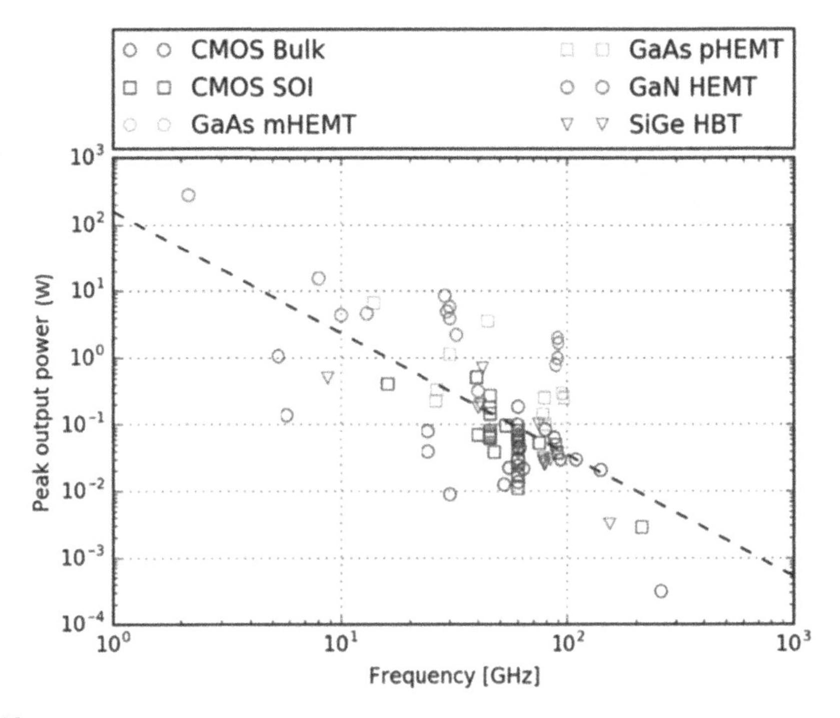

FIGURE 11.23

Peak output power of published power amplifier designs versus frequency for various semiconductor technologies. The dashed line illustrates the Johnson power scaling slope (−20 dB per decade).

Thus at mm-wave frequencies the available output power is limited by semiconductor technologies and the efficiency degrades with frequency.

The peak saturated power does not consider many aspects in a practical implementation and the achievable average power needs, in general to be roughly 10dB lower compared to saturated peak power in order to fulfill the required modulation quality (EVM), ACLR, or necessary linearization range and bandwidth. Considering the nonlinear behavior of the PA in terms of so-called amplitude modulation (AM)-AM/AM-phase modulation (PM) properties that describe the gain/phase characteristics of the PA versus input power, respectively, significant power back-off is necessary to achieve the linearity needed to meet the 3GPP and regulatory requirements. For FR2 with large array sizes with many transceivers, complex linearization algorithms are of limited use due to large signal bandwidths and power consumption of such algorithms. However, less complex algorithms can be used to reduce the power back-off levels to a lesser degree.

Fig. 11.25 shows output power and PAE as a function of ACLR, which is a measure for linearity of the PA, for CMOS and GaN type PAs operating at 30 GHz.

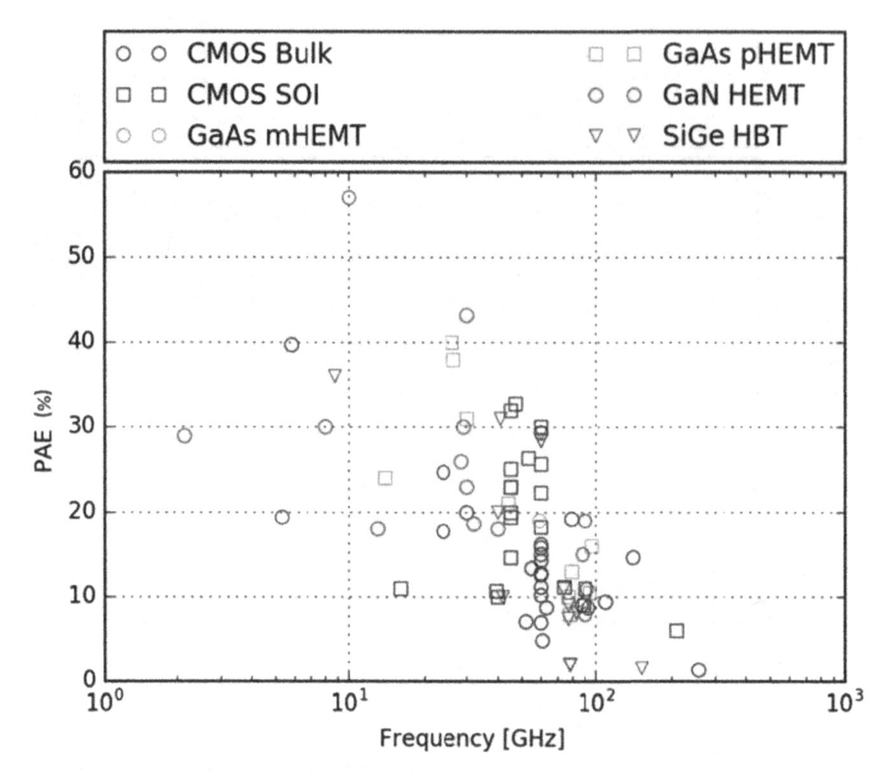

FIGURE 11.24

Saturated power-added efficiency versus frequency for published power amplifier designs using various semiconductor technologies.

Reproduced by permission of © 3GPP. © 2017. 3GPP™ TSs and TRs are the property of ARIB, ATIS, CCSA, ETSI, TSDSI, TTA, and TTC who jointly own the copyright in them. They are subjected to no further modifications and are therefore provided to you "as is" for information purposes only. Further use is strictly prohibited.

As shown in Fig. 11.25, overly strict requirements on linearity would result in poor output power and low efficiency. Considering the thermal behaviors in the small area/volume needed for mm-wave AAS base station array, the complex relation between linearity, PAE, and output power, heat dissipation will be challenging to handle. Even though the achievable output power for GaN technology is significantly higher than for CMOS and even though the PAE for GaN technology is slightly better than that of CMOS, thermal challenges due to higher power density for GaN in a size-limited design will still be significant.

As GaN technology provides higher output power compared to CMOS, the subarray antenna topology for a given fixed radiated power is different. The number of GaN transmitters need to be lower (to exploit the high-power capability), which will necessitate a larger number of radiating elements per subarray. Since the radiated pattern from a fixed subarray cannot be steered, a larger

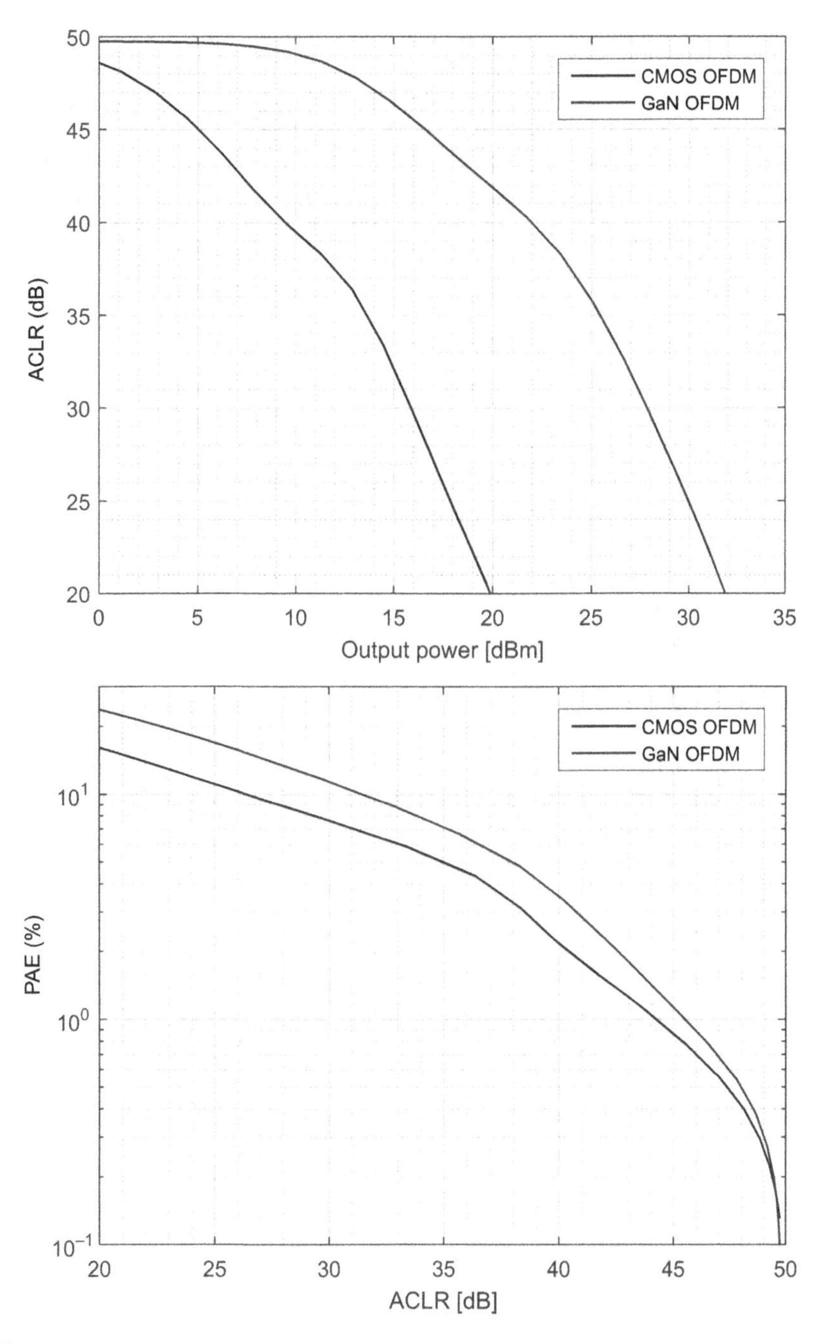

FIGURE 11.25

ACLR versus output power and PAE versus ACLR for CMOS and GaN power amplifiers ~30 GHz using OFDM-modulated input signal.

number of radiating elements per subarray may pose some restrictions in the beamforming capability.

Taking these considerations into account, as well as the results of co-existence scenarios that investigated the needed adjacent channel interference ratio, which is BS ACLR combined with UE ACS levels needed to ensure good co-existence performance for mm-wave systems, the base station ACLR requirement for FR2 was set at 26–28 dB. This requirement ensures good co-existence whilst enabling reasonable power-efficient transceiver designs. It is noteworthy that the base station ACLR is set to 45 dB for FR1.

11.6.2.2 Frequency generation and phase noise

In any communication system operating at higher frequencies, there is a need to up- and downconvert the frequency by using a *local oscillator* (LO) in the transceiver design. In any LO, *phase noise* (PN) is generated. PN is a description of signal stability in the frequency domain and can affect the system performance. On the transmitter side, PN contributes as an impairment to the signal quality. This is expressed as EVM and can result in restrictions on the maximum achievable SINR and consequently on the possibility to use higher modulation orders.

On the receiver side when a strong interferer is present close to the wanted signal, the interferer will not be significantly attenuated by the filter before the mixer, used for up and downconversion. The effect that PN is superimposed on the interferer because of so-called reciprocal mixing consequently leads to increased noise in the receiver and degrades sensitivity or SINR as a result. The separation between the interferer and the wanted signal is an important parameter for the level of degradation. The closer the interferer to the wanted signal, the higher the degradation (Fig. 11.26).

For mm-waves, the achievable PN performance affects the selection of numerology and SCS. The current 3GPP specifications allow for 60 and 120 kHz SCS for PDSCH, while 120 and 240 kHz for transmission of SSB.

PN has the following characteristics that are important to consider:

- PN increases by 6 dB every time when *frequency* doubles.
- PN is inversely proportional to signal strength of the LO signal, P_s.
- PN is inversely proportional to the square of the loaded quality factor of the resonator, Q.
- $1/f$ noise upconversion gives rise to close-to-carrier PN increase (at small offset).

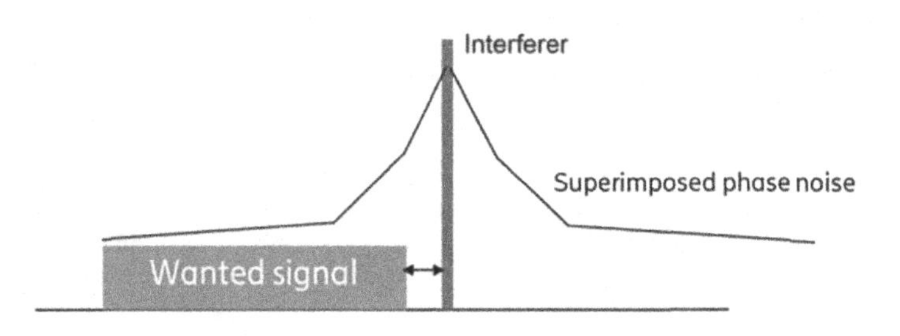

FIGURE 11.26

Reciprocal mixing of phase noise in the receiver.

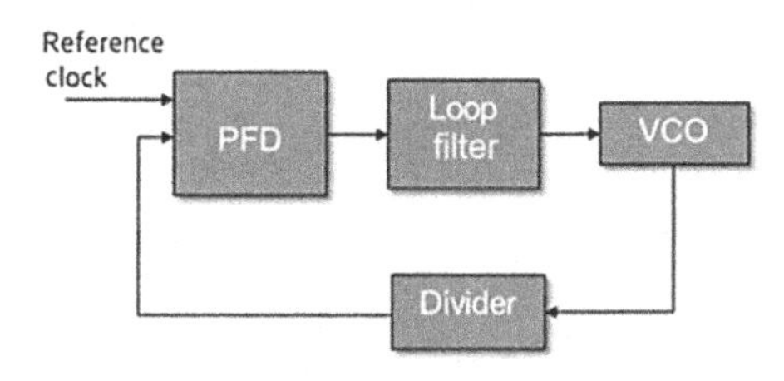

FIGURE 11.27

Block diagram of a phase-locked loop (PLL).

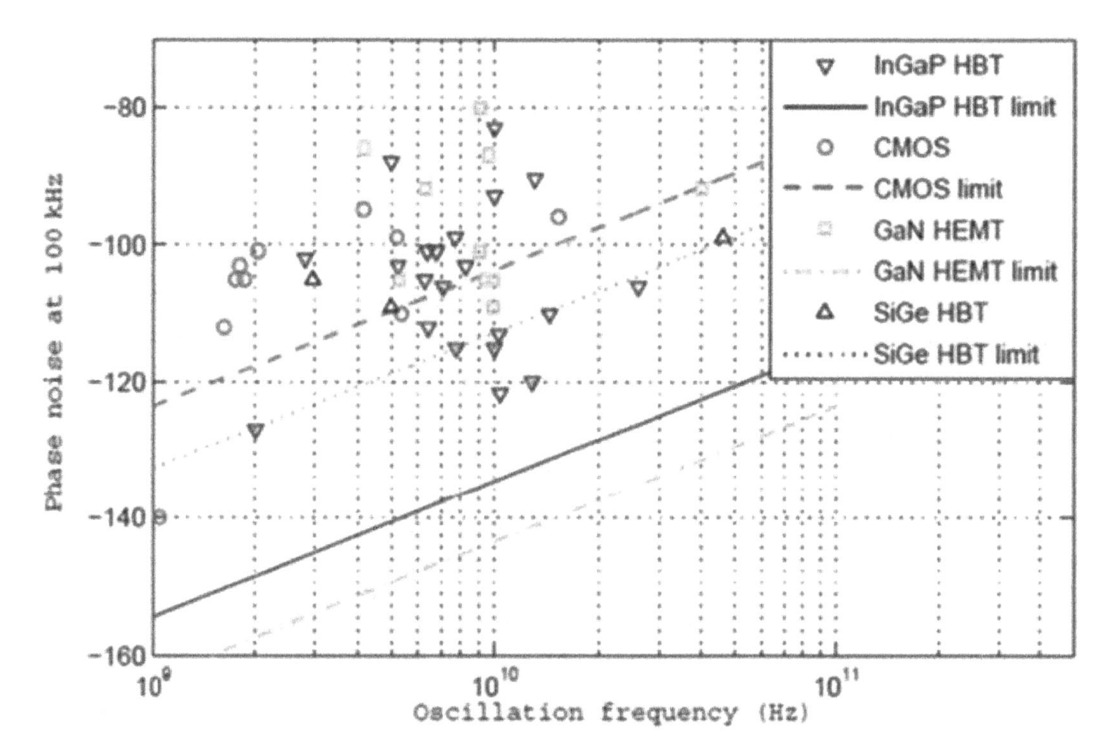

FIGURE 11.28

Phase noise performance at an offset of 100 kHz versus oscillation frequency for a number of published VCO designs using different technologies.

1/*f* noise or flicker noise is found in electronic devices and arises due to fluctuations associated with voltage or current. The magnitude of 1/*f* noise decreases with frequency and therefore it is referred to as 1/*f* noise.

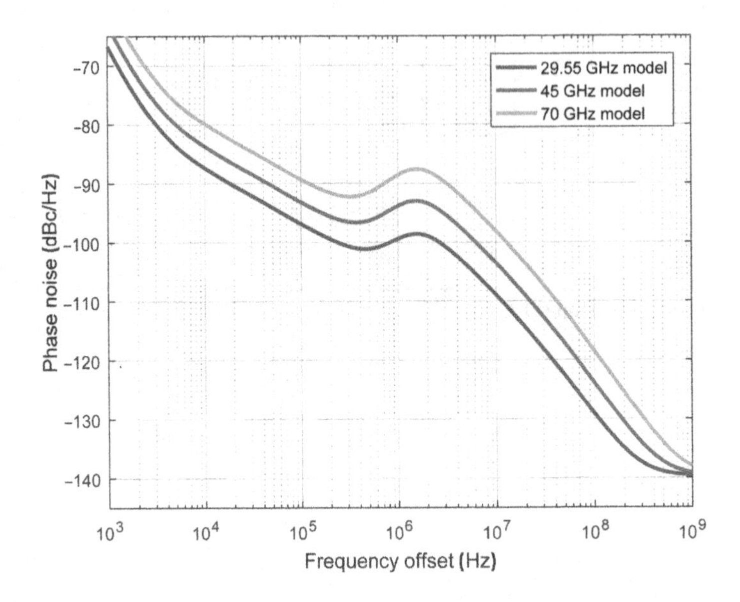

FIGURE 11.29

Phase noise models for different frequencies.

A common approach to suppress the PN is to use a phase-locked loop (PLL), as shown in Fig. 11.27, in which the voltage-controlled oscillator (VCO) is locked to a highly stable, low-PN reference clock.

The total PN of the PLL output is composed of different contributions from the VCO outside the loop bandwidth and the reference oscillator inside the loop. A significant noise contribution is also added by the phase/frequency detector and the frequency divider. The PN level and behavior depend on its frequency offset from the LO and rolls-off with increasing offsets.

PN performance versus oscillator frequency for different technologies is shown in Fig. 11.28 [10].

To examine the impact of PN, it is useful to construct a model of the degradation. The behaviour of an example PN model versus frequency offset for ~30, 45, and 70 GHz frequencies is given in Fig. 11.29.

The PN will degrade by ~20 dB when going from 3 to 30 GHz, assuming a constant power budget for generating the LO. To combat PN in 5G, the physical layer standard has incorporated the phase tracking reference signal (PT-RS) to facilitate common phase error (CPE) compensation in the receiver and to partially mitigate the increased PN for mm-wave frequency bands. Despite the CPE compensation in the receiver, PN still highly affects the signal quality in particular for higher-order modulations, the implementation of, for example, 256-QAM may be challenging.

It is worth mentioning that the architecture for LO generation in an AAS base station, and in particular for mm-wave frequency bands can also affect the design and performance over the air. A

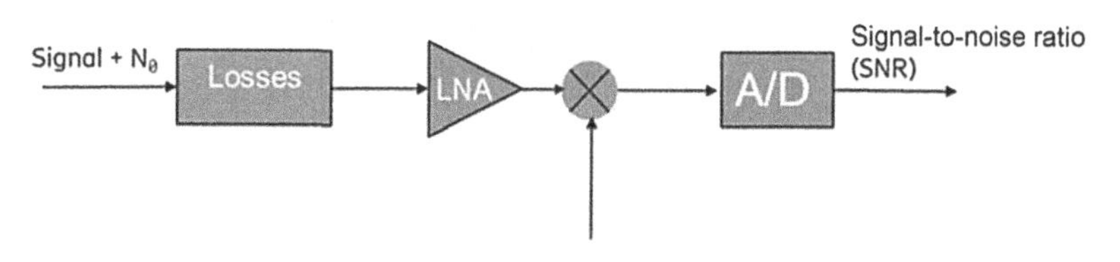

FIGURE 11.30

Simplified receiver model.

distributed LO architecture with LO generation implemented in each individual subarray will result in partial uncorrelated PN contribution over the air. In such architectures the performance of each LO may be poor, because the LO PN is related to the power consumption and the fact that power consumptions and dissipation must be limited due to thermal and power efficiency concerns. On the other hand, if individual PN contributions from the distributed LOs are uncorrelated, then a beamforming gain in EVM, as described in Section 11.7.2, may occur. With a centralized architecture, in which a single LO generator is used, there is no beamforming-related performance benefit over the air as PN will be correlated. However, in this case a LO with better performance at the expense of higher power consumption can be used.

11.6.2.3 Millimeter-wave receiver aspects

The mm-wave receiver performance, in the same way as for the transmitter, degrades over frequency. Due to the need for highly integrated and compact building practice for mm-wave AAS base stations with many transceivers, other often complex dependencies are essential to consider to be able to handle the power consumption, heat dissipation, and consequent thermal challenges.

One of the most important receiver performance parameters is noise figure (NF). NF is a measure of how SNR is degraded due to components and building blocks in the receiver chain. NF is not only determined by the LNA but also other components and building blocks in the receiver chain. For example, the insertion losses from, for example, filters, isolators, and signal routing before the LNA will linearly degrade the NF.

To assess the NF for mm-wave frequency bands, a simplified receiver model based on cascaded front-end (FE), analog/RF receiver (RX), and ADC can be used to derive typical NF values (see Fig. 11.30). Note that the intention with the model is not to replace a rigorous analysis but rather depict the main parameter interdependencies which, if evaluated still give quite realistic indicative values.

The impact of various impairments arising in the cascaded receiver chain can be calculated by considering Friis formula [11]. The Friis formula is used to calculate the total noise factor of a multistage LNA and can also be used to arrive at typical overall NF values for mm-wave frequency ranges.

In summary, when considering the overall typical NF some fundamental principles should be considered.

The NF is not only determined by the LNA, but also bandwidth, linearity, and dynamic range requirements will play an important role.

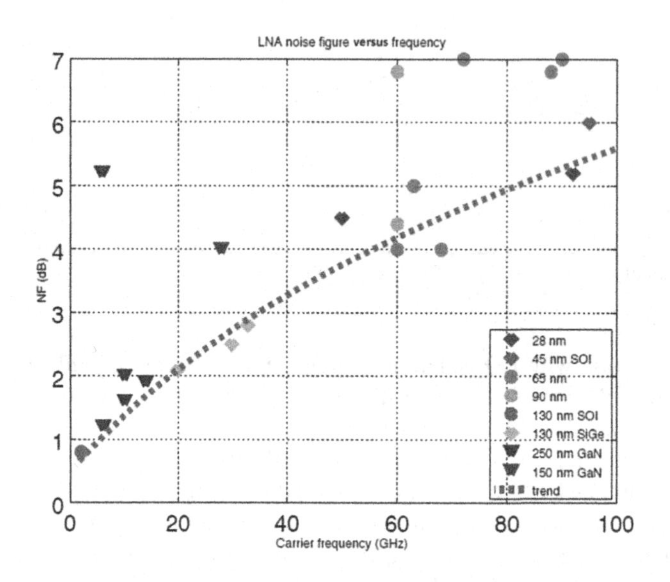

FIGURE 11.31

Published LNA-only noise figure for different technologies over frequency.

A full receiver chain all the way up to the radiating elements must be addressed as all parts in the chain will contribute to the overall receiver performance including switch, routing, filter losses, etc.

NF will trade against other parameters such as cost, power consumption, and heat dissipation. As an example, in the context of NF, it might be possible to reduce the noise contribution from an ADC by using more bits, but this will have a significant implication in terms of power consumption and heat dissipation. A single added bit to an ADC will result in a four times higher power consumption. The ADC power consumption is proportional to BW 2 $\cdot$DRADC, where DRADC is the ADC dynamic range.

There is also a trade-off between NF and linearity of the receiver, which is typically measured using a metric called input third-order intercept point (IIP3). The IIP3 can be improved at the expense of a slightly higher NF.

Considering recent state-of-the-art LNA NF publications, Fig. 11.31 presents the LNA-only NF for different semiconductor technologies over frequency.

It is evident that the LNA noise figure as expected degrades with frequency. In addition to published data for LNAs, also considering other impairments affecting the overall noise figure, for example, noise coming from mixers, antenna switches, signal routing, and due to filter losses, the typical noise figure for mm-wave frequency ranges around 30 and 45 GHz considering the simplified receiver model will be around ~10 and ~12 dB, respectively. It should be noted that the noise figure for FR1 frequencies is assumed to be 5 dB.

11.6.2.4 Millimeter-wave filtering aspects

RF filters have always been embedded in the analog front end (FE) in communication systems to either mitigate harmful interference generated in the transmitter or to create enough protection for the receiver against high level of interference. For FDD bands, the filters (usually duplex filters) are used to mitigate out-of-band unwanted emissions arising from, for example, nonlinearities in the transmitter chain, resulting in intermodulation products, harmonics generation, etc. In the receiver chain, filters are deployed to protect the receiver from the high-power leakage from the transmitter and to suppress high power out-of-band interferer at adjacent or other frequency bands (either cellular or noncellular). For TDD bands, a bandpass filter is used.

For any AAS base station, the implementation of filters is problematic, since there are many transceivers, and each transceiver must have a separate filter. Focusing on mm-wave frequency bands the implementation of filters becomes even more challenging, as discrete mm-wave filters are generally too bulky and pose challenges related to the integration into compact array structures. Fig. 11.32 illustrates two principal approaches to the integration, one in which filters are embedded inside a printed circuit board (PCB) carrying antenna elements on one side and RF circuits on the other side, and one in which filters are embedded in the antenna.

For AAS, RF filters in front of the PA or LNA, in conjunction with radiating elements, must provide high performance within the passband, particularly low insertion loss. Insertion loss directly affects the achievable EIRP and OTA sensitivity. For filters after the LNA or before the PA with the purpose of suppressing I/Q leakage or unwanted mixing products the performance is less critical.

Sharp RF filtering at mm-wave frequency bands in array applications tends to give unacceptably high insertion loss. Ideally a combination of low insertion loss and steep slopes is desired. In practice, a trade-off must be made between these two goals. Insertion loss (expressed in dB) is roughly inversely proportional to the relative bandwidth (bandwidth/center frequency). This implies, firstly, that insertion loss is higher, the higher we go in frequency except if the bandwidth is increased in proportion to the center frequency. Secondly, it implies that, for a fixed center frequency, the wider the passband the lower the insertion loss gets. The drawback in widening the passband is that it leads to a less steep slope (dB/GHz) in the attenuation outside the passband. The trade-off between insertion loss and steepness is illustrated by the simulated results in Fig. 11.33, which compares two bandpass filters based on ideal (but lossy) LC resonators, one with 500 MHz 3-dB bandwidth and one with 4×500 MHz 3-dB bandwidth. To increase the slope for a given relative bandwidth one can attempt to increase the order of the filter (number of resonators), but that will lead to

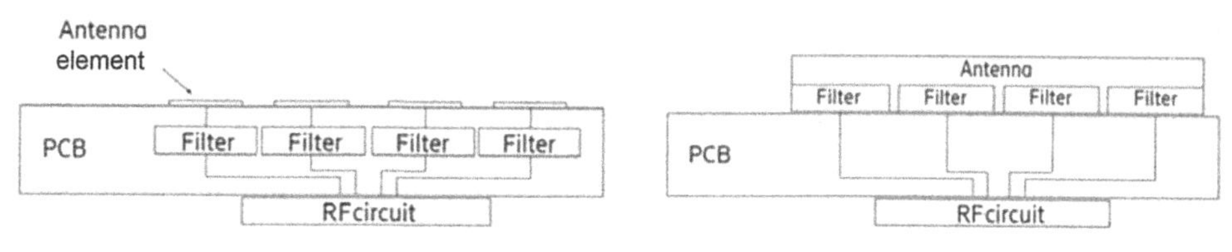

FIGURE 11.32

Illustration of two possible approaches for integrating filters on each antenna branch in a compact millimeter-wave array antenna system, with filters embedded in a PCB (left) and filters embedded in the antenna (right).

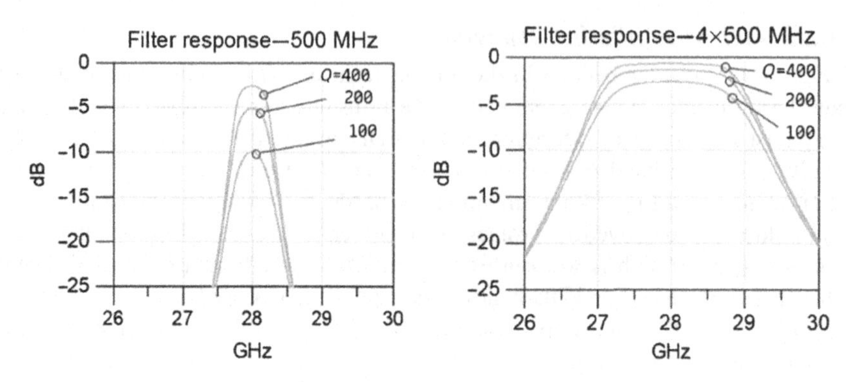

FIGURE 11.33

Filter response (S21) of a third-order bandpass filter, for different Q-values in the resonators, for a bandwidth of 500 MHz (left) and 4 × 500 MHz (right).

increased insertion loss (roughly proportional to the number of resonators) and a larger total volume.

To significantly improve both in terms of insertion loss and steep slope one should improve the Q-value. This is a fundamental property of the resonators in the filter, and it is highly dependent on the choice of filter technology. For PCB technologies Q-values can range from around 50 up to around 150, while for low-temperature cofired ceramic (LTCC) technology it is possible to reach up to around 200. There are technologies that provide a Q-value of 400 or even higher, for example, metal cavity filter and dielectric resonator filters, but those are generally difficult to integrate into a dense antenna array configuration and therefore require continued research. The influence of Q-value on insertion loss is illustrated in the examples in Fig. 11.33, where each trace corresponds to a certain Q-value (100, 200, and 400).

Frequency precision is also a critical aspect of any filter technology, and it must be considered when comparing alternatives, and when setting margins in a design. Variations in manufacturing processes along with temperature drift will give small errors in physical dimensions, and in material properties in the filter, which in turn will shift the band edge of a filter such that unwanted signals tend to pass and/or wanted signals tend to be blocked. It is reasonable to assume that individual tuning of filters resonators, which is done in many applications, is not feasible for array antennas operating at mm-wave frequencies due to the large amount filters, the small size, and the need for compact integration into an array without any test ports. Variations are therefore larger than for low-frequency applications and for single branch applications.

As an example, consider the filter response shown in Fig. 11.32. This is the result of a simulation of a fifth-order bandpass strip-line filter implemented in PCB technology, using high-performance microwave dielectrics to achieve a Q-value of 120 at 28 GHz. The filter was designed to cover 5G band n257, with 3 dB insertion loss, and to provide 20 dB suppression with a guard band of 2 GHz. We note that the filter provides enough suppression in the earth exploration satelite services (EESS) band at 23.6–24.0 GHz. The traces give the transmission response (S21) and the input matching (S11), respectively. Suggested requirement limits are indicated by dotted lines. Design margins were added to allow for variations in manufacturing. A Monte-Carlo simulation

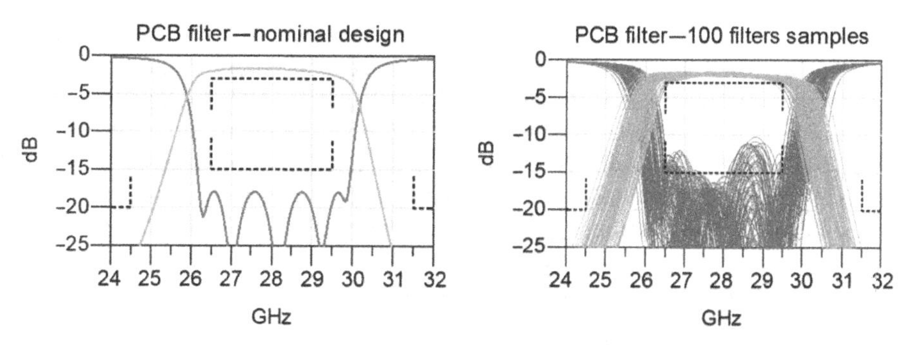

FIGURE 11.34

Estimated performance of a fifth-order bandpass filter implemented in PCB technology, for nominal design (left) and for 100 random samples.

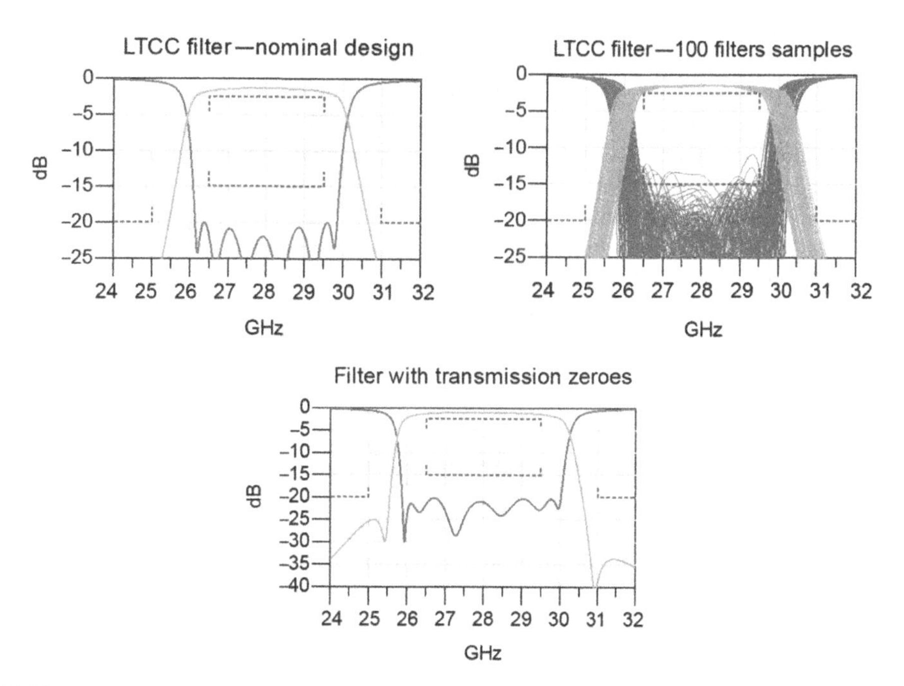

FIGURE 11.35

Estimated response of a sixth-order bandpass filter implemented in LTCC technology, for nominal design (upper left) and for 100 samples with random errors due to variations in manufacturing (upper right), and an example of nominal response with transmission zeroes (lower middle).

with realistic tolerance assumptions (1.5% in dielectric constant, 8% in line width, and 7% in dielectric thickness), for 100 random filter samples, produced the variation shown in the right-hand side of Fig. 11.34. Even though a relatively large guard band of 2 GHz to provide 20 dB suppression was assumed, some of the filter samples turned out to violate the requirement. Such violations can be accepted under certain conditions related to the randomness in production.

Filters based on LTCC technology have potential to provide better precision and better Q-value than filters based on PCB technology, but they are challenging to integrate into the antenna, and will give a major impact on the total cost, considering the large number of filters. A rough estimation of the achievable performance of a sixth-order bandpass filter implemented in LTCC technology providing a Q-value of 200 is shown in Fig. 11.35. The filter was designed to cover 5G band n257, with 2.5 dB insertion loss, and to provide 20 dB suppression with a guard band of ~1–1.5 GHz, with proper margin. In Fig. 11.35, a rough estimation of the possible impact of temperature drift and manufacturing variations assuming frequency variations up to 300 MHz, for 100 random filter samples, is made.

To make the slope steeper one can also consider adding transmission zeroes in the filter design, as exemplified in Fig. 11.35 (lower middle). However, transmission zeroes tend to increase the sensitivity to errors, and it is not clearly beneficial for filters that are not tuned.

As was shown in Figs. 11.34 and 11.35, manufacturing tolerances and temperature drift have a significant impact on performance and should be considered when designing mm-wave filters. Typically, tolerances and temperature variations for filter technologies discussed in this chapter can result in a frequency shift up to 300–500 MHz depending on implementation. This shift must be accommodated within the guard band.

From the above it is clear that there are limitations in performance and capability of mm-wave filters suitable for mm-wave AAS BS in FR2. This has been carefully considered within 3GPP standardization and regulatory bodies when deriving unwanted emissions requirements.

11.7 POSSIBLE RADIO DESIGN BENEFIT FROM MASSIVE MULTIPLE-INPUT MULTIPLE-OUTPUT

Within the wider research community, there has been interest during recent years in the potential for massive MIMO to be built using a large number of low-cost radios with reduced performance, known as "dirty RF." The beamforming gains of massive MIMO have the potential to compensate for reduced radio performance. Such an approach has obvious benefits in some cases, but not for all requirements. This section briefly reviews the potential for such approaches within the context of the full range of standardized radio requirements and regulatory constraints that the BS needs to meet. From an AAS base station testing perspective, the OTA testing in a chamber adds additional implication and limitations that also imply restriction of possible gains (e.g., it is not possible to create a test scenario with multiple layers and users in a multi-path fading environment that could be used to indicate performance in such circumstances).

This section considers the implications of massive MIMO for some important requirements that drive design considerations.

11.7.1 UNWANTED EMISSION

As outlined in Section 11.5.2, the unwanted emissions requirements are defined as TRP. The reasoning, as also outlined in Section 11.5.2, is that for in-band emissions, the statistical impact of the unwanted emissions on adjacent systems is generally correlated to TRP, regardless of the antenna pattern. The basic reason for this is that unwanted emissions impact victims that are generally not in the main beam direction, and the statistical impact of the emissions depends on average radiated power.

The basic mechanism for mitigating reduced radio performance is that unwanted effects do not combine coherently, whereas the wanted signal does. However, for unwanted emissions, noncoherent combining of emissions power may spatially whiten the spatial pattern of unwanted emissions but does not reduce the total emissions power. As described in Section 11.5.2, the emissions pattern will impact the pattern of distortion in a neighbor network, but will not lead to a large impact on the reduction in total capacity of a neighbor network.

Reduced PA performance in the radios will lead to an increase in adjacent channel leakage. Reduction of quantization bits will lead to wideband noise, which will also contribute to unwanted emissions. The fact that there is no spatial effect that reduces total emissions implies that, viewed from the context of in-band emissions requirements based on TRP as a metric, massive MIMO does not mitigate the need for suitable radio performance.

Out-of-band unwanted emissions are likely to be more spatially whitened. The exact pattern will depend on the out-of-band antenna characteristics. As with in-band, the position of an out-of-band victim will not correlate with the in-band energy direction. For out-of-band emissions, there is also no obvious means to reduce the total emissions or impact of the emissions due to massive MIMO.

Co-location-related emissions relate to co-existence between base stations that are close to one another, potentially in the near field of the array. Such emissions are also not mitigated by massive MIMO beamforming effects.

In addition both from standard and regulation perspective, all AAS base station transmitter requirements and in particular unwanted emission–related requirements shall be measured when base station is configured to its declared maximum power capabilities both in terms of EIRP and TRP, which would imply that statistical gains in relation to traffic variations occurring during operation that could in theory improve the typical level of unwanted emissions cannot be considered for compliance to standard and regulatory requirements.

11.7.2 TRANSMIT SIGNAL QUALITY

The transmit signal quality requirements consist of EVM, frequency error, and time alignment error.

EVM is the ratio of distortion to wanted signal. Typically, different EVM requirements are defined for different modulation schemes. The wanted signal experiences beamforming gain. The distortion may not experience coherent beamforming gain, depending on the dominant source of distortion. Thus there exists potential for mitigating radio distortion due to beamforming of massive MIMO for in-band EVM, provided that the distortion sources are not correlated between transmitters. The EVM is used as a metric to capture all impairments within the channel including PA nonlinearities, PN as well as distortion induced by peak reduction or clipping algorithms. The unclipped waveforms of WCDMA, LTE and NR typically have very high PAPR (~10 dB) in downlink. Peak reduction algorithms are needed to reduce the PAPR so that transmitter can be dimensioned for reasonable power efficiency. The peak reduction algorithms that are generally based on a nonlinear time-domain operation generate correlated distortion. The distortion from peak reduction is usually the largest component of the EVM. 3GPP specifies EVM for a single-user and beam. The distortion caused by clipping will combine coherently and thus not benefit from any gains. Also, the distortion induced by PA nonlinearities in the channel is likely to be fairly correlated. Other impairments such as PN assuming that the AAS base station has a distributed frequency generation would benefit from massive MIMO gain. Thus the majority of the EVM is made up of distortion components that add coherently.

If the EVM is evaluated considering more users, beams, or layers, the distortion arising from the correlated components may be more whitened than in the single-user case. However, this does not correspond to the requirements in the standards, which are based on single-user, single-layer transmission. In any case, it is not possible with current OTA testing setups to construct and decorrelate MIMO signals in fading channels, and so even if the specifications would be written differently, conformance testing with multiple layers would not be feasible.

Further development and evolution in OTA testing and OTA chamber types allowing for testing supporting on multiple users/layers and also fading environment might enable additional gain mechanisms to be exploited in the future.

The time alignment error relates to the timing accuracy between different reference signals, as seen by the UE. In some cases, when single layer is operated or virtual antennas at the UE are created by transmitting different beams through all transmitters, the time alignment error may be irrelevant. If different reference signals are transmitted by different array panels, the time alignment error will be relevant and must be met. If there is a large number of transmitters for each panel, there may be some potential for statistical improvement of the per-radio timing error, which also depends on the frequency and clock distribution in the AAS base station architecture.

11.7.3 RECEIVER REQUIREMENTS

For the basic sensitivity requirement, it is important to take into account that the composite OTA sensitivity is determined by the total array aperture and not the number of individual receivers. Consider an array with N antenna elements with element gain $G_{element}$, as illustrated in Fig. 11.36.

If the entire array would be passive combined and passed to the input of a single receiver with noise factor NF, then the OTA sensitivity would be:

$$EIS_{OTA} = -174 + 10\log_{10}(B) + NF - G_{element} - 10\log_{10}(N), \tag{11.3}$$

where B is the signal bandwidth.

At the opposite extreme, the array might contain N individual receivers. If the noise factor of each receiver is NF_{RX}, then the OTA sensitivity per receiver would be:

$$EIS_{OTA,RX} = -174 + 10\log_{10}(B) + NF_{RX} - G_{element} \tag{11.4}$$

Where $EIS_{OTA,RX}$ is the OTA sensitivity of each individual receiver and NF_{RX} is the noise factor of each individual receiver.

And the composite OTA sensitivity:

$$\begin{aligned} EIS_{OTA} &= EIS_{OTA,RX} - 10\log_{10}(N) \\ &= -174 + 10\log_{10}(B) + NF_{RX} - G_{element} - 10\log_{10}(N) \end{aligned} \tag{11.5}$$

Comparing (11.3) and (11.5), it can be seen that the relationship between the OTA sensitivity and the receiver NF is independent of the number of receivers (but does relate to the array size, or total number of elements). Thus there is no gain from massive MIMO in respect of receiver sensitivity given fixed array size.

Other requirements, such as blocking test the dynamic range that can be tolerated by the receiver without going into saturation. Since there is no statistical gain for receivers in saturation

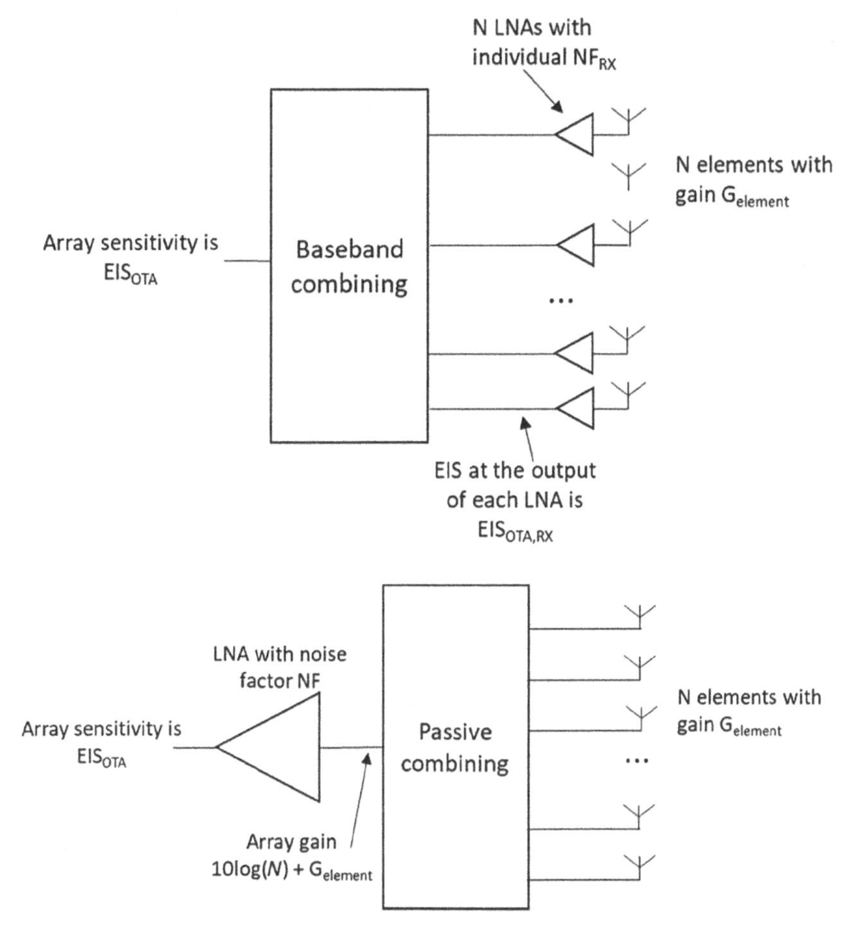

FIGURE 11.36

Example of AAS versus non-AAS receiver combining architecture.

(because signals received through saturated receivers due to blocking would be corrupted), blocking requirements will also limit any potential gain, for example, reducing the ADC bits that is achievable whilst maintaining full 3GPP receiver performance.

11.8 RADIO FREQUENCY ELECTROMAGNETIC FIELD EXPOSURE FROM ADVANCED ANTENNA SYSTEM BASE STATIONS

Like all radio equipment, AAS base stations need to comply with applicable standards and regulations related to human exposure to RF EMF. International standardization organizations such as IEC (International Electrotechnical Commission) and IEEE provide methods to be used to determine that the RF EMF exposure in the vicinity of base stations is in compliance with limits set by national and international authorities. Since AAS base stations apply beamforming and might achieve a larger EIRP in some directions compared to legacy equipment, the instantaneous EMF levels may be higher. However, due to the spatial distribution of UE, the time-averaged RF EMF exposure, which is relevant for regulatory compliance, is significantly lower and similar to levels from non-AAS base stations.

11.8.1 LIMITS ON HUMAN EXPOSURE TO RADIO FREQUENCY ELECTROMAGNETIC FIELDS

The International Commission on Nonionizing Radiation Protection (ICNIRP), which is a formally recognized expert group for the World Health Organization (WHO), develops exposure guidelines based on reviews of the available scientific literature about effects on human health of RF EMF [12]. The recommended limits have been established to offer protection against all identified health effects and have been set with wide safety margins. Most countries globally have adopted the ICNIRP guidelines in their national standards and regulations. For base stations, the relevant limits are expressed in terms of electric field strength (V/m), magnetic field strength (A/m), or incident plane wave equivalent power density (W/m^2) and are to be evaluated in free space. Table 11.4 shows the ICNIRP power density limits in the frequency range from 400 MHz to 300 GHz. Up to 2000 MHz the limits are frequency-dependent, meaning, for example, that in the 700, 900, and 1800 MHz bands used for mobile communications the limits are 3.5, 4.5, and 9 W/m^2, respectively. The power density is to be averaged over a specified time period, which is 6 minutes up to 10 GHz (Table 11.4).

Table 11.4 ICNIRP RF EMF Exposure Limits (Reference Levels) for the General Public in the Frequency Range From 400 MHz to 300 GHz.

Frequency, f (MHz)	Equivalent Plane Wave Power Density (W/m^2)
400–2000	$f/200$
2000–300,000	10

11.8.2 ASSESSING COMPLIANCE OF ADVANCED ANTENNA SYSTEM BASE STATIONS WITH RADIO FREQUENCY ELECTROMAGNETIC FIELD EXPOSURE LIMITS

The full beam steering range in both azimuth and elevation must be considered when assessing compliance of AAS base stations with the RF EMF exposure limits. If the antenna gain is known for all possible beam steering directions, a combined antenna radiation pattern can be constructed from the envelope of the maximum gain values of the individual beams. Using this combined gain $G_{env}(\theta,\phi)$ and the antenna input power P, the power density S at a distance d from the antenna can be determined using the spherical far-field formula

$$S = \frac{PG_{env}(\theta, \phi)}{4\pi d^2} \tag{11.6}$$

This equation and the applicable power density limit can be used to determine a three-dimensional EMF compliance boundary (exclusion zone) for the AAS base station. This will, however, be very conservative if the maximum power is applied, since the time-averaged power or EIRP (e.g., over 6 minutes) is lower due to the spatial distribution of the power in different directions. Studies have been published, which show that for a typical 8×8 massive MIMO array antenna the time-averaged power (or EIRP) contributing to RF EMF exposure in a certain direction is less than 25% of the instantaneous maximum power [13]. How to apply this so-called actual maximum power for AAS base stations is described in international standards and recommendations from IEC and ITU [14–16]. In the case that a site in addition to the AAS base station also contains equipment covering other frequency bands, the total RF EMF exposure from all the different bands must be considered when assessing compliance and determining EMF exclusion zones. How to do this is prescribed in international standards and national regulations.

11.9 OVER-THE-AIR TESTING

This section provides an overview different types of OTA test facilities. In addition, a short background around measurement uncertainties as well as test tolerances is given. The theory and some background around TRP measurement, which is needed for assessing most unwanted emission–related requirements, are also described in this section.

11.9.1 TYPES OF OVER-THE-AIR TEST FACILITIES

A number of practical approaches exist for base station OTA testing.

The most obvious approach to base station OTA testing is an indoor, screened anechoic chamber, as schematically illustrated in Fig. 11.37. The screening ensures that energy is not radiated to the outside world, and absorbers within the chamber remove reflections. In general, anechoic chambers are built such that within a so-called "quiet zone" reflections are acceptably low. The base station is mounted within the quiet zone on a rotating fixture, such that the BS can be rotated and illuminated in different directions. A test antenna is located some distance from the base station such that measurements can be made, or stimulus signals transmitted. The pathloss from the test

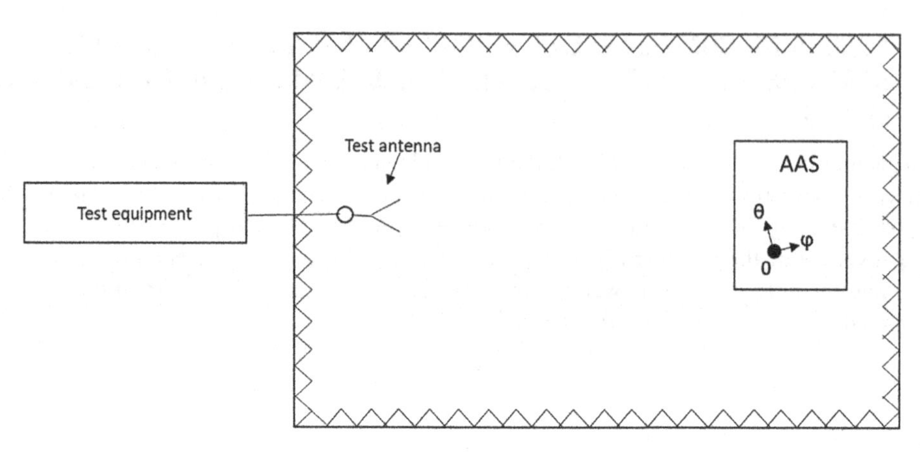

FIGURE 11.37

Schematic illustration of direct far-field chamber.

antenna connector to the base station position as well as the test instruments is carefully calibrated such that during testing, the strength of signals arriving at or leaving the base station is well known.

For testing of directional requirements and receiver requirements, the distance from the BS antenna to the test antenna must fulfill the so-called far-field criteria. In the far field, the waves arriving at the BS or test antenna are planar and as a result the beam patterns are correct. The far-field distance d_{FF} is calculated as:

$$d_{FF} \sim \frac{2D^2}{\lambda}. \tag{11.7}$$

Where D is the diagonal of antenna aperture size and λ is the wavelength. A more detailed description of the concepts of near field and far field is provided in Section 3.3.

A key drawback of the anechoic chamber is that the far-field distance can become large with reasonable antenna apertures. Large chambers are expensive and difficult to locate, for example, the far-field distance for an antenna with $D=1.5$ m at 2 GHz is $\sim$30 m. Large chamber sizes imply significantly increased cost and logistical challenges in locating chambers within labs. Moreover, large far-field distances imply large pathloss within the chamber. Large pathlosses can become challenging when attempting to measure low emissions levels or to achieve sufficient power for testing receiver requirements.

An evolution of the anechoic chamber approach is the so-called compact antenna test range (CATR). A CATR is an anechoic chamber containing a curved mirror as illustrated in Fig. 11.38. The curvature of the mirror is such that, when transmitting to the base station spherical waves emitted from the test antenna are reflected from the RF mirror / reflector as planar waves, which achieve the far-field condition at the base station receiver. In the reverse direction, the mirror focuses the base station planar waves such that the signal measured at the test antenna is the same as the signal that would be achieved in the far field. By means of focusing the reflected waves in this manner, far-field conditions can be achieved in a much smaller chamber (hence the name).

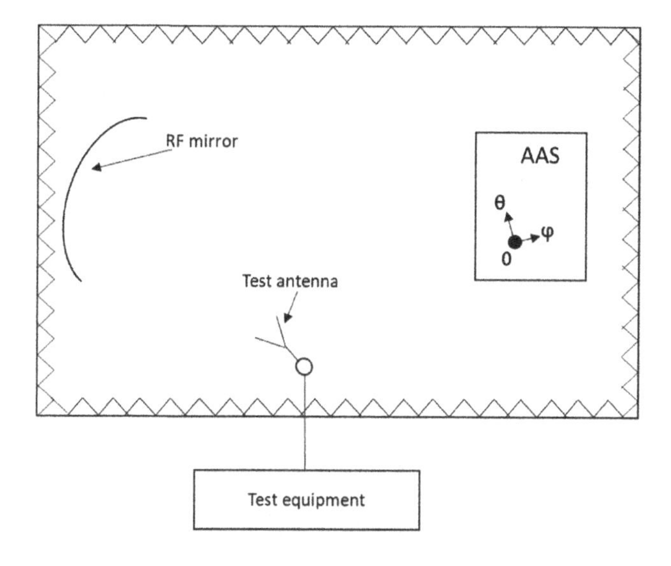

FIGURE 11.38

Schematic illustration of compact antenna test range.

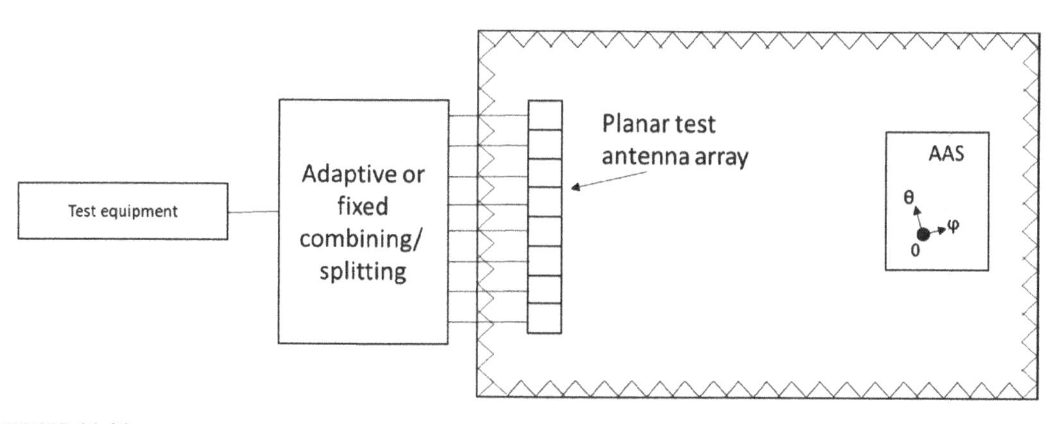

FIGURE 11.39

Schematic illustration of planar wave chamber.

The pathloss is influenced by the reflection of the RF mirror but is independent of the aperture size (as long as the base station fits within the quiet zone, that is, zone in which the reflections are negligible). The CATR is a particularly attractive versatile tool for in-band testing.

An alternative approach to reducing the physical size of the chamber is achieved by means of a so-called planar wave chamber (PWC) as depicted in Fig. 11.39. A PWC is in some ways similar to a CATR with the exception that the RF mirror and test antenna are replaced by an antenna array. The signals to and from the elements of the antenna array are subject to specific beamforming that

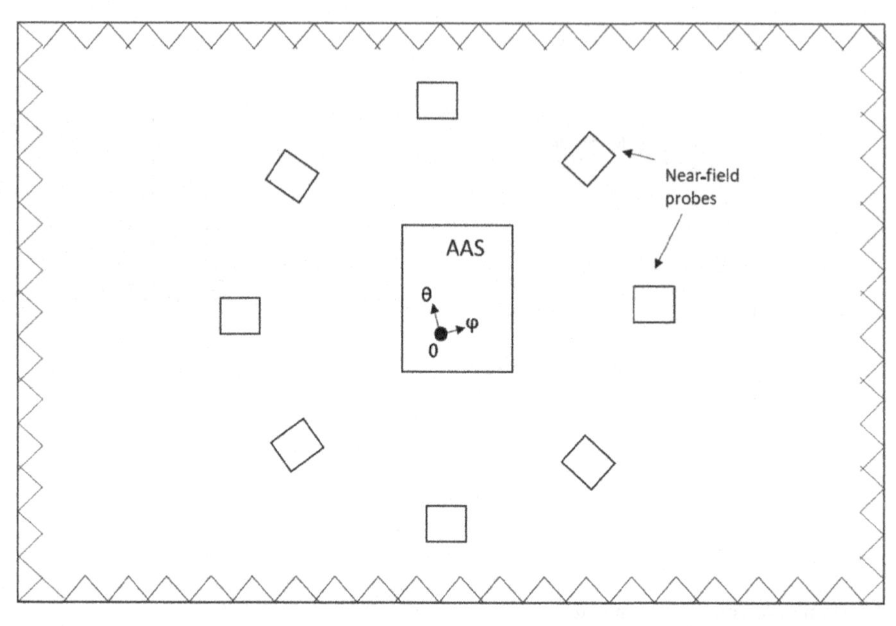

FIGURE 11.40

Schematic illustration of near-field measurement array.

performs the same function as the focusing of the parabolic RF mirror in a CATR. The array can create a planar wave such that far-field conditions are created toward the base station within a much smaller environment.

The indoor anechoic chamber, CATR, and PWC are all based around the concept of creating far-field wave propagation such that directional measurements that exist in the antenna far-field can be measured. A different class of measurement environment instead makes measurements close to the antenna, in the near field (see Fig. 11.40). The near-field radiation pattern differs from that of the far field, since the waves are not planar in the near field. For measuring TRP, measurements can be performed in the near field because the parameter of importance is the total power, and total power can still be captured in the summation. A more detailed description of TRP measurements is provided in Section 11.9.3. Advantages of measuring TRP in the near field are that the measurement chamber can be smaller, and that the pathloss between the base station and test antenna is reduced. Reduced pathloss is of importance when measuring low-power unwanted emissions levels.

For transmitter directional measurements, it is possible to take amplitude and phase measurements around the sphere of the base station and perform a mathematical transformation to calculate the far-field radiation pattern from the near-field measurements. To make such a transformation, a phase reference is needed either from within the base station or a conveniently placed reference antenna. In the receiver direction, mathematically speaking the inverse can be performed; the base station can be illuminated from around the sphere with precalculated amplitude and phase that correspond to a wanted far-field pattern. However, unless the base station is simultaneously illuminated in all directions in the sphere, near-field receiver measurements require amplitude and phase

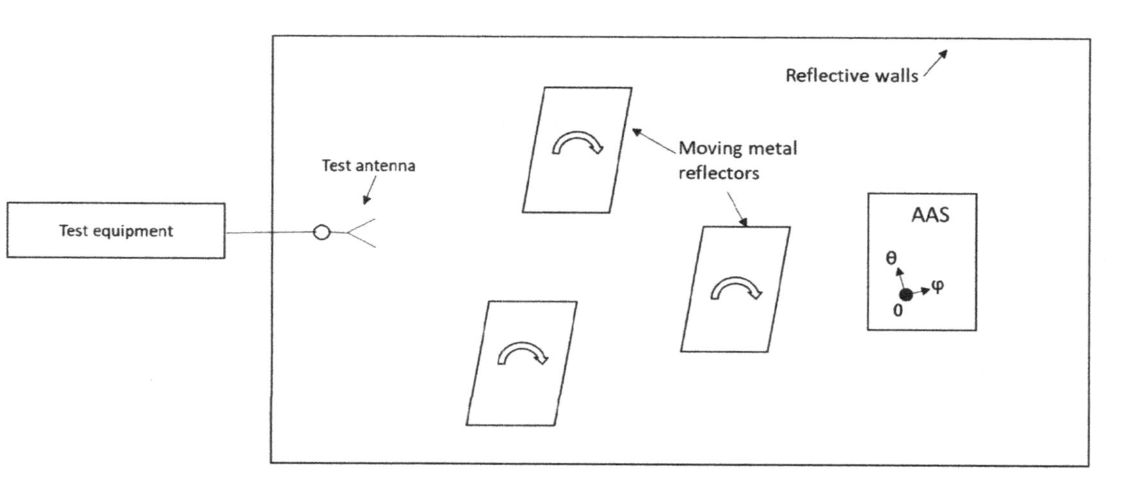

FIGURE 11.41

Schematic illustration of reverberation chamber.

mappings from the base station. Also, the mathematical process for calculating the near-field pattern assumes that the base station performs only linear processing on the received signal whereas most common receiver algorithms, such as MMSE receiver using channel estimates are not linear. Moreover, the near-field calculation cannot relate received power to throughput in the manner specified by the 3GPP requirements. For these reasons, near-field test setups involving near-field to far-field transformations are unsuitable for assessing receiver requirements.

In contrast to the anechoic chamber methods discussed above, a further approach to measurement is based around a measurement chamber that is shielded but deliberately non-anechoic, the so-called reverberation chamber (RC) as illustrated in Fig. 11.41. Mechanically moving metallic plates inside the chamber cause the pattern of reflections to shift with time. Due to the shifting reflection patterns, a static test antenna measures an average power pattern around a test base station over time, and so the RC is very suitable for measuring TRP. The main challenge is for measuring low emissions levels, where PIM within the chamber must be carefully measured. The RC is very suitable for measuring TRP but cannot measure directional requirements.

All the above facilities are only able to measure over a limited frequency range, depending on the design. Spurious emissions and out-of-band blocking requirements are defined over a frequency range from 30 MHz to 12.75 GHz, or more than 50 GHz for mm-wave systems. The above kinds of facilities do not cover such a wide range, in particular considering the large far-field distances relating to the lower end of the range. EMC testing has been performed over larger frequency ranges for many years using EMC chambers. EMC testing is also testing of electromagnetic emissions and susceptibility, although generally not from devices deliberately designed to radiate radio waves, in order to ensure that electronic devices do not create and are not susceptible to electromagnetic pollution to the radio environment. EMC chambers can also be applied for the measurement of spurious emissions and out-of-band blocking for AAS base stations. The main drawbacks of such facilities in this context are that the associated measurement uncertainties are quite high (in the order of several dB) and the positioning apparatus for the device under test may be limited in their degrees of freedom of movement, which limits the amount of points around the sphere that can be measured and increases TRP estimation error. In general, at frequency offsets far from the carrier frequency, the spurious emissions and out-of-band blocking requirements are met with a large margin, and thus high uncertainty and estimation errors can be tolerated.

When considering which kind of test facility to use, factors to be considered include whether the facility is appropriate for the requirement, cost, size, test time, link budget, and measurement uncertainty. Measurement uncertainty is discussed further in Subsection 11.8.2. To obtain reasonable measurement uncertainty, careful design and build of the test setup and careful calibration are needed. The pathloss must be sufficiently low so that for TX tests, the signal level to be measured is well above the noise floor of the test system and in the receive direction the test equipment can generate sufficient power to achieve the needed absolute power levels at the base station antennas under test.

It should be noted that for some requirements, for example, spurious emission, which is applicable over very large frequency range, multiple measurement methods and chambers may be needed due to the limitations and capabilities of chamber types. As an example, the spurious emission applies from 30 MHz and beyond which for FR2 could be extended to 100 GHz due to some regional regulations. For the lower range, for example, 30 MHz to 1 GHz, for not demanding excessive chamber size, EMC chambers are likely the best chamber type for verification of certain range, while for remainder of the frequency range other chamber types would be more suitable and provide lower measurement uncertainties.

11.9.2 OVER-THE-AIR TESTING IN STANDARD AND MEASUREMENT UNCERTAINTIES

The conformance part of the 3GPP RAN4 specifications specifies an acceptable test measurement uncertainty for each individual requirement. The overall Measurement Uncertainty (MU) for a requirement consists of a combination of individual uncertainty factors associated with different components of the test process. For the non-AAS type of base stations, the measurement uncertainties consist of contributions from measurement instruments together with complex and automated test systems with many switches, filters, possible PA, and LNA embedded into the test systems. The test setups enable testing of various requirements with different characteristics and levels, as well as large frequency ranges within a reasonable test time.

The overall measurement uncertainty is the accuracy to which the requirement metric can be measured with 95% confidence. Although 3GPP specifies acceptable measurement uncertainties, it is allowable to design tests with differing uncertainty provided that the uncertainty associated with the test setup is known and, should the test system uncertainty be larger than the 3GPP acceptable uncertainty, the test level is correspondingly tightened.

In addition to acceptable measurement uncertainties, the standard also captures a test tolerance for each individual requirement. Zero test tolerance implies that the tested requirement pass criterion is the same as the requirement level. Zero test tolerance de facto requires base station margins to the core requirement to be equal to or greater than measurement uncertainties; thus in practice a stricter level should be designed into the BS requirement than the pure requirement. The regulatory requirements in different regions usually specify zero test tolerance requirement on conformance tests.

For some requirements, the test tolerance is equal to the measurement uncertainty; this implies that the pass criteria for the test requirement are increased from the core requirement by the measurement uncertainty.

For OTA testing, the measurement uncertainty must, in addition to the contributions from measurement instruments, include contributions for test system filter banks, switch boxes and also the contributions from the chamber. Some examples of chamber-related measurement uncertainties are positioning and pointing misalignment between the AAS base station under test and the reference antenna, polarization mismatch between the AAS BS and the receiving antenna, impedance mismatch in the receiver chain and toward measurement instruments, measurement antenna frequency variation, measurement system dynamic range uncertainty, RF leakage, etc.

It is obvious that OTA testing is much more complicated compared to conducted testing and due to additional measurement uncertainties induced by the chambers, in many cases, the measurement uncertainties will be higher for OTA testing. This in practice implies that zero test tolerance requirements impose a more stringent implementation on the base station that is the case if conducted testing is used. The measurement uncertainties for OTA testing can also be different depending on the chamber types described in Section 11.8.1.

In addition, for OTA testing, due to propagation loss over the air in the chamber and the need to perform some measurements in the far field, the dynamic range of the chamber, and minimum detectable levels of test equipment, it is in some cases challenging for certain requirements to receive sufficient power for measurement. In some cases, inclusion of extra PAs and/or LNAs is necessitated in the test system to ensure enough interferer levels for receiver testing and reduce the total test system noise floor for the transmitter testing, respectively.

11.9.3 TOTAL RADIATED POWER THEORY AND METHODS

Section 11.5.2 outlined how many of the unwanted emission–related requirements, as well as the BS output power requirement has been defined in terms of the TRP of the base station. This necessitates a method for measuring the total power radiated in all directions from the BS enclosure. Since the in-band and spurious emissions requirements are defined at a large number of discrete frequencies, it is necessary to identify test methodologies that are sufficiently accurate whilst not time-consuming.

TRP can be defined as an integral of power over all directions in the sphere around the base station:

$$TRP = \int_{\theta=0}^{\pi} \int_{\varphi=0}^{2\pi} R(\theta, \varphi)\sin\theta d\theta d\varphi \tag{11.8}$$

where $R(\theta, \varphi)$ is the radiation intensity in direction (θ, φ).

Due to the principle of conservation of energy, the radial distance from the base station at which TRP is defined and measured is not of importance. The TRP can be measured in the far field or alternatively be measured in the near field of the antenna. What is of importance is that all of the energy from the radiated E-field is assessed.

Two fundamental approaches to measuring TRP have been identified. One of the approaches is to use an RC, discussed in Section 11.8.1. By means of varying the propagation paths within the chamber, the RC in effect measures radiation in all directions over a period of time. The RC can be an effective means of measuring TRP but is subject to limitations relating to the frequency range supported by the chamber and PIM.

The second fundamental approach is to mount the BS in an OTA test chamber and repeatedly measure the radiated power in different directions. Since moving of test antennas within a measurement facility is not straightforward, typically the base station itself is rotated in both axes whilst measuring power, in order that different directions with respect to the base station point toward the test antenna at different points in time.

Since test equipment needs a finite amount of time to accumulate and process sufficient samples to make an accurate power measurement, the BS cannot be rotated and measured continuously. Instead, the BS must be measured from a discrete number of directions, and the measurements in each direction summed to produce a TRP result:

$$\mathrm{TRP_{est}} = \frac{\pi}{2MN} \sum_{n=1}^{N-1} \sum_{m=1}^{M-1} \mathrm{EIRP}(\theta_n, \varphi_m)\sin(\theta_n), \tag{11.9}$$

where N and M are the number of samples in the θ and φ dimensions, respectively.

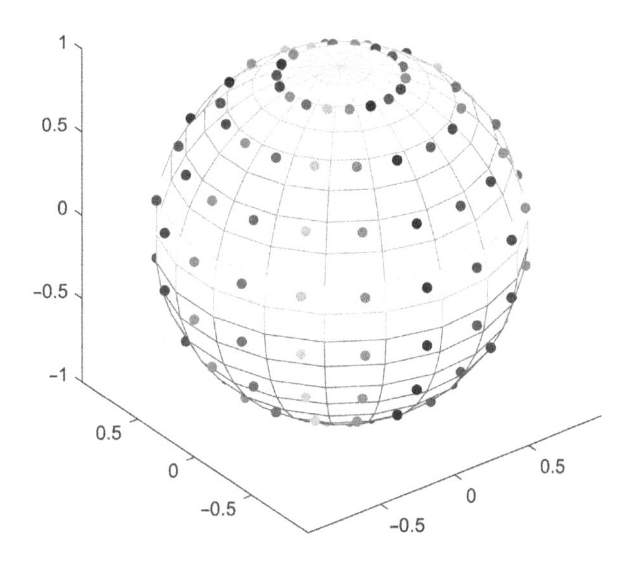

FIGURE 11.42

Illustration of measurement points for equal angle TRP grid [17].

Two aspects are of importance when considering the range of angles from which to make TRP measurements. The first is the grid of positions from which measurements are made. During 3GPP standardization, a number of different grids were considered. The first of these was the so-called continuous angle grid. In this simple grid, the measurement steps are at an evenly distributed set of angles in both the θ and φ axes. Although the angles are continuous, this approach leads to measurement points becoming closer together in space at the "poles" of the sphere (see Fig. 11.42).

It can be shown that in order to measure TRP accurately, the measurement points must be spread out by no more than λ/D in each axis. D is the antenna dimension in the axis in question and will be discussed later in this section.

Alternative grids considered in 3GPP include the "equal area" grid; this grid is defined such that the area of the sphere around each sample point is equal (i.e., sampling points do not cluster toward the "poles" of the sphere as is the case with the equal step size grid) and the Fibonacci grid, in which measurement points are arranged in a spiral.

As an alternative to measuring over a sphere, measurements may be made at equally spaced angles around the circles formed by a number of orthogonal cuts of the sphere. The number of cuts will relate to the amount of error in estimating the TRP. For known array types, it is possible to reduce the TRP error between cuts using a technique known as "pattern matching," which is elaborated in more detail in Ref. [17]. If the transmitter signal sources are not well correlated between elements, the benefit of the pattern matching approach will be low.

When measuring around the sphere, measuring at angles spaced by λ/D is sufficient to obtain an accurate TRP estimate. However, the large number of implied measurement points is too time-consuming for practical purposes and so methods to reduce the number of measurement points must be identified.

A first consideration on reducing the number of measurement points involves examining the D parameter in more detail. D is the antenna dimension in the axis in question. Typical antennas consist of a number of elements arranged in a uniform grid along the two axes. If the elements transmit coherently, then D is the accumulated size of all the elements. However, if the elements do not transmit coherently, then the resulting beamwidth is much larger. For noncoherent elements, D is the element size. With a smaller D, the number of measurement points is reduced.

With a beamforming system, in principle the transmitter radiation pattern should be held constant during TRP measurement. However, measurement time can be reduced if, during each individual measurement, the transmitter beam is swept in space. The beam must be swept over the same range of angles for each measurement. Measuring a swept beam has the effect of averaging out the measured radiation pattern, which allows for D to be reduced. Whether this approach is effective depends on the relationship between the reduced number of measurement directions and the increased time required for measurement in each individual direction due to beam sweeping.

The amount of measurement points can be reduced further if there is knowledge of the radiation pattern from the base station. This is likely to be the case for in-band measurements and in particular measurement of power on the wanted carrier. If the directions in which most power is likely to be radiated are known, then measurements outside of those directions can be skipped without significant impact to the measurement accuracy. In general, there exists a trade-off between the number of measurement points and the measurement accuracy. 3GPP has allowed for 0.75 dB in the measurement uncertainty budget for TRP measurements to account for the additional inaccuracy created by a reduced number of measurement points. The exact reduction depends on the points that are removed with respect to the radiation pattern if the radiation pattern is known.

For spurious emissions, the radiation pattern is in general not known, because the phase characteristics and antenna characteristics differ significantly from in-band. In general, spurious emissions are spread out in space (which implies potentially a low number of required measurement points). However, certain types of unwanted emissions, such as harmonics of the LO, can produce distinctly directional narrowband emissions spurs. Unfortunately, the direction and potentially frequency location of such spurs are not known in advance.

Typically, at high-frequency offsets from the operating band, base stations meet the spurious emissions requirements with a large margin. The large margin enables larger measurement uncertainty to be tolerated by means of tightening the test requirement in relation to the larger MU. This enables a reduction in the number of measurement points for spurious emissions.

The orthogonal cuts method is of particular interest for measuring spurious emissions. This is because measuring over two or three cuts with 15-degree spacing of the measurement samples replicates well-known methods and test chamber setups for EMC measurements, enabling EMC measurement facilities to be reused for spurious emissions measurements. EMC facilities have a larger measurement uncertainty than other types of chamber, and the orthogonal cuts approach will introduce greater TRP error than denser sampling, but the additional uncertainties and underestimation risk can be taken into account when setting the test requirement such that passing of base stations that exceed the TRP limit does not occur. This requires additional margin to the requirement to be available at the base station, which is generally not a major issue at spurious frequencies that are significantly far away in frequency for the band under test.

In summary, TRP can be assessed using a variety of methods ranging from RC to spherical measurement with techniques to reduce the number of measurement samples (such as beam sweeping) to measurements on orthogonal cuts. The best approach depends on the particular requirement to be assessed, considering the likelihood of correlation between the measured radiation sources (i.e., antenna elements) and the needed precision for TRP measurement. It is likely that different techniques will be applied for different requirements and parts of the frequency domain.

11.10 SUMMARY

Deployment and operation of cellular systems require carefully deviced radio requirement and conformance specifications. The specifications must protect the value of operators' investments in spectrum by managing unwanted emissions and incoming interference, ensuring that devices and networks behave in a predictable manner and ensuring a basic level of efficiency in using the spectrum, balanced against the cost and complexity of meeting the requirements. Due to the nature of the radio spectrum and industry, in which multiple stakeholders share spectrum and network nodes and components come from many independent vendors, standardized requirements are needed. Radio and conformance specifications are provided by 3GPP and regulators. The 3GPP requirements ensure good protection, whilst the regulatory requirements are a legal basis for deploying systems.

Over the first couple of decades of the cellular industry, a mature and well-known base station RF specification became established. The introduction of massive MIMO and 5G required a paradigm shift in the approach to requirement and conformance specification assessment because of the impact of dynamic beamforming on propagation characteristics and the need for OTA testing. OTA testing is needed both to provide complete system testing of integrated radio/antennas and to facilitate integrated array hardware without antenna connectors

A key challenge in the standardization process is to provide requirements that provide acceptable levels of protection and performance for all types of base station architecture whilst not restricting implementation flexibility by specifying requirements that are only feasible for certain types of system. 3GPP has put extensive effort into defining "over-the-air" requirements for AAS as whole system, implementation agnostic for this purpose.

5G NR technology is divided into systems that operate below 7.125 GHz and systems that operate in mm-wave. For below 7.125 GHz operation, NR and LTE AAS base stations may be deployed into existing bands and must co-exist with non-AAS based networks. For this reason, for FR1 the focus in standardization has been to create OTA requirements that are adapted to provide the same amount of performance and protection as was the case with the legacy specifications. For mm-wave, on the other hand, not only is the spectrum new, but the underlying technology characteristics are also radically different to FR1. Thus for mm-wave an investigation of technology fundamentals on aspects such as PA component behavior, PN modeling, filter technologies, receiver architecture and NF, and other things was needed. Based on this investigation, new requirements for mm-wave systems were created.

OTA testing requires investment in test chambers and facilities and developments of methods for accurate measurement. OTA testing is complex and more capital and time intensive than traditional testing methods, and deep expertise is needed to get the testing right. 3GPP RAN4 has investigated different types of facilities in terms of their suitability for assessing conformance to different types of requirement and the accuracy achievable in each case. Recommendations for measurement uncertainty and test tolerance are also provided in the specifications.

The process of creating radio performance specifications for massive MIMO and 5G systems has taken time in 3GPP. The first study into AAS characteristics began in 3GPP RAN4 in 2012, and the 5G specification conformance descriptions were completed at the end of 2018. The specifications are now finalized, paving the way for regulation and launch of massive MIMO systems.

REFERENCES

[1] 3GPP TS 38.104, NR; Base station (BS) radio transmission and reception.
[2] 3GPP TS 38.141, NR; Base station (BS) conformance testing.
[3] 3GPP TS 38.101, NR; User equipment (UE) radio transmission and reception.
[4] 3GPP TS 38.113, NR; Base station (BS) electromagnetic compatibility (EMC).
[5] 3GPP TS 38.133, NR; Requirements for support of radio resource management.
[6] 3GPP TS 37.104, NR, E-UTRA, UTRA and GSM/edge; multi-standard radio (MSR) base station (BS) transmission and reception.
[7] 3GPP TS 37.105, Active antenna system (AAS) base station (BS) transmission and reception.
[8] 3GPP TR 37.842, Evolved universal terrestrial radio access (E-UTRA) and universal terrestrial radio access (UTRA) radio frequency (RF) requirement background for active antenna system (AAS) base station (BS).
[9] Ericsson, R4–164226, On mm-wave technologies for NR, 3GPP TSG-RAN WG4 #79, 23–27 May 2016, Nanjing, China.
[10] 3GPP TR 38.803, Study on new radio access technology; Radio frequency (RF) and co-existence aspects.
[11] H.T. Friis, Noise figures of radio receivers, Proc. IRE 7 (1944) 419–422.
[12] ICNIRP, Guidelines for limiting exposure to time-varying electric, magnetic, and electromagnetic fields (up to 300 GHz), International Commission on Non-Ionizing Radiation Protection (ICNIRP), Health Physics 74 (1998) 494–522.
[13] B. Thors, A. Furuskär, D. Colombi, C. Törnevik, Time-averaged realistic maximum power levels for the assessment of radio frequency exposure for 5G radio base stations using massive MIMO, IEEE Access 5 (2017) 19711–19719.
[14] IEC TR 62669 Ed.2, Case studies supporting IEC 62232 – determination of RF field strength and SAR in the vicinity of radiocommunication base stations for the purpose of evaluating human exposure, 2019.

[15] ITU-T K-series Supplement 16, Electromagnetic field compliance assessments for 5G wireless networks, 2018.
[16] IEC 62232:2017, Determination of RF field strength, power density and SAR in the vicinity of radiocommunication base stations for the purpose of evaluating human exposure, 2017.
[17] 3GPP TR 37.843, Evolved universal terrestrial radio access (E-UTRA) and universal terrestrial radio access (UTRA); Radio frequency (RF) requirement background for active antenna system (AAS) base station (BS) radiated requirement.

CHAPTER

ARCHITECTURE AND IMPLEMENTATION ASPECTS

12

12.1 INTRODUCTION

Previous chapters explored the theoretical basis of array technology, beamforming algorithms, the possibilities enabled by the 3GPP specification, and some deployment considerations. For successful product development, apart from these issues, practical implementation considerations are of key importance. Chapter 11 considered the challenges posed on the design of the radio transmitters and receivers in the light of the radio frequency (RF) compliance requirements defined by 3GPP and regulators.

The required array processing leads to very large volumes of computational processing and associated large data flows with advanced antenna systems (AAS) and between AAS and other hardware units. The amount of data and computation is interrelated with the multi-antenna techniques employed. The sheer volume of data computation and data transfer, necessitated by large arrays operating on wideband signals, requires careful definition and dimensioning of the overall radio base station (RBS) architecture and the internal interfaces that challenge the traditional lower layer split (LLS) between the baseband processing and the radio unit. Also, an additional functional subdivision will be described—the higher layer split (HLS), which is geared toward cloud computing.

There is a trade-off between practical limitations at the radio site installation, such as size, weight, and the number of physical units allowed, and the beamforming performance of the AAS system. This trade-off becomes even more challenging as the number of frequency bands increases (see Section 14.4).

Implementation of beamforming can be performed in the analog or digital domain, as described in Chapter 7. The approach to the physical realization of beamforming is driven by several technological and deployment considerations.

Multibranch transmitter and multibranch receiver operation mean that array system issues, such as phase and amplitude misalignment and mitigation of coupling impacts to radio operation, need to be addressed.

In this chapter, the top-level base station architecture is outlined and placed in the context of the overall Radio Access Network (RAN) architecture. Based on this, the key drivers that shape the design decisions on beamforming implementation and further to the base station architecture are elaborated. Examination of the determining aspects of overall system design brings out the factors which currently tend to lead to different architecture and implementation solutions for mid-band and high-band AAS systems.

Advanced Antenna Systems for 5G Network Deployments. DOI: https://doi.org/10.1016/B978-0-12-820046-9.00012-5

12.2 5G RADIO ACCESS NETWORK ARCHITECTURE

12.2.1 2G, 3G, AND 4G RADIO ACCESS NETWORK ARCHITECTURE

In this section, a high-level overview of the RAN architectures of 2G, 3G, and 4G systems is given (also, see Section 2.1.2, network history).

For the 2G Global System for Mobile Communications (GSM) mobile telecommunications system [1], the cellular networks were based on a relatively fixed functional architecture and network functionality was in general geographically localized. The radio-related functionality and processing were located in the base transceiver station (BTS), whilst higher layer processing was located in the base station controller (BSC) (see Fig. 12.1). Backhaul to and from radio sites was not time critical given the nature of supported usage scenarios and traffic.

The base stations were predominately cabinet-based, with radio units located within a cabinet at the foot of the mast (Fig. 12.2).

In early 3G UMTS terrestrial radio access network (UTRAN) [2], radio transmission and reception functionality were located within NodeBs, whilst radio resource management and higher layer RAN user plane processing (as well as soft handover combining) took place in the radio network controller (RNC) (see Fig. 12.3). The RNC was interfaced with a core network, which was separate from the RAN. The Iub interface between the NodeBs and RNC involved significant latency, in particular, for control procedures. Resource allocation needed to be decided at the RNC and

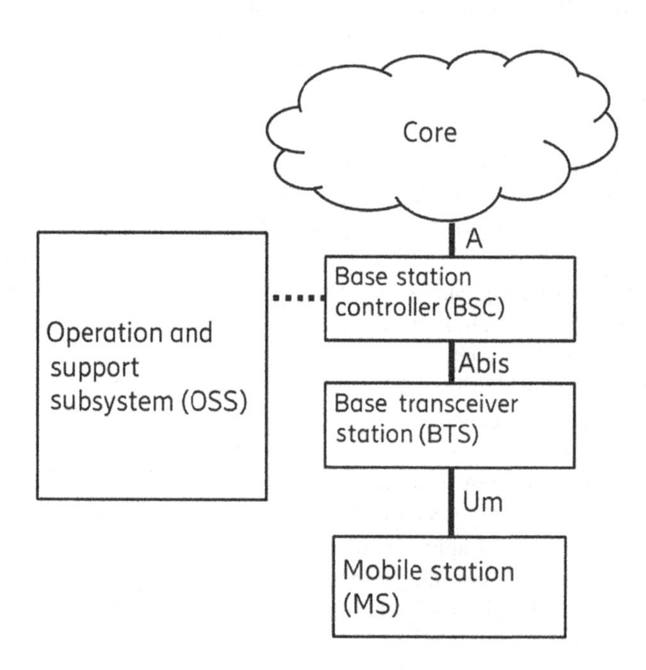

FIGURE 12.1

2G GSM RAN architecture. Note that mobile station is denoted as user equipment in later RAN architectures and elsewhere in this book.

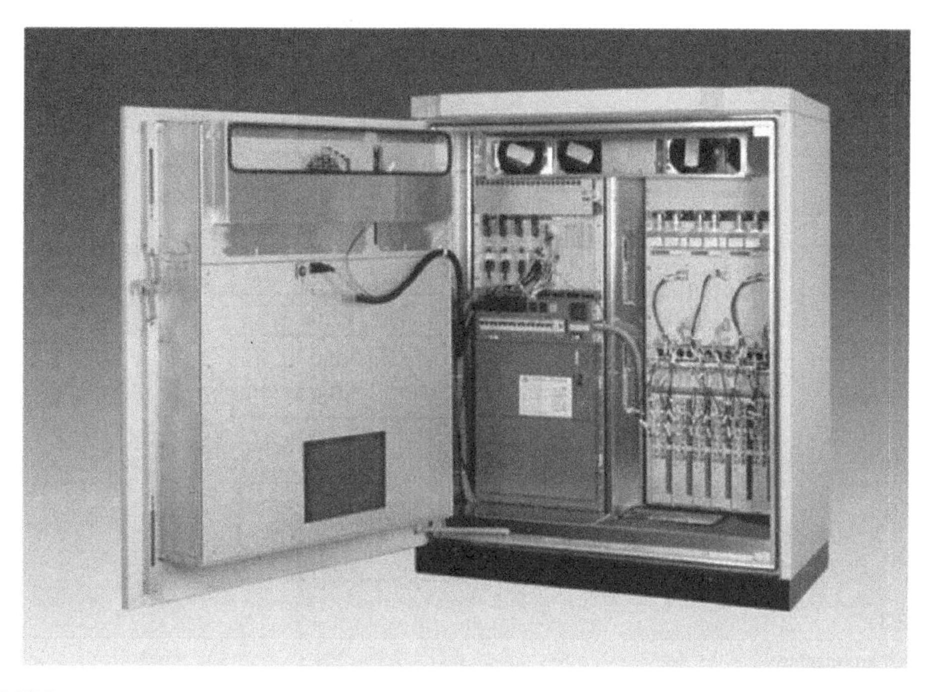

FIGURE 12.2

2G Outdoor GSM radio base station example. Upper left corner: power supply and transport/control unit. Bottom left corner: battery compartment and optionally more transport equipment. Upper right corner: combiner and filter units. Lower right corner: transceiver units (TRX) containing radio and signal processing. Air conditioning equipment in the door.

indicated to both the NodeB (via Iub) and user equipment (UE) (via radio resource control (RRC) signaling). Initially, even user plane related procedures such as modulation/coding rate selection were made at the RNC. The centralized approach had advantages in processing efficiency, but it proved to be disadvantageous as the resource allocation could not respond to radio and traffic conditions at the speed at which the conditions changed. During 3GPP releases 5–6, with introduction of high-speed downlink packet access (HSDPA) and high-speed uplink packet access (HSUPA) [3], some aspects of radio resource management and user plane control were migrated from the centralized RNC to the NodeB since fast management of these resources closer to the air interface increased efficiency, for example, hybrid automatic repeat request (HARQ) was introduced.

Early 3G base stations were built like 2G base stations, as cabinet-based solutions. In 2003, the Common Public Radio Interface (CPRI) specification [4] was released, and equipment vendors added main-remote solutions. The CPRI interface defines the interface between the main unit containing the baseband processing and a remote unit containing the radio (see Fig. 12.4). The main usage of the main-remote concept was to be able to add extra outdoor radios unit to sites where the cabinets were already full, that is, when extra sectors (e.g., at street level) or extra frequency bands were added. If the radio units were placed close to the antenna, also the feeder loss was reduced, improving the link budget both in receive and transmit direction. The outdoor radio units were also

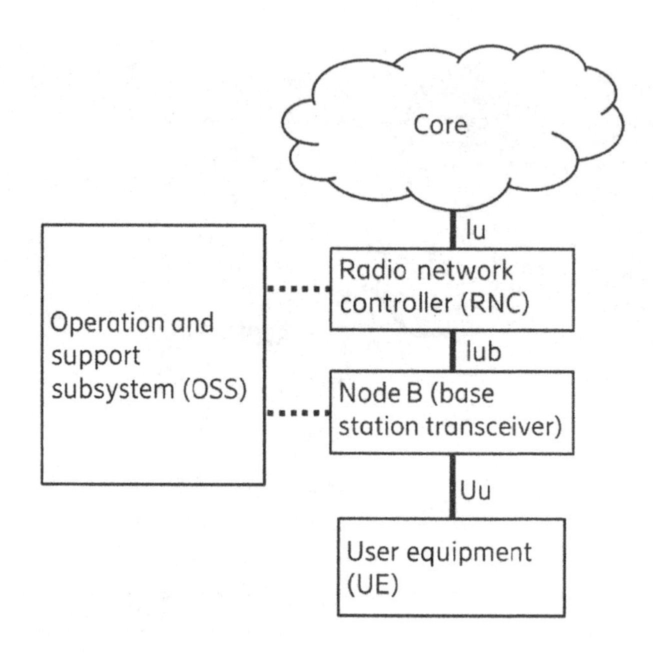

FIGURE 12.3

3G UTRAN architecture.

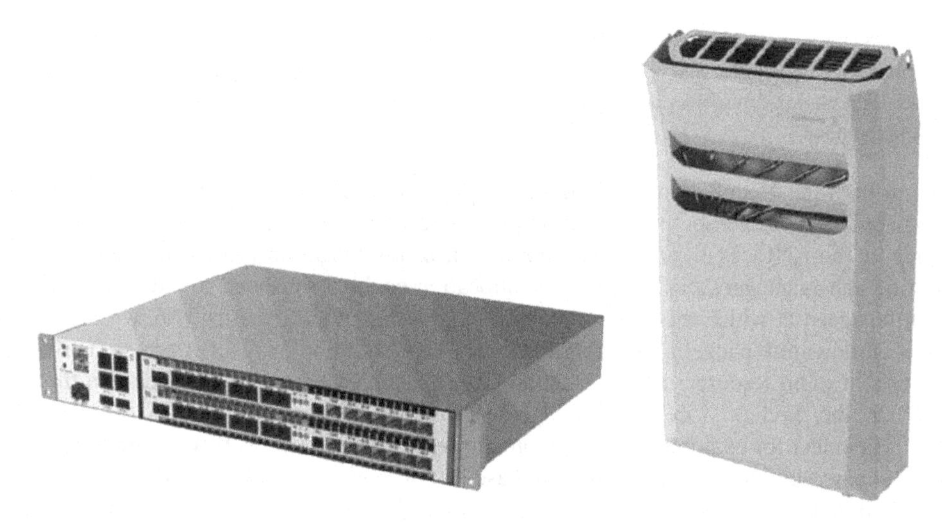

FIGURE 12.4

Radio base station based on the main-remote concept. The main unit (to the left) implements the control processing signal, thus processing and termination of both backhaul and *fronthaul* (CPRI) for multiple (up to tens of) remote radio units (to the right). The remote radio unit implements the radio receiver, transmitter, filtering, and control of antenna near equipment (e.g., tilting or power supplies). The interface in-between is called the *fronthaul* interface, typically based on CPRI. The main units also contain control of close-by support systems (power, fans, and alarm ports) and a router/switch for other *radio site* equipment.

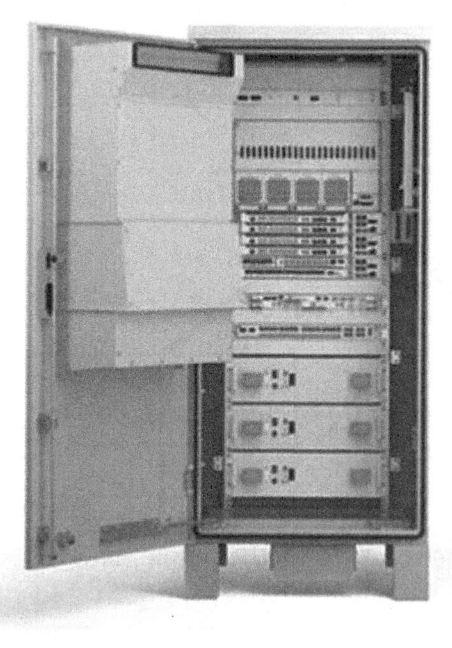

FIGURE 12.5

Main unit placement examples. To the left, a typical installation where the main unit is placed together with the power supplies and batteries. To the right, main unit is centralized in a baseband hotel. A few racks can easily serve hundred *radio sites*.

integrated with the antennas to create so-called active antennas, that is, antennas with integrated radio functionality such as power amplifiers (PA) and receivers. Some installations were made with centralized main units, co-locating from a few to many tens of units, in so-called baseband hotels, see Fig. 12.5, further reducing the size of equipment at the radio sites.

In 4G E-UTRAN [5], the whole RAN is collapsed into an eNodeB, creating a highly optimized vertically integrated node, with an X2 interface in between eNodeBs for coordination information exchange and for hand over (see Fig. 12.6). 4G E-UTRAN have evolved since 3GPP release 8 to incorporate increasingly complex air interface mechanisms, such as advanced multiple-input multiple-output (MIMO) and coordinated multipoint (CoMP) operation, see Sections 6.6.2 and 8.3.4, as well as diverse possibilities for aggregating data flows across different parts of the RAN, such as carrier aggregation and dual connectivity. The services supported in 4G have evolved from mobile broadband to other services such as the Internet of things (IoT), ultra-reliable low latency communication (URLLC), etc. (see Section 2.1.6).

The evolution of base station solutions has continued in line with the main-remote trends that started with 3G UTRAN, and traditional cabinet solutions are becoming very rare, and are often repurposed to house baseband for an area (e.g., functioning as a small baseband hotel).

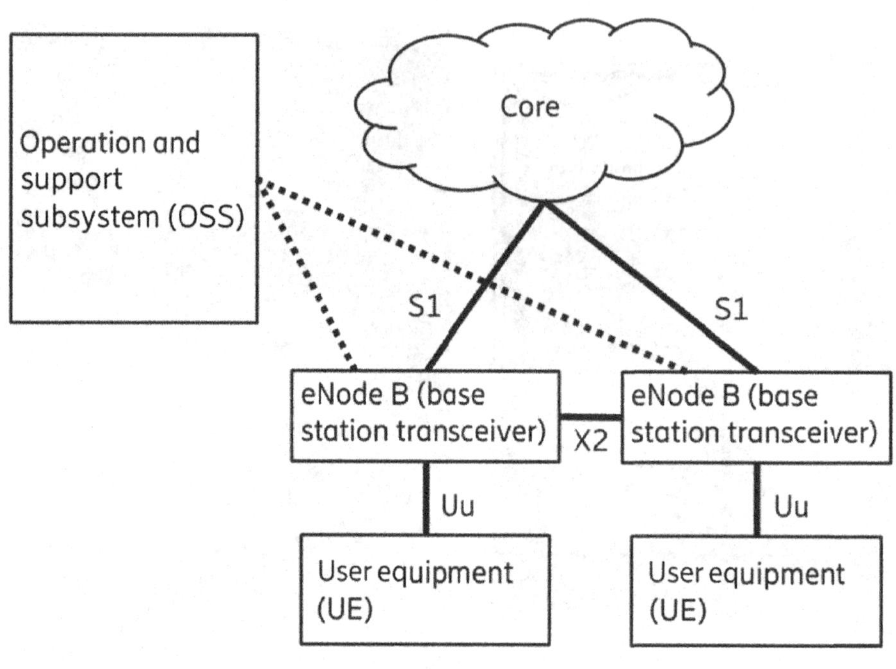

FIGURE 12.6

4G E-UTRAN architecture.

A major difference when introducing 4G E-UTRAN, as opposed to when introducing 2G or 3G, was the spectrum refarming. 2G and 3G were both introduced on new frequency bands dedicated for that purpose, whereas 4G was also added to existing spectrum, evicting 2G and 3G carriers. This drove the need for multistandard equipment, baseband units, and radio units which could simultaneously run 2G, 3G, and 4G.

The huge success of 4G around the world has also opened up many more purpose-optimized base station deployment architectures, for example, femtos and picos (low output power base stations), specialized distributed indoor systems, etc. (see Section 2.1.4).

For modern RANs, there are two deployment options that are crystalized as the main options: the distributed RAN (DRAN), where signal processing is co-located with the radios at the radio site, optimizing for backhaul cost and node integrity, and the centralized RAN (CRAN) where signal processing for multiple radio sites are co-located. To limit the fronthaul cost, the CRAN requires dark fiber (an unused fiber already installed, ready to be used) between the radio sites and the CRAN site and is, therefore, within a few kilometers from the radio site. A CRAN site typically serves 10–100 radio sites. This is very similar to the baseband hotels deployed in 3G, but the ambition on pooling of the CRAN hardware and coordination between the CRAN cells has increased to motivate the more expensive fiber infrastructure.

In each of the deployment options, further optimization of both hardware and software has continued. The distributed architectures with high constraints on power consumption have improved their fundamental processing engines (e.g., time-critical digital signal processing and coding) and the

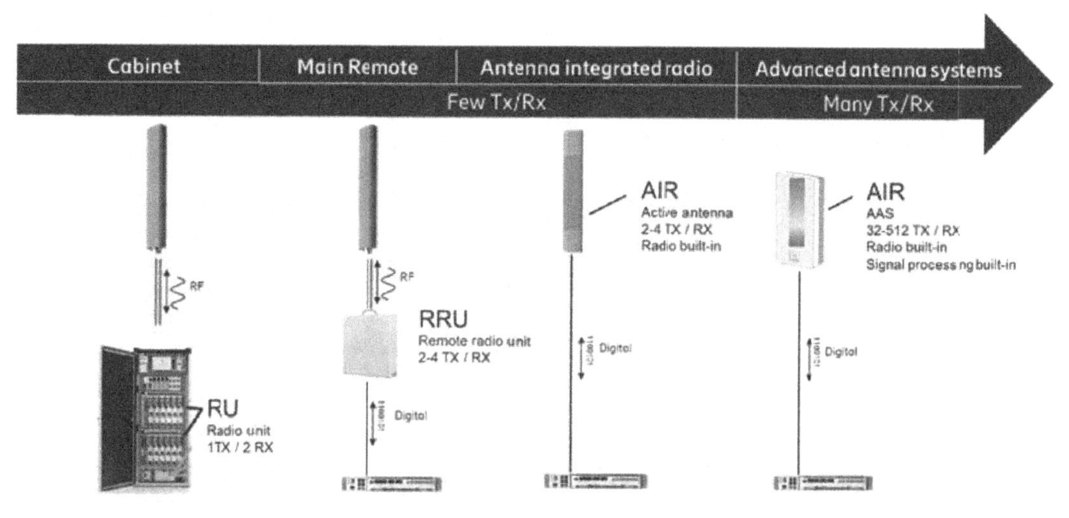

FIGURE 12.7

Evolution of radio placement. From radio units (RU), via remote radio units (RRU), to antenna integrated radios (AIR).As the number of transmitters and receivers has increased, the radio is moving closer to the antenna and has become more integrated with the antenna.

centralized baseband has explored the technological advancements in the cloud computing systems. These technological evolutions enable a much more tailored functional distribution of RAN computation between a generic core and an edge, tailored to operators' specific scenarios regarding deployment, services, geography, backhaul and fronthaul possibilities, etc. With the MIMO support in 4G E-UTRAN, the radio base station (RBS) has added support for more antennas in each sector. 2G and 3G used one transmitter (Tx) and two receivers (Rx) for the majority of the installations, but 4G used two Tx for MIMO from the beginning, and from four to eight Tx and Rx are not uncommon. This has put more focus on the amount and length of the RF feeders, pushing the radio unit closer and closer to the antenna. For AAS, the radio has to be integrated with the antenna (Fig. 12.7).

12.2.2 RADIO SITE INSTALLATION

The radio site is the physical location where the radio waves are transmitted and received between the base station and the UE. Over the years, the complexity of typical radio site installations has been increasing.

The complexity of the site installation increases as more "boxes" such as specific radios are added. For example, there is an increase in the cabling between the boxes, an increase in power supply requirements, and added dimensioning requirements on towers and other mounting devices to support all of these boxes.

The driving forces for this increased complexity are an increase in the available frequency bands for mobile communication, the trend to use antenna near radio units, and, in some cases, the use of higher order sectorization to increase the capacity of the radio site (Figs. 12.8 and 12.9).

FIGURE 12.8

A very tidy four-band rooftop radio site with microwave backhaul. One multiport antenna for each sector, with remote radio units that are placed close by to minimize feeder length and feeder loss. Each remote radio unit can have two, four, or eight radio branches, and single or multiple bands. The antenna can serve all bands using separate ports and wideband elements. The upper right sector also has a microwave backhaul unit. The power system and batteries are located in the cabinet to the bottom right. The cabinet also houses the main unit with the signal processing, control processing, radio site control, and transport network termination.

FIGURE 12.9

More site examples. Every radio site has its own challenges. When adding new AAS on an existing site, finding the mast space is crucial.

Examples of different products that can be installed at the radio site are as follows:

- antennas, including fixed tilt mount devices (mechanical tilt) and Remote Electrical Tilt (RET) devices (electrical tilt)
- low-noise amplifiers
- remote radio units
- baseband units
- power supply units with battery backup
- transport network equipment
- lightning protection.

The radio sites can be divided into different site types, depending on the physical deployment scenarios, which are given as follows.

- *Macro sites* are used in diverse scenarios, ranging from dense urban and suburban to rural areas. Typically, with rooftop and tower installations, with both many sectors and many frequency bands, especially in densely populated areas;
- *Street sites* are primarily used in denser urban areas, with below-rooftop installation, for example, on walls or poles (see also Section 14.5.1). The output power is typically lower than that of macro sites;
- *Indoor sites* usually consist of completely different types of products, with very low output power, and with distributed antenna systems.

From an AAS point of view, macro sites are of high priority, since extending coverage of higher frequency bands is one of the key properties of AAS solutions compared with non-AAS solutions. Street sites and indoor sites are also of interest, especially to provide high-capacity solutions and to provide coverage for high-band frequencies. Indoor sites will also be important for new use cases such as factories.

Each site type has its own constraints. For macro sites, the AAS is typically added to an existing radio site, where the antenna space on the mast may be already crowded. For mast integrity (wind load, carried weight) reasons or for appearance reasons, the AAS may compete with adding more antennas on existing bands. In extreme cases, adding AAS to a radio site may require modernizing the existing antenna system to a multiport antenna or other highly integrated solutions. Also, mast reinforcements can be necessary. The wind load requirements may also restrict the size of the AAS, and the appearance and shape of the AAS, both impacting the beamforming performance.

For street sites and indoor sites, the physical volume and appearance are crucial factors. This may limit the number of antenna elements that can be supported and as a result the achievable AAS performance. This is more an issue with mid-band AAS than high-band, where the size is smaller.

Installing an AAS on a radio site can lead to increased requirements on many RAN components, such as an improved power system, increased baseband capacity and more fronthaul transport.

Another aspect to consider is the purpose of the radio site, that is, the intended coverage and served capacity. A high-capacity macro site in a dense urban high-rise area with few indoor systems benefits from fine-grained vertical domain beamforming and multi-user MIMO, whereas a macro site in a suburban area may perform equally well with less flexibility (see Section 14.3). The less flexible AAS solution will typically have lower weight and power consumption, but could still have the same wind load impact on the mast, since the even less flexible solution needs the same antenna area to give the wanted coverage.

12.2.3 FUNCTIONALITY

The RAN consists of many functional areas, which are given as follows:

- the hardware equipment placed on the radio site and a potential central site
  - baseband hardware
  - radio hardware
  - control processing hardware
  - transport hardware, including routers, switches, wavelength division multiplexing (WDM), etc.
  - antenna hardware, including RET motors, Tower Mounted low-noise Amplifiers (TMA), etc.
  - power and power back up hardware
  - environmental hardware (fans, air conditioning, shelters, etc.)
- the data plane software
  - 2G, 3G, 4G, and 5G Radio Access Technologies (RAT), implementing the 3GPP protocol layers and vendor proprietary features and algorithms
  - transport
- control plane software
  - 2G, 3G, 4G, and 5G RAT
- management software
  - equipment management
  - self-organizing network software.

The traditional base station architecture originating from 2G, where no operation and maintenance (O&M) interface from the base station site was exposed to the management system, has resulted in all of these functions being implemented in the base station software. For example, there is no direct connection from the O&M system to the radio units. Instead O&M of the radio units are handled via the baseband units. This adds complexity to the software, but the advantage is the high automation and joint optimizations (e.g., automation of fail-overs, automated fault handling between domains, joint power efficiency optimizations, etc.). From an operating expense (OPEX) point of view, it is advantageous if the AAS is integrated into the same base station framework. From a performance point of view, it is advantageous if the AAS shares the same RAT software stack as the bands and sectors it is expected to interact with (e.g., interactions such as carrier aggregation between bands, and heterogeneous networks where close coordination between nodes is wanted).

12.2.4 INTRODUCING ADVANCED ANTENNA SYSTEM IN 5G RADIO ACCESS NETWORK

Section 12.2 has so far described the evolution of RAN architectures, how typical radio site installations look like, and what kinds of functionality are implemented in the different parts of the RAN. For each generation, the RAN becomes more and more powerful, but also more and more complex.

This is the environment where the new AAS products need to fit in efficiently in order to provide good cost-benefit advantages to mobile network operators. The following sections will describe the most important aspects that need to be considered when introducing AAS products in 5G.

12.2.4.1 5G architecture impacts

When designing the 5G RAN architecture, two new technologies have put additional requirements on the solution, namely, cloud computing and AAS.

With cloud computing, the vision is to centralize as much as possible of the RAT processing for several reasons, which are given as follows:

- utilizing commercial off-the-shelf (COTS) hardware, such as x86 servers. Hardware that is homogenous with other applications, such as the core network;
- utilizing scaling and pooling concepts developed in the cloud industry, including orchestration of hardware and applications; and
- having a centralized anchor point for mobility between standards (4G + 5G) and heterogeneous deployments (e.g., mid-band on some radio sites and high-band on others).

The AAS, on the other hand, primarily drives a significant increase in compute power and (internal) data transfers, to achieve the desired performance boost.

These two technologies lead to the need and possibility to introduce new functional splits for the 5G RAN architecture.

12.2.4.2 Split options

3GPP has studied different subdivisions of the RAT protocol layers, the so-called "split options" [6] (see Fig. 12.10). The further to the right in the figure, the higher the bit rates of the interface become and the more specialized 3GPP processing is required. More specialized 3GPP processing (the AAS itself being to the furthest right) leads also to more dependencies between RAN algorithms and hardware, such as highly optimized processing engines, hardware generation knowledge,

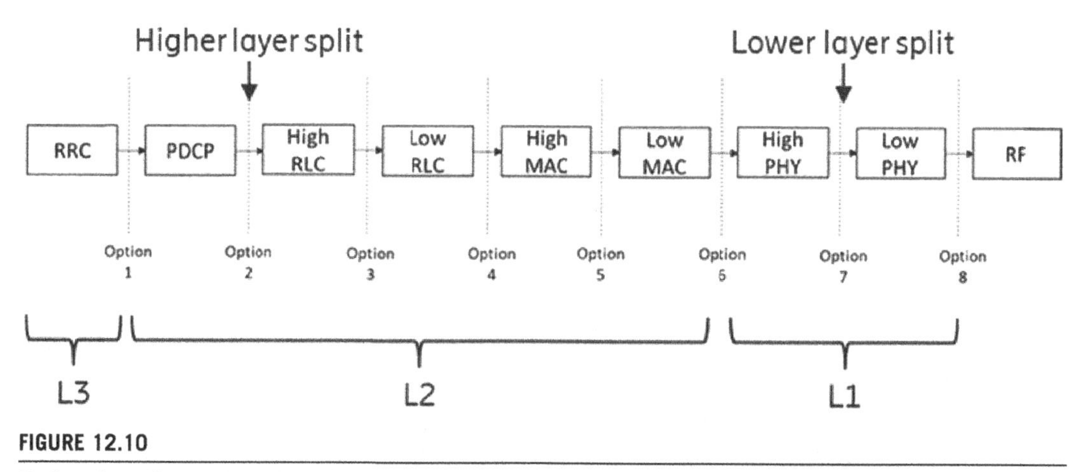

FIGURE 12.10

Options for splits considered by 3GPP. Of these, only Option 2 has been standardized in 3GPP.

product variants, power save functions, etc. Finally, any split to the right of Option 5 has very tight latency requirements as the HARQ processing is in the low media access control (MAC).

The RRC layer is the control plane of the RAT, managing mobility, setup of bearers, admission control, etc. This layer operates on a per UE basis and on a tens of ms rate.

The packet data convergence protocol (PDCP) tunnels bearer data to and from the core network. It ensures ordered delivery, security, and buffering of a bearer. It operates on bearer level with an internet protocol (IP) packet rate.

The radio link control (RLC) layer splits/merges packets to better fit the air interface, and to ensure delivery using RLC retransmissions. It operates on bearer level on a transmission time interval (TTI) rate (0.1–1 ms).

The MAC layer allocates the air interface resources to different UEs and different bearers. It operates on multicell level and multi-user/multi-bearer level, and on a TTI rate (0.1–1 ms).

The physical layer (PHY) contains the signal processing associated with the air interface (encoding/decoding, HARQ, scrambling, modulation/demodulation, equalizing, beamforming, etc.). This operates on multi-antenna level and multi-user level, and on a symbol to TTI rate (0.01–1 ms).

Sometimes, the PHY layer is denoted L1; PDCP, RLC, and MAC is denoted L2; and RRC is denoted L3 or control plane.

The split studies in 3GPP further subdivide each layer into a higher and lower part to enable discussions on function level, and not only on the aggregate level.

GSM implements implemented Option 6. Wideband code division multiple access (WCDMA) implements Option 6 for circuit-switched channels and Option 4 for HSDPA and HSUPA. Long-term evolution (LTE) has self-contained eNodeBs without any standardized split.

12.2.5 HIGHER LAYER SPLIT

In 3GPP release 15, the so-called HLS has been defined, corresponding to Option 2 in Fig. 12.10. This interface is denoted F1 by 3GPP [7]. This split enables RRC and PDCP to be located differently to the lower protocols. The control plane and user plane interfaces are called F1-C and F1-U. This allows for a central control plane covering multiple radio sites and multiple frequency bands. It also allows for a central core network termination point covering a large geographical area and supporting dual connectivity with 4G.

The main advantages with building an AAS product based on the HLS, see Fig. 12.11, are given as follows:

- optimization of the complete user plane stack to the right of the Option 2 to enable a very power-efficient product;
- utilizing that the HLS has low requirements in latency and is very bitrate efficient. It is also very transport network friendly and normal leased line IP transport can be used; and
- support for Integrated access backhaul, using the RAT also for backhauling (see Section 2.1.1).

The corresponding main drawbacks are as follows:

- carrier aggregation and L1 CoMP functions are more difficult, as they would require additional interfaces between units (they need information exchange to the right of Option 2) (see Section 6.6);

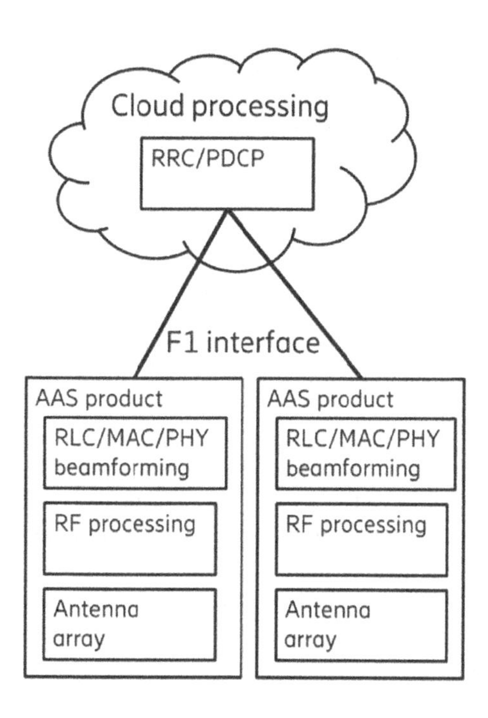

FIGURE 12.11

Example of AAS product architecture with higher layer split, using the 3GPP interface F1.

- significant baseband processing of the AAS drives size and power consumption. This may compete with other equipment on precious radio site space; and
- pooling of baseband processing hardware between sectors/bands that belong to different AAS is not possible.

An AAS based on the HLS enables a very flexible installation but has constraints on coordination features such as carrier aggregation and multitransmission point (multi-TRP) transmission/reception.

12.2.6 LOWER LAYER SPLIT

To combat the drawbacks of using the HLS as an interface to the AAS, a traditional main-remote architecture can be used, where a baseband unit implements the "main" unit and the AAS implements the "remote" unit. The interface between the two units is referred to as the lower layer split, or fronthaul.

The legacy fronthaul solution is the CPRI specification [4], Option 8 in Fig. 12.10. The CPRI specification was initially intended as a short distance interface between baseband and radio units, in the order of a few meters, with very tough requirements on time synchronization, but not so strict requirements on reducing the bit rates. Overtime, the CPRI interface has been extended to much larger distances, many kilometers, and it has then also become more essential to reduce the bit rates, in order to reduce the fronthaul cost.

With the introduction of AAS products and massive MIMO, the use of many antennas would lead to extremely high fronthaul bitrates if the CPRI interface is used. Especially for 5G RAN systems that have access to new frequency bands with larger bandwidths.

In order to reduce the fronthaul bitrate for AAS products, a new functional split between the baseband unit and the radio unit is needed.

12.2.6.1 Functional split

To reduce the fronthaul bitrates, there are different variants of Option 7 (see Fig. 12.12). The cuts D1 and D2 are two examples of functional splits for the downlink and cuts U1 and U2 are two examples for the uplink. The key property of these different functional splits is that they all have moved *beamforming* functionality from the baseband unit to the radio unit. This gives a large reduction of fronthaul bitrate.

In the downlink, there are three major blocks expanding the bit rates, namely coding, modulation, and beamforming. The more of these blocks that are placed in the radio unit, the lower the bit rate of the fronthaul interface will be. The choice between the two splits D1 and D2 is therefore a trade-off between how much reduction of fronthaul bit rates that is needed and how much of signal processing is acceptable to put in the radio unit.

The amount of reduction of data rate that can be achieved by moving these blocks to the radio unit depends on the actual use case. Roughly speaking, coding expands the data rate by a factor

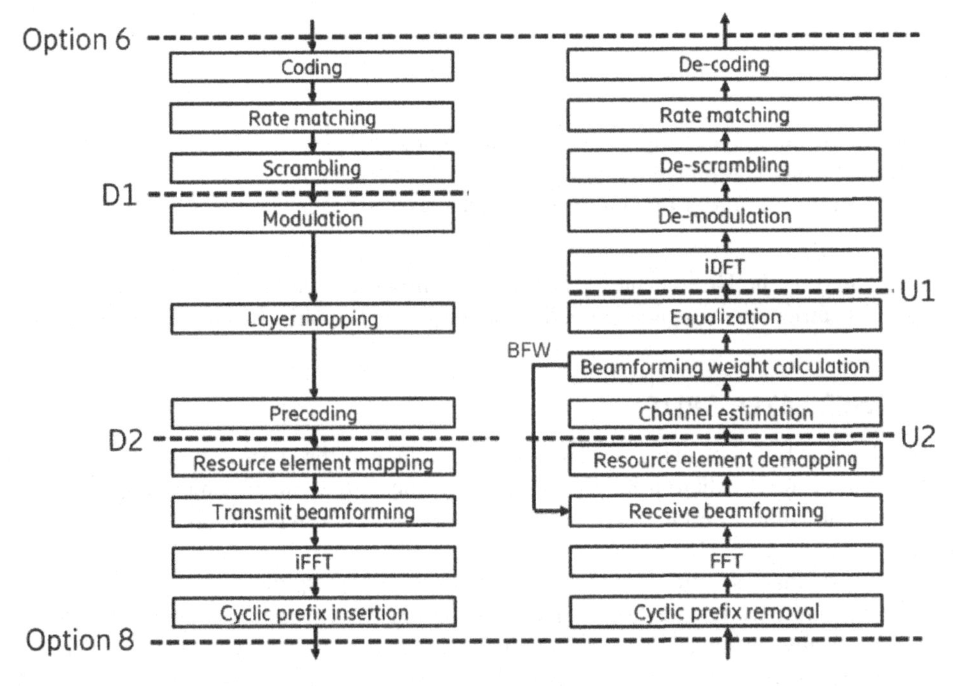

FIGURE 12.12

Different variants of the lower layer split, the so-called Option 7. BFW, beamforming weights.

Table 12.1 Bitrates for Different Functional Splits.

Bitrates (Gbps)	Control (to Radio Unit)	Downlink User Data (to Radio Unit)	Uplink User Data (From Radio Unit)
Split 6	6.2	8.8	3.3
Split D1 and U1	6.2	8.8	13.2
Split D2 and U2	6.2	26.4	13.2
Split 8 (CPRI)	0	177	177

one to three (since the code rate is 1/3), modulation by a factor of three (the ratio between IQ data bits and information bits) and beamforming a factor of four or more (the ratio between the number of radio branches and the number of layers) (see Table 12.1).

The main advantage of split Option 8 (the CPRI interface) is that it is very independent of the details of the RAN standard. A lot of improvements and changes to the standard can be made without impacting the interface and the components in the radio unit. The more blocks that are moved to the radio unit, the larger the impact will be on the radio unit due to future improvements and changes to the standards.

However, the standardization of the fundamental physical layer building blocks is typically very stable, and modern components, such as digital signal processor (DSP) based system-on-chip (SOC) or field-programmable gate arrays (FPGA) can be updated, and therefore this is not a major concern. The amount of processing in each block in the downlink is also rather limited, without any high complexity signal processing.

The conclusion for the downlink is, therefore, that one should at least move the beamformer and the modulator to the radio unit, that is, the split D1 gives a good balance between fronthaul bitrate reduction and increase in radio unit complexity.

In the uplink, Fig. 12.12 shows two examples of the Option 7 split, U1 and U2, in addition to the Option 6 and Option 8 splits. The three major blocks reducing the bit rates are beamforming, demodulation, and decoding. The more of these blocks that are placed in the radio unit, the lower the bit rate of the fronthaul interface will be (see Table 12.1).

Compared to the downlink blocks, the uplink blocks have some very different properties. On the one hand, the complexity of channel estimation and beamforming weight calculation can be large, which could indicate that it would be better to not place these in the radio unit. On the other hand, the performance will be better if the channel estimation and the beamforming weight calculations are in the radio unit. In Ref. [8], this performance difference is shown by the comparison between the schemes called "DMRS-based prefiltering" and "SRS-based prefiltering." For the case with low signal-to-noise ratio (SNR), the difference in required SNR for a given throughput can be as large as 4 dB, which corresponds to a significant loss of coverage.

The conclusion for the uplink is therefore that if both low fronthaul bitrates and good performance are important, then beamforming, channel estimation, and beamforming weight calculation should be in the radio unit. If performance can be sacrificed, then it is enough to only have the beamforming processing in the radio unit.

Assuming an AAS product with 64 antenna branches and 100 MHz bandwidth [corresponding to 273 physical resource blocks (PRB)], supporting 16 layers of user data in downlink and 8 layers in uplink, Table 12.1 shows the downlink control, downlink user data, and uplink user data bitrates for different splits, given the following assumptions:

- Downlink data: 256-QAM (8 bits/symbol), peak rate of 8 Gbps with 16 layers. Split 6 and D1 sends unmodulated data with 10% protocol overhead. Split D2 sends modulated data with 24-bit samples/8-bit symbol plus 10% protocol overhead. This gives 8.8 Gbps for split 6 and D1, and 26.4 Gbps for split D2;
- Uplink data: 64-QAM (6 bits/symbol), peak rate 3 Gbps with 8 layers. Split 6 sends unmodulated data with 10% protocol overhead. Split U1 and U2 sends modulated data with 24-bit sample/6-bit symbol plus 10% protocol overhead. This gives 3.3 Gbps in split 6 and 13.2 Gbps in split U1 and U2;
- Control: The control data for both uplink and downlink is sent to the radio, thus adding to the downlink user data. The peak occurs when sending beam weights for downlink transmission in a TTI, in parallel with sending downlink data, since both signals are time-critical and any delay adds to the total end-user delay. Assuming 16 layers, frequency varying beam with 1 beam per 4 PRB, each beam expressed as 64 complex-valued samples of 16 bit each, plus 10% protocol overhead, this gives 16 layers $\times$ 273/4 beams/layer $\times$ 64 $\times$ 16 bit/beam $\times$ 1.1 = 1.2 Mbit to transfer. The acceptable time to allocate to this transfer is assumed to be 0.2 ms leading to 6.2 Gbps;
- CPRI: Assume the 100 MHz signal is generated from a 115.2 Msps iFFT, each sample expressed as 24 bits. This gives a bit rate of 2.7 Gbps/radio branch. With 64 branches, this gives 177 Gbps.

The benefit of transferring per-layer data over the fronthaul with the D1/D2/U1/U2 splits, comparing with transferring per-antenna data over the CPRI split, is clearly seen. Also, the benefit will be even larger for designs with small number of layers and even larger number of radio branches. These splits scale much better than the CPRI split, since the fronthaul data rate scales with the actual user data, and not the number of radio branches.

12.2.6.2 Lower layer split standardization

3GPP performed a study [9] on the lower layer split. However, it was concluded without a specific recommendation on what split to be used, due to difficulties in defining a single split that will work well for the large varieties of functionality and product implementations that the 3GPP specification aims to support.

Later, the CPRI Forum released the eCPRI specification [4], an evolution of the CPRI specification that includes support for lower layer splits. The focus of the eCPRI specification is to add support for packet-based Ethernet transport, (see Section 12.2.6.3). The eCPRI specification enables the use of any kind of lower layer split.

The eCPRI specification is an *open standardized interface*, meaning that any vendor can develop a product according to the specification. That does not necessarily mean that products from different vendors can interoperate. This is also the case for the CPRI specification, which has been successful in driving an industry eco-system for the connection between the baseband and the radio units as small form-factor pluggable transceivers (SFPs) supporting the CPRI line rates and fronthaul solutions also based on coarse wavelength division multiplexing (CWDM) and dense wavelength division multiplexing (DWDM) technologies.

The O-RAN Alliance [10] is developing an *open multivendor interface* variant of the eCPRI specification that aims to achieve interoperability between different vendors. Supporting multivendor interfaces has both benefits and drawbacks. It can give more flexibility, but at a higher development and maintenance cost, slower innovation speed, and performance trade-offs.

12.2.6.3 Evolved Common Public Radio Interface Ethernet transport

Even though a lot of the focus on the lower layer split is on reducing the fronthaul bitrates for AAS products, one more significant improvement of the eCPRI standard compared with CPRI is the support of Ethernet transport.

Ethernet enables the use of standard Ethernet-based equipment, instead of very specialized CPRI equipment. Even though point-to-point connections are the first step, there is a future path toward multipoint-to-multipoint connections using a switch fronthaul with Ethernet switches.

The high-level difference between the CPRI and the eCPRI transport solution is illustrated in Fig. 12.13. In the CPRI case, MAC (layer 2) and PHY (layer 1) are both CPRI specific, leading to CPRI specific equipment. eCPRI, on the other hand, supports Ethernet-switched or IP-routed fronthaul networks as the Transport Network Layer.

A drawback with the CPRI interface has been that it is a constant bitrate service, sending radio samples also when no traffic is sent on the air interface. The new split is designed so that the fronthaul bit rates scale with the air interface traffic. This allows the fronthaul to be oversubscribed in the same way as a traditional backhaul interface, that is, the peak capacity of the fronthaul network can be less than the total peak capacity of the radio units.

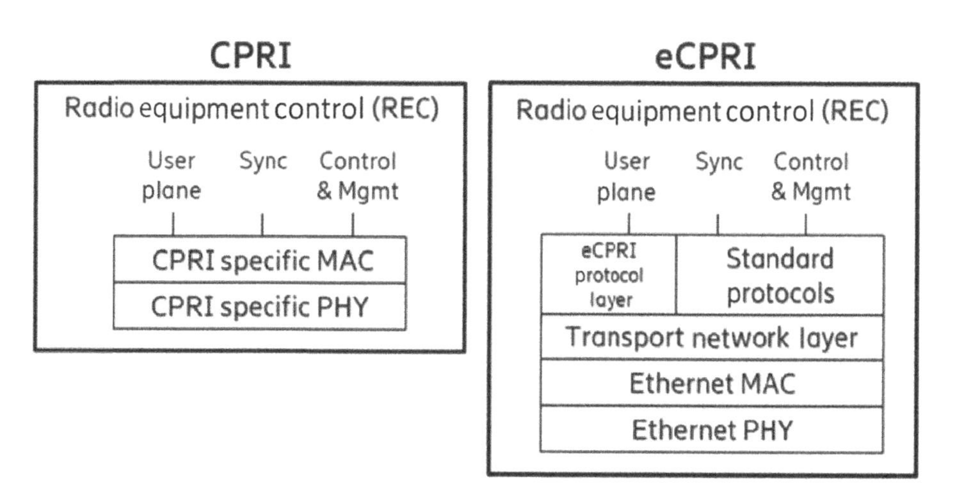

FIGURE 12.13

Architecture of CPRI and eCPRI.

12.3 5G RADIO BASE STATION IMPLEMENTATION IMPACTS

This section describes some of the implementation challenges for an RBS with an AAS. As can be seen in Fig. 12.14, illustrating a digital beamforming AAS, the AAS consists of a set of radio transceivers and a few central functions.

The primary part of the digital beamforming AAS is the radio block, which is instantiated multiple times. The radio block consists of a digital front end (DFE) consisting of digital channel filters and crest factor reduction (CFR), a transmitter/receiver block (TRX) consisting of analog-to-digital converters (A/D) and digital-to-analog converters (D/A), low-power analog circuitry (Tx and Rx), high-power transmit amplifiers (PA), low-noise receive amplifiers (LNA), RF bandpass filters, and finally a distribution network (not shown) to a set of antenna elements forming a subarray. The amount of TRX, PA, etc., depends on the number of individual RF branches integrated into the same radio block. Typically, four or eight, or even more branches can be integrated together, sharing DFE and other chips (e.g., quad D/A converters or antenna elements with dual polarization). If a radio block consists of eight branches and eight of such blocks are integrated into the same AAS, then the total amount of branches would be 64.

The AAS also has a few central parts, which are given as follows:

- The Central Digital Processing which implements the external fronthaul interface and the distribution of digital signals to the radio blocks. The digital processing steps of the radio are distributed between the Central Digital Processing and the DFE of each radio block. The split is dependent on, for instance, whether the algorithms to run can be distributed. Typically, the Central Digital Processing contains the beamforming and the control processor;

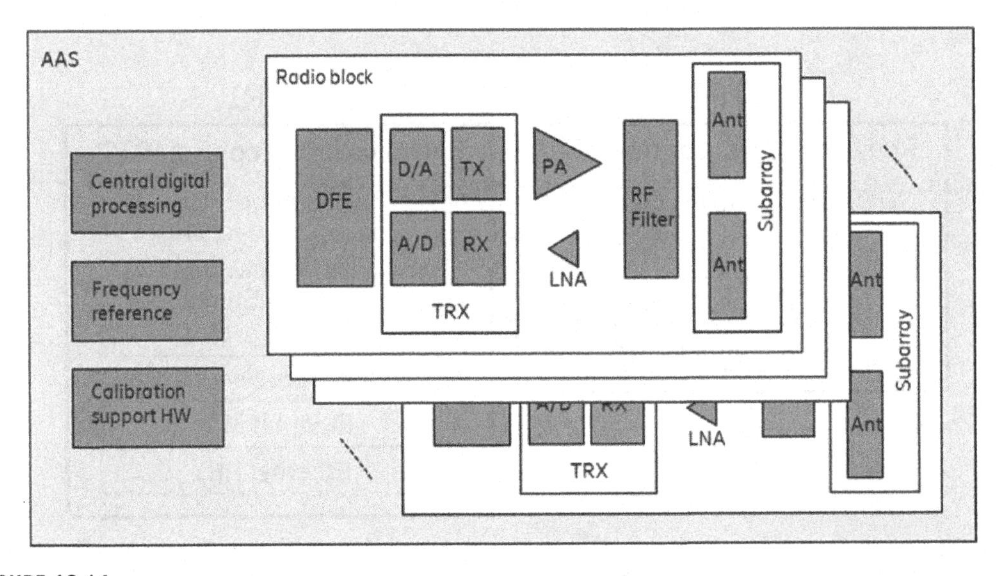

FIGURE 12.14

Sketch of digital beamforming AAS hardware.

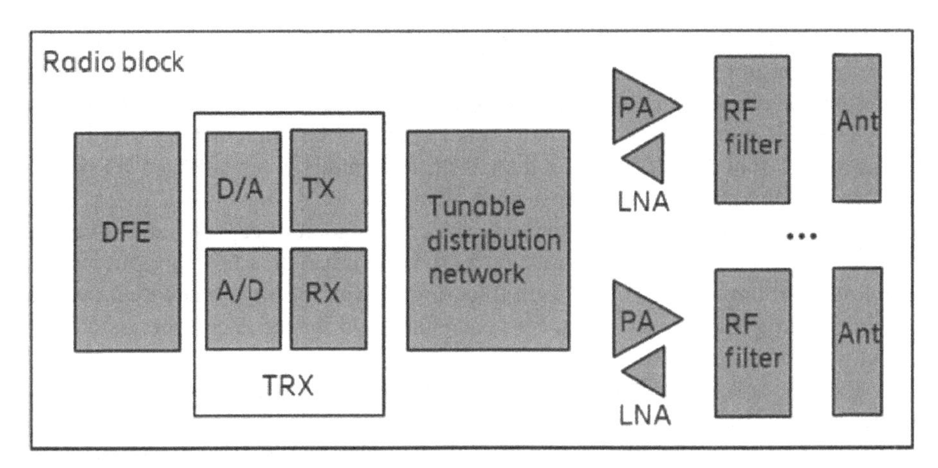

FIGURE 12.15

Radio block of an analog beamforming AAS.

- The Frequency Reference. One and the same frequency source is necessary to have a coherent combination of the RF signals, that is, to perform beamforming. The challenge lies primarily in having low phase noise and a stable distribution network;
- The Calibration Support hardware. The radio blocks gain and phase characteristics vary with time and frequency, and need continuous supervision and calibration. Depending on the method, central calibration hardware, such as coupling networks and dedicated transmitters and receivers may be needed.

Fig. 12.15 shows the radio block of an analog beamforming AAS. The AAS may contain one radio block or more, depending on the number of independent radio branches.

The radio block of an analog beamforming AAS has a dynamically tunable distribution network at RF creating different phases and gain variants of the same RF signal to different antennas. This creates one beam for the whole RF signal, as opposed to the full digital beamforming which can simultaneously direct different subcarriers within the RF signals to different beams. The Tunable Distribution Network can be implemented together with the PA and LNA in an analog front-end chip.

A combination of digital and analog beamforming can be implemented in an AAS. Different options exist, such as having multiple analog beamforming radio blocks with central functions.

A general challenge for all AAS is to match the cost and size with the targeted application. The optimal product for an RBS depends on, for example, the coverage area, the number and type of subscribers and their traffic model. The larger coverage area the larger physical antenna, and the more simultaneous users the higher flexibility is needed.

12.3.1 DIGITAL VERSUS ANALOG BEAMFORMING

In terms of beamforming potential, there is an important difference between frequency-domain beamforming and time-domain beamforming (see also Section 7.3). Frequency-domain beamforming involves the possibility to have different beams for different parts of a frequency carrier, typically different beams per resource block. Frequency-domain beamforming needs to be implemented with digital beamforming.

For time-domain beamforming, the same beam is applied for the whole frequency carrier and is typically implemented by analog beamforming, but could also be implemented with digital beamforming. That is, the functionality of time-domain beamforming is not affected by the choice of using analog or digital components for the implementation.

The mid-band AAS is typically deployed to gain a combination of capacity and coverage, where thousands of UEs are to be covered in, for example, a dense urban area from a rooftop deployment. It is then beneficial to have capable AAS with fully digital beamforming, many radio branches, and subarrays, in order to get higher signal power levels and good interference suppression. Fig. 12.16 shows a block scheme of a mid-band AAS RBS.

For time-division duplexing (TDD) mid-band frequencies, between 2 and 7 GHz, the number of subarray ports needed to achieve similar coverage as existing deployments are typically around 16, 32, or 64. The frequency bandwidths are typically in the order of 100 MHz, which means that it is technically feasible to implement full frequency-domain digital beamforming. This enables the AAS products to fully explore the spatial properties of the radio channel in order to maximize the spectral efficiency.

Beamforming is performed in the digital domain with frequency-domain beamforming, and there is one set of AD/DA converters, PA, RF receiver, and RF filter per subarray port in the antenna array.

The high-band AAS often handles very large bandwidths and is typically added for hot spot capacity as a supplement to mid-band. It serves tens to hundreds of users, using up to in the order of 1 GHz bandwidth. To get good coverage, it needs to be equipped with hundreds of antenna elements, or even thousands (see Section 7.2.2). Adding frequency-domain digital beamforming over such a large array and such wide bandwidth would give very large digital signal processing need, which in addition to high cost, also would drive the power consumption, and therefore indirectly

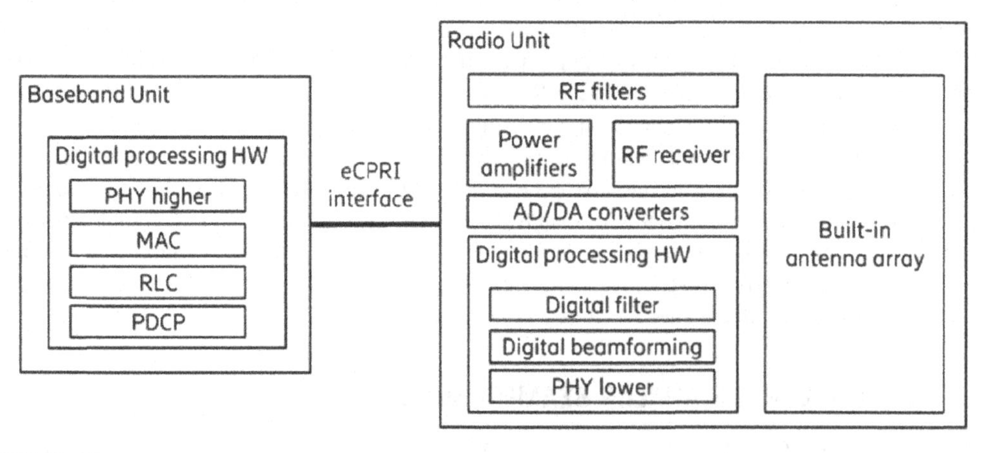

FIGURE 12.16

Mid-band AAS RBS block scheme.

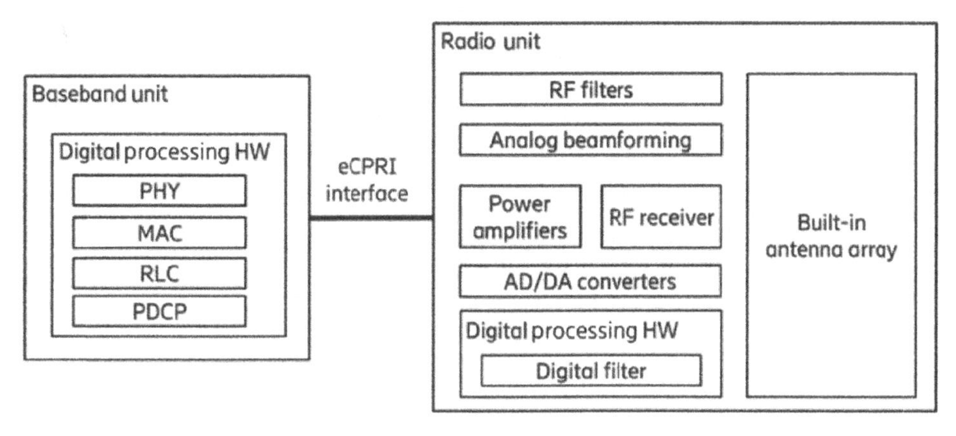

FIGURE 12.17

High-band AAS RBS block scheme.

the size of the box, and the size of the box is essential to acquire radio sites. Instead, time-domain analog beamforming is being used in the initial high-band AAS products (see Fig. 12.17).

Fortunately, the very short TTI at high-band allows for time multiplexing users at a very high rate, making spatial multiplexing within one TTI less important.

The high-band AAS RBS has more of the digital processing in the baseband unit, the whole PHY, compared with a mid-band AAS RBS, and with the use of analog time-domain beamforming, the number of AD/DA converters, PAs, and RF receivers no longer need to be as many as the subarray ports in the antenna array. Instead, it is enough to have as many as the number of layers that should be transmitted/received. For some realizations, the number of PAs and RF receivers is as many as the subarray ports, that is, it is possible to swap the order of analog beamforming and the PAs/RF receivers.

In the future, however, digital implementations of time-domain beamforming are likely to be developed. This is facilitated by taking advantage of the efficiency improvements that come with technology trends for digital processing.

12.3.2 EXAMPLE OF BEAMFORMING AT HIGH-BAND

The first NR AAS systems deployed at high-band frequencies are based on time-domain analog beamforming. A typical implementation, see Fig. 12.18, can consist of a DFE and a TRX that create an RF signal which is split into a number of antenna RF paths in the Tunable Distribution Network. Each path is amplified, and a phase shift is applied in order to steer the beam in the desired direction.

The size of the array depends on how much equivalent isotropic radiated power (EIRP) is needed for downlink coverage, but also on what array gain that is needed for the uplink. The EIRP is generated by a combination of the output power per branch and the gain of the antenna array. Assuming one PA per branch, a doubling of the array size would increase the EIRP with 6 dB, where 3 dB is from the doubling of the total output power and 3 dB is the increased antenna gain. In the uplink, on the other hand, a doubling of the array size only provides 3 dB gain from the increased antenna gain. Many of the early deployed radio units contain arrays with more than 256 elements.

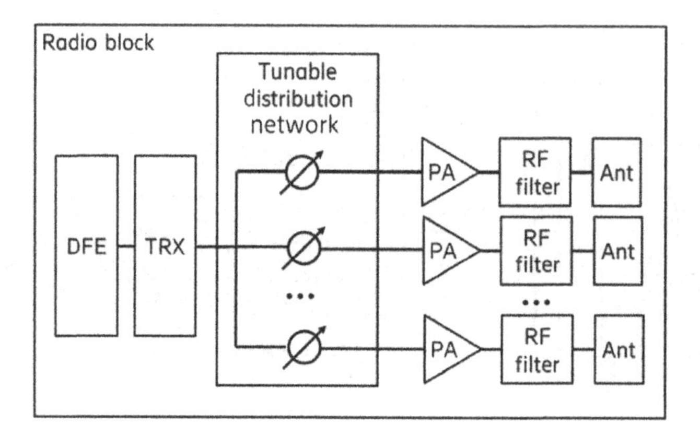

FIGURE 12.18

Time-domain analog beamforming.

A common implementation of the phase shift is to use digitally controlled phase shifters that can generate 2^B different phase values. For example, say that $B = 6$ bits are used to control the phase shifter. This would mean that a phase resolution of 360/64 = 5.6 degrees per branch can be achieved. If the phase of each branch is controlled by 6 bits, this would require a wide digital interface. With the numbers in this example, 6×256 bits are needed to set the appropriate beam per time unit. Further, assume that the beam should be updated every slot, which is once every 125 μs for a subcarrier spacing of 120 kHz. This would require an interface capacity of more than 12 Mbps. Assuming further that the amplitude of each branch should be controlled as well. For simplicity, it can be assumed that the amplitude, or power, per branch also can be controlled with 6 bits, this would increase the needed interface capacity to 25 Mbps. This would not be a problem with a state-of-the-art digital implementation; however, the phase shifters and PAs are implemented in a process developed for analog design, and hence the efficiency of digital processing may be very poor. To overcome this, a fixed set of phase and amplitude values can be indexed in a table to form a beam weight table. This table can consist of a fairly low number of beamforming vectors and hence the interface capacity can drastically be reduced. In practice, the beam table should consist of enough beams to span the wanted horizontal and vertical angular range. This can be in the order of 64–256 entries and hence only 6–8 bits per scheduling instance is needed to index a beam, which gives an interface capacity of 48–64 kbps.

If a digital implementation of the beamforming functionality would be possible, a full frequency-domain beamforming would consume hundreds of watts, given the number of antennas (256–1024) and the bandwidth (400–800 MHz), even with a state-of-the-art digital process. Another bottleneck in a digital design is the large bitrates that are created in the system. To see this, a simple but illustrative example of a typical transmitter chain is studied. The transmitter chain and its components are illustrated in Fig. 12.19.

Assume a system providing 4×100 MHz carriers. Further assuming a subcarrier spacing of 120 kHz, meaning that 66 PRBs per carrier are delivered by the physical layer at point A in the figure. There are 12 subcarriers per PRB and 14 Orthogonal Frequency-Division Multiplexing (OFDM) symbols per slot. Each slot is 125 μs, hence there will be 8000 slots per second. Spatial multiplexing might

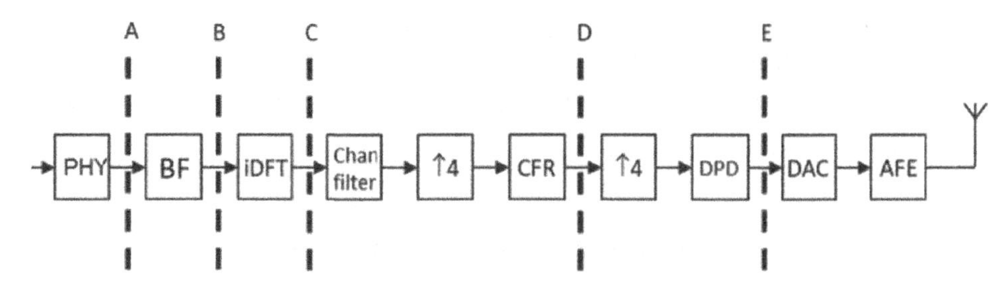

FIGURE 12.19

Transmitter chain for digital beamforming. Starting from the left with PHY layer processing up to beamforming (BF), inverse discrete Fourier transform (iDFT), channel filter, four-times upsampling, crest factor reduction (CFR), four-times upsampling, digital predistortion (DPD), digital-to-analog conversion (DAC), and analog front-end (AFE).

be less efficient for high-band systems, but it can be assumed that four layers should be supported by the system. All this would require $66 \times 12 \times 14 \times 8000 \times 4 = 355$ Msymbols/s. Typically, each physical layer symbol is represented by an IQ sample of about 20 bits. Moreover, in this example, four 100 MHz carriers are assumed. Then the physical layer would generate $355 \times 4 \times 20 = 28.4$ Gbps in this example.

The next step in the transmitter chain is beamforming. The four layers delivered by the physical layer should then be beamformed and transmitted over the antenna branches. If the four layers should be expanded to 256 antenna ports, a staggering $28.4 \times 256/4 = 1.8$ Tbps will be generated in the interface B. Note that the layers are added together in the beamformer, so that the bitrate numbers at interface B and later are independent of the number of layers.

A typical transmitter chain then consists of channelization filters which make sure that the generated OFDM signal confines to the spectral mask. The digital filter should also smooth the discontinuities between OFDM symbols as those can eventually generate out-of-band emissions. The transmitter chain may also include CFR and algorithms for the linearization of the PAs (see Section 11.3.1.1). The CFR algorithm usually requires oversampling of the signal to be effective and to, for example, combat peak regrowth [11]. Similarly, a typical digital predistortion (DPD) algorithm used to linearize the PA will require further upsampling of the signal to combat intermodulation terms generated by the nonlinearities of the PA.

Assume 120 Msamples-per-seconds/carrier/branch through the channel filter, hence a slight oversampling relative to the 100 MHz carrier bandwidth, where each IQ sample is represented by 20 bits. A typical upsampling factor for CFR would be 4, meaning that the CFR algorithm works on $4 \times 120 \times 20 = 9.6$ Gbps per branch per carrier. With 4 carriers and 256 branches, the total bandwidth would be 9.8 Tbps at point D in the figure. The DPD algorithm might require a further upsampling with, say, a factor of 4, hence the total bandwidth for the data into the DACs, point E in the figure, is in the order of 40 Tbps.

The cost and power consumption of a high-band system using full digital beamforming become very high: the central digital processing with beamforming itself consumes hundreds of watts, the radio block will have one chain per antenna both for DFE, converters, and TRX, and interface rates will go up. The potential benefit in capacity needs to be weighed against such a cost and size increase.

The best way of capturing more traffic is to further increase the coverage area, and that is achieved by further increasing the antenna area. This in turn leads to an even higher computational need and instantiation of radio blocks, thus even higher cost and size. The deployments where frequency-domain beamforming can be motivated instead of time-domain beamforming on high-band is thus few. Instead, different hybrid solutions will likely be developed balancing cost versus gain.

12.3.3 ANTENNA ELEMENT AND SUBARRAY IMPLEMENTATION

For traditional low-band and mid-band sector antennas, vertical-only subarrays with sizes of 8–16 elements have been used. For a given total antenna size, a larger subarray is typically reducing the cost of the product as the number of radio chains is reduced. Also, the complexity of the beam-forming algorithms is reduced if the number of radio chains is reduced.

The size of the subarray of an AAS product is set by two main factors, which are given as follows:

- The largest subarray that is acceptable given the angular spread of the UEs at the radio site, as will be discussed in Section 14.3;
- The smallest number of radio chains that are needed to accomplish the wanted resolution of the beam shapes or to get the total wanted output power.

The trade-offs are different for mid-band and high-band.

For mid-band AAS placed on rooftop or masts, the angular spread of the UEs allows for subarrays of 2–4 dual-polarized antenna elements in the urban area, and even more in the suburban areas. The number of subarrays is kept low since cost per radio chain is still significant given the tough RF purity requirements.

For high-band AAS, the size of the subarray is very much determined by the losses in the connecting material between the antenna element and the front-end components of the radio. For printed circuit board (PCB) based antennas distribution, the loss grows quickly, and the subarray is limited to 2–4 antenna elements. Alternatives with waveguides allow for larger subarrays.

The EIRP of an AAS is determined by the total number of subarrays, the subarray size and the output power of the radio chain feeding the subarray. For mid-band, the output power of the radio chain can be chosen fairly arbitrarily, and the subarray size and number of subarrays can be chosen based on the user distribution, allowed unit size, and wanted received performance. For high-band, the output power per radio chain is very limited and to achieve the wanted EIRP, more radio chains than needed to achieve the beam resolution or steering range may be required. The high signal distribution loss adds to this and drives many radio chains. This explains why mid-band typically uses 16–64 radio chains whereas high-band uses 256–1024.

The high-band subarray size can be increased, and thus the number of radio chains reduced, when lower loss distribution network is used, for example, waveguides with slits antennas. Such solutions get a restricted steering angle but could be used in some installations.

Fig. 12.20 shows subarrays for mid-band and Fig. 12.21 shows subarrays for high-band. The mid-band subarray is based on two dual-polarized patch antennas. Each subarray has two RF connectors, one per polarization, placed on the top or bottom of the PCB. Sixteen subarrays are combined on one PCB, and the array shown contains 4 PCBs, giving a total of 64 subarrays. Each PCB also contains traces for coupling the signal back to the antenna calibration block and to an RF test

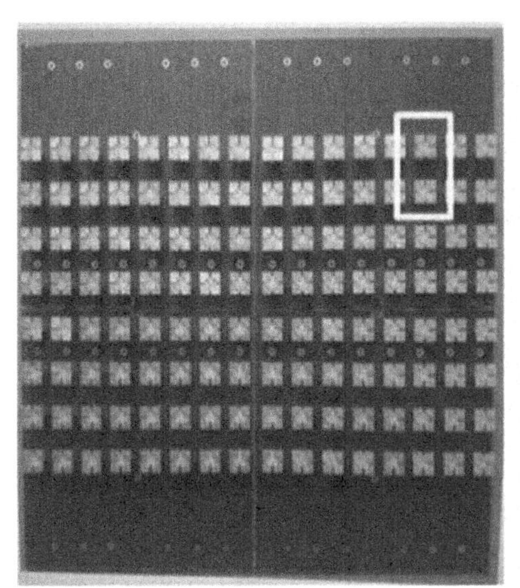

FIGURE 12.20

Example of an antenna array with 2×1 subarrays for mid-band. One of the subarrays is indicated by the yellow box. The picture to the left shows the antenna array from the front and the picture to the right shows the antenna array from the back.

FIGURE 12.21

Example of antenna array with 2×1 subarrays for high-band. The yellow box indicates one of the subarrays. The picture to the left shows the antenna array from the front and the picture to the right shows the antenna array from the back. 28 GHz phased antenna array module (PAAM), a collaboration between Ericsson AB and IBM T. J. Watson Research Center.

port. The antenna calibration signals for the 16 subarrays are combined on the PCB as can be seen in the horizontal space between the subarrays on each PCB.

The high-band subarray is mounted together with other subarrays on a joint PCB, with the radio front-end chip mounted on the back to keep distribution loss low. The shown radio front-end chip contains the front end of 16 subarrays, giving a total of 32 dual-polarized subarrays of size 2×1 on the PCB. Outside the active array is a row/column of passive elements to shape the overall pattern.

12.3.4 RADIO SITE IMPACTS AND CONSTRAINTS

Considering the already complex radio site installations, it is not always an easy straight-forward task to add AAS products to an existing radio site. The benefit and performance of AAS products come inherently from using many antenna elements (which should be organized in suitably sized subarrays), which means that the size has to be somewhat large, at least for sub-6 GHz.

There is, therefore, a trade-off between the number of branches, the power consumption, the complexity of beamforming algorithms, and the size and weight of the AAS product. This will be thoroughly discussed in Chapters 13 and 14. How large size and weight that are acceptable differs a lot between different mobile network operators; some can accept larger size and weight, but for others that would be a significant challenge. Even if the size and weight are reasonable, there might also be limitations on the number of units that can be installed at a radio site.

In order to facilitate new products on an already busy radio site, one can design products supporting multi-band operation, thereby avoiding the need to have one physical box per frequency band.

Another possibility is to also add passive antennas to the AAS product. That is, for some frequency bands, the AAS product provides RF ports to which external radio units can be connected (see Fig. 12.22). In this way, existing passive antennas can be removed, and the new AAS product will both provide the passive antennas for the existing radio units and provide radio functionality and the antennas for the new frequency bands, and in this way provide more frequency bands without adding more physical units to the radio site.

12.3.5 PHASE NOISE AT HIGH-BAND

The transceiver in the radio unit performs up and down conversion to/from the carrier frequency. As part of this conversion, a local oscillator (LO) is used as a frequency source. Due to inaccuracies in the LO, and in the frequency conversion, so-called phase noise will be generated (see Section 11.6.2.2).

As the carrier frequency becomes higher, the phase noise becomes more challenging. 3GPP is handling this by increasing the subcarrier spacing and by adding phase tracking reference symbols (PT-RS). Eventually, as the carrier frequency becomes higher and higher, this will not be enough, and more advanced functionality as high-order modulation and multi-user MIMO will be less feasible.

Phase noise is, therefore, an important aspect to consider when designing high-band standards and high-band radio unit products.

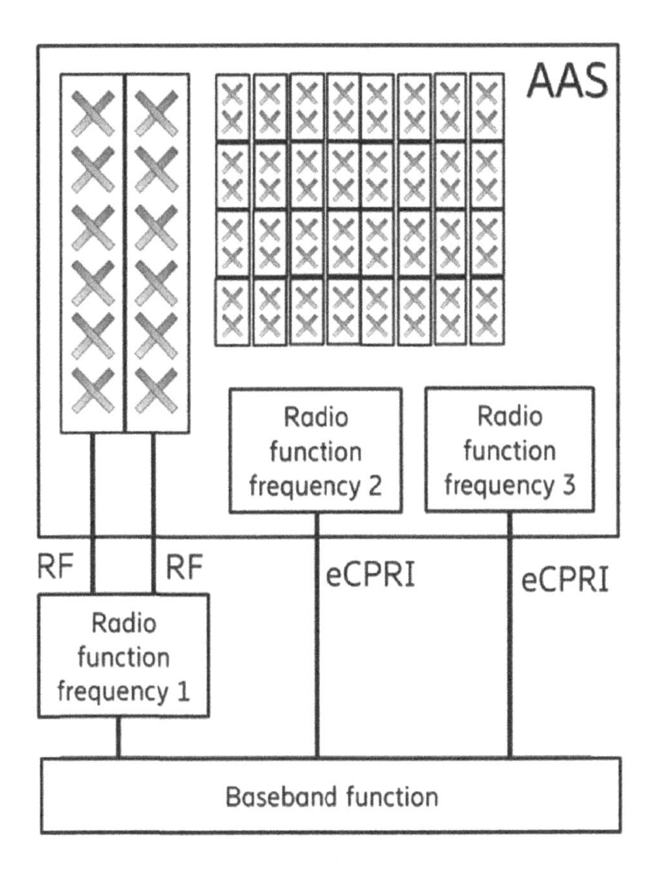

FIGURE 12.22

Example of AAS product with both multiple active antennas and passive antennas. In this example, the different sized antenna elements are places physically separated, but there are also techniques to interleave antenna elements to reduce the size.

12.3.6 ALIGNMENT REQUIREMENTS AND CALIBRATION

For AAS base stations capable of beamforming, the transmitted signals from the antenna array need to be aligned, which requires the phase and amplitude difference between transmitters to be controlled. This applies both to solutions where the beamforming weights are applied in the digital domain and to solutions where they are applied in the analog domain. If the alignment between transmitters is incorrect, then the beamforming experienced over the air will deviate from the intended beamforming pattern set by the beamforming weights. Similarly, in the receive direction, alignment between receivers is needed.

There are three types of alignment requirements, which are given as follows:

- transmitter alignment
- receiver alignment
- receiver-to-transmitter alignment.

The transmitter alignment is important for broadcast signals/channels, as well as downlink user-specific channels when beamforming is based on UE measurement feedback. The receiver alignment is important for the reception of uplink channels, especially if the receiver algorithms use knowledge of the antenna array structure. Finally, the receiver-to-transmitter alignment is important for downlink user-specific channel when beamforming is based on uplink measurements, that is, so-called reciprocity-based beamforming.

The reciprocity can be either large-scale reciprocity (e.g., directions), applicable to both frequency-division duplexing (FDD) and TDD systems, or fast fading reciprocity, applicable to TDD systems only (see Section 6.4.2). However, it is important to be aware of that interference is not reciprocal, that is, the interference experienced by the UE is not the same as what the base station can measure in the uplink.

Fig. 12.23 shows a simplified structure of a TDD transmitter and receiver. The switch selects whether a signal should be transmitted or received. In this example, there are only two antennas, but this can be generalized to an arbitrary number of antennas. The alignment requirements are requirements based on similarities between pairs of antennas.

Compared with an ideal device, a real implementation has various error sources which contribute to misalignment. These error sources can be divided into three categories, errors affecting both transmission and reception, errors that only affect reception and errors that only affect transmission.

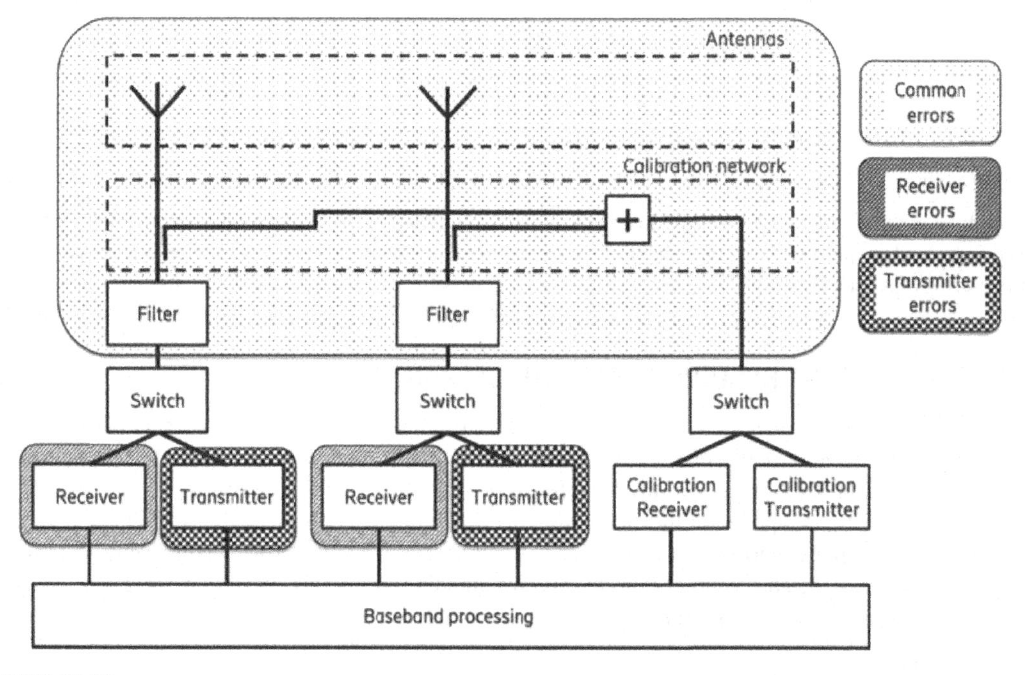

FIGURE 12.23

Alignment error types and calibration.

The requirement on transmitter alignment is therefore related to the sum of the common errors and the transmitter errors. The receiver alignment requirement is related to the sum of the common errors and the receiver errors. The receiver-to-transmitted alignment requirement is related to the sum of the receiver errors and the transmitter errors, but importantly, not to the common errors.

The method of achieving alignment is called calibration. In Fig. 12.23, one possible calibration solution is illustrated, which is typically used for mid-band AAS products. This solution consists of a dedicated transmitter and receiver that are used to inject a special measurement signal using a calibration network. The calibration network consists of couplers that leak a small part of the signals going between the filters and the antennas, and combiners that add/split the signal between the couplers and the calibration receiver/transmitter.

When performing transmitter calibration, the baseband processing block adds special measurement signals for each antenna. These signals are then leaked via the calibration network to the calibration receiver. The difference between the measurement signals for each antenna is then analyzed by the baseband processing block, which then can compensate for any measured misalignment.

When performing receiver calibration, the calibration transmitter is generating a measurement signal, which is leaked back to each of the receivers via the calibration network. The difference between the measurement signal for each antenna is then analyzed by the baseband processing block, which then can compensate for any measured misalignment.

Here it is important to note that the feedback loop used in this solution does not measure any misalignment in the antennas or in the calibration network itself. The antennas and the calibration network, therefore, need to be manufactured with high quality in order to fulfill the overall alignment requirements.

For mm-wave AAS base stations, the inclusion of coupler as part of the calibration network is challenging, and thus other schemes such as mutual coupling or even accurate array knowledge (empirical) are used.

Calibration may encompass not only phase alignment but, in some cases, also amplitude alignment. Alignment must be ensured between each and every pair of subarray ports. Amplitude alignment is less critical than phase alignment with regards to beamforming gain and pointing directions. The numerical value of the pointing error is generally much smaller than the numerical value of the phase error between the subarrays, due to statistical averaging of the errors between multiple subarray ports.

Amplitude alignment becomes more important when nullforming is needed. The requirements on phase alignments also become tougher when nullforming is needed, for example, when multi-user MIMO is implemented. Requirements on side-lobe suppression might also lead to tougher alignment requirements.

Similar to downlink transmission, the uplink receiver combining weights are essentially phase shifts that correspond to the direction, and consequently for some receiver architectures, calibration is important to ensure low sidelobes, correct pointing direction, and correct null steering to mitigate interference. For receiver architectures with a generic MMSE-based receiver, uplink alignment is less important, but still, there will be improvements from good alignment, especially if the receiver is using some knowledge of the antenna array structure.

For TDD reciprocity-based beamforming, there is a need to also ensure the relative phase and amplitude alignment between the receiver and transmitter, since the reciprocity-based operation in the base station first estimates the uplink channel and uses the uplink channel for downlink transmission.

Thus, for reciprocity-based beamforming, for any pair of subarrays, the relative phase differences between the receiver branches need to match the relative phase differences between the transmit branches for all possible pairs. The phase differences between any two receive antennas do not need to be the same as long as the phase differences are the same as the corresponding phase differences of the transmitters. If there is an error at reception, it is then compensated for at transmission.

As mentioned, the alignment accuracy requirement depends on the type of beamforming. The requirements are tougher for multi-user MIMO (where nulls must be formed) compared with single-user MIMO (where only a main lobe toward the intended UE is needed). Assuming a TDD system, using reciprocity-based beamforming for multi-user MIMO, one numerical example of phase alignment requirements could be the following:

- transmitter alignment within 25 degrees
- receiver alignment within 25 degrees
- receiver-to-transmitter alignment within 15 degrees.

Here it is important to point out the importance of having these three independent requirements. It would, for example, be possible to replace those three requirements with only a transmitter and a receiver alignment requirement. However, in that case, those two requirements become unnecessary tough. Assuming the worst-case scenario, the transmitter and receiver alignment requirement would need to be half each of the receiver-to-transmitter alignment requirements. That is, the alternative set of phase alignment requirements would be the following:

- transmitter alignment within 7.5 degrees
- receiver alignment within 7.5 degrees.

This is a much tougher requirement, giving a more costly solution, without any significant performance gain. It is therefore important to have three different requirements, to match the three different error sources, in order to enable cost-efficient solutions.

12.3.7 MUTUAL COUPLING

In an AAS base station, the antenna array will have the antenna elements placed close to each other in order to efficiently use the area. When antenna elements are placed close to each other, there will be some mutual coupling from surrounding elements (see Fig. 12.24). Coupling causes an element to not only radiate its intended waveform, but also the waveform of the elements that are coupled.

Mutual coupling affects the antenna array efficiency and beam characteristics depending on beamforming solution and beam direction. The impact of mutual coupling on the gain pattern of 3×3 uniform rectangular array with $0.5\ \lambda$ separations is visualized in Fig. 12.25. The figure illustrates the radiation pattern from individual radiating elements, which can be seen to be distorted due to the coupling. The mutual coupling also results in challenges for the PA because the coupled signals are reflected back into the PA.

Elements in the middle of the array experience the largest amount of mutual coupling due to being surrounded by eight adjacent antenna elements as shown in Fig. 12.24 (right), whilst elements at the array edges experience the lowest amount of coupling.

For AAS base stations where linearization is required due to stringent emission requirements, for example, for sub-6 GHz, the mutual coupling can potentially be a serious issue because

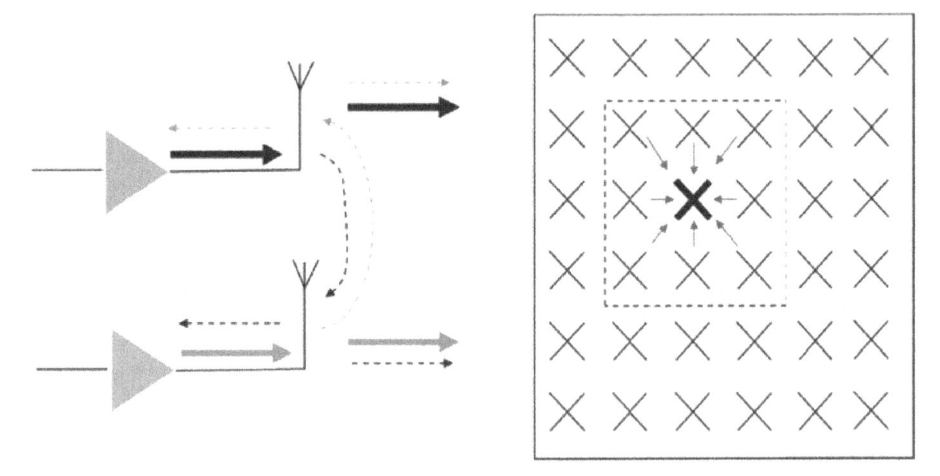

FIGURE 12.24

The thick arrows represent the wanted radiation, the narrow arrows represent the mutual coupling (left picture). Mutual coupling between neighboring antenna elements (right picture). The energy due to mutual coupling both goes back into the neighboring power amplifier (PA) and radiates out from the neighboring antenna element.

intermodulation between the wanted signal and coupled signals can occur, leading to increased unwanted emissions. Such arrays may require isolators/circulators with high isolation after the PA or more sophisticated DPD algorithms taking care of intermodulation induced by mutual coupling. For mm-wave AAS base stations, the emission requirements are less stringent compared to sub-6 GHz and isolators/circulators are difficult to integrate. Thus, accurate knowledge of the mutual coupling needs to be considered in the design.

Mutual coupling within an array is thus a serious issue which can give rise both to affected radiation characteristics as well as increased PA emissions. Mutual coupling may be dealt with by careful design and also in some cases algorithmically (e.g., with advanced DPD or beamforming). The existence of mutual coupling leads to further challenges for AAS base stations compared to non-AAS base stations.

Therefore, as a conclusion, the design of the array and the RF parts of the radio unit should be done so that the amount of mutual coupling will be low enough to avoid more advanced designs of the digital algorithms such as DPD and the beamforming itself. That is, given this, *all the existing theory and algorithms can be applied without needing to take any mutual coupling effect into account*. Note that this typically requires the antenna element distance to be 0.5λ or larger.

12.4 SUMMARY

In this chapter, a brief overview of the evolution of RAN architectures has been given and two new functional splits for the 5G RAN architecture have been illustrated. The higher layer split, that

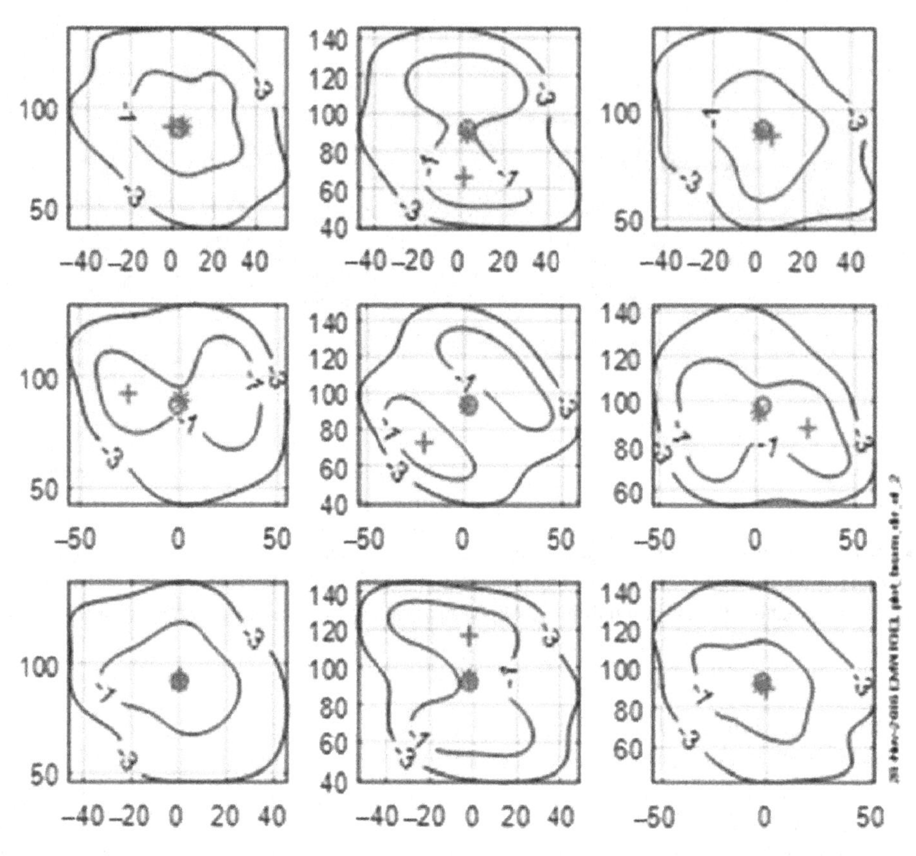

FIGURE 12.25

1 dB and 3 dB contours of the antenna element gain pattern foreach antenna element in a 3 × 3 rectangular array. The x-axis shows the horizonal angle from -50 to +50 degrees (0 degree being the middle of the main lobe), and the y-axis shows the vertical angle from +40 to +140 degrees (90 degrees being the middle of the main lobe).

provides a new split between dedicated hardware and cloud implementation, and the lower layer split which drastically reduces the fronthaul bitrates for AAS products.

The lower layer split is a key part of an efficient solution for AAS. To make a split in the middle of the physical layer is non-trivial and a lot of care has to be taken when handling the trade-offs between fronthaul bitrates, complexity, and performance. Different product types will also benefit from different splits. Especially it is shown that mid-band AAS product benefits from moving a significant part of the uplink baseband processing to the radio unit, to obtain a good trade-off between performance, fronthaul bitrate and complexity.

From a practical point of view, the installation of AAS products at the radio site is a real challenge for many mobile network operators. The masts and towers are already crowded, and to add more units might not be possible. This drives the need to develop both multi-band products and AAS products with a combination of active and passive antennas.

A few specific technical challenges with AAS are highlighted, phase noise at high-band, mutual coupling between antenna elements, and the different requirements on phase and amplitude alignment. When determining the alignment requirements it is important to take into account what type of beamforming should be supported and to formulate independent requirements for the three alignment requirements, receiver alignment, transmitter alignment, and the receiver-to-transmitter alignment.

REFERENCES

[1] GSM technical specification; Digital cellular telecommunications system (Phase 2 +); General description of a GSM Public Land Mobile Network (PLMN), ETSI, GSM TS 01.02.

[2] 3rd Generation Partnership Project, Technical Specification Group Radio Access Network; UTRAN Overall Description, 3GPP, 3GPP TS 25.401.

[3] HSDPA and HSUPA. <https://www.3gpp.org/technologies/keywords-acronyms/99-hspa>.

[4] CPRI and eCPRI Specifications. <http://www.cpri.info>.

[5] 3rd Generation Partnership Project, Technical Specification Group Radio Access Network; Evolved Universal Terrestrial Radio Access Network (E-UTRAN); Architecture Description, 3GPP, 3GPP TS 36.401.

[6] 3rd Generation Partnership Project, Technical Specification Group Radio Access Network; Study on New Radio Access Technology: Radio Access Architecture and Interfaces, 3GPP, 3GPP TR 38.801.

[7] 3rd Generation Partnership Project, Technical Specification Group Radio Access Network; NG-RAN; F1 General Aspects and Principles, 3GPP, 3GPP TS 38.470.

[8] W. Chang, T. Xie, F. Zhou, J. Tian, X. Zhang, A prefiltering C-RAN architecture with compressed link data rate in massive MIMO, in: IEEE 83rd Vehicular Technology Conference (VTC Spring), 2016.

[9] 3rd Generation Partnership Project, Technical Specification Group Radio Access Network; Study on CU-DU Lower Layer Split for NR, 3GPP, 3GPP TR 38.816.

[10] O-RAN Alliance. <https://www.o-ran.org>.

[11] G. Wunder, R.F.H. Fischer, H. Boche, S. Litsyn, J-S No, The PAPR problem in OFDM transmission, IEEE Signal Proc. Mag. 30 (2013) 130–144.

CHAPTER

PERFORMANCE OF MULTI-ANTENNA FEATURES AND CONFIGURATIONS

13

The objective of this chapter is to illustrate the radio network performance of multi-antenna features and configurations. The aim is to build on earlier chapters and visualize previously discussed concepts and results based on radio network simulations. Focus is on macro deployments with *mobile broadband* (MBB) as the main use case.

There are many aspects and dependencies that affect the performance of a multi-antenna or *advanced antenna system* (AAS) configuration, which makes it difficult to isolate and consider one aspect at a time. The spatial channel characteristics (see Section 3.5.3.2) are fundamental when discussing AAS performance and there is a close connection between the deployment scenario and desirable AAS characteristics, that is, different scenarios call for different antenna configurations and beamforming techniques to maximize the network performance.

This chapter focuses on how the antenna layout and the choice of multi-antenna algorithm impact the network performance, and the closely related discussion on how the deployment scenario impacts performance is more thoroughly treated in Chapter 14.

It should be highlighted that performance numbers and gains presented in the following text should be seen as indicative and representing trends rather than absolute truths. Evidently, the results of performance evaluations depend on many aspects, such as traffic models, propagation models, algorithm assumptions, antenna assumptions, system assumptions, and simulation methodology assumptions (see Section 13.2 and references therein). When considering simulation results, the underlying assumptions should always be scrutinized as these can strongly impact the results and conclusions. Given the discussion in Section 6.7.4, the aim is to use as relevant and realistic assumptions as possible in line with the industry standard.

A short outline of the chapter with key takeaways is given in Section 13.1. References to later sections where the details are given are also included. Each subsection will then start with a short overview and recap of relevant parts of previous chapters, followed by the main results. Additional details are provided in later subsections. This will hopefully help guiding the reader in the sense that the general reader might only need the overview, while the curious reader might want to dig into the details.

13.1 OUTLINE AND SUMMARY

The aim of this chapter is to illustrate how the *antenna configuration* and *multi-antenna algorithm* dimensions affect the network performance.

Advanced Antenna Systems for 5G Network Deployments. DOI: https://doi.org/10.1016/B978-0-12-820046-9.00013-7

By adopting the *array of subarray* (AoSA) antenna structure introduced in Section 4.6.1 and illustrated in Fig. 13.1, the antenna array consists of several subarrays, where the subarray is assumed to be the smallest dynamically controllable entity, with two radio chains per subarray, one for each polarization. Key AAS configuration aspects affecting beamforming performance given the AoSA layout are as follows:

- the subarray size;
- the number of subarrays; and
- the AoSA layout (horizontal and/or vertical stacking).

These aspects will be frequently discussed throughout this chapter, and more detailed descriptions, illustrations, and assumptions of the AoSA structure follow in Section 13.2.2.

As discussed in Sections 6.1 and 6.4, key aspects affecting the choice of beamforming algorithm for different UEs and physical channels include the required spatial coverage of beamformed signals and the availability of *channel state information* (CSI). Beamforming is herein broadly classified as *cell specific* or *UE specific*. Beamforming strives to increase the signal-to-interference-and-noise ratio (SINR) of desired signals, which can lead to increased coverage, capacity, and end-user throughput. UE-specific spatial multiplexing can also make better use of large SINRs by dividing the power over several layers, either to one UE (referred to as SU-MIMO) or to several UEs (referred to as MU-MIMO) as explained in Section 6.3. The impact on network performance from different AAS techniques is another central theme in this chapter. Considered algorithms are described in Section 13.2.3.

A detailed description of radio network simulation methodologies is outside the scope of this book, but for completeness, a few simulation aspects, models, and assumptions are discussed in Section 13.2. Note also that a brief introduction to network simulations was given in Section 2.2.4.

The outline of the chapter is summarized in Fig. 13.2 with more details and key takeaways following in Sections 13.1.1–13.1.5. The sections in this chapter can be read rather independently although Sections 13.3–13.5 are closely related. Note also that Sections 13.2.2–13.2.4 are fundamental in understanding notations and concepts discussed elsewhere in the chapter.

13.1.1 CELL-SPECIFIC BEAMFORMING

Cell-specific beamforming used to define the area where cell-wide signals are transmitted comes in many flavors ranging from site planning and tilt adjustments to more advanced automatic optimization of the cell-defining beam(s) in both the vertical and the horizontal domains as described in Section 6.2.

Cell-specific beamforming is typically used in the following situations:

- for signals requiring cell-wide coverage, that is, for signals that need to cover the entire cell. Typical examples of such signals for long-term evolution (LTE) are cell-specific reference signal (CRS), primary/secondary synchronization signal (P/S-SS), and physical broadcast channel (PBCH). Although new radio (NR) strives to minimize always-on signals and there are no common reference signals, there is still a need to transmit something that defines the cell. The synchronization channel will typically serve as a cell-defining signal for NR and will need cell-wide coverage. Note also that NR facilitates the so-called beam sweeping (see

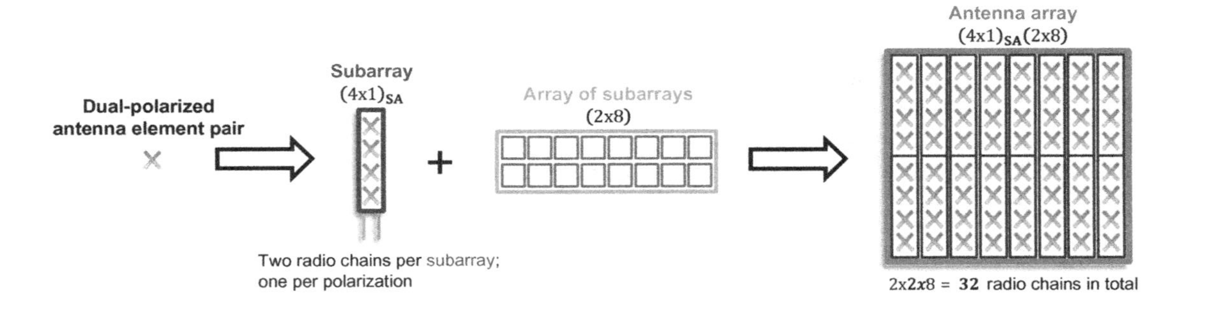

FIGURE 13.1

Illustration of a 32T $(4 \times 1)_{SA}(2 \times 8)$ antenna array constructed by stacking two subarrays vertically and eight subarrays horizontally, where each subarray consists of four dual-polarized antenna element pairs and has two radio chains. Thus, in total $2 \times 2 \times 8 = 32$ radio chains in this example.

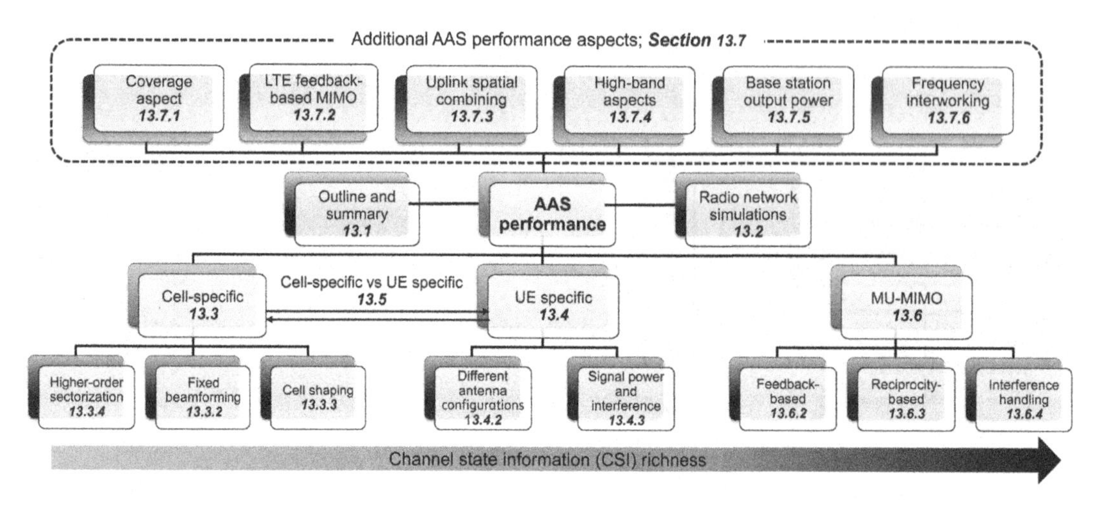

FIGURE 13.2

Overview of the chapter structure with section topics and associated section numbers.

Section 9.3.6) that implies that cell-defining signals may only need to cover parts of the cell at a time;

- whenever up-to-date CSI is not available, for example, in early steps of the initial access procedure. This may also happen for short burst transmissions in connected mode when sessions are so short that they have finalized before updated CSI has been obtained.

Cell-specific beamforming, discussed in Section 13.3, is about beamforming at least the cell-defining signals. Note, though, that in many cases, all the signals associated to the cell/point are beamformed in the same way, which is especially true for the common case of single-column classic antennas where all signals go through the same beam. Cell-specific beamforming encompasses the class of semi-static beamforming concepts discussed in Section 6.2 and can be broadly categorized as *fixed beamforming* and *cell shaping*.

13.1.1.1 Fixed beamforming

Fixed beamforming, discussed in Section 13.3.2, builds on "manual" setting of the cell-defining beam(s), for example, by moving a site or changing the tilt. These processes are typically performed very rarely, hence the name fixed beamforming.

Tilt (see Section 2.3.3.1) is an important tool for changing the cell shape and thereby achieving a good trade-off between maximizing the cell coverage and reducing the inter-cell interference.

13.1.1.2 Cell shaping

Cell shaping, illustrated in Section 13.3.3, builds on that an algorithm automatically adapts the cell-defining beam(s) on, in practice, a slow basis (see Section 6.2).

Cell shaping is often most beneficial in non-homogeneous deployment scenarios where proper site planning is difficult.

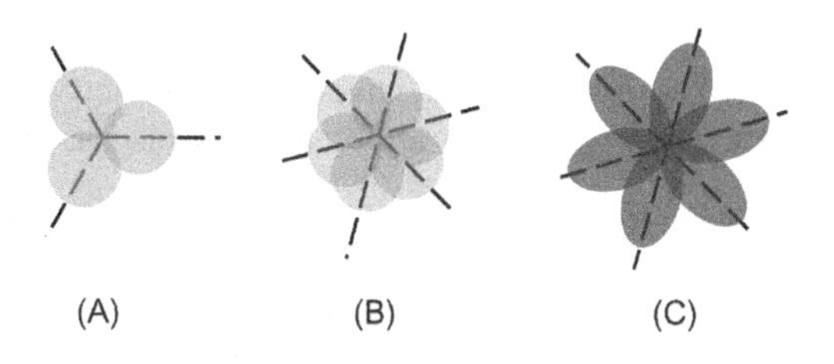

FIGURE 13.3

Illustration of sectorization. (A) Three-sector site based on classical 65 degree sector antennas, (B) six-sector site based on 65 degree antennas, and (C) six-sector site based on narrower 30 degree antennas.

13.1.1.3 Higher-order sectorization

Sectorization, that is, splitting an existing cell into two or more cells is discussed in Section 13.3.4 and illustrated in Fig. 13.3. Performance of sectorization is scenario dependent and works best in high load scenarios with good coverage. Hence, sectorization is generally said to be a capacity feature and not a coverage feature.

The main benefit of sectorization is a multiplexing gain, which allows UEs in the different cells to use the same time frequency resources. This is like MU-MIMO with fixed beams and two independently running schedulers.

The efficiency of sectorization depends on the isolation between the created sectors, which relates to propagation and antenna characteristics. A large antenna (in the dimension of sectorization) can create more isolated, well-defined sector beams and can result in less angular spread in the channel that often result in less inter-cell interference.

The capacity gain of sectorization where each three-sector site is split into a six-sector site based on antenna reuse and fixed total power shared between all additional sectors depends highly on the AAS capability relative to the baseline reference, as well as the scenario, but ranges from roughly 10%–40% in examples shown in Section 13.3.4.

13.1.2 UE-SPECIFIC BEAMFORMING

UE-specific beamforming, which is a form of channel-dependent beamforming, was thoroughly discussed in Section 6.1, and is characterized by:

- dedicated AAS processing tailored for one UE (SU-MIMO) or a group of UEs (MU-MIMO). The latter requires up-to-date fast-fading CSI, while the former is more robust with respect to CSI impairments (see Sections 6.4.1.1, 6.4.1.2, 6.4.2, and 6.7.2 for more details);
- beamforming strives to increase the SINR of desired signals, which can lead to increased coverage, capacity, and end-user throughput. Spatial processing can also make better use of large SINRs by dividing the power over several layers, either to one UE (SU-MIMO) or to several UEs (MU-MIMO) (see Sections 6.1 and 6.3 for further details);
- CSI is often classified as feedback-based or reciprocity-based and as large-scale-based or fast fading-based CSI (see Section 6.4 for a more thorough discussion).

Section 13.4 focuses on the performance of UE-specific beamforming for different beamforming algorithms and AAS configurations. A summary is given below for three main cases. Note also that the deployment scenario aspects of beamforming are thoroughly discussed in Chapter 14.

- *Case I:* Impact of adding antenna columns (increasing the antenna area and the number of subarrays)—although subject to diminishing returns, doubling the number of antenna columns gives typically SU-MIMO capacity gains in the order of 30%–50% in urban macro scenarios. The performance of generalized beamforming can be better than classic beamforming at least in scenarios with significant channel angular spread and good enough coverage (see Section 13.4.2.3 for more details);
- *Case II:* how to stack subarrays given a fixed number of subarrays (radio chains)—it is typically most beneficial to spend the degrees of freedom (radio chains) in the dimension where UEs are mostly spread. For many low-rise deployment scenarios with a rather large ISD, UEs are uniformly spread over the entire sector in the horizontal domain but confined to a small area in the vertical domain. In these cases, horizontal stacking of as many antenna columns (subarrays) as possible gives the best performance. Still, some vertical domain beamforming can be good to provide robustness against incorrect subarray tilt or channel angular spread in the vertical domain. In deployment scenarios with significant UE and channel angular spread in the vertical domain, on the other hand, vertical and horizontal domain UE-specific beamforming offer similar gains (see Section 13.4.2.2 for more details);
- *Case III:* impact of subarray size for a fixed antenna area—whether there are any significant performance gains by reducing the subarray size and thereby getting more vertical domain beamforming depend strongly on the deployment scenario. In a rural or low-rise urban macro scenario, less vertical domain beamforming is needed as the angular UE distribution as seen from the antenna is narrow and biased toward angles close to the horizon for which a fixed vertical beam generated from the subarray is adequate. The subarrays can hence be relatively large and still match the angular UE distribution in the vertical domain. In deployment scenarios with significant UE and channel angular spread in the vertical domain, however, there are generally significant gains by stacking several small subarrays compared to using fewer and larger subarrays (see Section 13.4.2.1 for more details).

13.1.3 RELATIONS BETWEEN CELL-SPECIFIC AND UE-SPECIFIC BEAMFORMING

Relations between cell-specific and UE-specific beamforming are addressed in Section 13.5. UE-specific beamforming relies on adapting the beamforming weights associated with each subarray for each UE individually, while cell-specific beamforming applies to cell-wide signals and is the same for all UEs connected to the cell.

For a fixed number of antenna elements (fixed antenna area), UE-specific beamforming tends to provide larger gains than cell-specific beamforming, that is, more radio chains with smaller subarrays is preferred over less radio chains with larger optimized subarrays. For a fixed number of radio chains (subarrays), on the other hand, it often holds that the gains increase by using larger subarrays, assuming that the subarray size matches the UE distribution reasonably well and that the static or semi-static subarray beam is chosen appropriately.

Furthermore, as will be shown in Section 13.5.3, it can be beneficial that beams used for cell-defining signals match the envelope of all beams used for UE-specific beamforming, that is, match the subarray antenna pattern. On the other hand, cell-wide signals need to cover the entire cell and in coverage-limited scenarios this might mean that the beamforming gain of cell-defining signals needs to be increased. One potential solution to this is to apply beam sweeping (see Section 9.3.6 for beam-based operation in NR).

13.1.4 MULTI-USER MIMO

Multi-user MIMO (MU-MIMO) performance gains are illustrated in Section 13.6, covering both feedback-based (Section 13.6.2) and reciprocity-based MU-MIMO (Sections 13.6.3 and 13.6.4). Note that while Section 13.6.2 primarily illustrates NR performance, more details regarding LTE feedback-based MIMO are given in Section 13.7.2.

MU-MIMO is thoroughly discussed in Section 6.3 and has the potential to bring significant network performance benefits under favorable conditions. For MU-MIMO to outperform SU-MIMO, there need to exist compatible UEs that have accurate fast fading-based CSI and high enough SINRs, so that SU-MIMO performance starts to saturate, and it thereby becomes beneficial to share the SINR among these UEs (cf. Section 6.3.4).

Some observations related to the performance of MU-MIMO are as follows:

- NR Type II CSI feedback gives good MU-MIMO gains in the order of 30%–40% gain over Type I SU-MIMO. In fact, NR Type II and reciprocity-based MU-MIMO exhibit similar performance. Hence, MU-MIMO is not only about reciprocity;
- reciprocity-based MU-MIMO with coordinated beamforming (CBF) to reduce inter-cell interference can offer significant capacity gains in interference-limited deployment scenarios;
- the number of required MU-MIMO layers is generally few in realistic deployments. Nevertheless, MU-MIMO gains and the required number of layers depend strongly on assumptions such as deployment scenario, traffic characteristics, and network load.

13.1.4.1 Illustration of MU-MIMO performance

To conclude this sub section, an example comparing the performance of different MU-MIMO schemes is given (see also Chapter 8 and Chapter 9 for a description of standard related aspects of LTE and NR).

Fig. 13.4 illustrates MU-MIMO capacity gains for a 32T $(2\times1)_{SA}(4\times4)$ AAS in the Third Generation Partnership Project (3GPP) urban microchannel model (UMi) scenario [1]. As discussed in Section 6.7.6, this 32T antenna configuration with four columns could perhaps be viewed as corresponding to an 8T in the massive MIMO literature studies.

Relative to LTE release 13 SU-MIMO, the capacity gains for LTE release 13 MU-MIMO, LTE release 14 advanced CSI, and NR Type II are roughly 10%, 20%, and 30%, respectively.

It is also evident that the potential of fast-fading reciprocity-based MU-MIMO is large, but with non-ideal and estimated CSI, including link adaptation and sounding reference signal (SRS) processing, gains drop. Furthermore, by comparing zero forcing (ZF) with CBF, where the latter includes inter-cell interference-awareness, it is seen that taking inter-cell interference into consideration is

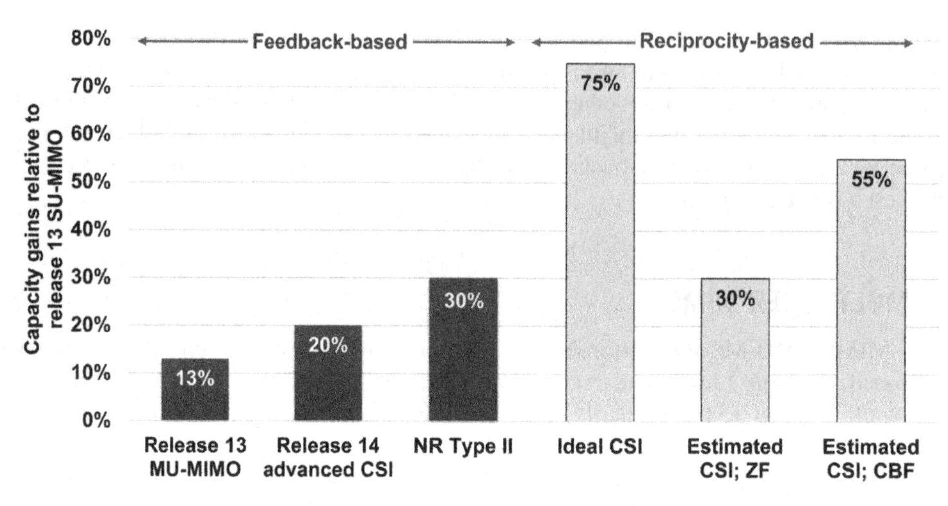

FIGURE 13.4

Illustration of MU-MIMO capacity gains relative to 3GPP Release 13 SU-MIMO for 32T $(2 \times 1)_{SA}(4 \times 4)$ AAS in the 3GPP UMi scenario.

important to reap the benefits of reciprocity-based MU-MIMO in an interference-limited scenario like 3GPP UMi.

It is also clear that the potential of NR Type II MU-MIMO is encouraging with gains in the same order as reciprocity-based MU-MIMO.

MU-MIMO gains are sensitive to assumptions regarding the scenario, traffic, AAS configuration and algorithm choices. These results assume equal buffer with relatively large data packet sizes compared with reality, hence in that respect optimistic MU-MIMO gains.

13.1.5 ADDITIONAL AAS PERFORMANCE ASPECTS

13.1.5.1 Coverage aspects of beamforming

One of the main benefits of AAS and beamforming is the improved coverage that makes it easier to reuse the existing site grid for deployments on higher frequency bands. This is illustrated in Fig. 13.5 and discussed in more detail in Section 13.7.1. One of the main challenges with deploying higher frequency bands using time division duplex (TDD) is uplink coverage that suffers from limited UE output power and downlink heavy slot formats.

Section 13.7.1 will also discuss how coverage affects different beamforming methods, and it is illustrated that a mix of classic beamforming for SU-MIMO (typically feedback-based) and fast fading-based generalized beamforming with MU-MIMO (reciprocity-based or feedback-based) is key to achieve a high-performing network.

13.1.5.2 LTE FDD feedback-based MIMO

There are two main tracks that can be pursued to improve the performance of LTE *frequency division duplex* (FDD) massive MIMO deployments, either higher-order sectorization or feedback-

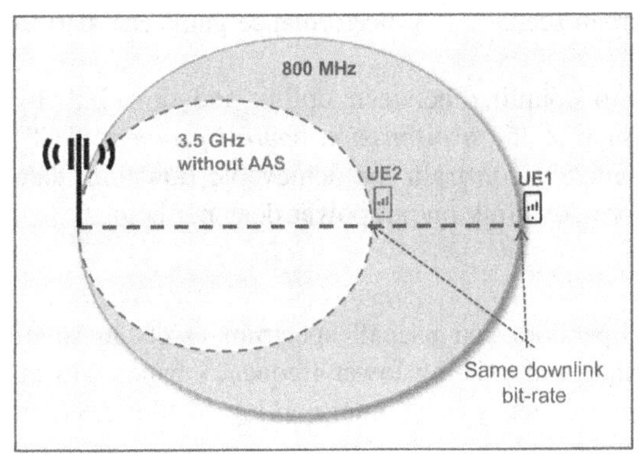

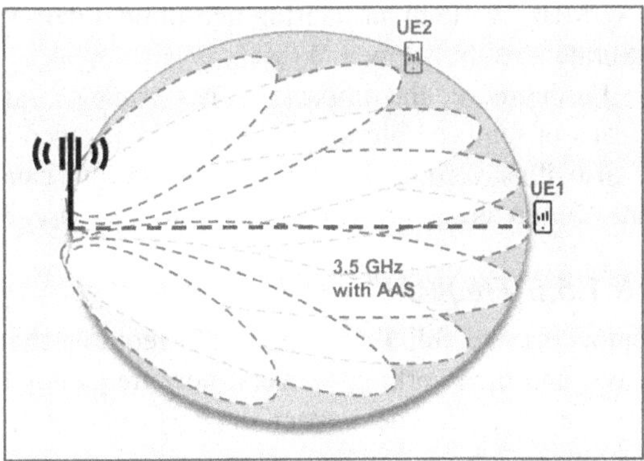

FIGURE 13.5

Illustration of coverage, defined as the minimum cell-edge downlink data rate, for an 800 MHz deployment without AAS and a 3.5 GHz deployment without AAS (*left*) and with AAS (*right*).

based CSI based on standardized transmission modes. The latter track has generally a better performance potential but suffers from the strong dependency on the UE capabilities, for example, which transmission modes are supported.

The performance of LTE feedback-based MIMO is illustrated in Section 13.7.2.

13.1.5.3 Uplink spatial receiver combining

The performance of uplink spatial receiver combining is considered in Section 13.7.3.

Uplink receive and downlink transmit beamforming considerations are similar, that is, they exhibit similar AAS characteristics and dependencies. A slight difference is that maximizing the antenna area is often more important for the uplink than for the downlink. The reason being that coverage tends to be more challenging for the uplink than for the downlink due to less output power of UEs than that of base stations. Although highly scenario dependent, the uplink is often more power limited than interference limited, while the downlink is typically more interference limited than power limited.

13.1.5.4 High-band aspects

Beamforming algorithms and AAS considerations are generally frequency agnostic. There are, however, a few characteristics that make beamforming more challenging at higher frequency bands. This is thoroughly described in Chapter 7. Section 13.7.4 illustrates some of these aspects. More specifically, the coverage challenge of higher frequency bands and performance implications of time-domain beamforming are illustrated.

13.1.5.5 Base station output power

The impact of base station output power on the AAS performance is illustrated in Section 13.7.5. More power tends to improve the performance but how much is scenario dependent. In a power-limited urban scenario, there are significant gains by increasing the output power well above

1 W/MHz, while in an interference-limited dense urban scenario, the performance gains can start to saturate already below 1 W/MHz.

Furthermore, the impact on performance due to coupling between uplink and downlink by means of required bitrates for transport protocols such as the *transmission control protocol* (TCP) is also illustrated. It is seen that the uplink can severely constrain the achievable downlink data rates for coverage-limited UEs and, in that case, more downlink output power does not help.

13.1.5.6 Frequency-band interworking

Unlocking the full potential of 5G requires that operators can use all spectrum assets in smart ways, and interworking between new frequency bands and existing lower frequency bands will be key for success. The impact on performance from frequency-band interworking is illustrated in Section 13.7.6.

Considering the example given in Section 13.7.6, it is seen that adding AAS on mid-band on existing sites can provide substantial capacity gains, in the order of 5–10 times, relative to existing LTE low-bands. The main reasons being more spectrum, use of AAS, and better UE capability (4 RX instead of 2 RX). Including mm-wave bands can improve the capacity by an additional ~30%. The exact numbers depend on many factors such as deployment scenario characteristics and the bandwidths of existing and new bands.

13.2 RADIO NETWORK SIMULATIONS

13.2.1 OVERVIEW

Radio network performance evaluations by means of simulations and trial measurements are key to judge what benefits an AAS solution can provide. This chapter focuses on AAS performance aspects by means of radio network simulations. These evaluations assess the performance on network level, capturing scenario characteristics, deployment, and propagation effects, as well as impacts of traffic load and interference by modeling relevant parts of the network including radio links from multiple base stations and UEs that interact in a dynamic fashion.

Typical inputs to a radio network simulator are scenario characteristics including the environmental topology (buildings, foliage, etc.), deployment aspects such as site and antenna models, system assumptions like bandwidth and power, traffic patterns, and the traffic/UE distribution. The network simulator then processes all of this information together with relevant scheduling models, single link performance abstraction models, and propagation models to produce desired outputs/results. Needless to say, there is a strong correlation between assumptions and obtained results.

Assessing the AAS network performance is a multi-dimensional problem where results depend on numerous factors that often interact in an intricate manner. It is impossible to capture all aspects in a simulator, but it is important to understand what parts need to be modeled in greater detail and what parts can be ignored or modeled with less details in order to get accurate enough results. This depends on the intention with the evaluation. For example, to compare different AAS products and features relative to each other and relative to non-AAS products, it is vital to capture the spatial domain in the simulator models. More specifically, it is necessary to understand how beamforming affects the own signal from the serving cell as well as the inter-cell interference characteristics.

Table 13.1 Overview of Default Simulator Parameter Assumptions Used in the Network Simulations.

Simulator Parameter		Settings
Carrier frequency		3.5 GHz
Bandwidth		100 MHz
Duplex		Synchronized TDD with 70% downlink
BS parameters	Output power	53 dBm (200 W or 2 W/MHz)
UE parameters	Output power	23 dBm
	Number of Tx antennas	1
	Number of Rx antennas	2
BS antenna parameters	Element characteristics	7 dBi gain with 90 and 70 degrees horizontal and vertical half-power beam width, respectively
	Subarray tilt	Generally optimized for each configuration Homogeneous scenarios use the same tilt in all cells, while nonhomogenous scenarios employ site-specific tilt
	Antenna array notation	$(M_v \times 1)_{SA}$ $(N_v \times N_h)$, where M_v denotes the number of dual-polarized element pairs in the subarray and N_v and N_h represent the number of vertically and horizontally stacked subarrays, respectively
Scenario		See Section 13.2.6
Cell selection		Generally based on a beam shape, which is similar to the shape of the subarray pattern
Traffic		Equal buffer (file download), uniformly distributed (see Section 13.2.5)

A detailed description of radio network simulation methodologies is outside the scope of this book, but for completeness a few simulation aspects, models, and assumptions are briefly discussed in Sections 13.2.2–13.2.7. Note also that a brief introduction to network simulations was given in Section 2.2.4.

Table 13.1 provides an overview of some of the baseline key assumptions and simulator settings used in the network performance evaluations following in later sections.

13.2.2 ANTENNA CONFIGURATIONS

Antenna configurations use the so-called AoSA structure introduced in Section 4.6.1, where the subarray is the smallest dynamically controllable entity with two radio chains, one per polarization. Antenna configurations are denoted by $(M_v \times M_h)_{SA}(N_v \times N_h)$, where M_v and M_h denote the number of vertically and horizontally stacked dual-polarized antenna element pairs in the subarray, respectively, and N_v and N_h represent the number of vertically and horizontally stacked subarrays, respectively. The intention with the subscript SA on the first parentheses is to help the reader remembering what dimensions correspond to the subarray and AoSA.

There are in total $2M_vM_hN_vN_h$ antenna elements and $2N_vN_h$ radio chains (two per subarray, one for each polarization). An illustration of the AoSA framework was given in Fig. 13.1. Note that throughout this chapter $M_h = 1$.

The number of radio chains is often denoted by T when discussing transmission and R when discussing reception. For example, a $(4 \times 1)_{SA}(2 \times 8)$ AAS configuration has $2 \times 2 \times 8 = 32$ T:s and 32 R:s, that is, 32 radio chains.

Antenna elements are modeled according to the 3GPP model [1] with 7 dBi gain and 90 and 70 degree horizontal and vertical half-power beam width (HPBW) (see Section 3.3.7.2 for the definition of HPBW).

Note also that definitions of vertical, elevation, and azimuth angles are given in Appendix 1.

13.2.2.1 Antenna array and choice of subarray

A key consideration when analyzing AAS performance based on the AoSA structure is the choice of subarrays in relation to the scenario and spatial UE distribution. The angular UE distribution in the vertical dimension for several scenarios is illustrated in Section 13.2.6.

Consider the AoSA antenna structure illustrated in Fig. 13.1. This antenna array setup facilitates both high antenna gain and adaptability. By adjusting the subarray tilt and the beamforming weights associated with each subarray, the associated beam shape can be adjusted. This is the baseline for both cell-specific and UE-specific beamforming discussed in Sections 13.3 and 13.4.

There are a few main aspects to consider when discussing the choice of subarray in relation to the scenario and angular UE distribution in the vertical domain:

- As discussed in Section 4.6.2, the subarray determines the angular coverage area for UE-specific beamforming, that is, the area where UE-specific beamforming works efficiently without any significant gain drop. This is illustrated in Fig. 13.6, showing how partitioning of an antenna of fixed size (eight vertically stacked dual-polarized element pairs) into subarrays of different size affects the antenna gain pattern. If the 8×1 subarray has an angular coverage area of α, it follows typically that 4×1 and 2×1 subarrays have an angular coverage area of 2α and 4α, respectively. Hence, for typical propagation channels with little angular spread in the vertical domain, it is enough that the width of the main lobe of the subarray matches the cell's angular UE distribution in the vertical domain, or rather is not too narrow in relation to the distribution;
- Some AAS related processing is done per subarray; hence, the subarray size may affect the resulting quality. For example, uplink channel estimates may get better quality by using larger subarrays. This is, however, an area subject to innovative signal processing, and several tricks exist to improve the performance in this respect. In theory, for a fixed antenna size, it is always better from a conceptual performance perspective to use many radio chains based on small subarrays compared to having fewer radio chains associated with larger subarrays. A larger subarray can always be constructed from many small ones via signal processing. Nevertheless, achieving optimal performance relies on good algorithm adaptation and no implementation losses;
- A third aspect concerns the process of associating UEs with a serving cell. This is commonly done by measuring the quality of a cell-specific signal and associate each UE with the strongest one. This cell-specific signal is generally sent via cell-specific

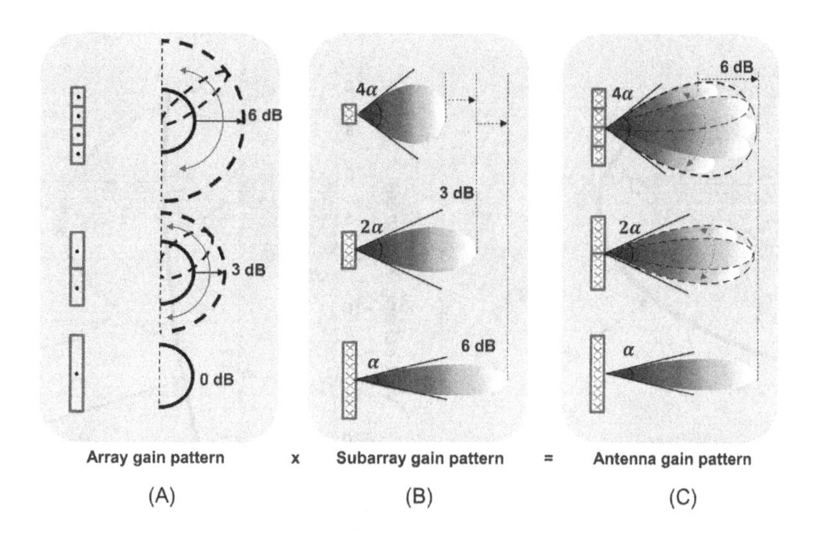

FIGURE 13.6

Illustration of how the antenna gain pattern (C) depends on the array factor (A) and subarray (B) for a fixed 8×1 antenna. (Upper part) $(2 \times 1)_{SA}(4 \times 1)$, (middle part) $(4 \times 1)_{SA}(2 \times 1)$, and (lower part) $(8 \times 1)_{SA}(1 \times 1)$ AoSA configurations.

beamforming as will be discussed in Section 13.3. In simulations, it is common that the cell-specific beam used for associating UEs with a serving cell is matched to the subarray shape. This means that the beam shape used for cell association will match the envelope of all the beams used for UE-specific beamforming. This will be further discussed in Section 13.5.

Fig. 13.7 shows the geometrical angular UE distribution in the vertical domain relative to the serving base station antenna together with the antenna gain patterns associated with the subarray and the traffic beams (discrete Fourier transform-based with two-times oversampling, cf. Section 6.3.3.5) for different antenna configurations in the 3GPP UMa scenario [1] (see Section 13.2.6.1 for a more thorough discussion of geometrical angular UE distribution).

In the left part of Fig. 13.7, a $(4 \times 1)_{SA}(1 \times 1)$ antenna configuration is used, that is, a subarray with four elements per polarization and no UE-specific beamforming. Hence, there is no adaptability in this case, and the fixed beam shape from the 4×1 subarray needs to cover all UEs sufficiently well. In the right part of Fig. 13.7, a $(1 \times 1)_{SA}(4 \times 1)$ antenna configuration is used, that is, a subarray with one element per polarization and UE-specific beamforming based on eight different beams (two-times oversampling) in the vertical domain. In this case, there is a good adaptability, and a wide angular range in the vertical domain can be covered with good gain. Note that the total antenna area 4×1 is the same in both cases.

From Fig. 13.7 it is seen that the width of the main lobe in the vertical domain, hence the coverage area, of the 4×1 subarray fits the 3GPP UMa scenario rather well, while the width of the main lobe in the vertical domain of the 1×1 subarray is unnecessarily wide. Furthermore, the $(1 \times 1)_{SA}(4 \times 1)$ configuration results in many unused traffic beams; essentially only one of

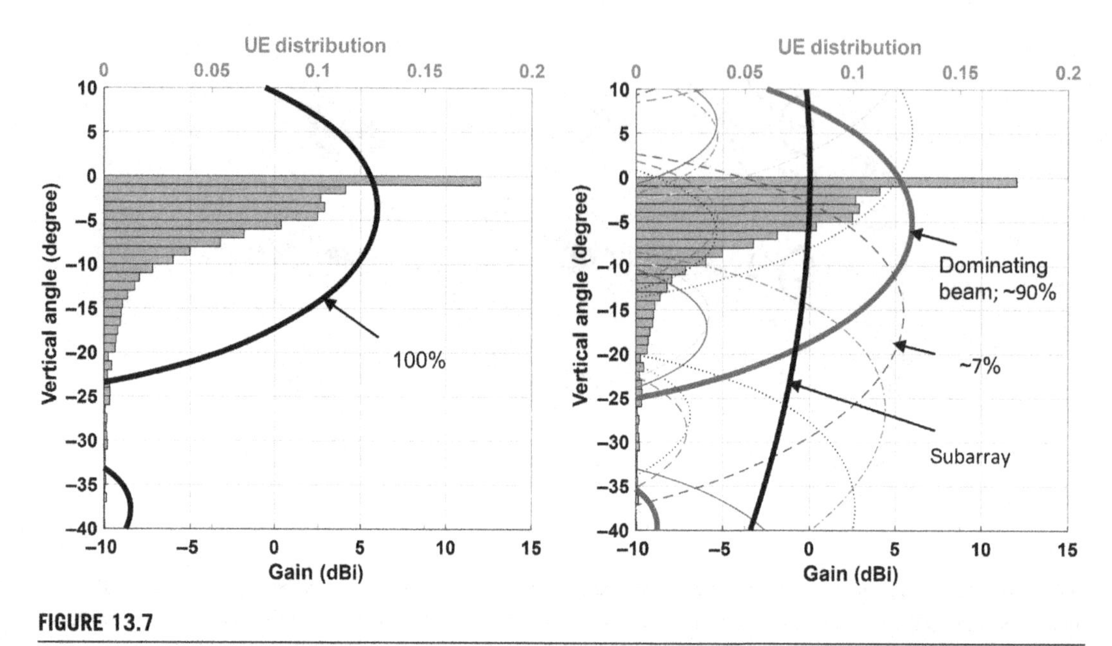

FIGURE 13.7

Geometrical angular distribution of UEs in the vertical domain (*bar plot*), subarray diagram and traffic beams based on a DFT codebook (*lines*) for the 3GPP UMa scenario. $(4 \times 1)_{SA}(1 \times 1)$ antenna configuration (*left*) and $(1 \times 1)_{SA}(4 \times 1)$ antenna configuration (*right*). Percentage numbers represent how often an associated beam is used to serve UEs in a simulation based on grid-of-beams (GoB). The y-axis represents vertical angles and the x-axis shows antenna gain (lower) and angular UE distribution (upper).

the eight beams is used. The dominating UE direction is close to the horizon, which makes sense as the cell size is significant, and there is also above rooftop propagation. Clearly, the $(1 \times 1)_{SA}(4 \times 1)$ array configuration behaves almost as the $(4 \times 1)_{SA}(1 \times 1)$ array configuration. Consequently, by considering only vertical domain UE-specific beamforming, it is likely that a 2T system based on a 4×1 subarray would perform almost as good as an 8T system based on 1×1 subarrays for this scenario. Note, though, that if horizontal domain UE-specific beamforming would be considered, that is, if subarrays were stacked in the horizontal domain instead of the vertical domain, then 8T would perform considerably better than 2T. This is illustrated in Section 13.4.2.

Fig. 13.8 corresponds to Fig. 13.7, but for the dense urban high-rise scenario (see Section 13.2.6.1). Here the main lobe width and therefore the angular coverage area of the 4×1 subarray are too small and would result in significant gain drops (and poor beamforming resolution) for a 4×1 antenna array. The width of the main lobe of the 1×1 subarray, on the other hand, captures the angular distribution of UEs in the vertical domain well, and the angular resolution provided by the $(1 \times 1)_{SA}(4 \times 1)$ antenna configuration would result in that an 8T $(1 \times 1)_{SA}(4 \times 1)$ configuration performs significantly better than a 2T $(4 \times 1)_{SA}(1 \times 1)$ configuration in this dense urban high-rise scenario. This will be further illustrated in Section 14.3.

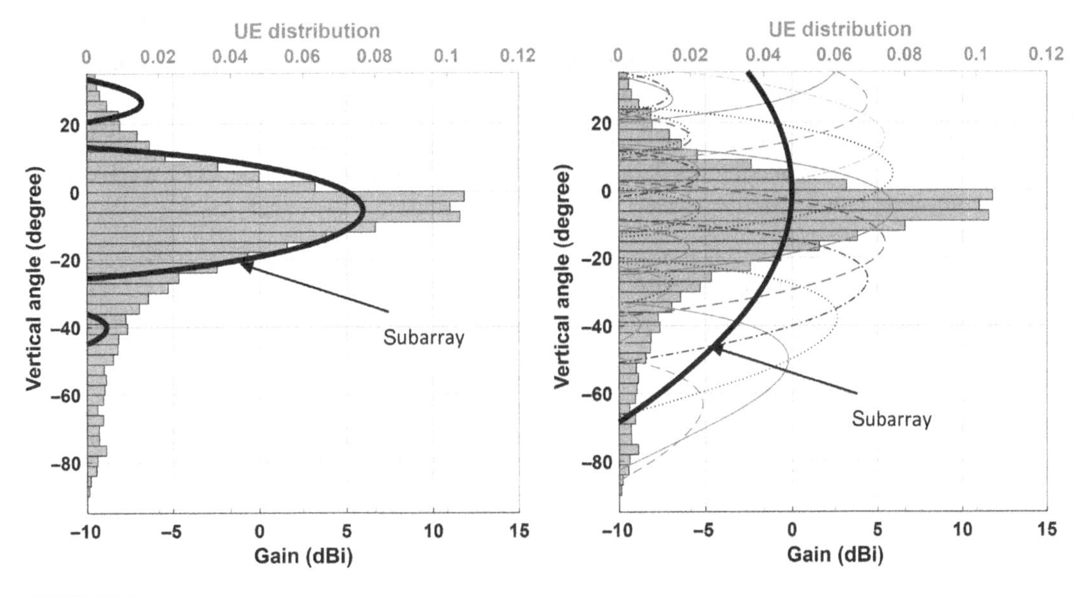

FIGURE 13.8

Geometrical angular distribution of UEs in the vertical domain (*bar plot*), subarray diagram, and traffic beams based on two-times oversampled DFT codebook (*lines*) for the dense urban high-rise scenario. $(4 \times 1)_{SA}(1 \times 1)$ antenna configuration (*left*) and $(1 \times 1)_{SA}(4 \times 1)$ antenna configuration (*right*). The *y*-axis represents vertical angles and the *x*-axis shows antenna gain (lower) and UE distribution (upper).

13.2.3 MULTI-ANTENNA ALGORITHMS

As discussed in Chapter 6, there exist numerous multi-antenna or AAS algorithms, but they can all be classified into relatively few main categories. To keep it simple, two main UE-specific beamforming algorithm types are considered in the performance evaluations to follow; *large-scale wideband* precoding and *fast-fading frequency-selective* precoding, where the former belongs to the category of *classic beamformers* (see Section 6.1.1.1) and the latter belongs to the group of *generalized beamformers* (see Section 6.1.1.2).

13.2.3.1 Large-scale wideband precoding via grid-of-beams

The grid-of-beams (GoB) algorithm uses wideband precoding similar to LTE and NR DFT-based codebooks that mainly target large-scale channel properties, although precoding over polarization is also performed.

Some of the characteristics of the GoB precoding used in the subsequent sections are as follows:

- captures long-term correlation properties of the channel;
- uses coarse wideband precoding, that is, a single precoder (beam) covering the entire frequency band;

- CSI is feedback-based, in which the UE estimates the preferred precoder and feeds it back to the base station. Note, however, that large-scale reciprocity-based wideband CSI obtained via SRS or DM-RS can also be used.

The GoB algorithm employs a DFT-based codebook with an oversampling factor of two to reduce straddling loss (cf. Section 6.3.3.5), where the beam that maximizes the UE's received signal (signal-to-noise ratio (SNR)) is chosen.

13.2.3.2 Fast-fading frequency-selective precoding

Some of the characteristics of fast-fading frequency-selective precoding used in subsequent sections include:

- coherent processing of fast-fading properties of the channel;
- uses frequency-selective precoding, where different parts of the frequency band may use different precoders;
- CSI is typically reciprocity-based, in which uplink sounding is often employed to estimate the channel used to derive the precoder. Note, however, that advanced feedback-based CSI will also be considered.

Fast-fading frequency-selective precoding is primarily exemplified by means of a reciprocity-based *maximum ratio transmission* (MRT) algorithm, where the precoding vector is given by the Hermitian transpose of the channel matrix normalized to take a sum power constraint over the subarray ports into account (see Sections 6.1.4 and 6.3.3.1). For MRT, the uplink sounding quality will impact how well the channel can be estimated, hence also impact the quality of the downlink precoding. That is, the downlink precoding gain will depend on the quality of the uplink sounding signal.

SU-MIMO with maximum two layers per UE is assumed unless otherwise stated. MU-MIMO will be discussed in Section 13.6 and will employ regularized ZF (see Section 6.3.3.3) or interference-awareness by means of CBF (see Section 6.6.2).

13.2.4 PERFORMANCE METRICS

Three main performance metrics used when evaluating AAS are *coverage*, *capacity*, and UE *throughput*. These metrics are defined in Section 2.2.3, and a summary is given below.

Capacity gains are based on the cell-edge (5:th) UE percentile where the reference case has a throughput requirement of 10 Mbps in the downlink. UE throughput gains are read at the same reference capacity point, that is, where the reference case has a cell-edge data rate of 10 Mbps (see Fig. 13.9 for an illustration).

At low network load (served traffic in Fig. 13.9 is low), there is no queuing effects and little interference, which means that the UE throughput is essentially dictated by the UEs' individual link quality. Note that the interreference situation at low load depends on the system and channel under study. For example, LTE has always-on signals like CRS that typically will interfere with physical downlink shared channel (PDSCH) at low load, which is not the case for NR. At high network load (served traffic in Fig. 13.9 is high), queuing effects and increased interference will substantially reduce the achievable UE throughput.

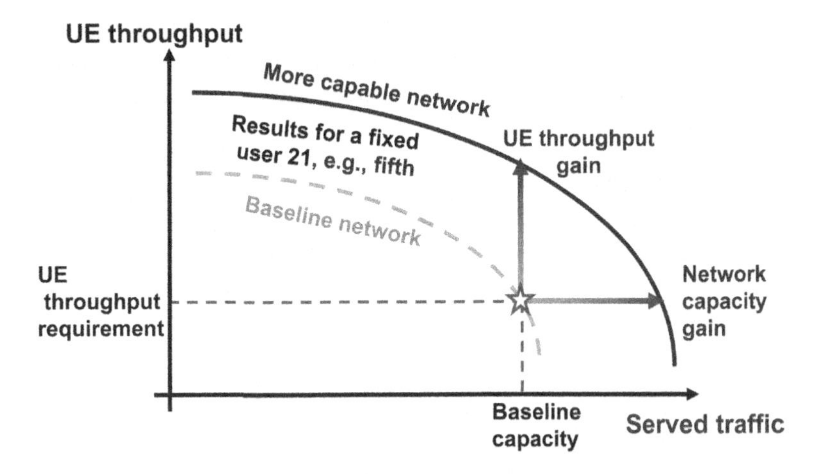

FIGURE 13.9

Illustration of capacity and UE throughput gains for a more capable network relative to a baseline network. The capacity gain is obtained by considering the cell-edge performance (5:th UE percentile) and is measured from the x-axis relative to a specific load or throughput requirement for the baseline network. UE throughput gains are measured from the y-axis relative to the same baseline point. UE throughput gains are often given for the 5:th (cell-edge) and 50:th UE percentiles.

13.2.5 TRAFFIC CHARACTERISTICS

The traffic characteristics can have a profound impact on the radio network performance; hence, it is important to understand the driving mechanisms and ensure that relevant models are used. As discussed in Section 2.2.2, MBB traffic, which dominates in today's mobile networks, is very bursty and consists of many small data packets. Each UE experiences repeated bursts consisting of short packages followed by a longer period of silence. Therefore, the number of connected UEs is typically very large, but the number of simultaneously active UEs is small. Several traffic models have been developed over the years, but two commonly used ones are the so-called *full buffer* and *non-full buffer* described in the following subsections.

Almost all evaluations in this chapter use an equal buffer traffic assumption, which is a special case of the non-full buffer model. The equal buffer traffic model is a lot more realistic than full buffer traffic, but in reality, even more complicated and heterogenous non-full buffer traffic patterns are encountered (see also the discussion in Section 6.7.4.1 where it is shown how MU-MIMO gains shrink by assuming non-full buffer).

13.2.5.1 Full buffer model

Full buffer refers to a case when there is a fixed number of UEs in the network whose data buffers are constantly full (see Section 2.2.4.2 for an illustration). In this case, all UEs are transmitting all the time, which results in very high resource utilization. Full buffer simulations are not very realistic and create a static environment where the network behavior is rather stable and predictable, which, for example, does not stress the dynamic behavior of network algorithms. Thus, it is difficult to judge whether Algorithm A performs better than Algorithm B. For full buffer evaluations, the system behavior in terms of capacity is typically dominated by the performance of "good" UEs, that is, the UEs having good link quality.

13.2.5.2 Non-full buffer model

A more realistic and commonly used traffic model is the non-full buffer or file transfer model, often realized by using an equal buffer where all UEs have an equal amount of data (fixed file size) to transmit. In this case, UEs enter the system dynamically, transmit their data files, and then leave the system. Alternatively, or in parallel, UEs stay inactive in the system until their buffers are filled with data, in which case they become active and transmit the data, and then once again go back to being inactive (see Section 2.2.4.2 for an illustration). The time it takes to serve a UE depends among other things on the scheduling policy, bandwidth, file size, network load, and the UE's link quality.

Network performance in terms of capacity is generally dictated by cell-edge UEs as they need longer time to complete their file transfer due to poor link quality. The non-full buffer model is dynamic in nature and has an inherent fairness in the sense that serving the cell-edge UEs quickly improves performance for the entire network.

The model is sensitive to overloading. Hence, admission control becomes essential to avoid system instability, that is, to avoid that, more traffic is offered than can be served. It is important to control both the number of UEs entering the system (blocking) and to remove UEs with too poor link quality (dropping) as these UEs will need to stay active for a very long time resulting in queue build-up.

13.2.6 DEPLOYMENT CHARACTERISTICS

The network deployment characteristics, including building properties, site and antenna types, and ISDs, have a profound impact on the AAS simulation results. Two common methodologies are to use *homogeneous* and more realistic *non-homogeneous* deployments.

Homogeneous deployments use often three-sector hexagonal network models with a fixed ISD and antennas deployed at a fixed height (see Fig. 13.10). Furthermore, statistical models are often used to capture environmental and propagation effects (see Section 3.6.3).

Non-homogeneous models can take different forms but are often based on map data, real network deployment data including buildings, roads, foliage, and clutter objects, and site-specific propagation models (see Section 3.6.4). These models are often calibrated by means of trial measurements (real network measurements) to end up at a correct operating point in terms of path loss.

In both cases, it is important to capture the spatial behavior in the model when evaluating AAS performance. Classical deployment and propagation models were two dimensional (2D) in the sense that only horizontal directions and distances between UEs and base stations were taken into consideration. This was often sufficient for classical antenna evaluations. However, for AAS evaluations, three-dimensional models, taking also the vertical domain into consideration, are required.

One key attribute for AAS evaluations is the angular UE distribution relative to the serving base station antenna. For example, this may have a significant impact on the usefulness of vertical domain beamforming as is further discussed in Sections 13.4 and 14.3. Similarly, it is important

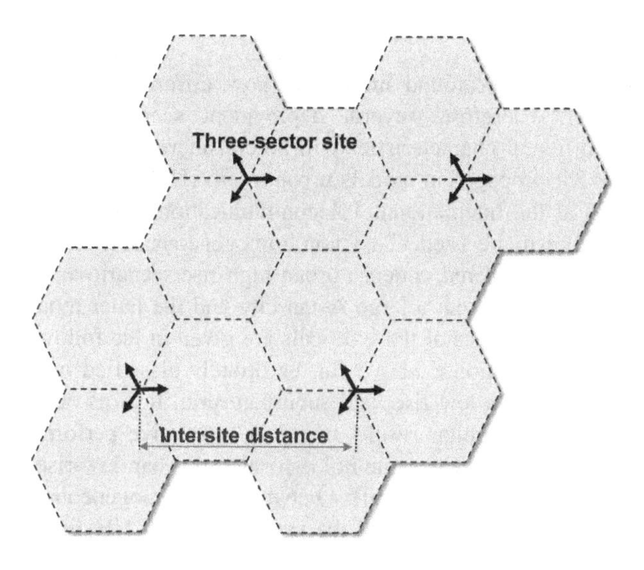

FIGURE 13.10

Three-sector sites placed in a regular hexagonal deployment with a fixed ISD.

whether antennas are above rooftop or below rooftop as this can affect the propagation characteristics, for example, the angular spread of the channel (see Section 3.5.3.3).

There is in general no right or wrong when it comes to using homogeneous or non-homogeneous models. Some aspects are easier to analyze and understand using homogeneous models, while other aspects are better observed using non-homogeneous models. However, some conclusions may differ by using one or the other approach.

As the name suggests, homogeneous models exhibit a homogeneous environment that can create artificial results. For example, having all base stations at the same height tends to make results very sensitive to directing "beams" toward the horizon as that will create excessive interference toward neighboring base stations. Hence, beamforming codebooks that avoid the horizon tend to be beneficial in homogeneous scenarios. For non-homogeneous scenarios with base stations at different heights, this problem is less pronounced.

Non-homogeneous models, on the other hand, require some experience and care when evaluating and interpreting results. For example, it is easy to end up with cells with very different load conditions that can result in that one or a few cells totally dictate the performance of the entire network. With a better deployment, results can then change significantly. Note, though, that many real deployments do suffer from quite significant load variations between cells.

Non-homogeneous models are particularly interesting in dense scenarios due to their inherent non-uniform characteristics and rich environmental effects, while in more uniform scenarios (e.g., suburban or rural), they do not provide much additional advantage over homogeneous models. It is also important to realize that by using a very detailed model over a specific scenario, results are essentially only valid for that scenario. Hence, it is often common practice to make non-homogeneous scenarios a bit more uniform in order to capture the performance of a specific class of scenarios.

13.2.6.1 Deployment scenarios

The deployment scenario has a profound impact on how different AAS configurations perform in relation to each other. Therefore, several deployment scenarios, both homogeneous and non-homogeneous, with different characteristics will be considered in the performance evaluations. The homogeneous 3GPP NR propagation models urban micro (UMi), urban macro (UMa), and rural macro (RMa) [1], as well as the International Telecommunications Union (ITU) propagation model suburban macro (SMa) [2] will be used. Two non-homogeneous scenarios based on site-specific propagation models are also considered: a dense urban high-rise scenario and an urban low-rise scenario, where the former could represent a large Asian city and the latter models a typical European main capital city. Some characteristics of these models are given in the following subsections.

All deployment scenarios mentioned above can be broadly classified into three different types: dense urban high-rise, dense urban low-rise, and suburban/rural. It turns out that dense urban high-rise and 3GPP UMi give rather similar results in terms of relative performance when evaluating AAS configurations and algorithms. The same holds for dense urban low-rise and 3GPP UMa. This might feel counterintuitive as many aspects differ between the homogeneous and non-homogeneous scenarios, but it really emphasizes how central the vertical angular UE distribution is to the beamforming behavior of the network. This is what essentially decides an appropriate subarray size and leads to similar conclusions for the urban high-rise as in 3GPP UMi and for the heterogeneous urban low-rise as in 3GPP UMa, even though many details between the scenarios are different.

13.2.6.1.1 Homogeneous scenarios

The 3GPP/ITU models are described in detail in Ref. [1,2], but for completeness, some of the key characteristics are summarized in Table 13.2 and Figs. 13.11 and 13.12. The different characteristics of the scenarios will impact performance. For example, SMa tends to be a more power-limited scenario than the other scenarios due to larger ISD, and 3GPP UMi is generally a more interference-limited scenario than the other scenarios due to short ISDs and more cluttered environment.

From the geometrical angular distribution of UEs in the vertical domain relative to the location of the serving base station antenna (Fig. 13.12), it can be seen that all UEs in 3GPP UMa and ITU SMa are below the horizon (vertical angle is negative) and confined to a rather small range of vertical angles, while in 3GPP UMi, there is a mix of UEs below and above the horizon, and roughly 25% of the UEs are located above the serving base station antenna (see Appendix 1 for the definition of vertical angle).

Table 13.2 Some Key Characteristics of the 3GPP/ITU UMa, UMi, and SMa Scenarios.

Characteristics of 3GPP/ITU Scenarios [1,2]	Urban Macro	Urban Micro	Suburban Macro
Inter-site distance	500 m	200 m	1500 m
Base station antenna height	25 m	10 m	25 m
Base station antenna deployment	Above rooftop	Below rooftop	Above rooftop
Indoor probability	80%	80%	50%

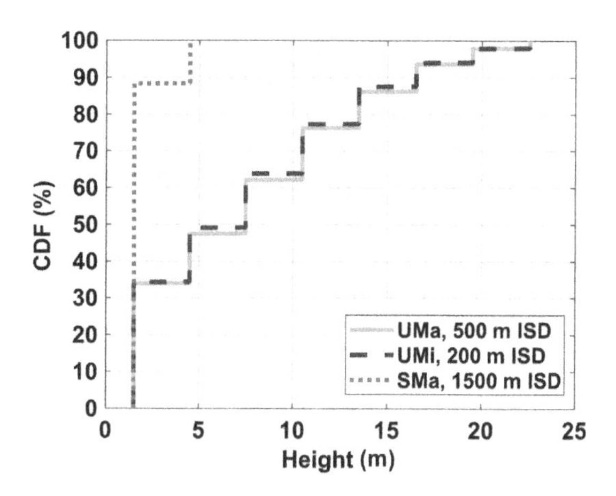

FIGURE 13.11

UE height distribution for the 3GPP/ITU UMa, UMi, and SMa scenarios.

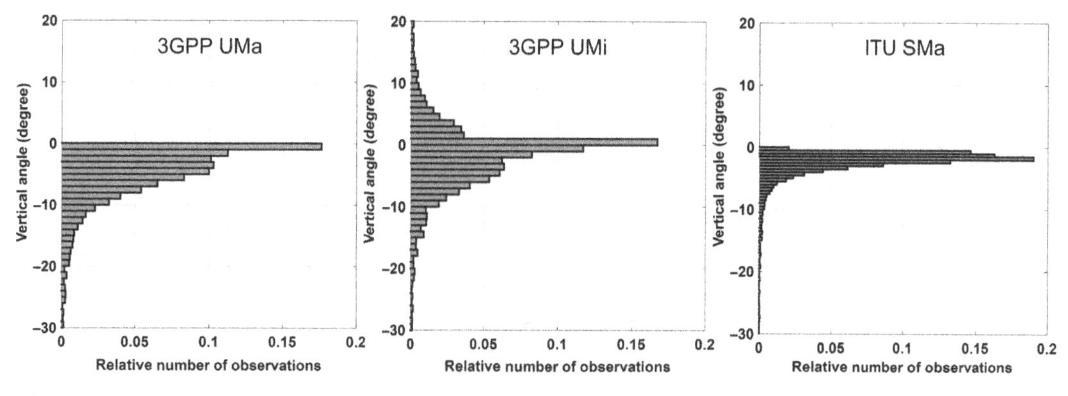

FIGURE 13.12

Geometrical angular UE distribution in the vertical domain relative to the location of the serving base station antenna for 3GPP UMa (*left*), 3GPP UMi (*middle*), and ITU SMa (*right*).

Geometrical angular distribution of UEs is a term referring to the distribution of angles in the vertical dimension corresponding to the Euclidian distance from the serving base station antenna to the served UE. Note that in practice, it is the radio channel that matters, and the angle of dominating radio channel as seen from the base station does not need to coincide with the geometrical angle due to channel angular spread stemming from scattering, over the rooftop propagation, reflections, etc. This is illustrated in Fig. 13.13, which indicates that the geometrical angular UE distribution overestimates the channel angular distribution in this example. Hence, the angle of departure of dominating channel paths would be a better measure, but a geometric angle analysis still provides valuable insights. Section 3.5.3.3 lists typical channel angular spreads for different scenarios.

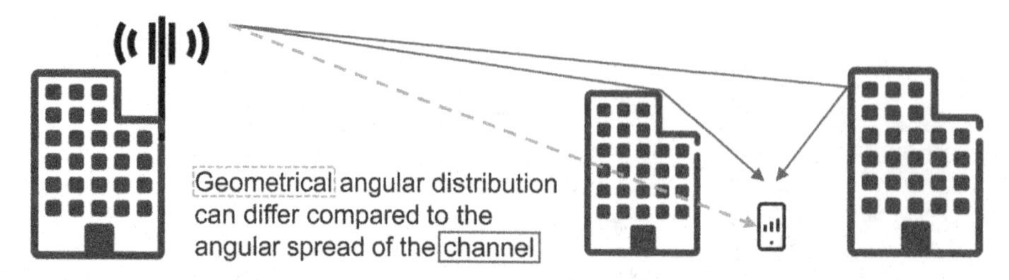

FIGURE 13.13

Example of above rooftop propagation illustrating that geometrical angles can differ from angles of the channel. The dashed line corresponds to the geometrical angle from the serving base station antenna to the UE, whereas the solid lines illustrate two possible propagation paths.

Table 13.3 Some Key Characteristics of the Non-homogeneous Dense Urban High-Rise and Low-rise Scenarios.

Characteristics of Non-homogeneous Scenarios	Urban High Rise	Urban Low Rise
Average inter-site distance	~350 m	~500 m
Indoor probability	80%	70%

13.2.6.1.2 Non-homogeneous scenarios

The non-homogeneous models use a three-sector hexagonal deployment (see Fig. 13.10) where the ISD varies slightly (see Table 13.3 for a summary of statistics). The distribution of UEs' height positions and base station antenna heights can be found in Fig. 13.14. The geometrical angular UE distribution in the vertical domain relative to the location of the serving base station antenna is illustrated in Fig. 13.15. The high-rise scenario has a much wider angular UE distribution in the vertical domain than the low-rise scenario.

It is also interesting to note that even though the topology and antenna deployments are different for the urban low-rise and the 3GPP UMa scenarios, the angular UE distributions in the vertical domain relative to serving base station are similar. A similar observation holds for urban high-rise and 3GPP UMi.

13.2.7 SYSTEM ASSUMPTIONS

There are many system assumptions that will impact the performance, for example, transmitter and receiver layer-1 algorithms, scheduling and radio resource management, and radio attributes such as output power and receiver sensitivity. A few aspects are discussed in what follows:

- the carrier frequency may affect coverage, system assumptions, and algorithm choices. High frequencies increase the Doppler, meaning that CSI becomes outdated faster for a fixed speed (see Section 3.5.3.4 for further details). Another aspect is that most operators have several frequency bands deployed that may impact the performance of a specific frequency

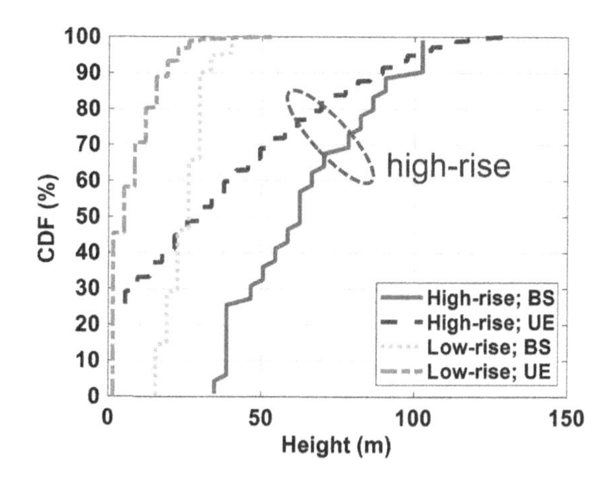

FIGURE 13.14

Height distributions of the base station antennas (BS) and UEs for the urban high-rise and low-rise scenarios.

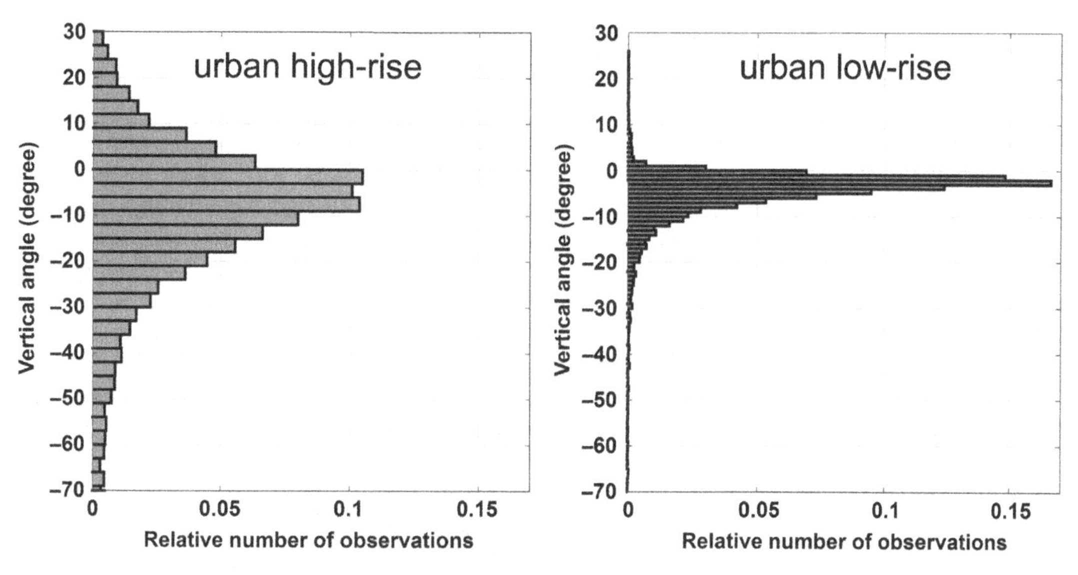

FIGURE 13.15

Angular distribution of UEs in the vertical domain relative to the location of the serving cell's base station antenna for the urban high-rise (*left*) and low-rise (*right*) scenarios.

band. For example, moving UEs with poor coverage to a lower frequency band can significantly improve the performance of a higher frequency band. Still, in most evaluations herein, a single carrier frequency is considered in order to isolate the performance impact of a feature or AAS configuration;

- increasing the system bandwidth leads generally to improved performance and can result in better frequency diversity. However, increasing the system bandwidth for a fixed output power also leads to worse power spectral density and potentially worse channel estimates. Similarly, it can affect coverage negatively and impact how efficiently reciprocity-based channel sounding can be done as thoroughly discussed in Sections 6.4.2.1 and 6.7.1.1;
- the output power and receiver sensitivity will affect coverage. At the base station side, it is generally feasible to increase the output power, while increasing the output power of a UE is often not possible;
- UE assumptions regarding the number of transmit and receive antennas are important. Employing more antennas increases the performance in general but using more receive antennas than transmit antennas (at the UE side) has implications on reciprocity-based sounding as discussed in Section 6.7.1.2;
- in real networks, there is in general a mix of UEs with different capabilities, and the network performance is often limited by the least capable UEs. Evaluations are here based on an assumption that 100% of the UEs support the studied features.

13.3 CELL-SPECIFIC BEAMFORMING

13.3.1 OVERVIEW

Cell-specific beamforming is about beamforming at least the cell-defining signals. In many cases, all the signals associated to the cell/point are beamformed in the same way, which is especially true for the common case of a single-column classic antenna where all signals go through the same beam weight. Cell-specific beamforming encompasses the class of static or semi-static beamforming concepts discussed in Section 6.2 and can broadly be categorized as *fixed beamforming* (see Section 13.3.1.2.1) and *cell shaping* (see Section 13.3.1.2.2).

A special case of cell-specific beamforming is higher-order sectorization, in which an existing sector is split into several sectors (see Section 13.3.1.2.3).

There is a tight connection between cell-specific beamforming, discussed in this section, and UE-specific beamforming, discussed in Section 13.4. This relation is the topic of Section 13.5.

13.3.1.1 Key characteristics

Cell-specific beamforming is used to define the area where cell-wide signals are transmitted. The benefits of cell-specific beamforming come from optimizing the beam(s) of cell-defining signals, resulting in

- increased isolation between cells, which decreases inter-cell interference;
- off-loading, that is, adapting the coverage area so that UEs in highly loaded cells get off-loaded to less loaded neighboring cells; and
- increased gain in received signal power.

It should be noted that the benefits of sectorization are slightly different. The main advantage is a multiplexing gain, where, similar to MU-MIMO, UEs in different sectors can use the same resources. This comes, however, at the expense of decreased cell isolation between some cells.

Achieving optimal cell-specific beamforming is often a complex procedure as there are many factors and dependencies that affect the outcome. For example, the optimal beam used for defining one cell depends on the beams of neighboring cells. Hence, cell-specific beamforming contains an element of inter-cell coordination. It involves also partly conflicting objectives, for example, maximizing coverage (increase gain) and capacity (minimize inter-cell interference) and the need to balance both uplink and downlink performance.

NR introduces support for so-called beam sweeping, mainly targeting high-band (see Sections 7.4 and 9.3.6). This framework can also be useful at mid-band for cell-specific beamforming and to facilitate better interworking between cell-specific and UE-specific beamforming. This will be further illustrated in Section 13.5, where the relation between cell-specific beamforming and UE-specific beamforming will be discussed.

13.3.1.2 Outline

13.3.1.2.1 Fixed beamforming

Fixed beamforming has been a vital part of the network planning and antenna deployment strategies since the early days of modern communication and builds on "manual" setting of the cell-defining beam(s), for example, by moving a site or changing the tilt. These processes are typically performed very rarely, hence the name fixed beamforming.

The main tool associated with fixed beamforming is tilt that is used to shape the beam associated with cell-wide signals in the vertical domain. There is often a trade-off between coverage and capacity when deciding the best tilt value. This will be discussed in Section 13.3.2.1.

13.3.1.2.2 Cell shaping

Cell shaping builds on that an algorithm automatically adapts the cell-defining beam(s) based on network measurements and network performance indicators. In practice, the adaptation is performed on a slow basis (see Section 6.2.2).

Key elements that make cell shaping difficult are the problem of achieving good observability that can be used as input to cell shaping algorithms and to come up with good algorithms. For example, an algorithm that optimizes some performance metrics may negatively affect other metrics. One main problem related to observability is that UEs not yet connected to the network also need to be considered.

Cell-specific tilt, where different cells may have different tilt settings, becomes important in non-homogenous deployments. Deciding the optimal tilt in this case is significantly more involved compared to using a single tilt setting for the entire network. This will be illustrated in Section 13.3.3.

Cell shaping is generally viewed as a promising but rather difficult area to enhance the AAS network performance, and it is subject to extensive research in areas such as artificial intelligence and machine learning. Cell shaping is the topic of Section 13.3.3.

13.3.1.2.3 Higher-order sectorization

Splitting an existing sector into two or more sectors is referred to as higher-order sectorization. This is a traditional technique for boosting the network capacity. Sectorization can be viewed as a special case of cell-specific beamforming, especially when sectorization is done based on antenna reuse, that is, when the same antenna array handles all the additional sectors associated with the original sector.

Like fixed beamforming, sectorization can lead to increased signal power by directing the sectors toward where UEs are located and thereby increase the antenna gain of cell-wide signals. The main gain of sectorization comes, however, from multiplexing, where UEs in different cells can use the same resources, but this is at the expense of decreased isolation between some cells. Sectorization will be discussed in Section 13.3.4.

13.3.2 FIXED BEAMFORMING

Fixed beamforming is closely related to network and site planning, that is, it concerns where to place sites and how to direct antennas (vertically and horizontally). For traditional antennas, tilt is the main tool for matching the vertical antenna characteristics to the scenario. Both mechanical tilt and *remote electrical tilt* are used, and although both can be controlled in a semi-static fashion, they are typically set once at installation or when the site plan is updated.

For AAS, cell-specific beamforming consists of one or several of the following components:

- mechanical tilt, affecting the entire antenna panel;
- subarray tilt, affecting the fixed beam shape of the subarray; and
- digital setting of weights used for cell-wide signals.

Note that mechanical tilt and subarray adaptation affect the characteristics of both cell-specific beamforming and UE-specific beamforming, while the digital adaptation of the beamforming weights associated with cell-defining signals is in general done independently of the UE-specific beamforming weights.

Note also that fixed beamforming can be applied in both the vertical and the horizontal domains.

13.3.2.1 Antenna subarray tilt

Antenna subarray tilt is used to match the vertical antenna characteristics to the deployment scenario, that is, adapting the subarray antenna pattern in the vertical domain to the UE distribution (see Section 13.2.2.1). This involves some partly conflicting objectives:

- maximizing the gain towards all own-cell UEs;
- minimizing the gain towards UEs in adjacent cells (minimizing inter-cell interference).

On a high level, maximizing the gain means typically that the gain should match path loss variations with the main lobe toward served UEs with worse path loss (cell-edge UEs), while minimizing inter-cell interference entails avoiding transmitting power into neighboring cells, that is, more down-tilting.

For traditional non-AAS antennas, there exist guidelines for how to set the tilt. These recommendations are often used also to set the subarray tilt for AAS antennas. Two commonly used rule-of-thumbs are (see also Fig. 13.16):

- *coverage tilt*—aim toward the physical cell edge, that is, the main lobe should point toward the physical cell edge in order to maximize the signal power toward the cell-edge UEs;

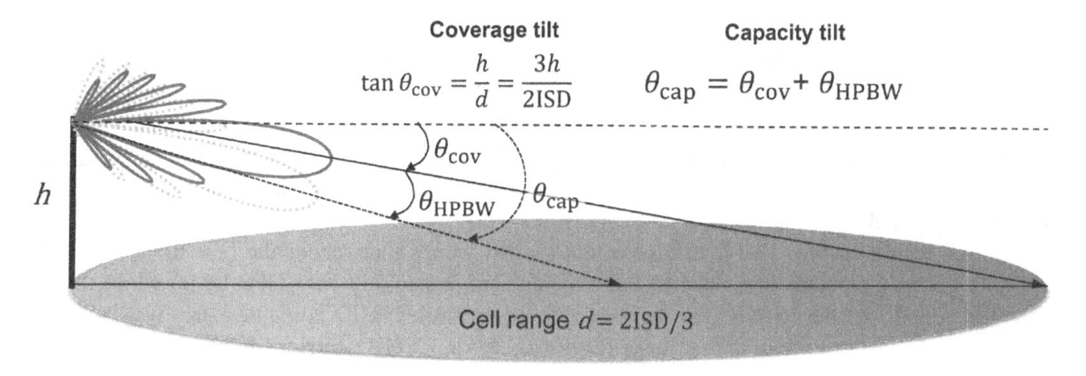

FIGURE 13.16

Illustration of coverage tilt θ_{cov} (solid line antenna gain pattern) and capacity tilt θ_{cap} (dotted line antenna gain pattern) for uniform hexagonal deployments, where h is the base station antenna height, ISD stands for inter-site distance and HPBW is the half-power beam width.

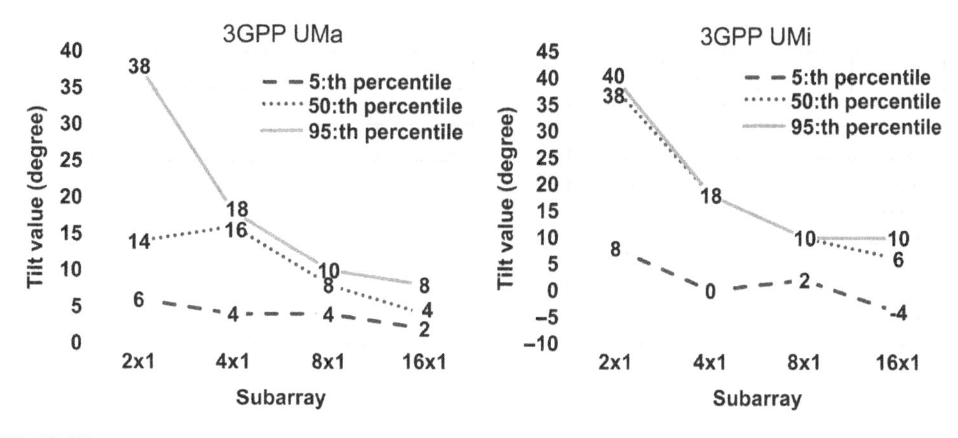

FIGURE 13.17

Subarray tilt values resulting in the best downlink SINR values for cell-edge (5:th), medium (50:th), and best (95:th) UE percentiles for different subarray sizes in the 3GPP UMa (*left*) and UMi (*right*) scenarios.

- *capacity tilt*—aim one HPBW below the physical cell edge, that is, the main lobe should point toward the cell edge minus on vertical HPBW. The objective of capacity tilt is to avoid spreading interference toward neighboring cells by placing a null close to the horizon. This is done at the expense of cell-edge coverage.

Fig. 13.17 shows optimized subarray tilt values rendering the best signal quality in terms of downlink SINR at high network load for cell-edge (5%), medium (50%), and best (95%) UE percentiles for different subarray sizes in the 3GPP UMa and UMi scenarios. Note that tilt is defined relative to the horizon (zenith angle is 90 degrees). Hence, no tilt corresponds to pointing the main lobe toward the horizon and a positive tilt means pointing the main lobe downwards.

Note that the coverage tilt for both 3GPP UMa and UMi is approximately 4 degrees as the antenna height relative to ISD is 0.05 for both cases, hence $\tan^{-1}(3 \times 0.05/2) \approx 4°$. As the capacity tilt is given by the coverage tilt plus the vertical HPBW, it will depend on the subarray size. Because the vertical HPBW of the elements in the subarrays is ~70 degrees, it follows according to Section 4.3.3.1 that the vertical HPBWs of subarrays of size 2×1, 4×1, 8×1, and 16×1 are roughly 35, 18, 9, and 4.5 degrees, respectively. The capacity tilt settings for these subarrays become then roughly 39, 22, 13, and 8.5 degrees, respectively.

From the results in Fig. 13.17, it is seen that the tilt values that render the best SINRs for cell-edge (5:th percentiles) UEs are close to the coverage tilt, especially for 3GPP UMa, while the tilt values that render the best SINRs for the best (95:th percentiles) UEs coincide fairly well with the capacity tilt. One thing to note is that these results are for a 3.5 GHz carrier frequency where coverage is more challenging compared to lower frequencies. By using a lower frequency with improved coverage, the capacity tilt setting often excels for all UE percentiles compared to the coverage tilt setting at high load in homogeneous scenarios like 3GPP UMa and UMi due to better inter-cell interference mitigation.

13.3.2.2 Sensitivity of antenna subarray tilt

Clearly, deployment characteristics, such as ISD, antenna heights, and antenna array layouts, will affect the appropriate tilt setting. However, there are also other aspects that affect the optimal tilt, such as traffic load, output power, and beamforming algorithm.

Fig. 13.18 illustrates the sensitivity of tilt by showing UE throughput gains associated with different subarray tilt settings relative to using no tilt (main lobe into the horizon) for the 5:th, 50:th, and 95:th UE percentiles, for two different traffic load points and two different subarray sizes in the 3GPP UMa scenario.

Evidently and as expected, small subarrays with wide main lobes are less sensitive to the subarray tilt setting compared to large subarrays with narrow main lobes. For 2×1 subarrays, the performance is insensitive to the tilt setting for all UE percentiles at all load points.

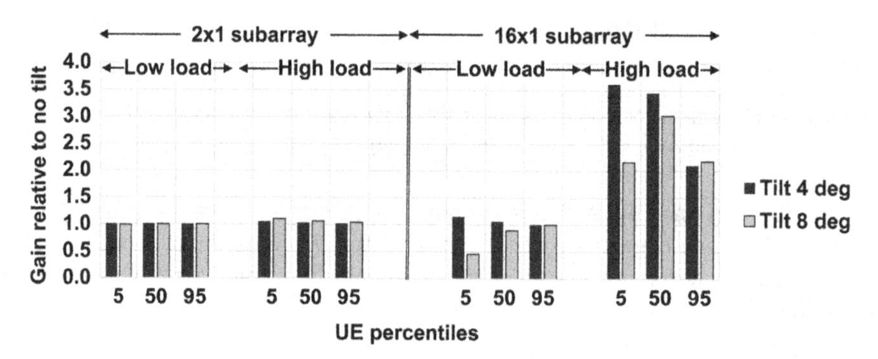

FIGURE 13.18

Illustration of sensitivity to tilt setting in 3GPP UMa. UE throughput gains for different tilt settings relative to no tilt (0 degree tilt) for 5:th, 50:th, and 95:th UE percentiles and low or high network load. Small 2×1 subarrays (*left*) and large 16×1 subarrays (*right*).

For 16 × 1 subarrays, on the other hand, the tilt setting is much more sensitive. Too much tilt is detrimental at low load, especially for the 5:th UE percentile, where 8 degree tilt performs much worse than no tilt or 4 degree tilt. At high load, some tilt is clearly required as no tilt performs much worse than 4 or 8 degree tilt.

13.3.2.3 Inter-cell interference handling via tilt

As discussed in previous subsections, tilt is an important tool to match the transmission to the deployment scenario by means of trading coverage versus interference suppression. A key factor for all deployments is how to handle inter-cell interference. Although important also for UE-specific beamforming (as discussed in Section 6.1.7 and illustrated in Section 13.4), this is particularly important for cell-wide signals using fixed beamforming in interference demanding deployment scenarios. The problem is more severe for always-on-like signals and especially for signals that have fixed positions in the resource grid and hence might collide between cells. Non-shifted CRS for LTE and synchronization signals in NR are examples of such signals.

As seen in the previous subsection, the interference issue can be improved by increasing the cell isolation between neighboring cells, for example, by using more down-tilt. It is, however, a non-trivial task to design good cell shapes that achieve high antenna gain and good cell isolation. This is illustrated in Fig. 13.19, showing the antenna gain pattern as a function of the zenith angle for a 10-element antenna with 13 degree down-tilt (capacity-based tilt). In the left part of the figure, DFT-based beam weights are used, while in the middle part of the figure "optimized" beam weights are employed. It is seen that the DFT-based beam has a rather strong side lobe pointing just below

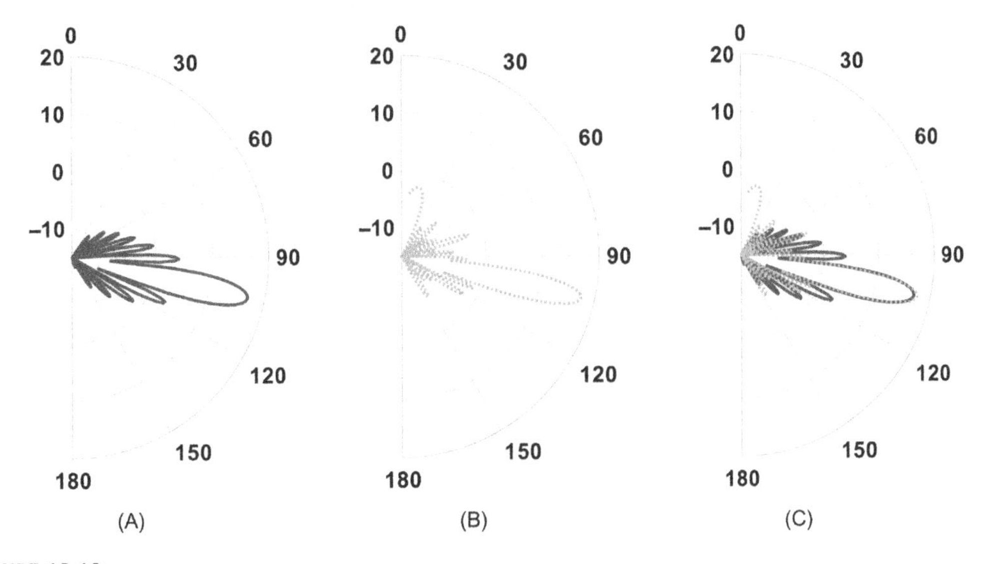

FIGURE 13.19

Illustration of the antenna gain pattern as a function of the zenith angle at 0 degree azimuth angle for a 10-element antenna with 13 degree down-tilt. (A) DFT-based beam weights, (B) "optimized" beam weights, and (C) both (A) + (B) in the same figure.

the horizon, which can lead to strong inter-cell interference. The optimized beam is designed to maximize gain in the pointing direction while suppressing side lobes above the main lobe (close to the horizon). It is indeed seen that the optimized beam has much less side lobe levels close to the horizon, which is good from an inter-cell interference perspective.

13.3.3 CELL SHAPING

This section will illustrate the potential of cell shaping by means of cell-specific subarray tilt, where an algorithm automatically optimizes the subarray tilt setting for each cell in the network.

Figs. 13.20 and 13.21 show downlink and uplink capacity gains, respectively, in the urban low- and high-rise scenarios for different subarrays with and without cell shaping. Cell shaping consists of cell-specific subarray tilt optimization where the optimal tilt of each cell is determined jointly. The optimization is done sequentially over all cells in a number of iterations, and the criterion to optimize is based on the experienced downlink UE throughput.

The case without cell shaping corresponds to applying mechanical tilt according to the capacity tilt rule-of-thumb discussed in Section 13.3.2.1 for an 8×1 subarray. The same mechanical tilt is then applied to all subarray configurations. Note also that results include only cell-specific beamforming, that is, no UE-specific beamforming is used (see Section 13.5 for a discussion on the relation between cell-specific and UE-specific beamforming).

By considering the downlink results (Fig. 13.20), it is apparent that the usefulness of cell shaping based on cell-specific tilt optimization is much larger in the high-rise deployment scenario than in the low-rise deployment scenario. It is evident that in the high-rise scenario with large angular UE spread in the vertical domain, careful cell planning is needed to get gains from larger subarrays. The low-rise scenario, where UEs are more homogeneously spread and confined to a rather narrow

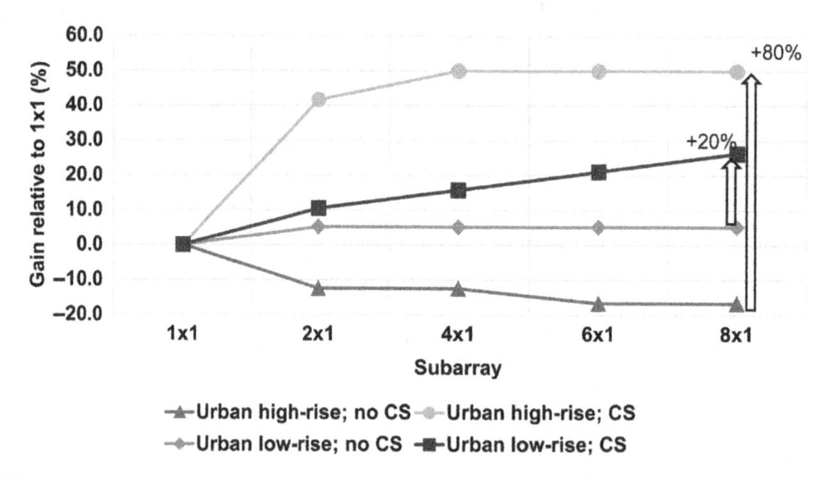

FIGURE 13.20

Downlink capacity gain in percent relative to a 1×1 subarray in the urban low- and high-rise scenarios for different subarrays with and without (no) cell shaping (CS).

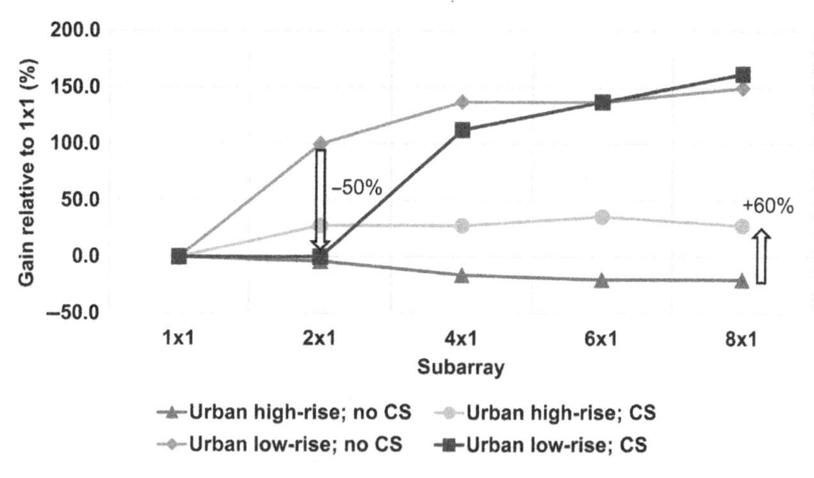

FIGURE 13.21

Uplink capacity gain in percent relative to 1×1 subarray in the urban low- and high-rise scenarios for different subarrays with and without (no) cell shaping (CS).

angle in the vertical domain, slightly below the antenna, is more robust with respect to cell planning.

In this example, it follows that the gains from cell shaping with 8×1 subarrays in urban low- and high-rise scenarios are roughly 20% and 80%, respectively. Also, although larger subarrays increase the antenna gain, it follows that using too large subarrays without cell shaping may incur losses, at least in the high-rise scenario. Note, though, that the tilt value obtained from the capacity-based rule-of-thumb applied to an 8×1 subarray, which is used to set the mechanical tilt for all cells without cell shaping, might be non-optimal, especially in non-homogenous scenarios.

The trends for the uplink results (Fig. 13.21) for the high-rise scenario are very similar to the downlink results, while the low-rise scenario exhibits very different uplink results compared to the downlink results. The main reason for the difference is that coverage is more challenging in the uplink compared to the downlink and the urban low-rise scenario has worse coverage compared to the urban high-rise scenario. Larger subarrays provide increased subarray gain and better coverage. Thus, uplink performance tends to increase significantly by increasing the subarray size in the low-rise scenario.

Somewhat counterintuitively, it is seen that cell shaping generally results in worse uplink performance for the low-rise scenario. The reason is that the cell shaping optimization criterion only considers the downlink performance. As the downlink results tend to be more interference limited than power limited in these dense urban scenarios, the cell shaping will generally result in down-tilting (or nulls close to the horizon). However, as the uplink has worse coverage compared to the downlink, too much down-tilting will decrease the uplink performance, especially for smaller subarrays with less gain. This also highlights the problem that different tilt strategies can be optimal for uplink and downlink.

Fig. 13.22 shows the distribution of tilt values in the network after applying the tilt optimization algorithm. Evidently, there is a large spread of tilt values in the high-rise scenario and a much

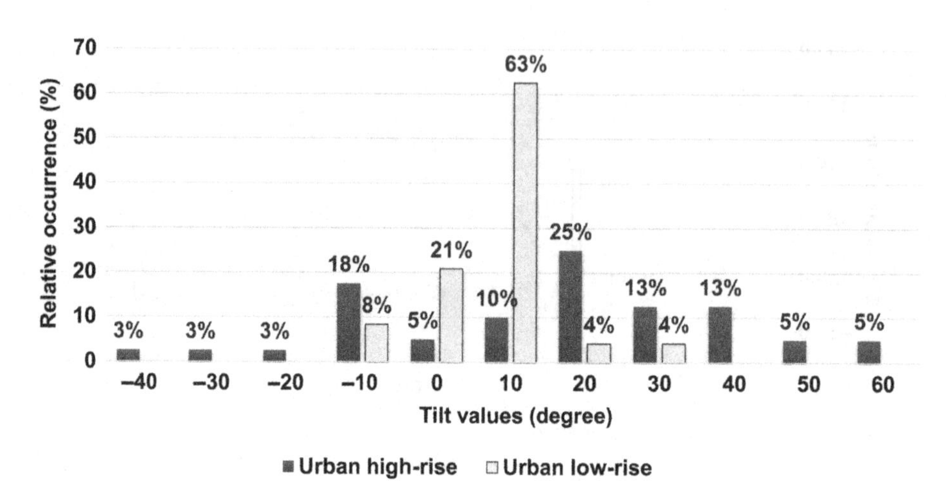

FIGURE 13.22

Distribution of used tilt values over the entire network for the 8 × 1 subarray case.

smaller spread in the low-rise scenario. Note, though, that these results do not reveal much about the sensitivity. For example, the criterion function can have many local optima with widely different tilt distributions. This effect is more likely to occur in the more non-homogeneous high-rise scenario compared to the more homogeneous low-rise scenario.

13.3.4 HIGHER-ORDER SECTORIZATION

Three-sector sites, where each site consists of three sectors and each sector consists of one cell, are very common deployments in cellular networks, but other constellations exist as well. A brief discussion regarding the site and cell concepts is found in Section 2.2.1.

13.3.4.1 Key characteristics

Splitting an existing cell into two or more cells is referred to as (higher-order) sectorization (see Fig. 13.3). The main benefit of this is a multiplexing gain, which allows UEs in the different cells to use the same time frequency resources. This is like MU-MIMO with fixed beams and two independently running schedulers. Sectorization leads to a reduction in the per-cell load, which means reduced queuing delays and increased capacity. Like for cell shaping, sectorization can lead to increased signal power by directing the sectors toward where UEs are located and thereby increase the antenna gain of cell-wide signals. These effects mainly bring gains at high load. For low load, there are in fact often losses by sectorization as more cells result in more common channels (at least for LTE) and thereby increased interference. For LTE, the main source of low-load interference comes from CRS:es that are transmitted continuously.

Sectorization can be done with or without antenna reuse. Sectorization without antenna reuse means essentially that a separate antenna (and radio) is deployed for each additional sector, while sectorization with antenna reuse means that the same antenna array (and radio) handles the original sector and all additional sectors. Note that sectorization with antenna reuse implies that the available power is split between all the associated sectors, which means that sectorization with antenna reuse is less beneficial in coverage-limited scenarios due to reduced power per cell.

The efficiency of sectorization depends on the isolation between the created sectors, which relates to the antenna characteristics. A large antenna (in the dimension of sectorization) can create more isolated and well-defined sector beams.

Sectorization without antenna reuse can employ antennas tailored to the deployment, for example, reduced width of the main lobe in the domain of sectorization to ensure better cell isolation. This fact together with twice the output power implies that sectorization without antenna reuse typically excels over sectorization with antenna reuse.

13.3.4.2 Performance results

The following example considers sectorization via antenna reuse for LTE with two radio chains per sector (one per polarization) for UE-specific beamforming in the 3GPP UMa and UMi scenarios. Fig. 13.23 shows capacity gains from horizontal and/or vertical sectorization for different antenna configurations relative to the baseline case of a single sector with an 8×1 antenna column. Note that the output power is split equally in all sectors. This means that the maximum power per cell for sectorization with two or four additional sectors is ½ or ¼ of the baseline case without sectorization. 2V denotes vertical sectorization where each sector is split into two vertical sectors and 2H denotes horizontal sectorization where each sector is split into two horizontal sectors. For vertical sectorization, one sector is down-tilted according to capacity-based tilt guidelines, while the other sector is up-tilted such that the side lobes of the up-tilted sector cover the nulls of the down-tilted sector. The implementation of vertical sectorization based on antenna reuse is illustrated in Fig. 13.24.

The results indicate that vertical sectorization based on an 8×1 antenna and horizontal sectorization based on an 8×2 antenna (8×1, 2V vs 8×2, 2H) provide similar gains. The likely reason is that vertical sectorization with an 8×1 antenna provides significantly better cell isolation than horizontal sectorization with an 8×2 antenna due to different antenna sizes in the dimension of sectorization. There are eight elements in the vertical dimension and only two elements in the horizontal dimension. Note also that the 8×2 configuration has twice the antenna gain as that of the 8×1 configuration as the 8×2 antenna configuration is twice as large as the 8×1 antenna configuration. Hence, it seems as the increased antenna gain with 2H 8×2 compensates the worse cell isolation relative to 2V 8×1.

Horizontal sectorization based on an 8×4 antenna compared to an 8×2 antenna (8×4, 2H vs 8×2, 2H) brings an additional capacity gain of 10% units. The gain is attributed to increased antenna gain and better sector isolation from the larger antenna.

Adding two vertical sectors on top of the two horizontal sectors (2H2V compared to 2H) adds roughly 20% and 40% additional capacity in 3GPP UMa and UMi, respectively. Hence, combining horizontal and vertical sectorization brings significant additional gains, especially in the 3GPP UMi scenario. Sectorization based on antenna reuse works best in scenarios with higher SINR levels (e.g., smaller cells) as then the higher SINRs can be shared among multiplexed layers/UEs.

The performance of sectorization is scenario dependent and works best in interference-limited scenarios with good coverage. Sectorization is generally said to be a capacity feature and not a coverage feature. Sectorization is closely related to cell-specific beamforming, and the two should

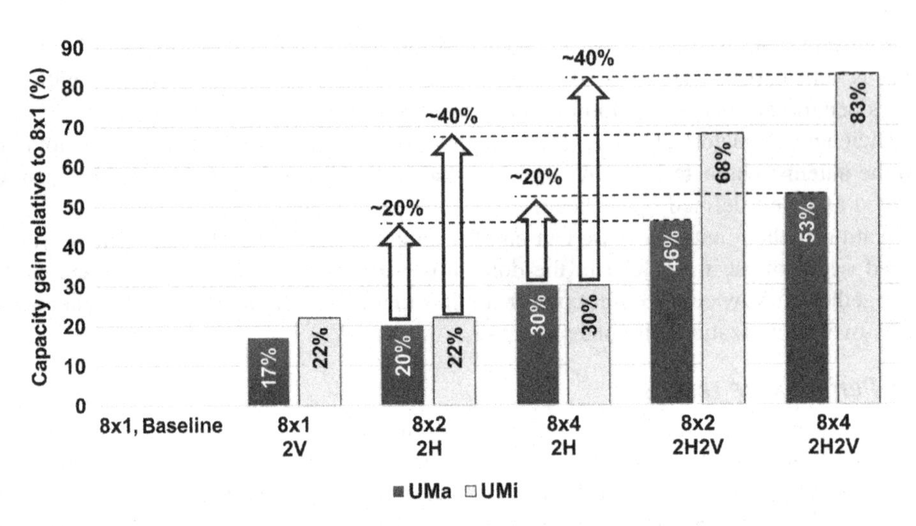

FIGURE 13.23

Capacity gains (%) from sectorization relative to 2T 8 × 1 in 3GPP UMa and UMi. n and m in $n \times m$ denote the number of vertical and horizontal antenna elements, respectively. 2H and 2V represent horizontal and vertical sectorization, respectively.

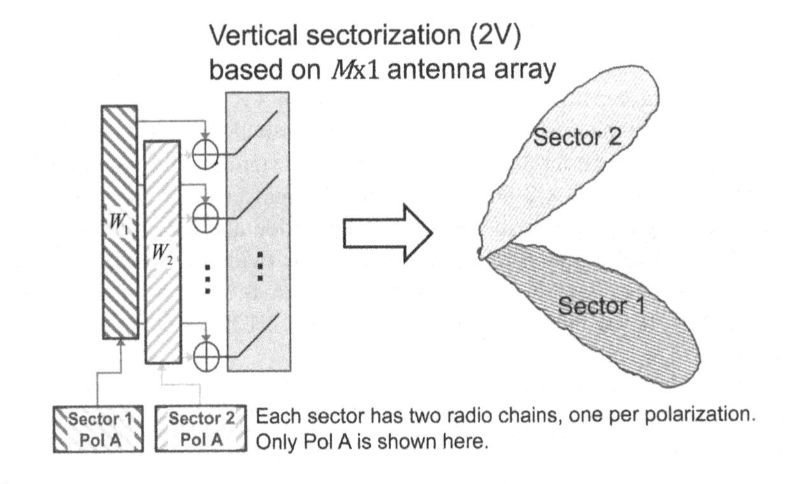

FIGURE 13.24

Illustration of implementation of vertical sectorization based on antenna reuse. Sectors 1 and 2 are created based on beamforming vectors W_1 and W_2, respectively, where W_1 creates down-tilt and W_2 generates up-tilt. Note that all M antenna elements are used to form both sectors. Only one of the two polarizations, Pol A, is illustrated in the figure.

preferably be combined to realize the full potential of sectorization. When combining UE-specific beamforming and sectorization, UE-specific beamforming should be explored before sectorization, that is, UE-specific beamforming brings more gain than sectorization.

13.4 UE-SPECIFIC BEAMFORMING (SU-MIMO)

13.4.1 OVERVIEW

The focus in this section is on *UE-specific transmit beamforming* and how different types of algorithms, antenna configurations, and deployment characteristics affect performance.

13.4.1.1 Key characteristics

As discussed in Section 6.1, UE-specific transmit beamforming relies on adapting the transmitted signal based on spatial and temporal characteristics of the radio channel. This process is often referred to as *precoding*, in which the signal is mapped to the radio chains, or to each polarization of all subarrays, and where the mapping adjusts the phase and amplitude of the signal individually for each radio chain.

Unless otherwise stated, two main types of UE-specific beamforming are considered, namely:

- *classic beamforming* illustrated by means of a GoB algorithm. This scheme relies on *feedback-based large-scale wideband precoding*, in which the UE estimates the preferred precoder and feeds it back to the base station;
- *generalized beamforming* exemplified by means of an MRT algorithm. This scheme uses *reciprocity-based fast-fading frequency-selective precoding*, in which uplink sounding is employed to estimate the channel used to derive the precoder.

See Section 13.2.3 for further algorithm details.

The AoSA antenna setup, introduced in Section 4.6 and discussed in Section 13.2.2, is adopted in this section. Given this antenna structure, the following aspects are key in relation to UE-specific beamforming:

- the *subarray size* defines the angular coverage area for UE-specific beamforming, that is, the region where beams can be directed without any significant gain drop. It can be discussed what is meant by "significant gain drop", but a common choice is to consider the 3 dB main lobe width (or HPBW) of the subarray gain pattern (see Fig. 13.6 for an illustration);
- the *number of subarrays*, or radio chains (number of T:s), and *how they are stacked*. A wide angular coverage area in the vertical domain requires many small subarrays to get high beamforming gain. Furthermore, using many subarrays and/or increasing the spacing between the subarrays lead to narrow beams, which can reduce the negative impact of interference. The preferred stacking of subarrays is closely related to the angular UE distribution.

Another related important antenna attribute is the *antenna gain* that relates to the total antenna area. Note, though, that given a uniform AoSA structure (uniform element separation), the number of subarrays together with the subarray size gives the total antenna area. Similarly, the antenna area and the subarray size yield the number of subarrays or the antenna area together with the number of subarrays gives the subarray size.

13.4.1.2 Outline

How the network level performance is affected by different flavors of SU-MIMO is the topic of Section 13.4.2. Here the performance impact of choosing different AAS configurations will be illustrated. More specifically, results from having a fixed antenna area (fixed number of antenna elements) with different number of subarrays are discussed in Section 13.4.2.1, the impact of different antenna layouts is illustrated in Section 13.4.2.2, and the performance impact by increasing the antenna area (the number of elements) is provided in Section 13.4.2.3. Finally, Section 13.4.3 will give some additional insights on the performance of UE-specific beamforming by illustrating how the AAS configuration affects the signal power and interference.

13.4.2 IMPACT OF DIFFERENT ANTENNA CONFIGURATIONS

In this subsection, the performance of different antenna configurations in the 3GPP UMa scenario is considered. To show how the AAS configuration affects performance, three different cases are considered:

1. different number of subarrays (or radio chains) for fixed antenna dimensions. Consequently, using many subarrays (radio chains) results in small subarrays and vice versa (see Fig. 13.25). In this case, the (maximum) antenna gain is fixed, while the vertical angular coverage area and the number of subarrays in the vertical domain are varying;
2. different antenna layouts for a fixed number of subarrays (or radio chains) of fixed size. The subarrays are consequently stacked in different ways. This corresponds to a deployment scenario with a fixed angular coverage area and a fixed antenna area but with different antenna shapes. This case highlights the need for horizontal or vertical domain UE-specific beamforming (see Fig. 13.25);
3. increasing the antenna area and the number of subarrays (or radio chains) by means of adding antenna columns or rows (see Fig. 13.26). This case shows the benefit of increased beamforming gain and flexibility in the horizontal or vertical domain for a fixed angular coverage area.

In Cases (1) and (2), the number of antenna elements is kept constant. As a result, the (maximum) antenna gain is fixed.

Recall that antenna configurations are denoted as $(M_v \times 1)_{SA}(N_v \times N_h)$, where M_v denotes the number of vertically stacked dual-polarized antenna element pairs in the subarray, and N_v and N_h represent the number of vertically and horizontally stacked subarrays, respectively. Thus, the number of radio chains (also referred to as number of T:s) is given by $2N_vN_h$.

13.4.2.1 Impact of subarray size for a fixed antenna area

The performance gains of MRT and GoB for different antenna configurations given a fixed 16×4 antenna array in the 3GPP UMa scenario can be found in Fig. 13.27. The number of antenna columns is fixed to four, while the subarray size varies between 16 and 1. That is, the following

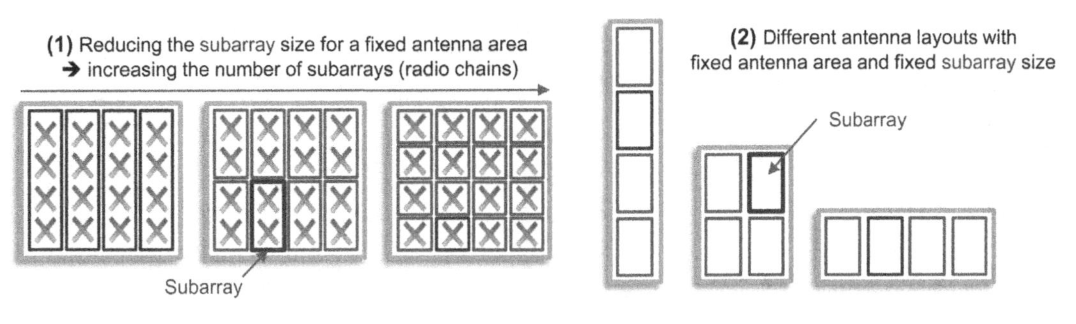

FIGURE 13.25

Illustration of Cases (1) and (2).

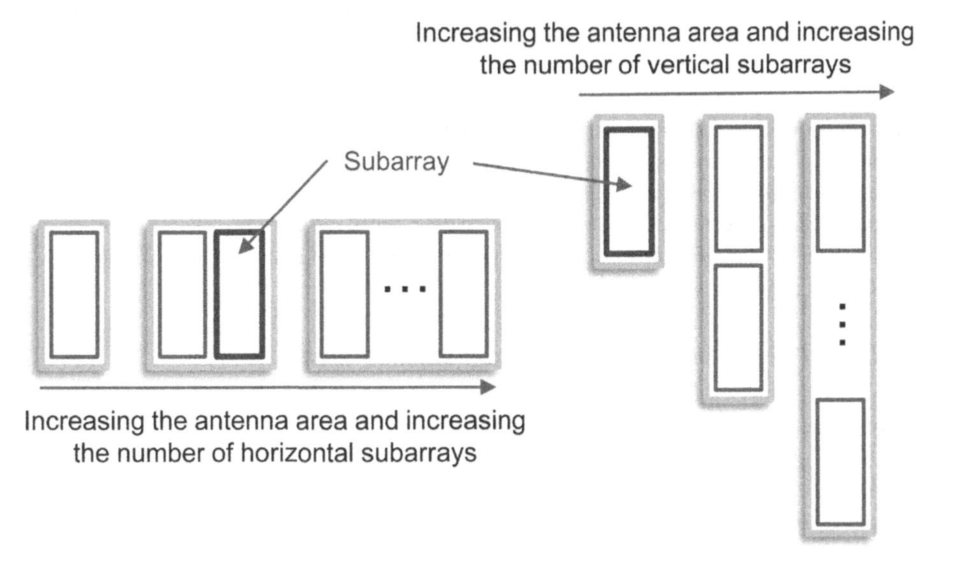

FIGURE 13.26

Illustration of Case (3).

configurations are considered: 8T with 16×1 subarray, 16T with 8×1 subarray, 32T with 4×1 subarray, 64T with 2×1 subarray, and 128T with 1×1 subarray.

The capacity is very similar for 32T, 64T, and 128T and slightly worse for 16T and 8T for both GoB and MRT. Hence, a subarray of size 4×1 or less is a good choice given a fixed 16×4 antenna area as this size matches the angular UE distribution in the vertical domain of the 3GPP UMa scenario. This result holds for other number of antenna columns as well.

Similar trends are observed for the UE throughput results as well. Note, though, that 8T with a 16×1 subarray is inferior relative to baseline 32T with a 4×1 subarray. Also, a bit surprisingly,

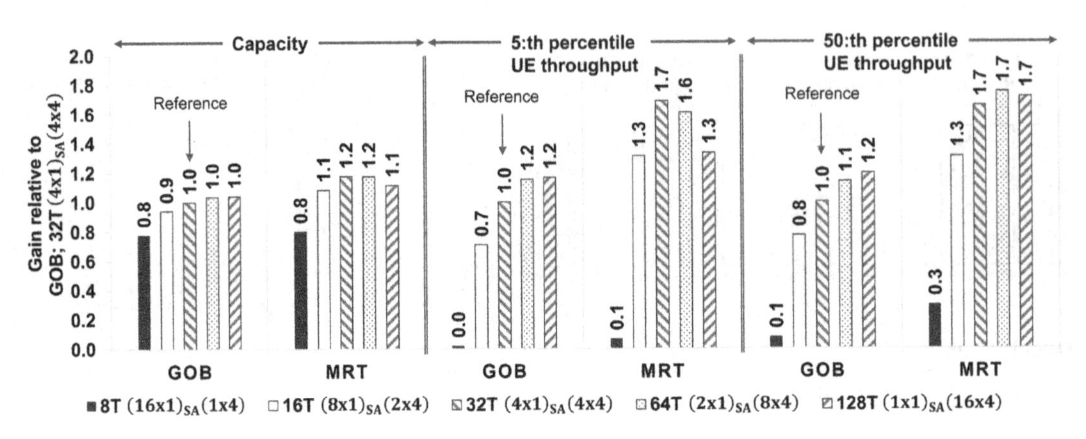

FIGURE 13.27

Capacity, 5:th percentile, and 50:th percentile UE throughput gains for GoB and MRT with different antenna configurations relative to 32T $(4\times1)_{SA}(4\times4)$ GoB in the 3GPP UMa scenario.

the 5:th percentile UE throughput of MRT starts to degrade by increasing the number of radio chains beyond 32. The main reason for this is that small subarrays may suffer from poor sounding quality, hence reduced beamforming gain. Nevertheless, in practice using a better adaptive MRT implementation, the performance should not be worse with more radio chains. For example, the 128T $(1\times1)_{SA}(16\times4)$ configuration can always be made to behave like the 32T $(4\times1)_{SA}(4\times4)$ configuration (see also discussion in Section 13.2.2.1).

13.4.2.2 Impact of antenna layout for a fixed antenna area

The performance gains of MRT and GoB for different antenna layouts (wide or tall) given a fixed 4×1 subarray and a fixed number of 32 radio chains (or 16 subarrays) in the 3GPP UMa scenario can be found in Fig. 13.28.

The horizontal antenna layout clearly excels over the vertical antenna layout. The reason is attributed to the homogeneous 3GPP UMa scenario with relatively large ISD, base station antennas above rooftop, and a narrow angular UE distribution in the vertical domain. Cells are essentially much wider than taller. A wider antenna layout will result in a more uniform spread of interference in the horizontal domain (see Section 13.4.3 for a detailed discussion). The importance of this is most apparent for MRT, in which case the beamforming gain is very similar irrespectively of the antenna layout; hence, the main performance difference between the configurations can be attributed to the interference characteristics.

13.4.2.3 Impact of increased antenna area and number of radio chains

Fig. 13.29 illustrates the performance gains for GoB and MRT by stacking 4×1 subarrays horizontally (making a wide antenna). Significant gains by increasing the antenna area and the number of radio chains are observed for all metrics. For GoB, the capacity is seen to increase by on average roughly 30% for each doubling of the number of radio chains (and antenna area), while the corresponding figure for MRT is roughly 40%. The gains are subject to diminishing returns, especially

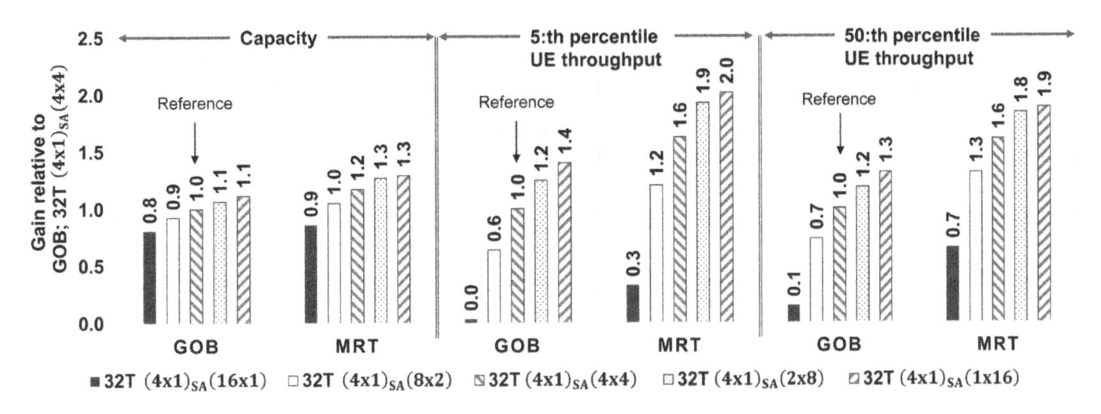

FIGURE 13.28

Capacity, 5:th percentile, and 50:th percentile UE throughput gains for GoB and MRT with different antenna configurations relative to GoB with $(4 \times 1)_{SA}(4 \times 4)$ in the 3GPP UMa scenario.

for GoB, due to effects such as channel angular spread. That is, for GoB, the capacity gain by doubling the number of antenna columns is 45% for 2T, 40% for 4T, 30% for 8T, and 20% for 16T. For MRT, the performance gain is more stable. Needless to say, for large SNRs, gains start to saturate unless SNRs are shared between several UEs, that is, unless MU-MIMO and not only SU-MIMO is considered (see Section 6.3).

Fig. 13.30 shows the performance gains for GoB and MRT by stacking 4×1 subarrays vertically (making a very tall antenna). As for horizontal stacking, significant gains by increasing the number of radio chains and consequently the antenna area are observed for all metrics. GoB and MRT show rather similar performance with capacity increasing by roughly 20%–30% for each doubling of the number of radio chains (and antenna area).

By comparing capacity gains for horizontal and vertical antenna configurations, see Fig. 13.31, it is evident that horizontal UE-specific beamforming excels in this homogeneous 3GPP UMa scenario with relatively large ISDs, base station antennas above rooftop, and a narrow angular UE distribution in the vertical domain.

It is interesting to see that GoB and MRT perform rather similar for most configurations. The most notable exception is the 32T case and especially for the horizontal layout. The main reason for this can be attributed to the channel angular spread that may affect performance when traffic beams become narrow (recall from Section 4.3.3.1 that the main lobe width is roughly inversely proportional to the size). The channel angular spread is substantially larger in the horizontal domain than in the vertical domain, and MRT can take advantage of the angular spread while GoB is sensitive to angular spread.

Vertical domain beamforming tends to be less robust than horizontal domain beamforming in the sense that performance impact is less predictable and more sensitive to, for example, the subarray size and subarray tilt. As many UE-specific beamforming algorithms do not take inter-cell interference explicitly into consideration, it is important to handle interference aspects in some

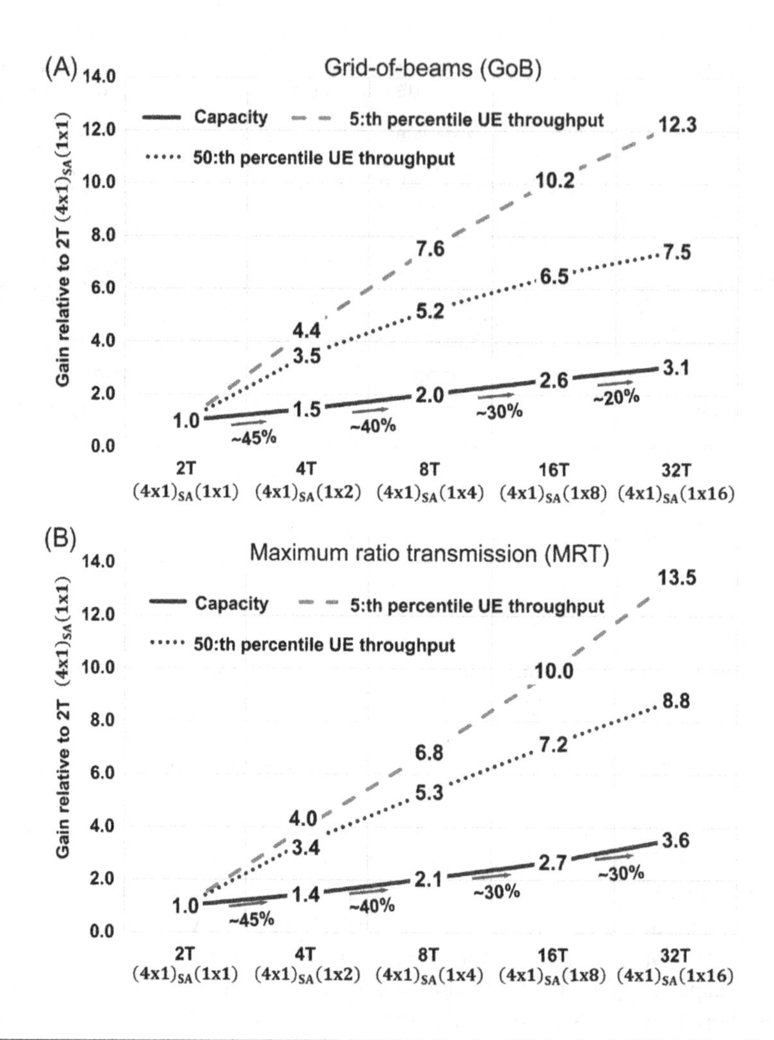

FIGURE 13.29

Capacity and UE throughput gains for horizontal antenna layouts. GoB (A) and MRT (B). Percentage values correspond to the capacity increase by doubling the number of radio chains (and antenna area).

other way. Different approaches for considering interference aspects can be envisioned, including, for example,

1. careful network planning and proper cell-specific beamforming can give significant system performance improvements (see Section 13.3.2);
2. imposing restrictions on the precoder codebook, for example, excluding "beams" creating excessive interference.

How important interference handling is and what approach is best depend on scenario characteristics, for example, is the network performance primarily limited in terms of coverage or capacity.

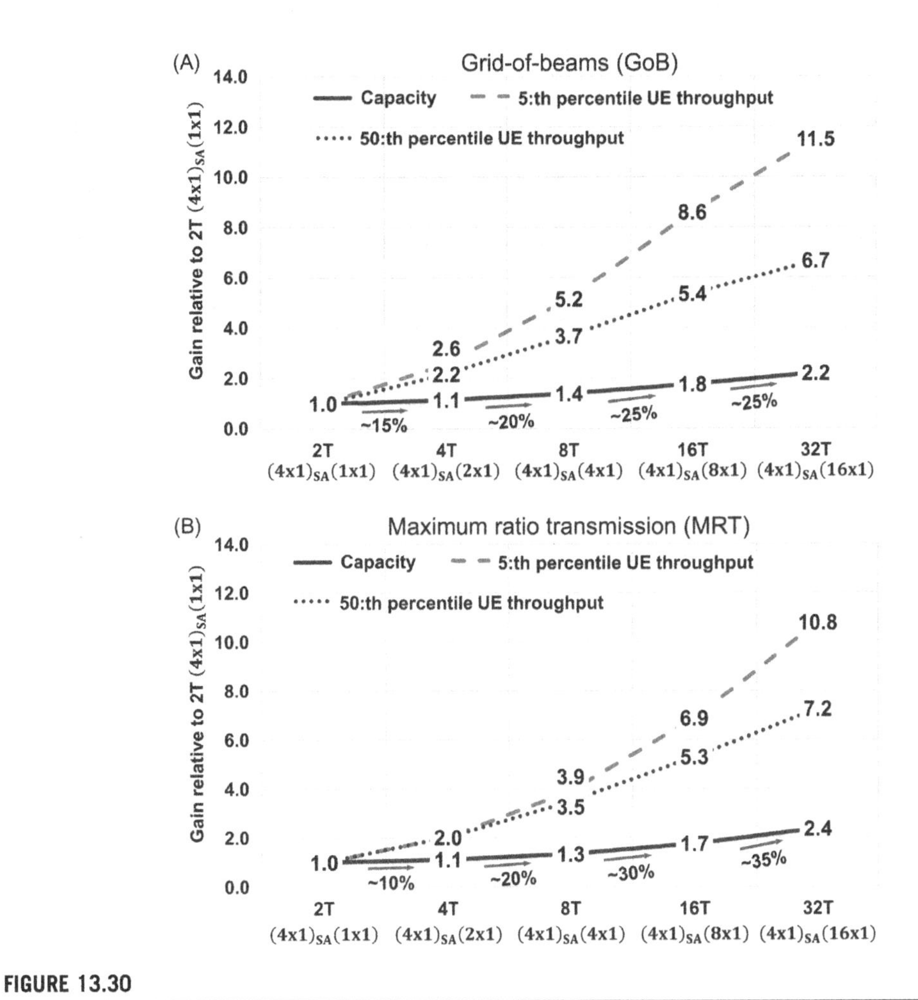

FIGURE 13.30

Capacity and UE throughput gains for vertical antenna layouts with 4×1 subarrays. GoB (A) and MRT (B). Percentage values correspond to the capacity increase by doubling the number of radio chains (and antenna area).

13.4.3 BEAMFORMING EFFECTS ON SIGNAL POWER AND INTERFERENCE

Performance for a specific UE (link) is a function of the achievable SINR, that is, the received signal power in relation to noise and interference from other transmissions. The own signal strength is improved by increased beamforming gain, which in turn depends on the antenna configuration, channel characteristics, and the effectiveness of the beamforming algorithm. While the instantaneous interference in a specific "direction" obviously increases as the beamforming gain increases, the characteristics of the average spatial interference is more involved and relate to aspects such as angular UE distribution, subarray size, and array layout. This was discussed in Section 6.1.7 and will be further illustrated below for MRT and GoB.

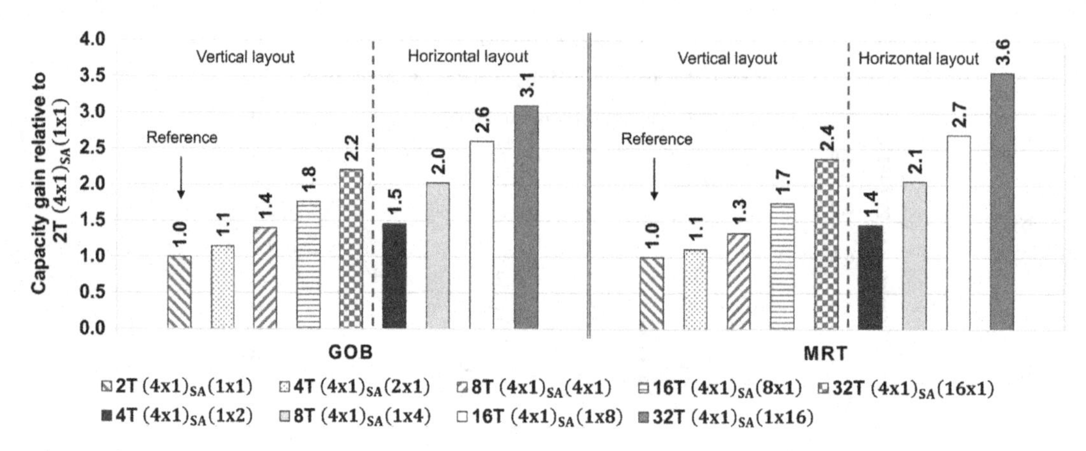

FIGURE 13.31

Capacity gains for GoB and MRT with a fixed 4×1 subarray and different horizontal and vertical antenna layouts relative to $(4 \times 1)_{SA}(1 \times 1)$. Note that the reference 2T case is without UE-specific beamforming; hence, GoB and MRT coincide.

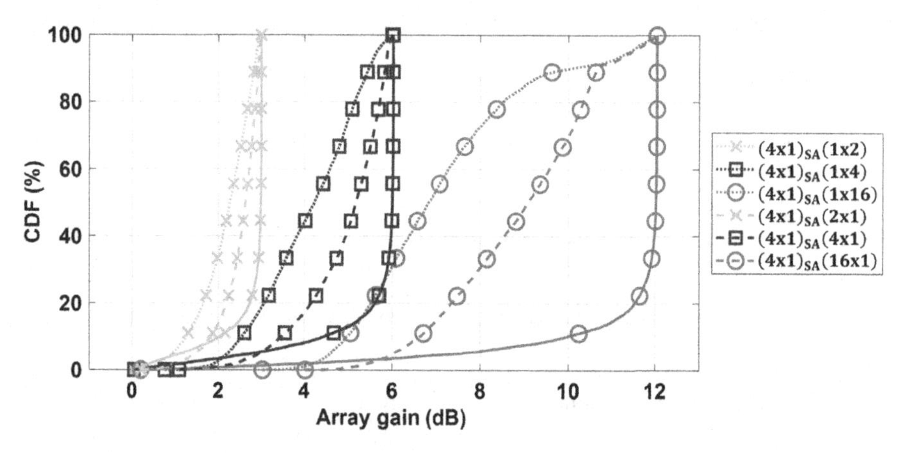

FIGURE 13.32

CDF of the array part of the beamforming gain for different antenna layouts. Dotted and dashed lines correspond to horizontal and vertical antenna layouts for GoB, respectively, and solid lines correspond to MRT. Note that for MRT, the gain is equal for horizontal and vertical antenna layouts, for example, array gains for $(4 \times 1)_{SA}(1 \times 4)$ and $(4 \times 1)_{SA}(4 \times 1)$ coincide.

13.4.3.1 Beamforming gain

Fig. 13.32 shows the distribution of the array part of the beamforming gain for GoB and MRT with different horizontal and vertical antenna layouts with a fixed 4×1 subarray in the 3GPP UMa scenario. The array part of the beamforming gain refers to the gain of using an AoSA compared to using a single subarray. It is important to understand that the beamforming gain depends not only on the antenna array layout but also on additional aspects like the choice of beamforming algorithm and channel properties like angular spread (see also Section 6.1.6.1).

For GoB, the array part of the beamforming gain is larger for vertical antenna layouts than for horizontal antenna layouts. The main reason being that angular spread is much larger in the horizontal domain than in the vertical domain. This effect becomes more pronounced for large arrays that give narrow GoB beams, which can be seen from that the difference in gain between GoB and MRT increases for large arrays.

In contrast to GoB, MRT is robust against angular spread, but performance is degraded when uplink sounding quality becomes poor. Hence, MRT gives close to maximum gain for most UEs in this 3GPP UMa scenario with relatively good coverage, but there are some UEs having poor coverage, for which the MRT gain rapidly deteriorates. As uplink sounding is done per subarray, this effect becomes most noticeable for small 1×1 subarrays, in which the uplink sounding quality is essentially reduced by 6 dB compared to 4×1 subarrays. Note, however, that for a fixed antenna size, a larger subarray can be constructed from smaller subarrays by means of signal processing. Hence, the problem discussed here can be alleviated (see also discussion in Section 13.2.2.1).

The distribution of the array part of the beamforming gain for MRT using 1×1 and 8×1 subarrays can be found in Fig. 13.33, where it is seen that the percentage of UEs having poor uplink sounding is much larger for 1×1 subarrays than for 8×1 subarrays, and depending on the array layout, roughly 15%–20% of the UEs employing 1×1 subarrays experience better beamforming gain with GoB than with MRT.

The larger the subarray, the narrower are the GoB beams. Hence, due to angular spread, an 8×1 subarray shows larger losses for GoB relative to MRT than a 4×1 subarray, which in turn shows larger losses than a 1×1 subarray.

The array part of the beamforming gain for MRT is insensitive to the array layout, which can be explained by the balloon analogy, that is, it does not matter how the balloon is squeezed (flat in

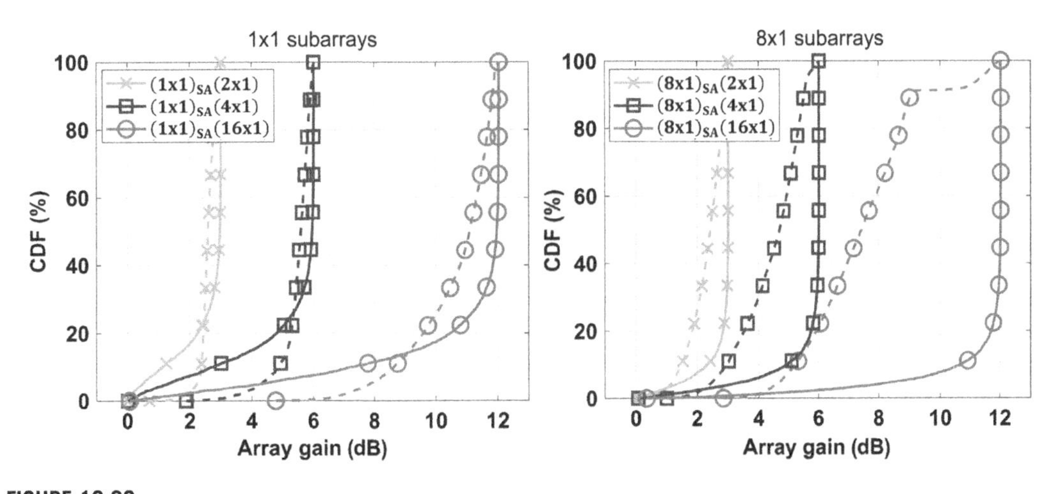

FIGURE 13.33

CDF of the array part of the beamforming gain for different vertical antenna layouts with 1×1 subarrays (*left*) and 8×1 subarrays (*right*). Dashed lines correspond to GoB, while solid lines correspond to MRT.

one or another direction), the maximum peak is still the same. A second effect that makes MRT perform better than GoB is that the channel is dispersive or frequency selective, which negatively impacts GoB but not MRT as GoB uses wideband precoding, whereas MRT uses frequency-selective precoding.

13.4.3.2 Interference characteristics

Fig. 13.34 shows the distribution of interference for full buffer MRT evaluations with 1×1 subarrays and three different antenna arrays: a 1×1, a horizontal 1×8, and a vertical 8×1. The 1×1 antenna array illustrates the impact on interference if the fact that interference is indeed beamformed, as for the 8×1 and 1×8 arrays, would be neglected. Thus, for the 1×1 antenna array, the interference looks like coming from a single subarray and experiences fixed beamforming.

The reason for considering full buffer evaluations here and not equal buffer is to ensure high and comparable load (interference) between different configurations and to remove any so-called turbo effects occurring in equal buffer simulations. The term turbo effect comes from that one effect might amplify or impact another effect. For example, in equal buffer simulations, removing poor UEs helps the good UEs as the poor UEs consume a lot of the system resources.

For the pure horizontal array, $(1 \times 1)_{SA}(1 \times 8)$, interference characteristics are similar to the $(1 \times 1)_{SA}(1 \times 1)$ configuration, whereas for the pure vertical array, $(1 \times 1)_{SA}(8 \times 1)$, the interference has increased compared to the $(1 \times 1)_{SA}(1 \times 1)$ configuration. Ignoring that interference is beamformed in the latter case would overestimate performance, that is, for the vertical array, beamforming increases the experienced interference, while for the horizontal array, beamforming has little impact on the interference distribution. The main reason for the difference between horizontal and vertical antenna layouts is because of the angular UE distribution that is larger in the horizontal domain compared to the vertical domain.

In 3GPP UMa, UEs are uniformly distributed in the horizontal domain, while confined to a small angular area in the vertical domain. Hence, beamformed interference is spread uniformly in

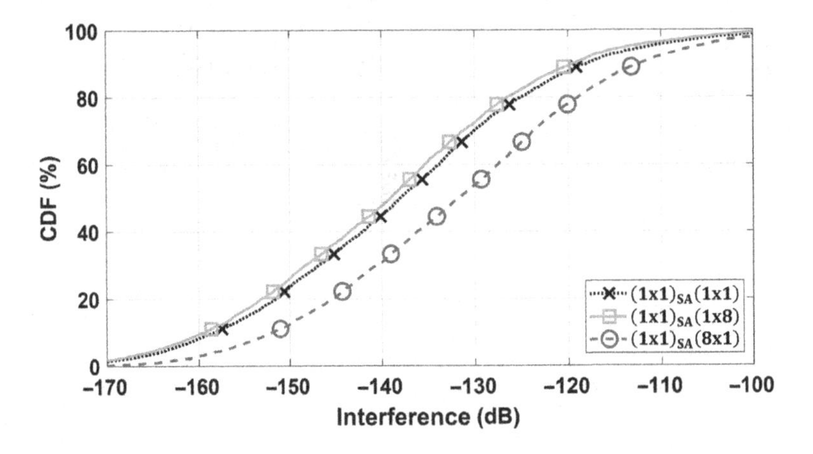

FIGURE 13.34

CDF of interference for different array configurations in 3GPP UMa based on full buffer simulations.

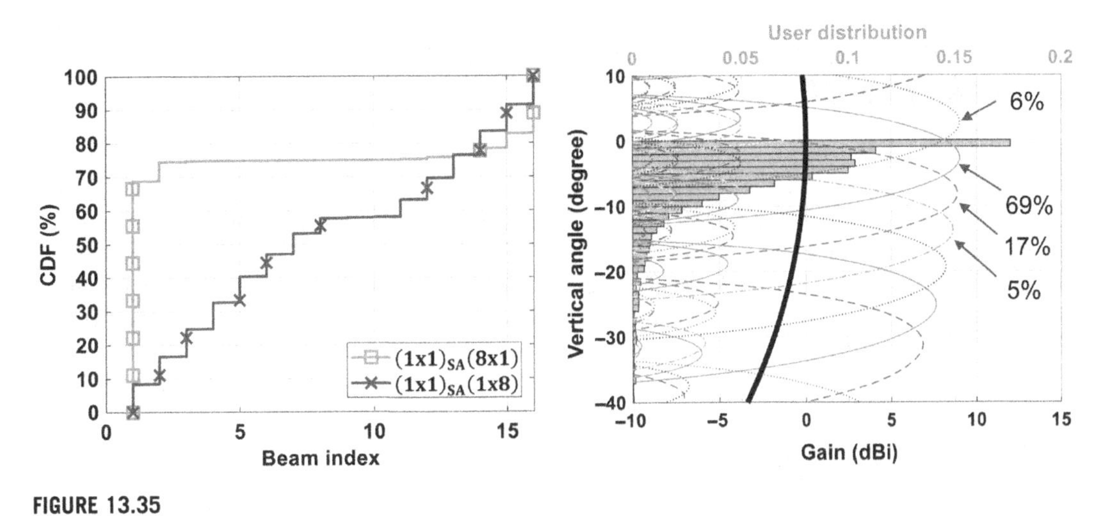

FIGURE 13.35

Illustration of used GoB beam indexes for wide and tall configurations (*left*) and GoB in relation to the angular distribution of UEs in the vertical domain for a tall $(1 \times 1)_{SA}(8 \times 1)$ configuration (*right*). Geometrical distribution of UEs in the vertical domain (bar plot), subarray diagram (solid thick line), and traffic beams based on GoB codebook (solid narrow lines). Percentage numbers represent how often different beams are used to serve UEs and correspond to the beam indexes to the left.

the vertical domain, keeping the average interference constant, whereas beamformed interference in the vertical domain is spread in a small angular region; hence, the average and instantaneous beamformed interference becomes similar. Note that the total transmitted power is the same for all cases, but different configurations distribute the power differently in space.

The results of Fig. 13.34 are further motivated by Fig. 13.35, where the left figure shows the used beam indexes (the precoding matrix indicators) with GoB for the wide and tall antenna configurations. Note that a codebook with oversampling two is used, hence in total 16 codewords (beams) to choose from. Evidently, with the wide configuration $(1 \times 1)_{SA}(1 \times 8)$, all 16 beams in the codebook are used with rather equal probabilities, whereas for the tall configuration $(1 \times 1)_{SA}(8 \times 1)$, one beam clearly dominates; beam index 1 is used in 69% of the transmissions. This further strengthens the argumentation that, with the wide configuration, interference is transmitted rather uniformly, while with the tall configuration, interference is transmitted in few directions. The right part of Fig. 13.35 shows the GoB beams for the $(1 \times 1)_{SA}(8 \times 1)$ configuration in relation to the geometrical angular UE distribution and illustrates that the used beams match fairly well the angular UE distribution in the vertical domain.

Other aspects impacting the interference characteristics include the channel angular spread and the beamforming method. Angular spread is typically significantly larger in the horizontal domain than in the vertical domain, and more angular spread has similar effect as more UE spread, that is, spreading interference wider. Classic beamforming (like GoB) typically spreads the power in a more well-defined beam, whereas generalized beamforming (e.g., MRT) spreads power in all directions with channel support, making the power distribution more scattered.

13.5 RELATION BETWEEN CELL-SPECIFIC AND UE-SPECIFIC BEAMFORMING

13.5.1 OVERVIEW

There is a tight connection between cell-specific beamforming, discussed in Section 13.3, and UE-specific beamforming, discussed in Section 13.4, that will be illustrated in this section.

Cell-specific beamforming will determine the coverage area for cell-wide signals like cell-selection signals and therefore impact which cell a specific UE is connecting to. At the same time, for UE-specific beamforming based on the AoSA antenna structure, it is the subarray that determines the angular coverage area, that is, the area where UE-specific beamforming works well without any significant gain drops. In other words, the subarray determines the envelope associated with all beams used for UE-specific beamforming.

For a fixed antenna array with limited degrees of freedom, it generally holds that UE-specific beamforming gives better performance than cell-specific beamforming. This will be illustrated in Section 13.5.2.

When the envelope of beams used for cell association does not match the envelope of beams used for data beamforming, a UE may connect to a non-optimal serving cell since the strongest cell as measured on cell-defining signals may not at all be the strongest cell for data beamforming. Thus, to secure performance improvements of UE-specific beamforming, signals used for cell association should be representative of the cell's ability to serve UEs in those directions, otherwise a performance loss may occur. This will be illustrated in Section 13.5.3. Furthermore, the impact of beam sweeping will be considered.

13.5.2 COMBINING CELL-SPECIFIC AND UE-SPECIFIC BEAMFORMING

In this section, the relation between cell-specific and UE-specific beamforming will be addressed. More specifically, capacity gains as a function of the total number of antenna elements and the number of radio chains (or subarrays) will be illustrated. The level of UE-specific beamforming (SU-MIMO based on GoB) is given by the number of radio chains, while cell-specific beamforming (cell shaping) is done by cell-specific subarray tilt optimization (see Section 13.3.2).

The subarray size is given by the number of antenna elements divided by the number of radio chains. Hence for a fixed number of antenna elements, there is a trade-off between the level of UE and cell-specific beamforming.

Fig. 13.36 shows the downlink capacity gain for different antenna configurations with both cell-specific and UE-specific beamforming in the urban high-rise scenario. For a fixed number of antenna elements, it is seen that UE-specific beamforming provides larger gains than cell-specific beamforming (gains increase by going in the y-direction). Similarly, for a fixed number of radio chains, it follows that a larger antenna (larger subarrays) with cell-specific beamforming is generally beneficial (gains tend to increase by going in the x-direction). However, although the antenna gain increases as the subarray size grows, it follows that making the subarrays too large can reduce the gains even with cell shaping. In this example, it is seen that a subarray with four elements seems to be a good choice. Too large subarrays result in too narrow angular coverage area in

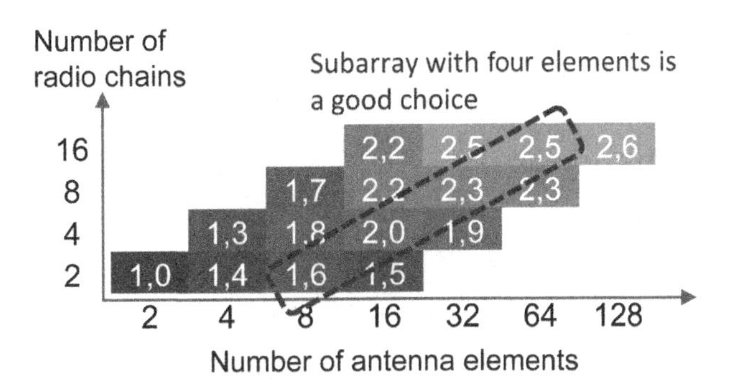

FIGURE 13.36

Downlink capacity gains for different number of radio chains and antenna elements relative to two radio chains with two antenna elements in the urban high-rise scenario. The subarray size is given by the number of antenna elements divided by the number of radio chains.

relation to the angular UE spread in the vertical domain. Note, though, that with full subarray weight adaptation, and not only tilt optimization, the loss with too large subarrays can be reduced.

From the first antidiagonal of the result matrix, where the number of radio chains equals the number of antenna elements, it is seen that the capacity gain by doubling the antenna area and the number of radio branches is roughly 30%, which is in line with results in Section 13.4.2.3. However, by increasing the subarray size, that is, considering an antidiagonal where the number of antenna elements is larger than the number of radio chains, the capacity gain of doubling the number of radio chains and antenna area tends to be (significantly) less than 30%. One reason for this is that the cell shaping and the UE-specific beamforming compete over the same limited antenna area and degrees of freedom. Hence, although the performance of combining multiple features exceeds the gain of each feature in isolation, it is not always true that the gains are fully additive. In general, feature gains tend to be additive when the antenna area and the degrees of freedom are not limiting. Another reason for reduced capacity gains is that gains saturate at large SNRs (see discussion in Section 6.3.4) unless SNRs can be shared between UEs as with MU-MIMO.

Fig. 13.37 corresponds to Fig. 13.36 but for the urban low-rise scenario. Most trends are similar for both scenarios, but cell-specific beamforming (cell shaping) seems to give less gain in the low-rise scenario, which is in line with the discussion in Section 13.3.3. Furthermore, it holds that larger subarrays are beneficial due to narrower angular UE distribution in the vertical domain.

13.5.3 CELL SHAPES AND BEAM SWEEPING

Fig. 13.38 illustrates the performance impact by considering different cell shapes and beam sweeping in the 3GPP UMa and UMi scenarios, respectively, for an 8×8 antenna array with 2×1 subarrays, that is, for a $(2 \times 1)_{SA}(4 \times 8)$ antenna configuration. In all cases, UE-specific beamforming is based on eight horizontal and four vertical subarrays (64 radio chains) with the angular coverage area determined by the subarray characteristics: horizontal and vertical HPBW of roughly 90 and 35 degrees, respectively.

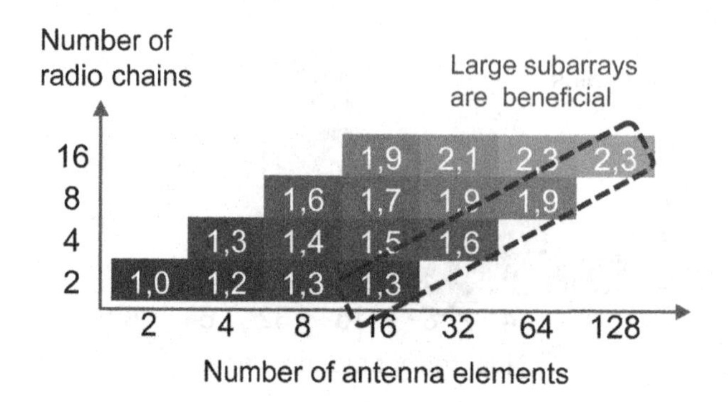

FIGURE 13.37

Downlink capacity gains for different number of radio chains and antenna elements relative to two radio chains with two antenna elements in the urban low-rise scenario. The subarray size is given by the number of antenna elements divided by the number of radio chains.

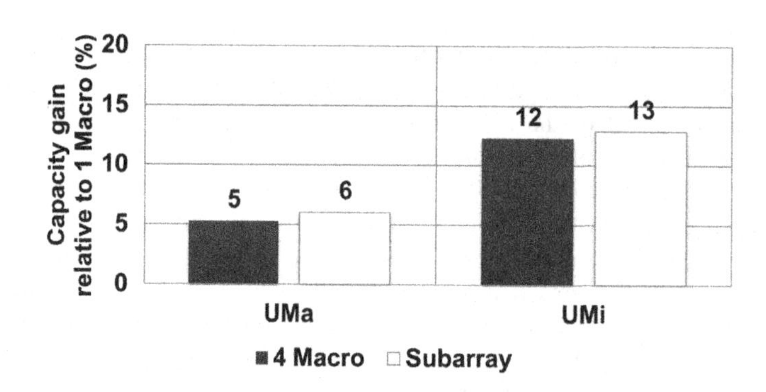

FIGURE 13.38

Examples of how cell-specific beamforming and beam sweeping affect PDSCH capacity in the 3GPP UMa and UMi scenario. Capacity gains of cell shapes "4 Macro" and "Subarray" relative to "1 Macro" are shown.

Three different cell shapes are considered:

1. "1 Macro"—one macro beam with roughly 65 degree horizontal HPBW and 10 degree vertical HPBW;
2. "4 Macro"—four macro beams spanning roughly 30 degree in the vertical domain, where each beam has roughly 90 degree horizontal HPBW and 10 degree vertical HPBW. In this case, beam sweeping in the vertical domain based on these four macro beams is applied, that is, each UE reports which one of the four beams that is preferred, and the network assigns each UE with the preferred beam or cell-part;
3. "Subarray"—one subarray beam with roughly 90 degree horizontal HPBW and 35 degree vertical HPBW.

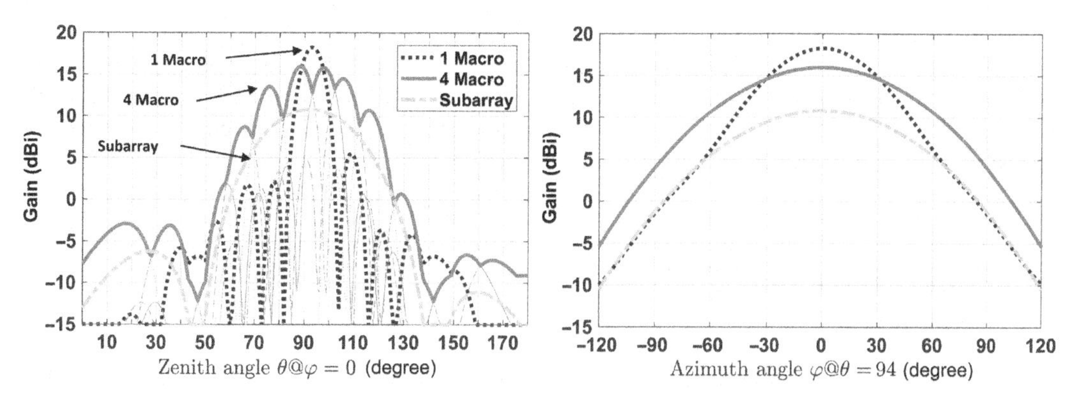

FIGURE 13.39

Illustration of cell shapes 1–3. Four degree subarray tilt is assumed. The four vertical beams for Case 2 (4 Macro) are shown with narrow lines and the corresponding envelope is represented by the thick solid line.

The cell shapes (CS) are illustrated in Fig. 13.39. CS "1 Macro" provides the highest peak gain but has a rather narrow main lobe width in the vertical domain. CS "Subarray" has less gain but a perfect match with respect to the angular coverage area of UE-specific beamforming. CS "4 Macro" achieves the main characteristics of both CS "1 Macro" and CS "Subarray", that is, considering the envelope of the four sub-beams, it follows that gain is rather close to CS "1 Macro" and main lobe width matches CS "Subarray" well.

Transmission of cell-wide common channels, like synchronization (SSB), as well as the procedure to connect a UE to a serving cell depends on the cell shape.

Fig. 13.39 shows that "4 Macro" and "Subarray" have much better PDSCH data performance compared to "1 Macro" in both scenarios, and especially in the 3GPP UMi scenario. Furthermore, "4 Macro" and "Subarray" exhibit similar performance in both scenarios. The reason for this is that for "1 Macro" there is a mismatch between the angular coverage areas of the macro cell shape used for cell association and the subarray used for UE-specific beamforming. This mismatch does not exist for "Subarray". For "1 Macro", the best cell-shaped cell might be inferior to another cell with respect to UE-specific beamforming based on the subarray's coverage area. This effect becomes more pronounced in the 3GPP UMi scenario compared to 3GPP UMa due to scenario characteristics; for example, is the scenario coverage- or capacity-limited and how wide is the angular UE distribution in the vertical dimension. By employing beam sweeping based on the narrow macro shape (4 Macro), the effective angular coverage area in the vertical domain becomes wider and similar to "Subarray", hence a good match between the shape of the cell-defining signal and the envelope of UE-specific beams.

One indicator of the mismatch discussed above is to consider how the cell association is affected by the cell shape. As "Subarray" gives a perfect match between the coverage areas associated with cell association and UE-specific beamforming, it is of interest to see how the cell association of other cell shapes relates to "Subarray". In this example, it follows that ~15% of the UEs choose a different serving cell with "1 Macro" compared to "Subarray" in the 3GPP UMi scenario. The corresponding number for "4 Macro" is only 2%.

Clearly there is a potential performance impact from the mismatch discussed above. However, it is important to realize that with proper settings this issue is significantly reduced. A wide cell shape and small subarrays are beneficial in scenarios with wide angular spread of UEs in the vertical domain. Such scenarios often have less ISDs, hence better coverage that reduces the need for narrow, high-gain cell-specific beams like the "1 Macro" in the example above.

13.6 MULTI-USER MIMO

13.6.1 OVERVIEW

Multi-user MIMO (MU-MIMO) is seen as a key capacity feature for AAS and this section will discuss different performance aspects of MU-MIMO.

13.6.1.1 Key characteristics

As thoroughly discussed in Section 6.7, there are several aspects that impact how effective MU-MIMO performs in relation to SU-MIMO. More specifically, for MU-MIMO to excel the following points need to be fulfilled:

- need to have sufficiently many simultaneously active UEs with data in their buffers to be able to find UEs to co-schedule. Hence, sufficiently high traffic load is required to see MU-MIMO gains. It is important to understand that the AAS itself does not generate any traffic;
- need to find co-scheduled UEs that can be separated, meaning that intra-cell interference can be kept sufficiently low. This step depends on many factors. A large spatial distribution of UEs, preferably in both the horizontal and vertical dimension, is beneficial. Channel conditions need to facilitate advanced MU-MIMO algorithms. The antenna configuration needs to have sufficiently many degrees of freedom (radio chains), meaning that the spatial filter that the antenna constitutes becomes flexible enough and has good enough spatial resolution. Finally, the scheduler needs to be dynamic and quickly adapt to latest CSI;
- need to have detailed and accurate (up-to-date) CSI to facilitate appropriate nullforming, co-scheduling and link adaptation. The requirement on CSI is thus much larger for MU-MIMO than for SU-MIMO. One problem with realistic MBB traffic is that most sessions consist of small data packets, hence a session has often ended before up-to-date CSI has been obtained;
- need to find compatible UEs that have high enough SINRs so that SU-MIMO gains start to saturate, and it thereby becomes beneficial to share the SINR among several UEs. Note that available base station transmit power is shared among scheduled UEs/layers, hence each scheduled layer reduces the per-UE signal power.

Accurate CSI is a key element for efficient MU-MIMO operation. CSI can be obtained by means of *reciprocity* or by *feedback.*

13.6.1.2 Outline

Feedback-based MU-MIMO based on fast-fading CSI will be discussed in Section 13.6.2.

Fast-fading reciprocity-based MU-MIMO will be covered in Section 13.6.3, and reciprocity-based MU-MIMO with explicit inter-cell interference handling is the subject of Section 13.6.4.

13.6.2 FEEDBACK-BASED MU-MIMO

13.6.2.1 Key characteristics

Codebook and CSI feedback designs for LTE before 3GPP release 14 targeted primarily SU-MIMO operation, where CSI based on rather coarse quantized channel information suffices. Applying these designs for MU-MIMO operation yields generally modest gains. Significant work was therefore put into 3GPP release 14 to enhance the CSI framework for MU-MIMO. The same design principles were also used to develop the advanced CSI feedback design for NR, referred to as Type II.

A few observations affecting these design principles were (see Section 8.3.2.4 for LTE, Section 9.3.5.5 for NR and Section 6.4.1.2.1 for further details):

- required feedback information can be significantly reduced by representing the channel in beam-space via a 2D-DFT transformation matrix. Feedback information related to the K strongest beams, where K can be (much) less than the channel matrix dimension, can in most cases give performance close to using full channel information;
- detailed frequency-selective channel information is key for good MU-MIMO performance, especially if the paired UEs' channels overlap. However, it is typically only phase information that needs to be frequency-selective. Power depends on the strength of the corresponding propagation paths in the channel, which is typically a large-scale channel property;
- relative power levels between beams need to be included in the codebook to get significant gains compared to the 3GPP release 13 DFT codebook. A relatively course quantization in the order of 2–3 bits of the relative power levels is typically enough. Nevertheless, as pointed out in the previous bullet, it often suffices to have wideband power information.

13.6.2.2 Feedback-based MU-MIMO in NR (Type II CSI)

As discussed in Section 9.3.5.5, the advanced CSI feedback design for NR, referred to as Type II, builds on the same principles as the 3GPP release 14 LTE advanced CSI framework. More specifically, the channel is represented in beam-space via a 2D-DFT transform, where the L strongest beams are selected. Phase information of each beam is given per sub-band and can be a direct quantization of the phase of the beams or based on a frequency-domain parameterization utilizing the correlation structure between beams. Amplitude information for each beam can be described wideband, per sub-band or both. Compared to LTE release 14, NR allows for, among other things, representing the channel with more than two beams, increased phase granularity, and enhanced amplitude weighting. A thorough discussion of the CSI feedback design can be found in Sections 8.3.2 and 9.3.5 for LTE and NR respectively, and also in Section 6.4.

Performance of the Type II CSI feedback approach described above for different number of beams (L), beam amplitudes represented by eight states, either wideband or a mix of wideband and sub-band, and phase information given by an 8-PSK alphabet per sub-band in 3GPP UMi with a 32T $(2 \times 1)_{SA}(4 \times 4)$ antenna configuration can be found in Fig. 13.40. Performance is relative to Type I SU-MIMO.

Relative to Type I SU-MIMO, Type II CSI feedback provides significant capacity gains in the order of 30%–40% depending on the CSI configuration. Similarly, the capacity gains of Type II CSI relative to Type I MU-MIMO range between roughly 15% and 20%.

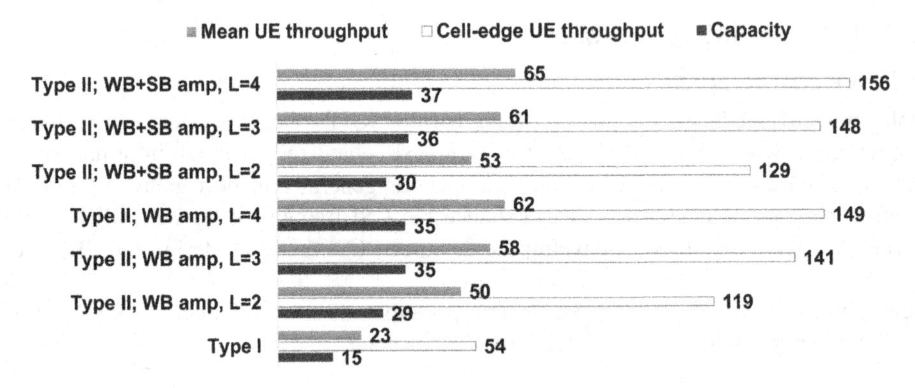

FIGURE 13.40

Illustration of NR Type I MU-MIMO and Type II MU-MIMO performance gains (%) relative to Type I SU-MIMO CSI feedback in 3GPP UMi with 32T. WB and SB amp stand for wideband and sub-band beam amplitude information and L denotes the number of beams.

Using more advanced Type II CSI feedback ($L=4$ with wideband and sub-band beam amplitude information) provides rather modest capacity gains in the order of 5% relative to using a more basic Type II CSI feedback setting ($L=2$ and wideband beam amplitude information). Furthermore, the number of feedback bits for Type II is in the order of 100:s and increases with, for example, the number of beams (see Section 9.3.5.5 for further details).

Fig. 13.41 illustrates the layer distribution for the different configurations discussed above at high network load and it is seen that all schemes have rather similar layer distribution and the maximum number of scheduled layers (with a percentage larger than 1%) is eight.

Fig. 13.42 shows how often MU-MIMO is being scheduled as a function of the resource utilization, and not so surprisingly it follows that MU-MIMO scheduling is rare at low load and then increases up to roughly 40% at very high resource utilization.

13.6.3 RECIPROCITY-BASED MU-MIMO

Reciprocity-based MU-MIMO obtains CSI information explicitly by means of uplink sounding. Hence, downlink channel information used to form the precoding matrix is obtained by estimating the uplink channel based on SRSes or other relevant uplink reference signals.

Performance gains of reciprocity-based MU-MIMO over SU-MIMO for different antenna port configurations can be found in Table 13.4. CSI is per sub-band (6 RBs) and CQI obtained via TM10 is used to estimate the interference-plus-noise variance. Sounding quality affects channel estimation accuracy and SRS is transmitted from both of the two UE antennas.

Using few radio chains gives no MU-MIMO gains, while significant gains are observed when employing more branches. Horizontal antenna layouts seem to excel over vertical configurations in 3GPP UMa, while in 3GPP UMi gains are more conditioned on the number of branches than the antenna layout.

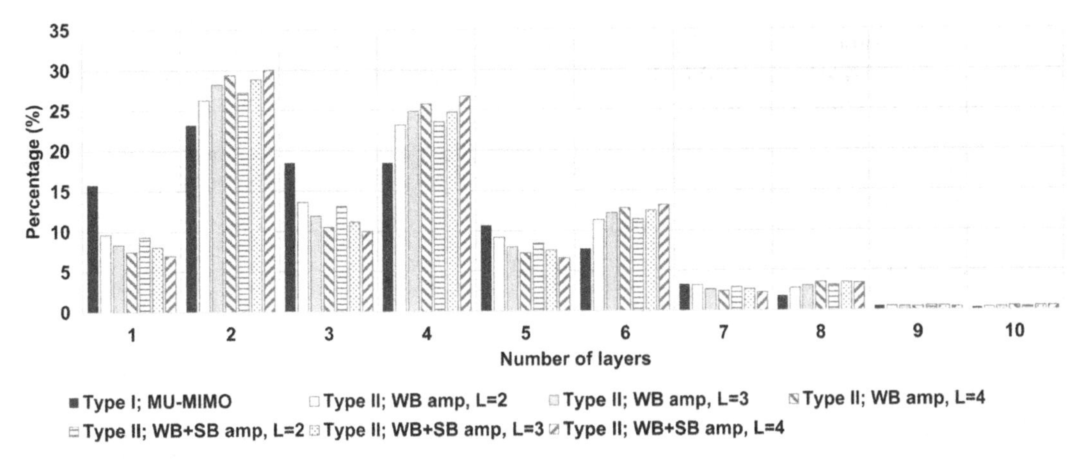

FIGURE 13.41

Illustration of layer distribution for different NR feedback-based configurations in the 3GPP UMi scenario with 32T.

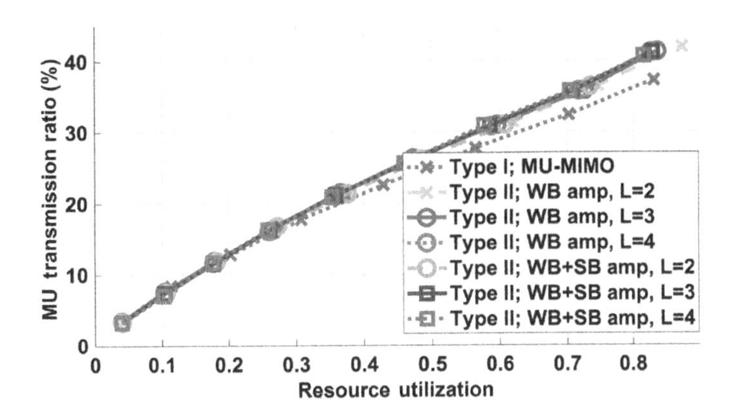

FIGURE 13.42

Illustration of the ration of MU-MIMO transmissions in relation to all transmissions for different NR feedback-based configurations in the 3GPP UMi scenario with 32T.

13.6.4 RECIPROCITY-BASED MU-MIMO WITH INTER-CELL INTERFERENCE HANDLING

Managing interference is key to improve network performance, especially at high load in interference-limited scenarios. Up to now it is mainly intra-cell interference that has been addressed by smart beamforming algorithms, for example, suppressing interference from multiple layers via ZF (cf. Section 6.3.3.3), while inter-cell interference has only been addressed implicitly via subarray tilting.

Table 13.4 Illustration of MU-MIMO Gains Over SU-MIMO Read at 50% Resource Utilization of SU-MIMO in 3GPP UMa and UMi. Note that 64T $(1 \times 1)_{SA}(1 \times 32)$ is a Special Case with a Very Wide Antenna.

	Capacity gain [%]		Cell-edge UE throughput gain [%]		Average UE throughput gain [%]	
	UMa	UMi	UMa	UMi	UMa	UMi
4T $(4 \times 1)_{SA}(2 \times 1)$	0		0		−7	
8T $(2 \times 1)_{SA}(4 \times 1)$		0		1		−4
16T $(1 \times 1)_{SA}(1 \times 8)$	18	17	45	39	8	11
16T $(4 \times 1)_{SA}(2 \times 4)$	9		16		1	
32T $(2 \times 1)_{SA}(4 \times 4)$		22		64		17
64T $(1 \times 1)_{SA}(1 \times 32)$	44	76	97	120	31	37

By taking inter-cell interference explicitly into consideration in scheduling and algorithm design, performance can be improved in interference demanding scenarios. This is achieved by considering the so-called *coordinated beamforming* (CBF) where transmissions from several cells are being coordinated. Note, however, that harvesting CBF gains is challenging as it puts additional complexity on the scheduler and relies on very accurate CSI from multiple UEs across several cells (see Section 6.6.2 for further details on CBF).

The following subsections will consider the performance of regularized ZF and CBF with either SU-MIMO or MU-MIMO based on fast-fading reciprocity. The coordination consists of taking inter-cell interference into account in scheduling decisions and precoder design. For all algorithms, the maximum number of layers per UE is two, which implicitly penalizes SU-MIMO compared to MU-MIMO.

13.6.4.1 Inter-cell interference-awareness

Figs. 13.43 and 13.44 show the capacity gain for different AAS configurations with MU-MIMO based on ZF or CBF in the dense urban high-rise scenario and the urban low-rise scenario, respectively.

In the dense urban high-rise scenario (Fig. 13.43), it is seen that interference-awareness increases the capacity substantially by roughly 45% for all antenna configurations. Furthermore, the capacity gain by doubling the number of radio chains is also substantial; in the order of 50-60%.

In the urban low-rise scenario (Fig. 13.44), the capacity gain by employing interference-awareness is roughly 10%, which is significantly less compared to the high-rise scenario. The capacity gain by doubling the width of the antenna is roughly 40% (going from 8T to 16T), but contrary to the high-rise scenario, the gain of increasing the number of radio chains by means of reducing the subarray size is very small.

Next, it is illustrated how different algorithm components affect the performance a bit more in detail. Fig. 13.45 shows the capacity gains of ZF and CBF with SU-MIMO and MU-MIMO for a 32T AAS in the dense urban high-rise scenario and the urban low-rise scenario.

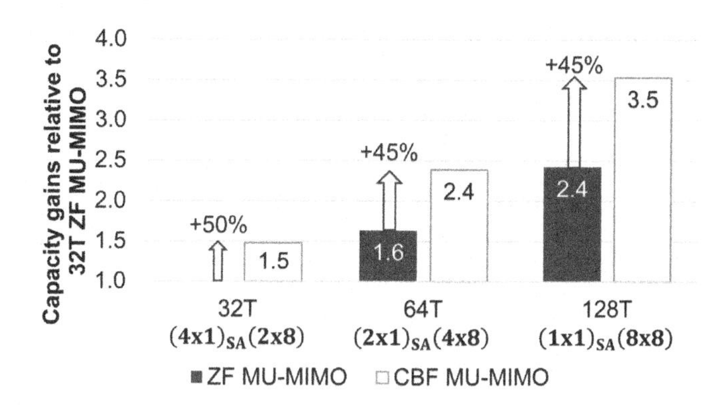

FIGURE 13.43

Capacity gains for zero-forcing (ZF) and coordinated beamforming (CBF) with MU-MIMO for different antenna configurations in the dense urban high-rise scenario.

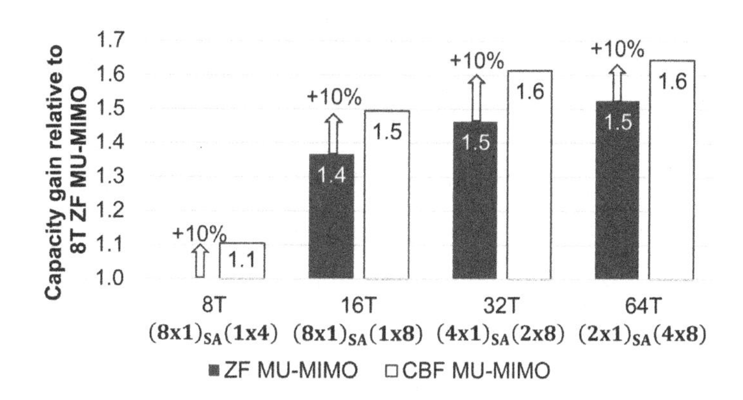

FIGURE 13.44

Capacity gains for zero-forcing (ZF) and coordinated beamforming (CBF) with MU-MIMO for different antenna configurations in the urban low-rise scenario.

In the high-rise scenario it follows that both MU-MIMO and interference-awareness bring rather similar gains relative to ZF SU-MIMO (40% and 30% respectively). However, combining MU-MIMO and interference-awareness offers significant gains; 90% relative to ZF SU-MIMO and 50% relative to ZF MU-MIMO.

In the low-rise scenario, on the other hand, both MU-MIMO and interference-awareness bring rather small gains and CBF MU-MIMO offers significantly less gain relative to ZF SU-MIMO compared to the high-rise scenario; 30% in the low-rise scenario compared to 90% in the high-rise scenario.

As a conclusion it is evident that offered gains from different beamforming algorithm components depend very much on the scenario and the AAS configuration. This will be further elaborated upon in Section 14.3.

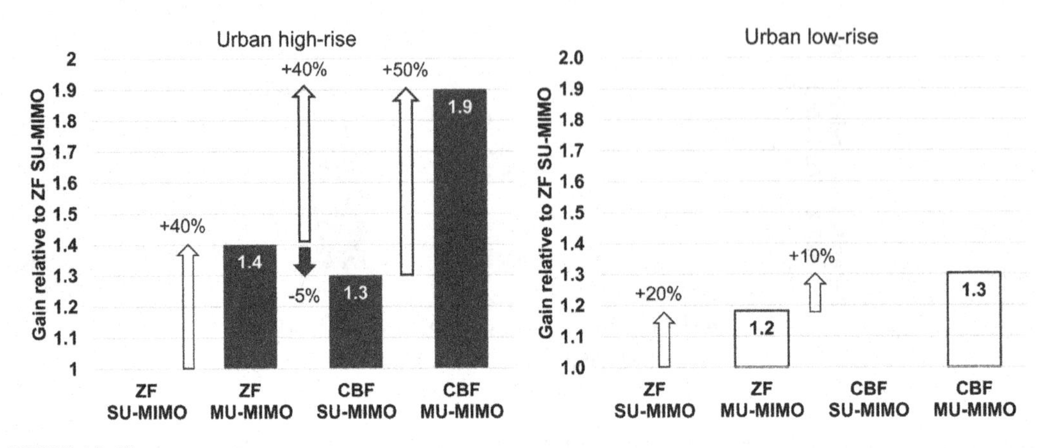

FIGURE 13.45

Capacity gains from different multi-antenna algorithms with 32T in the high-rise scenario (*left*) and the low-rise scenario (*right*). Note that the result for CBF SU-MIMO in the low-rise scenario is missing.

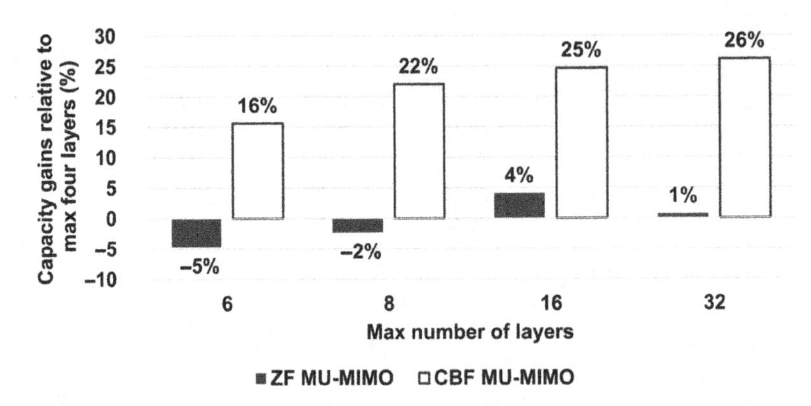

FIGURE 13.46

Capacity gains as a function of the maximum number of MU-MIMO layers for ZF and CBF in the dense urban high-rise scenario.

13.6.4.2 Number of MU-MIMO layers

Fig. 13.46 shows capacity gains as a function of the maximum number of scheduled layers in the dense urban high-rise scenario for a 64T $(2 \times 1)_{SA}(4 \times 8)$ AAS configuration and regularized ZF or CBF.

From the results it follows that a rather modest number of layers suffices. For ZF MU-MIMO the performance seems to be rather insensitive to the number of layers and for CBF, the MU-MIMO network gains seem to saturate at maximum eight layers. This is further illustrated in Fig. 13.47 showing the layer distribution over the entire network for CBF when the maximum number of layers is capped at 32. Relatively few layers are mostly scheduled. There are, however, a

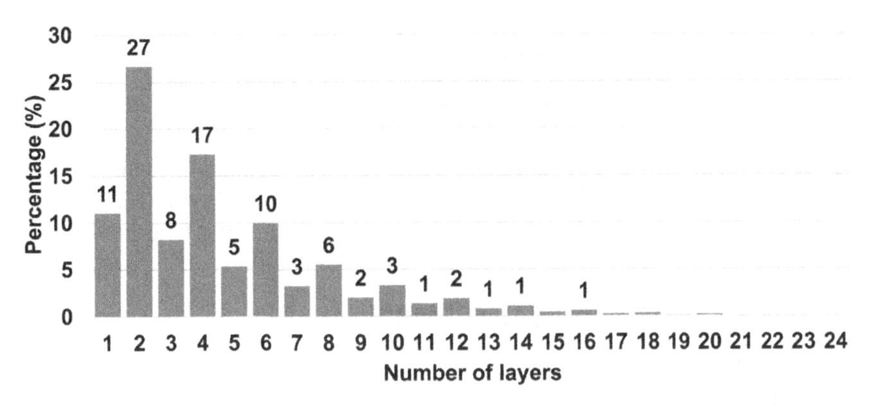

FIGURE 13.47

Percentage of scheduled number of layers at high load when using coordinated beamforming (CBF) in the dense urban high-rise scenario.

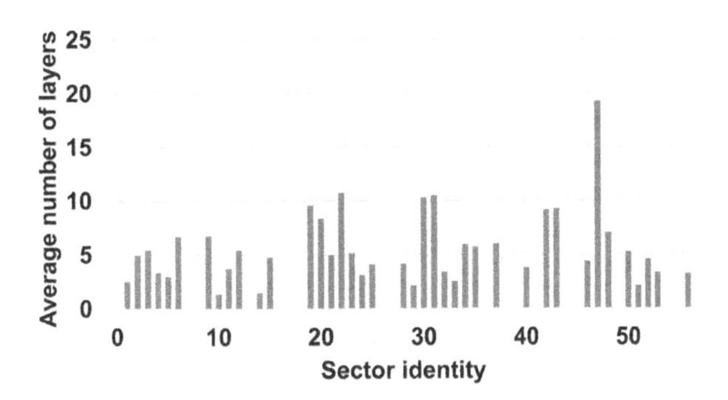

FIGURE 13.48

Average number of scheduled layers for each sector at high load.

few occasions when more layers are employed, but this has marginal impact on overall network performance.

As discussed previously, there are many requirements that need to be fulfilled to reap the benefits of MU-MIMO. One of the most critical condition is that the traffic situation needs to facilitate many simultaneously active UEs. In this deployment scenario, the average number of active UEs per sector at high load is ~1.2, which clearly indicates that many layers are generally not needed.

Nevertheless, as the scenario is inhomogeneous, there are sectors that are severely strained with high traffic demand and a lot of active UEs, and therefore can make use of more layers. This is illustrated in Fig. 13.48, showing the number of average scheduled layers per sector at high load. There is one sector (Sector 47) that on average schedules almost 20 layers. Nevertheless, as shown before this has marginal impact on overall network performance and demonstrates that better network planning to off-load this cell would be beneficial rather than being an indication that dimensioning for many layers is needed.

13.7 ADDITIONAL AAS PERFORMANCE ASPECTS

This section will illustrate the radio network performance impact of several additional AAS aspects. Coverage aspects of beamforming will be considered in Section 13.7.1, LTE feedback-based MIMO is the topic of Section 13.7.2 and uplink spatial receiver combining will be discussed in Section 13.7.3. Thereafter follows illustrations of high-band aspects in Section 13.7.4, impact of base station output power in Section 13.7.5, and finally frequency-band interworking in Section 13.7.6.

13.7.1 COVERAGE ASPECTS OF BEAMFORMING

13.7.1.1 Key characteristics

13.7.1.1.1 Reuse of existing sites when deploying higher frequency bands

One of the main benefits of AAS and beamforming is the improved coverage that often makes it easier to reuse the existing site grid when deploying higher frequency bands (see Fig. 13.5 for an illustration). How feasible it is to reuse the existing site grid depends on many aspects, such as the deployed frequency-band, product choices, and architectural options.

Coverage becomes generally worse as the frequency increases (see Sections 3.4.2 and 7.2). Evidently, a more capable product can combat coverage challenges better and an architecture that facilitates efficient frequency-band interworking reduces the need for full coverage on all bands. One of the main challenges with deploying higher frequency bands using TDD is uplink coverage that suffers from limited UE output power and downlink heavy slot formats.

A simple illustration of how coverage of a mid-band deployment with or without AAS relates to an existing low-band is given in Section 13.7.1.2.1.

13.7.1.1.2 Coverage of different types of beamforming

As illustrated in previous sections, advanced beamforming with MU-MIMO requires detailed and accurate channel knowledge. Both feedback-based and reciprocity-based methods can be used to obtain the fast-fading frequency-selective CSI required for MU-MIMO operation. Although the performance potential of reciprocity-based beamforming generally exceeds feedback-based beamforming as full channel information can be explicitly obtained in the former case, both approaches often exhibit performance results that closely match in many situations. There are, however, many challenges with obtaining detailed and accurate CSI.

Reciprocity-based MU-MIMO requires sufficient coverage of the uplink sounding, which is challenging in practice due to limited output power of most UEs. As discussed in Section 6.7.1, larger bandwidths, higher frequency bands and more receive antennas than transmit antennas at the UE side make uplink sounding particularly demanding. The SRS quality requirement to achieve good fast-fading reciprocity-based MU-MIMO operation can for a typical mid-band deployment with 100 MHz bandwidth correspond to an instantaneous PUSCH data rate in the order of several Mbps.

Feedback-based MU-MIMO operation, for example, based on NR Type II CSI, is also challenging from a coverage perspective. Detailed and accurate frequency-selective CSI corresponds to many feedback bits that results in worse uplink coverage. However, a clever quantization of the channel, as with NR Type II CSI, results often in less information bits compared to striving for full channel knowledge as is often the case for fast-fading reciprocity-based operation.

Better coverage is achieved by employing beamforming algorithms that require less detailed channel knowledge, for example, long-term wideband SU-MIMO. Performance of SU-MIMO using feedback-based or reciprocity-based CSI is generally similar although feedback-based CSI tends to be slightly more robust. As discussed in Section 9.3.5, Type I beamforming using wideband feedback, requires 10s of feedback bits, while Type II beamforming needs 100s of feedback bits.

The trade-off between coverage and capacity for beamforming can be seen as a question of classic beamforming for SU-MIMO (often feedback-based but can be reciprocity-based) versus fast fading–based generalized beamforming for MU-MIMO (reciprocity-based or feedback-based).

Another important aspect affecting the choice of beamforming is the time it takes to acquire the required CSI. Course CSI costs less in terms of capacity and stays up-to-date over a longer time-period, hence it is often configured to be obtained periodically and therefore readily available when it is time to transmit. Request for detailed and more accurate CSI, on the other hand, often needs to be triggered, thus it takes time to acquire. For small data packets this might result in that the data session, transmitted based on course CSI, has finalized before up-to-date detailed and accurate CSI has been obtained.

Although both feedback- and reciprocity-based operation rely on network and UE support of standardized features, feedback-based operation has often a stronger dependency and it is, for example, uncertain when UEs will support NR type II CSI feedback or how many CSI-RS ports for NR Type I initial UEs will handle.

The coverage challenge of obtaining detailed and accurate CSI is illustrated in Section 13.7.1.2.2, and a potential solution by means of so-called hybrid beamforming, where the type of beamforming is adapted to the needs and the coverage situation, is illustrated in Section 13.7.1.2.3.

13.7.1.2 Performance results

13.7.1.2.1 Illustration of the coverage challenge of reusing existing sites

Fig. 13.49 shows a link budget (see Section 2.2.4) illustration of data coverage for an 800 MHz FDD deployment without beamforming and a 3.5 GHz TDD deployment with and without beamforming. The site grid is assumed to be dimensioned for minimum 200 kbps uplink at 800 MHz, and coverage is given in terms of required ISD relative to this dimensioning point. Hence, the x-axis in the figure shows the ISD, relative to the ISD that supports 200 kbps uplink at 800 MHz, required to achieve a certain throughput. Typical values of UE and base station parameters, such as bandwidth, output power and noise figure, are assumed and a 3GPP non-line-of-sight UMa propagation model is used. Furthermore, a 7 dB SNR gain due to beamforming and TDD with a 70% downlink slot allocation are assumed.

For example, to satisfy a downlink data rate coverage requirement of 10 Mbps, a 3.5 GHz deployment without beamforming needs to reduce the ISD by 30% compared to the 800 MHz deployment, but with beamforming at 3.5 GHz no densification is needed as the gap to 800 MHz is closed. Note that no frequency-band interworking is assumed.

It is also evident that uplink coverage is challenging at higher bands employing TDD even though beamforming is used. The main reason for the worse uplink at 3.5 GHz compared to 800 MHz is the TDD penalty of using only 30% of the slots for uplink transmission that severely hampers the throughput and calls for frequency-interwork solutions (see Section 13.7.6).

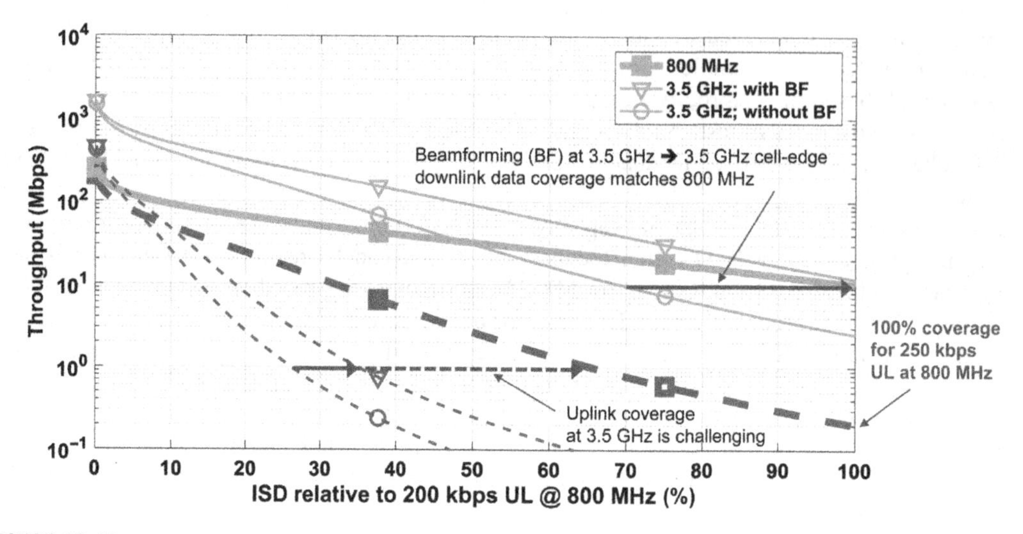

FIGURE 13.49

Link budget illustration of coverage of 3.5 GHz deployment with and without beamforming relative to 200 kbps uplink at 800 MHz. Solid lines and dashed lines represent downlink and uplink, respectively.

13.7.1.2.2 The coverage challenge of detailed channel-state information

The challenge of SRS coverage for reciprocity-based MU-MIMO is illustrated in Fig. 13.50 showing cell-edge UE throughput as a function of load for different antenna configurations with large-scale feedback-based beamforming (GoB) and fast-fading reciprocity-based beamforming (ZF) in a rural macro scenario. GoB clearly outperforms ZF and while performance of GoB is insensitive to the number of radio chains given a fixed antenna area, ZF performance increases with increased subarray size even though this means that less radio chains are used. The reason is that in this example the SRS quality depends on the subarray size and a larger subarray translates often into better SRS quality.

13.7.1.2.3 Hybrid beamforming

One approach to maximize beamforming performance is to employ a *hybrid approach*, where UEs with good coverage use MU-MIMO based on fast-fading frequency-selective CSI while UEs with worse coverage use SU-MIMO based on long-term CSI.

For example, UEs with good coverage can benefit from MU-MIMO gains by means of reciprocity or Type II advanced CSI feedback. UEs that start experiencing uplink quality problems can switch to SU-MIMO based on Type I frequency-selective beamforming, and finally UEs with poor uplink coverage can revert to SU-MIMO based on Type I large-scale wideband beamforming. Note, that large-scale wideband beamforming based on reciprocity or feedback has similar coverage so in principle reciprocity can be used instead of feedback also for SU-MIMO. Needless to say, both the network and the UEs need to support a feature to be able to enjoy its benefits.

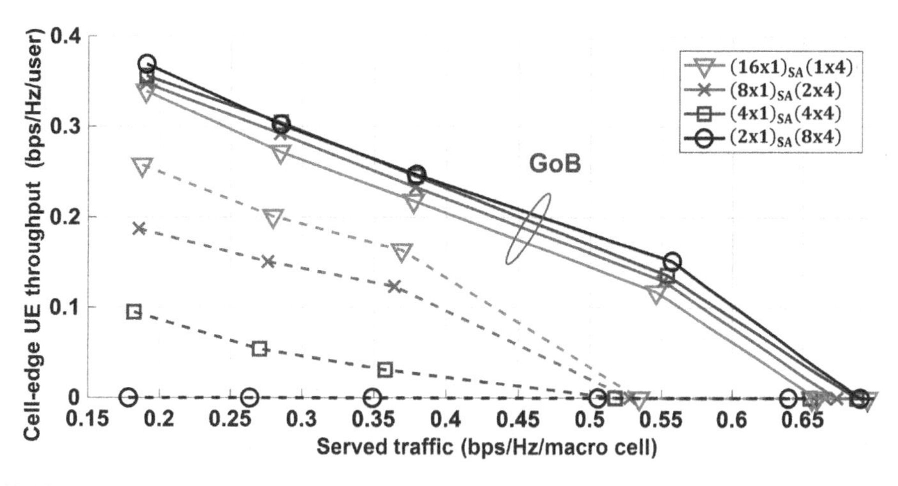

FIGURE 13.50

Cell-edge UE throughput as a function of load for feedback-based GoB (*solid*) and fast-fading reciprocity-based zero forcing (*dashed*) with different antenna configurations in a coverage demanding rural macro scenario.

By conveying uplink feedback on a lower frequency band with better coverage, using carrier aggregation, feedback-based beamforming can be used even though the uplink on the frequency band of interest is out of coverage. This observation provides a strong argument why frequency-band interworking is essential for good network performance when deploying on higher frequency bands with worse coverage (see also discussion in Section 13.7.6).

Fig. 13.51 provides an illustration of the coverage regions for different beamforming techniques. The illustration should be seen as indicative where values come from a link budget analysis of a typical 100 MHz mid-band AAS deployment. Type I wideband (WB), Type I frequency-selective (FS) and Type II advanced CSI are assumed to require 10, 100, and 500 feedback bits, respectively, and fast-fading reciprocity-based beamforming is assumed to require an SRS SNR of at least -15 dB over 10 MHz. The latter requirement is generally rather optimistic for achieving good MU-MIMO performance. The SNR requirement is often higher than -15 dB and the instantaneously sounded bandwidth often needs to be larger than 10 MHz to get up-to-date CSI over the entire bandwidth (see Section 6.7.1.1 for a more detailed discussion).

In this example, it is seen that Type II advanced CSI offers roughly 4 dB better coverage than fast-fading reciprocity-based beamforming, and Type I frequency-selective and wideband beamforming provide an additional 4 dB and 10.5 dB coverage relative to Type II.

With low-band interworking and carrier aggregation, uplink feedback can be conveyed on a low-band; hence all feedback-based beamforming types can be used as long as the low-band has enough coverage.

13.7.2 LTE FEEDBACK-BASED MIMO

This section will illustrate the performance of LTE feedback-based MIMO based on different 3GPP releases for different antenna and algorithm configurations.

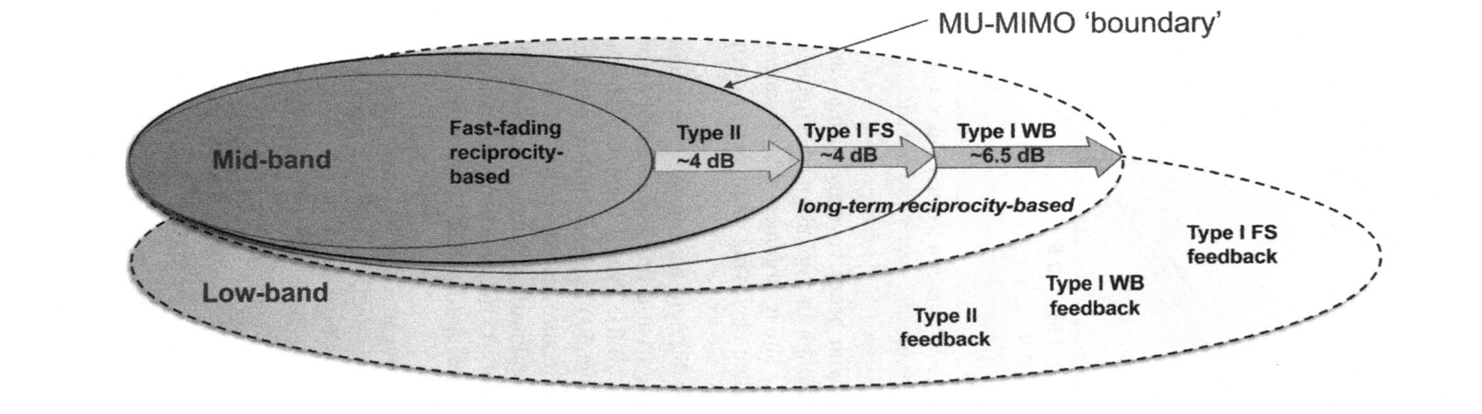

FIGURE 13.51

Illustration of coverage differences in dB for different beamforming techniques. FS and WB stand for frequency-selective and wideband, respectively.

13.7.2.1 Key characteristics

There are essentially two tracks that can be pursued to improve performance of LTE FDD massive MIMO deployments, either higher-order sectorization or feedback-based CSI based on standardized transmission modes. The latter track has generally a better performance potential but suffers from a strong dependency on the UE capabilities, for example, which transmission modes that are supported.

Higher-order sectorization was discussed in Section 13.3.4 and the feedback-based track will be illustrated in Section 13.7.2.2.

FDD deployments cannot rely on fast-fading reciprocity for acquiring CSI for MU-MIMO operation, and need to resort to advanced CSI feedback. The evolution of MIMO in LTE from 3GPP release 8 to 14 is thoroughly described in Section 8.2.

Even with the introduction of NR, it is important to understand that it will take a long time for the penetration of NR capable UEs to reach a substantial level, hence LTE needs to handle the increased capacity demand for many years to come, and here FDD massive MIMO is one important instrument.

13.7.2.2 Performance results

Fig. 13.52 shows downlink LTE FDD feedback-based MIMO performance gains for different number of radio chains and 3GPP releases and configurations relative to SU-MIMO with two radio chains (2T). Results are for the 3GPP UMa scenario at 2 GHz, where 2T, 8T, 16T and 32T are based on $(8\times1)_{SA}(1\times1)$, $(8\times1)_{SA}(1\times4)$, $(4\times1)_{SA}(2\times4)$ and $(4\times1)_{SA}(2\times8)$ antenna configurations, respectively (see Fig. 13.53). Thus, 8T and 16T have an antenna area that is four times the 2T case and 32T has an antenna area that is eight times the 2T case.

Evidently, increasing the antenna area is a recipe for boosting performance. For example, for SU-MIMO there is 80% capacity gain by going from 2T to 8T and an additional 30% gain from 8T to 32T, but very limited gain from 16T over 8T. Due to the scenario characteristics of 3GPP UMa

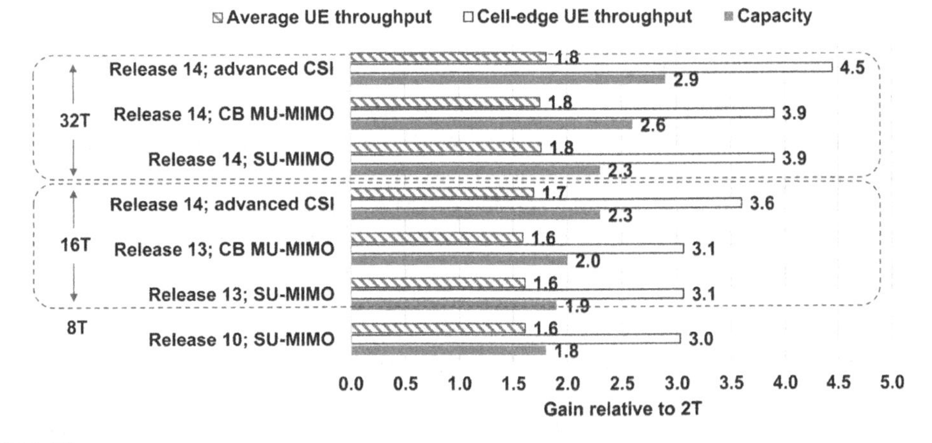

FIGURE 13.52

LTE FDD feedback-based MIMO performance gain for different number of radio chains and 3GPP releases and configurations relative to 2T SU-MIMO in the 3GPP UMa scenario.

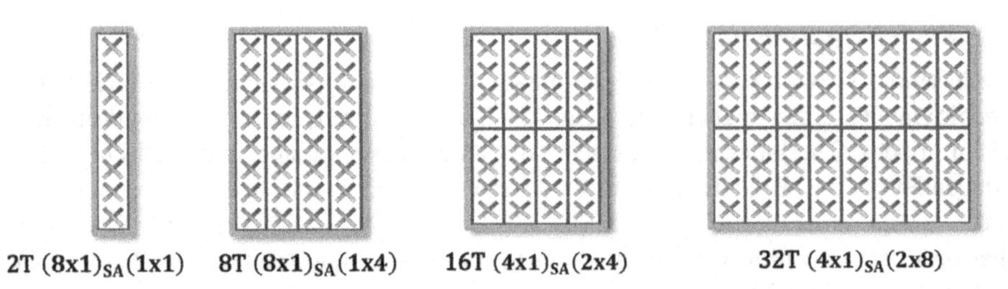

FIGURE 13.53

Illustration of the antenna configurations used in the results shown in Fig. 13.52.

with little angular spread of UEs in the vertical domain, it makes sense that 16T does not offer much gains over 8T (see also related discussion in Section 13.4.2).

Codebook and CSI feedback designs for early LTE releases (before release 14) targeted primarily SU-MIMO operation, where CSI based on rather coarse quantized channel information suffices. Applying these designs for MU-MIMO operation yields therefore rather modest gains.

The gain of MU-MIMO with more advanced CSI schemes is rather modest with few radio chains. Nevertheless, for 16T and 32T, capacity gains of 5%–20% and 10%–25% are observed from more advanced beamforming over SU-MIMO.

13.7.3 UPLINK SPATIAL RECEIVER COMBINING

As discussed in Section 6.3.3, transmitter precoding and receiver combining algorithms and characteristics are very similar. There are, however, aspects that differ between uplink and downlink that impact how effective different beamforming schemes are in uplink and downlink.

13.7.3.1 Key characteristics

For typical MBB cases, the output power of a UE is significantly less than of a base station, resulting in that the uplink tends to be coverage limited while the downlink is more interference limited. Hence, for uplink performance it is important to maximize the antenna area and it is often more beneficial using the AAS to increase the received power than suppressing interference.

UEs have often few (even one) transmit radio chains, hence SU-MIMO with many layers is uncommon in the uplink, at least for MBB.

For MMB, the traffic volume is much larger in the downlink than in the uplink. Hence, even though downlink heavy slot formats are often used for TDD deployments, which hamper the uplink data rates, the need for capacity features such as MU-MIMO is often larger in downlink than in uplink.

MU-MIMO relies on good CSI. As long as there is sufficient coverage, CSI for uplink receiver beamforming is generally more robust and has better quality compared to downlink transmit beamforming. Also, uplink MU-MIMO does not suffer from the problem of sharing the downlink power between layers.

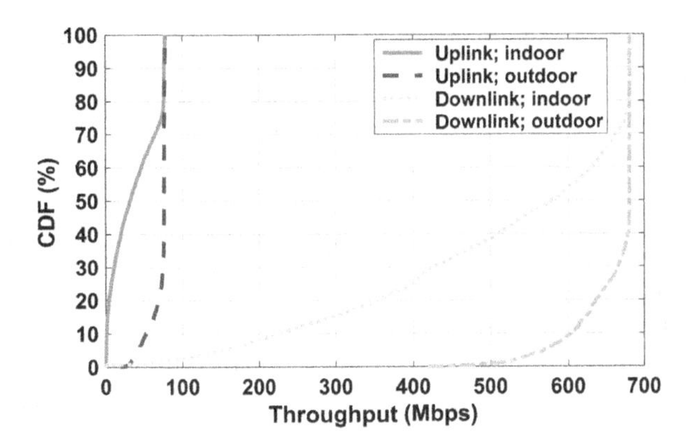

FIGURE 13.54

CDFs of average throughput at low load in the 3GPP UMa scenario with 16T and 16R.

13.7.3.2 Performance results

13.7.3.2.1 Coverage

An illustration of the difference in data rate coverage between uplink and downlink can be found in Fig. 13.54, showing a CDF of average downlink and uplink data rates at low load in the 3GPP UMa scenario. The frequency is 3.5 GHz, the bandwidth is 100 MHz and 75% of the slots are allocated for downlink transmissions. The results are separated into indoor and outdoor UEs, demonstrating that outdoor-to-indoor coverage is challenging. These results also indicate the importance of frequency-band interworking.

A substantial part of the UEs has very good data rates in both uplink and downlink. There are, however, a significant fraction of UEs having very poor uplink data rates, but still very good downlink data rates. Conveying the uplink of these UEs on a lower band with better coverage can improve the overall network performance (see Section 13.7.6.1).

13.7.3.2.2 Antenna configurations

Figs. 13.55–13.57 show uplink performance gains for different antenna configurations at 3.5 GHz in 3GPP UMa and ITU SMa, respectively, using interference rejection combining (IRC) and MU-MIMO with maximum eight layers. Fig. 13.55 shows results with a fixed 16×4 antenna area and different subarray sizes, hence different number of receive chains, Fig. 13.56 illustrates the impact of the subarray size with a fixed number of receive chains and Fig. 13.57 depicts the effects of adding antenna columns, hence increasing both the antenna area and the number of receive chains.

In both the 3GPP UMa and ITU SMa scenarios there are many UEs that suffer from very poor uplink coverage. These UEs will to a large extent dictate the network performance as they will dominate the resource utilization due to very poor link qualities. Hence, keeping these UEs will distort the evaluation and makes it difficult to make a fair comparison between different antenna configurations. Thus, UEs with poor coverage are dropped, which can be motivated by that these UEs should be served by a lower frequency band anyway. Roughly 10% and 20% of the UEs are dropped in 3GPP UMa and ITU SMa, respectively.

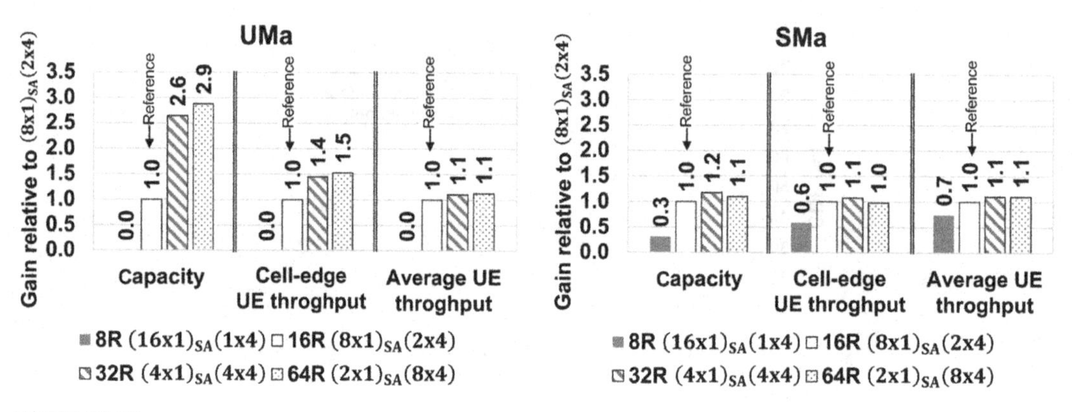

FIGURE 13.55

Illustration of uplink performance for different antenna configurations with a fixed 16 × 4 antenna area in the 3GPP UMa and ITU SMa scenarios. Coverage-based subarray tilt is employed (see Section 13.3.2.1).

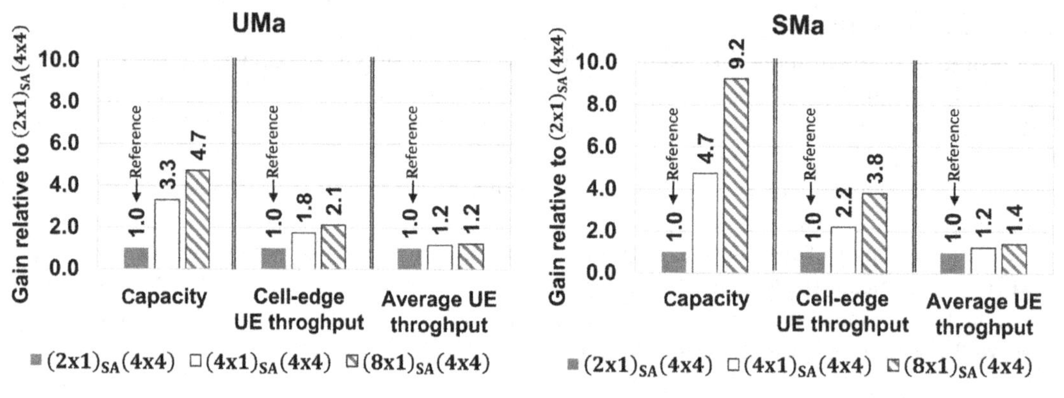

FIGURE 13.56

Illustration of uplink performance for different antenna configurations with 32 receive branches in the 3GPP UMa and ITU SMa scenarios. Coverage-based subarray tilt is employed (see Section 13.3.2.1).

The uplink results are generally well aligned with corresponding downlink results seen in Section 13.4.2. It is seen that for a fixed 16 × 4 antenna area, some degree of vertical domain beamforming is beneficial in 3GPP UMa, while for ITU SMa the performance is very insensitive to the level of vertical domain beamforming. In all cases the subarray size needs to fit the scenario, and the 16 × 1 subarray is too large, especially for the 3GPP UMa deployment scenario with smaller ISD and less flat characteristic than the ITU SMa deployment scenario. Hence, subarray sizes smaller than or equal to four and eight seem to be good choices for 3GPP UMa and ITU SMa, respectively.

For fixed number of radio chains, there is significant gain by increasing the subarray size, in particularly for ITU SMa that has worse coverage compared to 3GPP UMa.

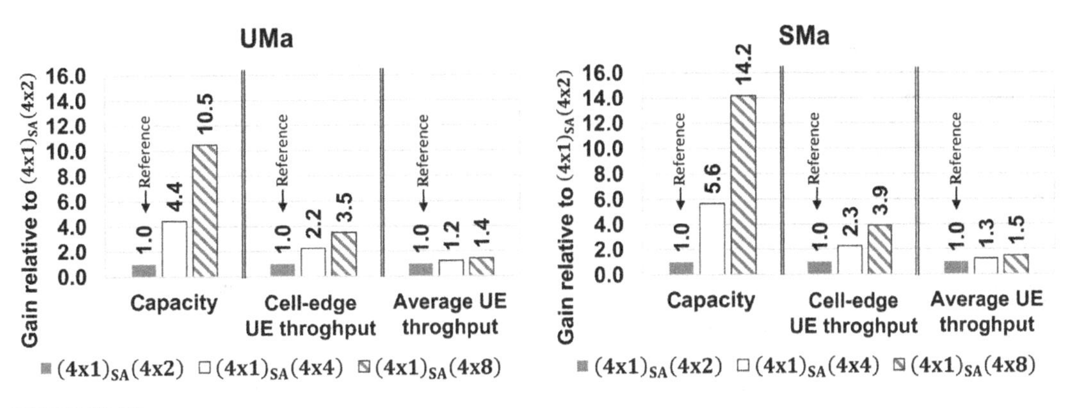

FIGURE 13.57

Illustration of uplink performance as a function of increased number of receive branches and increased antenna area (adding antenna "columns") in the 3GPP UMa and ITU SMa scenarios. Coverage-based subarray tilt is employed (see Section 13.3.2.1).

Increasing the antenna area and the number of branches (adding antenna columns), provide very large gains, especially for cell-edge UEs and in terms of capacity.

13.7.3.2.3 Impact of number of layers

Uplink performance as a function of the maximum allowed number of scheduled layers for a $(4 \times 1)_{SA}(4 \times 4)$ antenna configuration in the 3GPP UMa scenario can be found in Fig. 13.58. In this case the MU-MIMO gain saturates somewhere between four and eight layers, which is well aligned with related downlink results (see, e.g., Section 13.6.4.2).

13.7.4 HIGH-BAND ASPECTS

Multi-antenna techniques are essentially frequency agnostic, that is, the same AAS principles and techniques apply irrespectively of the deployed frequency band. However, as discussed in Section 7.2.2, the frequency has a profound impact on propagation properties and antenna characteristics that impact the design and systemization of the AAS. These effects affect also the expected performance of different AAS configurations.

13.7.4.1 Key characteristics

A short recap of high-band aspects impacting the performance of AAS at high-band follows next, while further description can be found in Sections 7.2 and 7.3, and 3GPP details are described in Section 9.3.6:

- coverage is generally more challenging at higher frequencies due to output power (or EIRP) limitations and increased propagation losses that typically cannot be fully compensated by an increased antenna area;

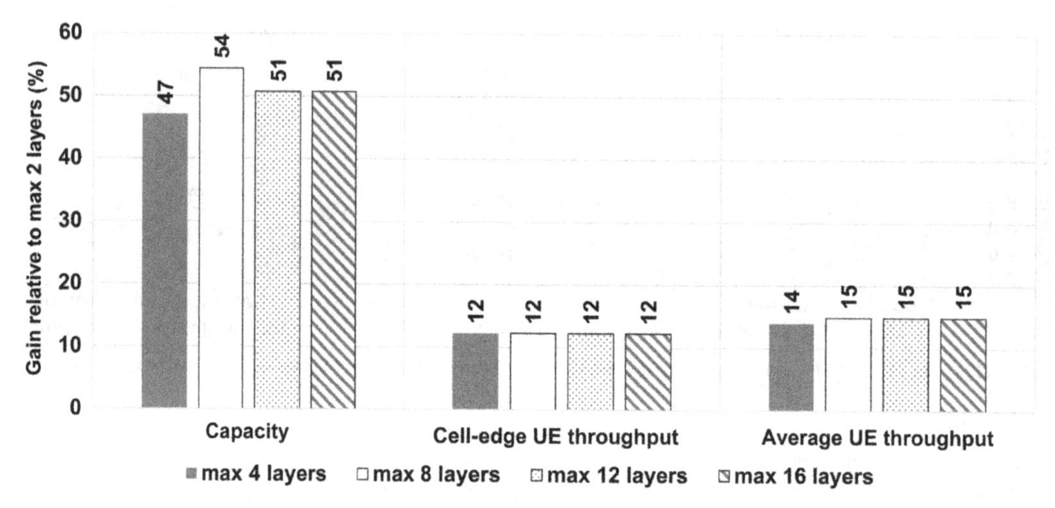

FIGURE 13.58

Illustration of uplink performance as a function of the maximum allowed number of scheduled layers for a $(4 \times 1)_{SA}(4 \times 4)$ antenna configuration in the 3GPP UMa scenario. Coverage-based subarray tilt is employed (see Section 13.3.2.1).

- products targeting higher frequency bands incur building practice constraints on, for example, output power, as components get smaller and a tight integration is needed. Also, cooling and heat dissipation become more challenging. Regulatory and co-existence aspects can also impose limits on the output power or rather the maximum EIRP (see Section 11.5.1). As high-band products typically need to have very high antenna gain (not the least from an uplink receiver perspective), a limit on the maximum EIRP can lead to a rather strict requirement on the maximum allowed output power;
- there are several aspects of the propagation channel that are highly frequency dependent such as foliage, diffraction and wall loss (the severity depends on the material). This leads to phenomena where the signal strength can be severely reduced or even completely blocked;
- as discussed in Sections 7.2 and 3.4.2, the antenna's ability to receive power decreases with increased frequencies given a fixed antenna gain. However, as the size of the antenna elements also gets smaller by moving up in frequency, more elements can fit a given area, leading to increased antenna gain. In fact, for a fixed physical antenna area that is fully utilized, the antennas ability to receive power is independent of the frequency. Hence, in order to keep the effectiveness of the antenna when going up in frequency, high-gain beams that become very narrow are required.
- high-gain antennas with very narrow beams are more sensitive to angular spread, which can affect performance at higher bands;
- fast-fading reciprocity-based beamforming becomes challenging at high-band due to worse sounding coverage, especially as the bandwidth can be very large;
- beam sweeping can become essential due to challenging coverage and AAS building practices. For example, a fully analog beamforming implementation together with

requirements on high-gain beams makes beam sweeping a necessity. On a high level, there is a need to have procedures that facilitate observability (measurements to decide the best way to transmit and receive), consistency between involved nodes (all involved entities need to know how other related entities transmit and receive) and recovery when something goes wrong (see Section 7.4.2 for more details);

- there is generally a substantial amount of bandwidth available at high-band. Hence, the potential for very high data rates is large. However, due to the coverage challenge, interworking with lower bands will be key to get satisfactory network performance. Whenever the high-band experiences sufficient coverage, valuable lower bands can be off-loaded, and great UE throughput can be achieved, and whenever the high-band is out of coverage, lower bands need to take care of the traffic;
- high-bands can work well in special deployments like *integrated access backhaul* (IAB) [3] or *fixed wireless access* (FWA) [4] due to vast availability of bandwidth, hence capacity, and improved coverage due to the characteristics of the deployment scenario, for example, less or no mobility and more line-of-sight propagation.

13.7.4.2 Performance results

Fig. 13.59 illustrates the average downlink UE throughput in 3GPP UMa at 28 GHz with 400 MHz bandwidth, 3:1 downlink versus uplink TDD allocation, a $(4\times 1)_{SA}(4\times 24)$ antenna configuration and GoB beamforming. Due to the vast amount of available bandwidth, throughputs well beyond 1 Gbps can be achieved in the downlink. It is also apparent that outdoor-to-indoor coverage is challenging; roughly 50% and 70% of the indoor UEs do not have downlink and uplink coverage, respectively, when the ISD is 500 m. Outdoor coverage is, on the other hand, rather good. It is also seen that high-band deployments need to rely on rather small ISDs to have sufficiently good coverage for all UEs in the network. This speaks in favor of using macro high-band deployments as local capacity boosters and for local off-loading of valuable low-bands. High-band can also provide a

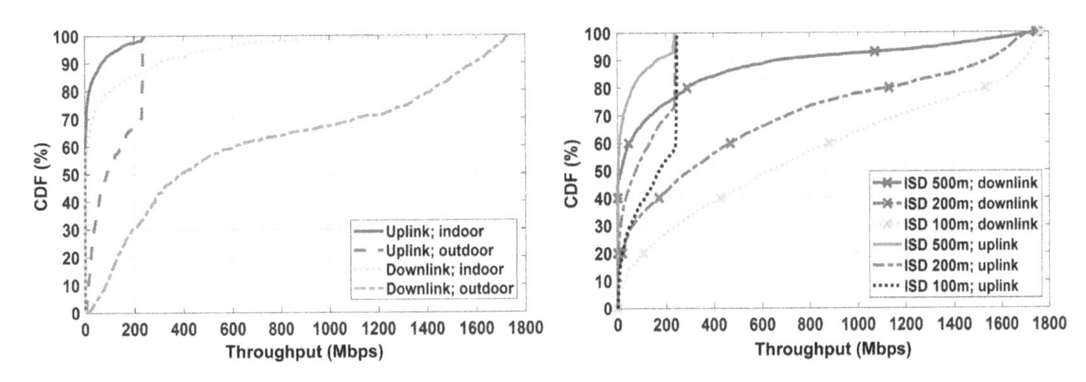

FIGURE 13.59

Illustration of average uplink and downlink UE throughput CDFs at low load in 3GPP UMa, 400 MHz bandwidth and 28 GHz carrier frequency. CDF conditioned on indoor or outdoor UEs with 500 m ISD (*left*) and CDF of all UEs at different ISDs (*right*).

high capacity solution to special deployments such as integrated access backhaul and fixed wireless access.

13.7.4.2.1 Time-domain beamforming

The discussion in previous sections has primarily addressed frequency-domain beamforming, which implies that different frequency regions (sub-bands) can use independent beamforming, that is, several UEs and/or physical channels can be frequency multiplexed and independently employ wideband (covering the entire allocated bandwidth) or frequency-selective beamforming.

As discussed in Section 7.3, time-domain beamforming, on the other hand, means that the entire frequency region needs to use the same precoder, or beam, that is, only one beam can be used per time-instance. Hence, all signals transmitted at a given time-instance must use the same beam, which makes it much more challenging to frequency multiplex UEs and/or different physical channels. Needless to say, by employing several parallel time-domain beamformers, where each beamformer is associated with different parts of the system bandwidth, different beams for different parts of the spectrum can still be used. This is, however, costly and typically not considered for the first mm-wave analog AAS implementations.

Strict time-domain beamforming, meaning strictly one beam per time-instance, can only use wideband beamforming, and, for instance, frequency-selective beamforming or MU-MIMO can in general not be used. Pure time-domain beamforming imposes scheduling restrictions that can reduce the potential capacity benefits offered by the large bandwidths at mm-wave. As only one beam can be formed per time-instance, and each beam is typically very narrow in order to maximize the gain, it is very difficult to find several UEs with data in their buffers that would like to be served by the same beam. An alternative would be to create a wider beam that covers several UEs, but such a beam would have much lower gain. Hence, in each time-instance the entire bandwidth is tied-up by one UE, irrespectively of the actual bandwidth-need. This can significantly reduce the potential capacity, especially as many small packets or signaling messages are commonly sent (see traffic discussion in Section 2.2.2).

To illustrate this, consider an example where only PDSCH is considered and the system has 100 MHz bandwidth, a $(2 \times 1)_{SA}(4 \times 24)$ AAS configuration, data packets of size 100 kB or 1.5 kB, and beamforming according to:

- frequency-domain beamforming (FD-BF) is allowed, meaning that UEs can be frequency multiplexed, each with a separate beam;
- time-domain beamforming (TD-BF) is used, meaning that only one beam per time-instance is allowed. Each beam belongs to a two-times oversampled DFT-based codebook. Hence, in total $2 \times 4 \times 2 \times 24 = 384$ narrow beams, which makes it unlikely to find two co-scheduled UEs that want to be served by the same beam.

Figs. 13.60 and 13.61 show time and frequency utilization as a function of load for packet sizes of 1.5 kB and 100 kB, respectively. The scenario is 3GPP UMa with only outdoor UEs to get good coverage. For the small 1.5 kB packet size, large differences in time and frequency utilization are observed for the two cases. In particular, it is interesting to see how the frequency utilization starts at roughly 10% for both cases at low load and then increases as the load increases for FD-BF, while for TD-BF it saturates at around 20%.

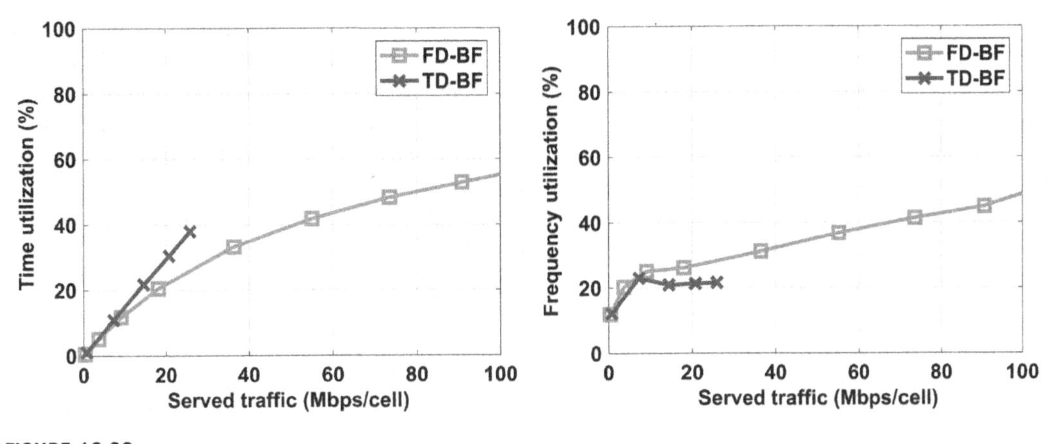

FIGURE 13.60

Average time (*left*) and frequency (*right*) utilization as functions of load for FD-BF and TD-BF, respectively. 3GPP UMa scenario with good coverage and packet size of 1.5 kB.

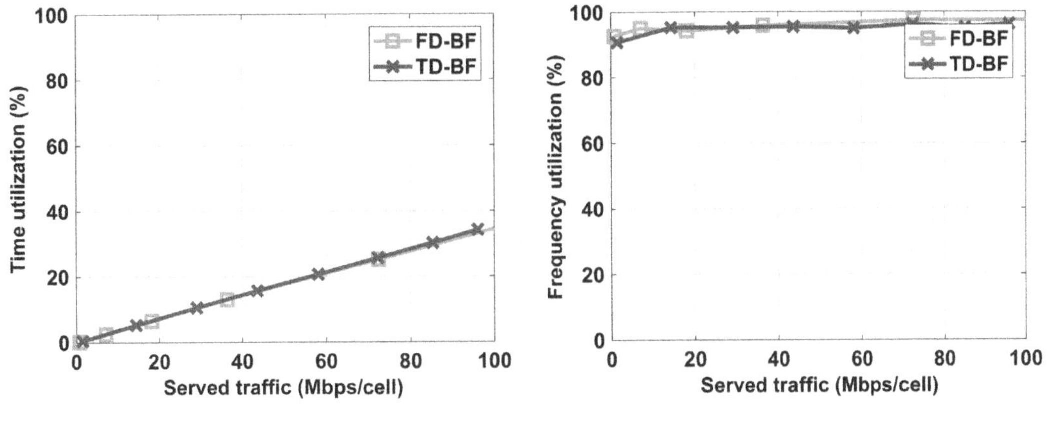

FIGURE 13.61

Time (*left*) and frequency (*right*) utilization as functions of load for FD-BF and TD-BF, respectively. 3GPP UMa scenario with good coverage and packet size of 100 kB.

The 10% frequency utilization at low load can be explained as follows. 100 MHz bandwidth with 120 kHz subcarrier spacing gives 66 resource blocks, where each resource block contains 12 subcarriers. Furthermore, 11 out of the 14 symbols per slot are assumed to be allocated for data, while the other symbols contain control and reference signals. Finally, 256 QAM (6 bits/symbol) and two SU-MIMO layers are assumed. This means that the maximum number of bytes (1B = 8 bits) per slot becomes

$$66 \times 12 \times 11 \times 6 \times 2/8 \approx 13.1 \text{ kB/slot}$$

As the data packet size is 1.5 kB and 1.5 kB/13 kB = ~11%, it follows that by using roughly 10% of the available resource blocks, the 1.5 kB data packet fits into one slot (or TTI). Note that the calculation above is rather ideal and in practice will the chosen modulation and coding scheme (MCS) depend on the link quality. Hence, UEs with a worse link quality require typically more resources due to more robust MCS choices. The reason for the saturation for TD-BF is that essentially no frequency multiplexing of UEs can be obtained.

For the large 100 kB packet size, on the other hand, all cases exhibit very similar results with respect to time and frequency utilization. This is reasonable as the 100 kB packet fills up several TTIs with full resource block utilization; hence there is less gain by allowing frequency multiplexing of UEs. Note, though, that this example considers only PDSCH transmissions with a fixed packet size. In practice, there will be many small packets, for example, ACK/NACK messages, downlink control information and RLC status messages that will severely hamper TD-BF.

The differences for the two cases depending on packet size are further illustrated in Fig. 13.62, showing the capacity gain for FD-BF relative to TD-BF. For the large 100 kB packet size, FD-BF and TD-BF provide similar capacity, whereas for the small 1.5 kB packet, FD-BF provides ~7 times more capacity than TD-BF.

Evidently, the relation between the bandwidth and the packet size is a key parameter when determining how detrimental scheduling restrictions as the one discussed above are.

13.7.5 BASE STATION OUTPUT POWER

13.7.5.1 Key characteristics

The optimal choice of base station output power level is an intricate question. From a pure performance point of view, more power is generally better although the gains are scenario dependent and subject to diminishing returns. In virtually all scenarios, there will be UEs with poor coverage that benefit from more output power. Still, more base station output power will not help the uplink and there needs in general to be a certain uplink data rate to sustain the downlink.

Performance is one driving factor that dictates the choice of output power, but there are many other aspects that are equally or even more important, such as regulatory requirements, cost and energy performance.

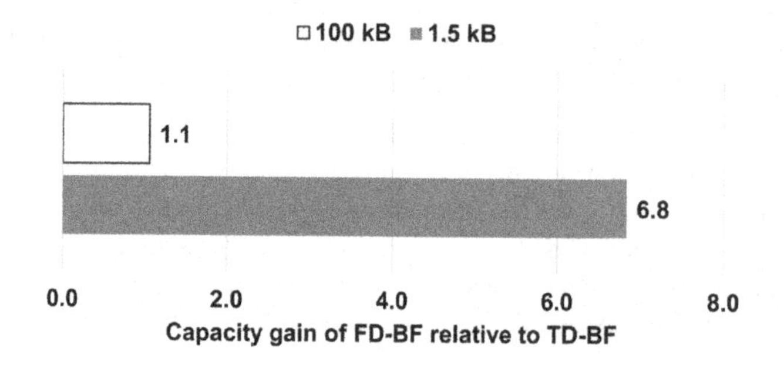

FIGURE 13.62

Capacity gains for FD-BF relative to TD-BF for two different packet sizes: 1.5 kB and 100 kB.

The objective of this section is to illustrate the impact of output power on AAS performance. Single-band 3.5 GHz with 100 MHz bandwidth and a 3:1 downlink versus uplink TDD ratio is considered. In many higher-band deployments, the downlink data rates are substantially higher than the uplink data rates due to downlink heavy TDD configurations and uplink coverage limitations. This raises the question whether the uplink can support the downlink in terms of required bitrates for transport protocols such as the TCP.

The TCP functionality is rather involved and depends on many factors including the traffic characteristics. However, realistic TCP feedback (acknowledgment) bitrates today seem to range from 1% to 10% of the data rate, essentially meaning that the uplink data rate needs to be larger than 1%–10% of the downlink data rate to sustain the downlink. For example, a downlink data rate of 100 Mbps would require 1–10 Mbps in the uplink to sustain a 1%–10% TCP feedback rate. A very simple model of the TCP impact ignoring any dynamic effects, would be to let the effective TCP "throttled" downlink data rate $r_{\mathrm{DL,eff}}$ equal

$$r_{\mathrm{DL,eff}} = \min(r_{\mathrm{DL}}, \alpha r_{\mathrm{UL}}) \tag{13.1}$$

where r_{DL} and r_{UL} denote the unconstrained downlink and uplink data rates, respectively, and α stands for the TCP ratio which is in the range of 100 to 10.

13.7.5.2 Performance results

Figs. 13.63 and 13.64 show some key downlink performance metrics as a function of the base station output power in the 3GPP UMa and UMi scenarios, respectively, with an $(2\times1)_{\mathrm{SA}}(4\times4)$ antenna configuration. Both unconstrained and TCP constrained with $\alpha = 50$ (2% uplink acknowledgment data rate requirement) downlink performances are considered.

In the 3GPP UMa scenario without TCP restrictions, there are significant gains by increasing the output power beyond 1 W/MHz, especially for the 5:th percentile UE throughput. Furthermore, it is evident that the uplink would limit the downlink with a 2% TCP ACK rate. Taking the simple TCP model (13.1) into consideration, the downlink performance is very similar for all output power

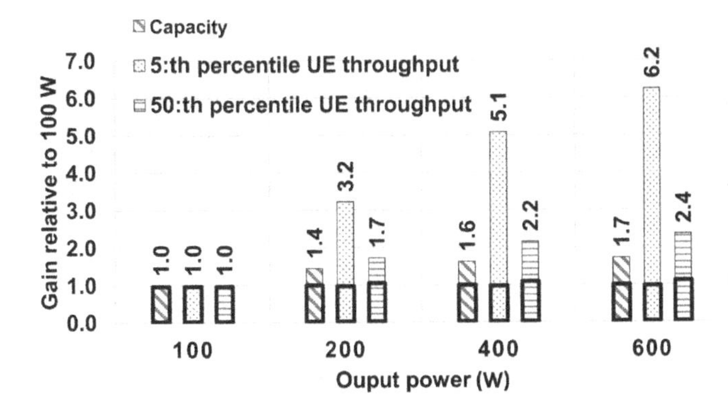

FIGURE 13.63

Downlink performance gains as a function of the output power in the 3GPP UMa scenario. Filled bars and numbers correspond to unthrottled results, while black unfilled bars represent TCP throttled results.

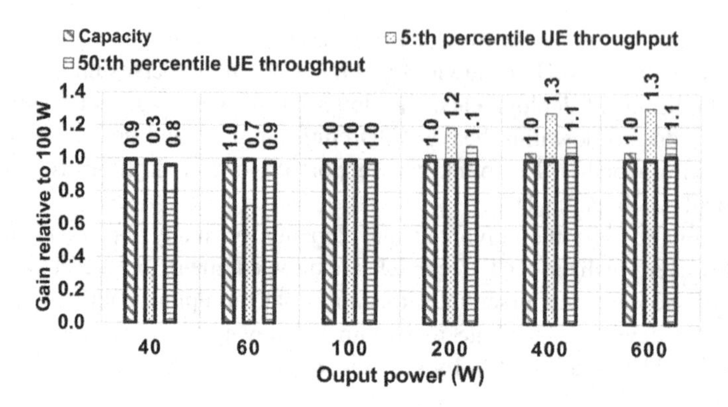

FIGURE 13.64

Downlink performance gains as a function of the output power in the 3GPP UMi scenario. Filled bars and numbers correspond to unthrottled results, while black unfilled bars represent TCP throttled results.

levels. The exception is a small increase in the 50:th percentile UE throughput from increased output power.

In the more interference-limited (defined as a scenario where more power does not lead to higher capacity) 3GPP UMi scenario, the capacity gain by adding more output power saturates below 1 W/MHz. For the 5:th percentile UE throughput, the corresponding saturation occurs around 2 W/MHz when not taking TCP restrictions into consideration. Also, in this case there is an impact from the TCP protocol on the achievable downlink data rate due to uplink limitations.

13.7.6 FREQUENCY-BAND INTERWORKING

13.7.6.1 Key characteristics

As seen in previous sections, 5G mid-band and mm-wave deployments have the potential to provide superior UE experience with very high UE throughput and substantial capacity gains due to vast availability of spectrum and AAS technologies facilitating massive MIMO. However, to fully unlock the potential of 5G will require operators to use all spectrum assets in smart ways and interworking with existing lower frequency bands will be key for success. There are a few observations that point in this direction, for example:

- mid-band and mm-wave TDD deployments may experience poor uplink coverage, particularly in specific environments. It will therefore be essential to be able to move the uplink for UEs with poor coverage to a lower (FDD) band;
- it will take time to reach a significant NR UE penetration. Thus, at the same time as NR is being introduced and LTE gradually migrated to NR, it might be necessary to also evolve LTE to satisfy the growing capacity need before NR has reached a significant UE base. In this process, it will be important to have tools that enable efficient sharing of existing spectrum assets between technologies.

13.7.6.2 Performance results

13.7.6.2.1 Frequency-band interworking

Fig. 13.65 illustrates cell-edge UE improvements of introducing NR at mid-band and high-band together with existing LTE low-bands using *E-UTRAN New Radio* (EN) *Dual Connectivity* (DC) in a dense urban scenario. In this example, LTE 800 and LTE 2.6 employ FDD with 10 MHz and 20 MHz bandwidth, respectively, 2T2R base station antennas and 1T2R UEs. NR at 3.5 GHz and 26 GHz has 75 MHz and 150 MHz effective bandwidth, respectively, assuming a 3:1 TDD slot format. Furthermore, at 3.5 GHz, UEs employ 1T4R antennas facilitating much better receiver processing and the base stations are equipped with an $(2 \times 1)_{SA}(4 \times 8)$ AAS with UE-specific beamforming and MU-MIMO. At 26 GHz, UEs are equipped with 2T2R based on analog beamforming with 14 dBi maximum directivity, and base station antennas use 2T2R analog beamforming with 30 dBi maximum directivity.

Evidently, mid-band NR at 3.5 GHz has the potential to substantially improve both the UE throughput experience and the capacity due to increased bandwidth, better UE capabilities and AAS facilitating beamforming and MU-MIMO. Adding NR 3.5 to existing LTE bands provides roughly 8 times more capacity and uses 1.5 times more bandwidth compared to the LTE bands, hence NR 3.5 is roughly 5 times more capacity efficient than the LTE low-bands in this example.

Although NR at 26 GHz has a significant amount of bandwidth, it follows that adding NR at 26 GHz on top of LTE 0.8 + LTE 2.6 + NR 3.5, gives only 30 percent increased capacity, hence the 26 GHz spectrum is much less capacity efficient than the NR 3.5 spectrum. One of the main benefits of adding NR at 26 GHz in macro deployments comes from off-loading the valuable lower bands, that is, by letting NR at 26 GHz serve the good UEs that have coverage on 26 GHz, more resources become available at the lower bands, including NR at 3.5 GHz, for serving the worst UEs.

Furthermore, coverage at high-band is rather binary, meaning that many UEs do not have coverage and cannot directly benefit from the high-band, but once you get coverage you enjoy great improvements due to the vast amount of spectrum. This is illustrated in Fig. 13.66 showing the 50:th

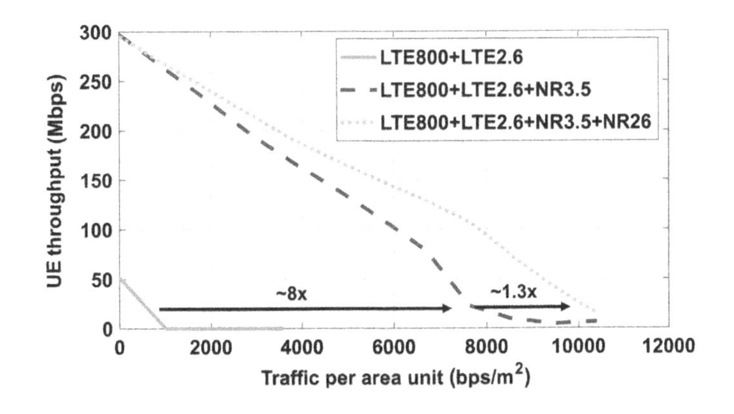

FIGURE 13.65

Illustration of downlink cell-edge (5:th percentile) UE performance gains from LTE + NR interworking in a dense urban macro scenario.

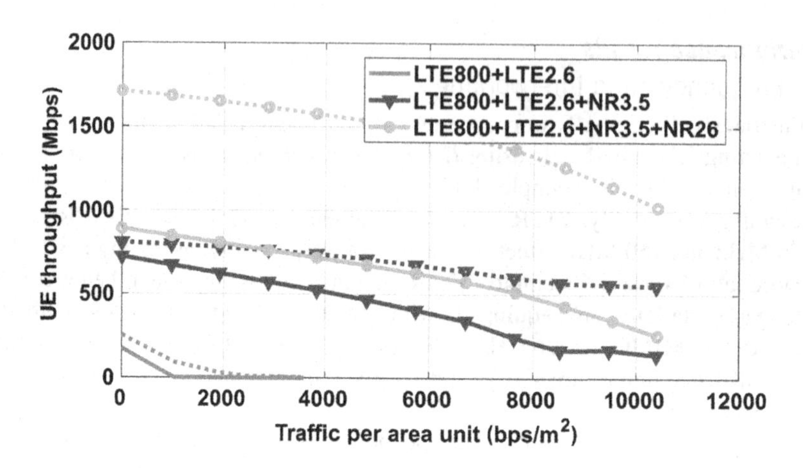

FIGURE 13.66

Illustration of downlink UE throughput gains from LTE + NR interworking for 50:th (*solid line*) and 95:th (*dotted line*) UE percentiles in a dense urban macro scenario.

and 95:th percentile UE performance. Evidently, the best UEs do get a substantial UE throughput performance boost from NR at 26 GHz.

It is important to understand that the combined LTE + NR performance assumes full penetration of NR capable UEs, and until this point has been reached, the network performance will be somewhere in between LTE and LTE + NR. Similarly, it is important to understand that coverage for NR will be important from day one, whereas the capacity demand will increase gradually over time.

13.8 SUMMARY

Based on assumptions following the industry standard, this chapter has illustrated how the choice of antenna configuration and multi-antenna algorithm affects the radio network performance.

At the time of writing, AAS are being deployed in both commercial networks and trial setups all over the world, and results from these AAS deployments show gains compared to classical (non-AAS) deployments in terms of coverage, capacity and end-user throughput and those gains are often in good agreement with simulation results shown in this chapter. AAS deployments with many radio chains and advanced multi-antenna techniques are still at an early stage, and learnings from these deployments will lead to improved AAS products and better AAS models used for radio network evaluations.

A detailed summary of the chapter is given in Section 13.1, but some of the key takeaways are the following:

- when considering simulation results, the underlying assumptions should always be scrutinized as these can strongly impact the results and conclusions. Assumptions do matter;

- the preferred choice of the antenna array layout is typically closely related to the angular UE distribution and channel angular spread in the horizontal and vertical dimensions. For example, given realistic channels with little angular spread in the vertical domain, the subarray size should typically match the angular UE distribution in the vertical domain;
- SU-MIMO can provide substantial performance benefits. Although subject to diminishing returns, doubling the number of antenna columns (increasing the antenna area and the number of radio chains) gives typically SU-MIMO capacity gains in the order of 30%–50% in urban macro scenarios. Generalized beamforming for MU-MIMO performs better than classic beamforming at high load;
- MU-MIMO is a capacity feature that can bring significant gains under favorable conditions, but in many practical cases, gains can be rather modest. NR Type II feedback-based CSI and reciprocity-based MU-MIMO have often similar performance. Hence, MU-MIMO is not only about reciprocity;
- one of the main benefits of AAS and beamforming is the improved coverage that often makes it easier to reuse the existing site grid when deploying higher frequency bands. When deploying a new frequency band, coverage is often important from day one while capacity demands gradually increase over time;
- coverage challenges imply that a mix of classic beamforming for SU-MIMO and fast fading-based generalized beamforming with MU-MIMO is key to achieve a high-performing network;
- frequency-interworking solutions are vital components to facilitate a smooth evolution of the network considering aspects such as coverage, legacy impact, spectrum, and capacity.

REFERENCES

[1] 3GPP, 3rd Generation Partnership Project; Technical Specification Group Radio Access Network; Study on channel model for frequencies from 0.5 to 100 GHz (Release 15), 3GPP TR 38.901.

[2] ITU-R, Guidelines for evaluation of radio interface technologies for IMT-Advanced, Report ITU-R M.2135-1.

[3] 3GPP, NR; Study on integrated access and backhaul, 3GPP TR 38.874.

[4] Ericsson, Fixed Wireless Access handbook. <https://ericsson.com>

CHAPTER

ADVANCED ANTENNA SYSTEM IN NETWORK DEPLOYMENTS 14

14.1 INTRODUCTION

One of the most important aspects to understand in terms of advanced antenna system (AAS) network deployments is that different scenarios call for different AAS solutions, and there is not a one-size-fits-all AAS product. What AAS configuration to deploy where depends on many factors including offered performance, site constraints, cost considerations, and strategic evolution aspects such as expected penetration of terminal capabilities, spectrum acquisitions, and regulatory requirements.

The objective of this chapter is to provide further insights into the role of AAS in network deployments by discussing some of the key considerations mentioned above.

To put AAS into perspective in terms of the overall network evolution strategy, a few illustrative case examples discussing potential network evolution steps for typical network operators are also included.

14.1.1 OUTLINE

The outline of the chapter is as follows.

The considered deployment scenarios are first described in Section 14.2, followed by Section 14.3 illustrating the performance of different AAS configurations in the different scenarios. Section 14.4 explains how site and multi-band techniques together with regulatory aspects influence the choice of AAS.

Thereafter follows a few examples of operator network evolution strategies in different markets in Section 14.5.

Finally, Section 14.6 discusses the importance of a total cost of ownership (TCO) perspective when the choice of what AAS to deploy is considered, before concluding the chapter in Section 14.7.

14.2 DEPLOYMENT SCENARIOS

The deployment scenario has a profound impact on the preferred choice of AAS from a performance perspective.

Four different scenarios are considered in this chapter (see Fig. 14.1): dense urban high-rise, dense urban low-rise, suburban/rural, and fixed wireless access (FWA) scenarios. Note, however, that many deployments in practice consist of a mix of deployment types, that is, deployments are typically inhomogeneous.

Advanced Antenna Systems for 5G Network Deployments. DOI: https://doi.org/10.1016/B978-0-12-820046-9.00014-9

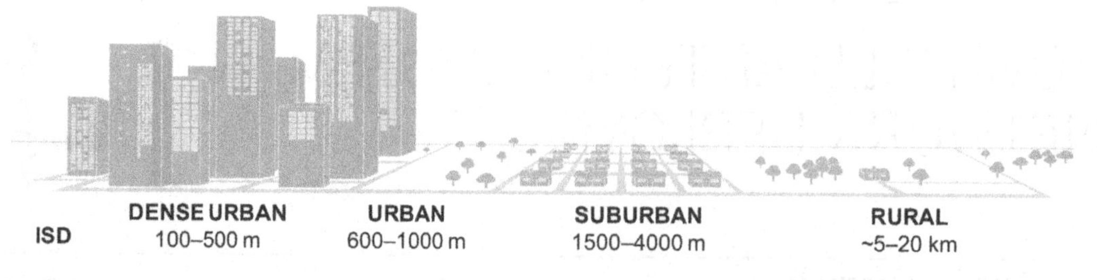

FIGURE 14.1

Illustration of different deployment scenarios with the typical inter-site distance (ISD) indicated.

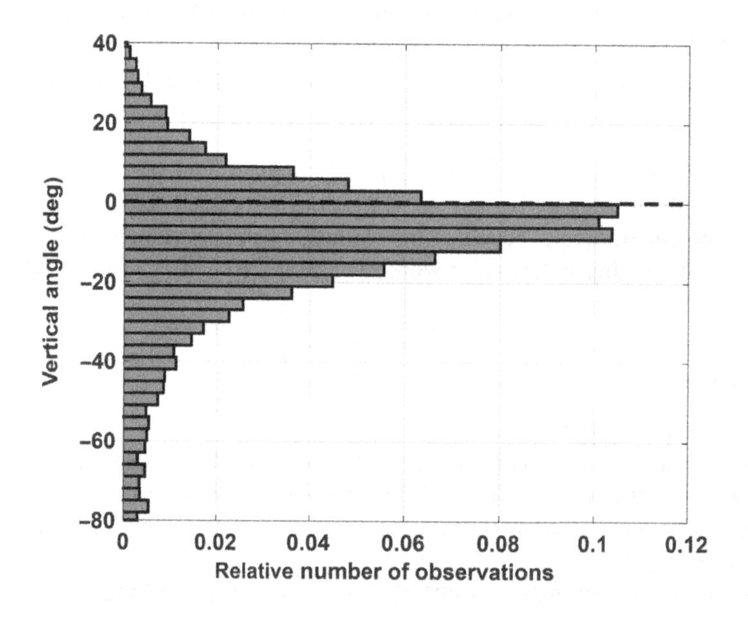

FIGURE 14.2

Illustration of geometrical angular UE distribution in the vertical domain relative to the location of the serving base station antenna for the dense urban high-rise scenario where 0 degree is toward the horizon.

Note also that most performance evaluations in Chapter 13, were done in scenarios corresponding to (dense) urban low-rise and dense urban high-rise and a few were done in suburban deployments.

14.2.1 DENSE URBAN HIGH-RISE MACRO SCENARIO

The dense urban high-rise macro scenario is characterized by high-rise buildings, short inter-site distances (ISDs) of 100–400 m, large traffic volume per area unit, and high subscriber density. User equipment (UEs) are located on different floors in high-rise buildings as well as outside, leading to a wide angular spread of UEs in both the vertical and the horizontal dimensions (see Fig. 14.2). Macro antennas are deployed both above and below rooftop with UEs both below and above them.

Most of the UEs are assumed to be indoor (~ 80%) in high-rise buildings. It should, however, be noted that many modern high-rise buildings have indoor solutions and do not need to be served by outdoor macro deployments. This makes the scenario having more low-rise characteristics with UEs typically well below the antennas. Nevertheless, the propagation environment will still experience (significant) angular spread in the vertical domain due to short ISDs and the cluttered environment.

14.2.2 DENSE URBAN LOW-RISE MACRO SCENARIO

The dense urban low-rise scenario represents many of the larger cities around the world, including the outskirts of many high-rise cities. Base stations are typically deployed on rooftops, with ISDs of a few hundred meters. This together with low-rise buildings means that most UEs are below the antennas creating a rather narrow angular UE distribution in the vertical domain (see Fig. 14.3). Compared to the dense urban high-rise scenario, traffic per area unit is lower. There is generally a mix of building types, which creates a rich multi-path propagation environment. Most UEs (~ 80%) are assumed to be indoor.

Coverage will be challenging for several UEs, especially at higher frequency bands employing time division duplex (TDD) and for deep indoor UEs, hence selecting as large antenna area as possible is important for improving uplink cell-edge data rates.

14.2.3 SUBURBAN/RURAL MACRO SCENARIO

Rural or suburban macro scenarios are characterized by rooftop or tower-mounted base stations with ISDs ranging from one to several kilometers, low or medium population density, and very small angular UE spread in the vertical domain (see Fig. 14.4).

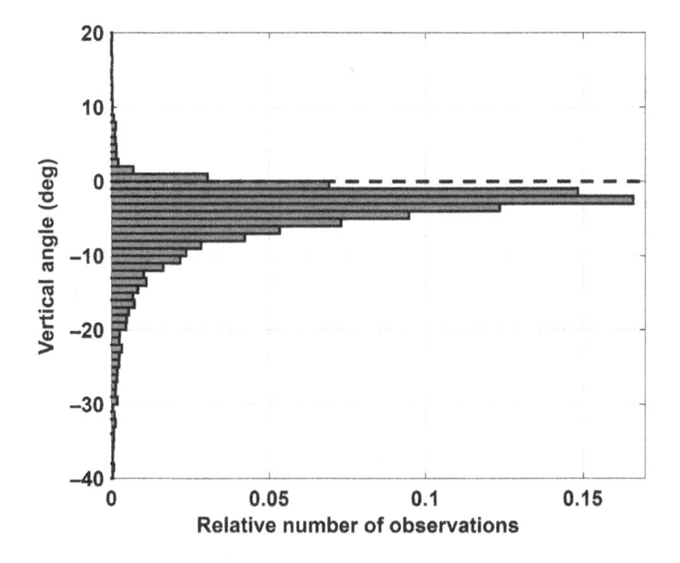

FIGURE 14.3

Illustration of geometrical angular UE distribution in the vertical domain relative to the location of the serving base station antenna for the dense urban low-rise scenario where 0 degree is toward the horizon.

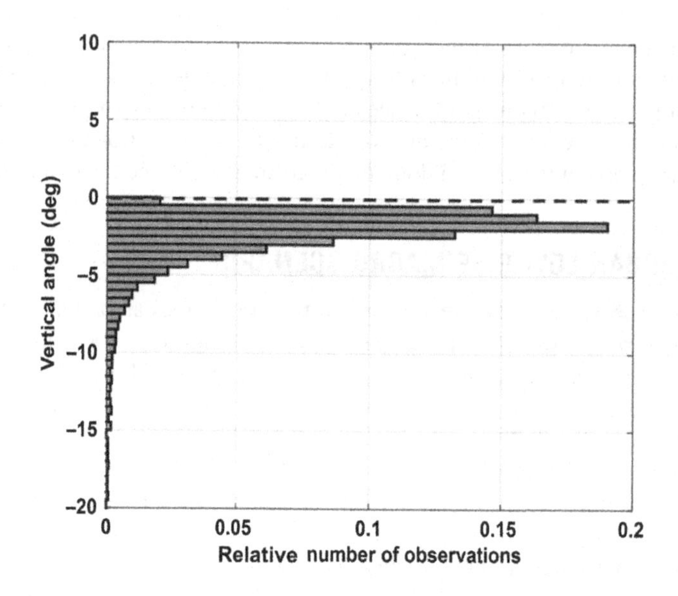

FIGURE 14.4

Illustration geometrical angular UE distribution in the vertical domain relative to the location of the serving base station antenna for the suburban macro scenario where 0 degree is toward the horizon.

Coverage is the main challenge due to very large ISDs, which favors large antennas. However, as the cells can become very large, there can also be a capacity demand that calls for multi-antenna deployments.

14.2.4 FIXED WIRELESS ACCESS

Fixed wireless access (FWA) was briefly discussed in Section 2.1.6.1, and it has slightly different characteristics compared to mobile broadband (MBB), for example, often higher capacity demand, no mobility as the *customer premises equipment* (CPE) is installed in households, and better link quality. For example, FWA deployments with outdoor CPE often experience a high degree of line-of-sight propagation, leading to an improved and predictable link quality.

While an indoor CPE is comparable with an ordinary MBB terminal in terms of capabilities and performance, an outdoor rooftop mounted CPE performs significantly better by using a high-gain directional antenna and potentially higher output power. A correctly installed outdoor CPE has its directional antenna directed toward the antenna of the serving cell, which leads to a very stable link quality with lower path loss. This also creates a multiple-user multiple-input multiple-output (MU-MIMO) "friendly" environment as data packets are potentially larger and the signal-to-interference-and-noise ratio levels typically become so high that it is beneficial to share power between CPEs.

Note that FWA is not 'a' scenario like urban or suburban as discussed for MBB in previous subsections. FWA can obviously be deployed in different environments, for example, urban, suburban, or rural (Fig. 14.5).

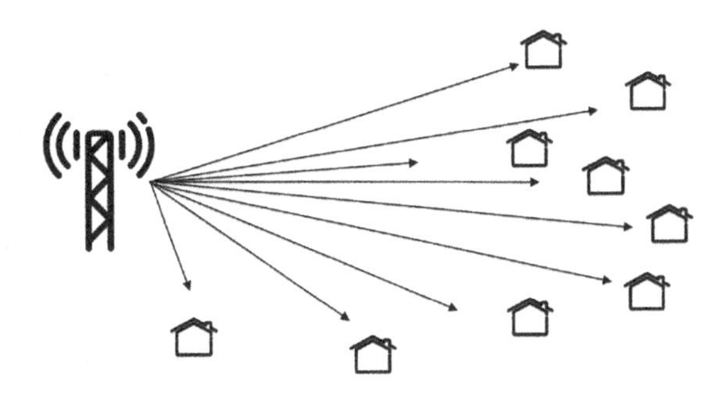

FIGURE 14.5

Illustration of FWA from one base station antenna to many households.

14.3 MULTI-ANTENNA PERFORMANCE IN MACRO NETWORK DEPLOYMENTS

There is a clear scenario dependency on the performance benefits of different multi-antenna solutions. The objective of this section is to discuss how performance in different scenarios depends on key multi-antenna characteristics, such as algorithm choice and antenna configurations.

Two algorithm types are considered:

- *feedback-based* SU-MIMO (single-user multiple-input multiple-output) by means of classic *large-scale wideband* precoding based on grid-of-beams—referred to as GoB;
- reciprocity-based MU-MIMO by means of generalized *fast-fading frequency-selective* precoding—referred to as zero forcing (ZF).

The ZF and GoB algorithms are described in Sections 6.3.3.3 and 6.3.3.4, respectively. How to select features from the (3GPP) toolbox to support feedback-based and reciprocity-based operation is described in Chapter 10, End-to-End Features.

The array of subarrays antenna configuration, described in Section 4.6.1, is adopted in this section. An illustration of typical antenna configurations used in the following subsections can be found in Fig. 14.6. Recall that antenna configurations are denoted by $(M_v \times M_h)_{SA}(N_v \times N_h)$, where M_v and M_h denote the number of vertically and horizontally stacked dual-polarized antenna element pairs in the subarray, respectively, and N_v and N_h represent the number of vertically and horizontally stacked subarrays, respectively.

As was thoroughly discussed in Chapter 13, three key aspects of the antenna configuration that affect performance are as follows:

- the impact of *maximum array gain*, which relates to the antenna area. The importance of this aspect depends on how good coverage the scenario offers, which in turn depends on ISD, line of sight probability, outdoor-to-indoor penetration losses, etc.;
- the *angular coverage area for UE specific beamforming*, that is, the area where UE specific beamforming works efficiently without any significant gain drop, which relates to the subarray size in relation to the angular UE distribution in the vertical domain;

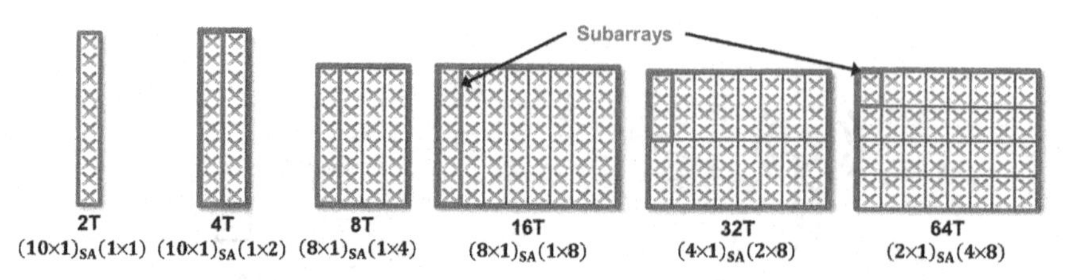

FIGURE 14.6

Illustration of different antenna configurations based on subarrays and arrays of subarrays with number of radio chains ranging from 2 to 64. Note that 16T, 32T, and 64T have the same total antenna area 8 × 8.

- the *number of subarrays* (number of Ts) and *how they are stacked to form the two-dimensional array*. A wide angular coverage area in the vertical domain requires many small subarrays to get high array gain. Furthermore, using many subarrays leads to narrow beams which can reduce the negative impact of interference. The preferred stacking of subarrays is closely related to the UE distribution.

Note also that given the array of subarrays antenna layout, the three aspects above are closely related. For example, the subarray size together with the number of subarrays gives the total antenna area, which is related to the antenna gain.

A detailed discussion of the network performance for each deployment scenario described in Section 14.2 is given in Sections 14.3.1–14.3.4, followed by a summary in Section 14.3.5. Note also that Chapter 13 provides an in-depth description of how different AAS configurations and algorithm considerations affect performance. Assumptions for results presented in the following sections are discussed in Section 13.2.

14.3.1 DENSE URBAN HIGH-RISE

The dense urban high-rise scenario is described in Section 14.2.1. Note that performance evaluations are done here without indoor solutions, hence indoor UEs are connected to an outdoor macro.

14.3.1.1 Performance

Capacity and 5:th percentile UE throughput results for different antenna configurations in the dense urban high-rise scenario are shown in Figs. 14.7 and 14.9, with associated antenna configuration displayed in Figs. 14.8 and 14.10, respectively. It should be noted that some assumptions may provide optimistic (capacity) gains, for example, a relatively large data packet size and limited algorithm impairments.

From the results, the following observations can be made in this scenario:

- Fig. 14.7 shows that for a fixed antenna area, there are significant gains by reducing the subarray size as it enabled more efficient vertical domain beamforming (more radio chains in the vertical domain);
    - For an 8 × 8 antenna array, a 64-branch AAS based on 2 × 1 subarrays using ZF provides almost four times more capacity than a 16-branch AAS based on 8 × 1 subarrays as the

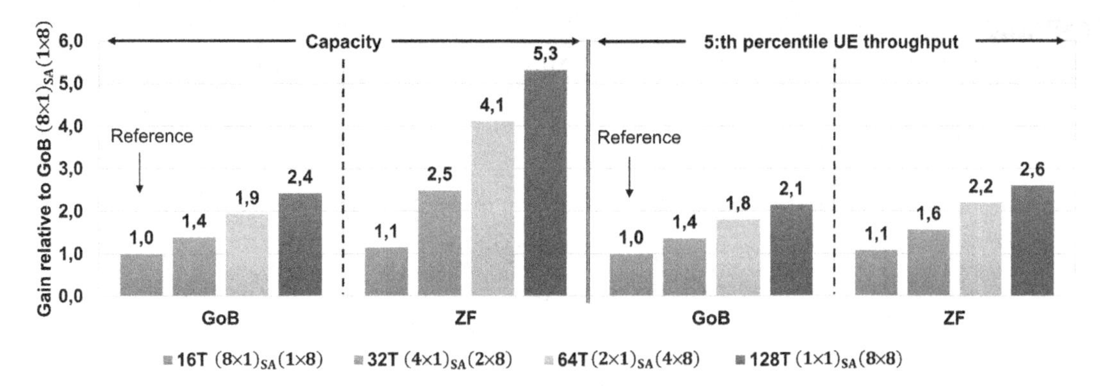

FIGURE 14.7

Illustration of ZF and GoB performance gains for various metrics and antenna configurations for a mid-band deployment in a dense urban high-rise scenario. Note that even though subarray layout differs, all results are with an 8×8 antenna area (see Fig. 14.8).

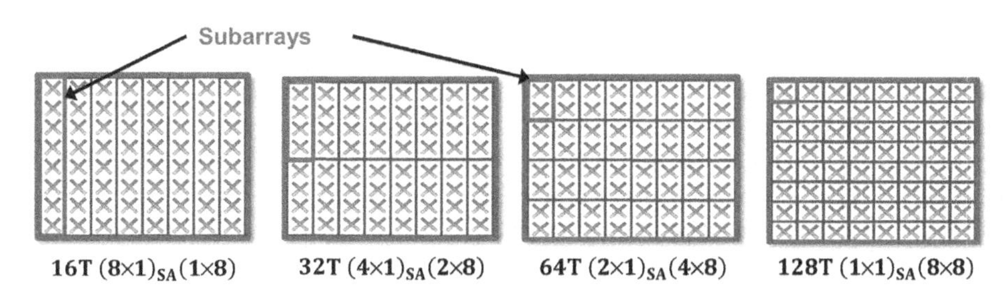

FIGURE 14.8

Illustration of AAS configurations used in Fig. 14.7.

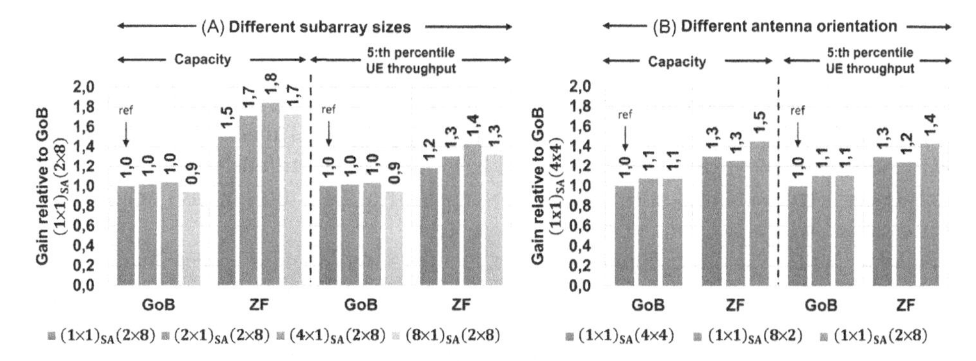

FIGURE 14.9

Illustration of ZF and GoB performance gains for various metrics and antenna configurations for a mid-band deployment in a dense urban high-rise scenario. All cases use 32 radio chains with (A) different subarray sizes and (B) different array of subarrays layouts (see Fig. 14.10).

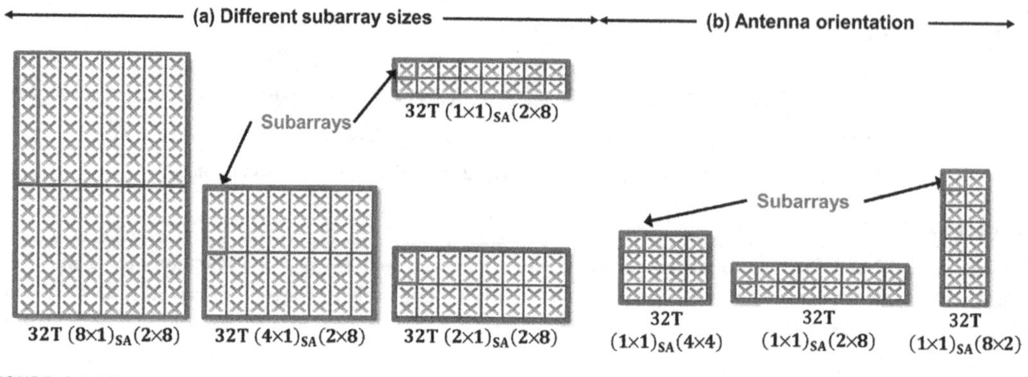

FIGURE 14.10

Illustration of the 32T AAS configurations used in Fig. 14.9.

former can utilize vertical domain beamforming while the latter cannot. The corresponding number for the GoB case is approximately 2.5 times more capacity;
  - Reciprocity-based fast-fading frequency-selective MU-MIMO (ZF) is superior to feedback-based large-scale wideband beamforming (GoB);
- Fig. 14.9A shows that for a fixed number of radio chains, a subarray with four elements gives a good trade-off between gain and angular coverage area;
- Fig. 14.9B shows that stacking subarrays horizontally or vertically provides similar gains, although horizontal stacking is slightly preferred. Due to the distribution of UEs in this scenario, "cells" are almost as tall as wide.

14.3.1.2 Analysis

The significant spread of UEs in the vertical domain in combination with the small ISD creates a situation where cell defining beam(s) should be wide and a rather large angular coverage area for UE–specific beamforming is needed, meaning that subarrays need to be small enough. Together with the high site density, this also leads to a situation where signals from interfering base stations are strong at the UE, and severe interference problems may occur. This can be counteracted from the network side by employing interference-aware beamforming algorithms that become very effective in conjunction with an AAS with many radio chains. The UE side can also employ advanced interference suppression algorithms to reduce the problem.

The detailed multi-path propagation environment with large angular spread favors frequency-selective beamforming.

The good coverage and large angular spread of UEs mean that the potential for fast-fading reciprocity-based beamforming (ZF) and MU-MIMO with a relatively large number of scheduled UEs is high.

An AAS where the antenna is partitioned into many small subarrays stacked both horizontally and vertically excels in this scenario. Hence, having many radio chains controlling all the small subarrays and advanced beamforming algorithms pay off in terms of cost-versus-performance.

14.3.2 DENSE URBAN LOW-RISE

The dense urban low-rise scenario is described in Section 14.2.2.

14.3.2.1 Performance results

Capacity and fifth percentile UE throughput results for different antenna configurations in the dense urban low-rise scenario are shown in Figs. 14.11 and 14.13, with associated antenna configurations in Figs. 14.12 and 14.14, respectively. From the results it follows that in this scenario,

- Fig. 14.11 shows that using 4 × 1 subarrays, that is, the $(4 \times 1)_{SA}(2 \times 8)$ case with 32 radio chains, provides the best performance for a fixed 8 × 8 antenna. That is, there is gain from

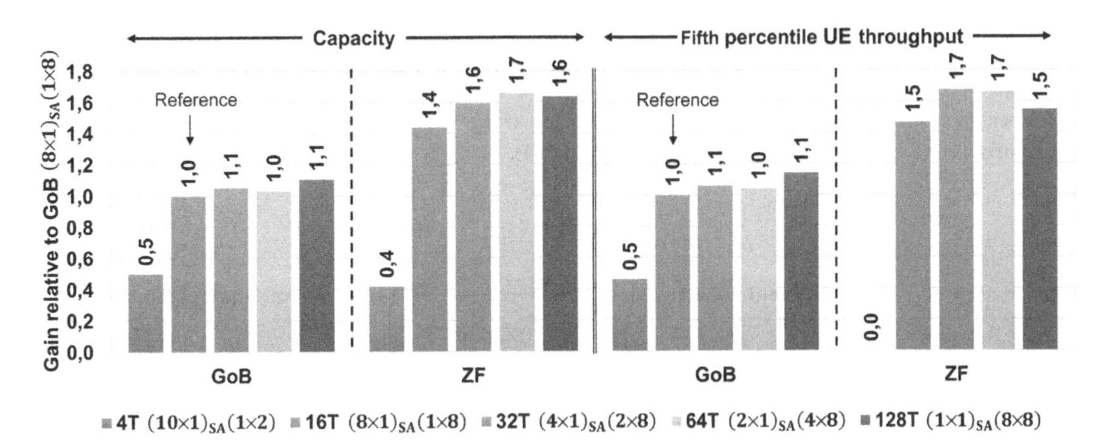

FIGURE 14.11

Illustration of ZF and GoB performance gains relative to 16T for various metrics and antenna configurations for a mid-band deployment in a dense urban low-rise scenario. Note that except baseline 4T with a 10 × 2 antenna array, all results are with a fixed 8 × 8 antenna area (see Fig. 14.12).

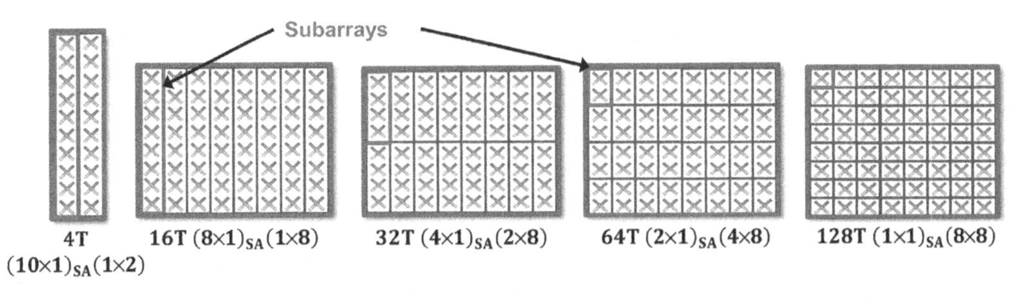

FIGURE 14.12

Illustration of antenna configurations used in Fig. 14.11.

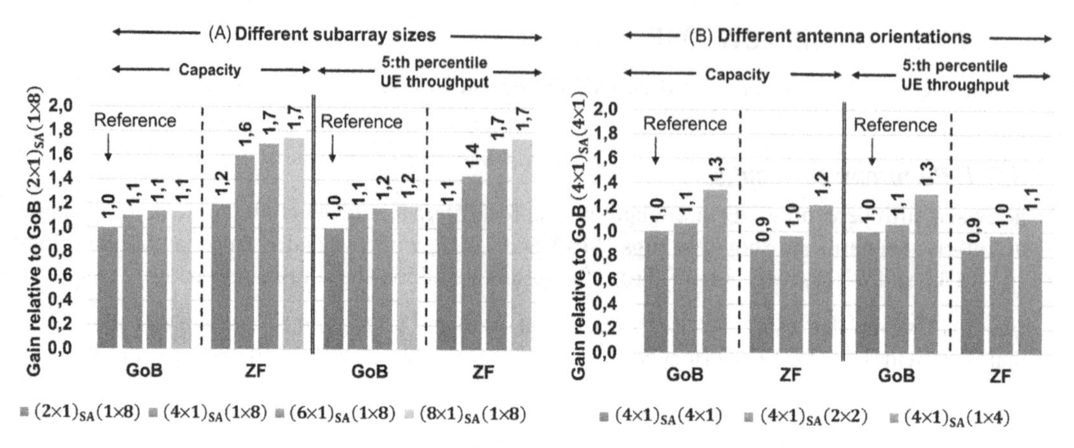

FIGURE 14.13

Illustration of ZF and GoB performance gains for various metrics and antenna configurations for a mid-band deployment in a dense urban low-rise scenario. All cases in (A) use 16T with different subarray sizes and all cases in (B) use 8T with different port layouts (see Fig. 14.14).

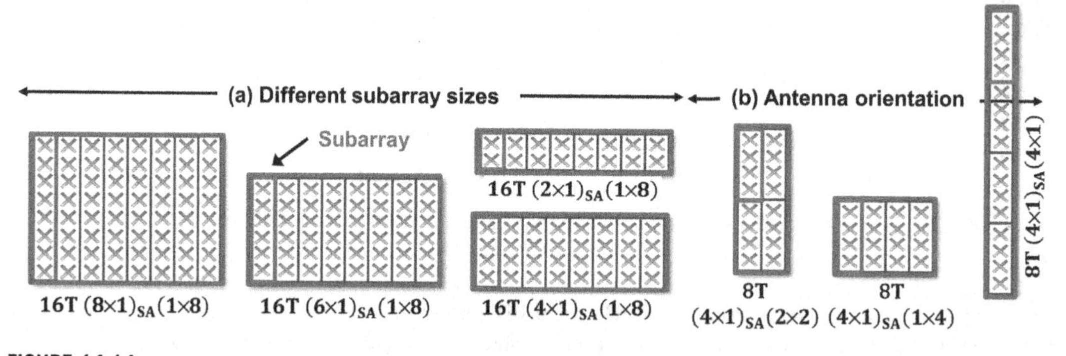

FIGURE 14.14

Illustration of antenna configurations used in Fig. 14.13.

vertical domain beamforming but much less compared to the high-rise scenario. Thus, horizontal domain beamforming is clearly favored over vertical domain beamforming;

- Similarly, frequency-selective MU-MIMO (ZF) that adapts to fast fading provides significant gains over large-scale wideband GoB, but the difference between ZF and GoB is reduced compared to the high-rise scenario;
- An AAS with a fixed 8×8 antenna size using 32T provides roughly two to three times capacity for GoB and ZF, respectively, relative to a 4T 10×2 sector antenna, thus significant gains by deploying AAS also in the dense urban low-rise scenario;

- Fig. 14.13A shows that for a fixed number of radio chains, relatively large subarrays are favored due to the increased gain (larger area) and less need for a large angular coverage area in the vertical domain;

- Fig. 14.13B shows that wide antennas perform better than tall antennas as the UE angular spread is significant in the horizontal domain, while more confined to a small area in the vertical domain.

14.3.2.2 Analysis

The mix of building types in this scenario creates a detailed multi-path propagation environment rendering frequency-selective and interfere-aware transmission techniques useful. Coverage will be challenging for several UEs, especially for higher frequency bands employing TDD and for indoor UEs, hence maximizing the antenna area is important for improving uplink cell-edge data rates.

Due to larger ISDs and decreased UE angular spread in the vertical domain (lower buildings), the vertical angular coverage area for UE specific beamforming can be decreased compared to dense urban high-rise. Hence, larger subarrays can be used and there is less gain to expected from vertical domain beamforming. Using larger subarrays for a given antenna area means that fewer radio chains are required. Horizontal domain beamforming is, however, still a very effective feature that provides large gains.

Fast-fading reciprocity-based beamforming schemes will work for many UEs, but there will be UEs with poor coverage that need to rely on techniques such as feedback-based beamforming. MU-MIMO is also beneficial at high network load due to the multi-path propagation environment, relatively good link qualities, and UE-pairing opportunities.

For the antenna configurations considered herein, a good trade-off between complexity and performance is obtained with an AAS with between 16 and 32 radio chains.

14.3.3 SUBURBAN/RURAL MACRO

The suburban/rural macro scenario is described in Section 14.2.3.

14.3.3.1 Performance results

Capacity and fifth percentile UE throughput results for different antenna configurations in the 3GPP suburban scenario are shown in Fig. 14.15, with the associated antenna configurations illustrated in Fig. 14.16. From the results it follows that

- cell-edge UEs using fast-fading reciprocity-based beamforming suffer from poor sounding coverage, especially when employing small subarrays;
- large subarrays with no or little vertical domain beamforming suffice, meaning that network-level performance based on GoB with 64T $(2 \times 1)_{SA}(8 \times 4)$ is only marginally better than 8T $(16 \times 1)_{SA}(1 \times 4)$ in this case.

  Needless to say, the performance of antenna arrays based on large subarrays is sensitive to appropriate tilt settings. It should also be highlighted that results in this section are based on a fixed 16×4 antenna with different subarray sizes, hence it is only the usefulness of vertical domain beamforming that is considered. Horizontal domain beamforming is always beneficial, so adding more antenna columns would improve performance. This also affects how many radio chains that are useful. With a fixed 16×4 antenna array, it follows that all configurations, 8T–64T, perform rather similar, especially in terms of capacity. For a fixed 16×8 antenna array, on the other hand, 16T–64T would perform significantly better than 8T.

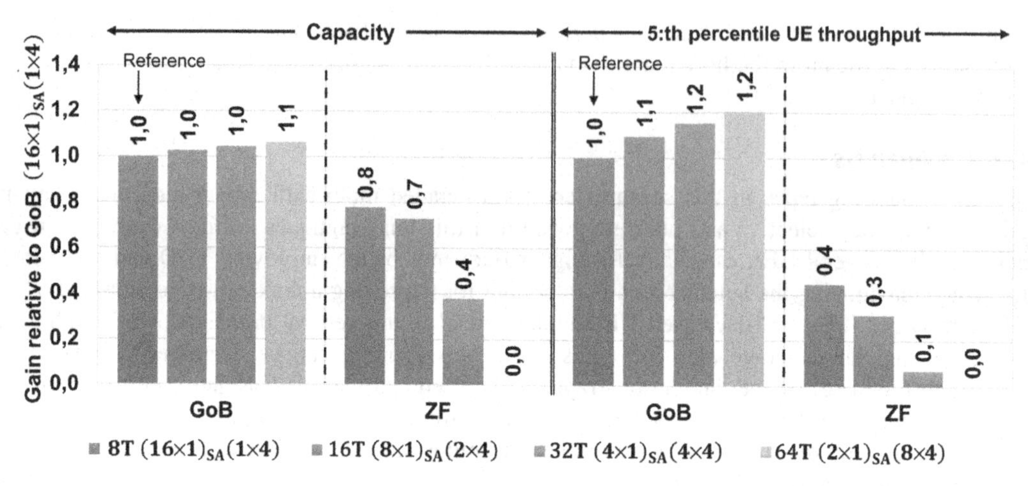

FIGURE 14.15

Illustration of ZF and GoB performance gains for various metrics and antenna configurations for a mid-band deployment in the 3GPP suburban scenario. All results are based on a fixed 16×4 antenna area (see Fig. 14.16).

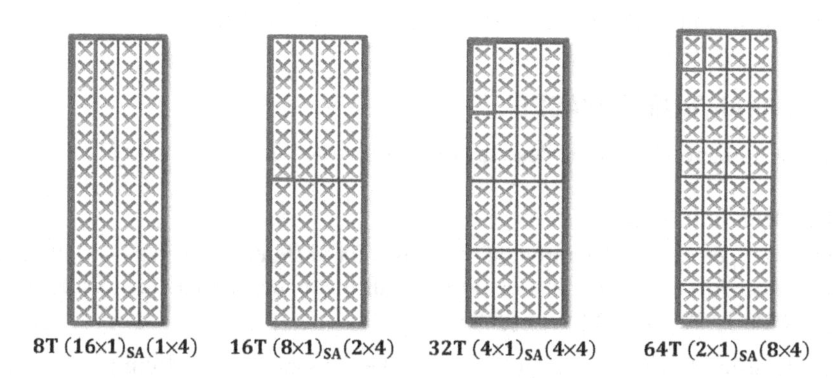

FIGURE 14.16

Illustration of antenna configurations used in Fig. 14.15.

14.3.3.2 Analysis

This scenario calls for an antenna solution with a large antenna area and the ability to support horizontal domain beamforming. Vertical domain beamforming, however, does not provide any significant gains as the UE spread in the vertical domain is low. Therefore, large vertical subarrays with small vertical coverage areas are appropriate as this maximizes gain while reducing the need for many radio chains. Note, though, that the ability to tune the narrow angular coverage area in the vertical domain via subarray tilt and/or mechanical tilt is often a must.

Fast-fading reciprocity-based beamforming is supported for a smaller fraction of UEs than in other scenarios, and MU-MIMO gains are limited.

For the antenna configurations considered herein, a good trade-off between complexity and performance is an antenna configuration with 8–16 radio chains.

14.3.4 FIXED WIRELESS ACCESS

FWA is described in Section 14.2.4.

Figs. 14.17 and 14.18 illustrate FWA performance gains in a typical suburban deployment at 3.5 GHz (TDD) with 1000 m and 2000 m ISD, respectively. The antenna configurations can be found in Fig. 14.19. Performance is given in terms of relative number of subscribing households for a downlink average busy hour data consumption of 540 GB/month and minimum downlink and uplink data rates of 10 Mbps and 1 Mbps, respectively.

Indoor CPEs are placed randomly within the buildings and use omni antennas with 0 dBi gain, while outdoor CPEs are placed at rooftop and employ directional antennas with ~14 dBi. Outdoor CPEs have their antennas directed toward the serving base station antenna. Note that the three last cases shown in the figures employ the same fixed 8 × 8 antenna area, whereas the other cases have 8 × 1, 8 × 2, and 8 × 4 antenna arrays. Note also that MU-MIMO, when configured, is only used for the downlink and not for the uplink in these results.

As opposed to typical MBB scenarios, MU-MIMO provides significant gains already for relatively few radio chains, especially with rooftop mounted CPEs, which leads to increased likelihood of line-of-sight propagation. The MU-MIMO gain over SU-MIMO ranges from ~30% to ~70% for the 8T $(8 \times 1)_{SA}(1 \times 4)$ setup considered here. Furthermore, due to the special characteristics of the FWA deployment with very limited CPE angular spread (and channel angular spread) in the vertical domain, large subarrays can be used and there are very limited gains of beamforming in the vertical domain.

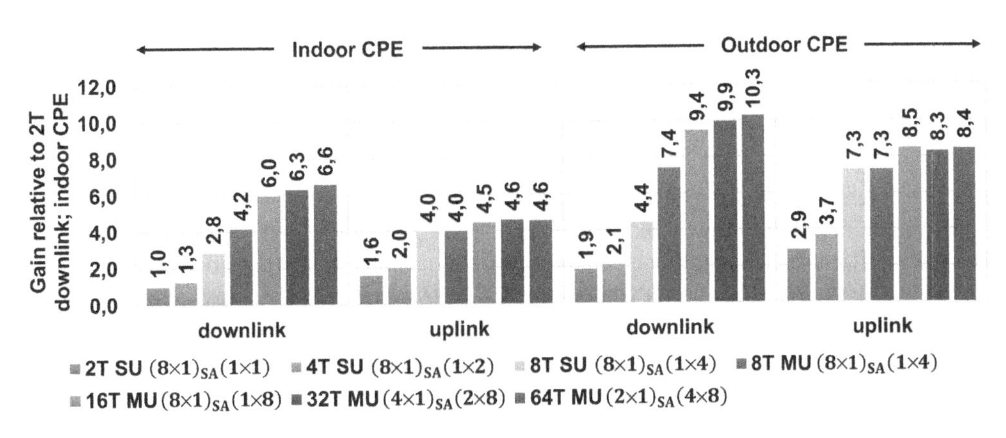

FIGURE 14.17

Relative number of subscribing households given a downlink average busy hour data consumption of 540 GB/month and minimum downlink and uplink data rates of 10 Mbps and 1 Mbps, respectively, in a typical suburban scenario with 1000 m ISD. SU and MU stand for SU-MIMO and MU-MIMO, respectively, both employing ZF.

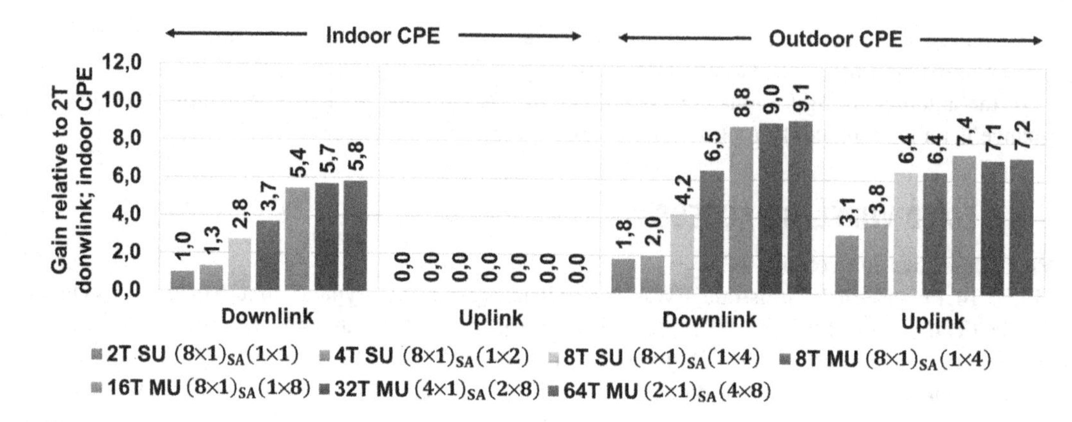

FIGURE 14.18

Relative number of subscribing households given a downlink average busy hour data consumption of 540 GB/month and minimum downlink and uplink data rates of 10 Mbps and 1 Mbps, respectively, in a typical suburban scenario with 2000 m ISD. SU and MU stand for SU-MIMO and MU-MIMO, respectively, both employing ZF. Note that uplink with indoor CPEs cannot fulfill the FWA requirement of 1 Mbps.

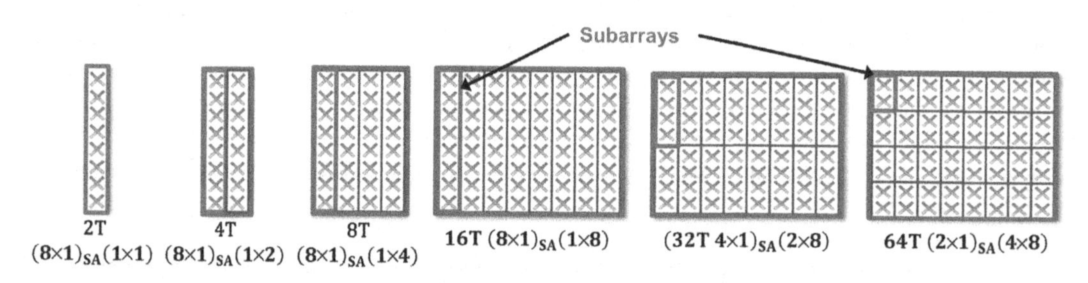

FIGURE 14.19

Illustration of antenna configurations used in Figs. 14.17 and 14.18.

It is also interesting to see that opposed to typical MBB deployments, the uplink is in many cases less strained than the downlink. For typical mid-band TDD MBB deployments, achieving good uplink performance is significantly more challenging than achieving good downlink performance, especially for cell-edge users, while for FWA, it is rather the other way around. The exception is the case of indoor CPEs with 2000 m ISD, in which case the uplink cannot any longer fulfill the FWA requirement of 1 Mbps for any of the antenna configurations.

14.3.5 SUMMARY AND DEPLOYMENT RECOMMENDATION

A summary of some key AAS attributes in different scenarios is given in Table 14.1. Keep in mind that results depend on numerous assumptions and should be therefore be seen as indicative.

Table 14.1 Summary of Some Key AAS Characteristics in Different Scenarios.

		Dense urban High-Rise	Urban Low-Rise	Rural/Suburban	Fixed Wireless
Key characteristics		Large vertical spread Interference limited	Medium vertical spread Interference limited	Small vertical spread Coverage limited	Small vertical spread
Subarray layout	For fixed antenna area (8×8)	Small (1×1) or (2×1)	Medium (4×1)	Large (8×1)	Large (8×1)
	For fixed number of subarrays	Wide angular coverage area (2×1) or (4×1)	Medium angular coverage area (4×1) to (8×1)	Small angular coverage area (8×1) to (16×1)	Small angular coverage area (8×1) to (16×1)
Algorithm considerations		Interference-awareness Reciprocity / feedback	Interference-awareness Reciprocity / feedback	Feedback	Interference-awareness Reciprocity / feedback
Antenna orientation (array of subarray layout)		Horizontal ~ vertical	Horizontal > vertical	Horizontal >> vertical	Horizontal >> vertical
Deployment recommendation for an 8×8 antenna		64 TR, 2×1 subarrays	32 TR, 4×1 subarrays	16 TR, 8×1 subarrays	16 TR, 8×1 subarrays

14.3.5.1 Dense urban high-rise versus rural

As an example for how different scenarios favor different AAS solutions, consider the two diverse scenarios, dense urban high-rise and rural (see Sections 14.3.1 and 14.3.3 for descriptions).

Dense urban high-rise represents a capacity driven and interference demanding scenario where AAS with advanced beamforming and MU-MIMO excels. Frequency-selective and interference-aware UE specific beamforming is key, and many horizontally and vertically stacked subarrays (many radio branches) help utilizing the spatial richness of the scenario.

Rural, on the other hand, represents a coverage demanding scenario where the antenna area is key and where horizontal domain beamforming is superior to vertical domain beamforming. Hence, horizontal stacking of as many large subarrays as possible is preferred from a performance point of view.

An illustration of the choice of subarray in these two scenarios is shown in Figs. 14.20 and 14.21. Recall that the AAS configurations were described in Section 14.3 and illustrated in Fig. 14.6.

As the angular UE distribution in the vertical domain is large in the dense urban high-rise scenario, small enough subarrays are required in order to have sufficient angular coverage area for UE specific beamforming.

The rural scenario, on the other hand, has large ISDs and less tall buildings, hence the required vertical coverage area is small. This means that larger subarrays fit the scenario well, and using small subarrays with more radio chains, which typically means increased cost, do not add much in terms of performance.

14.3.5.2 Capacity gains

Fig. 14.22 illustrates the capacity gain by using different antenna configurations relative to 2T for different MBB scenarios. Recall that each doubling of the number of T:s from 2T to 16T, that is, going from 2T to 4T, 4T to 8T, and 8T to 16T, results essentially in a doubling of the number of antenna columns (see Fig. 14.6). Some of the conclusions are as follows:

- from a capacity point of view, each doubling of the number of antenna columns yields roughly 30%–50% gain using SU-MIMO and up to ~100% gain using MU-MIMO in MU-MIMO friendly cases such as dense urban high-rise with large data packet size. Note though that gains are typically subject to diminishing returns;

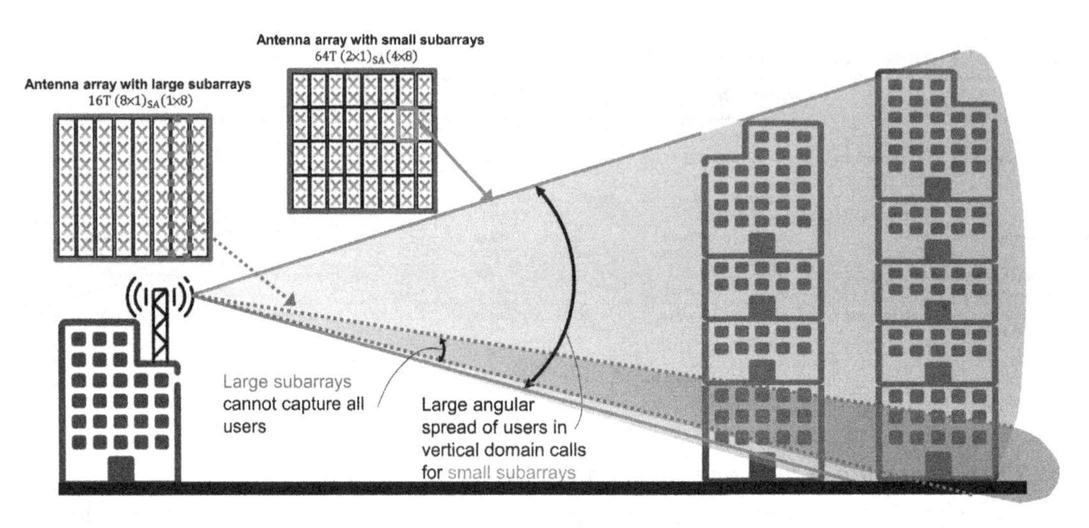

FIGURE 14.20

Illustration of the vertical coverage area in the high-rise scenario. Small subarrays are required to cover the deployment. Vertical domain beamforming provides significant benefits.

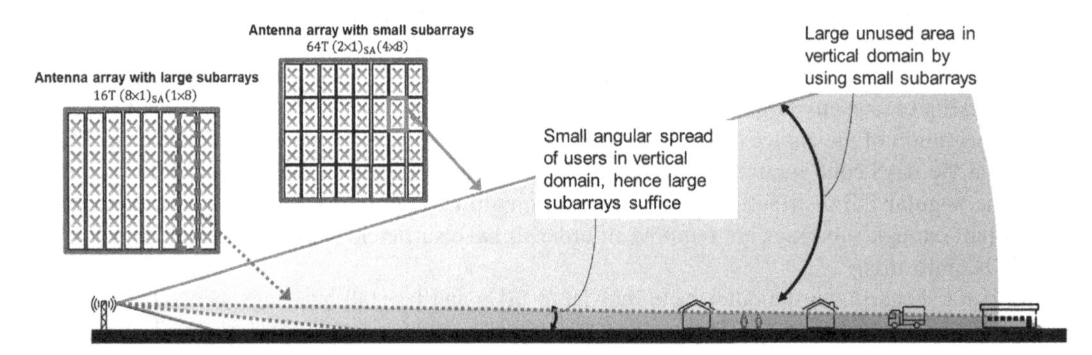

FIGURE 14.21

Illustration of the vertical coverage area in the suburban/rural scenario. Large subarrays have better gain and fit the scenario better than small subarrays. Vertical domain beamforming provides small gains.

- the gain of vertical domain beamforming is highly scenario dependent and ranges from almost zero for suburban/rural up to 100% for dense urban high-rise;
- results in Fig. 14.22 clearly demonstrate that the gains from a specific antenna configuration depend on the scenario. Given the antenna configurations considered here, the gain in suburban/rural saturates at 16T, that is, there is no gain of dividing the 8×1 antenna columns into smaller subarrays. The corresponding sweet spot for low-rise is 32T, that is, there are gains by splitting the antenna columns in half into 4×1 subarrays. In the high-rise scenario, there is no saturation, that is, the best performance is achieved by letting the subarray equal a single dual-polarized antenna element pair.

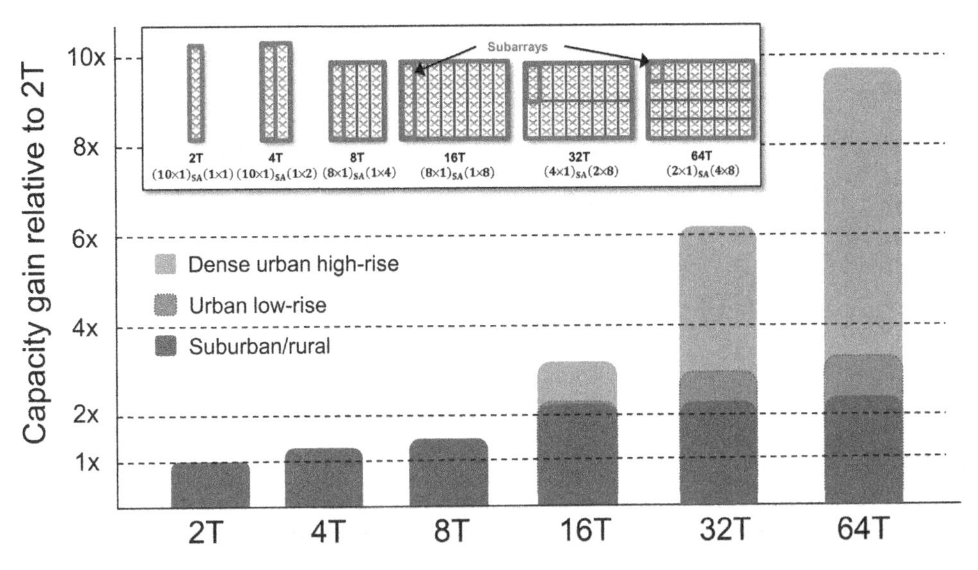

FIGURE 14.22

Illustration of capacity gains from various AAS configurations in different MBB scenarios. 2T, 4T, and 8T use SU-MIMO while 16T to 64T employ MU-MIMO. Antenna configurations are illustrated in Fig. 14.6. Note that results are indicative and some assumptions, for example, data packet size and algorithm impairments, are rather optimistic, especially for MU-MIMO performance.

It needs to be highlighted that gains in this comparison come from several sources, such as increased antenna area and more radio chains with associated SU-MIMO and MU-MIMO gains. In particular, the dense urban high-rise deployment scenario without in-building solutions and large data packet sizes is a perfect example where advanced multi-antenna solutions excel and can offer substantial gains.

14.3.5.3 Number of layers

Fig. 14.23 shows the distribution of number of scheduled layers as seen from the scheduler. For most MBB scenarios, rather few layers are typically scheduled, while for FWA deployments, especially with rooftop mounted CPEs, more layers are used. UEs have mostly two receive antennas and often use rank 2, hence the number of simultaneously scheduled UEs is less than the number of layers (often by a factor two). Nevertheless, dimensioning for more layers does not automatically translate into improved network performance (see Section 13.6.4).

14.4 DEPLOYMENT CONSIDERATIONS

Site deployments in real networks are often associated with constraints, which have an impact on the way new technologies are adopted in practice. Some of the most critical practical deployment constraints will be discussed in what follows.

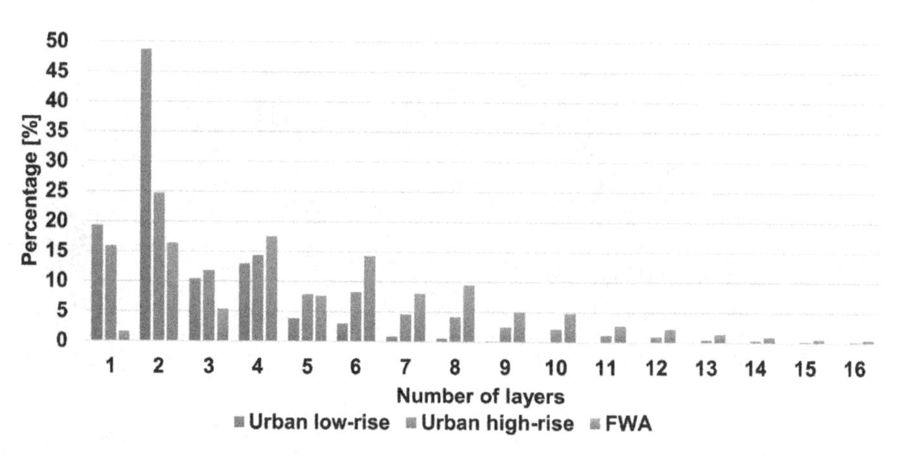

FIGURE 14.23

Illustration of the distribution of number of simultaneously scheduled layers for three different scenarios. All results are based on 32T with a 2 × 8 array of subarrays. FWA uses rooftop mounted directional CPEs.

14.4.1 SITE ASPECTS

Analysts estimate a global physical cell site count in the order of seven million today, a number that is expected to grow over time. The site density varies significantly between regions and countries, and each site (type) has its own requirements and constraints.

Adding new (macro) sites is a costly and a cumbersome process. Hence, there is a desire to reuse existing sites to as large extent as possible. At the same time, there is limited space for the deployment of antennas and auxiliary equipment at existing sites, and different operators, technologies (2G, 3G, 4G, and 5G), and features compete in many cases over the same limited physical site space.

Different markets and scenarios impose different deployment constraints. Some general trends/observations/remarks are as follows.

In many markets, there are visibility constraints that limit the number of antenna panels per operator and site to roughly two or three, which puts high requirements on the capability of each panel (see Section 14.4.2).

Furthermore, for each antenna array, there are strict requirements on size and weight dictated by construction constraints (e.g., each site is built to support a maximum weight or wind load) and regulated in site rental contracts. There are several aspects affecting the physical size and weight of an antenna panel, including

- The antenna array layout and frequency are fundamental aspects that impact the physical size. An illustration of this is given in Fig. 14.24. In this case, for a fixed antenna width of at most 0.4 m, it becomes very challenging to apply horizontal domain beamforming for low frequency bands as the number of antenna columns is limited to one or maybe two in most wide area deployments. Note that this does not prevent splitting up the antenna column(s) into smaller subarrays, thereby enabling vertical domain beamforming. However, as discussed in

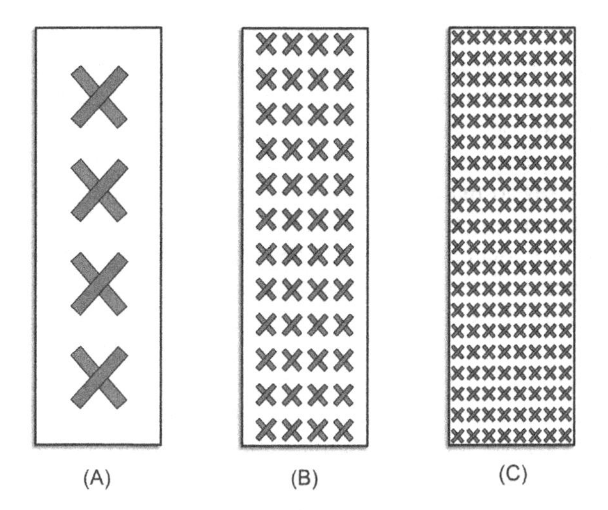

FIGURE 14.24

Illustration of the number of antenna elements with 0.5λ and 0.65λ horizontal and vertical element separation that fits a 1.3×0.4 m^2 antenna at (A) 700 MHz, (B) 1800 MHz, and (C) 3.0 GHz. Here λ denotes the wavelength that is roughly 43 cm, 17 cm, and 10 cm at 700 MHz, 1800 MHz, and 3.0 GHz, respectively.

Section 14.3, while horizontal domain beamforming is very useful to improve performance, the benefit of vertical domain beamforming is highly scenario dependent. Allowing the width of the antenna to grow beyond the typical ~0.4 m would significantly improve the beamforming potential also for lower frequency bands. This will, however, increase the visual site impact and wind load but might be acceptable in special deployments where the additional capacity is sought. Moving up in frequency makes advanced beamforming much more feasible and, as discussed in Section 7.2, sometimes even a necessity to maintain coverage;

- An active antenna with the radio integrated into the antenna has higher weight and larger size than a passive antenna where the antenna, radio, and baseband are separate physical entities;
- An active antenna requires cooling, which adds to the weight and size. The more output power the more cooling, hence increased size and weight;
- Increasing the number of frequency bands and the number of T:s and R:s supported by a single antenna panel increases generally the size and weight (see discussion in Section 14.4.2).

Today's antennas typically follow "standardized" sizes that reflect operators' deployment constraints. For example, fixed lengths of 1.3, 2.0, and 2.6 m and a width less than 0.4 m are common values.

14.4.2 MULTI-BAND

A key aspect affecting site deployment is the number of frequency bands that an operator is using. An operator has today typically several frequency bands deployed where each band requires its own radio(s) with associated filters. Whether different frequencies require separate antennas depends on the antenna technology and the frequency range.

An antenna supports a certain frequency range in which it works according to its specifications with respect to gain, main lobe width, etc. As an example, a typical lower mid-band macro antenna can today handle frequencies between roughly 1700 and 2700 MHz. However, deploying frequencies with large separations, for example, 800 and 2700 MHz, requires separate antennas, hence more equipment at site.

Multi-band support is key to manage the site deployment constraints. To construct such equipment is challenging from an engineering perspective, for example, building radios and antennas that cover a large range of frequency bands, combining AAS and conventional antenna systems in the same panel and to get desired performance on each band without degradation. Most passive antennas today have support for several bands and several ports. Examples of techniques used in multi-band antennas in order to minimize the impact on size are diplexing of radio chains and interleaved antenna elements. A diplexer is a passive device that performs frequency multiplexing, that is, it takes two input signals occupying different frequency bands and outputs a third signal consisting of both input signals (see Fig. 14.25 for an illustration of duplexing and interleaving).

Multiple frequency bands and advanced antenna technologies compete over the same antenna panel area, which is constrained as discussed in Section 14.4.1. By using advanced antenna building practices, including duplexing and interleaving, the situation is improved. Still, a choice whether to deploy several frequency bands or many T:s and R:s (radio chains) is often needed. This is illustrated in Fig. 14.26A, showing a passive 12 ports antenna with two diplexed low-band frequencies with 2 ports each and 8 ports dedicated to lower mid-band. These eight ports can be used to deploy four different lower mid-band frequencies with two ports each or to deploy one lower mid-band frequency with eight ports, or any mix in between. Fig. 14.26B illustrates an example of an antenna supporting 12 passive radio chains and an active AAS.

Needless to say, there is a trade-off between maximizing network performance and satisfying all site constraints, and to get cost-effective products, some performance might need to be sacrificed to comply with all other constraints. Already now there is a trend that more and more

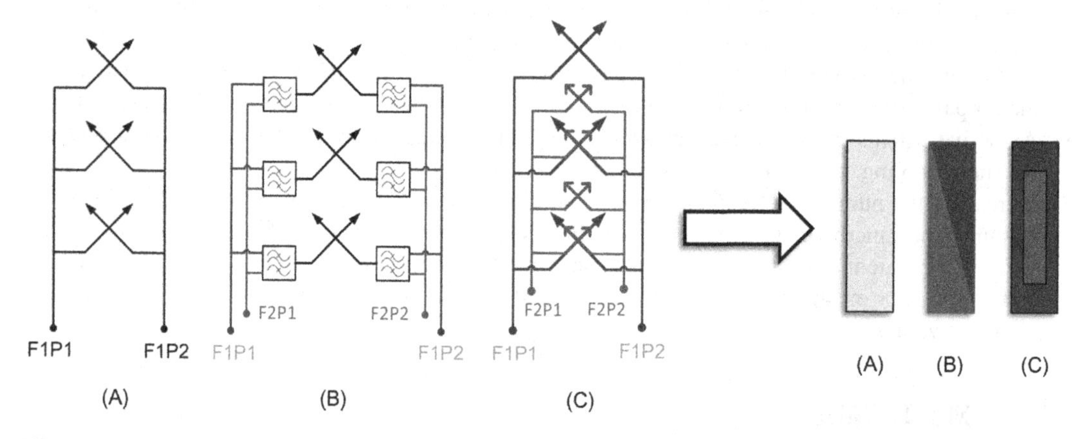

FIGURE 14.25

Illustration of (A) single-band elements, (B) multi-band elements based on diplexing, and (C) multi-band elements based on interleaving. Example of implementation on the left and schematic view on the right. F and P stand for frequency and polarization, respectively.

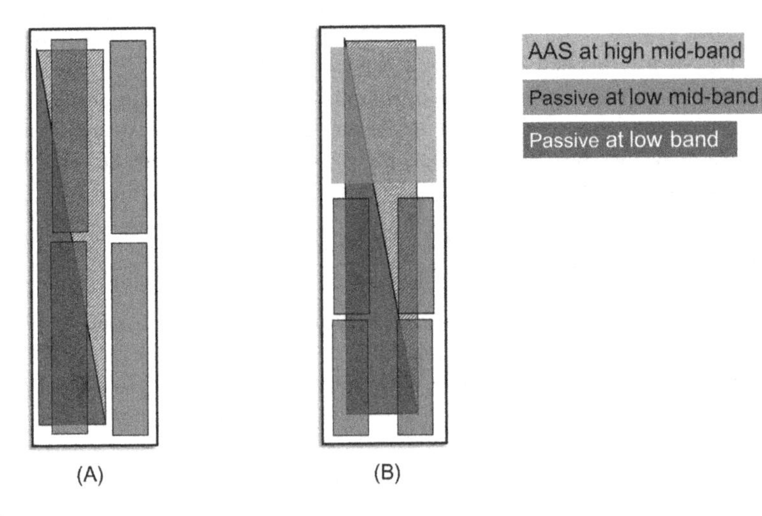

FIGURE 14.26

Illustration of (A) hexa-band, or 12 ports, antenna and (B) mix of AAS and passive antenna with support for 12 radio chains. Diplexing is used for low-band.

single-band antenna installations are replaced with only one or a few multi-band antenna systems. The benefit of this is that the total cost, size, and weight for one multi-band antenna are less than those for all the single-band antennas. The drawback is that performance tends to be worse for multi-band products compared to single-band products, but technology advances will hopefully reduce the performance gap.

14.4.3 DEVICE CAPABILITIES

History has demonstrated the importance of getting the standard right from the beginning, in the sense that advanced features standardized in later releases may never be implemented. One example of a feature that could boost the performance of advanced AAS techniques significantly in long term evolution (LTE) is TM10. However, since TM10 has very limited terminal penetration, it is not useful in practice.

As soon as the first deployment is launched, there will be legacy devices that need to be supported for a long time. This also makes it harder to introduce new advanced functionality with sufficient uptake to really impact the network performance and therefore motivate the investments. With the introduction of new radio (NR), there is an opportunity to ensure that good AAS technology becomes mainstream and supported by most of the devices from the start (see Section 10.1.3 for a discussion on mandatory and optional UE capabilities for NR that are relevant for AAS). For example, many NR AAS features, for example, Type II CSI (channel-state information) feedback or antenna switching for reciprocity, are mandatory with capability signaling (see Section 10.1.3). Hence, it is important to have a dialogue between device and network vendors to ensure that these features are being implemented and that interoperability device testing for these features is available. At the same time, legacy devices supporting older technologies will dominate the device fleet in networks for many years, hence legacy technologies need to evolve to handle the increased capacity need before NR has reached a substantial device penetration.

The device capability in the network has a profound impact on an operator's evolution strategy as it determines how many of the operator's subscribers will benefit from an upgrade. Network vendors, UE vendors, and operators need to align their respective strategies at some level to ensure well-performing networks and sound businesses.

14.4.4 REGULATORY CONSTRAINTS

Different regions of the world have different regulatory requirements that need to be fulfilled. For example, there may be limits on the radio emissions from a site toward the public, which may give implications on where antennas can be located, how much beamforming gain can be allowed, and what output powers can be used. Such requirements may also influence the choice of site solution (see also Section 11.8).

14.5 EXAMPLES OF OPERATOR NETWORK EVOLUTIONS

The role of AAS in the general network evolution will be highlighted in the sections below. The technique will be put in perspective and as will be seen, AAS will be one important component in the network evolution, but not the only one. AAS must fit together with existing radio solutions for the foreseeable future.

The use of AAS in MBB-only deployments as well as when FWA and MBB are codeployed will be described for some different US and European deployment cases. Note that mature Asian deployments are on a high level similar to the European case in terms of site density and spectrum holdings.

There is a general need for the evolution of mobile operator's radio network due to increasing MBB traffic demand and UE experience expectations (see Chapter 2: Network Deployment and Evolution).

The mobile network area capacity depends basically on two main factors, namely

- number of sites per km^2;
- amount of deployed spectrum per site or sector.

These two key factors are often different in different parts of the world due to regional spectrum strategies but also as operator deployment strategies can be different. Therefore, it is useful to study network evolution from a regional perspective, that is, taking the operator's site and spectrum assets into account. It is also beneficial to study the network evolution over time as decisions taken will have an impact or have dependencies of the evolution choices or steps in the coming years.

14.5.1 US OPERATOR EVOLUTION FOR MOBILE BROADBAND

14.5.1.1 Spectrum assets

In the United States, typical operator spectrum assets are

- low-bands below 1 GHz: ~20 MHz frequency division duplex (FDD) in two to three bands below 1 GHz. For example, it can be allocations in the 600, 700, and 850 MHz bands;

- mid-bands such as 1–2 GHz bands. A total of ~40 MHz FDD is not uncommon in bands such as 1900 MHz (personal communications service), 2100 MHz (advanced wireless service) and 2300 MHz (wireless communications service).

This means that about 60 MHz FDD is available for 2G, 3G, and 4G for most operators. There are exceptions where, for example, a large allocation of mid-band spectrum is available today that can be used for both 4G and 5G. The description here will exclude this case as it is also more similar to the European situation (see Section 14.5.2).

There is also a possibility to use the 5 GHz Wi-Fi band for 4G License Assisted Access (LAA), that is, 4G is coexisting with Wi-Fi under the same conditions [1]. This is a way to achieve very high peak rates in the order of 1 Gbps by aggregating 60–100 MHz.

The initial 5G deployments will use

- mm-wave bands in the 24–39 GHz bands. Allocations of 200–1000 MHz per operator are not uncommon.

There is also a TDD band in the 3.5 GHz band, parts of citizens broadband radio system (CBRS) [2] that will be auctioned in 2020. Due to sharing with other services/incumbents, this band will have relatively low output power allowed (maximum 47 dBm EIRP per 10 MHz). There are also two different types of allocations: *Priority Access License*, which is like a licensed type of band, and *General Authorized Access* (GAA), which will be used as an unlicensed band (see Fig. 14.27).

The CBRS is intended for 4G initially but is also planned to be used for 5G later.

14.5.1.2 Site assets

With regard to macro site assets, the US operators typically have sparser macro site grids compared to other operators, for example, those in Europe and Asia. This can be due to several reasons. The first is that analog and digital systems were deployed to offer mainly outdoor and in-car voice services along roads using low frequency band such as 850 MHz. Indoor coverage was not the main focus for voice as buildings often already used fixed line telephony. Charging for voice services was also different in the United States compared to other countries; the receiving party was charged for a voice call. One can suspect that therefore phones were turned off when a subscriber was reachable by a fixed phone. As a result, a sparse macro site grid was enough for voice services

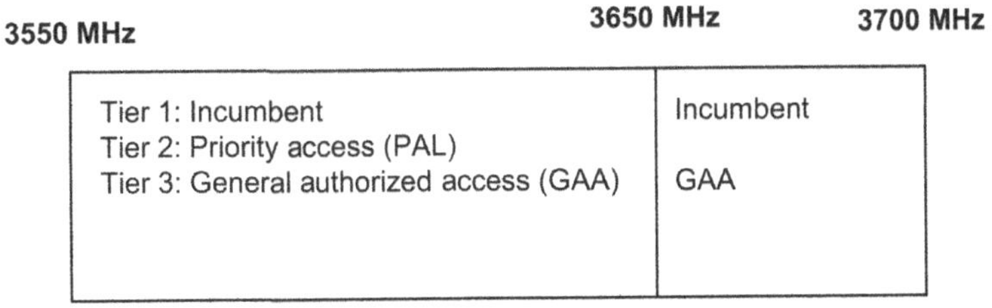

3550 MHz – 3650 MHz	3650 MHz – 3700 MHz
Tier 1: Incumbent Tier 2: Priority access (PAL) Tier 3: General authorized access (GAA)	Incumbent GAA

FIGURE 14.27

The CBRS spectrum band that is shared between different users (tiers) where incumbent has the highest priority and GAA the lowest priority.

outdoor in populated areas. The 850 MHz spectrum band allocations (e.g., 2 × 12.5 MHz) allowed for introducing 3G (e.g., using 5 MHz) along with 2G without need for more sites. Moreover, the 700 MHz low-band was introduced with 4G in the United States, which further made it possible to offer smartphone services using existing site grids.

The smartphone traffic and service offerings have been driven by the US Silicon Valley from companies like Apple, Google, and Facebook. This has resulted in the largest monthly traffic per average smartphone compared with other regions in the world (see, e.g., [3]). There is also a high penetration of premium smartphones such as iPhone supporting advanced radio capabilities such as LAA.

14.5.1.3 US urban network evolution

The sparse macro site grid and relatively high volume of traffic have resulted in that the US operators have deployed many frequency bands per site. Around 2019, more or less, all their available bands are deployed at macro sites in dense urban areas. This means 3–4 bands or around 50 MHz FDD for 4G and a smaller allocation for 2G/3G in the low-bands to maintain coverage. In some areas, street or small cells have started to be utilized in order to add capacity. The existing spectrum assets are hence more or less fully utilized at macro sites, and options to add AAS in existing bands or new sites can be considered.

A driver for introduction of small/street sites at this point is also that the LAA 5 GHz band for gigabit speeds is starting to be available in UE. This band with limited output power needs to be deployed close to the users in order to be effective and is therefore a good fit to street/small cell sites. There are also reports that macro rooftop site rentals are becoming high in dense urban areas and that operators have started to look at alternatives to macro sites. There are initiatives from, for example, FCC to streamline the small cell deployment process by, for example, limiting the approval time and yearly site rental for small/street cell type of sites. There are millions of poles (power or lamp posts) in the United States that can be utilized for cellular network deployments.

To cope with the increasing 4G smartphone traffic, street sites and/or 4G FDD AAS can be used to improve capacity in dense urban areas. The 4G demand increase is expected to continue in the near future, that is, 2020–2021, before 5G offload is widely available. 5G has started to be introduced during 2019, but the effective capacity offload will be small due to low penetration of 5G capable smartphones the initial years. 5G is deployed in the mm-wave bands (24–39 GHz) at start, and it may take some years for smartphones with attractive stand-by times and form factors to enter the market.

The mm-wave will use AAS technologies in order to achieve high output power and compensate the high frequency propagation losses. It is expected that existing street sites can be reused for mm-wave as the coverage is limited for this high frequency band, and deployment close to the users is beneficial similar to LAA. Reusing existing sites is beneficial from a TCO perspective. Typically, backhaul and site rental dominate the street site TCO [4].

Selected macro sites are also expected to be useful for mm-wave deployment. For example, a macro site with line of sight to an outdoor hotspot, for example, busy square, would be useful to deploy.

The mm-wave frequencies have challenging propagation properties, and it may be difficult to cover outdoor areas when a lot of foliage is present. Moreover, covering indoor areas, especially buildings with thermal efficient windows, is problematic at mm-waves. Therefore, interworking through carrier aggregation or dual connectivity with FDD bands is recommended. Dynamic spectrum sharing has been introduced in the standards as a way for NR to coexist with LTE in the same carrier. This can be applied in the 4G mid- and low-bands to provide 5G coverage. However, 5G capacity benefits from the sharing may be limited in dense urban areas in cases where 4G bands are more or less fully loaded in this period due to strong 4G traffic growth and limited 4G spectrum.

It is expected that 4G smartphone traffic continues to grow when 5G has been introduced. Data demand per phone is increasing as mentioned before. The 3.5 GHz CBRS band can therefore be deployed for 4G (initially) in order to reduce the need for further addition of street sites to handle the growing 4G traffic but also to create headroom in the 4G bands for 5G when dynamic sharing is applied. The CBRS band is a good fit to existing street/small cell sites due to the relatively low allowed output power.

An attractive option to street site densification is to deploy 4G based AAS, for example, 16T, in the mid-FDD bands to improve 4G macro capacity and hence minimize the number of new street sites. Street sites are still needed in order to provide LTE gigabit services using LAA.

Eventually 4G traffic demand is likely to flatten out when more and more smartphones will become 5G capable. This will increase the 5G traffic demand and need for more 5G capacity. Majority of smartphone traffic is generated indoor, and mm-wave outdoor deployment has difficulties in providing good indoor offload. Massive densification of outdoor sites is not expected to be practical due to high loss buildings in US dense urban areas. Indoor deployment is of course an alternative for high-rise and large footprint buildings. Another solution to this would be to deploy a potential new upcoming TDD frequency band around 4 GHz. The 3.7–4.2 GHz band is being considered in the United States with large spectrum allocations in the order of 100 MHz. AAS 16T–64T would typically be targeted for this band depending on the environment (see Section 14.3). The band could be available in the 2021–22 time frame and would provide a good 5G traffic offload if deployed from existing macro sites using AAS with MU-MIMO assuming high power transmissions are allowed (and not low powered as the CBRS band). The use of AAS provides good coverage and high capacity from existing macro sites as outlined in Section 2.3.4. Further, the narrow 850 MHz low-band could also be refarmed from 3G to 5G to be used as a 5G anchor band to support deep indoor subscribers from outdoor sites at this time.

The 4 GHz capacity relief may be temporary due to the relatively sparse macro site grid assuming a strong 5G uptake in this region. Eventually the 4 GHz band will also need to be deployed at street sites. AAS is still an option for street deployment of this band. A supplementary action is also to refarm 4G spectrum bands to 5G at this time perspective.

Fig. 14.28 shows an illustration of the described network evolution. It shows the frequency bands deployed per site type (macro and street) over time as well as what technology is used per band, including AAS.

In summary, there are many types of outdoor sites (macro and street) and multiple bands per site. AAS technologies would be applied for 5G in mm-wave and 4 GHz bands for both macro and street sites. There is also an option to apply AAS in the FDD mid-bands to further enhance the 4G macro capacity if possible from a site deployment perspective, for example, size constrains. Multi-band support for an AAS product would be a way to mitigate the problem.

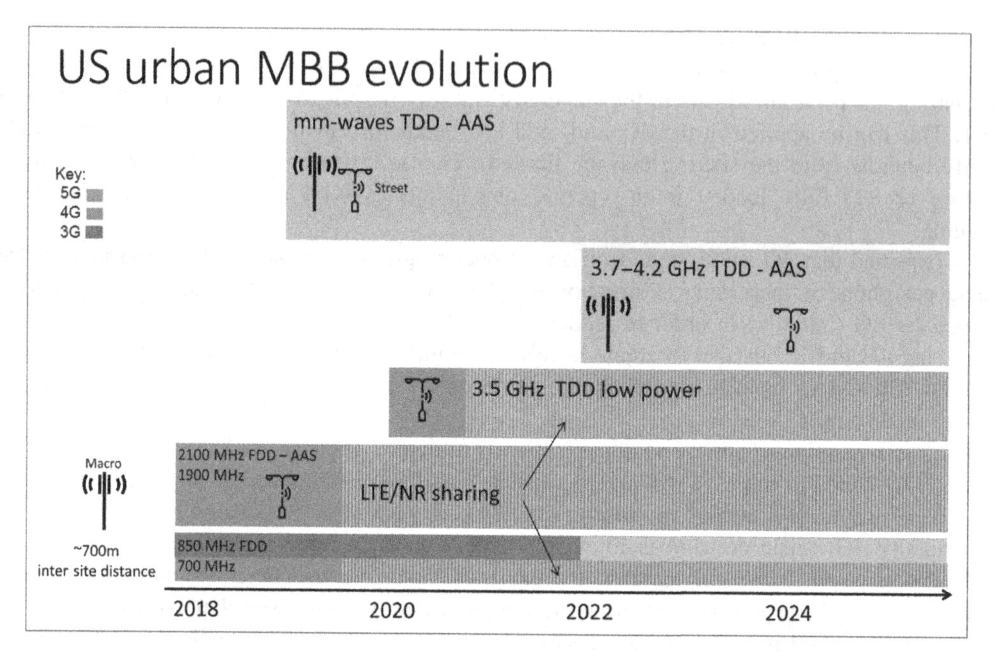

FIGURE 14.28

Illustration of dense urban network evolution showing bands, AAS, and radio access technologies deployed over time on a per site level to meet MBB demand. The radio network solution consists of both macro and smaller street sites.

14.5.1.4 US suburban network evolution

There are fewer sites in suburban areas due to lower subscriber density than in urban areas. An ISD of $\sim$2000 m or larger is common in the US densest suburban areas, that is, highly populated residential areas with around $\sim$3000 subscribers per km^2.

There can be up to three bands deployed at macro sites for 4G in 2019, for example, 700 MHz FDD, 1900 MHz FDD, and 2100 MHz FDD. With 3G still deployed in 850 MHz, there is not much capacity headroom to meet the growing 4G smartphone traffic.

Site densification is the conventional solution for more 4G capacity when there is spectrum exhaustion. Densification through macro tower additions is not popular in residential areas, and zoning requirements limit the antenna height. Instead, street sites may be a solution that has become acceptable and already used by some operators. Similar to urban areas, AAS for 4G FDD in 1900–2100 MHz FDD is a potential complement to street site densification that will limit the number of new street sites.

In suburban areas, street site deployment will be at a height that is often below the average tree height. 5G based on mm-wave is sensitive to foliage, and mm-wave only is likely not a cost-efficient solution for general wide area MBB coverage in residential areas with heavy foliage.

A solution for the initial 5G coverage is to apply LTE/NR dynamic sharing in the low and mid-FDD bands in order to provide similar peak speeds through carrier aggregation. Street site densification can provide capacity headroom for both 4G and 5G until the upcoming 4 GHz band targeted for AAS becomes available. 4 GHz is introduced on existing macro sites with AAS for coverage reasons (as described in Section 2.3.4). 16T AAS is a good option here (see Section 14.3). Further densification with street sites is prevented because of the AAS and wide 4 GHz band.

Refarming 850 MHz from 3G to 5G as in urban areas is a way to create a 5G anchor band in order to achieve range expansion for 4 GHz through carrier aggregation.

In summary, the dense suburban network evolution may include street site densification and mid-band FDD AAS upgrades for 4G to provide more capacity but also initial 5G services through LTE/NR sharing in low and mid-FDD bands using carrier aggregation. 4G FDD AAS is a means to avoid street sites. The new 4 GHz TDD band expected to be suitable for wide area 5G coverage from macro sites using AAS technologies will be a good solution mid-term wise, but additions of this band at existing street sites may be needed long term in densest suburban areas as the macro site grid is sparse. The suburban network evolution is illustrated in Fig. 14.29.

14.5.1.5 US rural network evolution

Due to deployment costs, there is seldom full area coverage in rural and wilderness areas. Instead, major roads and villages are typically covered, and there can be coverage holes, that is, blank spots,

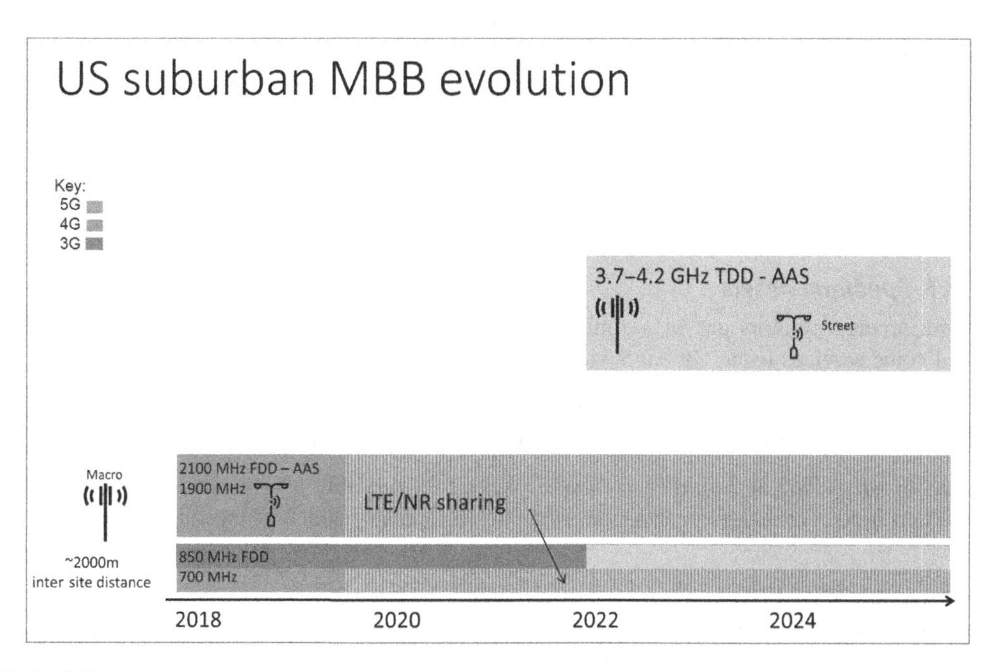

FIGURE 14.29

Illustration of suburban network evolution showing bands, AAS, and technologies access deployed over time on a per site level to meet MBB demand. The radio network solution consists of both macro and smaller street sites.

between sites in many cases. It is very common that mobile operators are targeting population coverage only.

The traffic demand in rural areas is low, and the need for network evolution to support higher capacity is smaller here. At some point, new technologies must be supported, especially if new services are built on this. Adding bands for supporting higher user experience targets, needs, however, to be considered, especially due to competition.

Today, sparse suburban and rural deployments in the United States consist usually of two deployed bands for LTE, for example, 700 MHz low-band as a coverage anchor and 1900 MHz mid-band for carrier aggregation to provide higher throughputs. Such a deployment would fulfill the capacity needs for LTE up to 2025, given current forecasts in case of population densities of ~50–100 subscribers per km^2.

When 5G is introduced in the same area, there is a possibility to complement this deployment to improve performance, and there are at least two options. The first option (option 1) is to add undeployed low and mid-bands, for example, 700–850 MHz and 2100 MHz that are commonly available and use LTE/NR sharing with carrier aggregation so that the investment would benefit both LTE and NR users. LTE/NR sharing is also an attractive solution for offering 5G services initially when 5G smartphone penetration is small, for getting basic 5G coverage. The other option (option 2) is to deploy a new 4 GHz AAS band for NR that can provide good coverage and higher user throughput performance from one single radio since the 4 GHz mid-band has a large bandwidth.

If FWA is to be offered on top of MBB, 4 GHz with AAS would be a good solution. The focus on cost is high in rural and will likely determine what solution would be most attractive in the end. 16T AAS is a good option for this deployment (see Sections 14.2–14.4). Competition is also a factor that will determine what strategy to choose. Fig. 14.30 shows the network evolution for this case.

14.5.2 EUROPEAN NETWORK EVOLUTION FOR MOBILE BROADBAND

14.5.2.1 Spectrum assets

In Europe, many operators are in a similar situation in terms of spectrum assets. The operators launched voice services using Global System for Mobile Communications (GSM) at

- low-band 900 MHz to provide coverage and later added;
- mid-band 1800 MHz for GSM capacity, while 3G was introduced at 2100 MHz.

Later 3G was added at 900 MHz to boost coverage in rural but also in dense urban, that is, to provide deep indoor coverage. A few new bands were typically added at the same time as 4G was launched:

- low-band 800 MHz FDD as a coverage band; and
- mid-band 2600 MHz both FDD and TDD as capacity bands. There are also some upcoming mid-bands such as the unpaired 1400 MHz supplementary downlink (SDL).

For 5G,

- 3.5 GHz TDD is being considered as the main band but also 2300 MHz TDD;

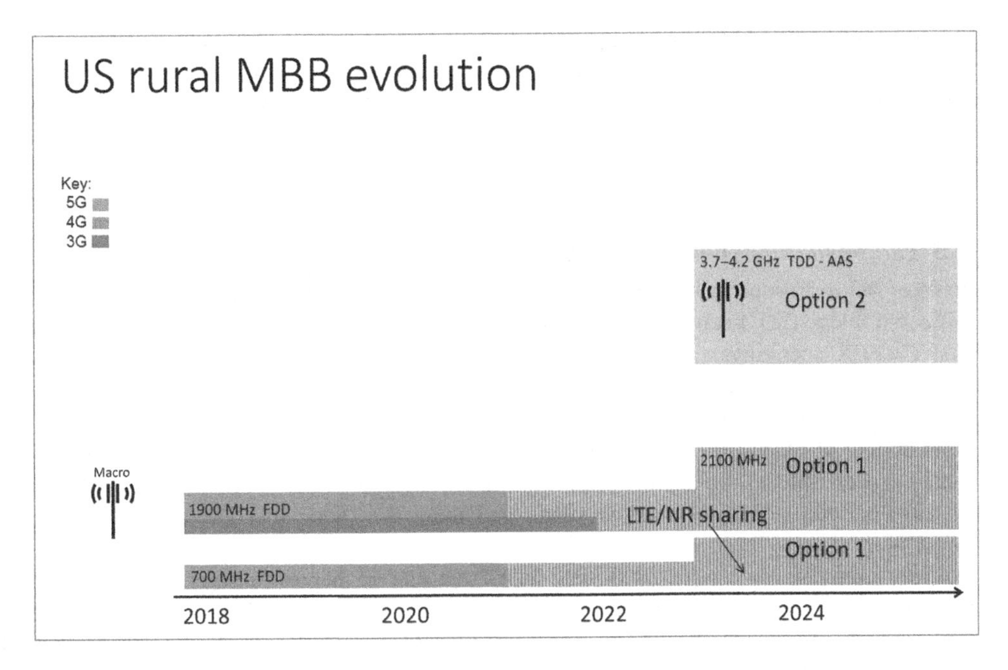

FIGURE 14.30

Illustration of rural network evolution showing bands, AAS, and radio access technologies deployed over time on a per site level to meet MBB demand.

- 700 MHz FDD as a coverage band (although the bandwidth is typically small, for example, 10 MHz FDD);
- 26 GHz based mm-wave is also studied to be allocated as a band for extreme capacity.

The initial idea for the 3.5 GHz band was to have an operator allocation of 100 MHz, but in some market, this may be problematic initially due to other services (e.g., radar) using the band today.

At the end of 2019, some 5G 3.5 GHz auctions in Europe have resulted in bandwidths of 20–130 MHz TDD or approximately 80 MHz in average per operator in some European markets.

It is not uncommon that European operators have a total spectrum asset of more than 100 MHz FDD or close to that (in FDD and TDD bands) before the introduction of 5G that will come with almost the same amount of TDD spectrum initially.

14.5.2.2 Site assets

The focus on indoor coverage for 2G was more apparent than in the United States, which resulted in a denser macro site grid generally. This was further emphasized with the introduction of 3G at 2100 MHz. Many 3G spectrum licenses came with coverage commitments, or coverage was a way to compete in application beauty contents. Still there are differences between the site density of different operators in Europe due to different coverage strategies.

The large amount of spectrum per operator in combination with a generally dense macro site grid puts European operators in a good situation to provide high mobile network capacity. The situation with large assets of FDD spectrum and dense macro site grids is similar for many Asian operators. Hence, the evolution of Asian operators is expected to be comparable to that of European operators on a high level.

14.5.2.3 European urban network evolution

The uptake of 4G in Europe was much slower than in the United States due to several reasons, for example, economical (3G licenses were costly in many European countries, which reduced the operators' CAPEX spendings) and not as intense 4G launch competition. The European macro deployments are quite capable, that is, dense and spectrum rich, and the traffic demand per smartphone in Europe is lower than in the United States according to Section 2.2.1.

A slow uptake of 5G in Europe relative to the United States is not unlikely to happen due to, for example, lower handset replacement rate than in the United States. This means that 4G traffic demand will continue to grow for some years, and there is a need to continue to deploy more capacity for 4G.

A possible option is to deploy 2600 MHz TDD that can be deployed early. The band is supported in many UEs today and will hence be beneficial from start, but the band can also be targeted for AAS deployment as described later. The already deployed 900 and 1800–2100 MHz (only 2100 MHz is illustrated below) will also be refarmed from 2G and 3G to 4G to provide better coverage and capacity.

An alternative to 2600 MHz TDD would be to deploy, for example, 1400 MHz SDL early due to better deep indoor coverage properties. The handset support of the 1400 MHz band is however not as good as for the 2600 MHz TDD band. 1400 MHz is a downlink only band and requires carrier aggregation operation with another band, for example, 800 MHz. The coverage of 2600 MHz TDD is worse, but this can be compensated with multi-antenna techniques or using low-band carrier aggregation as the SDL band. Hence, an attractive option can be to deploy 2600 MHz TDD using AAS techniques in order to avoid the additional deployment of the 1400 MHz band for 4G.

The European operators will typically try to acquire a large amount of TDD spectrum in the 3.5 GHz band intended for 5G services. There are also upcoming auctions of the 700 MHz FDD band that can be deployed (if acquired) as a 5G anchor band for deep indoor coverage. Interworking through carrier aggregation between, for example, 700 MHz and 3.5 GHz is beneficial as seen in Section 13.7.6. Some European operators have, however, quite large allocations in 900 MHz band, and this could be used for 5G as an alternative to the 700 MHz band.

Fig. 14.31 shows an illustration of the described network evolution for the European operator case. It shows the frequency bands deployed per macro site over time and the radio access technology used per band.

In summary, there are many bands per macro site. AAS is applied for the 3.5 GHz band to provide sufficient coverage (see Section 13.7.1 and [5]). The 3.5 GHz TDD spectrum together with AAS techniques such as MU-MIMO provides high capacity from the existing macro sites (see Section 2.3.4). AAS 32T–64T solutions would be deployed depending on the traffic distribution (see Section 14.3). In addition, there are some 3G/4G bands that can be refarmed to 5G when 5G is supported by most UE.

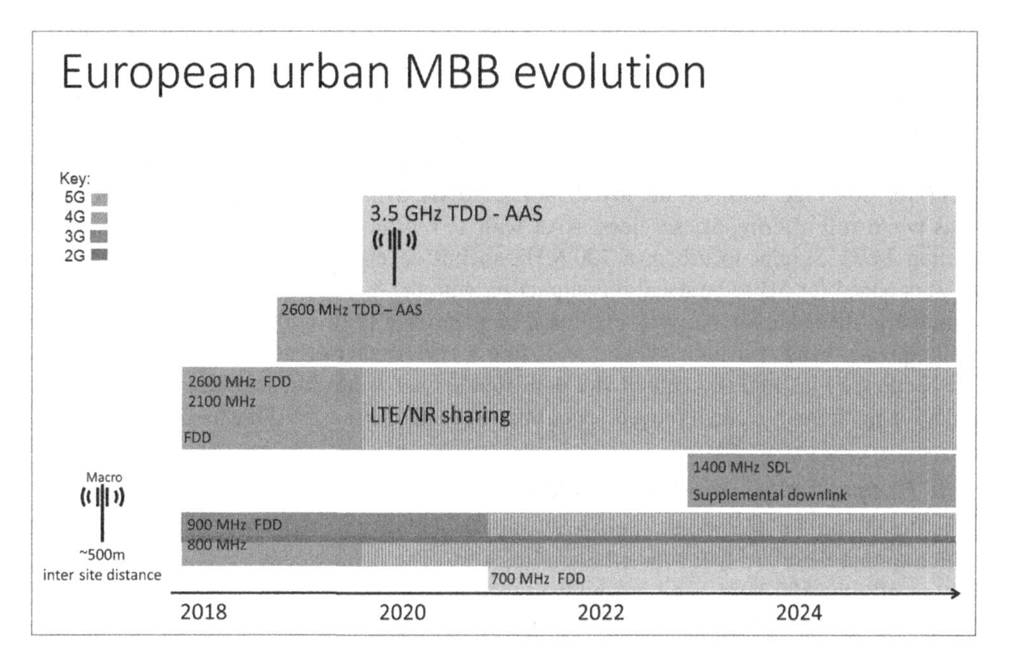

FIGURE 14.31

Illustration of European dense urban network evolution showing bands, AAS, and radio access technologies deployed over time on a per site level to meet MBB demand.

As a result, there is a limited site densification need foreseen generally (locally there can be areas where additional sites are needed) because of the AAS and new wide 3.5 GHz band (as described in Section 2.3.4). This means that macro site and backhaul infrastructure investments can be reused. A challenge is to fit all bands to the existing sites, that is, multi-band solutions are even more important. There is also an option to use 4G AAS in, for example, 2600 MHz, TDD band to reduce the need of deploying additional 4G bands. Adding street sites can be a solution to apply if the challenge with too many bands at sites becomes too difficult in practice.

14.5.2.4 European suburban network evolution

A deployment with ISD of about 1000 m is quite common in suburban and residential areas in Europe. Looking at one deployment situation, 800 and 1800–2100 MHz bands (only 2100 MHz is illustrated below) are currently deployed to provide coverage and capacity for 4G in suburban areas.

For 2G and 3G services, 900 MHz is typically used. There are hence undeployed bands available, for example, 1400 MHz SDL and 2600 MHz FDD and TDD.

Similar to the urban situation, a slow uptake of 5G is expected. This means that 4G must continue to evolve. Deploying additional 4G bands may be required within some years. Apart from refarming 900 MHz from 3G to 4G to increase the low-band bandwidth, there is the choice of using, for example, 2600 MHz FDD or 1400 MHz SDL. From an antenna system point of view,

2600 MHz may be preferred as it is close to 2100 MHz. The SDL 1400 MHz is in between the 2 GHz and the low-band 800–900 MHz.

The suburban 5G deployment will depend on the rollout strategy. Similar to the US scenario, it is possible to use LTE/NR dynamic sharing in mid- and low-FDD bands. There is capacity for both 5G and 4G since 4G is boosted using both refarmed 900 and 2600 MHz. Eventually the 3.5 GHz with AAS for coverage must be deployed, but LTE/NR dynamic sharing provides an option to delay this while still offering 5G services. AAS with 16T is an attractive solution here for 3.5 GHz (see Section 14.3). Similar to urban, a 700 MHz anchor coverage band (or alternatively a part of 900 MHz) is added for 5G to be used in conjunction with the 3.5 GHz band.

In summary, the suburban network evolution is macro site centric, that is, refarming and adding existing frequency bands for both 4G and 5G using LTE/NR sharing as well as using AAS for the 3.5 GHz band (see Fig. 14.32). Limited site densification needs are foreseen due to rather good site density and large amount of undeployed spectrum, especially when 3.5 GHz has been added.

14.5.2.5 European rural network evolution

Similar to the US rural deployments, European rural deployments consist of a minimal band configuration due to that there are few subscribers in the area. The deployment focus is on coverage, and for 4G, typically the 800 MHz band is deployed with a 10 MHz carrier, which is a common allocation in Europe. 2G and 3G may share the 900 MHz band typically. Given the good allocation of

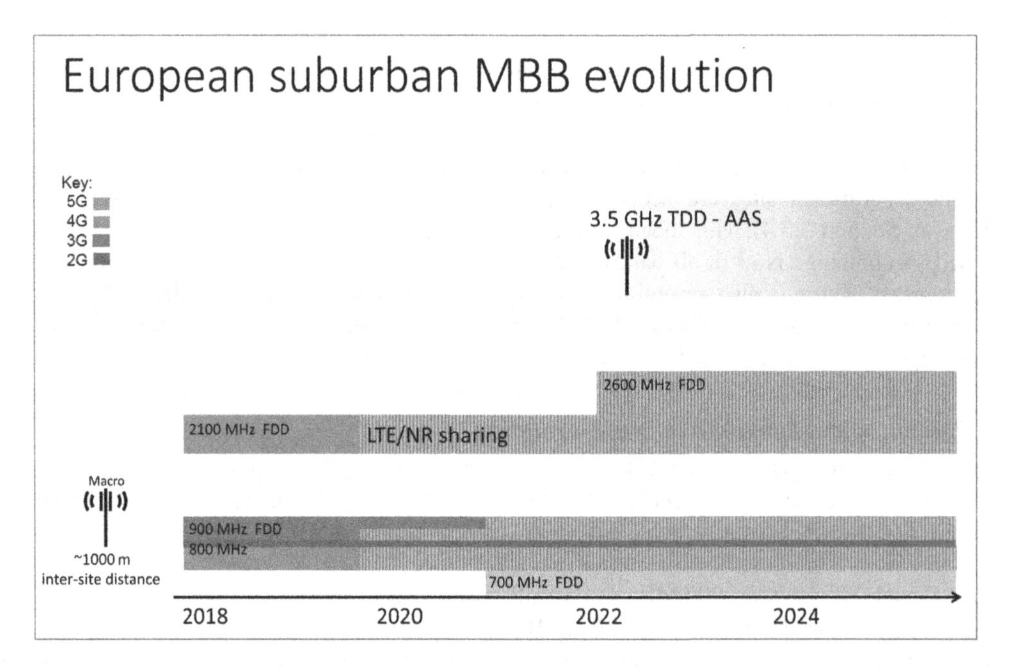

FIGURE 14.32

Illustration of European suburban network evolution showing bands, AAS, and radio access technologies deployed over time on a per site level to meet MBB demand.

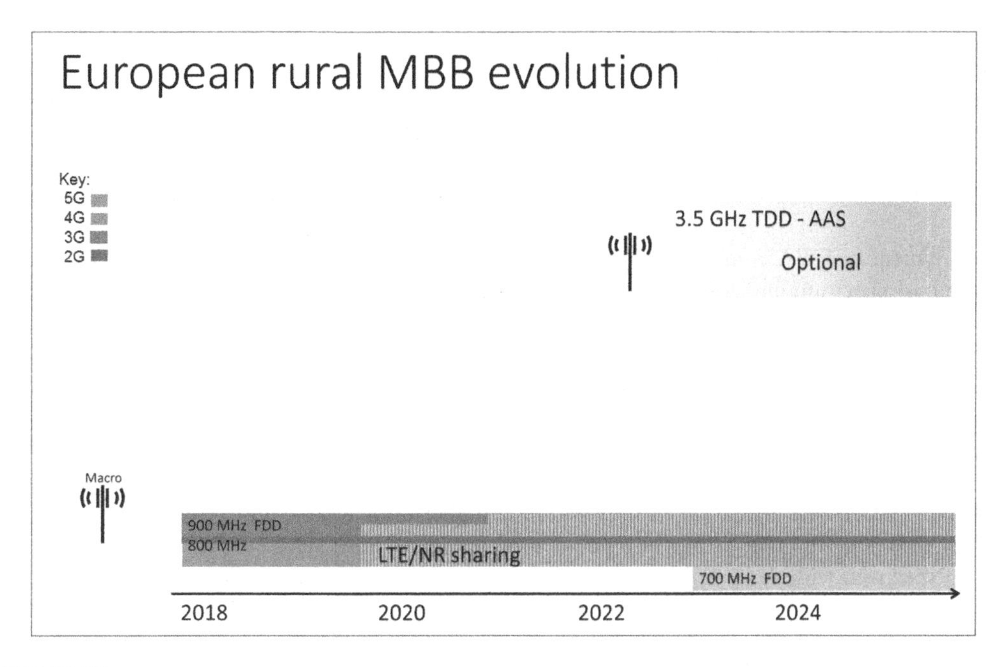

FIGURE 14.33

Illustration of European rural network evolution showing bands, AAS, and radio access technologies deployed over time on a per site level to meet MBB demand.

low-band spectrum in Europe, including the 700 MHz as an upcoming 5G coverage band, the evolution of rural deployment is quite straight forward (see Fig. 14.33).

Approximately 30 MHz low-band FDD would be available long term for proving good coverage and capacity for MBB-like services. 5G based on a 3.5 GHz AAS solution, for example, 16T, could be an option if very high user experience levels are desired or additional FWA services are to be offered in rural areas. The AAS deployment cost will have a large impact on the final solution.

14.5.3 DEPLOYING FIXED WIRELESS ACCESS ON TOP OF MOBILE BROADBAND

The MBB demand per smartphone is currently in the order of 10 GB per month (average ~0.045 Mbps in busy hour) in 2019 (see Section 2.2.1). For FWA, the demand per household is expected to be much higher. It is not uncommon that there are several persons in a household sharing the broadband access, and there is also a trend to use fixed broadband connection for video delivery or over-the-top TV.

In Ref. [6], a data volume estimation of 200–300 GB per household is assumed in a fixed wireless case where terrestrial or satellite services are used for linear TV. This means 20–30 times higher demand in case of FWA relative to an average MBB smartphone user.

The case with over-the-top TV and video included in a fiber substitute FWA offering would result in even higher demands, for example, more than 1000 GB/month or 7–8 Mbps demand per household in the busy hour for a 2017 US scenario [7]. Assuming a suburban household density of 1000 households per km^2 and a desired market share of 30%, the busy hour area demand for an FWA service would be more than 2 Gbps/km^2 (8 Mbps × 1000 households × 0.3 market share) in the latter case, which is much larger than typical MBB in dense urban areas today. The MBB demand would be 40 Mbps/km^2 (0.045 Mbps × 1000 households × 3 persons × 0.3 market share) in the same area, assuming three persons per household. FWA is 50 times higher than MBB in this area.

Deployment of FWA would hence need to support a very high area capacity. Deploying a large bandwidth of spectrum and AAS are two key capacity enablers for FWA. The largest bandwidth is available in the high frequency bands above 3 GHz where also AAS would be utilized from a coverage perspective. High capacity with AAS can be achieved with MU-MIMO.

Today most mobile spectrum is owned by operators using it for MBB services, and the upcoming new wide bands for 5G are expected to be used for MBB also given the large uptake of MBB traffic. Sharing spectrum between FWA and MBB is hence a natural way to go [6], especially considering the deployment and spectrum costs. Adding spectrum for FWA, which can also be used for MBB, improves the efficiency. It implies that a higher utilization and thereby higher capacity is possible. This is due to that a "bigger pipe" is used compared to several smaller individual "pipes." Splitting spectrum with dedicated bands for MBB and FWA results in an efficiency loss. Moreover, the site infrastructure such as tower and backhaul can be reused when FWA is introduced. The investment made for FWA can be reused by MBB services in the FWA nonbusy hours and vice versa.

Deployment needs for an FWA and MBB codeployed case with an FWA offering without over-the-top TV for a European country town were analyzed in Ref. [6]. The solution involves adding undeployed 2.6 GHz band and 3.5 GHz using multi-antenna solution (8T) with MU-MIMO to cope with both MBB and FWA demands with up to 60% market share.

A similar analysis for MBB and FWA with over-the-top TV for an American suburb was presented in Ref. [3]. Fiber like performance was targeted for FWA. To meet FWA and MBB expectations in this scenario, the deployment solution included adding 3.5 GHz and mm-wave bands at existing and some new street sites. MU-MIMO is an important component for the 3.5 GHz band.

In summary, the high capacity demand for FWA from households' video usage will need large spectrum bandwidths and AAS or multi-antenna technologies and features.

14.6 AAS SELECTION TAKING COST EFFICIENCY INTO ACCOUNT

As discussed in Chapter 2, an operator's strategy concerning how to evolve the network is based on business decisions and as discussed in the previous section the evolvement is either

- Stepwise: The operator has coverage and capacity in an area that needs to be evolved;
- One-swap mode: The operator is shifting from one technology to another like from 2G/3G to 4G and now toward 5G or is installing new equipment for new spectrum.

AAS type of radios are efficient tools for both types of expansions as they offer increased capacity, coverage, and end-user throughput. One of the main benefits of installing new AAS equipment for the operator is easier access to new technology and spectrum at higher frequency bands due to better coverage. The main drawback is the higher hardware cost per product. Cost efficiency is therefore a fundamental factor when setting network deployment strategies, and the overarching goal for operators is usually to shift *operational expenses* (OPEX) cost into *capital expenditures* (CAPEX) cost. Hence, when and where to deploy AAS will be based on cost-versus-performance measures according to some chosen principles.

There are several ways of calculating cost-versus-performance. The method described below is based on cost-per-offered-capacity. This is one approach for measuring cost efficiency, and other relevant methods exist.

14.6.1 HARDWARE COST

A very simplified hardware cost model illustrated in Fig. 14.34 is adopted for the analysis to follow. The starting point of the line in the figure is a 2TR radio, and the slope of the line is highly dependent on the technology used at the time for the acquisition of radio equipment, the frequency spectrum used, the number of branches in the radio, etc. Over the years, the cost for a 2TR radio has been significantly reduced, and the cost for each radio chain increment has been decreasing. How this will evolve in the coming years is not the topic for this book, but one observed trend is that Moore's law [8] that has been valid for 50 years has started to saturate. As a consequence the tremendous growth in feature richness but continued decrease in both product cost and footprint will slow down, compared to the early days of telecom. Before elaborating on cost-per-offered-capacity, the methods for measuring cost and capacity are described.

14.6.2 TOTAL COST OF OWNERSHIP

Cost can be measured in several ways, for example, cost-per-AAS-radio, TCO-per-AAS-radio, TCO-per-site, and TCO-per-network. The reason for using TCO is that the real cost for an operator is so much more than only the hardware enabling coverage and capacity. The choice of solution will therefore be affected by the cost efficiency evaluation.

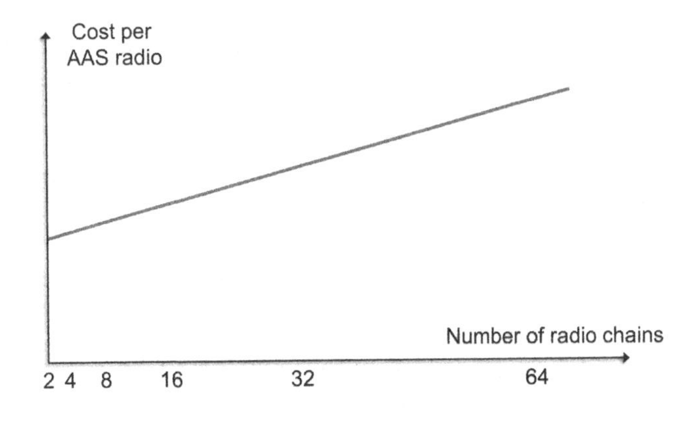

FIGURE 14.34

Simplified example of hardware cost as a function of the number of radio chains.

CAPEX typically includes

- Hardware cost of radio, baseband, antenna, and site equipment;
- Software license cost;
- Network rollout, that is, the total project cost to plan, procure, deploy, and enable a total functional site.

OPEX typically includes

- Power consumption;
- Site rental; if the operator is forced to expand the current grid, the site-rental cost will rise significantly. One objective for operators is therefore to keep the ISD and the current cell structure of the sites;
- Support cost;
- Transmission, that is, fronthaul between radio and baseband and backhaul between the baseband and the core network.

TCO is the total cost over an investment cycle, that is, TCO = Total CAPEX + Total OPEX over the investment cycle (typically around 6 years).

As shown in the cost per radio chain graph (see Fig. 14.34), an AAS will have higher CAPEX than a traditional 2TR radio. Furthermore, AAS will have OPEX depending on, for example, site rental and power consumption cost.

14.6.3 TOTAL COST OF OWNERSHIP AND NETWORK CAPACITY

A common approach in network design is to dimension the network to offer a certain required capacity in a specific geographical area. As shown in Fig. 14.22, the efficiency in terms of offered capacity for a specific radio type is highly dependent on the deployment scenario. This affects the performance versus cost analysis as will be illustrated by a simple example in what follows next.

Fig. 14.35 illustrates how TCO per-offered-capacity depends on the number of radio chains in an urban low-rise deployment scenario. Inputs to the analysis are the network capacity evaluations presented in Section 14.3 and Fig. 14.22 and the postulated cost model in Section 14.6.1. In this scenario, the 32T AAS comes out as the most cost-efficient choice compared to 16T or 64T. Similarly, for the 64T AAS, the cost increase is higher than the capacity increase. In other deployment scenarios, the outcome could change, that is, less or more radio chains become most cost efficient.

The aim with this strongly simplified example is to highlight the importance of the network view. A more capable radio product will inevitably cost more, but it can also offer better coverage, capacity, and end-user throughput that are essential for an efficient network evolution. Also, a more costly product can be more cost efficient in terms of TCO per-offered-capacity. Obviously, assumptions are very much simplified in this example; for example, it is assumed that the TCO and capacity are evaluated in a single-band scenario for a given site grid. An alternative to using AAS to increase the capacity is densification, as described in Section 2.3.3.2. This is however usually very costly and is hence considered as a second step. The TCO per-offered-capacity evaluation will then be considerably more complex and is beyond the scope of this book.

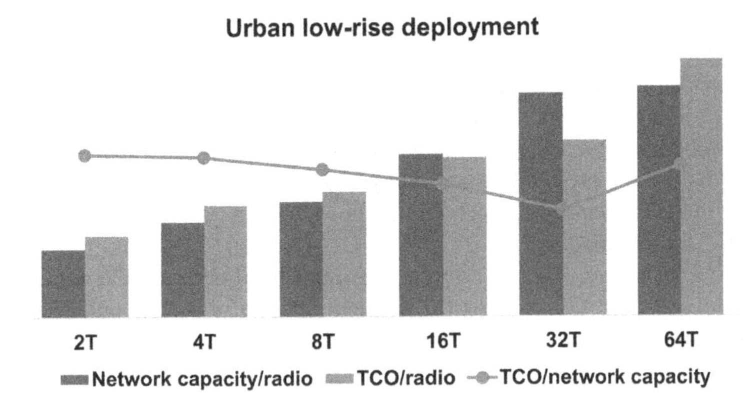

FIGURE 14.35

Shows that given the assumptions made in this example, the 32T has the best (lowest) TCO-per-offered-capacity ratio for the urban low-rise scenario.

14.7 SUMMARY

The mobile broadband deployments are today continuously being evolved toward higher capacity and higher user speeds, and this is expected to continue in the future. As a result, TDD frequency bands above 2.5–3 GHz are now being targeted for 5G, and AAS will have a major role in these bands especially in urban and suburban areas. Because of AAS beamforming techniques, the coverage from existing sites on these new high bands can easily match the coverage on existing bands. Interworking with low-bands through, for example, dual connectivity or carrier aggregation is important for uplink performance. In addition, high capacity is also provided with AAS, making AAS a tool for fully exploiting existing site investments and minimizing the need for new sites.

Upper mid-band deployments (e.g., 3.5 GHz) are very valuable on existing grids when used together with lower bands, but also mm-waves (e.g., 28 GHz) provide clear benefits in deployments with good enough coverage; for example, cases exhibiting line of sight to buildings, small cells, FWA, backhaul, outdoor-to-outdoor, and indoor-to-indoor.

AAS will also be important in order to handle 4G traffic growth before 5G offload is effective (initially when 5G handset penetration is small) while minimizing the need for costly new sites. For this purpose, AAS is expected to be introduced in FDD bands above 2 GHz. AAS will likely not appear in bands below 1 GHz as solutions will be too bulky and impractical to deploy.

There is a clear scenario dependency on the performance benefits of different AAS solutions. While horizontal domain beamforming is always beneficial, the benefit of vertical domain beamforming is highly scenario dependent and, in many cases, fixed beamforming in the vertical domain suffices. Another scenario dependent aspect is the benefits of advanced MU-MIMO. For example, dense urban high-rise with outdoor-to-indoor coverage (assuming no or little indoor systems) represents a capacity driven and interference demanding scenario where AAS with advanced beamforming and MU-MIMO can excel. Rural, on the other hand, represents a coverage demanding scenario where the antenna area is key, SU-MIMO is more important than MU-MIMO and where horizontal domain beamforming is superior to vertical domain beamforming.

In addition to performance considerations, deployment constraints and cost aspects also influence the network evolution strategies and specifically the choice of AAS solution. The cost-per-bit for different AAS products depends on the capacity need and the deployment scenario. The general trend is to go for multi-band products where one panel can replace several panels and hence have many deployment related advantages. Each site is, however, unique, and the site solution must be adapted to its specific conditions.

Solutions that might have a disruptive impact of multi-band deployments include the ability to construct multi-band products with support for wideband radios, wideband antennas, interleaving of active and passive antennas, combining of FDD and TDD, etc. in a cost-effective manner.

REFERENCES

[1] T. Cheng, License assisted access: operation principles. <https://www.ericsson.com/en/blog/2015/2/licensed-assisted-access-operation-principles> (accessed 25.09.19).

[2] Federal Communications Commission, 3.5 GHz band overview. <https://www.fcc.gov/wireless/bureau-divisions/mobility-division/35-ghz-band/35-ghz-band-overview>.

[3] Ericsson, Ericsson mobility report – November 2018. <https://www.ericsson.com/en/mobility-report/reports/november-2018/mobile-traffic-q3-2018> (accessed 17.06.19).

[4] 5G Americas, Small cell siting challenges and recommendations. 5G Americas and Small cell forum, August 2018. <http://www.5gamericas.org/files/9815/3547/3006/195_SC_siting_challenges_final.pdf> (accessed 17.06.19).

[5] F. Kronestedt et al. The advantages of combining 5G NR with LTE at existing sites. Ericsson Technology Review, November 2018. <https://www.ericsson.com/en/ericsson-technology-review/archive/2018/the-advantages-of-combining-5g-nr-with-lte> (accessed 17.06.19).

[6] H. Olofsson et al. Leveraging LTE and 5G for fixed wireless access. Ericsson Technology Review, August 2018. <https://www.ericsson.com/en/ericsson-technology-review/archive/2018/leveraging-lte-and-5g-nr-networks-for-fixed-wireless-access> (accessed 17.06.19).

[7] J. Wellen et al. Sustained throughput requirement for future residential broadband service. 2017 SCTE ISBE Cable-Tec Expo, Denver Co.

[8] Moores law on silicon cost and computer power. <https://en.wikipedia.org/wiki/Moore%27s_law> (accessed 25.09.19).

CHAPTER

SUMMARY AND OUTLOOK

15

15.1 SUMMARY

Multi-antenna technologies have been used in many industries for several decades. In mobile communication networks, multi-antenna technologies reached wider global spread with the introduction of 4G, where such technologies were standardized and implemented from the start. These technologies were gradually developed and became more advanced over time to evolve into what is here referred to advanced antenna systems (AAS), that is, highly integrated multi-antenna solutions with a large number of radio chains. In 5G, support for AAS is an intrinsic, standardized technology component. AAS is therefore expected to be commonly used as one key technology in mobile networks from now and in the future.

Coverage and capacity are the main drivers for AAS. Increasing coverage ensures efficient reuse of existing sites by providing additional coverage on new, and higher, frequency bands on mid-band (see Fig. 15.1) and to get maximum coverage for new sites on high-band. AAS can compensate for higher path loss and, as these bands typically use time division duplex (TDD), also the additional loss due to limited time for uplink transmission. AAS can also increase the benefits of new sites by providing better coverage than other solutions. The capability to provide additional capacity in high-traffic areas both for existing and new bands, see Section 2.3.4, is also a strong driver for deploying AAS.

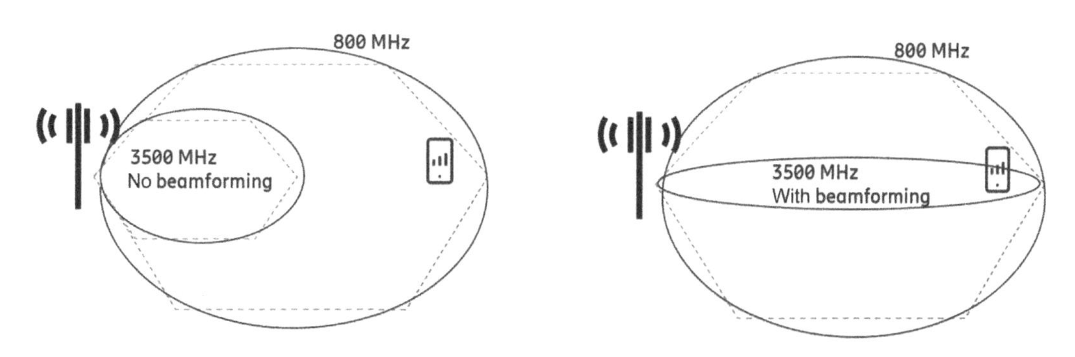

FIGURE 15.1

Illustration of coverage benefits from beamforming at 3.5 GHz relative to 800 MHz coverage.

Advanced Antenna Systems for 5G Network Deployments. DOI: https://doi.org/10.1016/B978-0-12-820046-9.00015-0

5G spectrum is available on both low/mid-band and high-band in most countries, and there is work ongoing to allocate even more spectrum for 5G use cases. AAS solutions will therefore be used in many segments of mobile networks around the world.

Introduction of AAS has also created a paradigm shift considering the definition and testing of AAS radio requirements where the traditional conductive approach has shifted to over-the-air (OTA) approach for definition and testing. The OTA approach has implied new metrics and considerations for requirements. Furthermore, sophisticated OTA testing methods using different OTA chamber types are needed.

In this book, a holistic view of AAS and their performance in real networks has been provided. To do so, the knowledge and experience from several areas in science and industry have been combined.

15.1.1 ADVANCED ANTENNA SYSTEM PERFORMANCE

Multi-antenna techniques can be applied both to transmitter and receiver sides for both TDD and frequency division duplex (FDD). Multi-antenna techniques can offer increased coverage, capacity, and end-user throughput, and although all these metrics can be improved simultaneously, a specific solution may be optimized to favor one of the metrics over the others.

The most efficient way to achieve higher spectral efficiency is to make use of beamforming (see illustration in Fig. 15.2 and related discussion in Section 6.1). Single-user MIMO (SU-MIMO), which makes use of beamforming, can provide substantial performance benefits and often results in capacity gains in the order of 30%–50% per doubling of the number of columns of the antenna

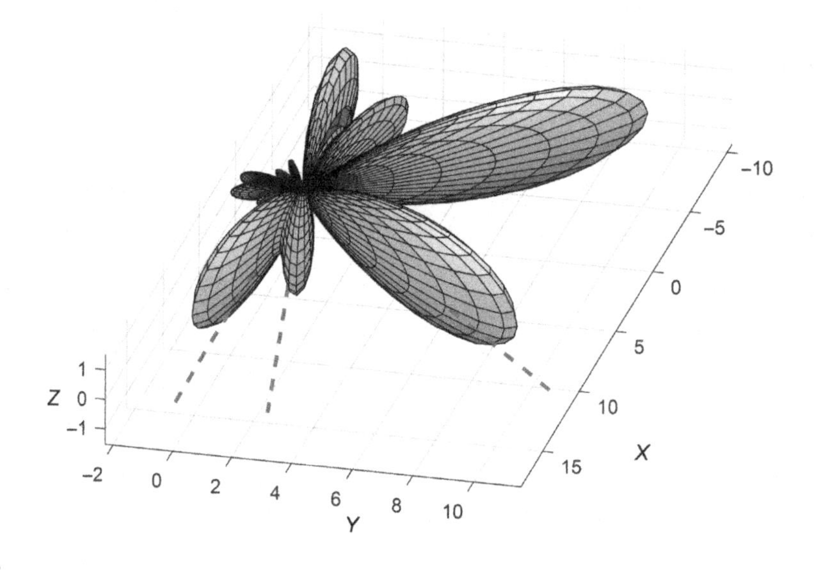

FIGURE 15.2

Beamforming example, where the beam shape is adapted to the channel, in this case three signal paths from the transmitter to the receiver, to provide optimum performance.

array in urban macro scenarios (see Section 13.4). Multiple user multiple-input multiple-output (MU-MIMO) is a capacity feature that can bring significant gains under favorable conditions (see Section 13.6).

The antenna array structure, that is, the number of antenna elements and the subarray partitioning, determines the degrees of freedom available for the supported software features and can have a profound impact on, for example, the angular coverage area or the number of supported MIMO layers. It is therefore beneficial to adapt the array structure to the given deployment, in order to optimize the AAS cost versus performance for a given form factor (see Fig. 15.3). Particularly, the level of vertical domain UE specific beamforming required has an impact on the subarray structure in the vertical dimension.

Scaling up the number of radio chains in AAS base stations gives rise to several challenges in terms of architecture, hardware, and software complexity. Different architectures and multi-antenna algorithms have been developed to embrace the deployment and usage scenarios foreseen by AAS.

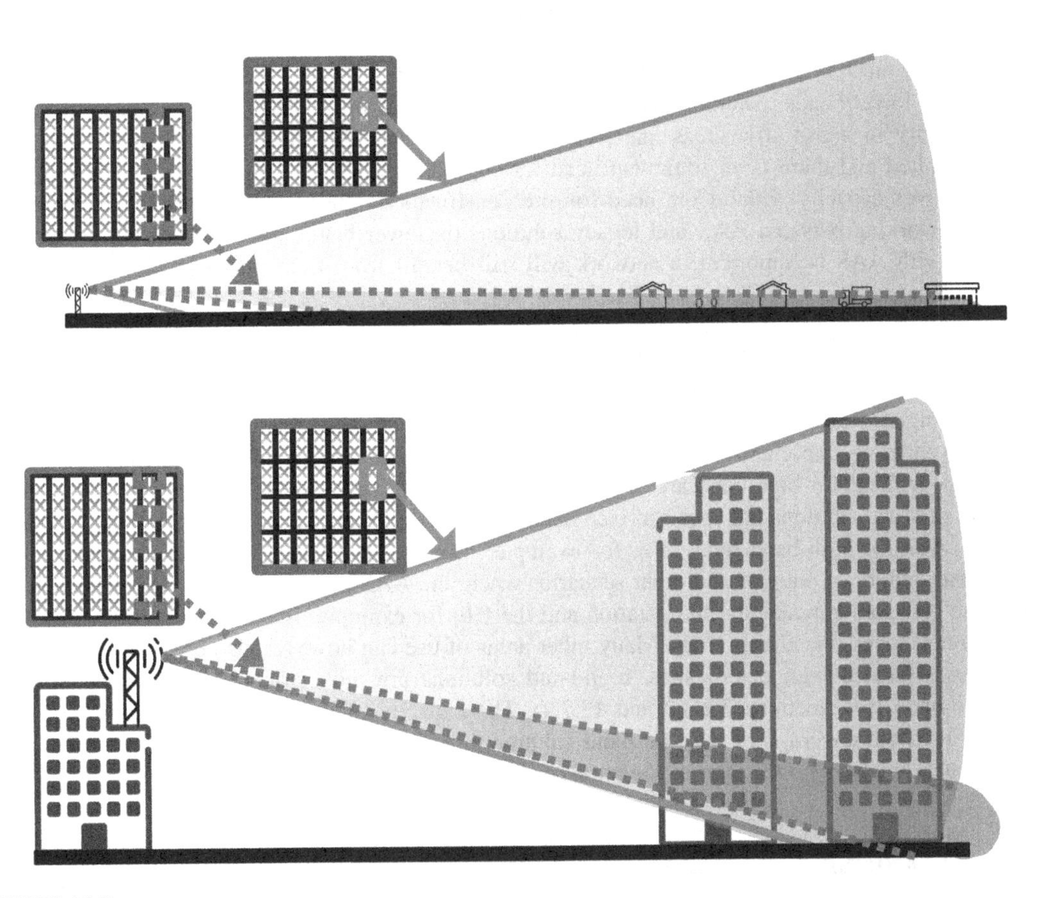

FIGURE 15.3

Examples on subarray structures and how they affect the coverage area.

15.1.2 BUSINESS VALUE AND USE CASES

The success of any new technology depends on how competitive it is with respect to business value compared to alternative solutions. AAS can offer higher performance than conventional non-AAS systems but comes at a higher cost as it is a more advanced product and contains more hardware and more sophisticated software solutions in each radio unit.

The industry interest in multiple-antenna solutions with a large number of antennas was relatively limited during the 2G, 3G, and early 4G deployment periods as other solutions with fewer radio chains were more attractive from business and deployment perspectives. With 5G, there are several new frequency bands defined both on mid-band and high-band and, as described above, AAS can bring benefits by enhancing coverage and by supporting higher capacity on both new and existing bands.

AAS facilitates efficient site reuse when adopting bands at higher frequency, which are usually TDD based. The alternative costs of adding new sites to provide coverage are substantial. AAS has therefore been an important technology component in order to make efficient use of the new 5G bands. For example, when deployments are available on 1.8, 2.1, or 2.6 GHz and a new deployment is considered on 3.5 GHz. AAS can then help to improve the coverage on 3.5 GHz to match the coverage of lower bands. Another important use case for AAS on mid-band is to enhance the network capacity in high-traffic areas, particularly in areas where the spectrum availability on mid-band is limited and there is an imminent need for capacity expansion. Using AAS on the existing sites improves capacity without the need for site densification. There are also significant benefits from interworking between AAS and legacy solutions on lower bands. Even though the improved coverage with AAS is important, a network will still benefit from tight interworking between the different bands.

On high-band frequencies, there are large chunks of spectrum available for mobile communication, and it is therefore in principle possible to support very high capacity on these bands. Coverage is however more challenging on high-band than on mid-band due to for example higher path loss, higher penetration loss, and lower output power levels (see Section 7.2.2). AAS is therefore an even more important tool to improve coverage for high-band than for mid-band. The preferred deployments for high-band are somewhat different from those for mid-band. Also, somewhat different product solutions are required (see Sections 12.3.1 and 12.3.2). Some of the most promising use cases for high-band today are, for example, mobile broadband (MBB) in relatively open areas either indoor or outdoor and other scenarios where the AAS transmitter and receiver can overcome the path loss between the base station and the UE, for example, fixed wireless access (FWA) or integrated access backhaul (IAB). Many other areas of use can however also be envisioned.

When colocated with macro sites, high-band solutions are most effective when interworking with mid-band (see Sections 7.2.3.2 and 13.7.4). The capacity of the high-band solution working together with the coverage of the mid-band solution provides a solution with overall performance that is much greater than the sum of the parts of the two solutions acting separately.

15.1.3 CLARIFICATION ON SELECTED ISSUES

Multi-antenna technologies are advanced solutions, and many factors contribute and interact to influence the network performance. It is therefore almost inevitable that there is some discrepancy in the industry understanding of certain aspects. Some highlights concerning ongoing discussion topics in the industry are reflected below.

15.1.3.1 Effectiveness of multiple user multiple-input multiple-output

MU-MIMO is a technique for increasing capacity (see Section 6.8). This can be shown both in simulations and in various test and trial setups. The effectiveness of MU-MIMO does however require that certain conditions are fulfilled, for example, high load and high signal-to-interference-and-noise ratio (SINR) for all users. When these preconditions are fully met, MU-MIMO is very efficient and very large performance improvements can be achieved. In most cases, these conditions are only partially fulfilled. In such scenarios, the performance improvements may still be significant, but lower than the impression often given by commonly used artificial demo setups and by the academic literature (see Section 6.8).

15.1.3.2 Effectiveness of vertical domain beamforming

Most UEs are spread in the horizontal dimension. Therefore, there is almost always a benefit of UE specific beamforming in the horizontal dimension. For the cases where the vertical spread of UEs is small, for example, in typical European cities and suburbs with predominantly low-rise buildings, the gains from UE specific beamforming in the vertical domain are limited. In dense urban areas with many high-rise buildings or, for other reasons, significant channel angular spread in the vertical domain, there may be significant benefits of UE specific beamforming also in the vertical domain, for example, when serving indoor UEs from outdoor base stations. Hence, the benefits of UE specific (horizontal and vertical) beamforming are only prominent if the UEs are spread accordingly (see Sections 13.4 and 14.3).

15.1.3.3 Beamforming: shaping the beam pattern

Traditionally, for example, in radar applications, beamforming is used to maximize the transmitted or received SINR in one direction in a propagation environment with predominantly free-space conditions and with little or no interference, for example, in surface-to-air, air-to-air, and space applications. Because of this, the main lobe is used for the purpose of transmission/reception, and the antenna side lobes are not deliberately used. As this was the dominating application for many years, this is referred to as classic beamforming (see Fig. 15.4A).

In mobile systems on the other hand, particularly in urban environments, there is usually mult-ipath propagation between the transmitter and the receiver. Hence, the transmitted signal will find its way to the receiver through many propagation paths. The concepts of main lobe and side lobes may therefore lose their meaning. In this case, the whole beam pattern, including all lobes, is adapted to match the actual instantaneous radio channel to optimize the criterion of choice, for example, maximize power or SINR at the receiver. This optimization process comprises shaping the beam in all directions, including also the possibility to form nulls in certain directions to reduce interference (see Section 6.1.1). To distinguish this significantly, a more comprehensive way of shaping the beam pattern, referred to as generalized beamforming, is used (see Fig. 15.4B).

A beam shape based analysis only describes the properties of the transmission (or reception) focusing on physical directions and not the properties of the resulting received signal propagating over a multi-path channel. The common use of such an analysis is therefore, although

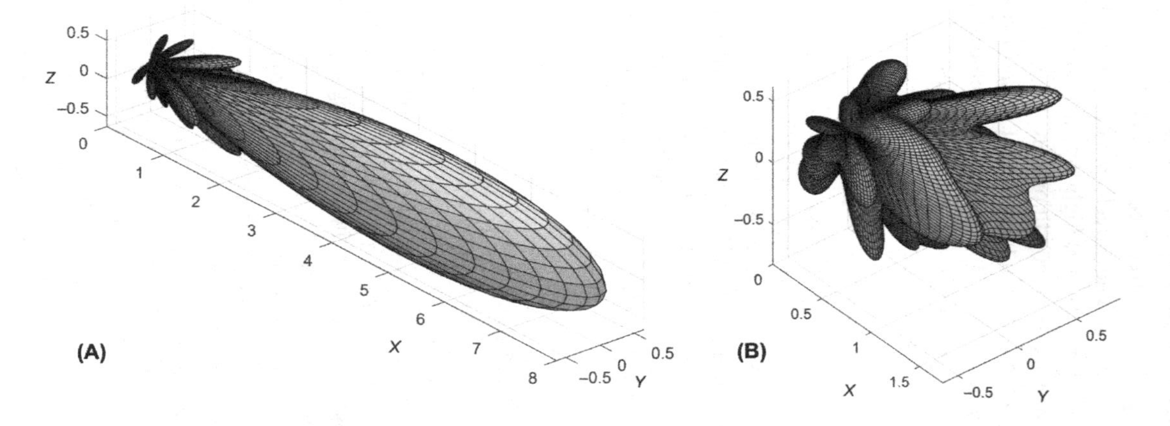

FIGURE 15.4

Illustration of beam gain pattern for (A) classical beamforming and (B) generalized beamforming.

useful for building intuition, often insufficient for providing a full understanding of multi-antenna schemes. A sufficient characterization of relevant signals is only obtained with free-space channels. In reality, radio channels are however rarely free-space channels, and some important effects may therefore not be properly reflected, in particular fast fading. Fast fading effects are best understood using a vector space description of the channel. This model properly captures the effect of multipath channels on the received signals and is a dominating view in the massive MIMO literature. The case of spatially uncorrelated Rayleigh fading is one example of when a beam shape based analysis completely loses its meaning which in contrast the vector space approach handles well. On the other hand, real physical radio channels are typically far from also this idealization, and therefore, a pure vector space approach without any connection to physical directions is also insufficient. A more powerful approach is therefore to use both analysis approaches in tandem, that is, a beam shape and physical direction based analysis together with the use of vector spaces, to obtain a deeper understanding of all relevant radio propagation effects.

15.2 OUTLOOK FOR THE FUTURE

So, how do AAS technologies and their deployment evolve from here? Many factors influence the evolution of AAS technology and its adoption in commercial networks, and this can play out in many ways. Some speculative ideas are however outlined below.

15.2.1 FIELD FEEDBACK AND ADAPTATION

For a long time, research of beamforming, multi-antenna technology, and standardization were ahead of the deployment. Hence, requirements on AAS were driven from a technology point of view rather than actual field performance. As the mobile operators are now starting to deploy AAS in large numbers, the requirements from deployments and new use cases are evolving together with AAS technology in a more interactive way. AAS capable products both on network side and UE side are deployed in large volumes in commercial networks, and the product capabilities, variations, and volumes are growing rapidly. It is now expected that experience will be gathered from real network deployments for several years to come. In this process, it is expected that there will be a focus shift toward the actual issues of most concern, for example, further enhancements of network performance in relevant scenarios, deployability, multi-band operation, UE support, and total cost of ownership (TCO). Academic research also has the opportunity to get useful feedback on the needs from the industry on the issues of concern that need research support. Hence, some research capabilities may also be directed to studies that integrate the findings from real systems with respect to, for example, channel models, traffic models, relevant deployment scenarios, product cost, and deployment restrictions.

15.2.2 COST EVOLUTION AND USE CASES

The continued success of AAS depends largely on the cost competitiveness compared to other solutions in commercial network deployments. The added business value of AAS with respect to coverage, capacity, and end-user performance has motivated more complex products with higher initial cost. The cost evolution of AAS will however likely be similar as for other product families. The cost of hardware will likely go down over time. The successively increased use of AAS drives volume of production which generally reduces cost. As cost per AAS radio providing a given performance level will therefore likely decrease, the application areas of cost-efficient AAS use will increase from what is seen today.

As long as the cost for acquisition of new sites is high, efficient reuse of existing sites will continue to be important and lead to a strong case for using AAS on new and higher frequency bands. Also, even though site reuse may be the most important factor, the cost versus performance per AAS radio deployed onsite is still relevant and is expected to continue to be so in a foreseeable future. Hence, a differentiation between AAS configurations and their beamforming capabilities is anticipated.

For mobile broadband, there is a potential of increased use of AAS in suburban and rural areas, since there are performance benefits expected there. For example, AAS could be used to a larger extent to provide coverage in rural areas where it may become competitive compared to other solutions, for example, to provide FWA.

In most countries, AAS will mainly be used on mid-band, both on TDD and FDD bands, since the mid-bands are the bands used by the operators to provide coverage in urban areas. The main use case in the near term will likely continue to be enhancing capacity on existing frequency bands and to enable sufficient coverage on new frequency bands. Non-AAS solutions, for example, 4T4R remote radio units, will likely continue to be competitive in many scenarios, particularly in areas that are not in imminent need for increased capacity and where coverage is sufficient with non-AAS solutions.

In some countries, the access to mid-band spectrum is limited. Large amounts of mm-wave spectrum may however be available, and may even be the only option, and is therefore attractive to exploit. A more aggressive use of mm-wave can then be expected. For the case those countries later will deploy mid-band, a large capacity boost can be expected when this early mm-wave network is combined with mid-bands as shown in Section 13.7.6.

The use of AAS in indoor environments may also increase. High-band solutions may be particularly useful in dedicated indoor environments since the available bandwidth is very large and the wall penetration loss reduces interference toward outdoor. In such environments, a very high capacity can be supported in limited indoor areas with very low interference between indoor and outdoor users. This may, for example, be an attractive solution for industrial use. There may, however, still be an advantage to combine high-band with mid-band in order to improve robustness.

Requirements from deployments including size and weight, multi-band/multitechnology solutions, etc., for different specific frequency bands increase over time and drive development of both product solutions and features. Multi-band solutions are already seen as very important as they provide benefits in terms of, for example, lower site rental, maintenance costs, and visual impact (see Section 14.4). The use of such solutions is expected to increase. This is a challenge as the number of frequency bands increases, and hence, there will inevitably be a diversification of the number of bands and band combinations both including AAS and non-AAS solutions. In order to accommodate such a variety of solutions, a higher degree of modularization of the products can be expected.

15.2.3 TECHNOLOGY EVOLUTION

15.2.3.1 Hardware technologies

The development of hardware technologies for processing, interfaces, capacity, and power efficiency are key enablers to reduce cost per transmitted/received bit. Particularly, the cost reduction per radio chain will benefit broader use of AAS features. New hardware technologies may improve the cost efficiency both on mid-band and high-band. The maturity of high-band solutions is, however, lower and hence there is likely more room for improvements. The general trend of making products smaller, lighter, and more capable will also continue.

It is expected that the communication industries will continue to look for and exploit new spectrum. This may include both low- and mid-bands as well as exploitation of new higher bands. Lower bands may be FDD or TDD. Higher bands are almost only TDD. New bands for AAS and new band combinations for AAS in combination with both other AAS and non-AAS solutions will likely be exploited. The TCO for the whole site will often benefit from reducing the number of hardware units and is hence expected to continue to be an important measure for evaluation. It should be noted that there are many important implementation challenges relating to AAS/non-AAS multi-band solutions.

New types of antennas may come into use. Larger and more capable antenna elements may compete with subarrays of simpler antenna elements in certain applications. New types of antenna materials may also be exploited.

The use of many lower quality radios, that is, "dirty radios," that has attracted a lot of attention for some time, will likely not come into commercial use. The industry, mainly through the work in 3GPP RAN4, has concluded that it is necessary to keep high requirements on the radios in order to ensure high performance (see Section 11.7).

15.2.3.2 Advanced antenna system solutions

Since UEs are typically spread more in the horizontal dimension than in the vertical dimension, large gains can be achieved using wide antenna arrays with many columns. In many countries, there is a reluctance to deploy wide antennas, often due to that the existing antennas have a certain width and that new antennas must have the same width. If this restriction can be reconsidered, higher performance gains can be achieved.

Up to now, rectangular array structures have been dominating. This will likely be the case also in the future due to that the spread of users is predominantly in the horizontal domain. Other array structures are however possible, for example, circular arrays, but have so far found limited use. Other related solutions, for example, coordinated usage of antennas distributed over geographical areas (e.g., CoMP) may also become more popular. This may be particularly useful in indoor environments with many obstacles, specifically for high-band where the diffraction is lower and the penetration loss is higher, and hence, cooperation between many transmission/reception points may be beneficial.

For mm-wave, there is a potential to reach higher performance and flexibility by implementing digital beamforming, as used for mid-band. For example, there is a possibility to implement time domain beamforming in the digital domain (see Section 7.3). It is also possible to implement frequency domain beamforming, as for mid-band. Furthermore, it is possible to also multiplex users onto different beams simultaneously.

Coverage may be improved for mm-wave solutions. The antenna area could be increased on both AAS and UE sides. This does however require development in some technology areas, for example, the techniques to distribute power to all antenna elements with limited power losses. An increase in the number of antenna elements and radio chains can also be expected. It should be noted that increasing the number of antenna elements tends to make the beams narrower, which gives a limited ability to match the radio channel even for modest levels of angular spread.

15.2.3.3 New standardized features

The 3GPP standards are extremely flexible, particularly for new radio. There is significant work to evaluate the optimum parameter settings in different environments, and here, the use of machine learning algorithms can be a useful tool. Hence, there is a large potential for additional performance improvements and refinements, particularly with respect to specific environments. There are many years of maturity and refinement before the AAS potential is exhausted even in existing deployments.

New 3GPP features can be foreseen, for example, new measurements to allow for machine learning, but there is currently no clear major new trend in the industry concerning new 3GPP AAS features to be exploited in the immediate future, except for the ones already mentioned.

The 3GPP technologies will however likely expand beyond the current frequency bands, even going above 100 GHz is being discussed. This may require a new waveform, and consequently, current MIMO features such as beam management need to be revisited and possibly adapted.

In order to deploy new features or more advanced features, there needs to be a sufficient fraction of UEs supporting these features. Therefore, a continued cooperation between the network and the UE suppliers is necessary for supporting new and advanced features.

15.2.4 SITE SOLUTIONS

In the near term, the radio sites in the macro networks will likely evolve gradually since the overall network structure and the use of MBB as a dominating use case will remain in foreseeable future. There are however signs that this may change. New options for site deployment, for example, street sites, are becoming more frequently used. New use cases, or use cases of renewed interest, are also appearing. Hence, it is quite possible that the requirements on networks and UE solutions also will evolve.

FWA is one example of a use case that has got a renewed interest during the last few years. The requirements from FWA are still developing. Due to the characteristics of FWA, that is, high data usage to a limited number of fixed installations, AAS may provide significant benefits, particularly for outdoor installation of the customer premises equipment to which the propagation conditions often are very good. IAB is another example of a site solution of growing interest, particularly where mm-wave is used for backhaul.

In addition to the classical macro network solutions, different types of new site solutions are starting to appear also in regular operator networks, for example, street sites. There are several reasons for this evolution, for example, the cost structure, availability, accessibility, and lead time to acquire street sites may be more attractive than for macro sites. In these scenarios, AAS can, due to their higher coverage and capacity, be used to limit the number of new sites, and it can also operate from nonoptimum site locations. AAS may therefore be a necessary feature since these solutions may need to cover the buildings in the vicinity, which may be of both various widths and heights, and hence, both horizontal and vertical beamforming may be beneficial.

Industries and authorities have an increased interest in private networks dedicated for companies or organizations. The requirements on network and UE solutions from these segments are very broad since the character of the industries and their communication needs are often very different. Many of these networks will operate in limited areas, either indoor or outdoor. Depending on the actual needs with respect to coverage, capacity, and reliability, AAS may be a part of their network solutions.

15.2.5 FINAL REMARK

AAS brings already benefits that other solutions cannot match. The range of actual and potential use cases is very high. AAS is deployed in real networks today in many thousands of sites and works in harmony with other solutions. AAS development will continue as part of an overall drive to tame the complexity of today's networks and support the increasing demands of end users and applications that drive today's and the future's connected world.

APPENDIX

MATHEMATICAL NOTATION AND CONCEPTS

1

The following mathematical notation is used unless stated explicitly otherwise.

A.1 COMPLEX NUMBERS

- Imaginary number $\sqrt{-1}$ is written: j
- Complex number: $z = x + jy = |z|(\cos\varphi + j\sin\varphi) = re^{j\varphi}$
  - Real part: $\mathrm{Re}\{z\} = x$
  - Imaginary part: $\mathrm{Im}\{z\} = y$
  - $r = |z| = \sqrt{x^2 + y^2}$ is the absolute value of z
  - Argument of z :$\arg\{z\} = \varphi$
- The complex conjugate of $z = x + jy$ is $z^c = x - jy$. It can also be written as z^* since z is a scalar (see complex conjugate transpose notation)

A.2 CONVENTIONS

- Variables/parameters are in italics: x, y.
- Operators are in normal text, words are in text: $\log(\bullet)$, $\sin(\bullet)$, $(\bullet)^{\mathrm{T}}$, SINR, Q_{MMSE}.
- Matrices are in uppercase italics and boldface: $\boldsymbol{A}$.
- Vectors are in lowercase italics and boldface: $\boldsymbol{x}$. There is, however, an exception to this in Chapter 3, Deformation/rotation in convected coordinate system, to adhere to a convention in physics on notation common for vector fields, which are uppercase italics and boldface: $\boldsymbol{E}(\boldsymbol{r}, t), \boldsymbol{B}(\boldsymbol{r}, t), \boldsymbol{S}(\boldsymbol{r}, t)$.
- When vectors are specified using their coefficients, for example, in a coordinate system or in an abstract vector space as used in linear algebra, they are always column vectors. In this case a row vector is denoted using a transpose: $\boldsymbol{h}^{\mathrm{T}}$. Example: $\boldsymbol{x} = [x_1 \quad x_2 \quad x_3]^{\mathrm{T}}$.

A.3 OPERATORS

- Element m of vector $\boldsymbol{x}$:$[\boldsymbol{x}]_m$. An alternative notation is $x_m \triangleq [\boldsymbol{x}]_m$
- Element on row m, column n of matrix $\boldsymbol{X}$: $[\boldsymbol{X}]_{m,n}$. An alternative notation is $X_{m,n} \triangleq [\boldsymbol{X}]_{m,n}$
- Kronecker product of two matrices $\boldsymbol{A}$ and $\boldsymbol{B}$: $\boldsymbol{A} \otimes \boldsymbol{B}$

- Hadamard product of two matrices $\boldsymbol{A}$ and $\boldsymbol{B}$: $\boldsymbol{A} \odot \boldsymbol{B}$. The Hadamard product is the element-wise product of two matrices with the same size, $[\boldsymbol{A} \odot \boldsymbol{B}]_{m,n} = [\boldsymbol{A}]_{m,n}[\boldsymbol{B}]_{m,n}$
- The base b logarithm is denoted by $\log_b(\bullet)$, for example, $\log_{10}(x)$. The natural logarithm is denoted by $\ln(\bullet)$
- Absolute value: $|x|$
- Complex conjugate: $(\bullet)^c$
- Transpose: $(\bullet)^T$
- Complex conjugate transpose: $(\bullet)^*$
- Scalar product of two vectors $\boldsymbol{a}$ and $\boldsymbol{b}$: $\boldsymbol{a} \cdot \boldsymbol{b} = \boldsymbol{a}^* \boldsymbol{b} = \sum_m ([\boldsymbol{a}]_m)^c [\boldsymbol{b}]_m$
- The norm of a vector $\boldsymbol{a}$: $\|\boldsymbol{a}\| = \sqrt{\boldsymbol{a} \cdot \boldsymbol{a}} = \sqrt{\boldsymbol{a}^* \boldsymbol{a}} = \sum_m |[\boldsymbol{a}]_m|^2$
- Unit vectors are denoted by a hat: $\hat{\boldsymbol{a}} = \frac{\boldsymbol{a}}{\|\boldsymbol{a}\|}$
- The vector (cross-product) of two three-dimensional (3D) vectors $\boldsymbol{a}$ and $\boldsymbol{b}$: $\boldsymbol{c} = \boldsymbol{a} \times \boldsymbol{b} = (a_2 b_3 - a_3 b_2)\hat{\boldsymbol{x}} + (a_3 b_1 - a_1 b_3)\hat{\boldsymbol{y}} + (a_1 b_2 - a_2 b_1)\hat{\boldsymbol{z}}$
- Proportional to: $\sim$
- Defined to be equal to: $\triangleq$

A.4 VECTOR FIELDS

In a vector field, each point in space is associated by a vector (compare Fig. A1). The electric field $\boldsymbol{E}$ and the magnetic field $\boldsymbol{B}$ are both vector fields in three dimensions. The *divergence* of a 3D

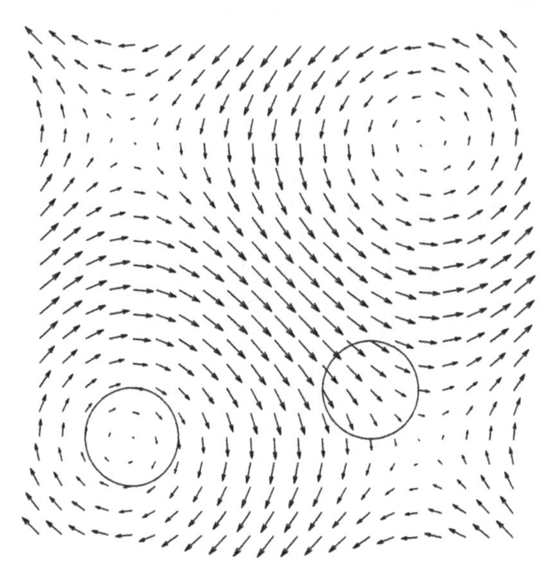

FIGURE A1

Illustration of a vector field in two dimensions, where the arrows represent the local strength and direction of the field. The divergence and the curl of the field represent local properties of the vector field, for example, the behavior in a small volume around each point in space, compare the two circles illustrating two local regions in two dimensions. Note that electromagnetic fields are three-dimensional fields.

vector field $\boldsymbol{F}$, which is denoted by $\nabla \cdot \boldsymbol{F}$, is a scalar field that is a measure of the local "flow" and whether it is outgoing (due to a source) or incoming (due to a sink). The *curl* of $\boldsymbol{F}$, denoted $\nabla \times \boldsymbol{F}$, is also a vector field, which measures the local rotation (strength and axis of rotation) at each point in space. Maxwell's equations describe how the electric field and the magnetic field are coupled, where the time derivative of one of the fields is proportional to the curl of the other, while the divergence of the electric field is proportional to the electric charge density. Since there are no magnetic charges, the divergence of the magnetic field is zero.

A.5 COORDINATE SYSTEMS

The right-handed Cartesian coordinate system (x, y, z) and the spherical coordinate system (r, θ, φ) where r, θ, and φ denote radial distance, zenith angle, and azimuth angle, respectively, with $r \in [0, \infty)$, $\theta \in [0, \pi]$, and $\varphi \in [0, 2\pi)$ are used in the book. These coordinates are usually collected into a column vector, for example, $\boldsymbol{r} = \begin{bmatrix} x & y & z \end{bmatrix}^{\mathrm{T}}$, to facilitate further computations and in fact all the vectors can be, and often must, be interpreted as a tuple of coordinates collected in a column vector. The relations between the Cartesian coordinates (x, y, z) and the spherical coordinates (r, θ, φ) are, as evident from Fig. A2,

$$\boldsymbol{r} = \begin{bmatrix} x & y & z \end{bmatrix}^{\mathrm{T}} = \begin{bmatrix} r\sin\theta\cos\varphi & r\sin\theta\sin\varphi & r\cos\theta \end{bmatrix}^{\mathrm{T}}$$

with

$$r = \sqrt{x^2 + y^2 + z^2}$$
$$\varphi = \tan^{-1}\frac{y}{x}$$
$$\theta = \cos^{-1}\frac{z}{\sqrt{x^2 + y^2 + z^2}}$$

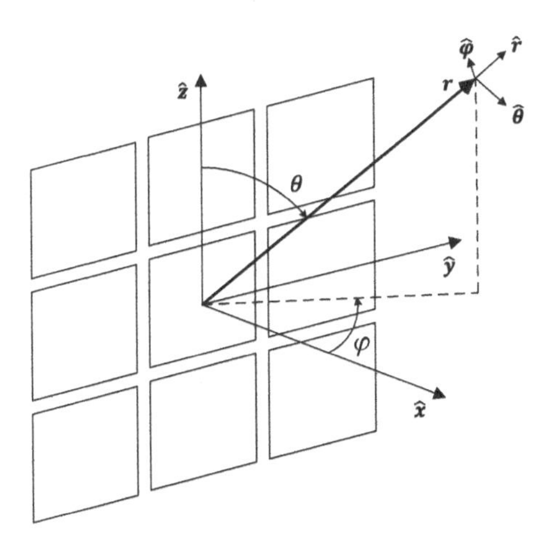

FIGURE A2

A spherical coordinate system placed. Relative an antenna array.

As seen in Fig. A2, the unit vector $\hat{\boldsymbol{r}}$ is parallel to vector $\boldsymbol{r}$. A plane perpendicular to the vector $\hat{\boldsymbol{r}}$ is spanned by the orthogonal unit vectors $\hat{\boldsymbol{\theta}}$ and $\hat{\varphi}$. Thus the unit vectors $\hat{\boldsymbol{\theta}}$, $\hat{\varphi}$,$\hat{\boldsymbol{r}}$ are all orthogonal. Expressions for the orthogonal basis vectors will be given toward the end of this section.

Furthermore, the term elevation angle, also referred to as vertical angle, will be used and is defined relative to the horizon and is positive when pointing upwards and negative when pointing downwards. Thus the elevation angle equals 90 degrees minus the zenith angle. Note that zenith and the horizon correspond to $\theta = 0$ degree and $\theta = 90$ degree, respectively.

Planar arrays are assumed to lie in the yz-plane with the x-axis pointing in the front direction of the array, that is, the boresight direction corresponding to $\theta = 90$ degree and $\varphi = 0$ degree in the spherical coordinate system, as shown in Fig. A2.

It is seen that a vector $\boldsymbol{r}$ from the origin to the point (x, y, z) can be specified as

$$\boldsymbol{r} = r\hat{\boldsymbol{r}} = x\hat{\boldsymbol{x}} + y\hat{\boldsymbol{y}} + z\hat{\boldsymbol{z}} = \begin{bmatrix} \hat{\boldsymbol{x}} & \hat{\boldsymbol{y}} & \hat{\boldsymbol{z}} \end{bmatrix} \begin{bmatrix} x \\ y \\ z \end{bmatrix}$$

where the vectors $\hat{\boldsymbol{x}}$,$\hat{\boldsymbol{y}}$,$\hat{\boldsymbol{z}}$,$\boldsymbol{r}$ all are represented as 3×1 column vectors. In Chapter 3, Deformation/rotation in convected coordinate system, electromagnetic vector fields are treated, where the field at a point in space $\boldsymbol{r}$ is represented by a vector $\boldsymbol{E}(\boldsymbol{r})$ as shown in Fig. A3. This vector field can similarly be specified as $\boldsymbol{E}(\boldsymbol{r}) = E_x\hat{\boldsymbol{x}} + E_y\hat{\boldsymbol{y}} + E_z\hat{\boldsymbol{z}}$ in Cartesian coordinates or

$$\boldsymbol{E}(\boldsymbol{r}) = E_r\hat{\boldsymbol{r}} + E_\theta\hat{\boldsymbol{\theta}} + E_\varphi\hat{\varphi} = \begin{bmatrix} \hat{\boldsymbol{r}} & \hat{\boldsymbol{\theta}} & \hat{\varphi} \end{bmatrix} \begin{bmatrix} E_r \\ E_\theta \\ E_\varphi \end{bmatrix}$$

in spherical coordinates where $\boldsymbol{E}(\boldsymbol{r})$, $\hat{\boldsymbol{r}}$,$\hat{\boldsymbol{\theta}}$, $\hat{\varphi}$ all are represented as 3×1 vectors.

For transformations between these coordinate systems, the following relations between the basis vectors $\hat{\boldsymbol{x}}, \hat{\boldsymbol{y}}, \hat{\boldsymbol{z}}$ and $\hat{\boldsymbol{r}}, \hat{\boldsymbol{\theta}}, \hat{\varphi}$ are useful

$$\begin{aligned} \hat{\boldsymbol{x}} &= \sin\theta\cos\varphi\hat{\boldsymbol{r}} + \cos\theta\cos\varphi\hat{\boldsymbol{\theta}} - \sin\varphi\hat{\varphi} \\ \hat{\boldsymbol{y}} &= \sin\theta\sin\varphi\hat{\boldsymbol{r}} + \cos\theta\sin\varphi\hat{\boldsymbol{\theta}} + \cos\varphi\hat{\varphi} \\ \hat{\boldsymbol{z}} &= \cos\theta\hat{\boldsymbol{r}} - \sin\theta\hat{\boldsymbol{\theta}} \end{aligned}$$

$$\begin{aligned} \hat{\boldsymbol{r}} &= \sin\theta\cos\varphi\hat{\boldsymbol{x}} + \sin\theta\sin\varphi\hat{\boldsymbol{y}} + \cos\theta\hat{\boldsymbol{z}} \\ \hat{\boldsymbol{\theta}} &= \cos\theta\cos\varphi\hat{\boldsymbol{x}} + \cos\theta\sin\varphi\hat{\boldsymbol{y}} - \sin\theta\hat{\boldsymbol{z}} \\ \hat{\varphi} &= -\sin\varphi\hat{\boldsymbol{x}} + \cos\varphi\hat{\boldsymbol{y}} \end{aligned}$$

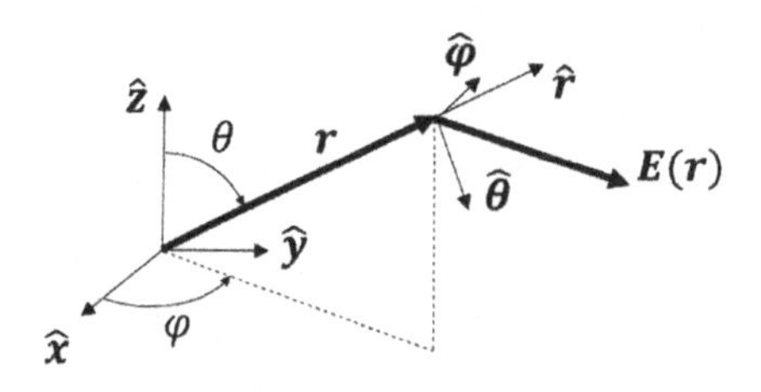

FIGURE A3

Cartesian and spherical coordinate system and a vector field $\boldsymbol{E}(\boldsymbol{r})$.

These relations can be expressed just as coordinates collected in 3×1 vectors without specifying the basis vectors, which makes the tacit assumption that there is an unspecified orthonormal basis common to one or several vectors in the expressions of interest. In particular for expressing the vectors $\hat{\boldsymbol{r}}$, $\hat{\boldsymbol{\theta}}$, $\hat{\boldsymbol{\varphi}}$, but also more generally, it is often convenient to tacitly assume, without loss of generality, that there is some global Cartesian coordinate system, and then only specify the coordinates within that system. In that coordinate system, the coordinates of $\hat{\boldsymbol{x}},\hat{\boldsymbol{y}},\hat{\boldsymbol{z}}$ can be taken to be

$$\hat{\boldsymbol{x}} = \begin{bmatrix} 1 & 0 & 0 \end{bmatrix}^{\mathrm{T}}$$

$$\hat{\boldsymbol{y}} = \begin{bmatrix} 0 & 1 & 0 \end{bmatrix}^{\mathrm{T}}$$

$$\hat{\boldsymbol{z}} = \begin{bmatrix} 0 & 0 & 1 \end{bmatrix}^{\mathrm{T}}$$

which means the above expressions for $\hat{\boldsymbol{r}},\hat{\boldsymbol{\theta}},\hat{\boldsymbol{\varphi}}$ simplifies to the coordinates

$$\hat{\boldsymbol{r}} = \begin{bmatrix} \sin\theta\cos\varphi & \sin\theta\sin\varphi & \cos\theta \end{bmatrix}^{\mathrm{T}}$$

$$\hat{\boldsymbol{\theta}} = \begin{bmatrix} \cos\theta\cos\varphi & \cos\theta\sin\varphi & -\sin\theta \end{bmatrix}^{\mathrm{T}}$$

$$\hat{\boldsymbol{\varphi}} = \begin{bmatrix} -\sin\varphi & \cos\varphi & 0 \end{bmatrix}^{\mathrm{T}}$$

The 3×1 electric field vector $\boldsymbol{E}(\boldsymbol{r})$ is in many cases lying in the 2D plane spanned by the basis vectors $\hat{\boldsymbol{\theta}}$ and $\hat{\boldsymbol{\varphi}}$. In other words,

$$\boldsymbol{E}(\boldsymbol{r}) = E_\theta\hat{\boldsymbol{\theta}} + E_\varphi\hat{\boldsymbol{\varphi}} = \begin{bmatrix} \hat{\boldsymbol{\theta}} & \hat{\boldsymbol{\varphi}} \end{bmatrix}\begin{bmatrix} E_\theta \\ E_\varphi \end{bmatrix}$$

where here $\hat{\boldsymbol{\theta}}$ and $\hat{\boldsymbol{\varphi}}$ are 3×1 vector of coordinates as above. Much of the computations can then be conveniently performed within that 2D basis instead of in the full 3D space. Thus, only the coordinates in the 2×1 vector $\underline{\boldsymbol{E}}(\boldsymbol{r}) \triangleq \begin{bmatrix} E_\theta & E_\varphi \end{bmatrix}$ are needed if the basis $\hat{\boldsymbol{\theta}}$ and $\hat{\boldsymbol{\varphi}}$ is known from the context. To clearly distinguish whether a vector is represented directly in 3D space or as coordinates in a 2D plane of that space, this book uses the notation $\underline{\boldsymbol{x}}$ to mean the 2×1 vector of coordinates corresponding to the 3×1 vector $\boldsymbol{x}$ (cf. Section 3.2.3). Vector $\underline{\boldsymbol{x}}$ may be referred to as the Jones representation of the vector $\boldsymbol{x}$. It turns out that also 3×3 matrices can have 2×2 Jones representations (see the footnote in Section 3.6).

A.6 LINEAR AND LOGARITHMIC UNITS

- dB (decibel) is used to express the ratio of one value relative to another on a logarithmic scale. If the ratio is P on a linear scale, then it is

$$10\log_{10} P$$

  on a logarithmic scale.
- dBi is the gain of an antenna relative to an isotropic antenna in dB scale.
- dBm is used to express power relative to 1 mW in dB scale.

Index

Note: The page numbers followed by "*f*" and "*t*" refer to figures and tables, respectively.

D

G

J

L

M

N

O

P

Q

R

S

T

U

V

W

X

Z

Printed in the United States
By Bookmasters